International Maritime Conference 2010

(Pacific 2010)

Sydney, Australia
27-29 January 2010

ISBN: 978-1-61738-016-7

Printed from e-media with permission by:

Curran Associates, Inc.
57 Morehouse Lane
Red Hook, NY 12571

Printed by Curran Associates, Inc. (2010)

For permission requests, please contact The Royal Institution of Naval Architects]
at the address below.

The Royal Institution of Naval Architects
10 Upper Belgrave Street
London SW1X 8BQ UK

Phone: 44 207 235 4622
Fax: 44 207 259 5912

publications@rina.org.uk

Additional copies of this publication are available from:

Curran Associates, Inc.
57 Morehouse Lane
Red Hook, NY 12571 USA
Phone: 845-758-0400
Fax: 845-758-2634
Email: curran@proceedings.com
Web: www.proceedings.com

TABLE OF CONTENTS

WELCOME

On behalf of the Organising Committee, the Royal Institution of Naval Architects, the Institute of Marine Engineering, Science and Technology and Engineers Australia, it is my pleasure to invite you to the Pacific 2010 International Maritime Conference. The Conference is being held in association with the RAN Sea Power Conference as part of the Pacific 2010 Maritime Congress.

Building on the success of Pacific 2008, our Program Committee has arranged an excellent program around the theme of the Conference; Maritime Industry - Challenges, Opportunities and Imperatives. A wide range of topics will be addressed by outstanding speakers from Australia and overseas.

I look forward to welcoming you to the Pacific 2010 International Maritime Conference.

JOHN C. JEREMY
Conference Chairman

PACIFIC 2010 INTERNATIONAL MARITIME CONFERENCE

DELEGATE LIST

27-29 January 2010

Sydney Convention & Exhibition Centre

ENGINEERS AUSTRALIA

Institute of Marine Engineering, Science & Technology

IMarEST

Pacific 2010
International Maritime Conference
27 - 29 January 2010
DELEGATE LIST

NAME	ORGANISATION	EMAIL
AUSTRALIA		
Ahmed, Mr Ziauddin	DMO	ziauddin.ahmed@defence.gov.au
Amey, Mr Paul	DSTO	paul.amey@dsto.defence.gov.au
Anderson, Dr Peter	ESSys	peter.anderson@essys.com.au
Anderson, Mr Ron F	Strategic Marine	marketing@strategicmarine.com
Andreasson, Mr Tommy	Deep Blue Tech	tommy.andreasson@deepbluetech.com.au
Angus, Mr Mike	Thales Underwater Systems	katrina.hardy@thalesgroup.com.au
Armstrong, Dr Tony	Austal Ships	tonya@austal.com
Atkinson, Mr Jack	ASC Pty Ltd	jack.atkinson@asc.com.au
Bailey, Mr Simon	Overseamarine	simoncbailey@yahoo.com.au
Baker, Mr Adrian	Deep Blue Tech	adrian.baker@asc.com.au
Barnes, Mr David	Aurecon	barnesdm@ap.aurecongroup.com
Beaton, Mr Alistair	Thales Australia	alistair.beaton@thalesgroup.com.au
Beer, Mr Jason	BAE Systems Maritime	jason.beer@baesystems.com
Benjamin, Mr John	Thales Australia	john.benjamin@thalesgroup.com.au
Beuster, Mr Gerhard	Ausnautics / Loop Group Enterprises	gerhard@ausnautics.com.au
Bixley, Mr William	Bixley Marine	wbixley@bigpond.net.au
Boden, Mr Peter	Department of Defence	peter.boden@defence.gov.au
Bondar, Mr Greg	Inst. of Marine Eng. Science & Technology	execdirector.anzspac@imarest.net
Bonner, Mr Dave	Babcock Integrated Technology Australia	paula.rowney@babcock.com.au
Bosustow, Mr George	Department of Defence	george.bosustow@defence.gov.au
Boulton, Mr Craig	ASO Marine Consultants Pty Ltd	craig.boulton@asomarine.com.au
Bourke, Mr Robert	Department of Defence	robert.bourke1@defence.gov.au
Branch, Mr Jonathan	Lloyd's Register Asia	sydney@lr.org
Brennan, Mr Rod	Defence Material Organisation	rod.brennan@defence.gov.au
Broadbent, Mr Adrian	Lloyd's Register Asia	sydney@lr.org
Buckland, Michael	DSTO Maritime Platforms Division	michael.buckland@dsto.defence.gov.au
Bumpstead, Mr Neil	Country Fire Authority Victoria	n.bumpstead@cfa.vic.gov.au
Burdekin, Ms Susan	ADFA	s.burdekin@adfa.edu.au
Burke, Mr Darren	Department of Defence	darren.burke@defence.gov.au
Burke, Mr Michael	Department of Finance and Deregulation	michael.burke@finance.gov.au
Cannon, Dr Stuart	Defence Science and Technology Organisation	stuart.cannon@dsto.defence.gov.au
Capizzi, Mr Vincent	BAE Systems Australia	vincent.capizzi@baesystems.com
Carlsson, Mr Roger	Deep Blue Tech	roger.carlsson@deepbluetech.com.au
Carson-Jackson, Ms Jillian	Australian Maritime Safety Authority	jillian.carson-jackson@amsa.gov.au
Chan, Dr Steve		stevechan@internode.on.net
Chandran, Mr Arvind	DSTO	arvind.chandran@dsto.defence.gov.au
Chowdhury, Eur Ing Maghfar	CME - DMO	maghfar.chowdhury@defence.gov.au
Clark, Dr Peter	DSTO	peter.clark@dsto.defence.gov.au
Coonan, Mr Darren	Department of Finance and Deregulation	darren.coonan@finance.gov.au
Corkindale, Mr Kevin	ASC Pty Ltd	kevin.corkindale@asc.com.au
Corniani, Mr Valerio	Gurit Australia	valerio.corniani@gurit..com
Dall'armi-Stoks, Dr Giuseppina	DSTO	leah.schwartz@defence.gov.au
Darby, Mr Mike	BAE Systems Australia	mike.darby@baesystems.com
Davies, Mr Paul	Hub Consulting Pty Ltd	paul.davies@hubconsulting.com.au
Davis, Mr John	Raytheon Australia Pty Ltd	John.davis@ausawd.com
Davis, Prof Michael	University of Tasmania	m.r.Davis@utas.edu.au
Dawson, Mr Edward	BMT Design & Technology	edawson@bmtdt.com.au
De Yong, Leo	DSTO	leo.deyong@dsto.defense.gov.au
Deeks, CDRE Mike	Raytheon Australia Pty Ltd	mdeeks@raytheon.com.au
Donovan, Mr Patrick	Babcock Integrated Technology Australia	paula.rowney@babcock.com.au
Dunn, Mr Matthew	Department of Defence	matthew.dunn5@defence.gov.au
Firth, Mr David	Gurit Australia	david.firth@gurit.com
Flapan, Mr Mori	National Marine Safety Committee Inc	mflapan@nmsc.gov.au

Pacific 2010
International Maritime Conference
27 - 29 January 2010
DELEGATE LIST

NAME	ORGANISATION	EMAIL
Fleck, LCDR Christopher	New Zealand Ministry of Defence	christopher.fleck@nzdf.mil.nz
Fletcher, Ms Lyn	Defence Science and Technology Organisation	lyn.fletcher@dsto.defence.gov.au
Foley, Mr Peter	Australian Transport Safety Bureau	peter.foley@atsb.gov.au
Fox, Mr George	Cubic Transportation Systems Australia Pty Ltd	
Gallacher, Mr John	AWD Alliance	john.gallacher@ausawd.com
Garrard, Mr Peter	ESSys	peter.garrard@essys.com.au
Gehling, Mr Rob	Australian Maritime Safety Authority	rob.gehling@amsa.gov.au
Georg, Mr Stephen	Department of Defence	stephen.georg@defence.gov.au
Glanville, Mr Jordan	BMT Design & Technology	jglanville@bmtdt.com.au
Goodin, Mr Peter	Australian Marine Technologies Pty Ltd	samantha.felgate@ausmarinetech.com.au
Grande, Mr Vincent	ASC Pty Ltd	vincent.grande@asc.com.au
Grech, Dr Michelle	Defence Science and Technology Organisation	michelle.grech@dsto.defence.gov.au
Griffin, Mr Martin	BAE Systems Australia	martin.griffin@baesystems.com
Griffiths, Mr Darryl	BMT Systems Maritime	darryl.griffiths2@baesystems.com
Grove, Mr Ken	Babcock Integrated Technology Australia	paula.rowney@babcock.com.au
Grover, Mrs Samantha	Department of Finance and Deregulation	samantha.grover@finance.gov.au
Haddad, Mr Peter	State Government of Victoria	peter.haddad@diird.vic.gov.au
Hamilton, Mr Steve	Nova Defence	steve.hamilton@novagroup.com.au
Hammond, Dr Lloyd	Maritime Platforms Division, DSTO	lloyd.hammond@dsto.defence.gov.au
Hargreaves, Mr Denzal	Det Norske Veritas	denzal.hargreaves@dnv.com
Harman, Mr Matthew	BAE Systems Australia	matthew.harman2@baesystems.com
Helmore, Mr Phillip	University of New South Wales	p.helmore@unsw.edu.au
Hodge, Miss Danielle	DMO - DSME	danielle.hodge@defence.gov.au
Holden, Mr Stephen	BAE Systems Australia	stephen.holden2@baesystems.com
Hooper, Mr David	BAE Systems Australia	david.hooper3@baesystems.com
Howlett, Mr Craig	Raytheon Australia Pty Ltd	chowlett@raytheon.com.au
Hughes, Mr Craig	ABS	crhughes@eagle.org
Ibal, Dr Gokhan	DSTO	gokhan.ibal@dsto.defence.gov.au
Jeremy, John	RINA	jcjeremy@ozemail.com.au
Johnson, Mr Colin	BAE Systems Australia	colin.johnson3@baesystems.com
Jones, RADM Peter	Department of Defence	peter.jones14@defence.gov.au
Kelkar, Ms Shruti	BAE Systems Australia	shruti.kelkar@baesystems.com
Kim, Mr Yong-Jin	BMT WBM	ykim@wbmpl.com.au
Kinsey, Mr Jeremy	Department of Defence	jeremy.kinsey@defence.gov.au
Kirkby, Mr Darren	BAE Systems Maritime	darren.kirkby@baesystems.com
Kitching, Mr Nick	BAE Systems Australia	nick.kitching@baesystems.com
Knevett, Mr Brent	Profile Ray and Berndtson	lamis@p-m-c.com.au
Knox, Ms Jennifer	Lightning Naval Architecture	navlight@bigpond.com
Kumaranayakam, Mr Kumar	Department of Defence	kumar.Kumaranayakam@defence.gov.au
Lahiff, Mr Simon	Deep Blue Tech	simon.lahiff@asc.com.au
Lane, Mr John	Symred Pty Ltd	symred@optusnet.com.au
Lawrence, Prof Norman	Australian Maritime College	n.lawrence@amc.edu.au
Leach, Mr Marcus	Department of Defence	graham.watson3@defence.gov.au
Lee, Mr Karkhee	Department of Defence	karkhee.lee@defence.gov.au
Lockley, Mr Patrick		pat.lockley@gmail.com
Loft, Mr Neil	BAE Systems Australia	neil.loft@baesystems.com
Longmuir, Mr Brett	Australian Marine Technologies	brett.longmuir@ausmarinetech.com.au
Ludlam, Mr Stephen	ASC Pty Ltd	vicki.highet@asc.com.au
Lyford, Mr Mark	BAE Systems Maritime	mark.lyford@baesystems.com
Macdonald, CMDR Ralph	Training Authority Maritime Warfare	ralph.macdonald@defence.gov.au
Macneill, CMDR Mike	Training Authority Maritime Warfare	mike.macneill@defence.gov.au
Makin, Mr Murray	Thales Australia	murray.makin@thalesgroup.com.au
Marshall, Mr Gary	BAE Systems Australia	gary.marshall2@baesystems.com
Mart, Dr Peter	DSTO Melbourne	peter.mart@dsto.defence.gov.au

Pacific 2010
International Maritime Conference
27 - 29 January 2010
DELEGATE LIST

NAME	ORGANISATION	EMAIL
Maxwell, Mr William	Thales Australia	william.maxwell@thalesgroup.com.au
McAuliffe, Mrs Catriona	Engineers Australia	cmcauliffe@engineersaustralia.org.au
McCormack, Mr Errol	Aviops Pty Ltd	aviops@bigpond.com
McLaughlin, Mr John	ASC Pty Ltd	john.mclaughlin@asc.com.au
Mechanicos, Mr Michael	Germanischer Lloyd Australia	michael.mechanicos@gl-group.com
Meldrum, Mr John	ANZAC SPO	john.meldrum@anzacspo.gov.au
Middleton, Mr Philip	BAE Systems Maritime	philip.middleton@baesystems.com
Miley, Mr Robert	Tasmanian Defence Initiative	robert.miley@development.tas.au
Milroy, Mr Ben	Frontline Australasia	bmilroy@frontline-aust.com.au
Mohamad Ayob, Mr Ahmad Faisal	ADFA	ahmad.mohamadayob@student.adfa.edu.au
Mole, CDRE Denis	Sydney Ferries	denis.mole@sydneyferries.nsw.gov.au
Moran, Mr Gary	Gryphon Technologies International	gmoran@gryphonlc.com
Morris, Mr Brett	Defence Science & Technology Organisation	brett.morris@dsto.defence.gov.au
Morton, CDRE Geoff	Goeffry Morton and Associates	davis_michael@bectech.com
Nandagopal, Dr Nanda	DSTO	nanda.nandagopal@dsto.defence.gov.au
Nayler, Mr Graeme	Nova Defence	graeme.nayler@novagroup.com.au
Neill, Dr Roger	Defence Science and Technology Organisation	roger.neill@dsto.defence.gov.au
Newman, Mr Andrew	Serco Australia Pty Ltd	andrew.newman@serco-ap.com.au
Nightingale, Mr Gerald	Department of Defence	gerald.nightingale@defence.gov.au
O'Donnell, Mr Edward	Electric Boat Australia	ed.odonnell@asc.com.au
O'Donnell, Mr Gary	Navy Certification and Safety	gary.odonnell@defence.gov.au
Ozerova, Ms Marina	DSTO	Marina.ozerova@dsto.defence.gov.au
Paragios, Mr George	Department of Defence	george.paragios@defence.gov.au
Parsons, Mr James	Microsoft	james.parsons@microsoft.com
Pedrina, Dr Lenore	DSTO	lenore.pedrina@dsto.defence.gov.au
Perez, Dr Tristan	The University of Newcastle	Tristan.Perez@newcastle.edu.au
Peters, Mr Noel	Det Norske Veritas (DNV)	noel.peters@dnv.com
Phelps, Mr Bernard	DSTO	bernard.phelps@dsto.defence.gov.au
Plant, Mr Paul	Deep Blue Tech	paul.plant@deepbluetech.com.au
Polglaze, Mr John	URS Australia	john_polglaze@urscorp.com
Pollet, Mr Robin	BAE Systems Australia	robin.pollet@baesystems.com
Porter, Mr Kevin	Lloyd's Register Asia	sydney@lr.org
Pottinger, Mr Kevin	DMO/TIEIO	kevin.pottinger@defence.gov.au
Poynter, Mr Michael	ASC Pty Ltd	mike.poynter@asc.com.au
Prince, Mr Michael	Australian Hydrographic Service	makprince@iprimus.com.au
Prusty, Prof Gangadhara	UNSW	g.prusty@unsw.edu.au
Rago, Mr Louis	Microsoft	louis.rago@microsoft.com
Rahman, Mr Tauhid	Det Norske Veritas	tauhid.rahman@dnv.com
Raju, Mr Raju	University of New South Wales	z3046119@student.unsw.edu.au
Ranmuthugala, Prof Dev	Australian Maritime College	dranmuth@amc.edu.au
Ratcliffe, Dr Max	Babcock Integrated Technology Australia	max.ratcliffe@babcock.com.au
Raymond, Mr Ian	Directorate of Navy Platform Systems	ian.raymond1@defence.gov.au
Rehborn, Mr Hans	Sigma Coachair Group	hrehborn@sigmacoachair.com
Renilson, Prof Martin	Renilson Marine Consulting Pty Ltd	martin@renilson-marine.com
Ridgway, Mr Simon	ASC Shipbuilding	simon.ridgway@ausawd.com
Riding, Mr Douglas	BAE Systems Australia	sue.woods@baesystems.com
Ryan, Mr Jeffrey	Australian Customs and Border Protection Service	jeffrey.ryan@customs.gov.au
Ryan, Dr Peter	Defence Science & Technology Organisation	peter.ryan@dsto.defence.gov.au
Salter, Mr Ian	Nova Defence	ian.salter@novagroup.com.au
Simonds, Mr Craig	BAE Systems Maritime	craig.simonds@baesystems.com
Skeen, Dr Michael	DSTO	michael.skeen@dsto.defence.gov.au
Skinner, CAPT Christopher	DISplay Pty Ltd	cjskinner@acslink.net.au
Slater, Mr Karl	BAE Systems Australia	karl.slater@baesystems.com
Smith, Dr Warren	ADFA	w.smith@adfa.edu.au

DELEGATE LIST

NAME	ORGANISATION	EMAIL
Spalj, Mr Dusko	Sydney Institute of Tafe	Dusko.Spalj@tafensw.edu.au
Speer, Mr Timothy	Austal	tims@austal.com
Stals, Dr John	EAABC Pty Ltd	sajeind@ozemail.com.au
Stevens, Mr Andrew	ASC AWD Shipbuilder Pty Ltd	andrew.stevens@dusawd.com
Stevens, Mr Matthew	DMO-CME	matthew.stevens10@defence.gov.au
Stokes, Mr Graham	Beca	graham.stokes@beca.com
Such, Mr David	Serco Australia Pty Ltd	david.such@serco-ap.com.au
Tait, Ms Samantha	BMT Design & Technology	stait@bmtdt.com.au
Taylor, Mr Graham	Taylortech Pty Ltd	graham.taylortech@gmail.com
Thomson, Mr Andrew	AMOG Consulting	jan_renshaw@amogconsulting.com
Toffoletto, Mr Robert	DSTO	robert.toffoletto@dsto.defence.gov.au
Tomaselli, Mr Richard	Beca	richard.tomaselli@beca.com
Torresan, Mr Hugh	Defence Science and Technology Organisation	hugh.torresan@dsto.defence.gov.au
Tu, Mr Ziming	DSTO	ziming.tu@dsto.defence.gov.au
Turner, Mr Terry	Defence Science & Technology Organisation	terry.turner@dsto.defence.gov.au
Vierow, Mr Michael	L-3 Nautronix	michael.vierow@l3com.com
Warren, Mr Mike	ASC Pty Ltd	mike.warren@asc.com.au
Washington, Mr Daniel	AMOG Consulting	jan_renshaw@amogconsulting.com
Watson, Mr Graham	Department of Defence	graham.watson3@defence.gov.au
Waugh, Mr Malcolm	Anzac Systems Program Office	mal.waugh@anzacspo.gov.au
Weiland, Mr Bruce	Drivetrain Power and Propulsion	bruce.weiland@drivetrainpower.com
Whiley, Mr Stuart	ASC Pty Ltd	stuart.whiley@asc.com.au
Wicklander, Mr Hans	Deep Blue Tech Pty Ltd	hans.wicklander@deepbluetech.com.au
Wilhelm, Mr Geoff	Centre for Maritime Engineering (CME) DMO	geoff.wilhelm@defence.gov.au
Williams, Mr Sean	Deep Blue Tech	sean.williams@deepbluetech.com.au
Williamson, Mr Matthew	BAE Systems Australia	matthew.williamson6@baesystems.com
Willis, Mr Mike	BAE Systems Australia	michael.willis2@baesystems.com
Wilson, Mr Bruce	Australian Customs and Border Protection Service	bruce.wilson@customs.gov.au
Wilson-Haffenden, Mr Sam	Australian Maritime College	wilsonsj@utac.edu.au
Woodhouse, Mr Adrian	Department of Defence	adrian.woodhouse@defence.gov.au
Young, Mr Kevin	APC Technology	kyoung@apctechnology.com.au
CANADA		
Mihova, Mrs Natasha	S.H.M. Marine International, Inc.	natasha.mihova@yahoo.ca
FRANCE		
Bolot, Mr Jean-Michel	DCNS	jean-michel.bolot@dcnsgroup.com
Dematte, Mr Patrick	DCNS	patrick.dematte@dcnsgroup.com
GERMANY		
Tigges, Mr Kay	Siemens Ag	kay.tigges@siemens.com
Von Drathen, Mr Arndt	MTU Friedrichshafen	jill.power@mtudda.com.au
INDIA		
Basra, Mr Gagandeep	Basra Brothers	gdsinghbasra@rediffmail.com
Talwar, Ms Sonia	Basra Brothers	sonia.talwar1@yahoo.com
KOREA, SOUTH		
Norton, Mr Paul	Converteam	paul.norton@converteam.com
MALAYSIA		
Abdul Rahim, Mr Rushdi	MiGHT	rushi@might.org.my
Aris, Ms Nor Azliza	MiGHT	azliza@might.org.my
Nyamekye, Mr Eric		mark994@hotmail.com
NETHERLANDS		
Wei, Ms Yan		y.wei@tudelft.nl
Wildenberg, Mr Albert	Thales Nederland	albert.wildenberg@nl.thalesgroup.com
NEW ZEALAND		
Arrell, Mr Scott	Thales New Zealand	scott.arrell@thalesgroup.co.nz

Pacific 2010
International Maritime Conference
27 - 29 January 2010
DELEGATE LIST

NAME	ORGANISATION	EMAIL
Fairweather, CMDR David	Royal New Zealand Navy	dave.fairweather@nzdf.mil.nz
Martin, Mr John	Royal New Zealand Navy	alan.martin@nzdf.mil.nz
Tucker, CAPT John	Royal New Zealand Navy	john.tucker@nzdf.mil.nz
Wishart, CMDR Kelvin	Royal New Zealand Navy	kelvin.wishart@nzdf.mil.nz
SINGAPORE		
Ham, Miss Wan Ling	Defence Science & Technology Agency	hwanling@dsta.gov.sg
SOUTH AFRICA		
Adedeji, Mr Kehinde Adewale		neyo_d@yahoo.com
TAIWAN		
Lien, Prof Chang-Hua	National Kaohsiung Marine University	chlien@mail.nkmu.edu.tw
Su, Dr Dong-Taur	National Kaohsiung Marine University	tonysu@mail.nkmu.edu.tw
Wu, Prof Jia-jang	National Kaohsiung Marine University	jjangwu@mail.nkmu.edu.tw
Yang, Dr Tsun-Lirng	National Kaohsiung Marine University	yang2006@mail.nkmu.edu.tw
THAILAND		
Adedeji, Mr Michael Edward		checklama@yahoo.com
UNITED KINGDOM		
Best, Mr Paul	Frazer-Nash Consultancy Ltd	p.best@fnc.co.uk
Blakeley, Mr Trevor	Royal Institution of Naval Architects	tblakeley@rina.org.uk
Brooking, Mr Charles	Frazer-Nash Consultancy	c.brooking@fnc.co.uk
Buckingham, Mr John	Bmt Defence Services Limited	johnb@bmtdsl.co.uk
Deacon, Mr Andrew	Defence Science & Technology Laboratory	adeacon@dstl.gov.uk
Guild, Rear Admiral Nigel	Institute of Marine Engineering, Science and	
Harris, Mr Philip	Frazer-Nash Consultancy Ltd	p.harris@fnc.co.uk
Morgan, Mr Eugene	BMT Defence Services Ltd	emorgan@bmtdsl.co.uk
Quilliam, Mr Roy	BMT Defence Services Ltd	rquilliam@bmtdsl.co.uk
Simpson, Mr Robert	Lloyd's Register EMEA	robert.simpson@lr.org
Tippett, Mr George	Conidia Bioscience Ltd	george@conidia.com
Twomey, Mr Bernard	Lloyd's Register EMEA	bernard.twomey@lr.org
USA		
Alvarez, Ms Kristen	Gryphon Technologies LC	kalvarez@gryphonlc.com
Anding, Mr Fred	Raytheon Missile Systems	fwanding@raytheon.com
Beckett, Mr Lloyd	Raytheon	lloyd@raytheon.com
Beitinger, CPO Craig	AMI International	cbeitinger@amiinter.com
Blanch, Mr Scott	ViaSat, Inc.	scott.blanch@viasat.com
Booth, CMDR Charlie	Cisco Systems, Inc.	chabooth@cisco.com
Bowler, I I I, Mr R. T. E.	Bath Iron Works	tom.bowler@biw.com
Carlin, Mr William	Argon ST, Inc.	deborah.richardson@argonst.com
Copsey, Mr Thomas	The Columbia Group	tcopsey@csc.com
Cumm, Mr Eugene	Northrop Grumman Corporation	gene.cumm@ngc.com
Davis, Ms Carol	Gryphon Technologies	cdavis@gryphonlc.com
Davis, Ms Cynthia	Raytheon Missile Systems	cynthia.a.davis@raytheon.com
Dean, Mr Dennis	Gryphon Technologies	commercial@dennis-dean.com
Deblasio, Mr Marc	Northrop Grumman Corp	marc.deblasio@ngc.com
Delpizzo, Mr Richard	ABS	rdelpizzo@eagle.org
Doughty, Mr Alan	Bath Iron Works	alan.doughty@gdbiw.com
Feeley, Mr Michael	Lockheed Martin	mike.feeley@lmco.com
Frick, RADM Michael	Bectech, Inc.	davis_michael@bectech.com
Gagnon, Mr Mark	General Dynamics Electric Boat	mgagnon@gdeb.com
Grant, Mr Steven D.	Naval Surface Warfare Center Panama City	steven.grant@navy.mil
Heil, Mr Bradford	General Dynamics/ Electric Boat	bheil@gdeb.com
High, Mr Jeffrey	Northrop Grumman	jeff.high@ngc.com
Houlihan, Ms Melissa	Lockheed Martin	melissa.evanskey@lmco.com
Jenkins, Mr Henry	Northrop Grumman Information Systems	henry.jenkins@ngc.com

Pacific 2010
International Maritime Conference
27 - 29 January 2010
DELEGATE LIST

NAME	ORGANISATION	EMAIL
Klinedinst, Mr David	General Dynamic Information Technology	david.klinedinst@gdit.com
Kocin, Mr Michael	ViaSat, Inc.	mike.kocin@viasat.com
Loraine Iii, Mr Jacques	Raytheon Space and Airborne Systems	jacques_b_loraine@raytheon.com
Lovgren, Mr Manuel	General Dynamics Information Technology	manny.lovgren@gdit.com
Luntz, Mr Mike	Lockheed Martin	mike.d.luntz@lmco.com
Mackett, Ms Janet	ViaSat, Inc.	janet.mackett@viasat.com
Martin, Mr Robert	Raytheon Company	ram@raytheon.com
Merritt, Dr Christopher	US Navy, Maritime Liaison Unit - Pacific	christopher.merritt@fe.navy.mil
Mitchell, Capt Anthony	USN, Undersea Weapons Program Office	anthony.e.mitchell@navy.mil
Moncada, Mr Louis	Raytheon Missile Systems	ljmoncada@raytheon.com
Morel, Mr David	Naval Undersea Warfare Center	colleen.beaver.ctr@navy.mil
Naqvi, Dr Waseem	Raytheon	waseem@raytheon.com
Nohara, Mr Steven	Electric Boat Corporation	snohara@gdeb.com
Padgett, Mr John B.	General Dynamics, Electric Boat	jpadgett@gdeb.com
Potter, Mr David	Northrop Grumman Corporation	kevin.king@ngc.com
Ritchey, Mr Matt	Raytheon IDS	ritchey@raytheon.com
Shields, Mr Ronald	Raytheon	rcshields@raytheon.com
Smith, Mr Richard	Raytheon Company	ricksmith@raytheon.com
Smith, Mr Thadeous	Raytheon	thadeous_smith@raytheon.com
Sutton, Mrs Rebecca	BecTech, Inc.	davis_michael@bectech.com
Sutton, RADM Robert	BecTech, Inc.	davis_michael@bectech.com
Tompkins, Mr Robert	Naval Undersea Warfare Center Division Newport	robert.tompkins@navy.mil
White, CAPT Richard	BecTech, Inc.	davis_michael@bectech.com

SPONSOR

We would like to extend our appreciation to our sponsor for their kind support:

Principal Sponsor
Satchel Sponsor, Lanyard Sponsor and Official Congress Program Sponsor

LOCKHEED MARTIN

What happens when you start getting wet feet? The revised standard for buoyancy and stability after flooding

Mori Flapan
National Marine Safety Committee Inc
mflapan@nmsc.gov.au

ABSTRACT

This paper describes the development of a new performance-based standard for buoyancy and stability after flooding on domestic commercial vessels. The standard specifies controls that reduce the consequences of flooding. Flooding is an event associated with by far the majority of vessel losses. The standard bridges the gap between relatively stringent international standards intended for full seagoing operations and standards applicable to small craft in sheltered conditions. Some of the key changes relative to current standards have been highlighted in the paper.

1. INTRODUCTION

The Australian domestic commercial vessel standard called the Uniform Shipping Laws Code[1] (USL Code) has been the subject of a progressive review by the National Marine Safety Committee[2] (NMSC). The objectives of the review include updating the standard to accommodate current and new technologies based on relevant international and national standards, and introduce a performance-based approach[3]. The revised standards are being published as the National Standard for Commercial Vessels[4]

The standards for subdivision and damaged stability contained in the USL Code were based on SOLAS provisions and US CFR 46[5] requirements from the 1970s. Apart from amendments to the criteria for damaged stability, there had been little change to the USL Code provisions since that time.

An analysis of 108 Australian commercial vessel losses over the last 15 years is given in Figure 1[6]. The requirements for buoyancy and stability after flooding work to directly reduce the consequence of the incidents that lead to a vessel being wrecked (i.e., grounding), being in collision and foundering (i.e., capsize, swamping or flooding). It also can play a role in reducing the consequences of fire by separating compartments and limiting the potential effects of flooding.

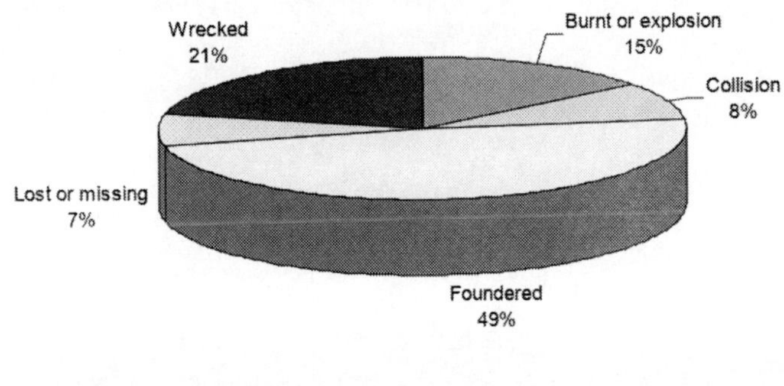

Figure 1—Analysis of 108 Australian vessel losses (1992-2007)

2. A MORE TAILORED APPROACH

The current USL Code requirements[7] for subdivision and damaged stability are largely centred around four grades of vessels: First they are differentiated by length; 35 metres and more or under 35 metres; then these are divided by the type of operation; passenger vessels and vessels not being passenger vessels. Within these groups are a few adjustments for length; but as a general observation, the grading of requirements tend to be relatively coarse. Application of the current USL Code standards frequently results in sub-optimal solutions, requiring the application of surveyor "interpretation" and discretion to provide a practical solution.

The revised standard aims to capture this surveyor experience within the standard and so avoid the need to exercise surveyor discretion that is frequently to some extent subjective and leads to less than consistent outcomes.

3. INTERNATIONAL STANDARDS SET THE UPPER BENCHMARK FOR NSCV

The buoyancy and stability after flooding requirements for passenger ships and large cargo ships engaged in international trade are contained in Chapter II-1 of SOLAS[8]. SOLAS is not applicable to fishing vessels or cargo ships of less than 500 gross tons. The flag state is given discretion to vary requirements for vessels engaged in international trade operating not more than 20 nautical miles from the coast. For domestic vessels, the new NSCV standard applies the IMO Regulations as the upper benchmark for ships engaged in domestic operations having levels of risk essentially similar to those engaged in international trade. Hence the NSCV refers to SOLAS requirements for passenger and cargo ships above a certain size that operate at sea. Because SOLAS is not applicable to fishing vessels, the buoyancy and stability after flooding requirements in the IMO Safety of Fishing Vessels Code[9] (SFV Code) form the upper benchmark for large seagoing domestic fishing vessels. The relevant SOLAS requirements are also applied to passenger vessels of smaller sizes engaged in unlimited domestic operations (Operational area A), and also certain passenger vessels carrying relatively large numbers of day passengers (more than 450) or berthed passengers (more than 36) in operations from 20 nautical miles up to 200 nautical miles from the coast.

SOLAS and SFV are also specified as alternative deemed-to-satisfy solutions for some of the remaining vessels; see Table 1 from the NSCV below. Thus a Class 1C passenger vessel 50 metres long carrying 200 day passengers could either apply SOLAS or the specific deemed-to-satisfy provisions contained in the NSCV. These alternative solutions have been provided so that a vessel imported into Australia that is built to international standards will be deemed-to-satisfy the NSCV.

The lower limit of application of the international standards is determined by the effective limits of their application. SFV for example, while nominally intended for fishing vessels 24 metres or more, only contains verifiable requirements for vessels 45 metres or more; the requirements below this length being applied at the discretion of the flag state. The NSCV requirements for fishing vessels between 24 metres and 45 metres represent the Australian interpretation of this discretion and so the interpretations of foreign states must be considered in the context of an equivalent solution. A key pillar that shapes the Australian interpretation is that differences in minimum achieved safety levels between fishing vessels and other non-passenger vessels should be eliminated[10].

Table 1 — Deemed-to-satisfy solutions for buoyancy and stability after flooding

Operational Area	Class 1 Passenger	Class 2 Non-passenger	Class 3 Fishing
A	SOLAS Chapter II-1	SOLAS Chapter II-1 if $L_m \geq 80$ m	Torremolinos SFV if $L_m \geq 45$ m
		or NSCV if $L_m < 80$ m	or NSCV if $L_m < 100$ m
B	SOLAS Chapter II-1	SOLAS Chapter II-1 if $L_m \geq 80$ m	Torremolinos SFV if $L_m \geq 45$ m
	or NSCV if $L_m < 60$ m (2)	or NSCV if $L_m < 80$m	or NSCV if $L_m < 100$ m
B20, C	SOLAS Chapter II-1	SOLAS Chapter II-1 if $L_m \geq 80$ m	Torremolinos SFV if $L_m \geq 45$ m
	or NSCV if $L_m < 80$ m	or NSCV if < 100 m	or NSCV if < 100 m
D, E	SOLAS Chapter II-1	SOLAS Chapter II-1 if $L_m \geq 80$ m	Torremolinos SFV if $L_m \geq 45$ m
	or NSCV	or NSCV	or NSCV

4. RECREATIONAL BOAT STANDARDS SET THE LOWER BENCHMARK

Recreational boat standards were not mandatory in Australia at the time the USL Code was first written in the 1970s. However, since the implementation of the Australian Builders Plate[11] in 2005, new recreational boats in Australia have been required to meet certain minimum standards for maximum number of persons, load capacity, engine power and, for vessels less than 6 m in length, buoyancy. The NSCV acknowledges that properly constructed recreational craft should suffice for use in the lowest risk categories for commercial service. Hence, the revision of the NSCV has used recreational boat standards as the lower benchmark for requirements, applicable to commercial vessels of lowest risk. The concepts of basic and level flotation used in the NSCV have been closely based on the US standards for recreational craft[12], with modifications to extend the application to vessels above 6 metres in length and to accommodate the wider range of configurations seen in commercial vessels. This should have the practical benefit of permitting the use of certain production recreational boats for certain applications. One significant departure however is the assumed mass of each person which has been adjusted from 63.5 kg to a minimum 80 kg per person to take into account important effects arising from changes in the physical characteristics of the community[13].

5. CATEGORIES OF FLOODING RISK BRIDGE THE GAP

Between the two extremes of IMO standards and recreation craft standards for buoyancy and stability after flooding; the new NSCV standard adopts a more comprehensive method of risk grading vessels to replace length as the prime driver of requirements. Four so-called flooding risk categories I, II, III and IV are defined taking account factors such as the Vessel Use (passenger, non-passenger, fishing), Area of Operation (Operational Areas A, B, C, etc), number of passengers on board and the number of berthed passengers on board. Table 2 from the standard shows how the Flooding Risk Category of a vessel is determined. This table also shows how the requirements specified in SOLAS are dovetailed into the NSCV requirements. The Flooding Risk Categories provide a framework that grades the requirements from those applied to recreational craft to the full SOLAS standard. This is illustrated in Figure 2. The estimated distribution of vessels into the Flooding Risk Categories is as follows: 45.6% in FLRC 1, 41.6% in FLRC II, 11.8% in FLRC III, 0.8% in FLRC IV and 0.2% are required to apply SOLAS[14]. By far the majority of vessels (87.2%) are expected to apply FLRC I or II.

Table 2 — Flooding Risk Categories of vessels

Operational area category (see NSCV Part B) **Vessel use category (see NSCV Part B)**	**Class A** Unlimited domestic operations	**Class B** Offshore operations	**Class B20** Offshore operations limited to 20 nm	**Class C** Restricted offshore operations	**Class D** Partially smooth waters	**Class E** Smooth waters
Class 1— Length of vessel		**< 60 m**	**< 80 m**	**< 80 m**	**All lengths**	**All lengths**
Class 1: 13 to 36 day passengers	SOLAS Ch. II-1	III	III	III	III	II
Class 1: 37 to 400 day passengers	SOLAS Ch. II-1	IV	IV	III	III	III
Class 1: 401 or more day passengers	SOLAS Ch. II-1	SOLAS Ch. II-1	IV	IV	III	III
Class 1: 13 to 36 berthed passengers	SOLAS Ch. II-1	IV	IV	III	III	III
Class 1: 37 or more berthed passengers	SOLAS Ch. II-1	SOLAS Ch. II-1	IV	IV	IV	III
Class 2— Length of vessel	**< 80 m**	**< 80 m**	**< 100 m**	**< 100 m**	**All lengths**	**All lengths**
Class 2: Nil passengers	II	II	II	I	I	I
Class 2: 1 to 12 passengers	III	II	II	II	I	I
Class 3— Length of vessel	**< 100 m**	**< 100 m**	**< 100 m**	**< 100 m**	**All lengths**	**All lengths**
Class 3	II	II	II	I	I	I

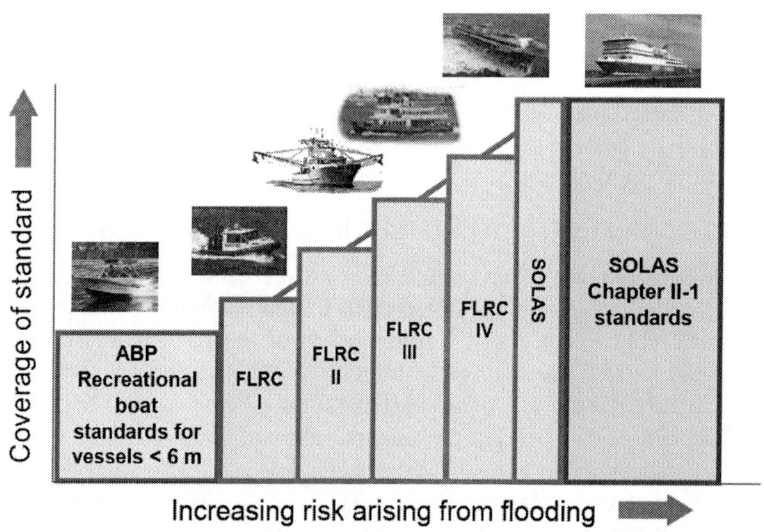

Figure 2—Illustration of how the Flooding Risk Categories provide a graded approach to the requirements for buoyancy and stability after flooding

6. OUTCOME RATHER THAN SOLUTION FOCUS

A significant difference between the USL Code provisions for subdivision and damage stability and the new NSCV standard "Buoyancy and stability after flooding" is the focus on hazards and safety outcomes rather than specific solutions. This is reflected in the name of the standard. There are now specific chapters that deal with the key hazards of swamping, local flooding, grounding and collision (both bow collision and side collision). The provisions within each chapter determine the application of the chapter to a particular vessel. Graded requirements are specified depending on the magnitude of relevant key risk parameters, including the flooding risk category. Thus, while a small vessel may be subject to the swamping chapter, the grounding and collision chapters may not apply. Similarly, the swamping chapter may not apply to certain vessels required to meet the side collision requirements of the collision chapter.

7. KEY REFORMS IN DEEMED-TO-SATISFY SOLUTIONS

7.1 Clear requirement for certain vessels to withstand swamping

Swamping is defined in the standard as the unintentional entrapment of sea or fresh water within spaces open to the weather such as found on open vessels, or in the wells or cockpits on vessels. Chapter 4 of the new standard specifies that the following vessels are required to withstand swamping: Any vessel of measured length less than 6 m; or a vessel of FRC II of measured length 6 m or more and less than 24 m. However it does not apply to vessels having buoyancy spaces in the form of watertight compartments protected by decks with closures to openings and rapid drainage compliant with the NSCV requirements for Watertight and Weathertight Integrity[15]. Two standards of compliance are specified: Basic flotation and level flotation corresponding to the definitions used in the Australian Builders Plate and ABYC Rule H-8 standards.

7.2 Concept of local flooding

Previous standards have professed an "all or nothing" approach to flooding. Either the vessel can survive flooding from a given extent of collision damage or it cannot. However, the reality of a one compartment standard that is commonly applied to domestic passenger vessels is that the probability of hitting and damaging the shell plating in way of a bulkhead is no less than that of damaging the equivalent length of shell plating between bulkheads. The probability of survival of a vessel designed to a one compartment standard in the event of collision may well be quite marginal depending upon the actual arrangement used. At the very least, this gives a false impression of the extent of protection.

The draft standard defines local flooding as the unintentional entry of sea or fresh water into a single buoyant space through an unintended opening, hole, crack or gap. This could include a limited amount of damage from grounding or collision, but does not purport to provide comprehensive levels of protection from a reasonably likely range of possible collision or grounding scenarios. Local flooding also includes flooding from failure of sea water systems, accidental flooding by persons, flooding from structural failure or corrosion, flooding past stern glands and vulnerable appendages.

The concept of local flooding provides a performance basis for the prescriptive requirements that specify bulkheads for vessels not required to survive flooding such as cargo and fishing vessels. These provisions are used to provide opportunities for reducing the consequences of flooding by providing time and protecting essential systems to try and prevent or limit flooding and for evacuation spaces and, if necessary, the vessel. Most importantly, the effect of the provisions is to modify the way that a vessel founders by preventing premature capsize. Provisions are provided for: Early warning of flooding (flooding detection systems), Separation of spaces liable to flooding, Measures to control the risks of local flooding causing premature capsize, Local flooding of enclosed Ro-Ro spaces, and Slack bilges in landing craft.

A challenge when adapting existing provisions to a performance-based framework is how to deal with apparent anomalies such as the acceptance of yachts without effective transverse subdivision.

Great care has to be taken to ensure that the new framework does not exclude hitherto acceptable arrangements. This has been dealt with by constraining the consequences of flooding by other means. Limits are placed on the power of machinery and the vessel must be of a type not prone to premature capsize when flooded; e.g., a vessel with a substantial ballast keel or of catamaran configuration.

7.3 Performance based options for protection from grounding

The USL Code requirements have focused on highly prescriptive requirements for double bottoms presented in such a way that they are hardly related to the hazard of grounding. Chapter 5 of the NSCV standard applies the approach used by the current version of SOLAS Chapter II-1 that provides the choice of either a prescriptive requirement without calculations or a more performance-based requirement determined using an assumed extent of bottom damage. This allows much greater flexibility of solution allowing the use of a double bottom, watertight compartments or a combination of both. A progressive approach to application has been adopted depending on Flooding Risk Category similar to the approaches used in the HSC Code 2000[16].

The chapter also includes damage scenarios involving vulnerable appendages. Changes to the design of keels on large yachts including the use of extreme fin keels and canting keels have meant that previous assumptions as to the robustness of ballast keels may be invalid. Instead of providing a degree of protection they may be a point of vulnerability. The provisions in the NSCV standard raise the issue and require appropriate consideration to be given to the risks of flooding.

7.4 Collision reforms

Two types of collision damage are considered. The first is bow collision, while the other is side collision. Bow collision incorporates the prescriptive requirements for the collision bulkhead, but with modifications that introduce performance-based concepts allowing for greater flexibility in the location of the collision bulkhead. One modification that could potentially affect some larger displacement vessels and vessels that operate at higher speeds is the introduction of a minimum distance from the bow for the collision bulkhead based on an energy coefficient based on the collision acceleration calculations in the HSC Code 2000[17]. It is unreasonable to expect that the degree of protection offered by a bulkhead located 5% from the bow will be the same for a vessel that travels twice as fast or displaces twice as much. However, this change should be more than offset by the much greater freedom the provisions provide.

Recent amendments to SOLAS Chapter II-1[18] have profoundly changed the basis on which passenger vessels are assessed for subdivision and damaged stability. Probabilistic methods are employed overlaid over a framework of deterministic criteria. The question of whether this new methodology should be applied to domestic commercial vessels was raised for public comment and was considered by the reference group formed to consider the public comments and steer the course of the revision. It was decided that a mandated direct use of the probabilistic method would be inappropriate for two reasons. The first was that its application to vessels of the type typically considered for the NSCV might not be valid because of differences in size and configuration. The second was the difficulties that would be faced by attempting to introduce the methodology into the industry and the burden of higher cost of analysis that would have to be borne by the industry. It was felt that the benefits would not warrant the burden.

Notwithstanding that the probabilistic methods have not been applied, a number of changes to the deterministic aspects of the new IMO standard have been incorporated into the new NSCV standard. The SOLAS provisions specify that vessels that carry 400 or more passengers are required to meet two compartment standard of subdivision[19]. Similar requirements have been incorporated into the NSCV to ensure that the highest risk vessels reflect community expectations as expressed within those International Standards. Not all the changes are more onerous. The NSCV standard incorporates the SOLAS extent of damage provisions which in some aspects represent a lesser requirement. Similarly, for vessels carrying 12 to 36 passengers, no side collision criteria are specified, these being replaced by local flooding requirements. For vessels carrying between 36 and 400 passengers, a one compartment standard is specified in such a way that it prevents incorporation of numerous bulkheads that could jeopardise the effectiveness of the outcome.

7.5 Reforms to compliance criteria, more flexibility but with increased analysis

Chapter 8 of the new standard contains the criteria for compliance after application of the flooding scenarios specified in Chapters 4 to 7. The changes reflect the almost wholesale adoption of CAD and computer analysis for the design of vessel since the original USL Code. The main change has been the removal of the traditional reliance on non-immersion of the margin line, the removal of subdivision length analysis and the removal of permissible length calculations[20]. These have been replaced by more performance-based criteria based on SOLAS and the HSC Code 2000. The NSCV criteria now consider flooded freeboard, equilibrium angle, and for passenger vessels analysis of the flooded G_fZ curve looking at the range of stability and the height of maximum G_fZ. Two alternative sets of criteria are provided for the analysis of the flooded G_fZ curve, one from SOLAS and the other from HSC Code 2000.

7.6 Limits to the use of hinged watertight doors

A major point of contention in the USL Code was the use of watertight doors. The USL Code required watertight doors to be the minimum number practicable and prohibited hinged watertight doors entirely on passenger vessels 35 metres and more. The Code was ambivalent as to the application of hinged watertight doors on smaller vessels; leaving it largely to the discretion of the Authority. The result was that there was no consistent interpretation of requirements. Issues sometimes arose when a vessel transferred between jurisdictions.

The NSCV standard clearly specifies the extent and circumstances to which manual hinged watertight doors will be deemed-to-satisfy. Their use depends on the Flooding Risk Category, their size, the distance between the lowest sill of the door and the bulkhead deck and the number of such doors proposed.

Table 3 — Applicable Types of watertight doors

		Type 1 **Sliding door, operated by power and by manual gear**	**Type 2** **Sliding door, operated by manual gear only**	**Type 3** **Hinged door**
Additional limits on application		None	Door sill not more than 4 metres below the bulkhead or freeboard deck	Door sill not more than 2.5 metres below the bulkhead or freeboard deck Maximum width 800 mm
Flooding Risk Category	I	Required where Type 2 or Type 3 cannot apply	Optional for vessels having no more than four watertight doors per hull	Optional for vessels having one or two watertight doors per hull
Flooding Risk Category	II	Required where Type 2 or Type 3 cannot apply	Optional for vessels having no more than four watertight doors per hull	Optional for vessels having one or two watertight doors per hull
Flooding Risk Category	III	Required where Type 2 or Type 3 cannot apply	Optional for vessels having no more than three watertight doors per hull	Optional for vessels having a single watertight door per hull
Flooding Risk Category	IV	Required where Type 2 cannot apply	Optional for vessels having a single watertight door per hull	Not applicable

A pragmatic yet cautious approach has been applied based loosely on the provisions contained in the US standards[21], see Table 4. There have been a number of incidents where watertight doors were found to be wanting at the time of need, both within Australia[22] and overseas[23]. Quite apart from the perennial problem of their being left open, there have been cases where the watertight door has been found incapable of providing an effective seal. Watertight doors should be installed sparingly, and only when absolutely necessary. Table 4 specifies the deemed-to-satisfy solution and does not preclude the proposal of an equivalent solution provided equivalent safety can be proven.

7.7 Collared vessels are covered

One of the most significant changes in the design of vessels over the last 30 years has been the development and increasingly widespread use of vessels employing collars. Collared vessels are defined[24] as a vessel with a rigid hull provided with one or more buoyant collars around the periphery of the vessel such that—

a) the buoyant volume of the collars constitute greater than one-fifth of the total intact buoyant volume of the vessel, and

b) inclusion of the buoyancy of the collars is essential for the vessel to meet the applicable intact or damaged stability criteria.

The collar may be inflatable, foam filled or of rigid construction. Collared boats include RIBs (rigid inflatable boats). Such vessels were never envisaged in the USL Code and were not easily accommodated within the existing provisions. An interim set of guidelines was developed in 2005 that provided for separate standards applicable to inflatables and RIBs[25]. While the guidelines accommodated the majority of such vessels, they were not able to provide solutions for collared vessels that were not inflatable, nor large collared vessels of length greater than 12 metres. Another issue is the inconsistency in the safety outcomes provided by the standards for inflatable collared vessels compared to other vessels engaged in similar service. The challenge has been to consider collared vessels within the same overall framework provided by the graded approach contained within the NSCV, while at the same time acknowledging the strengths and addressing any particular weaknesses associated with the wide range of collared vessel configurations.

An analysis was made of the requirements specified by a number of relevant national and international standards to compare these against one another and also against the NSCV requirements for buoyancy and stability after flooding applicable to conventional (non-collared) vessels, see Table 4. This was used to help identify the differences between the conventional requirements and those for inflatable collared vessels. The standards imply that inflatable collars have advantages over rigid air chambers in that the status in regard to the effectiveness of the chamber is manifest and any time, while rigid air chambers may have a latent defect that is never identified until the critical moment in an emergency. Hence, the proposal allows for a reliability assumption that only one inflatable chamber need be considered deflated for swamping or local flooding, compared to the assumption that two rigid air chambers should be considered ineffective.

The standards indicate that inflatable chambers are considered more vulnerable to damage in certain situations such as bow collision, side collisions and raking damage. The draft standard provides for additional requirements applicable to inflatable chambers to provide equivalent safety.

The majority of provisions in the draft standard apply to collared vessels in just the same way as to other vessels. However, a number of specific provisions are also included that address the particular strengths and weaknesses of collared types. These are listed as follows:

7.3.7 Special provisions for bow collision on collared vessels with inflatable collars.	*10.6.2 Number and disposition of chambers*
7.4.5 Special provisions for side damage on collared vessels with inflatable collars.	*10.6.3 Collar to hull connection*
7.5 Measures to reduce the consequences of raking damage on collared vessels with an inflatable collar	*10.6.4 Fittings for inflation and deflation*
8.6.3 Criteria for basic flotation	*10.6.5 Additional tests for inflatable collars*
8.6.4 Alternative criteria for basic flotation	*10.6.6 Means to repair and reinflate collars*
8.7.4 Criteria for level flotation	*C4.5 Volume of additional buoyancy*
10.6 Inflatable collars	*E4.3 Weights to simulate swamped conditions*
10.6.1 Collar materials	*H3.6 Vessels with inflatable collars*

8. IMPACT OF THE NEW STANDARD

The review of the NSCV Buoyancy and Stability after Flooding Standard has been driven by a number of key constraints. These are:

a) Incorporation of a performance-based approach that focuses on safety outcomes rather than solutions

b) Consistency with relevant national and international standards; and in particular the changes to relevant IMO standards in SOLAS Chapter II-1 and SFV since 1979; and standards applicable to recreational craft in Australia.

c) Changes in technology in vessels and the methods of designing and analyzing vessels.

d) A desire for any changes not to burden the industry unless the benefits can be justified.

The new standard has been drafted to find a balance between these constraints. Some of the changes are more onerous, however others are less so. It is envisaged that the net effect should be relatively neutral in terms of overall cost to the industry. For some higher flooding risk vessels, the increased standards for two-compartment flooding that are based on the IMO requirements will undoubtedly add to the costs for individual vessels, but with a benefit to their safety. For the lowest level of flooding risk, certain recreational craft standards will suffice increasing the chance that a recreational craft can be bought off the shelf for commercial use.

The revised criteria for deemed-to-satisfy solutions are much more performance driven being limited by safety outcomes rather than limits of past available methods of analysis. It is envisaged that this should give designers more room for flexibility when considering deemed-to-satisfy solutions. As new technologies develop and new methods of analysis become available, there is the option to depart from the deemed-to-satisfy criteria by developing equivalent solutions that provide equivalent safety.

Part of the potential benefit of the new standard will take time to realize as designers become familiar with the content and learn to adapt their designs to take advantage of its enhanced features.

9. ACKNOWLEDGEMENTS

The author gratefully acknowledges the significant contributions made to the development of this standard by members of the Buoyancy and Stability after Flooding Reference Group and the many stakeholders from both industry and government who provided input through their public comment. A process that encourages the expression, consideration and balancing of a range of views as part of this drafting of this standard has done much to enrich the outcome.

Table 4 — Comparison of various standards for inflatable collared vessels

Scenario	New NSCV Conventional Vessels	NMSC RIB Guidelines[25]	ISO 6185.3[26]	ISO 6185.4[27]	MCA MGN 280[28]	SOLAS Chapter III Rescue boats	NZ Maritime Rules 40A and 40C[29]	NSCV inflatable Collared vessels
Swamping	Application: Vessels < 6 m FLRC II vessels 6m to 24 m. Excludes vessels with rapid drainage (i.e. most vessels) Recesses & drainage arrangments	Min freeboard of well deck with pumping criteria. USL Code Sec.10 Appendix N with 2 largest chambers excluded. Inherent buoyancy not less than 60% Appendix N	Drainage requirements Basic (?) flotation with largest chamber deflated	Hull drainage Bilge System if integral hull/deck. Drainage of totally flooded deck areas in less than 3 minutes	SOLAS Rescue boat if > 20 nm Swamped condition + drainage	Min 0.17 m3 inflated buoyancy per person. Inherently buoyant material of Basic flotation plus 280N of buoyant force per person.	Supports all persons when swamped, prove drainage effective.	Application as per Conventional using Level & Basic but with only largest collar air chamber excluded if inflatable, two if not.
Local flooding	Application: All vessels Largest 2 air chambers excluded.	USL Code Sec.10 Appendix N with 2 largest chambers excluded. Inherent buoyancy not less than 60% Appendix N	Min 2 compts in Rigid portion of RIB. Basic (?) flotation with largest chamber deflated	Min 2 compts in rigid portion of RIB. Min 5 chambers in collar. Residual inflated buoyancy at least 50% of manufacturers rated max load capacity after failure of largest chamber.	SOLAS Rescue boat if > 20 nm Swamped condition + drainage	Inherently buoyant material. Basic flotation plus 280N of buoyant force per person.	10 to 25 bhp requires 2 to 3 chambers. > 25 bhp requires 3 to 4 chambers	Application: As per conventional Level & Basic but with only largest collar air chamber excluded if inflatable, two if not.
Grounding	If > 50 m Vulnerable appendages if draft > 4 metres	No specific requirement	No specific requirement	No specific requirement	No specific requirement	No specific requirement	No specific requirement	If > 50 m Vulnerable appendages if draft > 4 metres
Bow collision	Application >16 m Collision bhd requirement Analysis in certain circumstances	No specific requirement	No specific requirement	No specific requirement	SOLAS Rescue boat if > 20 nm Bow chambers deflated	+ve freeboard over entire periphery with:bow chamber deflated	If <12m or <= 12 pax then +ve freeboard over entire periphery with bow chamber deflated	If > 16 m (where applicable) As per conventional plus collar in way of both sides of stem deflated. If < 16 m collar in way of both sides of stem deflated

Scenario	New NSCV Conventional Vessels	NMSC RIB Guidelines[25]	ISO 6185.3[26]	ISO 6185.4[27]	MCA MGN 280[28]	SOLAS Chapter III Rescue boats	NZ Maritime Rules 40A and 40C[29]	NSCV inflatable Collared vessels
Side collision	Application > 36 pax	Class 1 vessels only, as per conventional	No specific requirement	No specific requirement	Not applicable as <13 pax	Side impact while launching		As per conventional but assume 2 largest chambers in inflatable collar in way of damage deflated
Raking damage	No requirement	No specific requirement	No specific requirement	No specific requirement	SOLAS Rescue boat if > 20 nm	+ve freeboard over entire periphery with: 2. Entire buoyancy along one side deflated 3. Entire buoyancy along one side & bow chamber deflated.	Conventional damaged stability including allowable angle, min range, min GZ and min area under curve.	Inflatable collar along one side damaged. Level or basic flotation. No requirement for offset load in level flotation

9. NOTES AND REFERENCES

[1] Australian Transport Council. *Uniform Shipping Laws Code.* 1991. 1993. Australian Government Printing Service. Canberra. http://www.nmsc.gov.au/commercial_vessels/index.php?MID=92&COMID=1&CID=92
[2] Home page at: http://www.nmsc.gov.au
[3] National Marine Safety Committee. *Strategic Plan.* 2008-13. NMSC Inc. Sydney. page12 http://www.nmsc.gov.au/standards_publications/index.php?MID=32&COMID=1&CID=32
[4] Australian Transport Council. *National Standard for Commercial Vessels* 2005. 2006. 2007. 2008. National Marine Safety Committee. Sydney. http://www.nmsc.gov.au/index.php?MID=16&CID=97
[5] US National Archives and Records Administration. *US Code of Federal Regulations 46.* Washington 2009 http://www.access.gpo.gov/cgi-bin/cfrassemble.cgi?title=200246
[6] Flapan, Mori. *Register of Australian and New Zealand Ships and Boats.* Unpublished. www.boatregister.net
[7] USL Code Subsections 5C Construction - Watertight Subdivision Of Passenger Vessels and 5D Construction - Watertight Subdivision Of Class 2 And Class 3 Vessels
[8] International Maritime Organization. *The International Convention for the Safety of Life at Sea (SOLAS),* London. Chapter II-1 Part B.
[9] International Maritime Organization. *The International Convention for the Safety of Fishing Vessels 1977 (The Torremolinos Convention) as adopted and amended by the 1993 Torremolinos Protocol*
[10] Flapan, Mori. *Fishing Vessel Safety—A New Approach*. Paper presented at the Ausmarine East Conference. 2003 Cairns.
[11] Australian Transport Council. *National Standard for the Australian Builders Plate for Recreational Boats*. Sydney. 2005
[12] American Boat and Yacht Council. *H-8 Buoyancy in the event of swamping.* July 2004.
[13] Flapan, Mori. *Improvements and updates to Intact Stability Standards applied to domestic vessels in Australia.* Presented to Pacific 2008 Conference. Sydney.
[14] National Marine Safety Committee. *Final Regulatory Impact Statement for Buoyancy and Stability after Flooding.* Sydney. 2009. (Unpublished at the date of writing).
[15] To be contained in NSCV Part C Subsection 2. This Section of the NSCV is currently under development.
[16] International Maritime Organization. *HSC 2000 Code - International Code of Safety for High-Speed Craft, 2000.* London. Clause 2.6
[17] HSC Code 2000 Clause 4.3
[18] International Maritime Organization. *Resolution MSC.216(82) Adoption Of Amendments To The International Convention For The Safety Of Life At Sea, 1974, As Amended.* London. 2006.
[19] *MSC 216(82)* Regulation 8
[20] Criteria devised to provide practical solutions when using manual calculation methods in the days prior to the widespread use of computer numerical analysis.
[21] *US Code of Federal Regulations title 46.* 170.250
[22] Sedco Helen 31st January 1970. Offshore supply vessel holed and sank with loss of 9 lives. WT door left open.
[23] The Ro-Ro passenger ship Express Samina that sank in Greece 26 Sept 2000 with the loss of 80 lives. Nine of the total 11 watertight doors had been left open. www.naval.ntua.gr/~sdl/Publications/Papers/SAMINA-PAPER.pdf .
[24] *NSCV Part C Subsection 6A—Intact stability criteria.* Clause 1.7 Definition
[25] National Marine Safety Committee. *Guidance circular for the survey of inflatable boats and RIBs of measured length less than 12 metres.* 2005.
[26] International Standards Organization. *ISO 6185-3:2001 - Inflatable boats -- Part 3: Boats with a maximum motor power rating of 15 kW and greater.* 2001
[27] International Standards Organization. *ISO 6185-4 - Inflatable boats -- Part 4: Boats with an overall length of between 8 m and 24 m and with a motor maximum power rating of 75 kW and greater*
[28] UK Marine and Coastguard Agency. *MGN 280(M) Small Vessels in Commercial Use for Sport or Pleasure, Workboats and Pilot Boats – Alternative Construction Standards.*
[29] Maritime New Zealand. *Maritime Rules Part 40A Design, Construction and Equipment - Non-SOLAS Passenger Ships* and *Part 40C Design, Construction and Equipment - Non-SOLAS Non Passenger Ships.*

Maritime and Marine Emergency Response
Integrated Emergency Management

Presenting Author: Neil Bumpstead

(Operations Manager, Country Fire Authority, Victoria, Australia)

INTRODUCTION:

Prevention, Preparedness, Response and Recovery are the four key factors to the management of all emergencies. Minimal prevention and preparedness to address the emergency management to 'all hazards, all agencies' approach to Maritime and Marine environments is evident in most regions of the world. The critical factor identified through research is the lack of communication and interaction from all parties and therefore, the lack of Emergency management preparedness and planning for maritime and marine emergencies.

Specifically in Australia, the development by emergency services, statutory authorities and industry in response to maritime or marine emergencies has not been identified, developed There is no specific identified national plan or state plans to include all stakeholders. The responsible authorities and emergency services have minimal understanding of each other's legislative responsibilities or jurisdiction, There is minimal interaction between the emergency service agencies to glean knowledge of each others capacity or capabilities.

This paper is to introduce agencies and industry to the need to develop strategic alliances and approaches in emergency management across Federal and State jurisdictions for all hazards and in all waters and to strengthen response to maritime and marine emergencies. Lessons learned from approaches in both the United Kingdom and United States of America are drawn on as an example of how high level strategic approach can be achieved.

Australian states have varied jurisdiction and legislative responsibilities for the response of their emergency service agencies to incidents in the marine environment.

Current arrangements for emergency services and authorities to attend and combat an incident or emergency is varied and in some cases not possible as this is dependent on the level of capability or pre planning by the responsible stakeholders.

BACKGROUND:
There is an integrated approach for response that is well established on state and federal level in search and rescue, however, the same cannot be said for fire, hazardous materials incidents or other types of emergencies. Currently response to emergency incidents in the maritime and marine environment is sporadic and at different levels of implementation and capability across Australian states. Every Australian state or territory's Emergency Services are aware of their legislative responsibility for maritime and marine emergencies however the detail of strategic direction or development of integrated management, response and unified command approach to the risk are not being developed or practiced across all Australian jurisdictions.

There is minimal consistency in the jurisdiction and legislative responsibility for Australian state's or territories' emergency services. The services responsible for the response do not have well established multi agency policies or procedures for "All of Hazards" approach for maritime or marine emergencies, however there are some 'specific event' response plans at state and national level. Several states have infrastructure and equipment to perform a response where others have agreements with marine bodies or agencies.

Australia's emergency services responding to a Marine Emergency have differing responsibility of who is the combating authority. For example Victoria's fire services legislative responsibility for emergency response is for "all inland waters and up to three nautical miles to sea". Beyond Victorias state waters Australian Maritime Safety Authority (AMSA) co-ordinates the National Search & Rescue Response and National Maritime Emergency Response Arrangements. It is time to consider the current response plans and support the development of additional arrangements which detail other marine emergencies of, Fire, Specialist Rescue, Hazardous Materials Incidents or other incidents along the Australian coastline and on inland waterways.

Current management of a Search and Rescue incident is detailed in the National Search and Rescue Response as state resources in state waters and then Federal resources and processes in Commonwealth waters. The issues to be addressed for our Emergency services is how they deliver response and the management in a multi agency "All Hazards" approach to an incident other than, or which may include, Search and Rescue and to be cognisant of the organisations jurisdiction and legislative responsibilities to respond.

Australian Emergency Services and Industry involved in maritime and marine emergency response need to focus on all areas of risk in conjunction with all aspects of who has the legislative responsibility and jurisdiction as the combating authority.

CURRENT SITUATION:

Marine response is sporadic and at differing levels across all of Australia's state and territorial waters. Incidental and ad hoc approaches to maritime and marine emergency response by emergency services are common.

This raises the question of, what is the ability to respond to all hazardous incidents with an integrated approach, including unified command, skilled operators, sufficient suitable appliances/vessels and understanding to operate within legislative compliance?

The opportunity exists for future development of strategic direction across state and federal government to establish a position and direction for the management of overall response, training and equipment. In general terms the states emergency services have varied jurisdiction and legislative powers for emergency response to any maritime and marine emergency in their state or territory waterways. Many of the reported maritime or marine emergencies will involve a response from numerous services and authorities. This response should be for "all of hazards" and include response from the relevant agencies responsible for the incident through legal jurisdiction and with agencies having the capacity and capability to respond and perform their role at the incident.

This is not the case with different agencies providing the service but within the realm of their own organization and as independents. It must be noted this is not done in a negative manner towards the other agencies, rather, an immediate reaction/response to an emergency by agencies capable of, or responsible for action.

The key areas of high potential have been identified as:

- Security and hazards associated with identified ports
- Increased number of passenger and vehicle ferries
- Increase in the number of commercial vessels in all major and numerous smaller ports
- Increase in recreational vessels being registered and widely used around Australia. (In excess of 700 thousand recreational vessels registered in Australia)
- Hire vessels off shore and on inland lakes and rivers
- All other risks of access and development on waterfront
- "Other" water access incidents including Swift Water Rescue and all water/marine access areas where incidents may occur.

Interagency operations have lacked direction of interagency policy & procedures to ensure a seamless operation. The solution is for an, integrated agencies and industry, all hazard approach to all emergencies or incidents. The development of policies, procedures and memorandum of understanding documents will be required to cover all agencies and authorities however the achievement of a unified command structure and an integrated response must identify a "point of contact" for response to all incidents.

RESEARCH OF INTERNATIONAL DEVELOPMENT:

The United Kingdom's Maritime and Coastguard Agency (MCA) has a statutory duty for the initiation and co-ordination of civil maritime Search and Rescue within the United Kingdom Search and Rescue Region (UKSRR). The MCA is consequently responsible for the initiation and co-ordination of fire fighting and rescue at sea. MCA identified the gap of Fire and Hazardous materials response and entered the agreement with the UK Fire Brigades who are experts in this field. This was the catalyst for the Maritime Incident Response Group (MIRG) plan. The value of the partnership provides the 'All of Hazards' multi agency response. The MCA entered into partnership with 15 coastal Fire & Rescue Services around the UK. The MCA Maritime Incident Response Group (MIRG) was instigated to address the issue of further improving maritime resilience and delivering a more formal approach to dealing with incidents of fire, chemical release and industrial accidents on vessels/structures 'at sea'.

The plans detail response to all maritime hazards including coastal, inland and the jurisdictional boundaries of the UK Maritime & Coastguard Agency. The Maritime Coast Guard Agency (MCA) and UK Fire Services have an established Memorandum of Understanding on the arrangements for Fire fighting, Chemical Hazards and Rescue on Vessels at Sea. This has been the base document for the development of the "Sea of Change" project that has evolved in the UK. The Seas of Change project saw the development of a project team of experienced fire service personnel and MCA specialists to create the package of Memorandum of understanding, Policy and Procedures for response, Training and all associated directional information for the services to align their training and response procedures. The simplicity of this plan sees the MCA as the 'one number call" portal for notification and initial responder. Under their agreement the MCA evaluate the emergency, activate the appropriate response and follow the MIRG plan to draw their response team together and respond to the emergency.

The US Marine Terrorism Response plan was developed by a project team that was tasked with the development and validation of a multi-agency response system and the national model plan to aid the safe and effective mobilization of local, state and federal resources to marine terrorism incidents. This project team recognised four (4) critical objectives.

1. Develop a plan that will enable responders to minimize the impact of marine terrorism on life, property and the environment.
2. Ensure the plan provides for a "tiered" response capability
3. Model the tools needed to achieve an enhanced and sustainable maritime terrorism response capability that is also appropriate to other marine emergencies.
4. Engage local and national agencies and maritime community representatives in the plan development process.

The United States of America Marine Terrorism Response (MTR) project was initiated to provide U.S. port communities with tools needed to enhance their ability to minimise the loss of life and property and the disruption to trade caused by marine terrorist incidents.

Marine Emergency Response to all hazards is the responsibility of the relevant agencies and their expertise in the search and rescue field is world renowned. Both the MIRG and MTR models have been based on an "All of Hazards" application and encompass fire, hazardous materials, and incidents including terrorism threat, waterborne response and access to water for all emergency services personnel.

Research into the current active models in the United Kingdom and the United States has included the investigation by Northern Europe and specifically France and Netherlands who have agreed to adopt the UK MIRG model by their government and industry to their services. This will ensure a holistic approach for the key waterways of the North Sea, English Channel and significant ports vital to their development and sustained survivability.
Significant research of the UK and USA emergency services at all levels showed the support of their state and federal governments, Statutory Authorities, Emergency Services and key industry bodies in the maritime industry.

Research and understanding of international models have provided some definitive proactive thought for a proposal to the current and future needs of Australian waters. The lack of an State National or pacific region, All of Hazard, Marine Emergency Response structure and plan has driven the main themes for proposed development on this topic.

HUMAN FACTORS AND INTERACTION/RELATIONSHIPS:

Many personnel from varying UK and US services state the relationships they have developed across the services, with government departments and the maritime industry have forged the working relationships at the management level. The feeling is that this has been enhanced markedly with the continual exercises and review of their planning process and networking. However it is fair to say that certain elements of some areas held a differing opinion of most regarding the value and worthiness of their developed models and plans. This is seen as a personality trait and lack of understanding of the strategic direction of the relevant services.

The UK and USA have developed and established Marine Response models and although very different they offered the opportunity to research their interagency procedures, operation and interoperability in an All Hazard approach to review the risk and consider dedicated response from Integrated Management to treat the risks identified.

OPPORTUNITIES/FINDINGS:

Australian emergency services, authorities and industry have a need to support the development of a model for maritime and marine response. Progress and development will only be limited by the opportunity to engage and evolve with all statutory authorities and industry stake holders. The benefits will be substantial for all involved with the interaction and sharing in all facets. The development of an All of Hazards approach and integrated agency response direction would be driven by the following points.

- Development and enhancement of partnerships with all state and federal service providers for integrated emergency response in the maritime field
- Provide strategic direction in an All Hazards approach in Marine Emergency Response
- Emergency Services, Provide input and direction for the development of Marine Emergency Response Policies and Procedures
- Development and sharing of Training information and techniques
- Provide direction for Equipment and responder resources

Relevant research and understanding of the inception of the international models and the opportunity to cross pollinate their development and knowledge into a local concept would be strongly supported by our international colleagues. Acceptance of and value adding to the UK (MIRG) and US (MTR) models will provide a solid reference base for the development of an Australian body to co-ordinate an inter agency all of hazards maritime and marine emergency management procedures and response by all stakeholders.

The future direction must provide a directional framework for all of hazards co-ordinated and collaborative approach to PPRR (Prevention, Preparedness, Response & Recovery). This must include the development of the basic foundation to interagency response and should endorse the universal Command & Control structure (ICS).

- Include programmed exercises, with valued debrief actions on outcomes.
- All agencies, authorities and industry to share experiences in generic emergency management terms.
- Understand the differences in emergency service, statutory authorities and industry structure and management.
- Provide the opportunity to interact with and discuss all levels of Marine Emergency Response and the development of Training regimes and Planning tools with wide ranging effect on fire and emergency service departments.

The international services identified have developed plans for maritime and marine response and a very high commitment level of interagency involvement required to function capably to perform their role.

Internationally developed models have been given a clear direction from their government and senior management for the development of a national model or plan to be developed.

Planning for the event and the outcomes of operational response are the driver for the maintenance of the models and there is significant direction to elaborate the supporting of the Australian known PPRR process which was utilised but not well documented in the UK model. The development of the US model is well detailed into four stages as Plan, Response, Field Ops (Field Operations), and Technology

RESEARCH OUTCOMES:

A proposal for the development of a plan for maritime and marine emergency management and response strategy for Australia to work in conjunction with current maritime emergency management agency's at state and national levels (AMSA) to accomplish the following critical objectives:

- Buy in at Federal and State government levels for the development of a strategic approach in both legislative requirements and proposal development.
- Development of a model/plan that will enable the management of responders to minimise the impact of maritime and marine incidents on life, property and the environment
- Engage local, national & international agencies and maritime representatives in the development of the plan
- Ensure the development of a "tiered" response capability for emergency agencies/authorities throughout the plan
- Model the tools needed to achieve an enhanced and sustainable marine emergency response capability that is applicable to all maritime emergencies
- Develop an endorsed national recognized training base for all operational response to cover all fields of maritime and marine emergency management and response
- Access the support of international and national emergency services and the maritime industry in maritime and marine emergency management and response
- Communicate with Industry and other services on their findings from research and development in training, procedures and equipment.

PROGRESS / DEVELOPMENT:

- Continual review of current and future best practice models of service delivery by authorities and organisations associated with the maritime and marine industry
- An established network of fire services and other Emergency services and agencies to interact and share maritime and marine emergency management and response skills and learning's based on similar experiences across many industries such as 'lessons learned centre's'.
- Consultation with emergency response authorities to program and develop a state/national maritime and marine emergency management framework

- This detail of an agreed plan for multi agency response is the key focus for the proposal to develop an Australian model that will reflect an integrated agencies all of hazards approach.
- The development of a co-operative agreement with national and international fire and rescue organisations to share skill, knowledge and research in maritime and marine emergency management and response.

Disclaimer: The opinions and recommendations of this paper are those of the author, not the position of the Country Fire Authority, Victoria.

Bibliography:

Australian Maritime Safety Authority, National Maritime Emergency Response Arrangements (NMERA), August 2006. (www.amsa.gov.au)

Australian Maritime Safety Authority, National Search and Rescue Response Arrangements. June 2004. (www.amsa.gov.au)

Australian Maritime Safety Authority, National Plan to Combat Pollution of the Sea by Oil and Other Noxious Hazardous Substances. May 2001. (www.amsa.gov.au)

Her Majesty's Fire Service Inspectorate, Maritime Resilience and the Role of the UK Coastal FRS's Circular, July 2006

Kettle, M. 2003. 'Ship Fires at Sea – The Fire Service Response". The 1st "National Fires On Ships" Conference – March 2003.

Marine Terrorism Response Plan (MTR) – National Model Edition (NME), December 2005 (www.mtr.org)

Marine Terrorism Response Plan (MTR) Project. MTR Plan Development Process April 2006. (www.mtr.org)

Maritime and Coastguard Agency, March 2007. Report of the "Sea Of Change" Fire fighting and Rescue at Sea Project 2003 – 2007. (www.mca.gov.uk)

Maritime and Coastguard Agency and the UK Fire and Rescue Service, Memorandum of Understanding Between The Maritime and Coastguard Agency and The Fire Service on the Arrangements For Fire fighting, Chemical Hazards, and Rescue on Vessels at Sea., (www.mcga.gov.**uk**)

MCA UK Fire and Rescue Services Maritime Incident Response Group (MIRG), SAR Operating Procedures, October 2005. (www.mcga.gov.**uk**)

MCA UK Fire and Rescue Services Maritime Incident Response Group (MIRG), Provision of Marine Training For Coastal Fire and Rescue Services. October 2005. (www.mcga.gov.**uk**)

Executive Summary:

Neil Bumpstead

Operations Manager: Country Fire Authority

Ph: **0417 334 318**

Email: **n.bumpstead@cfa.vic.gov.au**

Fixed Water Mist System Implementation within the RAN

I Raymond[1], B McNeice, and J Colquhoun
Directorate of Navy Platform Systems – RAN,
[1]Ian.Raymond1@defence.gov.au

Due to changes in international and national regulations and the naval practice of compartment re-entry to recover fire damaged spaces, the Royal Australian Navy (RAN) has investigated requirements for the implementation of water mist systems onboard RAN platforms, primarily to replace halogenated fire fighting agents. Several platforms within the RAN have been fitted with IMO compliant water mist systems in conjunction with existing gaseous fire fighting systems. These IMO based water mist systems have required additional redundancy and functionality to be added to overcome known and observed limitations of these systems when applied in the naval context.

The RAN has developed system requirements for the implementation of stand-alone water mist systems throughout a navy platform. The selection of the water mist nozzle can be achieved three ways; through an Empirical Method, Semi-Empirical Method, or Standard Test Method. The RAN has requirements for water mist system redundancy that exceeds the IMO requirements. These additional requirements address the risk posed to system availability from combat damage.

System testing of water mist systems on RAN platforms requires a full discharge test for the first-of-class platform, as well as after any significant system or compartment change. A controlled discharge test is required for commissioning of all platforms as well as regular functional testing of the system throughout the life of the platform.

The RAN is continuing to support the development of system requirements and testing regimes for the installation of water mist systems onboard current and future RAN platforms.

INTRODUCTION

Montreal protocol, UNEP (1999), and concerns with synthetic fire fighting agents has led the RAN away from halogenated fire fighting agents. In looking for a replacement the RAN was seeking a fire fighting agent which would align with navy fire fighting and recoverability practices and requirements.

Water mist is attractive to the RAN because it:

- Improves evacuation and rescue capability;
- Allows earlier safe activation of the fixed fire fighting system;
- Can be used to assist and protect fire fighting parties;
- Can control risk from fuel sprays;
- Facilitates thermal management of the compartment;
- Does not rely on the compartment being air tight for it to function;
- Presents no known environmental or personnel hazards; and
- Is rechargeable whilst at sea.

In the RAN water mist can be implemented in any of three configurations. These are:

- Combined Water Mist System – An IMO compliant or equivalent commercial based water mist system which is supplemented by a separate gaseous or foam system which meets RAN requirements;
- Hybrid Water Mist System – A water mist system that is conjoined with a gaseous fire fighting system, using the one distribution and nozzle system and satisfying all the RAN performance requirements. In such a system the straight water mist could be used for evacuation, fire control, and fire extinguishment of hydrocarbon fires, whilst the gaseous component extinguishes other classes of fires; and
- Water Mist System – A RAN approved stand-alone water mist system that can control, suppress and extinguish designated fires.

DIFFERENCE BETWEEN IMO AND RAN WATER MIST PERFORMANCE REQUIREMENTS

Like commercial marine operations, the RAN is responsible for the safety of its crews. Water mist systems have a clear benefit over other fixed fire fighting system agents as they can be used to assist evacuation and rescue of crew from within a compartment containing a fire. Additionally, the RAN expects to be able to restrict the spread of damage within a platform during a fire. This is especially important when conducting operations. Water mist is expected to restrict fire spread to adjacent compartments and have limited impact on components within the compartment post activation of the fixed fire fighting system.

The difference between IMO and RAN water mist systems result from the need to cover navy specific compartments that the IMO requirements do not cover in IMO (2007) and the performance and system requirements for navy machinery space arrangements. Navy specific compartments are discussed later. It should be noted water based systems for accommodation spaces, covered in IMO (2007), generally align with RAN requirements.

Navy machinery spaces are generally more compact and congested than large merchant vessels. This leads to vertical separation between the top of high fire risk components and the water mist nozzle being limited to somewhere in the order of a meter. This creates a disconnect between the IMO water-base testing standard, MSC Circ 1165, IMO (2007), which is based on much larger separations (cathedral type machinery spaces) thus allowing the water mist time to spread providing full coverage of the machinery.

In other words the application of IMO water mist system into naval machinery spaces includes technical risk with regards to the ability of the system to be able to perform its expected function. The technical risk is associated with MSC Circ 1165 permitting systems that have large nozzle separation and small nozzle angle (see Figure 1). In the test rig the separation between the top of the engine mock and the nozzles is such that the spray from adjacent nozzles will merge to form a continuous layer of mist that then propagates down. When these commercial water mist systems are fitted into a naval machinery space in accordance with IMO requirements it is possible to have areas of no mist coverage, such as on top of diesel engines.

Further to the system layout issues discussed above, the RAN has a number of constraints that differ from IMO requirements.

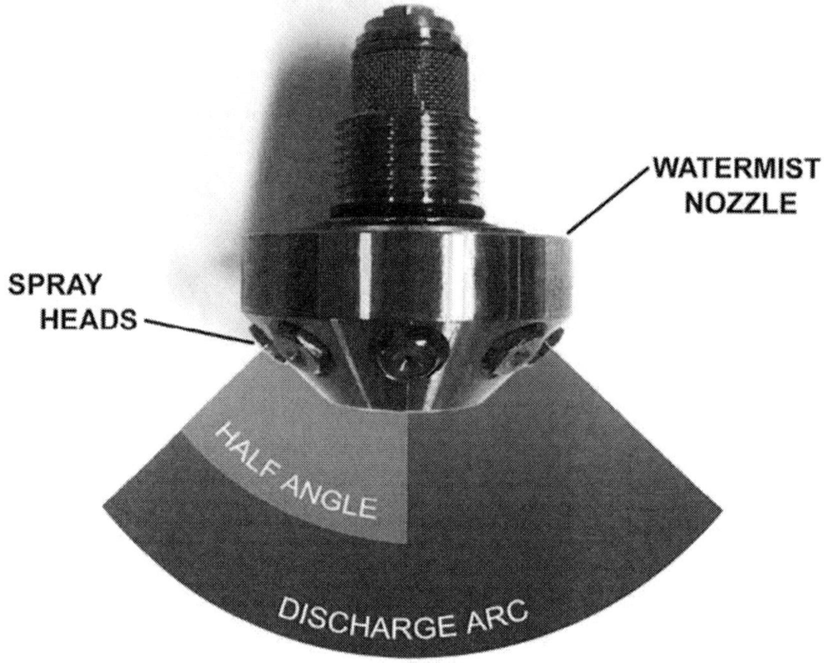

Figure 1 – Water Mist Nozzle

IMO sets a thermal management acceptance criterion of 100°C, whilst the RAN seeks to limit this to 80°C, as recommended by Williams *et. al.* (1999). This reduction in the maximum acceptable temperature is to:

- Reduce the likelihood of thermal damage to crew and equipment within the compartment; and
- Inhibit the formation of steam throughout the compartment away from the primary fire.

Under MSC 1165 the water mist system has 15 mins to extinguish the test scenario fires, whilst Williams *et. al.* (1999) recommend the maximum extinguishing time be 10 mins. This reduction in time period is to:

- Reduce the likelihood of injury to crew;
- Reduce the likelihood of additional damage to platform systems; and
- Improve the recoverability of the platform.

An outcome of the above is that water mist systems that meet these requirements have a greater water mist flux, 0.4 to 0.8 lpm/m^3, compared to IMO commercial water mist systems, flux is generally between 0.15 to 0.3 lpm/m^3.

The MSC 1165 permits door nozzles to be used in meeting the requirements of the tests. Although door nozzles may have a benefit in assisting fire fighting parties to access the compartment, these should not be used to meet the requirements of a total-flooding system.

Recently IMO has released MSC Circ 1272, water-based extinguishing fitted systems for Ro-Ro (Vehicle) deck. This standard is not supported for application in the RAN as the:

- Deck height is limited to 5m;
- Fuel load is Class-A instead of Class-B (n-heptane); and

- Obstructions used in the test are not of a level comparable to navy vehicle deck operations.

The level of assurance that the water mist system is available in all conditions of the platform, including loss of main and auxiliary power and certain combat damage, means that navy water mist systems require significantly higher survivability requirements when compared to IMO systems.

RAN STAND-ALONE WATER MIST REQUIREMENTS

The following overview of the RAN water mist requirements is based on DEF(AUST)5000 Vol7, Pt2 (2010).

Nozzle Approval

The outcome of nozzle approval is the nozzle type, the nozzle operating parameters and grid layout of the water mist system to meet the RAN acceptance criteria. There are three nozzle approval methods, these are:

- Empirical Method which requires the testing of the system in a mock-up of the space identical to the one for which it is intended to be installed and is not seen as being practical for the RAN in developing water mist systems for specific navy applications as every class of navy vessel has a different compartment layout and so a mock-up would be required for each class at a significant cost;
- Standard Test Method which is based on approved standards that are recognised and applied throughout the industry for approving water mist systems for a specific application or compartment configuration. This method is not seen as practical for all navy applications as there are currently no standard test methods available for some navy compartment types; and
- Semi-Empirical Method which is used where no standard test exists and is based on a navy specific test standard being developed. An example of a semi-empirical method is water mist nozzle approval for naval machinery spaces covered in Williams *et. al.* (1999).

The applicable standards for the Standard Test Methods are:

- Machinery Spaces

  There are no recognized standard test methods applicable to all machinery spaces for navy platforms. Therefore, if water mist is to be applied in support of a Fire Safety Study then a semi-empirical test would be required to prove the nozzle and nozzle parameters.

  For larger RAN platforms which have large 'cathedral' type machinery spaces the standard test for water mist nozzle approval in MSC Circ 1165 can be used.

  Where MSC Circ 1165 is permitted door nozzles can not be used to meet the MSC Circ 1165 requirements.

  Water mist shall not be used within bilge areas.

  High fire risk equipments, such as engines, are to be fitted within enclosures or protected by an additional appropriate fitted fire fighting system.

- Equipment Enclosures

  The standard test for water nozzle approval for use within generator enclosures is FM5560 (2005) Appendices B, E, or G, depending on size of the enclosure.

  The standard test for water nozzle approval for use within turbine enclosures is FM5560 (2005) Appendices C, F, or H, depending on size of the enclosure.

  When selecting a water mist systems to protect turbine enclosures, the thermal shock to the turbine needs to be considered/assessed.

- Accommodation, Office, and Workshops

  Water mist systems used to protect accommodation, office, and workshops with low to moderate fire risks must have passed either FM5560 (2005) Appendix I or IMO Resolution A800.

  For vessels with a platform shock requirement, individual thermally activated nozzles will require the removal of the glass bulb or fusible link if these components of the nozzle are not shock rated. In this situation the nozzle shall be treated as dry pipe nozzles.

- Flammable Liquid Storage Locker, Paint Lockers, and High Fire Risk Compartments

  There are no recognised commercial test methods for approving water mist systems for protecting Flammable Liquid Storage Lockers, Paint Lockers and High Fire Risk compartments. The RAN allows the use of water mist for this application provided the nozzles used in this system have passed the USN machinery space tests as described in DEF(AUST)5000 Vol 7 Pt 2 Annex D.

- Electrical and Electronic Compartments

  The water mist systems used for protecting compartments that house Electrical and or Electronic equipment shall be "equivalent to sprinklers". The water mist system must have passed either FM5560 (2005) Appendix I or the appropriate test in IMO Resolution A800. The appropriate IMO test shall be identified in the Fire Safety Study based on the fuel loading and configuration in the space.

  For vessels with a platform shock requirement, individual thermally activated nozzles will require the removal of the glass bulb or fusible link if these components of the nozzle are not shock rated. In this situation the nozzle shall be treated as dry pipe nozzles.

- Vehicle Storage Areas and Transit Paths

  There are no recognised standard test methods applicable to Vehicle Storage Areas and Transit Paths for navy platforms. IMO MSC/Circ. 1272 may be applicable depending on the conditions in the space, i.e. if the compartment and the fuel load align with this test standard. Therefore, if water mist is to be applied, in support of a Fire Safety Study a semi-empirical test would be required to prove the nozzle and nozzle parameters.

- Explosive Ordnance Stowage Compartments

  There are no recognized standard test methods applicable to Explosive Ordnance Stowage Compartments for navy platforms. Therefore, if water mist is to be applied, in support of a Fire Safety Study a semi-empirical test would be required to prove the nozzle and nozzle parameters.

- Access Cooling
  An access cooling system is designed to assist with the re-entry into a compartment containing a fire. This system shall consist of a limited number of (one or more) overhead nozzles installed just inside the access into the area. The nozzles used in this system must have passed the USN machinery space tests described in DEF(AUST)5000 Vol 7 Pt 2 Annex D.

- Galley
  Water mist systems used to protect galley cooking equipment must have passed ISO 15371.

Nozzle Layout

Water mist nozzles shall be laid out within a protected compartment such that the nozzles meet the nozzle parameters approved within the nozzle approval test. The nozzle layout (grid) shall be as uniform as practical, with a specific focus on high fire risk components. In designing the water mist grid, it shall be shown that adjacent spray patterns will merge before interference from high fire risk components.

Protected compartments that traverse multiple decks shall be configured such that the nozzle grid for each deck is fed by an individual section valve.

High fire risk compartments, such as machinery spaces, shall have two feeds from diagonally opposite ends of the compartment. Water mist shall not be fitted to bilges, unless it is a foam water mist / sprinkler system that has demonstrated performance for the role. Turbine enclosure shall have overhead nozzles as well as centerline bulkhead nozzles.

Water mist nozzle layout within flammable liquid storage lockers and paint lockers shall be installed in the overhead of each isle/passageway and on the bulkhead directed towards each isle/passageway.

Access cooling water mist systems shall have at least one nozzle within one meter of each door / hatch.

Water mist nozzle layout within galleys shall cover deep fat fryers, other fire risk cooking equipment, and at least 3 meters of the exhaust trunk beyond the galley.

Distribution System

For RAN platforms that have no combat role the water mist distribution system needs to only meet a commercial standard, such as that set out by the IMO. This type of system would consist of a single water mist pumping station, a single master control system, multiple water mist activation points, and a centerline distribution main or a single manifold.

The distribution system complexity and survivability increases as the combat capability of the platform increases. Therefore, for warships the water mist distribution system is expected to be either a vertically offset distribution main or a composite main. In either case the ability to isolate the main at each fire boundary is required, as well as cross connection in each fire zone.

Water Mist Pumping System

There shall be sufficient water mist pumping capacity to meet the most demanding load from a grouping of water mist pumps at one or more water mist pumping station/s. The most demanding load is determined from the potential fire scenarios and water mist system response to them. The water mist system response may include water mist hose reels as well as the fixed system. Following NFPA750 (2006), the most demanding load is increased by 10% to account for possible system changes and performance deterioration through the life of the system. As required by IMO (2007), an additional pump is to be included to account for the possibility of pump failure.

On non-combatant platforms there needs to be only a single water mist pumping station. This pumping station is to be in accordance with the requirements of IMO (2007).

For combatant platforms there needs to be multiple water mist pumping stations feeding into the dispersed distribution main such that the loss of 25% of the platform by length means there is sufficient pumping capacity such that the most demanding load can still be achieved. This can be achieved by either having two identical pumping stations at opposite ends of the platform, or distributed pumping stations spread throughout the platform.

Control Systems

Water mist systems may be either automatically or manually activated. Automatic activation may be achieved by using a separate detection system or individually thermally activated nozzles depending on the application. When a separate detection system is used, activation is to occur after two detectors have detected the fire. Automatically activated systems that use a separate detection system shall also be equipped with manual activation points. Manual activation points for water mist fire protection systems shall be included local to the protected compartment(s), local to each pump station, and in the MCR, DCC and ADCC or equivalent spaces.

The water mist fire protection system shall be equipped with an audible and visual alarm that must be able to be seen and heard at any point within the compartment noting the likely clutter and ambient noise expected in the compartment. Audible alarms shall be such that they can be heard when the occupant is wearing standard hearing protection equipment and the machinery combination producing the greatest noise is running. The water mist fire protection system shall be equipped with a visual alarm (flashing light) located at the access to each protected space to indicate that the system is activated. The water mist system shall be instrumented to determine if the system is functioning properly during discharge.

Each water mist storage tank shall be provided with local and remote tank level indicators. Remote indication shall be provided at DCC and ADCC at a minimum. Compressed gas cylinders where used in the water mist system shall be provided with local and remote level indicators.

For pump driven systems, alarms shall be provided to indicate the changeover from the potable water source to a seawater source. The first of these alarms shall activate five minutes prior to changeover and the second two and a half minutes prior to changeover. The two alarm format allows the crew to stop the seawater changeover and gives an indication of the time remaining prior to changeover.

The control system components shall not be fitted in high risk spaces and shall be accessible to the crew without the need of special equipment. The control system components on combatants shall be duplicated to eliminate any single-point of failure.

Survivability Requirements

For RAN combatants the water mist system shall not have any single points of failure. The design study needs to show that the water mist will be able to perform as required for all design incidents.

To assist in this the following requirements are specified:

- IP rating – Water mist system components are to have an IP rating of 56 if situated below the V-line or on the weather deck, otherwise they shall be rated to 44;
- Backup power – If the main and auxiliary power configuration on a combatant platform can not be shown to be reliable such that it can power the water mist system in all designed incidences then an additional diesel generator (air cooled) will need to be fitted away from the current generators to overcome this survivability limitation of the system. If a compressed gas system is to be used to power the water mist system, then the system is to have sufficient gas such that it can discharge water mist for the designed discharge period (i.e. 10 to 15 mins), the nozzles are to be approved for this type of discharge (i.e. gas powered), and uninterrupted power supply (UPS) is to be available for the water mist control systems;
- Casualty power - Emergency/casualty power supply arrangements is to maintain electrical power to critical equipment, including the water mist system; and
- Jumper - The water mist distribution mains shall have at least one jumper hydrant per fire zone that would permit water flow to be diverted through hoses to the next available jumper hydrant so damage sections can be by-passed. To support this capability sufficient water mist jumper hoses are is to be carried so that jumper hydrants separated by two fire zones can be connected.

SYSTEM ACCEPTANCE TESTING

System testing of fitted water mist systems is, at this stage, the only method to show that any fitted system will meet the RAN requirements as well as meet the system design requirements. This is partly due to the absence of any viable numerical modeling package for fire dynamics and water mist systems. Therefore the system acceptance testing needs to be thorough as it is critical for the RAN to have confidence in the systems capability.

First of Class and Commissioning Trials

For 'First of Class' the water mist testing is structured to fully test the system. For 'Commissioning' the water mist testing is structured to confirm that the correct components are fitted and that the system parameters can be meet. Both trials include the following:

- The water mist distribution system shall be pressure tested to one and a half (1.5) times the operational pressure prior to the installation of the water mist nozzles;
- Post-installation acceptance of all water mist fire protection systems and components shall consist of both visual inspections and functional testing;

- All water mist systems shall be visually inspected to ensure that the installed system meets all of the agreed parameters and that the nozzle types and locations meet the tested design requirements;
- The water mist systems shall be discharge tested to ensure that all system components operate properly;
- All other water mist protected compartments shall be discharged tested (socks or tubes to direct water flow from each nozzle to a bucket are acceptable); and
- During these discharge tests, all of the methods of activating the water mist fire protection systems shall be tested.
- During the discharge testing, the flow rate and operating pressure shall be confirmed to fall within the system design parameters. Any instruments used to confirm these parameters must be calibrated by a nationally recognized test facility (NATA or equivalent).; and
- The ingress protection (IP) of machinery and electrical systems within compartment(s) protected by water mist fire protection systems shall be confirmed.

In addition the 'First of Class' trials include:

- Discharge tests shall be conducted in all uniquely configured protected compartments to verify complete spray/mist coverage of the space;

Through Life

Regular testing of the water mist fire protection system shall be undertaken in accordance with the OEM requirements, relevant class society and the following RAN requirements:

- The pump maintenance shall follow the manufacturer's recommendations and additionally shall meet the following requirements:
    - The water mist pumps shall be started weekly from each control location;
    - The filters on the suction inlets to the water mist pumps shall be checked semi-annually and replaced annually;
    - The suction supply lines from the potable water reservoir and the pump location shall be designed for flushing or circulating flow;
    - The back-flow prevention device separating the pump suction manifold from the platform's potable water storage tanks shall be checked periodically and serviced annually;
    - The emergency (stand-by) power generation capacity shall be checked and a pump operating test utilizing the full power draw shall be carried out annually; and
    - The opening and closing of electrically operated sectional control valves shall be verified annually.
- The maintenance and replacement of nozzles shall follow the manufacturer's recommendations and at least meet the following requirements:
    - Nozzles shall be inspected and cleaned every six months;
    - Nozzles shall be replaced if damaged; and
    - Nozzles shall be replaced after being used in a fire.
- At least one water mist protected compartment discharge shall occur annually on all platforms. This discharge may be done with socks to minimise turn-around time and water damage, as the test is to confirm system pressure and flow rate. Within a five year time period all water mist protected compartments are to have had a test discharge.

FUTURE WORK

There is ongoing work in developing requirements and guidance for developing and updating the semi-empirical method for nozzle approval. Especially, for the approval of water mist nozzles in navy specific compartments.

Recently FM Global has released FM5580 (2009), standard for hybrid water mist / nitrogen systems. This offers a way forward in applying water mist systems for lifesaving and fire suppression, whilst the nitrogen system provides extinguishment. Developing such a system is likely to be easier than providing the required nozzle approval.

The ability to analysis / numerically model water mist systems and fire dynamics currently is very limited. Current capacity does not permit the design validation of a water mist system design. When this capability is improved the level of testing, nozzle approval and system acceptance, could be reduced.

CONCLUSION

Water mist system requirements for the RAN now exist and can be used to apply water mist systems as a stand-alone fixed fire fighting system. These requirements offer the ability to have a fixed water mist fire fighting system that has equivalent performance to Halon and similar gaseous / clean agents. In fact due to how water as a fire fighting agent works, it offers improved fire safety to the crew with the potential for faster recoverability operations. The disadvantage of water mist systems is nozzle approval and system design that currently leads to significant 'First of Class' testing.

REFERENCE

1. UNEP, 1999, "The Montreal Protocol on Substances That Deplete the Ozone Layer", Beijing, United Nations Environment Programme
2. IMO, 2007, "FSS Code: International Code for Fire Safety Systems", International Maritime Organisation, London, England
3. Williams, F.W., Back, G.G., DiNenno, P.J., Darwin, R.L., Hill, S.A., Havlovick, B.J., Toomey, T.A., Farley, J.P., and Hill, J.M., 1999, "Full-scale Machinery Space Water Mist Tests: Final Design Validation," NRL/MR/6180--99-8380, Naval Research Laboratory
4. Commonwealth of Australia, 2010 "RAN Fire Safety Systems", DEF(AUST)5000 Vol 7 Pt 2 Annex D
5. FM5560, 2005, "Approval Standard for Water Mist Systems", FM Approvals
6. NFPA750, 2006, "Standard on Water Mist Fire Protection System"
7. FM5580, 2009, "Approval Standard for Hybrid (Water and Inert Gas) Fire Extinguishing System", FM Approvals

Enhancing shipboard crew safety-related behaviour: Is Crew Resource Management training the answer?

Michelle Grech[1], Susan Burdekin[2], Jim Taylor[1], Derek Abraham-James[3]

[1]Maritime Platforms Division, Defence Science and Technology Organisation
[2]Australian Defence Force Academy (ADFA)
[3]Director Navy Safety Systems, Royal Australian Navy

ABSTRACT

One of the threats to safety on ships includes crew work behaviour and performance as influenced by their shipboard and environmental conditions and by the organisational context and culture in which they work. It is well known that accidents and incidents involve some form of human failure that incorporates errors, violations or flawed decisions. In most cases, these human errors occur despite operator technical competence. One way to potentially reduce and manage these human errors and threats effectively is to introduce error management intervention in the form of training. Within this context, Crew Resource Management (CRM) training is being increasingly seen as a fundamental part of the human error management philosophy. Although, Royal Australian Navy (RAN) crews are provided with highly effective technical and procedural training, there has been comparatively little focus on the non-technical skills that enable ship's crews to effectively manage shipboard hazards and errors in such a dynamic and demanding operational environment. CRM training which is a specialised form of applied human factors training is designed to provide crew with the awareness, knowledge, skills and attitudes to manage themselves and available resources more safely and effectively. It is increasingly seen as a capability enhancer in the improvement of performance and safety.

This paper will outline a project which is currently looking at evaluating the effectiveness of the implementation of CRM training programs for RAN crews (submarine and surface ships). The evaluation will also examine whether the effectiveness of CRM practices is contingent upon organisational safety climate.

INTRODUCTION

Background

The maritime domain's assertion that the majority of mishaps are due to *human error* (in some form) is a consistent and stable trend throughout all other safety critical industries [1]. Many researchers today view human error as emanating from *a family of problems of behaviour* which originate from various sources such as poor management, inadequate supervision, lack of communication as well as slips and lapses [2]. In the last decade, accidents, incidents and near-misses have increasingly highlighted the fact that the most competent of operators and systems are fallible. A review of maritime and Royal Australian Navy (RAN) incident and accident statistics indicated that human factors contributed in some way to the majority of all incidents and accidents. Common issues identified include such factors as failure to develop or maintain adequate situational awareness, decision errors, procedural failures, not applying training, design flaws or the application of equipment not fit for service. One issue that appears to prevail across incident reports is failure to adhere to standard operating procedures.

CRM: A Historical Perspective

In an effort to arrest these worrying trends the Directorate of Navy Safety Systems (DNAVSAFE) is considering the implementation of safety interventions that other high risk industries have employed. One initiative under consideration is CRM which is a behavioural safety program that has been developed as a training strategy to reduce the likelihood of human error, and to minimize its impact upon operators and operations if it does occur. CRM was introduced in the 1970's within the aviation industry when it was recognised that many aircraft incidents were due to human error, rather than technical failures. Examples of these errors included lack of team coordination, communication breakdown, leadership or decision making errors [3]. The maritime domain absorbed the CRM philosophy after recognizing the need for non-technical or resource management skills following a number of accidents in which lack of team communication and co-ordination played a major role. Between 1995 and 2008 alone, the Maritime Accident Investigation Branch in the UK reported five similar accidents in which lack of teamwork and communication were highlighted as major contributors. One such accident example is of the passenger ship *Pride of Canterbury* which grounded on a charted wreck in January 2008 while sheltering off the port of Dover in bad weather conditions [4]. At the time of the accident the vessel had 275 passengers and 101 crew on board. Fortunately, no injuries were reported, however the financial costs involved were huge with the ship sustaining major damage, including the total loss of her port controllable pitch propeller hub and a section of its propeller shaft [4]. The report cites that while the vessel was waiting for the port of Dover to reopen a breakdown in the bridge team management occurred, with the following failures identified:

- The crew did not develop a passage plan;
- Hazards were not identified and marked on the chart;
- Positions were only sporadically plotted and the paper chart was not consulted at the crucial time, even though the paper chart was the primary source of navigation;
- The officer of the watch was changed on an ad-hoc basis, and the handovers were not structured with the result that important information might have been lost;
- Navigation was conducted visually and by reference to an electronic navigational chart display (ENC). None of the bridge team had been trained in the use of ENC, and the settings were inappropriate such that key dangers would not have been displayed;
- Throughout the waiting period, there were a number of telephone calls to the bridge, principally regarding matters on the vehicle decks. These constant interruptions distracted the bridge team from their primary function of navigation and lookout.

This accident and others highlight the potential risks involved when breakdown in communications, team work and decision making occurs. In this regards the CRM philosophy is designed to address these issues.

SAFETY CULTURE AND CRM

Organisational Culture

Organisational culture and practices have been found to influence individual attitudes and behaviours towards work. Organisations develop norms, values and beliefs which are reflected in the strategies and attitudes of management towards concepts like for example, open communication, teamwork and training. Some accidents can be traced back to organisational policies and decisions, which can lead to human error types such as lack of situational awareness, high fatigue and unacceptable workload levels with serious potential implications [2]. The maritime domain has only very recently started addressing the influence of organisational culture on crew behaviour. One such initiative in this area is the Man/Ship Interface System (MASIS) project [5] initiated by the European Commission. This project aimed to assess practical tools and procedures capable of reducing the impact of human error in shipping. One of the main contributing factors identified as affecting human behaviour onboard ships is the *man/organisation interaction system*. Even with the most appropriate regulations, best designed equipment and the most qualified and trained personnel, no system can be better or safer than the system or organisation that binds them. Individual behaviour is therefore influenced by the organisation and one means of inducing optimum behaviour and performance is to develop a good safety culture.

Safety Culture

There has been a growing interest in recent years within the maritime domain in the concept of safety culture. Recent developments in the manning supply-demand dynamics in the maritime industry have resulted in increasing interest in and acknowledgement of the need to address human factor issues, not only at an individual level, but also at the team and organizational levels that influence behaviour. A submission by the UK to the International Maritime Organization (IMO) Maritime Safety Committee in 2003 defines safety culture as '*a culture in which there is considerable informed endeavour to reduce risks to the individual, ships and the marine environment to a level that is as low as is reasonably practicable*' [6]. The term '*safety culture*' has now been discussed in detail in a number of maritime casualty reports, namely the *Herald of Free Enterprise* casualty report [7]. The roll on/roll off passenger ship capsized in 1987, shortly after leaving Zeebrugge Harbour. 150 passengers and 38 crew members lost their lives with many others injured as a result of this accident. This report indicated that crew behaviour had a detrimental effect on the final outcome. Successive accident enquiries into this casualty indicated that crew behaviour that led to negative outcomes came about from a poor safety culture within the organisation resulting from, amongst other things; poor communication, unclear job responsibilities and management failures [8].

Some industries have developed widely varying techniques for addressing the uncertainties arising from human involvement, with varying degrees of success. The aviation domain for example has been particularly successful in this regard with its adoption of a non-punitive approach and also ensuring that safety data on errors is collected in order to identify where improvements need to be made. Within this context the maritime domain is following suit and has made considerable progress in the area of safety and error management. A number of strategies being increasingly adopted within the maritime domain include the introduction of formalised Safety Management Systems and CRM training.

Safety Management Systems: Progress within the Maritime Domain and the RAN

Safety culture is, of course, closely linked to the philosophy underlying Safety Management Systems. In July 1998, the maritime domain adopted a Safety Management System more commonly referred to as the International Safety Management (ISM) Code. This was made mandatory by incorporating it as a new chapter into the IMO International Convention for the Safety of Life at Sea (SOLAS) 1974. The ISM Code is widely regarded as one of the most important measures adopted by IMO, the question of whether or not it is working as intended is, however, still debatable. Nonetheless, there is some evidence to indicate that the ISM Code does work and the potential rewards may be worthwhile.

Within the RAN, DNAVSAFE developed Safety Management System - Navy (SMS-N) tailored specifically for the Navy. In addition, a number of initiatives were undertaken to evaluate the extent of safety culture within the organisation. In 2003, the RAN commissioned a baseline safety culture study and as a result of this, a number of safety programs were implemented. In 2006, a culture recalibration was conducted by the same company using a similar methodology. Although, the results demonstrated that safety culture had improved, there were a number of issues raised [9]. These included:

- The perception that a blame culture still exists within Navy;
- A lack of commitment to safety from the top;
- Poor safety related communications;
- The need to review safety training, timing and content; and
- The application of risk assessment requirements was confusing.

In 2004, the Defence Occupational Health and Safety Committee (DOHSC) endorsed a study of Defence-wide OHS behaviours. The results specifically relating to Navy found that, although OHS performance was generally good, OHS behavioural drivers such as attitudes, beliefs and behaviours were significantly lower than the average Defence-wide results. The authors of the report concluded that 'Navy personnel do not feel valued; do think frequently of leaving the service, and do not have very high levels of trust and confidence in their superiors and service leaders' [10, p.6]. The comments related to service leaders and superiors appear to be common to the Zeal (2008) report results concerning safety culture. In full knowledge of these findings DNAVSAFE is in the process of introducing a number of initiatives. The Navy Safety Culture Development Project is one such initiative, and it is suggested that CRM training can contribute to this project by tailoring its learning outcomes towards the very foundations on which to build a better safety culture. This has also been reflected in the recent Navy Strategic Plan for 2009-10 where it is stated that *'New Generation Navy is focused on ensuring the future health and enrichment of Navy through cultural, leadership and structural reforms that will permit Navy to deliver on future capability and embed signature management behaviours that reflect a focus on people, performance and professionalism. Only by doing this will Navy be able to meet its mission of fighting and winning in the maritime environment*" [14].

CRM TRAINING

Objective of CRM Training

The aim of CRM is to achieve more efficient use of the available resources, and to improve teamwork which often results in increased morale, safety and performance. CRM has evolved into a program of training based on a philosophy of threat and error management, where hazards to the operation are identified, the threats that they pose are managed, and error is avoided, trapped or mitigated. CRM training helps crewmembers gain greater influence over the human and situational factors that contribute to critical incidents. This can help optimise the benefits of multi-person crews to significantly improve safety. The aviation domain is now in their 5th generation program of CRM training. They have developed sound and tested theories, methods, principles, approaches, and content available to shape the design and delivery of CRM training [11]. The concepts of CRM have also become more diversified. For example, the notion of *resource* in aviation has gone past its initial reductionist meaning and encompasses all resources i.e. not only crew members themselves, but also the aircraft (and its functions), procedures, Air Traffic Control (ATC), ground staff and other aviation stakeholders. In the aviation domain *resource management* has also spread its wings to incorporate skills aspects, which go beyond the collective and interactive dimensions of the activity, and includes amongst others all individual cognitive aspects [12]. Thus CRM training deals more and more systematically – at the individual as well as the collective level – with topics such as: understanding and awareness of the situation, decision making, error management and confidence. Although, as with civil aviation CRM training has now become well-established in the maritime curricula, there remains a question mark about how effective such training is in improving safety and operational performance.

Evaluation of CRM Training

A number of research initiatives have focused on developing effective evaluation tools for the assessment of CRM training. An important outcome of CRM training evaluation is to ensure that the goals and the objective of the training are appropriate and are achieving the desired results. In addition it is important to ensure that the content and method of training used are compatible with the training goals and inform future improvements of such training. Formal evaluation of the effectiveness of these training workshops have generally been positive. In this regard there are a number of tools available that offer useful ways to capture, assess and diagnose CRM-related processes and outcomes. Most of these tools have been developed with the aviation domain in mind, hence some work is required to customize these for use within the maritime context. One tool worth mentioning here is the Kirkpatrick approach to training evaluation which has been used by researchers conducting meta-analyses across a range of industries [13]. The advantage of adopting this approach to evaluate the proposed CRM effectiveness is that the results can be compared to the results found in the previous CRM studies. This would give some measure of the impact that the CRM trial achieved within the maritime/Navy as compared to that of other industries.

Does CRM training work?

There are a number of studies that have evaluated CRM training. One of the most notable is a review carried out by Salas et al in 2001 and 2006 [15, 16]which looked at a large number of published articles (125) in order to evaluate the effectiveness of CRM training using the Kirkpatrick approach. The results of the first review [15] indicated that overall CRM training had a positive impact on attitudes in most domains surveyed (these were mainly commercial and military aviation). However, some variability in results came about with respect to the actual learning of trained competencies with some inconsistencies evident, with the result that its impact on safety could not be determined. The outcome from this first review suggested that collecting data at multiple levels may lead to easier interpretation and conclusiveness of findings. As a follow on study Salas et al., [16] conducted an extended evaluation following the same methodology as the previous study. This time, the review incorporated other domains of interest, such as the naval and maritime domain. An extract of this review containing the maritime references is included in Table 1. Although the results highlighted in Table 1 do contain positive findings, the results of the extended review in general indicated that further work needs to be done to improve the CRM training evaluation process.

Table 1: Maritime examples extract from summary of CRM evaluation efforts where training was provided from [15].

Source	Community	CRM Training Content	Type of Study/Data Collection	Findings
Burn et al., 2000	Royal Norwegian Naval Academy	Coping in critical situations Communications and decision making Teamwork	Experimental	Reactions: In general, reactions to training were positive Learning: No change in shared mental model scores. No difference in shared mental model scores between trained and untrained groups.
Byrdorf, 1998, as cited in O'Connor et al., 2003	Danish shipping/maritime (AP Moeller and MAERSK Compny)	Resource Management Assertiveness Communication Teamwork Stress Coping	Quasi-experimental	Organisational impact/results: Reduction in nautical and machinery casualties since training inception 4 years earlier. Insurance premiums decreased by 15%
Fonne & Fredriksen, 1995	Norwegian maritime	No specifications of skills taught	Quasi-experimental Self-report survey	Learning: Attitude changes for trainees overall very positive (e.g. more likely to inform other crew members of personal stress, more likely to reduce speed in adverse weather conditions). Attitude changes were stable or improved over time.

CRM PILOT STUDY WITHIN THE RAN

The Defence Science and Technology Organisation (DSTO) together with the RAN are currently evaluating whether CRM has an application in Navy. Submarine operations were selected as a pilot group for this study, mainly because submarine operations are conducted by a discrete group of professionals who work in an unforgiving, isolated, hostile environment where there is little margin for error. Earlier this year, the 2009 Defence White Paper was released [17]. One of the capability priorities announced for Force 2030 is the acquisition of twelve new future submarines. Acknowledging that the present submarine capability is degraded, the Australian Government has undertaken to 'respond decisively to deficiencies in the current availability of operational ready submarines by embarking on a major reform program to improve the availability of the Collins class fleet' [17, p71]. It was stated that 'these reforms will change how we attract, remunerate, train and manage the submarine workforce' [17, p71]. Targeted incentives for submariners including training were nominated as one reform. The Government's desire to identify training as an incentive provides Navy with an opportunity that should be embraced, and improved training is also a major component of New Generation Navy as indicated in the Navy Strategic Plan 2009-10 [14].

Recent public reports within Navy and specifically within the submarine workforce have highlighted cultural, OH&S and workplace issues. By focussing on the submarine environment, this review has identified that many characteristics of the operations in a submarine are similar to those of other specialisations where CRM has been successfully introduced. As an example, operations in the control room are similar to those in an air traffic control radar environment, and operations on the bridge and engine room are comparable with those involved in complex operations and other safety critical activities.

Unlike other high-risk industries however, the submarine is unique in some of its characteristics, particularly those associated with social settings on board. Therefore, based on the evidence presented in this report, it is highly likely that CRM can be successfully tailored to the submarine environment.

STUDY: CRM WITHIN NAVY

During a brief visit to HMAS Stirling in Western Australia in February 2009, a number of interviews were held with a representative sample of crew members from HMAS Waller. Participants included Watch Leaders, engineering and technical officers, and non commissioned crew. A series of questions were posed which aimed to gain an insight into present working conditions and issues. Furthermore, the interviewees were invited to put forward suggestions on how they would see the problems currently confronting them resolved. The crew members were then given an outline of the type of content that might be included in a CRM course tailored to submarine operations, and asked for their views on how such a course might benefit them.

RESULTS

Issues

Issues arising from the interviews included many items that have subsequently been reported in the public domain with efforts made to address them announced by Chief of Navy (CN). Issues such as, lack of internet access on boats and insufficient shore leave when alongside were clearly articulated. Whilst the senior Watch Leaders believed they already receive adequate training, all other crew members interviewed expressed great interest in the concept of a CRM course. During the research conducted for this report it became apparent that the results from this small sample in fact reflected the results from other navy cultural surveys.

Adapting CRM Training to Submarine Operations

In order to determine whether CRM might be able to be adapted to the submarine environment it was necessary to review the present submarine training syllabus. The submarine safety training documents reviewed revealed no evidence of CRM training content duplication. The submarine safety training syllabus content is focused on teaching essential duties in order to perform specific tasks, similar to vocational training. It contains instruction for the awareness of and compliance with procedures, rules and regulations, documentation and OH&S requirements. This content and training approach is quite different to the CRM content and approach.

Based on the mapping of CRM learning outcomes against the submarine safety training syllabus it appears that there is little duplication of the type of content and delivery method that a CRM course of training would provide. Consequently, a logical next step would be to trial a CRM course to a sample of participants from the submarine workforce. In order for CRM to be fully effective, it needs to be adapted to the specific working environment and carefully tailored to the target population [18]. Additionally, prior to the development of a trial course, it is suggested that baseline data be collected. This data will not only provide a measurement by which any change after the training intervention can be determined, but could also provide valuable information on what the content of the course should cover.

CONCLUSION

The successful outcome of CRM training within the aviation domain has led other industries, including the maritime domain to follow suit. The CRM training philosophy is anchored in threat and error management (TEM) where hazards are identified by those within the workplace and the subsequent risk that they pose is analysed, treated and managed by using a common risk management framework. CRM evaluation studies have shown large positive effects on the attitudes and behaviours of participants which contributes to a better safety culture. Furthermore, a business case which was prepared for the rail industry illustrated that the adoption of CRM training could save the industry up to US$33m per annum in reduced accident costs. As CRM has been successfully adapted to high risk industries, such as nuclear power, rail, medicine, and aviation, there is some considerable empirical support for the proposal that CRM could also be adapted to operations within the RAN. It is hence argued that CRM training in combination with the other planned safety initiatives will certainly be a positive intervention that would improve the submarine workforce situation. One element of CRM is 'workload management' which targets the development of strategies to create a more effective and efficient workplace. This element also focuses on how to achieve a better work-lifestyle balance. It should be noted here that Navy has already identified that CRM type training can influence the attitudes of individuals and teams. Attitude change is an essential step towards behavioural change and improved operational performance.

REFERENCES

1. Grech, M. R., Horberry, T. and Koester, T. (2008) *Human factors in the maritime domain*. Boca Raton, FL Taylor & Francis.
2. Reason, J. (1997) *Managing the risks of organisational accidents*. New York, Ashgate Publications Ltd.
3. Helmreich, R. L., et al. (1998) Culture, Error, and Crew Resource Management. In: Salas, E., Bowers, C. A.Edenn, E. (eds.) *Applying resource management in organizations: A guide for professionals*. Hillsdale, NJ, Erlbaum
4. Maritime Accident Investigation Branch (2009) *Report on the investigation into the grounding of Pride of Canterbury "The Downs" - off Deal, Kent, 31 January 2008*. 2/2009, Southampton,
5. European Commission (1998) *Man/Ship Interface System II (MASIS)*. [Final Report] Brussels, European Commission
6. International Maritime Organization (2003) *77th session of the Maritime Safety Committee, Agenda item 17*, London, International Maritime Organization,
7. Department of Transport, U. (1987) *MV Herald of Free Enterprize*. 8074, [Formal Investigation] Southampton,
8. Sheen, H. L. J. (1987) *Mv Herald of Free Enterprise*. Report of Court No.8074 Formal Investigation, [Her Majesty's Stationery] London,
9. Zeal (2008) *Safety Culture Recalibration*. . Canberra, Royal Australian Navy
10. Ellwood, M. and Lockyer, D. (2004) *Behavioural Baseline Research: Navy special report*. . Canberra, Defence OHS Program Office. Australian Government Department of Defence
11. Helmreich, R. L., Merritt, A. C. and Wilhelm, J. A. (1999) The evolution of Crew Resource Management training in commercial aviation. *International Journal of Aviation Psychology* 1 (9) 19-32

12. Helmreich, R. L. and Merritt, A. C. (2004) *Culture at work in aviation and medicine.* Hants, England, Ashgate
13. Kirkpatrick, D. L. and Kirkpatrick, J. D. (2006) *Evaluating training programs: the four levels* 3ed. San Francisco., Berrett-Koehler Publishers Inc.
14. Royal Australian Navy (2009) *Navy Strategic Plan 2009-2010*, Canberra
15. Salas, E., et al. (2001) Team Training in the skies: Does Crew Resource Management (CRM) training work? *Human Factors* 43 641-674
16. Salas, E., Wilson, K. A. and Burke, C. S. (2006) Does Crew Resource Management Training Work? An update, an extention, and some critical needs. *Human Factors* 48 (2) 392-412
17. Commonwealth of Australia (2009) *Defending Australia in the Asia Pacific Century: Force 2030* [Defence White Paper] Canberra,
18. Burdekin, S.G., (2000) *Crew Resource Management: Its role in the Australian Defence Force.* Directorate of Flying Safety – Focus 1/00. Canberra.

CAPABILITY ANALYSIS OF MODULAR MULTI-ROLE WARSHIPS FOR AUSTRALIA

Jordan Glanville
BMT Design & Technology, Melbourne Australia
JGlanville@bmtdt.com.au

ABSTRACT

A significant amount of research has been undertaken in the field of modularisation of naval vessel capabilities, which potentially offers significant procurement and operational cost benefits to owners and operators as well as increased fleet flexibility. The Australian Government is considering proposals for a single class of modular multi-role offshore combatants to conduct the tasks currently undertaken by four existing role specific fleets; patrol boats, mine countermeasure vessels (MCM), hydrographic survey and oceanographic environmental assessment vessels.

This paper considers the benefits and technical risks/issues associated with the application of modular technology in the design of a multi role platform and the degree to which the level of effectiveness of the individual role specific capabilities may be compromised. In order to inform the acquisition process and determine the cost effectiveness of the proposed approach, savings in procurement and through life cost must be compared with the capability implications and the Navy's associated role specific measures of effectiveness.

Building on the work undertaken for the *Venator* concept design, BMT analysed the differing requirements of the proposed capabilities, with specific reference to Australian operating conditions and existing fleet capabilities, in order to assess the suitability of modular technology to the intended roles. The key design characteristics of the platforms are considered along with the integration requirements of modern modular technologies that are deemed viable for use in the future combatant platform. An overview of the resulting cost capability analysis tool is provided, demonstrating how the benefits of modular multi-role platforms may be modelled to create cost benefit based comparisons.

INTRODUCTION

As modular/autonomous naval technology advances and fiscal factors become increasingly pertinent, Navies are looking for methods to consolidate fleets comprising several classes of vessels into fewer classes offering a wider range of capabilities. Modular capabilities have been utilised by various Navies around the world for several years as a method of achieving this, particularly within the European Union. The adoption of modular technology and design methodology has the proven ability to provide significant procurement and operational cost savings as well as increased fleet flexibility, if the implications are understood and it is applied correctly.

In a bid to reduce procurement and operational expenditure and increase fleet flexibility, the Royal Australian Navy (RAN) is considering proposals for a single class of modular multi-role offshore combatants, herein referred to as 'Future Offshore Combatant' or 'FOC', to conduct the tasks currently undertaken by role specific fleets. The widely varied capabilities proposed to be provided by the FOC require that a thorough analysis of both operational and cost implications is undertaken with specific reference to Australian operating conditions in order to verify the operational and financial implications. The purpose of this paper is to firstly outline the design factors that are critical to an Australian based modular fleet, then present an outline of a cost benefit and capability analysis completed to determine the probable financial and operational outcomes of the adoption of modular technology for an Australian based FOC fleet.

MODULAR TECHNOLOGY

Modular capabilities, also known as 'Mission Modularity', refer to the discretisation and modularisation of capability specific equipment so that they may be rapidly and easily exchanged in order to change the overall capability of a vessel. Modular capabilities typically include: weapons systems such as guns and missiles; auxiliary equipment such as cranes or loading equipment; or unmanned vehicles for mine countermeasure or anti-submarine warfare. There are several different forms of implementation of modular capability, which may be summarised as follows [1]:

Type I	Type II	Type III
Modular containers or other modular installations (modular plug and play space with minimal installation time).	Modular installations (plug and play, but with significant installation time).	Modular space potential (reconfigurable space available for capability specific equipment).

Existing modular vessel concepts use the three forms of modularity in varying degrees. The logic behind the choice for application of the different forms is best described by investigating their application in existing designs or concepts, as outlined below.

EXISTING VESSELS

Stanflex Concept (Type 1)

The Stanflex concept was developed in the early 1980's to replace an ageing Danish fleet of 22 small surface warships, comprising torpedo boats, patrol and minesweeper vessels. Budget constraints precluded the one for one replacement of the role specific fleets, therefore technological advances and novel design concepts were applied in order to produce the 'Standard Flex' concept. The concept relies on the modularization of capability specific equipment into standardised container interfaces that can be rapidly changed out from a common fixed platform. The fixed platform comprises; the hull, propulsion system, sensors common to all roles, and the command and the control and communication system (C^3 system). The result was a fleet of 14 vessels; the 'Stanflex 300' series or Flyvefisken class, which offer capabilities equivalent to or greater than the previous role specific fleets.

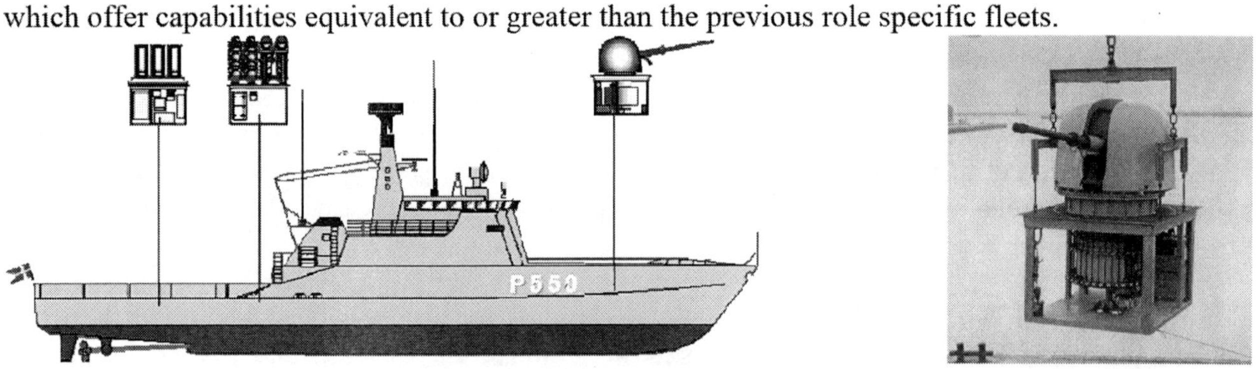

Figure 1 – STANFLEX Concept of the RDN

In their current form, the Flyvefisken Class vessels cover eight roles: Surveillance, Combat (ASuW), Mine laying, Anti Submarine Warfare (ASW), MCM, Hydrographic survey, Oceanographic survey and Environmental protection. The Stanflex concept has been adapted for use within other classes of the Royal Danish Navy (RDN) and it is envisaged that by 2013, almost all vessels within the Royal Dutch Navy (RDN) fleet will utilize modular capabilities. By utilizing the Stanflex concept across almost the entire fleet, greatly increased flexibility and redundancy is obtained through the interchangeability of modules between vessels.

BMT's Correspondence with the RDN has highlighted that the key areas in which advantages have been realised, as compared to the previous role specific fleets, are:

- the ability to offer increased operational flexibility for a small navy;
- allows the outfit of a vessel to be tailored to a specific operation;
- significantly increased fleet-wide redundancy, resulting in cheaper and easier maintenance and overhaul with less off-task time; and
- simpler and cheaper design and construction of the next generation of vessels due to ongoing development and utilization of similar technology.

The most significant risk highlighted by the RDN regarding the use of modular fleets is that of crew training and manning levels not being sufficient to enable full utilisation of fleet flexibility. This is an area which requires in depth analysis for an Australian based fleet.

MEKO (Type 1 and 2)

The MEKO concept of German shipyard Blohm & Voss (B&V) is undoubtedly the most widely employed example of modularity in ship design and construction. The concept relies on the strict modularisation of certain ship systems, with all components needed to run the system accommodated in a single module, with modules connected to power supply, HVAC, and the data network via standard interfaces. Over 64 MEKO vessels have been delivered to navies worldwide.

Figure 2 - B&V MEKO CONCEPT

The initial development of the MEKO concept was in an effort by B&V to utilize a standard ship platform and offer variable and customisable levels of outfit for different customers. The use of standardised and modularised systems for items such as weapons and machinery allows for significant savings in production costs by allowing system manufacture and testing to be completed on shore. In contrast, the Stanflex concept utilises modularity in an effort to provide an increased level of fleet capability with a reduced number of ships and weapons. Financial savings are predominantly gained through reduced through life cost due to the use of a common platform across multiple capabilities.

MOPCO (Type 2)

The Modular Platform Concept (MOPCO) concept of German shipyard Abeking & Rasmussen draws on years of experience in the development of minehunters, namely the Fast Minesweepers Class 343 and the Minehunters Class 332. The concept combines elements of the MEKO and Stanflex concepts, however on a larger scale of implementation. Common systems are combined into large ′Task Modules′ which are resiliently/shock mounted onto a common SWATH platform. The concept relies on a platform designed with MCM as the prime task. It is therefore optimised for properties such as shock resistance, low signature etc. Each task module contains all equipment required for the intended capability or task, such as MCM, accommodation or ship control. A conceptual design of a 45 metre vessel is shown below, comprising a low-signature and shock resistant platform built from Non-Magnetisable Steel and three, shock-acoustic mounted task modules. Role reassignment in this case is achieved by complete replacement of the MCM module.

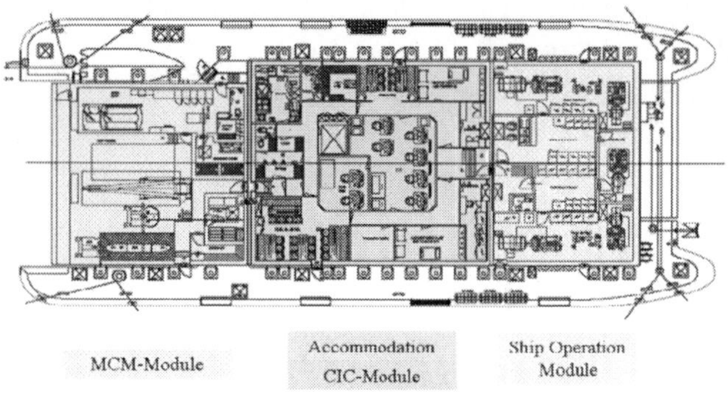

Figure 3 - MOPCO Concept - Abeking & Rasmussen

The MOPCO concept has been referenced as it shows the extent to which modularity may be implemented, i.e. basic platform with three large almost independent modules.

BMT Venator concept (Type 2 and 3)

The Venator Concept of the UK based BMT Defence Services is the result of an extensive research program into the requirements of a globally deployable platform and implementing best practice modular technology and design philosophy [3]. The concept is for a small surface combatant capable of being configured to perform the roles of MCM, maritime security, hydrographic and environmental assessment and patrol. The concept development focused on the optimisation of small hullforms for global deployment and maximising commonality within a relatively small hull envelope.

Figure 4 - BMT Venator conducting MCM with ASV - © BMT Defence Services 2008 credit DefenceImaging.com

In contrast to the Stanflex 300 series, the Venator concept relies heavily on the use of reconfigurable internal mission spaces and modularised off board vehicles for the conduct of varying activities, rather than fast interchangeable standardised units. The concept requires a relatively large reconfigurable ′garage′ area, incorporating a telescoping hangar arrangement with overhead gantry system and permanent stern ramp for off board vehicle launching. For tasks such as MCM, it is intended that the main vessel is stationed far behind the area of operations being swept for mines by off board UAVs, hence negating the requirement for shock resistance and low signature. This significantly affects the design of the platform, allowing optimisation from a structural and performance point of view rather than having to design for issues such as signature and shock resistance.

Summary - Modular Technology Advantages/Disadvantages

A number of benefits and disadvantages of the application of modular technology to ship fleets can be extracted from the aforementioned concepts and experiences.

Benefits	Disadvantages	Risks
-Possible fleet cost saving during both acquisition and through life due to single common platform; -High degree of commonality for stores and maintenance; -Increased fleet flexibility; -Ability to store major components/weapons shoreside during peacetime; -Reduced off-task time and redundancy; and -Ability to develop and apply modular technology fleet wide.	-Generally results in larger platform than a role specific vessel, with associated secondary effects, i.e. less efficient than a role specific vessel; -Functionality trade-offs are invariably required for certain capabilities as compared to role specific fleets; and -Complication of command and control as well as auxiliary systems leading to increased weight, complexity and initial cost.	-Potential inability to provide sufficient numbers of trained crew to allow for full utilisation of modular capabilities; and -Higher level of risk of in service failure due to increased system complexity and use of emerging technologies.

Whilst it is envisaged that the FOC may utilise all three types of modularity, albeit in different quantities, it is both beyond the scope of this investigation and too early a stage to perform a detailed analysis on the likely configuration of the FOC. However it is clear that detailed operational and requirements analysis is required in order to identify the ideal application of different forms of modularity.

DESIGN DRIVER ANALYSIS

It is necessary at this stage of investigation to identify the pertinent design drivers for a modular warship operating within the context of an Australian navy.

Existing Fleet

The specific roles and principal particulars of the existing RAN fleets considered in the FOC are outlined in Table 1, along with the key design drivers relating to their operation.

Class	Armidale Class Patrol	Huon Class Minehunter	Leeuwin Class Ocean Hydrographic Survey	Paluma Class Coastal Hydrographic Survey
Number in Class	14	6	2	4
Primary Roles	Border protection Vessel interception apprehension Surveillance	Mine countermeasure	Hydrographic survey	Littoral/coastal /shallow water hydrographic survey
Design Drivers	Range Pursuit/cruise speed Seakeeping	Magnetic and Acoustic signature control	Range Volume for equipment	Shallow draft
Length / Displacement	56.8m / 270 tons	52.5m / 720 tons	71.2m / 2,550 tons	36.6m / 320 tons
Speed (Cruise/Max)	12 / 25 knots	12 / 14 knots	9 / 14 knots	12 knots
Range	3,000 nm @12 knots	1,500 nm at 12 knots	18,000 nm at 9 knots	3,500 nm

Table 1 - Current RAN Fleet

It is clear that each role has very specific and independent functions with little overlap in capability. Each capability requires different levels of performance, therefore making the design of a common platform a complex and multi-variable problem.

Design Drivers for FOC

Almost all design drivers stem from the primary definition of the capabilities the system must deliver. It is therefore imperative that the implications of defining the required capabilities for a modular fleet are understood implicitly, as discussed further in 'Capability Analysis'. The relative level of importance of major design variables can be mapped to the objectives of the FOC as shown in Figure 5. The objective of 'minor warfare missions' has been included as it has been prescribed as an important requirement for the FOC and presents significantly increased operational requirements over basic patrol capabilities, such as increased armament, range and operability in higher sea states.

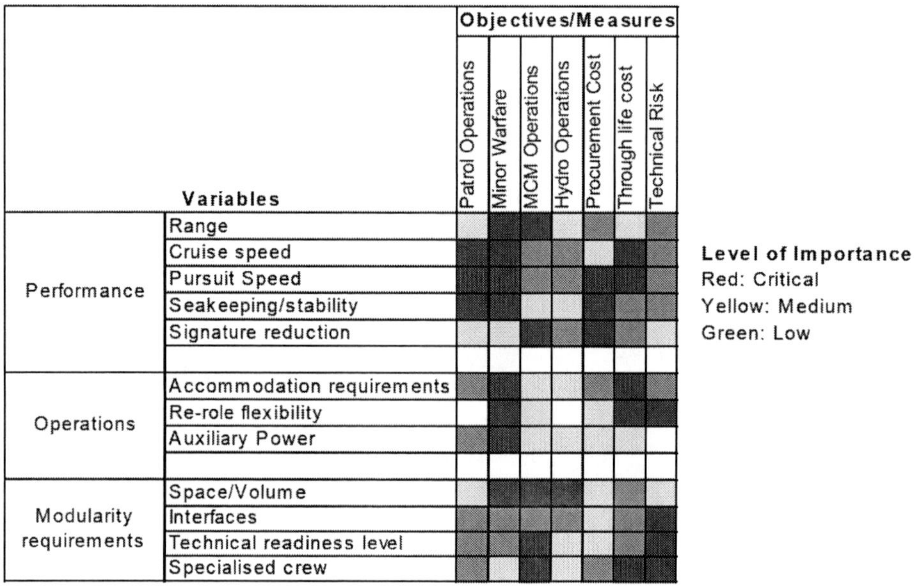

Figure 5 - FOC Design Drivers

It is clear that the objectives of minor warfare missions and patrol operations have the greatest effect on performance requirements of the platform, whilst MCM and hydrographic objectives represent the greatest area of technical risk along with level of re-role flexibility. The identification of these drivers is critical to the development of the cost model described below.

COST/CAPABILITY ANALYSIS

Introduction

In order to analyse both the financial and operational gains that are potentially offered by the adoption of modular technology, a two staged approach that concurrently analyses both the procurement/lifecycle costs of the vessel fleet and the operational effectiveness of the candidate solution is required. Analysis of the financial and operational effectiveness of the current RAN fleets is a relatively simple task, requiring consultation with fleet operators and analysis of historical operational data. However, the analysis of the operational effectiveness of a conceptual future fleet is much more difficult, with the complexity of such an analysis significantly increased through the introduction of modular capabilities.

For example in the context of the proposed Australian fleet, a platform with a maximum speed of 25 knots is required for patrol operations. However, this is somewhat excessive for a hydrographic survey vessel. Designing for the high pursuit speed will possibly lead to inefficient operation during utilisation in hydrographic survey operations, because of factors such as hullform and propulsion system selection. If the vessel spends the majority of its 30 year service life completing survey operations, cost benefits gained through the use of a common platform may be negated completely. However, the financial savings from the use of a common platform, or the reserve capacity for use in combatant roles during wartimes, may indeed outweigh the financial implications of the inefficiency. As a result, the analysis must include detailed performance evaluation and lifetime scenario modelling to determine the real world benefits of such technology. Integral to such an evaluation is a detailed assessment of the amount of time that each capability is required over the lifetime of the fleet.

In order to determine the applicability of modular technology to an Australian fleet, the anticipated level of effectiveness of the vessels in each role must also be compared with the levels offered by dedicated single role vessels. To effectively draw conclusions regarding the potential benefits, a value must be assigned to the relative capability loss or gain offered by the adoption of modular vessels. Such a process is outlined in the following section.

Capability Costing

A Mission Effectiveness Analysis that directly relates mission performance to cost, such as that described by Rains [2], is an effective method of analysing the combined effectiveness of a naval fleet `system´ made up of multiple vessel types. Such analysis allows a cost based comparison of various fleet combinations based on Measures of Effectiveness (MOEs). MOEs are generally derived from quantifiable mission performance variables, i.e. the parameters that directly enhance the ability to perform a mission, such as availability, armament, speed, number of vessels and total life cycle cost (TLCC). In their most basic form, the MOEs are developed based on a ratio of the general form of mission completed per dollar invested [2], for example Area Patrolled/Fleet Cost. MOEs are typically normalised over the life cycle of a vessel fleet in order to obtain a meaningful measurement of the effectiveness of a given combination of vessels. Factors such as vessel efficiency are taken into account through such an analysis due to increased TLCC, hence reducing the mission completed per dollar invested.

The introduction of modular capabilities to naval fleets adds another dimension to the derivation of MOEs, as fleet composition is inherently variable over the life cycle of the vessels, with potential variations in capability between peacetime and times of conflict. Some aspects of costing, such as procurement and TLCC, become significantly simpler to analyse when compared with multi class fleets, as the TLCC of modular vessels is relatively constant across the entire fleet due to a common platform. However other variables, such as strike effectiveness, are more difficult to quantify over the lifecycle of the fleet, as in times of conflict several vessels may undergo role reassignment to augment the capabilities of other vessels. There are also other factors which must be factored in such situations, such as time required for role change, which effectively lowers the mission effectiveness of re-roled vessels as compared with vessels already in the correct form of outfit.

As it is all but impossible to define the amount of time that vessels will be required to augment other capabilities of the fleet during their service life, i.e. in times of conflict or humanitarian aid, a probabilistic form of analysis is required. Further to this, such a model allows for scenario modelling to be undertaken in conjunction with a sensitivity analysis in order to understand the significant issues regarding cost implications and help derive decisions on such aspects.

Platform Costing

At this early stage of investigation, it is generally accepted that top down, parametric based cost assessment methodologies are most suitable for costing studies. A lack of design detail and information at this stage of assessment precludes the use of bottom up cost estimation methodologies. It is accepted that due to the complexity, innovation and adoption of new technologies involved with the application of modular capabilities, a bottom up cost estimation method would be able to provide greater resolution on potential cost savings.

Due to the lack of previous experience within Australia regarding modular installations, costs of interfaces for modular equipment have been estimated using simulated interface installations and costing information for similar systems. Further studies would include detailed cost assessment of modular kit through detailed discussions with suppliers. A simplified process by which the concept synthesis and costing analysis of conceptual platforms may be undertaken is shown in Figure 6.

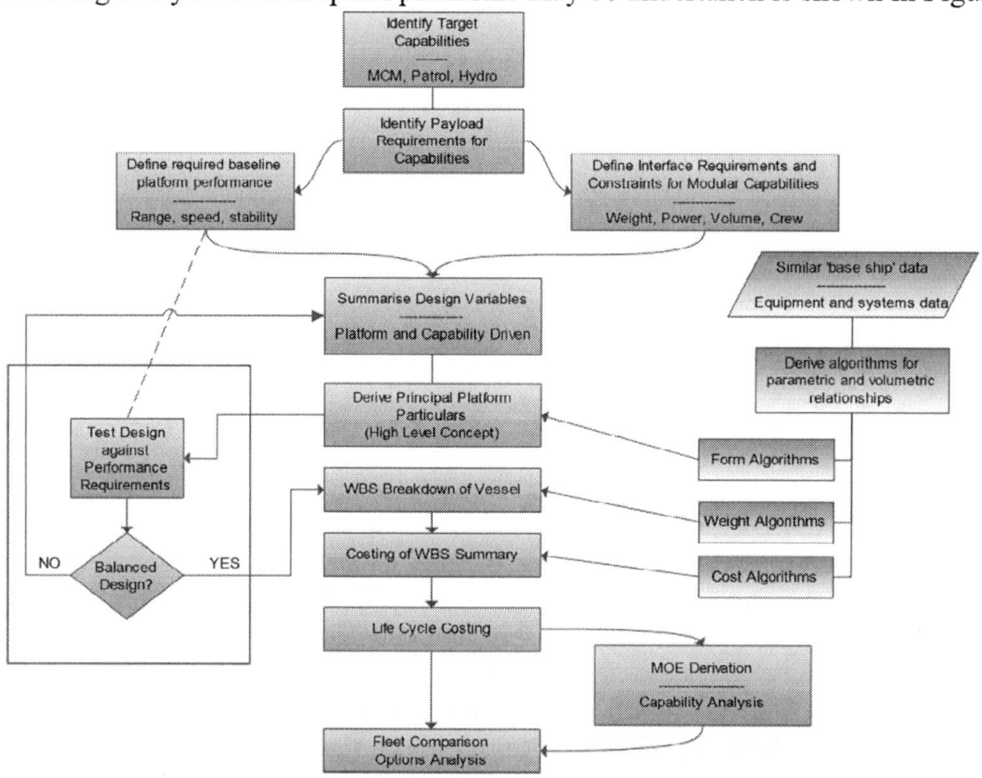

Figure 6 - Simplified Concept Synthesis and Costing Process

This process allows for high level concept designs to be synthesized and compared rapidly for comparison within the cost/capability model. Lifecycle costing is completed using BMT's established project costing processes.

Costing comparisons are based on platform costs only and do not take into account specific armament and capability specific equipment, rather the interface requirements to support such equipment and performance requirements of the platform to support operations. It is assumed that, based on existing examples of modularity and future concepts, armament for role specific and modular fleets will be similar with modifications of kit for modularisation.

SCENARIO ANALYSIS

BMT have developed and implemented a preliminary cost/capability analysis model utilising role specific MOEs and modifications to account for the application of modular technologies, including fleet flexibility. The analysis includes a cost assessment of the concept platforms previously described and incorporates complete life cycle costing. This analysis method was applied to varying fleet configurations, including an entirely role specific fleet, in order to derive informed conclusions about the application of modular technology to a FOC fleet.

MOEs have been based on the initial mission requirements outlined within the RAN 2009 Defence Whitepaper. Each MOE consists of several MOPs in order to gain a quantitative measure of how well the fleet meets the requirements of each capability. For example, the MOE relating to "The ability to conduct border protection missions" is derived from MOPs relating to coastal patrol area covered by fleet and intercept capability (mean time for a vessel to reach threat). These MOPs are normalised and combined to give five MOEs with values between 0 and 100. The MOEs are then combined to give an overall MOE (OMOE) through scenario modelling, involving weighting factors derived from relative importance of certain capabilities under differing scenarios and anticipated percentage of life spent in alternate scenarios. This process is described in Figure 7 below.

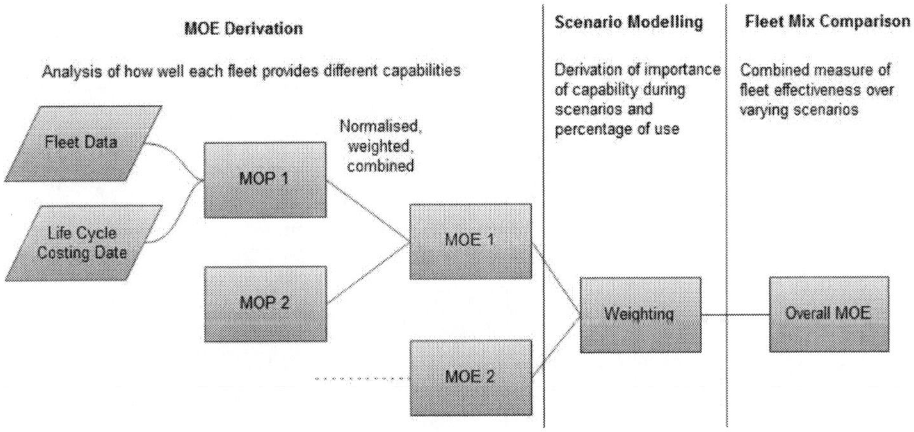

Figure 7 - MOE Derivation Process

The following variables were utilised in the analysis.

Fleet Composition	MOEs	Scenarios
Fleet A – Role specific fleets, as per current Fleet B – Entirely modular fleet, i.e. common platform across all capabilities Fleet C – Entirely modular fleet with no reroling requirement Fleet D – Role specific Patrol boat fleet and modular MCM and Hydro vessels.	Ability to: -MOE1) conduct offshore and littoral warfighting missions -MOE2) conduct border protection missions -MOE3) conduct long range counter terrorism missions -MOE4) provide hydrographic services -MOE5) conduct mine countermeasure missions	-Scenario 1) - Peacetime -Scenario 2) - Minor Conflict -Scenario 3) - Major Conflict

Table 2 - Capability Analysis Variables

The composition of Fleet C was selected in order to analyse the effect of the reroling requirement on the combined effectiveness of the fleet over it´s lifecycle. The composition of Fleet D was selected in order to analyse the effect of differing service speed and range requirements on fleet cost/effectiveness.

For simplicity, this paper assumes the application of the BMT Venator concept utilising offboard systems for the primary functions (MCM, hydro, patrol) for the derivation of cost comparisons. It is also assumed that the capabilities of both the SML and Hydrographic survey vessels may be completed by one state of outfit, as the use of offboard survey equipment negates shallow draft requirements.

A fleet with a total of 20 vessels was adopted for the analysis; 12 patrol, 4 MCM and 4 survey. Re-role flexibility was incorporated for Fleet B by varying states of outfit in different scenarios. Information regarding modular kit likely to be employed on a vessel such as the FOC has been gathered through discussions with Blohm & Voss, the RDN and Germanischer Lloyd.

Results

The results of the cost model analysis are presented in Table 3. It must be noted that estimates are purely indicative at the current point of investigation based on preliminary investigations and are presented for comparative purposes only.

	Fleet Acquisition Costs (Tendering, Design, Construction, NRE)	**Fleet Support Costs** 30 year service life (Maintenance, support)	**Fleet operating costs** 30 year service life (Consumables, crew)	**Fleet TLCC**
Fleet A- 12 Patrol, 4 MCM, 4 Survey	\$2.4 bn	\$1.3 bn	\$5.1 bn	\$8.8 bn
Fleet B- 20 Modular	\$2.7 bn	\$0.9 bn	\$5.6 bn	\$9.3 bn
Fleet C- 20 Modular – No re-role	\$2.7 bn	\$0.8 bn	\$5.6 bn	\$9.2 bn
Fleet D- 12 Patrol, 8 Modular MCM/Survey	\$2.3 bn	\$1.1 bn	\$5.3 bn	\$8.7 bn

Table 3 - Comparative Fleet Costs

The results indicate that significant savings in procurement and operational costs were not realised as initially expected. This was found to be due to the increased performance requirements of the Patrol/Warfare capabilities over the MCM and survey vessels. A platform capable of performing global warfare roles as intended by the RAN must be significantly larger than the required size of the MCM and survey vessels. Therefore 8 of the 20 vessels in the concept fleet were significantly larger than required. The follow on effects of this, such as increased fuel usage and increased baseline crew numbers, carry on through life and offset savings due to reduced design, maintenance and support costs. It was found that if patrol capabilities were kept similar to the existing level of capability, i.e. basic littoral patrol with no global warfare/deployment requirement, then significant financial savings could be achieved. It can therefore be concluded that fleet cost savings are achievable if the platforms are similar in baseline requirements for operability, volume etc. Whilst this does not show that financial savings are not achievable for the FOC, it is clear that further investigation of requirements definition must be undertaken.

The results of the cost/capability analysis are presented in Figure 8. The MOEs for each scenario are shown along with the normalised overall MOE.

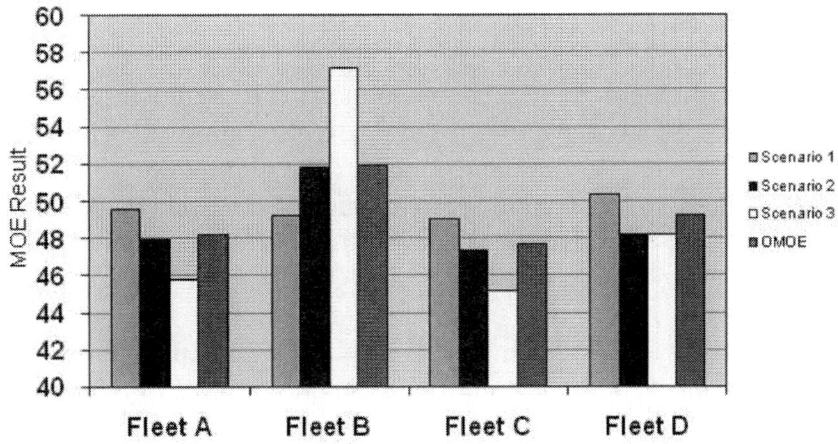

Figure 8 - Capability Analysis Results

The increase in fleet effectiveness as level of conflict increases is clearly evident for Fleet B. During normal operations, the increased TLCC of the modular fleets leads to a decreased MOE value compared with the role specific fleet. However when normalised over varying scenarios, it is found that Fleet B offers a 10% increase in capability per dollar invested over Fleet A and 12% over Fleet C. In physical terms this translates to an increased fleet capability equivalent to an extra vessel compared with a role specific fleet. It is therefore evident that for capability benefits to be realised for the FOC, re-role flexibility is of critical importance and must be investigated thoroughly.

To illustrate the importance of common performance requirements of the base platform, Figure 9 shows the OMOE value for fleets A and B as the ratio of platform size required for patrol/warfare operations to MCM/hydrographic survey operations increases, i.e. a ratio of 2 indicates that patrol requirements necessitate a platform with twice the displacement required for MCM/survey operations.

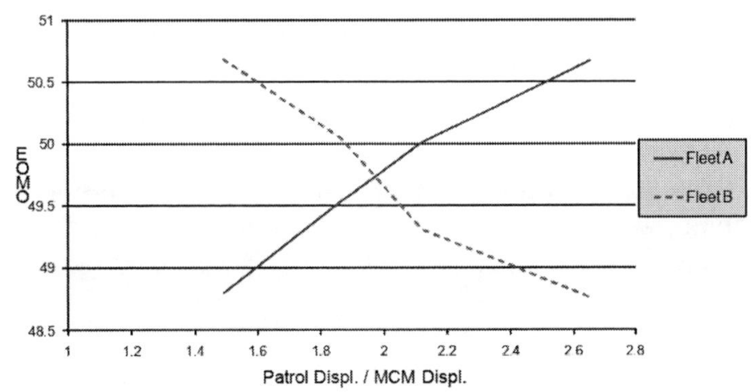

Figure 9 - Platform Variation Analysis

It is therefore evident that there are definable limits to how much variation in baseline performance requirements there may be between capabilities before there is a detrimental effect on overall cost based effectiveness of the fleet. This is a simplistic example, as similar relationships are found for other design factors such as service speeds.

Whilst it was found that the use of a role specific patrol vessel and common platform for MCM and hydrographic survey offered little capability benefits, it has the potential to offer competitive financial benefits and should be investigated in detail if it is determined that re-role flexibility is not feasible or advantageous.

SUMMARY AND RECOMMENDATIONS

The results of the study have shown that whilst the application of modular technology has the potential to provide operational and financial benefits to the RAN, it is imperative that all aspects of modular design are understood prior to CONOPS and requirements definition. There is no single 'correct solution' with regard to the extent of application of modularity. However through careful options analysis it is possible to make educated decisions regarding the most beneficial configuration of the FOC.

The development of an integrated costing and capability analysis model has allowed for the significant cost drivers and areas of highest risk to be identified such that they may be further investigated in order to refine the model as it develops the RAN specific solution. The study has highlighted that the issues having the most significant impact on the FOC design and effectiveness are those of re-role flexibility and performance requirements for the patrol/minor warfare missions as compared to MCM and hydrographic survey missions.

Recommendations for further work include:

- Further development of costing models and scenario analysis methods to increase validity of results;
- Development of bottom up cost estimating model to allow for comparison of novel design features such as multi-hulls and varied materials usage;
- Further investigation into the application of different forms and levels of modularity; and
- Study of risks highlighted through RDN experience regarding manning issues with direct relation to an Australian based fleet.

REFERENCES

1. MacKenzie, S.C. & Tuteja, R., (2006), "Modular Capabilities for the Canadian Navy's Single Class Surface Combatant – A Perspective on Flexibility", Defence R&D Canada CONTRACT REPORT DRDC -CR-2006-004
2. Rains, D.A., (1999), "Fleet Mix Mission Effectiveness Analysis", Naval Engineers Journal, January (1999), p. 65-81
3. Kimber, A. & Giles, W., "Minor Warship Roles - How technology is leading to a new vessel type", Pacific 2008 Proceedings

The U.S. Navy/ABS Service Life Assessment Program

RDML Thomas J. Eccles USN, Naval Sea Systems Command
Richard Delpizzo American Bureau of Shipping
Glenn Ashe American Bureau of Shipping
Sam Albrecht American Bureau of Shipping

Abstract

One of the most challenging aspects of fleet husbandry today for any Government ship owner is the proper assessment of where to spend maintenance resources to keep ships mission ready and safe for the personnel who sail them as well as for the operating environment. Budgets are becoming more and more constrained, personnel have less time to spread their attention over all the systems that comprise the total ship system, and operating tempos are increasing. As a result, it is imperative to identify and implement processes facilitating the ability to focus attention where it is most needed, and where it will provide the best return on investment.

The Naval Sea Systems Command and the American Bureau of Shipping (ABS) have recently partnered in a "Service Life Assessment Pilot" for the U.S. Navy introducing a structured, third-party periodic condition assessment process for selected systems on a number of classes of Navy ships. The intent is to determine areas where ABS can assist the Navy and share common knowledge and processes in identifying where maintenance action is most needed on ships, and recommend focused inspections whereby the continuing health of these systems may be tracked.

The intent of this paper is to describe the components that make up the Service Life Assessment Program and the potential benefits from application of ABS in-service classification practices applied to selected naval ship classes in the context of current Navy methods, processes, and procedures for inspection and maintenance.

Introduction

The pace of operations for the U.S. Navy over the last two decades has required a sustained commitment of many of the fleet's surface assets with reduced time and available budget for focused maintenance and modernization. In addition, the resources that have been applied to asset upgrade and maintenance have been constrained by the costs of the actual operations and the needs of the other services.

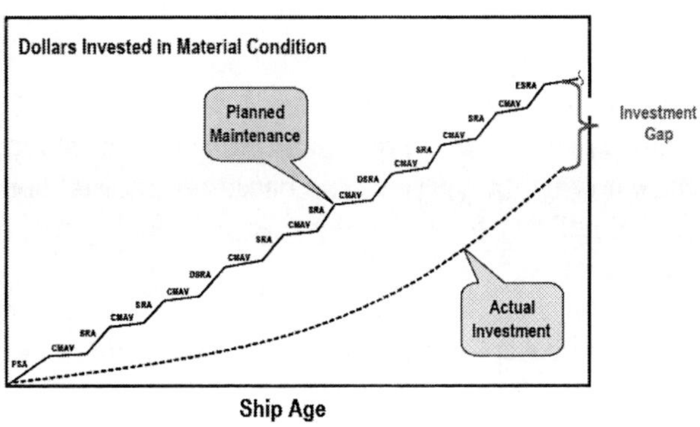

Figure 1: Maintenance Investment Gap

Concurrently, the requirement to eventually arrive at a 313 ship Navy has remained relatively constant, and a fundamental reality has emerged. In order to reach a 313 ship Navy, it is imperative that those assets already in service are maintained such that they reach their expected service life or, in some cases, be managed such that they reach service lives beyond those originally envisioned.

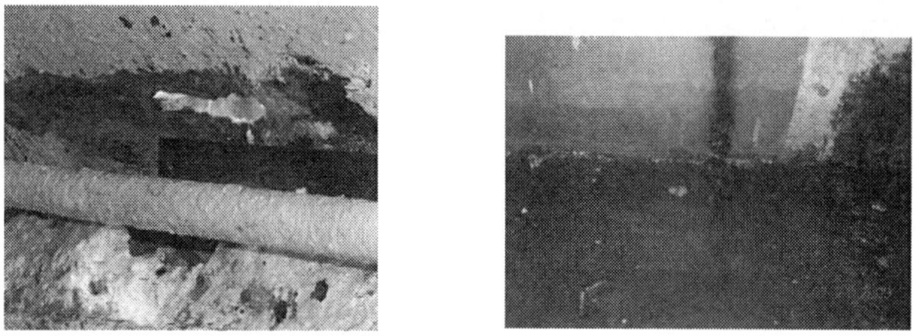

Figure 2: Typical Maintenance Concerns

This necessary emphasis on service life attainment recently galvanized a several pronged initiative on the part of Commander Naval Sea Systems Command (COMNAVSEA) to put in place programs that might make such a goal achievable. Such programs included the creation of the Surface Ship Life Cycle Management (SSLCM) Activity to provide centralized surface ship lifecycle management including; life cycle maintenance engineering, class maintenance planning and management closely aligned to the Fleet, Surface Warfare Enterprise (SWE), and NAVSEA's needs and priorities. SSLCM will help address the Navy's long term challenges of providing maintenance and support and integration services similar to programs already in place for aircraft carriers and submarines. The CLASSRONs were also charged with monitoring and managing some of these activities for the ships they represent.

As a desired supplement to these programs, COMNAVSEA and his Chief Engineer approached the American Bureau of Shipping based on their current support of U.S. Navy classification during acquisition and construction to ask the primary question:

Where can the U.S. Navy benefit from the existing ABS in-service survey process to help achieve expected service life?

In this context, the survey process under consideration is the normal ABS procedure for examination of a classed vessel. It is the process applied by ABS in the work it performs to assess the continuing compliance of over 10,000 vessels to the established set of technical criteria the vessels were constructed to. This pilot service life assessment also seeks to examine the potential to leverage the following:

- Combine advanced analysis with visual survey and tracking of observed deficiencies to better focus maintenance planning and provide improved material readiness.
- Project the current remaining service life based on material condition
- Establish a disciplined risk assessment process to identify and prioritize areas that do not meet technical requirements
- Determine capability to predict where critical or limiting material conditions may develop
- Provide an analytical basis for targeting investment and repair resources to achieve expected service life

In order to determine the effectiveness of this assessment approach, one ship from each of four different ship classes considered representative of typical maintenance issues were selected for the initial pilot efforts and include:

- USS GERMANTOWN (LSD-42),
- USS COLE (DDG-67),
- USS UNDERWOOD (FFG-36) and
- USS MOBILE BAY (CG - 53).

Key systems being examined to determine material condition of these ships included:

- Ship structure
- A/C cooling capacity
- Electrical plant health and capacity
- CHT system (piping and components)
- Firemain and chill water systems
- Ballast system

The information below describes the approach for this program, currently in progress, and summarizes some of the initial findings.

Background

The American Bureau of Shipping [ABS] is one of the world's prominent classification societies. It is a not-for-profit, independent technical organization that establishes and administers standards, known as Rules, for the design, construction, and periodic survey of ships and other marine structures. Classification is a process of certifying adherence to these Rules, and verifying continuing conformance throughout the vessel's service life. A classed vessel is, therefore, one that has been engineered and constructed to provide for both the safety of personnel and cargoes carried, as well as to protect against the operational environment the vessel is required to meet.

As the classification society of the United States, ABS has a unique relationship with the U.S. Government for vessels owned by the U.S. Government. While there is no requirement for U.S. Government-owned vessels to be classed, public law states that if a Government agency voluntarily chooses to class their ship, they must use ABS. Military Sealift Command, the Maritime Administration, and NOAA all choose as part of their ship management policy to build and maintain all of their ships in class with ABS. The Army Corps of Engineers and the Army Tank, Automotive and Armaments Command require in their vessel acquisition programs to have their ships built to ABS class. For many years the U.S. Navy has built many of its auxiliary and support vessels to ABS class.

Historically, governments have carried out parallel processes to the classification process on military vessels primarily through the use of internal empowered technical authorities. In order to address the challenges of technical authority in an austere environment, a two-fold joint decision was made by NAVSEA and Program Executive Officer Ships (PEO Ships) in February 2003. Results of this agreement included the Navy's preparation of specific rules for the design and construction of naval combatants and ABS publishing the rules officially as the Naval Vessel Rules (NVR). The rules developed, and subsequently authorized for use by the ABS Naval Technical Committee, are a set of rules that are Navy controlled. The second companion decision was to require ABS Classification for DDG 1000, all LCS Ships and non-nuclear future surface combatants.

ABS pursues its mission through classification of ships and other marine structures. In addition to classification, and surveys during construction, ABS provides life cycle surveys after construction.

Surveys After Construction: Though a new vessel may be granted ABS classification and thereby judged as having been designed and constructed in accordance with the applicable Rules, such status is not automatically retained throughout its service life. As the rigors of sea can be wearing on a vessel's hull and machinery, the society conducts periodic surveys to determine whether a vessel is being maintained in a condition whereby it remains in conformance with the applicable Rules and standards. As specified in the Rules, the owner must present the vessel to ABS on a periodic basis for survey of hull and machinery items. Also, should there be any reason to believe that an ABS classed vessel has sustained damage that may affect classification status; it is incumbent upon the owner to so inform the society. Upon request, ABS surveyors would then survey the vessel to determine whether it meets the ABS Rules and, if not, recommend appropriate repairs to maintain classification.

Periodic survey after construction is the core of the Service Life Assessment Pilot applied to the selected Navy ships.

Ongoing Navy Initiatives

In addition to the ABS Service Life Assessment Pilot described herein, several other initiatives have been completed or are currently in active development. In addition to the SSLCM Activity and CLASSRON activities previously described, focused effort has been applied to process and tool enhancements to help better define the scope of issues, provide means of tracking, and to improve current shipboard practices. Table 1 provides some of the planned and underway initiatives.

One of the primary challenges faced by the naval maintenance community is maintaining the structural integrity and combat/mission readiness of the fleet while concurrently extending their ships' class Expected Service Life (ESL). Corrosion and structural wastage continue to be significant factors in diminished structural integrity for Navy ships. Corrosion induced structural problems identified during inspections over the past several years have adversely impacted several ships ability to continue service. Historical and current Navy ship design philosophy does not include specific wastage allowances, and rather relies on proper upkeep and maintenance to allow for lighter weight vessels meeting speed, endurance and other performance requirements. As a result, Navy ships are less tolerant of structural wastage and degradation before remediation or assessment is required. Reduced manpower on ships, reduced maintenance budgets, and lack of required regular inspections over the past 10-15 years have resulted in declining material conditions. Further exacerbating the issue is the fact that Navy designs typically employ higher strength materials, allowing thinner structures that are again less tolerant of structural wastage. Additionally, naval designs must accommodate extreme loading associated with shock and weapons effects and other survivability threats not expected from commercial designs. These degraded material conditions have led to increased repair cost, readiness concerns, and potentially reduced service lives. However, service experience has confirmed that Navy designs have inherent robustness using naval design methodology, and using materials that have high material toughness and ductility, fatigue resistance, and the ability to withstand significant overload without brittle collapse failures.

Naval authorities involved in maintenance processes and their general functions are as follows:

Organization	Function
NAVSEA 21	Program Management
SSLCM	Life Cycle Management
NAVSEA 05	Engineering and Technical Authority
NAVSEA 04	Maintenance Activities and Scheduling
OPNAV	Funding
Fleet/Type Commander	Operational Requirements, and Availability Schedules

During the third quarter of FY09, a significant new initiative was implemented to gain a better understanding of the material condition of the ship hulls through mandatory execution of a standardized structural inspection process using Maintenance Requirement Cards (MRC) initiated via established Integrated Class Maintenance Plans (ICMP). Inspection periodicities, processes and procedures for inspections, maintenance, repairs and reporting have been established based on a systems engineering approach across platforms that focus on areas of high stress and fatigue that recognize structural members should not require structural repair when properly coated and maintained (i.e. structural coatings focused). See Figure 3 for the scheduled periodicity implemented as a result of the Structural Corrosion Working Group initiative for aircraft carriers and surface ships.

In addition, simplified analysis tools and updated technical guidance and requirements are being developed to empower the maintenance disciplines to quickly assess and determine the adequacy of the structure to carry the required loads.

Area	Status	Benefits
Rapid Cure Single Coat	Approved	Projected 25% of the coating application cost and an equivalent 56% of the coating application schedule for Ballast and Fuel Tanks and Voids
Stripe Coat Elimination	Approved	Eliminates 15% of coating application cost & equivalent 33% of coating application schedule for Ultra High Solids (Edge Retentive) Coatings
Shipboard Corrosion Assessment Training. SCAT	Approved	Training for assessors for consistent classification of coating and structural conditions. Free for government personnel
NSTM 631 Update	Approved	Paint products for best value surface preservation
NSI 009-32 Update	Approved	Universal coatings requirements for contracting
Corrosion Control and Assessment Maintenance Manual (CCAMM) Update	Approved	Effective guidance for inspection and assessment of tanks and helps prioritize tank work
Paperless QA	Approved	Provides real time adjudication of non-compliances Provides cost savings since data uploaded directly and greater data accuracy
Peel and Stick Non-Skid	Approved	Easily applied by ship's force, durable product
ICMP/Corrosion MRC	Approved	Provides structured approach to assessing critical ship surface coatings
Impressed Current Cathodic Protection (ICCP) Log Rev	Requirements Promulgated	Formal log review to ensure conditions and proper system operation
NSTM 100	Updating	Repository of requirements for structural assessment, maintenance and repair
NSTM 081	Updating	Formalize periodicity requirements for underwater hull inspection and cleaning
Service Life Assessment Pilot	Ongoing	Establishes service life assessment and material condition of ship
Structural Corrosion Analysis Tool Software	Under Consideration	Strength calculation software for the RMCs to easily execute determination of local repair/maintenance requirements for corrosion and holing
Thin Hull Criteria Development	Under Consideration	Booklet of minimum thickness requirements to be maintain for strength per ship configuration

Table 1: Representative Current and Proposed US Navy Maintenance Improvement Initiatives

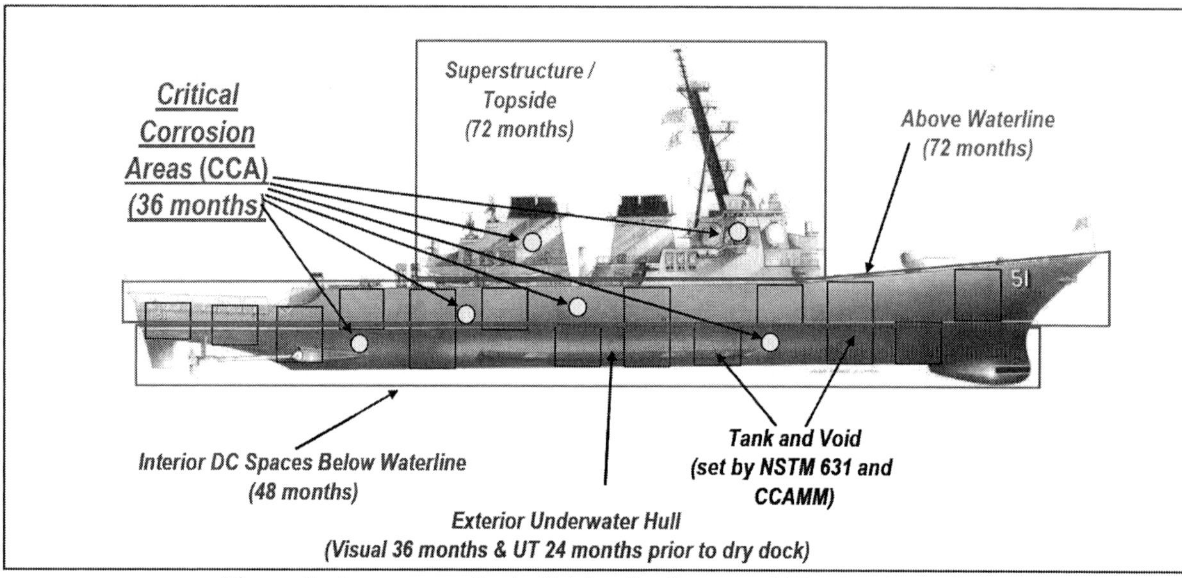

Figure 3: Assessment Periodicities for Structural/Coating Integrity

Another successful program currently employed by the Navy is the Tank and Void Condition Based Maintenance and Inspection program. The working group administering this program has developed periodicities for all tanks and voids based on gathered historical data on coating and anodic protection life, and historical records of known problem areas.

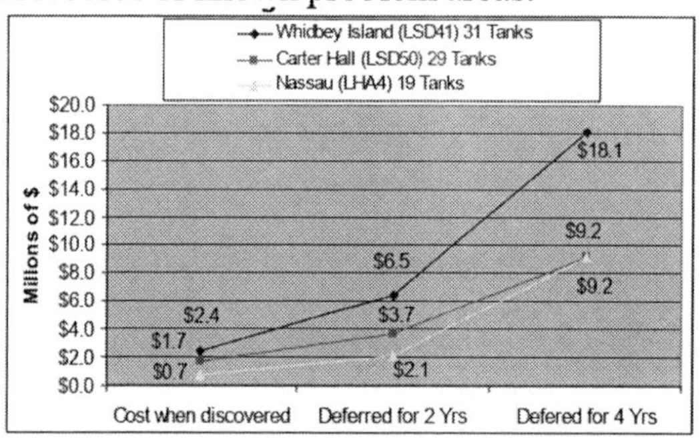

Figure 4: Tank Coating Deferrals Lessons Learned

Both of the NAVSEA working groups have aggressively refined the technical requirements and guidance for when and where inspections are to be performed, how the inspections are to be performed, methods and degree of repairs, procedures for how deferrals are to be prepared, and the required approval authority, and format for data gathering and reporting. The efforts from these working groups have been combined with data reporting using an existing and proven corrosion and assessment reporting database, Corrosion Control Information Management System (CCIMS), currently used in carrier, submarine and surface ship communities.

ABS Service Life Assessment Process

The objective of the pilot program is to establish technical insight into the current condition and life expectancy of key systems of a ship as follows:

1. Prepare the assessment and process model that will be applied to conduct the initial assessment of each individual ship
2. Conduct the initial assessment of each ship

3. Conduct periodic, continuing assessments of each ship to track the maintenance impact on the selected systems and provide a tool for targeting future maintenance
4. Integrate the process so that the assessment work performed on the ship minimizes intrusiveness and impact, and that the Navy retains Technical Authority over all decisions

The following paragraphs summarize the work envisioned for each of these steps.

ABS Process Description

The proposed ABS Service Life Assessment Program includes the following steps:

1. Build a Finite Element Model (Figure 5) for the as-built configuration of the ship, and apply the projected operational loadings envisioned for the ship to identify the following:

- Inherent as-built structural strength margins
- Inherent corrosion allowances
- Identify areas of high stress for specific inspection interest
- Predicted fatigue resistance for primary structural details and components.

2. Modify the as-built finite element model to represent the current condition of the structure based on ABS survey and hull thickness measurements, then exercise the modified model to assess remaining strength and fatigue capability.

3. Develop process check sheets listing the systems, equipment, and components to be surveyed.

4. Build the Hull Maintenance Model (Figure 6) facilitating the graphical representation of the structural history of each ship. The Hull Maintenance Model provides a detailed visual representation of the structure of the ship containing the information related to the structure in an easily obtainable format and allows one to track the structural history of the individual ship including member thicknesses, damage, repair and alteration.

Such models are being developed for each new construction naval combatant classed by ABS.

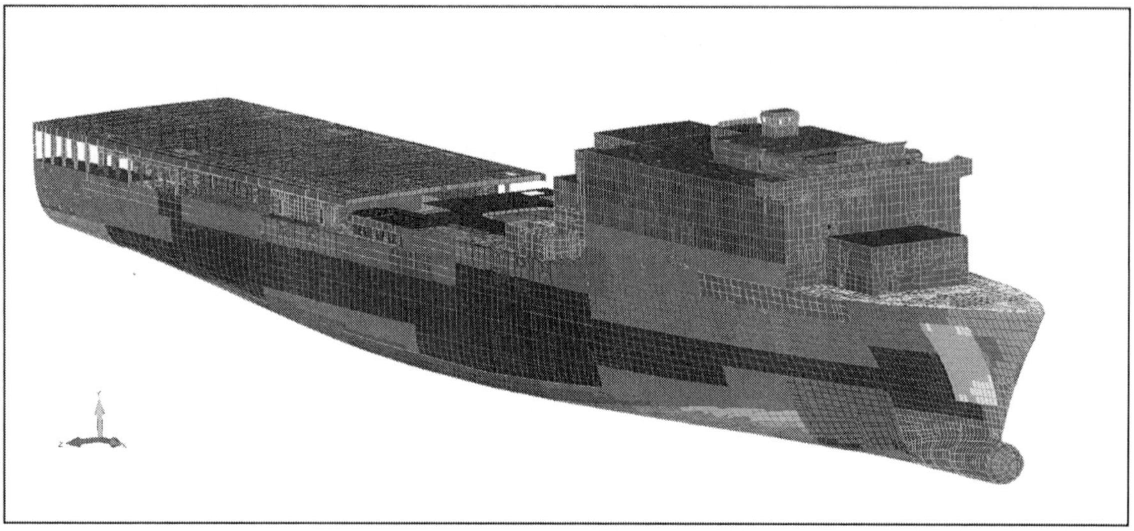

Figure 5: USS GERMANTOWN Finite Element Model

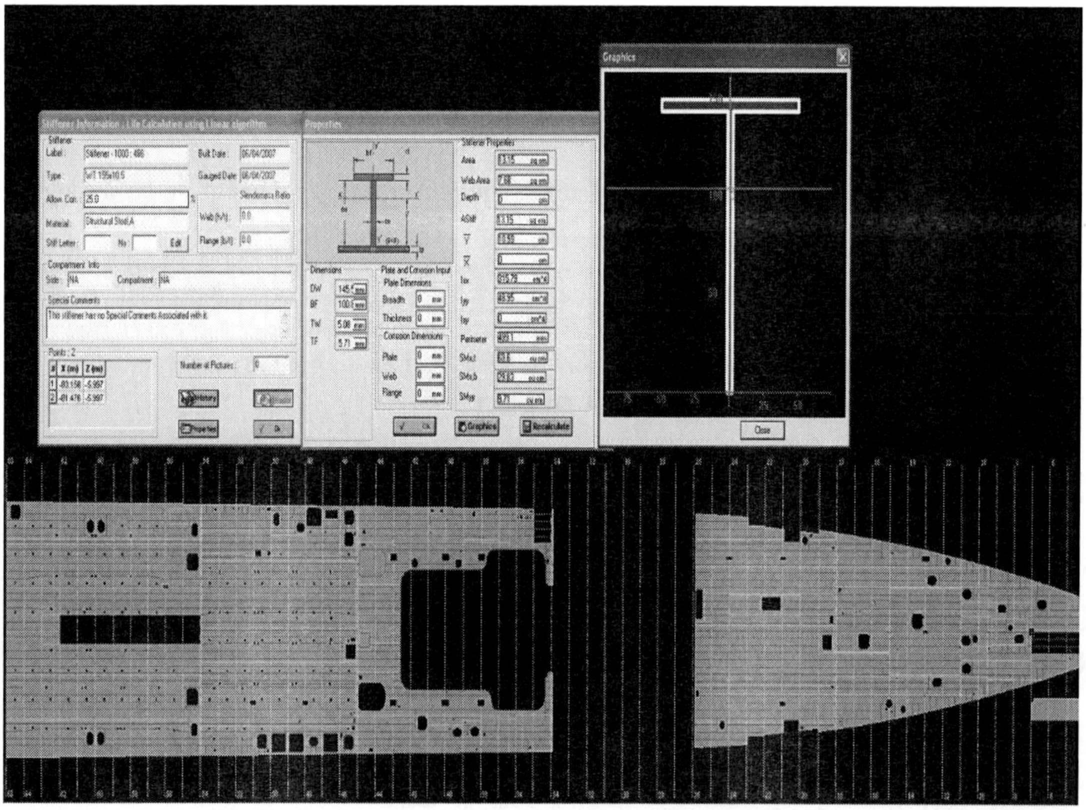

Figure 6: Hull Maintenance Model Details

5. ABS initial and continuing assessments of each ship to determine the current condition of each vessel. Using the check sheets to document survey findings and survey scope based on ABS Periodic Survey Rule requirements, an overall assessment is conducted for several ship systems that include:

- A/C Cooling Capacity
- Electric Plant Health & Capacity

- CHT System (Piping & Components)
- Seawater Systems
  - Firemain
  - Chill Water
- Ballast system
- Ship Structure
- Underwater hull condition (when available in dry dock)
- Internal close-up examination of spaces
- Assessment of key distributed mechanical and electrical system surveys based on design life requirements

More specifically, the structural inspection survey includes:

- Close-up examination of tanks, voids and cofferdams including coatings
- Thickness measurements of the Hull including:
  - Three (3) transverse sections within the amidships region
  - Internals of Forepeak and Afterpeak tanks
  - Hatch covers and coamings
  - Main deck
  - Wind and Water strakes, port and starboard, full length
  - All keel plates, full length
  - Shell plating in way of sea chests and overboards
  - Lowest strakes in way of decks of all transverse bulkheads in machinery spaces

Accomplishment of this survey establishes the initial baseline for the continuing assessment of the hull and selected distributed systems. The surveys are structured and recorded on standardized check sheets to maintain consistency of application and reporting, and provide a means to quickly judge relative system health.

Follow-on annual surveys for each ship are proposed, allowing for tracking of system health and will facilitate the assessment of the service life.

Upon survey completion, a complete report outlining the findings of the survey will be provided in a narrative ABS Survey Report to the Naval Technical Authority for review and concurrence.

Initial Results

The initial results of the service life assessment program have been encouraging. Full ship surveys were conducted, material conditions documented for the systems surveyed and are currently being evaluated in the context of service life and directed inspections. The fleet, ship's force, CLASSRONs and their command structure, and shore based technical support elements such as the RMCs provided outstanding support.

Over the course of the pilot efforts, the scope and focus evolved over time (and continues to evolve) by virtue of the vision of the affected stakeholders, lessons learned and expectations in implementation. Initial take-aways include the following:

- Program can facilitate an integrated assessment process supporting corporate control, visibility and decision making
- Meaningful system assessments require earlier implementation of surveys, review of PMS and application of analysis tools in conjunction with visual survey
- Reporting must include a systematic approved method to prioritize findings using risk-based condition assessments

- Need to integrate with other initiatives to maximize value and cost effectiveness, to take advantage of common effective processes, and properly capture and address findings by Navy activities
- Need a closed loop to apply lessons learned to technical criteria (NVR)

Follow-on annual surveys for each ship are proposed by ABS allowing for tracking of system health and will facilitate the assessment of expected ship service life. Surveys of the other systems have provided similar results. As all four ships are being surveyed during drydocking upkeep periods, much of the individual ship systems surveyed are already being addressed by the work package and are in good condition at the conclusion of the period. As with the hull, there have been a few areas discovered not in comparable condition.

Conclusion

The ABS Service Life Assessment Pilot has demonstrated an approach for consideration by the Technical Authority for further review and consideration.

Initial structural survey results, with the supporting hull thickness measurements, have shown hulls in varying degrees of fitness, thereby confirming the overall robustness of the naval hull structure platforms, while emphasizing the need for a sound inspection, maintenance, and repair program. The NAVSEA Technical Authorities will review and assess the ABS Pilot Program and formulate areas where integration of ABS methods can be supplemented with the current Navy processes.

The next steps involve completing the assessment of the ships currently involved in the pilot, better defining the content of the Service Life Assessment Pilot products with the applicable Technical Authority, and then working with activities such as SEA 21, SSLCM, and SEA 05 to integrate the Service Life Assessment process into the current Navy processes where appropriate.

Implications of the Defence White Paper 2009 for Naval Submarine Technology in Australia

Christopher J Skinner BSc(Eng) MEngSc MIET MIEAust MACS CPEng
DISplay Pty Ltd
cjskinner@acslink.net.au

1. INTRODUCTION

The Australian Defence White Paper 2009 [DWP] (Australian_Government 2009) has affirmed the commitment to an increase in the numbers and versatility of Australia's naval submarine force

However there has been little discussion of the implications for naval ship technology for Australia beyond the bipartisan commitment for the future submarine [FSM] force to be built in Australia. This is changing with the recent award of the consultancy contract for a Future Submarine Design Capability Study to the RAND Corporation, an organisation that has performed extensive reviews of nuclear submarine design and industrial capabilities for both the UK Royal Navy and the USNavy. (Raman, Murphy et al. 2005; Schank, Cook et al. 2005; Schank, Riposa et al. 2005; Schank, Arena et al. 2007)

Due recognition must be given for the attainment by ASC Ltd of Design Authority [DA] for the Collins Class submarine [CCSM]. This does not necessarily equate to the capability to undertake ab initio design for the FSM, except in conjunction with submarine designers of long standing proficiency. Further the onerous responsibilities inherent in the DA role have also been in the media recently with regard to CCSM availability.

There are many submarine technologies that extend well beyond those used in the CCSM, for example Special Forces deployment, support and recovery; air independent propulsion [AIP]; and the capability not only to operate submarine launched land-attack missiles [SLAM] but also to participate in their targeting process; and uninhabited vehicles, both airborne and undersea, which provide many capabilities to gather information for intelligence, surveillance and reconnaissance [ISR].

And then there is the ultimate naval submarine technology – nuclear propulsion – that will need to be addressed one day in the future. The total absence of any domestic nuclear engineering industry makes this option infeasible due to the significant extension of project timeframes that would result, even if the political will to adopt this option existed. It is notable in passing that the recently announced nominee for chief executive of ASC Ltd comes from a position with Rolls Royce submarine division and has extensive experience of nuclear submarine propulsion engineering.

This paper discusses the implications of these issues in the light of the DWP for the further development of naval submarine technology in Australia.

2 THE REQUIREMENT FOR FUTURE SUBMARINE CAPABILITY

The 2009 Defence White Paper (Australian_Government 2009) makes many references to naval submarine capability, and to strategic and other force structure issues that bear on submarine capability. This paper examines these references as a basis for discussion of the Future Submarine Capability [FSM] Project (SEA 1000)

2.1 The Approach Taken in this paper

A conventional approach to generate FSM Top Level Requirements [TLR] would be two-fold:

a. Extrapolation from world-wide submarine operational and technical experience over more than a century. And
b. Application of Australian lessons from the Oberon Class and its Submarine Weapons Upgrade Project [SWUP] and then the Collins Class Submarine [CCSM] development and corrective action programs.

However there is merit in taking a more varied approach. In the same way that systems architecture is understood to require several views or perspectives this also applies to the definition of TLR. Consequently this paper provides an additional view of the TLR for the FSM based on the statements and implications from the DWP. Additional commentary is provided where needed to infer additional requirements.

Coincidentally this paper also adverts to a recent publication by Derek Woolner, co-author of the 'The Collins Class Submarine Story…' (Yule and Woolner 2008) that also reflects the DWP in discussion of SEA 1000 in the light of the CCSM. Where relevant this paper refers to the Woolner paper (Woolner 2009)

2.2 DWP Chapter Headings

For convenience this discussion follows the sequence of the DWP chapter headings.

Chapter 1: Government Approach to Defence Planning

Inter alia, the DWP says that the Government 'plans to strengthen the foundations of Australia's defence... to meet the challenges of an uncertain future.' (DWP p15)

Discussion: Is the future any more uncertain than in the past? We could say there are some generally agreed bases for consideration of the future:

- The increasing global human population and the resulting pressure on sustainability of food and energy supplies
- The global financial crisis [GFC] and its aftermath.
- Continuing nuclear proliferation, notwithstanding the laudable efforts of many countries, including Australia, to resist this trend.
- Global climate change due partially to human exploitation of non-renewable resources. And
- . The finite availability of these same energy resources and hence the need to find alternatives.

In addition Australia is an island with an economic development strategy that is dependent on immigration and tangible and intangible maritime trade and commerce. Australia is overwhelmingly a maritime nation and therefore dependent on protection and sustainment of its maritime interests. The DWP acknowledges this as the fundamental basis of policy and strategy.

The timing and scope of the DWP are necessary and constructive, notwithstanding the lack of detail in some areas, notably funding.

The DWP says that it: '... explains the level of resources... to invest, and what the Government expects in return from Defence.'

Discussion: the tension between large financial investment projections and the Defence Reform Programme [DRP] is to be expected, and should encourage a constructive articulation process.

The DWP also says:

- 'Defence faces a series of organisational and resource challenges, including the recruiting and retention of skilled people.' No surprises there, however it should be noted that there has been recent concern regarding the sustainment of engineering and logistics knowledge and experience within Defence to perform the competent buyer role. This will be further exacerbated as the Strategic Reform Programme [SRP] places further pressure on human resources
- And calls for: 'effective management of Defence's supporting systems', including logistics, facilities, and information and communications technology assets.
- There has been a higher operational tempo since 2000, and the implications of this have not been fully appreciated in all areas. And in conclusion says:
- Defence demands: 'Decisions on significant aspects... including air combat and submarine forces.' Noble words indeed!

Chapter 2: Defence and National Security

This chapter says that national security is concerned with ensuring:

- Freedom from attack on Australia as a nation;
- Maintaining Australia's territorial integrity;
- Promoting Australia's political sovereignty;
- Preserving Australia's freedoms; and
- Sustaining Australia's economic prosperity.

The DWP discusses the future role of armed forces in differing contextual settings:

- Major war;
- Intra-state conflict;
- Non-state global actors perpetrating conflict and terror;
- Domestic security;
- Emergency response;
- Humanitarian assistance; and
- Disaster relief.

Discussion: These contextual settings are generic and entertain a very wide range of contingencies, not all of which can be predicted. Flexibility and the capacity to respond and adapt are always needed.

Chapters 3 to 5: Managing strategic risk; Australia's strategic outlook, and strategic interests

The most important discussion in chapters 3 and 4 is on strategic warning. The timescale involved in perception of clear warnings is generally less than the lead times for major capability acquisition, and this means that capability development is a flexible process that requires periodic review and adjustment. This is especially so in lengthy acquisition programs such as the FSM.

The possibility of strategic surprise can never be ruled out entirely and this risk requires:

1. The maintenance of versatile armed forces to provide an immediate response capability.
2. Development and maintenance of the necessary skill levels. And,
3. Provision of cadres for expansion over a longer period should that become necessary.

Discussion: The demand for competent and knowledgeable people will be especially high in times of strategic surprise. We should therefore keep up the skill levels and the best way to do that is to apply them in a useful manner such as a long-term project like the FSP.

The most significant feature of chapter 4 is the discussion of the rise of China as a global power. In passing it is significant to note that China possesses the largest number of submarines in the Asia-Pacific region (Howarth 2006)

Australia is committed to strategic stability in the Asia-Pacific region, based on the truism that geography is a determinant of strategy.

Chapters 6 and 7: Australian Defence Policy and Principal Defence Tasks

Australia's Defence policy addresses a number of key factors including:

- Defence against attack on Australia
- Defence force structure
- The Australia-USA alliance
- Coalition operations
- The primary operational environment.

Principal tasks for the ADF are defined in a framework of four decreasing priorities including:

- First priority: defence against direct attack on Australia. This has an important linkage with the third priority, as discussed below.
- Second priority: to contribute to stability and security in the South Pacific and East Timor. It is of note that this includes Fiji where China has been increasing its influence in the vacuum caused by the exclusion of Fiji from the Commonwealth and South Pacific Forum. The Australian intervention in East Timor in 1999 comes under this priority
- Third priority: contributing to military contingencies in the Asia-Pacific region. This would encompass the scenario discussed later in this paper. It is important to note that the DWP makes special note of retaliatory action that could arise because of Australia's contribution under this head.
- Fourth priority: contributing to military contingencies in support of global security (such as Afghanistan).

Discussion: The FSM contributes to all of these priority tasks and is especially suited to the higher priority tasks.

Chapter 8: Future Development of the ADF

This chapter covers the major directions of force development of the ADF, including maritime forces, strategic strike, force projection and joint operations. Important considerations in capability development are discussed including:

- weight and reach of force projection
- strategic capability advantage
- precision in the application of force
- networked forces
- operational flexibility
- survivability
- interoperability
- cost-effectiveness; and
- the need for leadership in the attainment of professional mastery.

Chapter 9: Capability priorities and Chapter 10: Preparedness

The discussion of priorities for ADF capabilities starts with maritime capabilities and goes on to discuss air power and strategic strike and other current and future capabilities. It is notable that the discussion of other capabilities is in generic terms whereas maritime priorities are stated explicitly by platform type, starting with submarines (rather than the capability for undersea warfare).

Chapter 10 discusses preparedness of permanent and reserve forces, the role of contractors and the process of national mobilisation, should it be needed at some future point.

Chapter 11: Alliances and Chapter 12: Defence Intelligence

The main discussion is of the Australian/US alliance and its importance for many aspects of Australia's defence; these include:

- technology transfer
- interoperability
- intellectual property rights and licensing; and
- combined operations.

Discussion: all of these aspects are very relevant for the FSM as they were all significant in the Collins Class submarine program and we are still living with the consequences.

The assumption of the responsibilities of parent navy and CCSM design authority has been onerous, and in many ways unanticipated. This may have been excusable for the CCSM but must not happen again. Other countries with longer experience of these roles ensure that full consideration is given from the very start of such a major project. In common with large resources and infrastructure projects in all fields an all of life approach is taken including the ultimate safe and sustainable disposal of the asset. This requires the capability development process to work from the perspective of the full cost of ownership including in service engineering, integrated logistic support, human resources and the operational availability to be delivered over the full life of the asset.

Chapter 12 covers strategic warning as it relies on intelligence, intelligence support for current operations, and intelligence support for capability development. Submarine forces are routinely employed to contribute some forms of intelligence information in any level of conflict or high peace.

Chapter 13: Defence Management and Reform

A major management programme has been the Defence Strategic Reform Programme [SRP] focussing on management of costs to deliver savings to be applied to capability development and sustainment. There are also efficiencies to be gained within the sustainment processes themselves.

Chapter 14: Defence People, Chapter 15: Support and Chapter 16: Procurement, Sustainment and Industry
This chapter discusses Defence's integrated workforce. For the FSM we can expect significant participation by CDG, DMO, DSTO, industry – domestic and international – as well as liaison through alliances.

Inter alia, chapter 15 discusses information management and communications capability and proposes reforms to deliver a secure and robust ICT capability that supports war fighting and business requirements. Defence has recently announced that its ICT needs will be market tested and there is much to the argument that much of Defence ICT needs can be met by advanced industry practice where cyber-security is just as much a challenge as it is for Defence.

Chapter 16 covers the international industrial context for the ADF, Australia's industrial capacity and the development of Defence industry policy.

Discussion: This has been the focus of recent policy development but this has been from the perspective of globalised supply chains and participation by Australian companies at second or lower tiers. This may not be adequate in time of emergency because it may not ensure the sustainment of skills at the higher levels of design analysis and synthesis needed for self-reliant responses to strategic surprise.

Chapter 17: Defence Science and Technology
Chapter 17 discusses the support provided by the Defence Science and Technology Organisation [DSTO] for current operations, as well as research for capability development and contingency planning.

The final Chapter 18: provides a brief statement on the funding required to deliver all the programmes and capabilities discussed earlier in the DWP. Discussion: The lack of detail has been widely deprecated and has not been offset by the complementary Defence Capability Plan which was itself abbreviated from a ten-year to a five-year horizon. This timeframe has since become the subject of a review.

3 A SCENARIO FOR DISCUSSION OF FSM CAPABILITY

For the purposes of this paper it is helpful to consider a plausible scenario for the employment of the Future Submarine capability. However this should be done in the light of the shortcomings of this scenario-based approach, namely:

- Such a scenario may be attacked by critics on the content of the scenario rather than the manner in which this straw-man has been used.
- Whatever scenario is presented, it can only ever be a single instance among an infinite range of possibilities, any one of which may be more likely and/or more insightful than that selected. This very real limitation is accepted for this paper but should be emphasised in any further work that the scenario has no special merit or rigorous basis – it is only for discussion of the central issues of this paper. And

- The final short-coming is that by defining a given scenario, this may actually influence the evolution of events such that the scenario is rendered invalid and hence the arguments advanced using the scenario for insight may also be invalidated. That possibility is unavoidable for this paper.

3.1 Scenario: Submarine Contribution to the Defence of Taiwan

China continues to grow in economic and geo-political power. The general principles of China's national maritime strategy provide for a three-phase expansion to encompass the first island chain, then the second island chain and finally a blue-water naval capability that can influence events wherever China's national interests are involved.(Howarth 2006)

The re-integration of Taiwan within China's sphere of influence is the most important part of the first phase expansion. It is also at the forefront of the geo-political confrontation of China and USA as the latter has guaranteed its support to the Government of Taiwan.

Meanwhile Australia adheres to a One China policy that recognises the primacy of China's sovereignty in land and maritime regions that are recognised internationally. This includes Taiwan.

USA also has its own interests to consider in its response to the expansion of China's maritime military capabilities. These include an expanded PLAN submarine force that is suitable for use against carrier battle groups, and the gradual expansion of a submarine nuclear deterrent force to a level that would threaten continental USA.(Erickson, Goldstein et al. 2007)

China on the other hand may decide at some point that it can achieve its reincorporation of Taiwan by military means with a sufficient probability of avoiding major conflict with USA or others. Should this conflict nevertheless occur, it is possible that Australia would be requested to assist in the defence of Taiwan, and this would most likely be based in the first instance on the contribution of submarine forces.

In that event, as the DWP points out, Australia could become the target of direct attack in retaliation for its involvement in the Taiwan theatre. This possibility has not arisen in conflicts since WW2 but would certainly be the case for a scenario such as described here. This would make direct attack on Australia a very real prospect that would need to be taken fully into account by the Australian Government. In particular Australian ASW force capabilities would be challenged.

I restate that this scenario is only for discussion purposes in this paper.

3.2 Determinants of Submarine Capability for Such a Scenario

What would be the determinants of capability that Australian submarine forces should possess for such a scenario? The submarine would need to possess a number of characteristics:

- Survivability, as a factor of many parameters, including low signatures, low indiscretion rates, mobility, endurance, effective countermeasures, shock resistance, navigation and hydrography
- Interoperability with other forces in a coalition that might include USA and Japan, for example.

- Effective command and control, communications, computers and intelligence access via networks both in space and terrestrial – above and underwater.
- Effective sensors to contribute to coalition ISR capability. These should include off-board sensors including UUV and UAV
- Effective weapons and effectors carried in sufficient numbers for use in strategic strike, engagement of general maritime forces and for ant-submarine warfare [ASW].
- The capability to insert and recover special-forces personnel and other specialised intelligence operators.
- Crew competence, and capacity, based on realistic training and rehearsal.

The FSM would be needed in sufficient numbers to support effective contribution to the coalition and at the same time to the defence of Australia and her national interests.

3.3 Submarine General Characteristics

The future submarine should therefore be of a sufficient size to provide for mobility and endurance and to support the operation and coordination of a variety of sensors, processors and effectors for a wide range of roles.

The roles may include the need for additional personnel to be carried either for supplementary roles or for external deployment. Consequently there will need to be additional personnel accommodations and expanded hotel services.

3.4 Naval Submarine Technology in Australia

The top-level requirements [TLR] for the FSM including operational sustainability, and the technologies that are needed to meet those requirements are beyond what has been done before – in some cases anywhere in the world. There will need to be careful evaluations of the risk profiles and essential further research and development needed for these options and then rigorous tradeoffs performed to optimise the design baseline.

All this is needed before locking into long-term construction contracts. Engineering work is needed – building operating prototypes and evaluating them under realistic environments and operating profiles. This is mandatory practice for experienced parent navies and we must learn from them for the FSM.

It can fairly be said that the most important implication of the DWP is that Australia will not find an existing deign that will meet requirements without substantial changes. As experienced engineers know, once the changes become significant, then it is a new ballgame!

4 LESSONS FROM THE COLLINS CLASS SUBMARINE PROJECT

A recent publication (Woolner 2009) from Derek Woolner, co-author of *The Collins Class Submarine Story* (Yule and Woolner 2008) examines the FSM from the perspective of the Collins Class Submarine [CCSM] project and makes a number of important observations, including the following that are pertinent to the DWP and the FSM:

- The need for a well-validated concept of operations document for the new submarine capability

- Acknowledgement that a submarine capability provides benefits in areas of information gathering for intelligence, surveillance and reconnaissance [ISR] and in strategic deterrence when carrying suitable long range strike weapons, provided the submarine can operate covertly (aka stealthy operation), as well as in unlimited warfare.
- The expected leverage of network-centric warfare [NCW] in the effective use of force assets including submarines. This is yet to be demonstrated in the case of submarines due to the acknowledged difficulty in reliable, high capacity communications when submerged.
- Advances in submarine propulsive technology have opened up the possibility that a future submarine will attain performance levels that provides flexible mobility that more closely attains that of nuclear propulsion.

Woolner also comments on the acquisition strategy for the FSM and again draw a number of lessons from the CCSM project, including:

- The careful identification of technical and commercial risks and use of effective measures to manage the risks, if necessary by judicious use of research programs to reduce uncertainty in technology or approaches for systems integration.
- The choice of appropriate contractual frameworks that permit the continuing access by the Commonwealth of Australia to the assets and expertise needed at any stage of the project to resolve issues that arise, and also allow for the inevitable changes in corporate ownership and expertise over the life of the project.
- The clear distinction among the roles of ab initio designer, the design authority [DA] for the life of type and the design approval authority [DAA] (DMO with DSTO and naval engineering resources on behalf of the Commonwealth)
- The essentiality for production to be undertaken by companies with appropriate experience
- A well understood process for management of design changes that will inevitably arise from evolution in operational and support requirements over the life of the class, and also from changes in design to resolve problems that may arise or to take advantage of advances in technology, where cost-effective.

5 CONCLUSIONS

1. The Defence White Paper provides a coherent basis for the Future Submarine project (SEA 1000), but will require further analysis in the light of a range of contingencies that can be envisaged within the future timeframe described in the DWP.
2. There is not yet any apparent justification for the upper range of mobility capabilities, such as available with nuclear propulsion. However this may be expected to change in the future.
3. The implications of the DWP for naval submarine technology in Australia are wide-ranging and will have significant effect on the definition of submarine requirements both for operational capability and for through-life-of type sustainability.
4. Finally and most substantially the financial requirements for the Future Submarine Capability are profoundly affected by the technological dimensions of the project, including the need to develop and sustain essential knowledge and competences to be available through the life of the submarines and their successors.

Acknowledgement

This paper was partly supported by the 2006 Maritime Advancement Australia Award from the Australian Naval Institute and the Australian National Centre for Ocean Resources and Security, sponsored by Booz and Company.

The insightful comments by several colleagues in the naval submarine operational and engineering domains are gratefully acknowledged.

6 BIBLIOGRAPHY

Australian_Government (2009). Defending Australia in the Asia Pacific Century: Force 2030. Defence White Paper 2009. D. o. Defence. Canberra, AGPS: 140.

Erickson, A. S., L. J. Goldstein, et al., Eds. (2007). China's Future Nuclear Submarine Force. Anapolis, Maryland, Naval Institute Press

Howarth, P. (2006). China's rising sea power : the PLA Navy's submarine challence / Peter Howarth. New York, NY :, Routledge.

Raman, R., R. Murphy, et al. (2005). Options for Initial Fuelling. Santa Monica, California, RAND Corporation.

Schank, J. F., M. V. Arena, et al. (2007). Sustaining U.S. Nuclear Submarine Design Capabilities. Santa Monica, California, RAND Corporation.

Schank, J. F., C. R. Cook, et al. (2005). Ministry of Defence Roles and Required Technical Resources, RAND Corporation.

Schank, J. F., J. Riposa, et al. (2005). Sustaining Design and Production Resources. Santa Monica, California, RAND Corporation.

Woolner, D. (2009). "Taking the Past to the Future: The Collins Submarine Project and SEA 1000." Security Challenges. Kokoda Foundation **5**(3): 47-71.

Yule, P. and D. Woolner (2008). The Collins Class Submarine Story. Steel, spies and spin. Melbourne, Cambridge University Press.

Hydrodynamic Design Optimization of a Hard Chine Planing Craft for Coastal Surveillance

A F Mohamad Ayob, T Ray and W F Smith
University of New South Wales @ Australian Defence Force Academy
Ahmad.MohamadAyob@student.adfa.edu.au

ABSTRACT

In this paper, an optimization framework for the design of planing craft with regards to resistance, stability and seakeeping consideration is presented. The framework incorporates a surface information retrieval module, geometry manipulation module, an optimizer and several accepted naval architectural estimation methods. The framework performs automatic surface information retrieval from offset data using B-splines and uses the resulting ship geometry for geometric variation through parametric transformation. A parametric transformation module that maintains the realistic hull form design is driven by an optimizer that further identifies the optimized candidate ship. Rough water resistance minimization is used as the objective function while stability and seakeeping criteria serve as constraints given the preliminary mission requirements. A single objective resistance minimization is carried out using Non-dominated Sorting Genetic Algorithm II (NSGA-II). In an attempt to further reduce the number of function evaluation required by NSGA-II a proposal to use surrogate assisted optimization is presented. The surrogate assisted evolutionary algorithm (referred here as EASDS) reduces the number of function evaluations. Since one can expect the optimum design to lie on constraint boundary, a proposal to use Infeasibility Driven Evolutionary Algorithm (IDEA) is also incorporated in this paper.

NOMENCLATURE

B	Beam (m)	P	Power (W)
$Disp$	Displacement (kg)	R_T	Total resistance (N)
Fn	Froude number	R_C	Calm water resistance (N)
GM	Metacentric height (m)	R_A	Added resistance (N)
$H_{1/3}$	Significant wave height (m)	T	Draft (m)
Ie	Half angle of entrance (degrees)	V	Speed (ms^{-1})
L	Length (m)	I_a	Impact acceleration (g)
C_b	Block coefficient	C_p	Prismatic coefficient

1 BACKGROUND

Optimization frameworks for the design of displacement crafts have been reported in literature such as works of Ray [1] and Ganesan [2] for the design of containerships, Smith [3] for design of frigates, Gammon [4] and Majumder , Akinturk and Calisal [5] for the design of fishing vessels. Such frameworks typically consider maximization of net present value or minimization of total resistance as a single objective optimization problem subject to constraints on stability, displacement etc. The optimization algorithms used in the abovementioned frameworks include simulated annealing [1], genetic algorithm[6] and particle swarm optimization [7].

Although a number of optimization frameworks for the design of displacement crafts have been successfully developed and used over the years, very little attention has been paid towards the design and development of optimization frameworks for planing crafts. The development of an optimization framework for the design of planing crafts is a non-trivial problem due to the lack of adequate data, experience and availability of estimation tools [8]. Some significant early works in this area include Almeter [9] on the resistance optimization of Soviet's BK and MBK planing hulls, Jons, Koelbel and Sheldon [10] on an integrated CAD process for the design of planing crafts, and more recently the efforts to minimize resistance of hard chine planing crafts by Mohamad Ayob, Ray and Smith [11]. In terms of the optimization methods used in above studies, proposals include evolutionary algorithms [11], systematic variations [9] and comprehensive design geometry library [10]. The formulations consider resistance minimization and improvements in the seakeeping performance.

Observations and a critical analysis of past approaches towards design of planing crafts highlight the following;

(a) The shape representation scheme employed should be able to handle one or more hard chine forms.
(b) The candidate hull forms should be intrinsically smooth [12].
(c) The underlying naval architectural estimation methods such as resistance estimation, estimation of seakeeping performance, etc. should be valid for the type of vessel under consideration.
(d) The optimization method used should be able to seamlessly deal with single and multi-objective optimization problems with constraints.
(e) Since the performance analyses of candidate designs are likely to be computationally expensive, the optimization method should have a high rate of convergence in order to be effective.

This paper presents an optimization framework for the design of planing crafts. The framework incorporates a surface information retrieval module, geometry manipulation module, an optimizer and several accepted naval architectural estimation methods. The framework performs automatic surface information retrieval from offset data using B-splines and uses the resulting ship geometry as the parent ship. The resistance estimation is based on Savitsky [13], while the seakeeping performance is computed using Savitsky-Brown [14] derived from Fridsma's experimental tank test [15]. The underlying optimization algorithm (NSGA-II) is a population-based, stochastic, zero-order form. The performance of the algorithm is further improved through a surrogate assisted optimization model [16] which is embedded with multiple surrogates such as ordinary response surface method (ORSM), normalized response surface method (RSM), ordinary radial basis function (RBF), ordinary radial basis function (ORBF) and kriging method (DACE). Further improvements in design are achieved through the use of Infeasibility Driven Evolutionary Algorithm (IDEA) [17] which is known to converge faster for constrained optimization problems.

The following sections provide an overview of the framework along with the necessary details of its components. The mathematical formulations of resistance minimization subject to constraints on seakeeping performance is presented for a well-studied coast guard craft by White and Savitsky [18]. The performance of NSGA-II, IDEA and EASDS is studied within the proposed framework.

2 THE COMPONENTS OF THE OPTIMIZATION FRAMEWORK

The optimization framework consists of: (a) geometry manipulation module, (b) estimation method for resistance and seakeeping and (c) optimizer. The next subsections will elaborate the components of the optimization framework summarized in Figure 1.

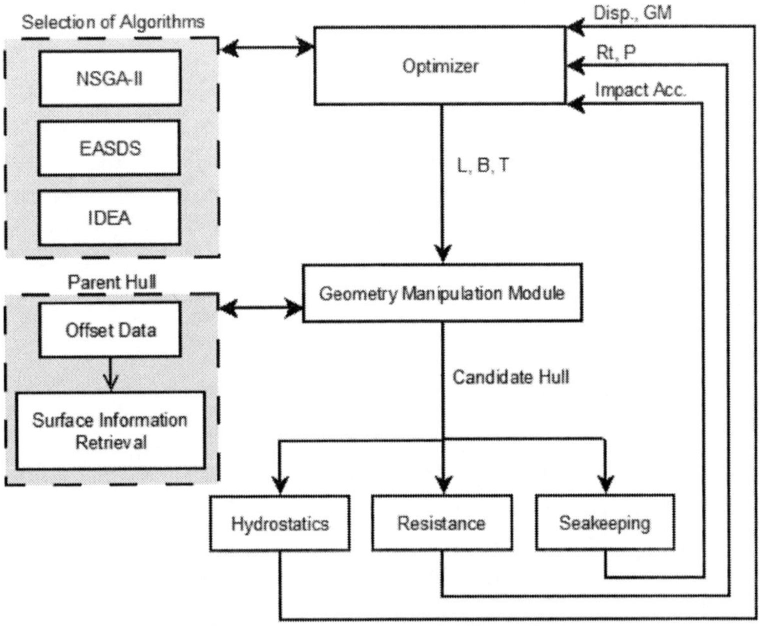

Figure 1: Framework Flowchart

2.1 Geometry Manipulation Module

In this paper, surface information retrieval module [11] and linear parametric transformation using NURBS surface offered by Maxsurf [12] based on Lackenby's method [19] is incorporated, due to its capability to preserve the fairness of the resulting surface and maintaining hull coefficients such as C_p and C_b of the parent hull. Hull form's length (L), beam (B) and draft (T) will serve as the input given by the optimizer during the optimization process.

2.2 Resistance Estimation Method

The resistance estimation of a planing craft is dictated by its speed coefficient. In the example (Savitsky and Brown) discussed later in the paper, the speed coefficient is more than 1.5 and hence the resulting dynamic lift forces cause a significant rise of the center of gravity, positive trim, emergence of the bow and separation of the flow from the hard chines. Also, for this value of speed coefficient, resistance estimation of Savitsky-Brown is valid.

The details of the coast guard vessel studied in this paper are presented in Table 1.

Table 1: Characteristics of a typical 110 ft. coast guard patrol boat

Displacement (kg)	118246.6
Length (m)	32.93
Beam (m)	7.52
Draft (m)	1.6
VCG (m)	2.7
GM	1.64
Speed (knots)	30
$C_v = V/\sqrt{(g*B_T)}$	1.797

2.3 Seakeeping Estimation Method

Based on the earlier discussion regarding the available seakeeping estimation methods, Savitsky-Brown [14] equations derived from Fridsma's [15] work are used in this paper. Irregular waves with energy spectrum of Pierson-Moskovitz characterized by significant wave height, $H_{1/3}$ have been used in the application. The details of the seakeeping equations (Added Resistance, Average Impact Acceleration at CG, and Average Impact Acceleration at Bow) governing the behaviour of planing craft at seaway can be found in [14].

2.4 Optimization Methods

In this framework, the underlying optimization algorithm used a population-based, zero-order, stochastic algorithm known as Non-dominated Sorting Genetic Algorithm II (NSGA-II). This particular algorithm is by far able to solve wide range of engineering problems, and for the details of the algorithm readers may refer to Deb *et. al* [20].

Since NSGA-II or any other form of evolutionary algorithm requires the evaluation of numerous candidate solutions prior to its convergence, their applicability is restricted for computationally expensive optimization problems. In order to contain the computational time within affordable limits, surrogates or approximations can be used. In this study an evolutionary algorithm with spatially distributed surrogates (EASDS) is used to illustrate the benefits offered by surrogate models. The complete details of EASDS are presented in [16].

Solutions to real-life constrained optimization problems are often likely to lie on constraint boundaries. NSGA-II and most other optimization algorithms intrinsically prefer a feasible solution over an infeasible solution during the search. However, some recent works suggest that effectively utilizing the marginally infeasible solutions during the search can expedite convergence rate. To this effect, Infeasibility Driven Evolutionary Algorithm (IDEA) [17] was suggested. In the presented studies, the performance of IDEA is also studied within the optimization framework.

3 NUMERICAL EXPERIMENTS

3.1 Optimization Problem

A typical coast guard planing craft based on White-Savitsky's [18] experiment is considered as shown in Figure 2. The hull form is constructed from a set of coordinate points (offset data) where the body plan is shown in Figure 2(a). B-splines surface information retrieval method [11] is applied to produce the ship surfaces as shown in Figure 2(b) that will be used on the later stage of the framework. The details of the ship are tabulated in Table 1.

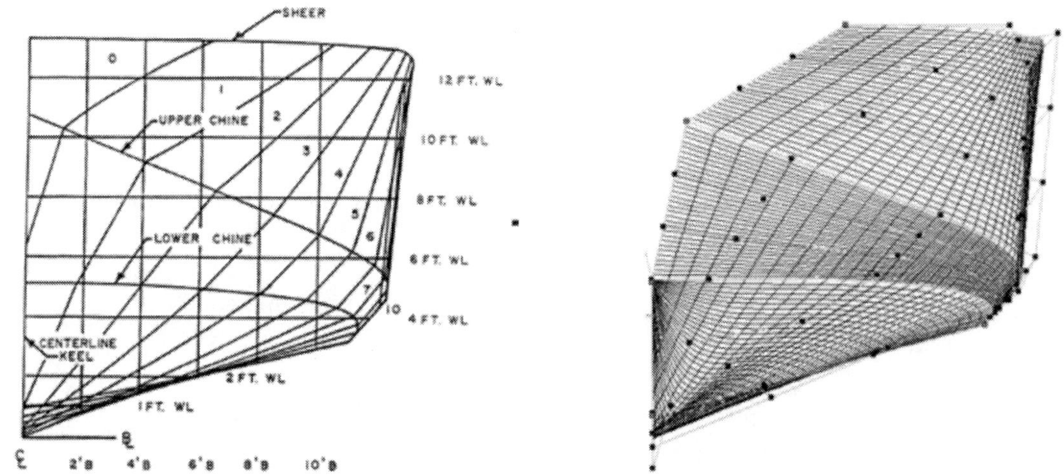

(a) Ship lines (b) B-splines surfaces

Figure 2: Body plan view of the parent hull

The optimization problem is described as identification of a coast guard patrol boat with minimum total resistance subject to the constraints on displacement, stability (transverse metacentric height) and impact acceleration corresponding to the operational sea-states. Seaway operational scenario with three sea-states conforming to Pierson-Moskovitz energy spectrum represented by significant wave height $H_{1/3}$ is chosen, as shown in Table 2. The objective functions and constraints are listed as below where subscripts 'B', 'I', 'T', 'C' and 'A' resemble basis hull, candidate hull, total resistance, calm water resistance and added resistance due to waves, respectively.

Minimize:

$f(1) = R_T$, where $R_T = R_C + R_A$

Design variables:

$29.93m < L < 35.93m \; (L_B = 32.93m)$

$6.51m < B < 8.51m \; (B_B = 7.51)$

$1.5m < T < 1.7m \; (T_B = 1.59m)$

Constraints:

$g(1) = Disp_I > Disp_B$

$g(2) = GM_I \geq GM_B$

$g(3) = 3.07 < L_I/V_I^{1/3} < 12.4$

$g(4) = 3.7^o < Ie_I < 28.6^o$

$g(5) = 2.52 < L_I/B_I < 18.28$

$g(6) = 1.7 < B_I/T_I < 9.8$

$g(7) = I_a < 1.5$

Table 2: Operational sea-states of the optimization formulation

Sea-state	$H_{1/3}$ (m)
Sea-state 3	1.34

Sea-state 4	2.32
Sea-state 5	3.62

3.2 Simulation Results

For each algorithm (NSGA-II, IDEA and EASDS), ten independent runs are done. A population size of 40 has been used for each algorithm. The crossover and mutation parameters used are listed in Table 3. The number of evaluations used by each algorithm is kept approximately equal for fair comparison. The surrogate models used are restricted to RSM, ORSM, RBF, ORBF and DACE. A training period of 3 and prediction error of 0.05 has been used for EASDS.

Table 3: Summary on the optimization parameters

Parameters	NSGA-II	EASDS	IDEA
Crossover Probability	1.0	1.0	1.0
Mutation Probability	0.1	0.1	0.1
Crossover distribution index	10	10	10
Mutation Distribution Index	20	20	20

The results across 10 runs for best, median and worst designs are tabulated in Table 4, 5 and 6 respectively. This information provides an indication on reliability of the three optimization algorithms used in this framework in perspective of resistance reduction. Percentage of minimized resistance is determined using the expression below:

$$\% \text{ of Minimized Resistance} = \frac{\text{Basis Hull Resistance - Optimized Hull Resistance}}{\text{Basis Hull Resistance}} \text{ x } 100\%$$

Table 4 shows the best design for various sea states across 10 optimization runs. It can be seen that for sea-state 3 and 4, IDEA and EASDS was able to perform better than NSGA-II. However for sea-state 5, IDEA is at par with EASDS while NSGA-II performs better than the two. In contrary for both sea-state 3 and 4, IDEA provides better best designs compared with EASDS. However, EASDS was able to build the surrogate model thus reducing the computational effort while performing better than NSGA-II for sea-state 3 and 4.

The values of median design across 10 runs are tabulated in Table 5. It can be observed that EASDS is able to provide better solution than IDEA and NSGA-II for sea-state 5 with less number of function evaluations and IDEA consistently reports better design solution than NSGA-II for all sea-states. The same scenario can be observed for the worst design across 10 runs as shown in Table 6. Even though EASDS is able to report better solution for sea-state 5, IDEA consistently outperformed NSGA-II in providing optimized design for all sea-states.

In this particular case study as presented in Table 4, 5 and 6, close observation on the results shown that both EASDS (with lesser number of function evaluations) and IDEA consistently reports better results compared with NSGA-II.

Table 4: Summary for the best design across ten runs

		NSGA-II	EASDS	IDEA
Sea State 3	Avg. Num. of Fun.Eval.	720	688.4	720
	% of Minimized Resistance	7.35	7.71	**7.74**
Sea State 4	Avg. Num. of Fun.Eval.	880	856.6	880
	% of Minimized Resistance	5.69	5.94	**6.11**
Sea State 5	Avg. Num. of Fun.Eval.	760	744.3	760
	% of Minimized Resistance	**4.19**	4.02	3.99

Table 5: Summary for the median design across ten runs

		NSGA-II	EASDS	IDEA
Sea State 3	Avg. Num. of Fun.Eval.	720	688.4	720
	% of Minimized Resistance	6.96	7.37	**7.39**
Sea State 4	Avg. Num. of Fun.Eval.	880	856.6	880
	% of Minimized Resistance	5.08	4.61	**5.53**
Sea State 5	Avg. Num. of Fun.Eval.	760	744.3	760
	% of Minimized Resistance	3.41	**3.89**	3.63

Table 6: Summary for the worst design across ten runs

		NSGA-II	EASDS	IDEA
Sea State 3	Avg. Num. of Fun.Eval.	720	688.4	720
	% of Minimized Resistance	5.59	5.81	**6.35**
Sea State 4	Avg. Num. of Fun.Eval.	880	856.6	880
	% of Minimized Resistance	2.90	3.88	**4.58**
Sea State 5	Avg. Num. of Fun.Eval.	760	744.3	760
	% of Minimized Resistance	2.51	**2.77**	2.71

3.3 Scenario Examples

The results presented in previous subsection are extended to provide an insight of the operational scenario to assist planing craft designers in choosing the overall best design with respect to the value of total resistance and impact acceleration in a particular sea-state. Each optimized candidate design in their respective sea-states is allowed to operate in all sea-states to visualize the design that will operate well in all conditions. The comparisons of performance are judged based on the value of total resistance and the impact acceleration at the given sea-state for each optimized hull.

Each optimized design produced using NSGA-II, in their respective sea-state 3, 4 and 5 is allowed to operate in all sea-states as shown in Figure 3. It can be seen that the candidate design optimized at sea-state 5 performs better as it has lesser total resistance than the candidate design optimized at sea-state 3 and 4 in all sea-states as shown in Figure 3(a), (b) and (c). On the contrary, the ship optimized by EASDS at sea-state 4 in Figure 4(a) and Figure 4(b) has outperformed both optimized candidate design in sea-state 3 and 4. If lower impact acceleration is favoured in sea-state 5 operation, the ship optimized at sea-state 4 presented in Figure 4(c) has an overall advantage for all operational scenarios.

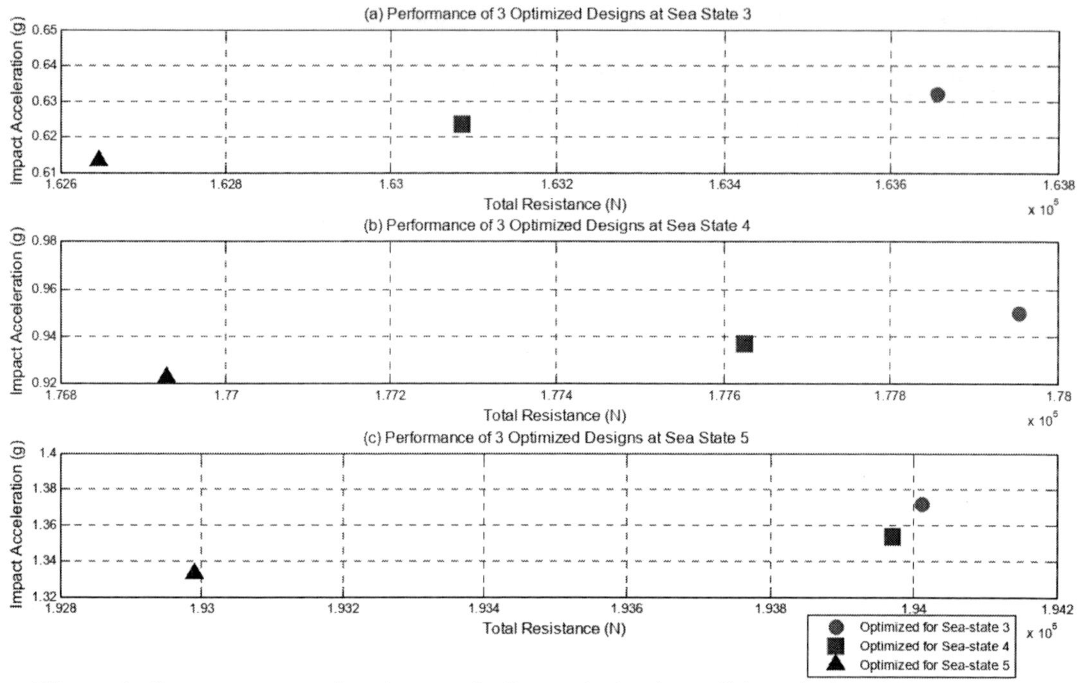

Figure 3: Sea-state operational scenario for optimized candidate design (NSGA-II)

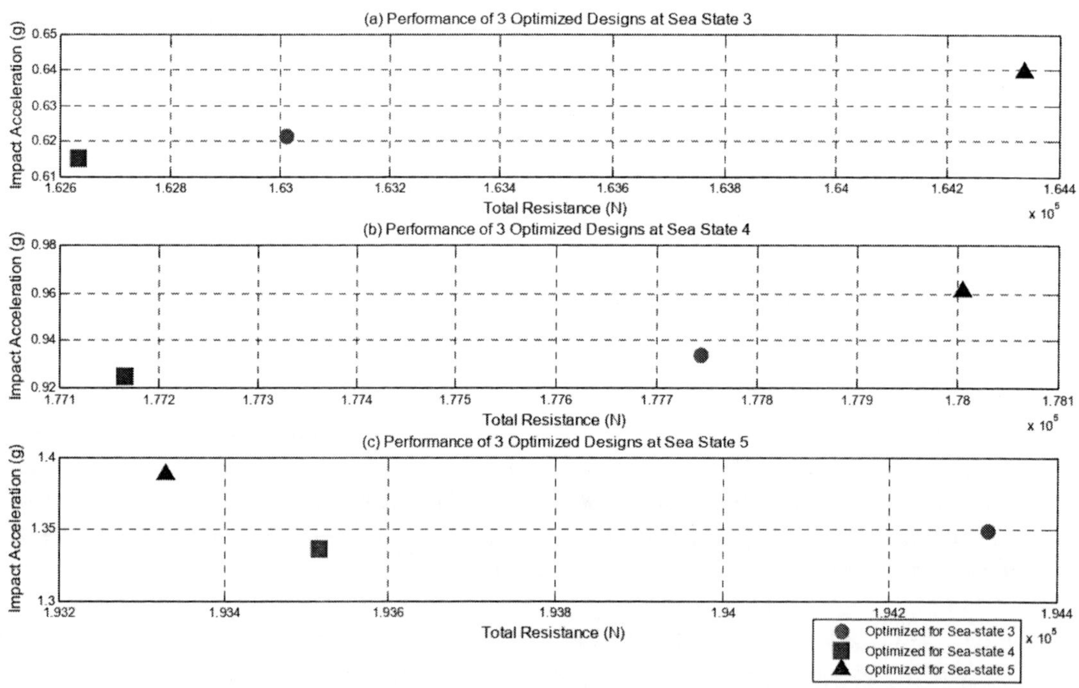

Figure 4: Sea-state operational scenario for optimized candidate design (EASDS)

Each optimized design produced using IDEA, in their respective sea-state 3, 4 and 5 is allowed to operate in sea-state 3 as shown in Figure 5(a). It can be seen that the optimum design for sea-state 4 performs better as it has lesser total resistance than the candidate design that is optimized at sea-state 3. However the candidate design that is optimized for sea-state 5 operates in sea-state 3 is having the largest resistance and impact acceleration compared with the two previous designs. The same pattern is observed for all optimized designs allowed to operate in sea-state 4 as shown in Figure 5(b). In Figure 5(c), the candidate design optimized in sea-state 5 has less resistance than the candidate design optimized in sea-state 3, while the candidate design optimized in sea-state 4 has less resistance and impact acceleration for all sea-state operational scenarios compared with the candidate design that is optimized at sea-state 3 and 5.

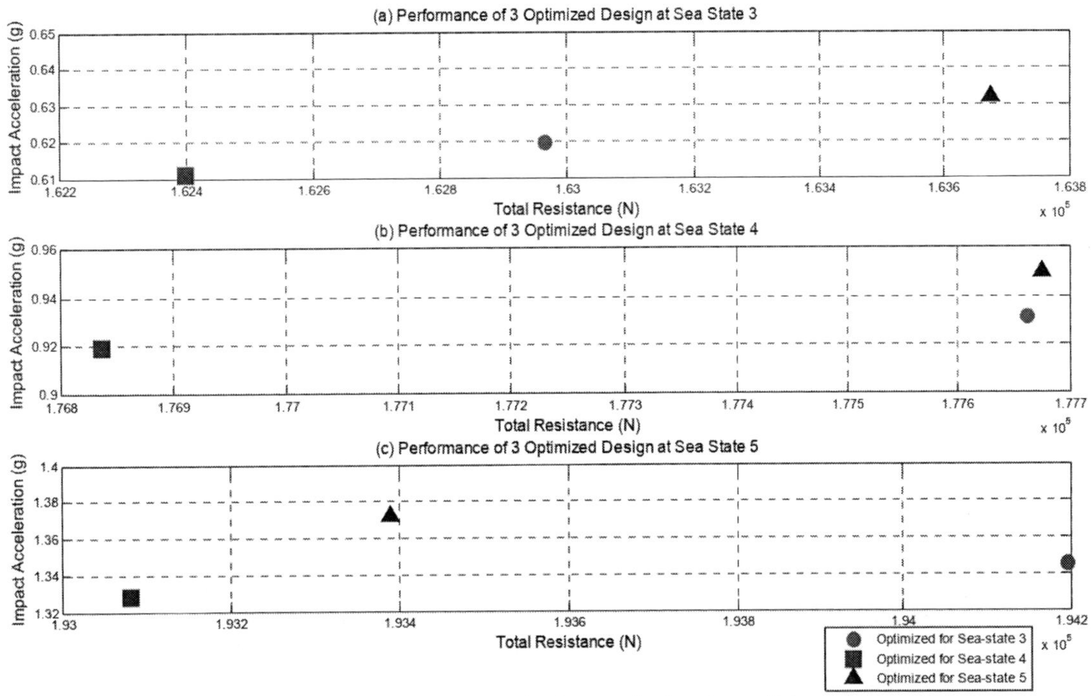

Figure 5: Sea-state operational scenario for optimized candidate design (IDEA)

4 SUMMARY AND CONCLUSIONS

A hydrodynamic design optimization framework for a hard chine planing craft has been presented in this paper. The proposed framework incorporates three evolutionary algorithms, namely NSGA-II, EASDS and IDEA and extends the previous work of Mohamad Ayob, Ray and Smith [11] by incorporating seakeeping consideration during optimization. Optimization problem of a typical coast guard planing craft in sea-state 3, 4 and 5 is formulated and it is seen that IDEA and EASDS consistently perform better than NSGA-II in providing the resistance-optimized candidate design. Examples of operational scenario are presented using the data of the optimized designs identified using abovementioned algorithms to assist planing craft designers in choosing the best overall design in terms of the value of total resistance and impact acceleration in a particular sea-state. The identification of candidate design that operates well in all sea-states using NSGA-II, EASDS and IDEA clearly highlights the benefit of the proposed framework for planing craft designers.

ACKNOWLEDGMENTS

Various people helped in the development of this project. Thanks are due to Dr. Patrick Couser and Mr. Andrew Mason from Formation Design Systems (FormSys) for providing the Maxsurf Suite and support. Also thanked are colleagues Mr. Amitay Isaacs and Mr. Hemant Kumar Singh for assisting in programming the methods.

REFERENCES

[1] T. Ray, "Preliminary Ship Design in the Multiple Criteria Decision Making Framework," Kharagpur, PhD Thesis, 1995.

[2] V. Ganesan, "A Model for Multidisciplinary Design Optimization of Containership," Masters Thesis, 1999.

[3] W. F. Smith, "The Modeling and Exploration of Ship Systems in the Early Stages of Decision-Based Design," PhD Thesis, 1992.

[4] M. A. Gammon, "Ship Hull Form Optimization by Evolutionary Algorithm," Istanbul, PhD. Thesis, 2004.

[5] M. Majumder, A. Akinturk, and S. M. Calisal, "Optimized Design of Small Craft," *Marine Technology,* vol. 39, pp. 67-76, April 2002.

[6] A. H. Day and L. J. Doctors, "Resistance Optimization of Displacement Vessels on the Basis of Principal Parameters," *Journal of Ship Research,* vol. 41, pp. 249–259, December 1997.

[7] A. Pinto, D. Peri, and E. F. Campana, "Multiobjective Optimization of a Containership using Deterministic Particle Swarm Optimization," *Journal of Ship Research,* pp. 217-228, 2007.

[8] D. Hendrix, S. Percival, and F. Noblesse, "Practical Hydrodynamic Optimization of Monohull," SNAME Annual Meeting, 2001.

[9] J. M. Almeter, "Resistance Prediction and Optimization of Low Deadrise, Hard Chine, Stepless Planing Hulls," in *Ship Technology and Research Symposium (STAR)*, 1995.

[10] O. P. Jons, J. Koelbel, and R. Sheldon, "A New Generation of High-Performance Planing Craft," *Naval Engineers Journal of Ship Research,* pp. 234-247, 1985.

[11] A. F. Mohamad Ayob, T. Ray, and W. Smith, "An Optimization Framework for the Design of Planing Craft," in *International Conference on Computer Applications in Shipbuilding 2009 (ICCAS09)*, Shanghai, 2009, pp. 181-186.

[12] *Maxsurf User Manual Version 13*: Formation Design Systems, 2007.

[13] D. Savitsky, "Hydrodynamic Design of Planing Hull," *Marine Technology,* vol. 32, October 1964.

[14] D. Savitsky and P. Ward Brown, "Procedures for Hydrodynamic Evaluation of Planing Hulls in Smooth and Rough Water," *Marine Technology,* vol. 13, pp. 381-400, October 1976.

[15] G. Fridsma, "A Systematic Study of the Rough-Water Performance of Planing Boats (Irregular Waves - Part II)," 1971.

[16] A. Isaacs, T. Ray, and W. Smith, "An Evolutionary Algorithm with Spatially Distributed Surrogates for Multiobjective Optimization," in *Progress in Artificial Life*, ed: Springer Berlin / Heidelberg, 2007, pp. 257-268.

[17] T. Ray, H. K. Singh, A. Isaacs, and W. Smith, "Infeasibility Driven Evolutionary Algorithm for Constrained Optimization," in *Constraint-Handling in Evolutionary Optimization*, ed: Springer Berlin / Heidelberg, 2009, pp. 145-165.

[18] J. A. White and D. Savitsky, "Seakeeping Predictions for U.S.C.G Hard Chine Patrol Boats," in *N.Y. Metropolitan Section of SNAME*, 1988.

[19] H. Lackenby, "On the Systematic Geometrical Variation of Ship Forms," *Trans. INA,* vol. 92, pp. 289-315, 1950.

[20] K. Deb, A. Pratap, S. Agarwal, and T. Meyarivan, "A Fast and Elitist Multiobjective Genetic Algorithm: NSGA-II," *IEEE Transactions on Evolutionary Computation,* vol. 6, pp. 182-197, 2002.

Advanced Anchor System for Submarines

John McLaughlin, © ASC Pty Ltd, john.mclaughlin@asc.com.au
Tommy Andreasson, ASC Pty Ltd, tommy.andreasson@deepbluetech.com.au
Dr Colin Kestell, The University of Adelaide, colin.kestell@adelaide.edu.au

ABSTRACT
With submarines increasingly operating in shallow waters, greater opportunities to make use of the anchoring system arise, compared to blue water operation. The challenge for the designer is to create an anchoring system that is reliable, is silent in operation, maintenance friendly, and most importantly having good holding performance. Additionally having the ability to use the anchoring system when the submarine is submerged would give improved mission capability in littoral waters. For this project some demanding requirements were set, including the need for low acoustic signature during release and retrieval while maintaining good retrieval and release times. Improvements in the reliability of stowage and integration of the total anchoring system into a submarine were other key drivers.

An advanced anchoring concept for submarines has been prepared using lessons from anchoring advances outside the naval industry. The concept design activities that were undertaken used both published data and verification by new testing of scale models. The final design was then analysed for estimated performance and integration of the total system into a submarine. The new conceptual design optimises key anchor performance parameters, such as holding power and weight and space budgets. The finished concept compares well to the chosen benchmark anchoring system and has potential of offering new levels of performance and reliability to military submarines.

BACKGROUND

Throughout naval history vessels have needed the ability to remain stationary relative to the seabed, which has been achieved principally by an anchoring system. This project has been limited to naval submarines, with their unique requirements of maximising submerged time and minimising detectable signatures. Increasingly operations are in shallow waters, and there may be a need to remain stationary to gather intelligence or deploy missions. Safety of the vessel is also a concern when close to obstructions or shoreline, with anchor failures on surface vessels having caused loss of ships [1][2].

Submarines have generally used modified surface vessel anchors, hence there may be opportunity to optimise an unique anchor design for submarine use, and incorporate advances made by modern anchors, [3][4][5]. According to published data submarines also use chain line [6][7], indicating there is opportunity for signature improvements.

The primary aim of this project was to improve the performance of submarine anchor systems while enabling the capability of anchoring when submerged. Potential areas for improvement being the ability to set and hold, minimise weight, signature, maximise operational speed, reliability of stowage and remote operation. The focus was on improving the areas of anchor and line design, with some consideration of ancillary components required for integration into a submarine.

The approach undertaken was to perform initial research and theoretical analysis and concept generation followed by some verification by calculation and by scale model testing. The result is a new concept for an anchoring system referred to as the Improved Submarine Anchor System, ISAS.

LITERATURE REVIEW

A significant amount of literature is available from the offshore oil industry, surface vessels, and leisure boat industry for anchor design. The largest body of recent research has been undertaken to support the offshore oil industry for anchoring and mooring of large offshore structures [8], and while different to ship anchoring, the research provides understanding on how anchors function.

Some key anchor performance drivers were highlighted and then used as guidelines for generating new concepts for submarine application, with the key anchoring system attributes used for this project listed in Table 1. Veering of an anchor refers to pulling the anchor in a different direction to the initial set, such as when the sea current direction changes. Scope is the ratio of the length of cable to water depth, with short scope meaning there will be a larger than zero angle of the line to the bottom.

Table 1. Key attributes for Anchoring

No.	Key Attribute
1	Remote operation for underwater functionality
2	Low Noise signature
3	Reliability of setting into ground
4	Holding power adequate, reliable, and weight efficient
5	Ease of retraction of anchor from the ground
6	Reliability of stowage
7	Holding power at veering horizontal load directions
8	Holding power at short scope
9	Time to deploy/retract
10	Space and weight

Each of these attributes can be further broken down into key performance features. For example the ability of an anchor to initially set is related to the anchor weight, which assists in the initial penetration [11], and the holding capacity is related to the fluke area [3]. References [12] and [13] list results for setting and holding power of full sized anchors in several locations and seabed types. Other important factors were found such as, low resistance to burial to allow for improved setting and deeper penetration, and the anchor stability to prevent breakout [3][19]. Having the correct fluke angle for the different soil types was another critical aspect [3] and single fluke designs appeared to perform better at larger angles to the seabed [16][17].

Many types of anchors are available with advantages depending on the seabed condition and application method, [10]. Large increases in holding capacity are possible with some modern anchors, [15], but for application to a submarine there are many practical issues to be resolved. These include noise reduction during operation, stowage issues, ease of use, retrieval and cleaning of the anchor. It was found there was little information on veering the angle of pull, as well as shortening the scope and retrieval of the anchor. However there is good information on anchor holding performance for some types of seabeds.

A claw shaped, twin shank, fixed fluke anchor, the Bruce TS, was selected as a current high performance anchor for comparison with a new submarine design, as good data was available, refer to Figure 1. It has the advantage of being able to turn 360°, when the load veers, without breaking out [19]. The flukes are fixed rather than pivoting, avoiding some issues with jammed flukes. Recent designs for the leisure boat industry as per Figure 2, are predominantly variations of the fixed fluke design [18].

Figure 1 Anchor Scale Model based on the Bruce TS

Figure 2 Typical plough shaped anchor on a small vessel

BENCHMARK SELECTION

Options for an existing submarine benchmark anchor are very limited, with standard surface vessel anchors not being suitable as they do not mate flush with the hull. The AC.14 is a hinged anchor with a version that is adapted for submarine use [6]. A similar submarine anchor is the AC.16 [9], and it was selected as the benchmark, as it had slightly better holding and veer performance when tested as part of this project. Figure 3 shows a simplified model of the AC.16 design, with stabilisers added to the outside of the crown for improved veer performance. From published data it appears chain is the most commonly used solution for the anchor line, therefore this was selected as the benchmark [6].

Figure 3 Benchmark Anchor Model, a simplified version of the AC.16 type anchor

CONCEPT DEVELOPMENT AND TESTING PROCEDURE

Hybrid anchor concepts were developed that combined features from existing anchors. Other existing high performing anchors were selected, then models of these were fabricated, tested and compared to the benchmark and some hybrid concepts. Scale model testing of anchors can be found in the literature [11][20][21], with modern techniques using a centrifuge to create representative pressures that would be experienced at depth.

As fairly good holding force data is available for various anchor sizes, the goal was to compare the test data of different models to get relative ratings rather than accurate scalable results. Therefore testing scale models at 1g is adequate to get this relative data [20][21]. Also for this reason accurate soil properties were not essential. The model anchors were tested in sand, clay and mud, with testing performed by pulling the anchors at various angles in the test beds and measuring the holding force. A total of 13 basic model anchors were made, and from these many variations were tested, trialling different stabilisers, positions, angles and proportions. Only a few of the relevant anchor results will be listed for clarity.

EXPERIMENTAL RESULTS

Figure 5 summarises the results from testing scale models of four anchors, the benchmark, the Bruce TS and two versions of hybrid anchors based on the twin shank fixed fluke type anchors, (Hybrid 1 and 5 per Figure 4).

Holding Power

For holding power, which is the holding force divided by the anchor weight, the benchmark anchor performs well in these tests, as do the Hybrid and the Bruce TS anchors. The hybrids performed the best for overall holding power across the soil types.

Setting
The ease of initial setting determined the ability to self right when landing on the seabed in the wrong orientation, and then digging into the seabed and staying imbedded. The benchmark didn't score well here, as it tended to roll to the surface when dragged and sometimes the flukes did not trip. Hybrid 1 self righted well, setting nearly every time, whereas Hybrid 5 did not in some soil types. Hybrid 5 was a larger and heavier version of Hybrid 1, with modifications to improve mating to the submarine. All anchors had difficulty setting in firm clay, with the literature indicating that difficulty in setting in hard clay is a common problem [14]. Heavier anchors have more likelihood of penetrating as the stiffness of the clay increases [11] [22], explaining the poor performance of the models.

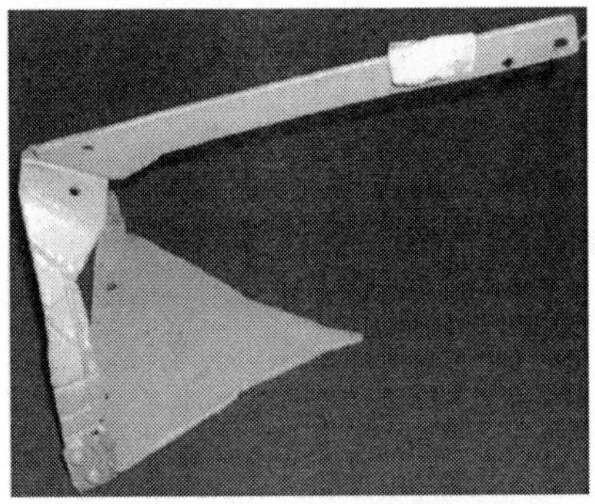

Figure 4 Model anchor, Hybrid 5, a strengthened version of Hybrid 1.

Veer
The benchmark anchor performed poorly for veering load, as it was unstable, and rolled to the surface when a load at ninety degrees was applied in sand. The hybrid anchors performed exceptionally in this test, being very stable as they were dragged.

Increasing Angles
The benchmark anchor did not hold as well when the angle was increased, reducing by about 50%, while the Hybrid 5 anchor only reduced in holding power by about 20%.

Other Attributes
Other key attributes were tested such as ease of break out, cleaning the anchor and fouling of the line. Break out is the force required to retract the anchor, with all anchors having reasonably low forces required. Sand and clay adhering to an anchor may impede the housing of the anchor back in the submarine. Such was the case for the benchmark anchor, which was found to jam with mud or sand in the open fluke position. The anchor is usually housed in the lower bow area, with a recess shaped to match the crown of the anchor. In this hard to access area, jammed flukes would present a serious problem.

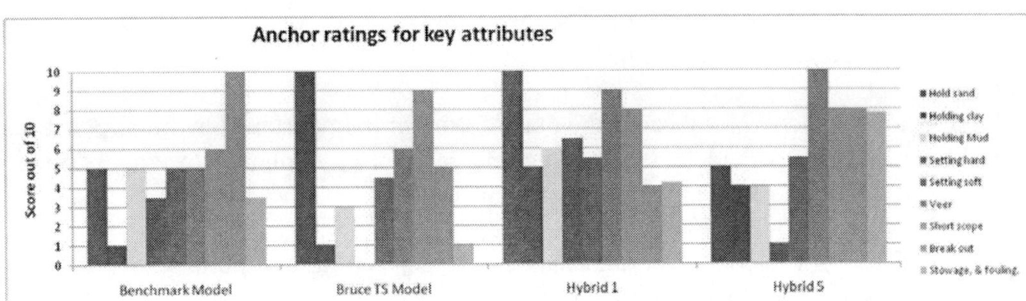

Figure 5 Anchor ratings for key attributes.

Comparison to Published Data

There was some significant potential for experimental errors in the testing, as the main aim was for comparison and not scalable data. Possible causes of error include the soil not being representative of actual sea bottom conditions and pressures, measurement errors, and build inaccuracies due to lack of accurate drawing data.

The curves from published data have been derived from experimental testing of anchors of various sizes [4]. The anchor design of Hybrid 5 has similarities with the Bruce TS and Stevpris anchors, which have identical scaling curves according to reference [4]. Three of the models are compared to the published data in Figure 6. The points for Hybrid 5 and the Bruce TS from model testing in this project are close to the Bruce TS published curve. Due to the small sample size caution should be taken in any conclusions made, however results are in the approximate range expected.

Using the data from Figure 6 and assuming that Hybrid 5 follows the curve for the Burce TS anchor, the weight can be obtained for the anchor required to achieve the maximum holding force for a conventionally sized submarine, giving an anchor weight of about 500kg.

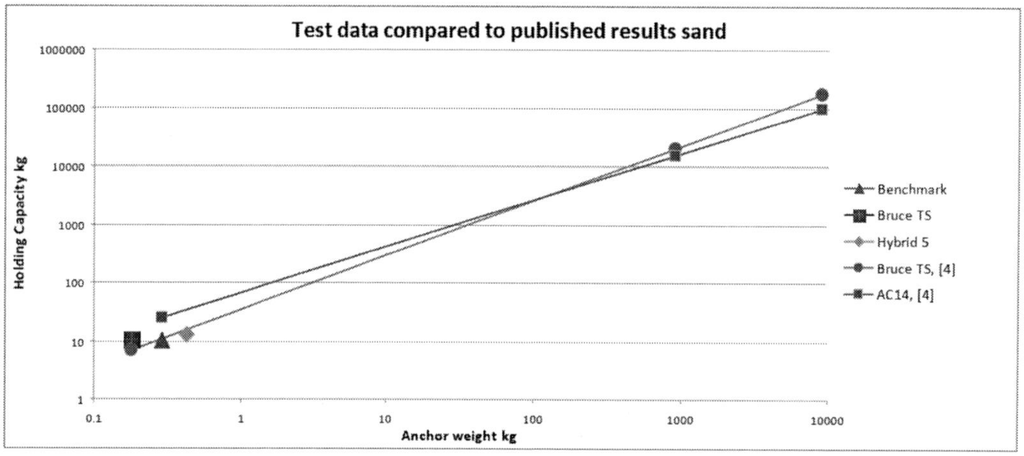

Figure 6 Comparison with published results, [4].

DISCUSSION

In the literature there are some general patterns that emerged such as from references [15] and [21] indicating that a 50% reduction in holding force can be expected with an angle of about 10°. Reduced angle testing with setting at about 10° agreed with this for the benchmark model and Bruce TS model. However for the fixed single fluke anchors such as Hybrids 1 and 5 the reduction was only 20%. This agrees well with reference [17] which indicated that for a 10° angle the single fluke anchor only looses 20% of its holding power, while others loose much more.

Reference [11] and [22] indicated the anchor weight is important for initial penetration of the fluke into the seabed to set the anchor. This may explain why some trouble was experienced with most anchors in harder surfaces.

The hybrid design rated well in all areas except for setting and holding in clay and mud. Further research could be undertaken in these areas however it is expected that a larger heavier anchor would improve the ratings for hard ground. Combining the features of

Hybrid 1 and Hybrid 5 should enable successful setting while being suitable for integration into the submarine.

Hybrid 5, refer to Figure 4 was the final model manufactured, and attempted to duplicate the successful features of Hybrid 1 while making it more suited to integration into a submarine. The central shank connects to the roll bar, creating less surface area for soil to adhere to. The small ribs close to the top end of the shank are for alignment with the submarine. Figure 7 is a drawing of a modified version of the Hybrid 5 to include the successful features from Hybrid 1. Further testing is required to optimise the shape for performance and strength.

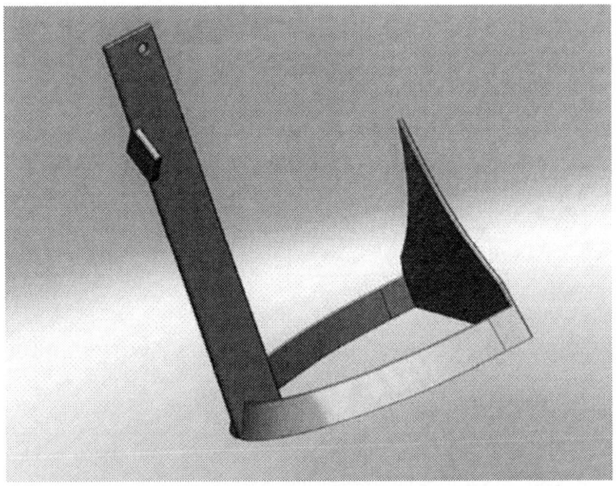

Figure 7 Improved Hybrid 5 CAD model.

As Hybrid 5 is adapted for use on a submarine, it is the basic design which is recommended as the initial concept solution. When features from the Hybrid 1 anchor are combined it provides a solution with superior anchoring performance in practically all areas compared to the Benchmark option.

ANCHOR LINE

The anchor line performs the critical function of pulling the anchor horizontally into the seabed. If higher angles to the seabed are used significantly lower holding power will result. Heavy chain line is generally used in anchoring which provides the horizontal pull at the anchor. Wire rope or synthetic cable are used to save weight and cost, especially for deep water offshore platform moorings, but their low weight make them poor for establishing a catenary curve and the horizontal pull [23]. Chain however emits excess levels of noise when being belayed and has high weight, making it a poor choice for a naval submarine. Figure 8 shows the catenary curves for the three types of equivalent strength cables with a 20kN load. This demonstrates clearly the advantage of chain, in that a much shorter length is required.

Another option to obtain the horizontal pull on the anchor is to use a sinker, that is, a weight that is lowered part way down the line, to increase the sag of the cable and assist with snatch loads. Adding a heavy weight to the cable ensures less cable is needed while still achieving a zero pull angle at the anchor. For wire rope a large effect is achieved by adding the weight, as shown in Figure 9, with the required length reduced by about 80%. This diagram shows a wire cable and synthetic line with the same weight sinker of 500kg added close to the anchor compared to a standard chain of similar strength, resulting in little difference between the various line lengths. However when the total weights of the lines are considered, chain is significantly heavier, as shown in Table 2. A line of synthetic material gives the lightest weight and when used with a sinker the project requirements of low weight, low signature and fast operation can be satisfied.

Table 2. Weight and length comparisons for different line types at 20kN load

Line Type	Total weight in air [kg]	Total Length [m]
Chain	10620	296
Wire + sinker	1114	361
Synthetic + sinker	587	380

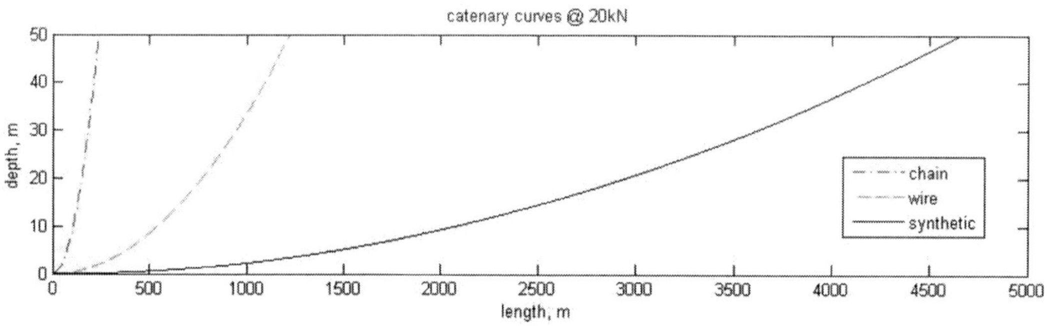

Figure 8 Chain, wire and synthetic cables, catenary curves with 0° angle to the seabed

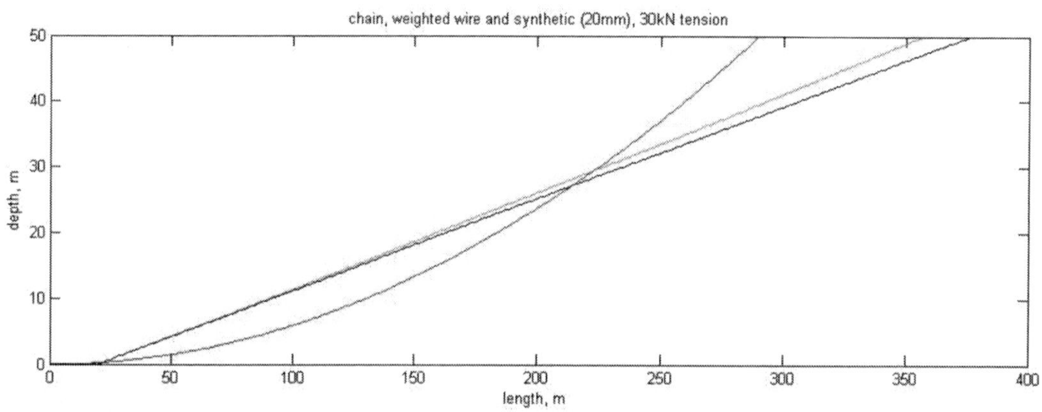

Figure 9 Three cable types, with 500kg weight, red: chain 7/8", green: 20mm wire cable with 500kg sinker and black: 20mm synthetic cable with a 500kg sinker.

ANCILLARY EQUIPMENT

Some equipment commonly used for anchoring system operation would be a winch or windlass, a motor or drive mechanism, a braking mechanism, and a control and monitoring system. The anchor and line also need to be stowed so as to form a flush surface with the submarine hull when not in use. The ISAS has been designed for no specific submarine and assuming space is available in the bow free flood area. The lower bow area would be used to stow the anchor and the upper bow for the winch and motor. The ancillary equipment was considered for the ISAS but limited analysis was undertaken and will thus not be discussed in detail.

INTEGRATED SOLUTION

The integrated solution requires further development but would include a sinker weight provided by a heavy mass fixed to a synthetic line a few metres from the anchor. The winch drum would be mounted high in the bow for easy access. The anchor would self orientate as it mates with a shaped hole provided in the lower bow.

The ISAS concept is compared to the benchmark total anchoring system in Table 3, which lists the expanded project requirements and rates each out of 10, with the higher rating option for each requirement highlighted green. Taking a simple average of the scores, the benchmark system scores 4.9 indicating that it is functional but opportunity for improvements are available. The ISAS scores much higher at 7.4 and has major advantages in signature reduction, weight reduction, stability of the anchor at varying angles of pull and reliability of stowage, which are vital areas in submarine performance.

Table 3 Benchmark compared to the ISAS (Improved Submarine Anchor System)

Requirement	Benchmark	ISAS	Comments
Improve anchor design			
Holding sand	5	5	Both perform acceptably, but not as well as some modern anchors
Holding clay	1	4	Low weight models performed poorly in clay
Holding Mud	5	4	Benchmark is marginally better, but didn't always set
Setting hard	3.5	6.5*	*Assuming Roll bar shape and features from Hybrid 1
Setting soft	5	9*	
Veer	5	10	Hybrid design is far superior in this area
Short scope	6	8	
Break out	10	8	Both performed well
Stowage & fouling	3.5	7.8	The hybrid is specifically designed for improvements in this area
Improve Horizontal Anchor Loading			
Weight of line required to hold 30kN	1	10	The chain performs well but weighs significantly more
Reliability of Stowage and cleaning			
Anchor	5	7	Hybrid has a self cleaning mechanism during housing
Line	9	9	Both systems are acceptable functionally
Automatic operation of anchoring system			
Ability to automate	9	9	ISAS automated features could be applied to benchmark system
Noise Reduction of Deployment and Retrieval of the Anchor			
Line deployment and retraction noise	1	10	Chain is extremely noisy
Anchor housing noise	5	4	Some minor additional rubbing from the hybrid anchor shank
Total (average score)	4.9	7.4	

CONCLUSIONS

Literature was reviewed with many potential concepts and design features collected. Research opportunities were shown to exist in several areas, including improved anchor design for veering, short scope, improving the horizontal loading, improved stowage and reduction in noise. A hybrid solution that had improved reliability and suitability for submarines was developed. A benchmark anchor was selected and scale models were tested and then compared.

It was found that improvements could be made with the anchor and the line, and this was the focus of the project, with less detail in the solutions for the stowage, control, and winching parts of the system. Methods of reducing the length of line required were developed with theoretical calculations performed to optimise the layout, with a view to integrating into a submarine.

The final ISAS concept has potential of offering significant advantages in reducing the total weight of the system, improving reliability of stowage and reducing the signature during operation. Future work is recommended to optimise the ISAS and a full scale prototype should be constructed to validate the system operation.

REFERENCES

[1] Buckle. A, (1974), Anchoring and Mooring Equipment on Ships, The Royal Institution of Naval Architects

[2] King. J, Ojo. A, (1983), Some Practical Aspects of Anchoring Large Ships, The Naval Architect, The Royal Institute of Naval Architects

[3] Vryhof Anchors, (2005), Vryhof Anchor Manual 2005, www.vryhof.com

[4] US DoD, (2005), Design: Mooring, Unified Facilities Criteria (UFC), UFC 4-159-03

[5] Rocker. K, (1985), Handbook for Marine Geotechnical Engineering, Deep Ocean Technology, Naval Civil Engineering Laboratory, Port Hueneme.

[6] Ministry of Defence, (2002), Requirements for Anchoring, Berthing, Towing and Securing to a Buoy, Defence Standard 02-117, Issue 2, Defence Procurement Agency, Glasgow

[7] Burcher, R, Rydill. L, (1994), Concepts in Submarine Design

[8] Sincock. P, Sondhi. N, (1993), Drag Anchors for Floating Systems, Offshore Technology Report, Health and Safety Executive, OTH 93 395, HSE Books

[9] http://www.waymarking.com/waymarks/WM28B6_Submarine_anchor_Greenwich, Accessed Oct 2009.

[10] U.S. Navy, (2008), Engineer's Handbook – U.S. Navy Salvage Engineer's Handbook Vol 1, S0300-A8-HBK-010, Appendix G, Anchoring Systems

[11] Towne. R, (1959), Mooring Anchors, The Society of Naval Architects and Marine Engineers, Nov 1959

[12] Taylor. R, (1981), Performance of Conventional Anchors, 13th Annual Offshore Technology Conference in Houston

[13] Taylor. R, (1982), Drag Embedment Anchor Tests in Sand and Mud, Report TN-1635, Naval Civil Engineering Laboratory

[14] http://www.ussailing.org/safety/anchor/results_13.htm, Accessed Sept 2008

[15] Dove. H, Ferris.G, (1960), Development of Anchors, The Royal Institution of Naval Architects, Mar 1960

[16] Ura.T, Yamamoto.Y, (1979), Stability considerations of anchor dropped in non-cohesive seabed, Applied Ocean Research

[17] Ura.T, Yamamoto.Y, (1981), Development of Absolutely-Roll-Stable Anchors, 13th Annual Offshore Technology Conference in Houston

[18] Hodges. T, Springer. B, (2006), Ultimate Holding Power, www.yatchingmonthly.com

[19] Vold. R.C, Eie. J, (1983), Anchor Holding Tests in the Norwegian Trench, Offshore Technology Conference, Texas

[20] Farrell. K, (1950), Improvements in Mooring Anchors, Institution of Naval Architects, Oct 1950

[21] Dove. H, (1950), Investigations on Model Anchors, Institution of Naval Architects, Mar 1950

[22] Craig. W, (1994), Size effects in anchor performance, Canadian Geotechnical Journal, vol 31

[23] Bexco, (2004), DeepRope Manual 2004, Polyester and Dyneema Mooring Ropes Manual

Vessel Identification & Positioning System (VIPS®) - The Future of Maritime Surveillance

Dr. David K.M. Chew
Stratech Systems Limited, Singapore
David_KM_Chew@StratechSystems.com

INTRODUCTION

The needs for tracking have been addressed for many years. Th Automatic Identification System (AIS) is widely deployed to provide ship location, movement as well as ship identification. Other than AIS, many coastal surveillance installations still use radar extensively for vessel tracking and vessel traffic management.

The incorporation of computer vision technology will further enhance the ability of vessel tracking and traffic management by providing visual information to the vessel traffic controller. Together with AIS and radar, the vessel's location, movement, and its declared information can be tracked and verified. The use of computer vision technology allows visual tracking, identification and verification of declared vessel information.

In this paper, a computer vision based approach is discussed for vessel identification. There are a few assumptions on the approach taken: vessel information is available from AIS, radar reports the location of vessels, and electro-optic sensors (EOS) are installed at strategic locations along the coast line to acquire visual imagery with the control of pan-tilt-zoom to obtain images of vessels showing the necessary details for its identification and verification. The computer vision approach combines vessel detection and identification as an integral process.

STRATECH VIPS®

Stratech's Vessel Identification and Positioning System (VIPS®) is an *intelligent* Vision based system which uses EOS with radar and AIS information to monitor the locations and identities of vessels. VIPS® runs on a unique combination of advanced image processing technology and intelligent video analysis algorithms. It can perform vessel measurement of key characteristics such as height, shape and size to validate against radar and AIS information as well as past information stored in the database.

VIPS® is a complete Vessel Traffic Management System that validates the identity of vessels and monitors whether they exceed the height allowed in the waterway. Other features include the ability to provide weightage calculation and threat analysis.

VIPS® IDENTIFICATION PROCESS

VIPS® learns the object to be detected and recognised. This is supported by a database of image samples of each type of vessel to be detected and identified (from Lloyd's Register and the Jane's publication on ships). A revised version of the Talbot Booth recognition system based on ship type, sequence and hull form is also included.

In addition, 3D human face modeling technique is used to build the vessel models. Vessels are modeled by a mathematical representation, which essentially builds a statistical template that captures the typical 3D structure and appearance of the vessels in the category. From the 3D models, 2D appearance image can be generated and used to verify the 2D appearance of a vessel captured by a camera. By projecting the 3D model through typical camera viewing parameters, a collection of 2D images can be obtained for modeling and the appearance manifold. This facilitates the identification of a vessel's pose.

From the identification of pose, the corresponding views of the vessels can be pulled from the database for vessel verification. The features used for the identification of a vessel are based on landmark based representation such as Scale Invariant Feature Transform (SIFT) features or the Maximally Stable Extremal Region (MSER). The principle is to represent a vessel by such landmarks found on the vessel. Such statistical representations will be learned in the visual model construction and vessel identification from the captured images.

The process of identification is more a verification process than a recognition process in the sense that radar and AIS information are available to provide the type of vessel and its location. Hence, vessel identification will make use of such information to zoom in to the 3D models from the corresponding category. This narrows down the search scope and speeds up the identification process.

The principle of the revised version of the Talbot Booth system is to describe a merchant vessel using features that should be visible at a distance where the name cannot be read. The three elements used in classification order are: ship type, sequence and hull form. There are 21 types of ships used, including tankers and combination carriers, liquefied gas carriers, gearless dry cargo vessels, dry cargo ships, refrigerated cargo ships, geared/gearless container ships, low-airdraught ships, Ro-Ro / Lo-Lo, vehicle carriers, passenger ships (including ferries) and specialised cargo ships.

Below are the scenarios for vessel identification depending on the availability of AIS information:

- **If AIS information is available:**

EOS are installed at strategic points, such as ports, harbours and strategic assets such as off-shore oil platforms, to monitor a specific waterway, supplemented by radar as well as AIS information. Once the vessels enter the Region of Interest, the EOS will detect the vessels and validate the vessel information against AIS using one-to-one matching. It will retrieve the corresponding 3D image from the database and compare it with the image captured to identify the vessel. The system will analyse the shape of hull and presence of superstructure, including their relative size, shape and position, to identify the vessel. It will also detect the vessel name on hull, and compare it with the name given by AIS. The user will be alerted if the identification result does not match the AIS information.

Figure 1: Vessel identification

- **If AIS information is not available:**

In the case that AIS information is unavailable, VIPS® will use one-to-many matching. This include filtering out some vessel types and classes based on the vessel size and moving speed; detecting the vessel name on hull and using it to identify the vessel, if possible; and displaying all possible vessel types and classes to let the user determine.

- **Identification of Local Vessels Without AIS**

For local vessels which are not legally required to have AIS, they can have a scaled-down version of AIS. This simplified version of tracking device uses GPS for positioning, and the position and identification information are transmitted to base station (on land) via GPRS/GSM. The device can be permanently mounted on board of vessels which frequent within the sea port, harbour craft which services the sea port, and patrol vessels which patrol the sea port, etc. The device allows perimeter surveillance of 360° coverage within 15 km or more from the sea port, depending on the GPRS coverage at the sea port. The device is able to detect vessels approaching the sea port and entering the user-configurable virtual sea zones.

Another beauty of VIPS® is that it can obtain live AIS information such as the arrival schedules of vessels. This allows VIPS® to "queue" the vessels according to their arrival time and pull out the corresponding AIS information (class, type, name etc.) before the vessels come in. There are more than 300,000 vessels certified with AIS in the database.

RULES OF ENGAGEMENT

VIPS® works on the principle of Rules of Engagement. The system will track and identify the vessels based on their threat level. Vessels with higher threat level will be identified first, and VIPS® will choose the most suitable EOS to capture the side view of the vessel for identification.

The factors for threat level calculation include: moving direction, moving speed, existence of AIS information, detection by radar, traversed zones and distance from user-defined restricted zones. VIPS® will alert the user if threat level is above a certain threshold.

VIPS® APPLICATION – VHMS-III

One example of VIPS® application is the third-generation Vessel Height Measurement System (VHMS-III) project with the Civil Aviation Authority of Singapore (CAAS) for surveillance of flight paths over water. The ability to provide vessel height surveillance is crucial for airports that have their take-off and landing flight paths over water and the sea to ensure minimum height clearance in the aircraft flight zone.

VHMS-III is an automated solution which integrates radar, AIS and vision-based height measurement. Using the VIPS® concept, the system detects vessels entering the traffic information area and automatically measures the vessel height, as well as validates against the declared height obtained from the AIS information. If the measured height exceeds the minimum clearance height, the system will alert the users at the operations centre.

VHMS-III employs the Rules of Engagement, which includes the vessel mode (lock and unlock), fence type and weight type. The weighting system enables the system to assign points to vessels passing through various locations along the waterway so as to prioritise the vessels for tracking and height measurement.

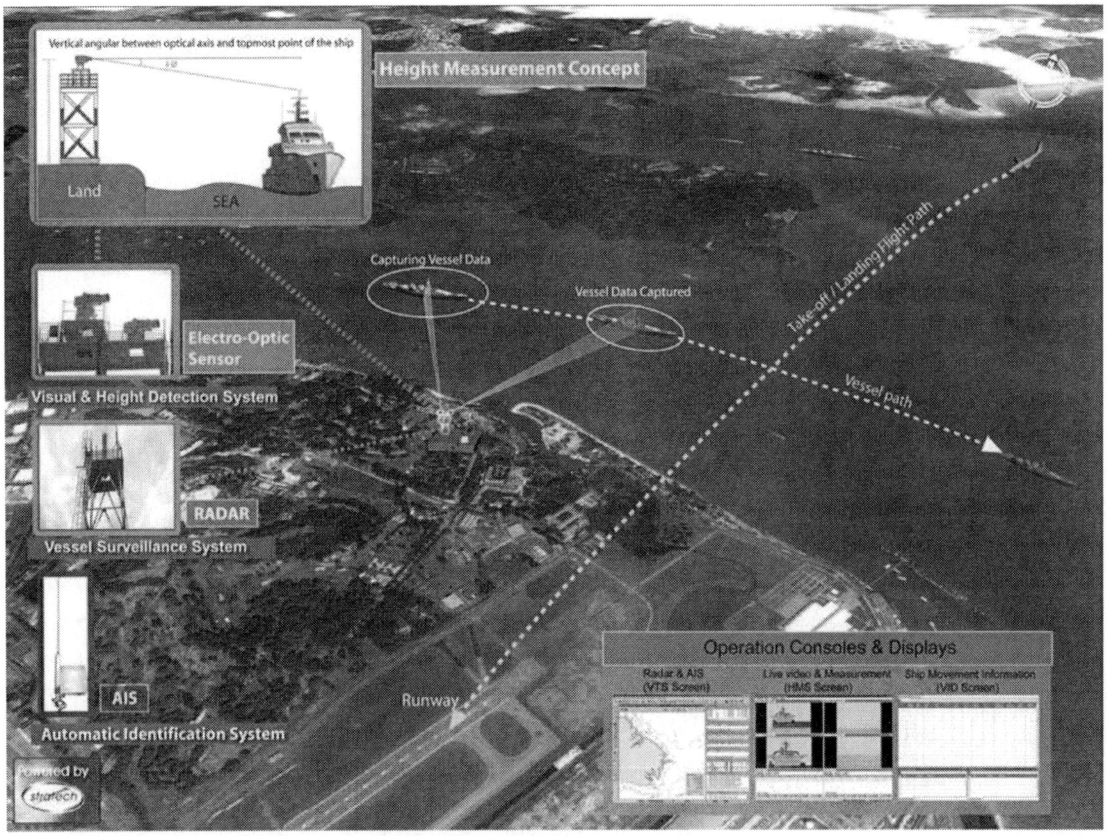

Figure 2: VHMS-III for surveillance of flight paths over water

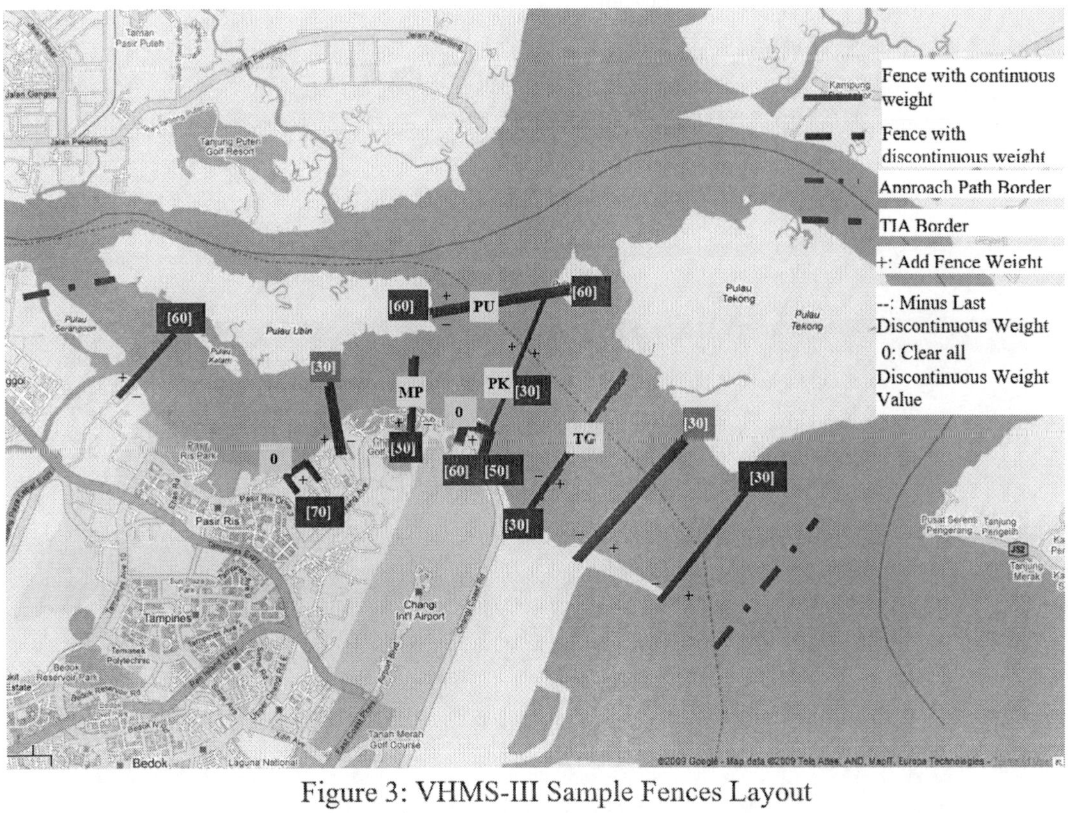

Figure 3: VHMS-III Sample Fences Layout

Prevention of Terrorist Attacks

Other than vessel height measurement, VIPS® can be used to counter terrorist attacks. One example is the U.S.S. Cole attack by the al-Qaeda terrorist organisation in October 2000. A small boat filled with explosives rammed into the vessel in Yemen, killing 17 American soldiers.

In response to this attack, Stratech has partnered the US Navy and Technical Support Working Group (TSWIG) of the US Department of Defense to help prevent recurrence of such incidents. As part of the partnership, Stratech will integrate its VIPS® to enhance port security through real-time situation awareness. VIPS® delivers automatic intruder detection, classification and tracking of vessels for coastal surveillance. It complements maritime radar system with the element of sight. Monitoring over large areas of the waterside, VIPS® identifies vessels and scours the waterway for all vessels entering its region of interest and tracks their movements. In addition, the system can be used to detect/capture activities such as smuggling or illegal immigration.

The system will also deliver vessel trajectory prediction capabilities so as to provide early warning of abnormal behaviour and possible collision of vessels. This strategic alliance will bring about a paradigm shift in maritime security as it addresses the critical issue of vessel identification through real-time validation.

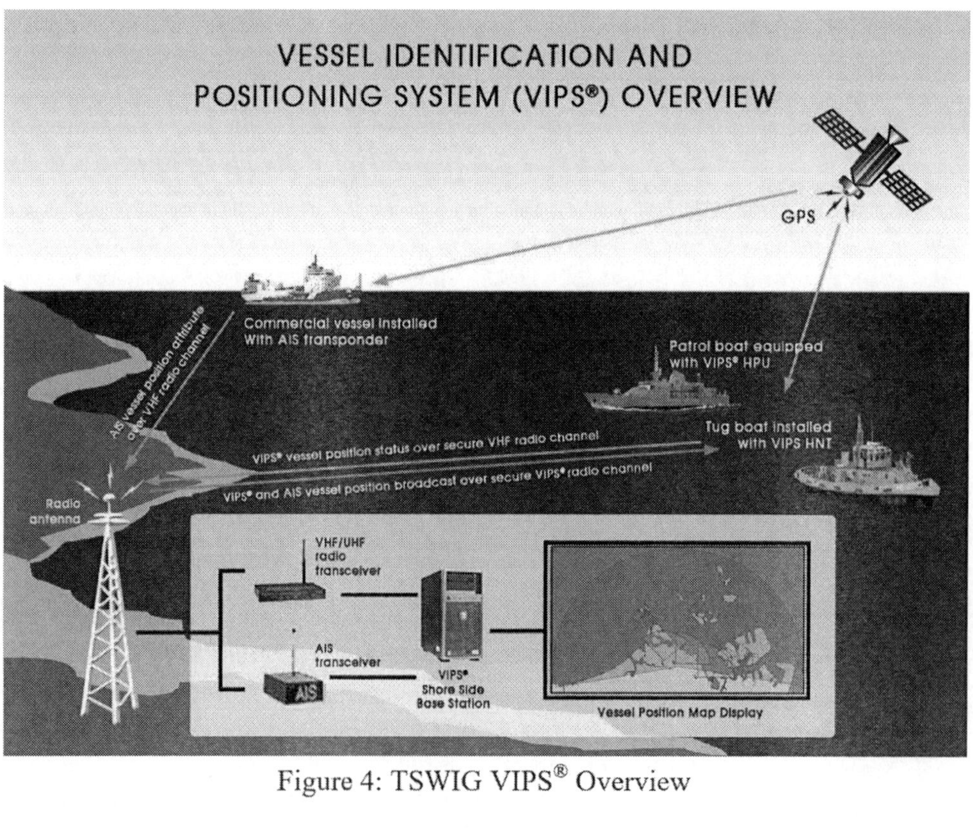

Figure 4: TSWIG VIPS® Overview

Protection of Strategic Assets

Another important application of VIPS® is the surveillance of off-shore oil platforms and nuclear plants – which are disasters waiting to happen. VIPS® is a cost-effective solution which can help to manage obvious points of such strategic assets.

VIPS® automatically detects and identifies suspicious activities over water and alerts strategic asset owners of impending sea borne attacks or incursions with visual data. All approaching vessels are tagged for identification and VIPS® integrates seamlessly with existing AIS and radar systems to provide centralised alarm management.

Anti-Piracy

Another application of VIPS® is anti-piracy. VIPS® automatically detects and identifies suspicious activities over water and alerts strategic asset owners of impending sea borne attacks or incursions with visual data.

VIPS® Anti-Piracy is a long-range vision detection system which is mounted onboard the vessel. It sees further than a human spotter in day or night in most weather conditions at sea. It employs *intelligent* Vision techniques to detect and track small vessels that may pose a collision or pirate threat, providing crucial response time for the ship's crew to take necessary action. It can be integrated with GPS transponder, radar, AIS and other navigational sensors.

COMMERCIAL POTENTIAL OF VIPS®

VIPS® has huge market potential. It can be applied to close to 10,000 sea ports and over 1,000 off-shore oil platforms worldwide. It can also be deployed at off-shore nuclear plants, refineries, harbours, fishery ports, hydro-electric dams and various military installations.

With its unique vessel identification capability, VIPS® is poised to bring about a paradigm shift in vessel traffic management.

Fitting Square Pegs into Ever-Changing Round Holes: Environmental Compliance Challenges in Modern Warship Design

John Polglaze
URS Australia
john_polglaze@urscorp.com

DISCLAIMER

The views presented in this paper are those of the author alone. No particular navy nor specific class of ship is considered in this paper, unless and only to the extent that such navy or ship may be specified.

ABSTRACT

Warship compliance with International Maritime Organization (IMO) marine environment protection requirements is entering into a new paradigm, where adherence to some elements of IMO regulations is becoming more onerous. By extension, compliance by warships with IMO requirements will become more divergent from the IMO caveat of "not impairing the operations or operational capabilities of such ships".

New, emergent and forecast IMO regulations address aspects such as ballast water, biofouling, greenhouse efficiency, protection of oil fuel tanks, and the selection of materials for ship structure and fittings. Some requirements can be accommodated in warships in their design, build and operations and may actually enhance their tactical profile, but many others cannot be readily addressed without penalty in combat capability, survivability and through-life support costs. The lifecycle of warships, from concept development through design, build and service-life is also incompatible with the development and implementation schedules of many IMO initiatives.

The period of straightforward navy compliance with IMO requirements is closing. Navies are at the threshold of an era when strict compliance with many IMO requirements will neither be practicable nor achievable, at least not without severe constraints and limitations on warship operational capability and effectiveness. Practicable application of IMO marine environment protection requirements should remain a primary policy goal for navies, but actual application will arguably need to be on a more balanced, judicious and less rigid basis if warships are to realise optimal operational capabilities.

INTRODUCTION

The international community, through national and regional arrangements but primarily at the global level through the United Nations (UN) based International Maritime Organization (IMO), has recognised the myriad threats that shipping presents to the environmental health and biodiversity of the world's oceans. Through a range of international instruments, the IMO has introduced controls on many aspects of ship design and operation, with the intention of prohibiting or limiting emissions of harmful substances, and modifying ship design to limit the risk of pollution. These rules are in a constant state of flux, as the IMO deals with emerging marine environment protection priorities or exploits the opportunities presented by new or improved technologies.

Warship observation of IMO marine environment protection measures is not mandatory. In many aspects, such compliance is relatively simple to achieve, unavoidable or otherwise makes good sense for reasons of operational capability and flexibility, through-life support costs or similar. Ease of compliance is not always the case, however, and warship adherence to IMO marine environment protection requirements can be seen as entering into a new paradigm, where observation of some elements of IMO regulations is becoming more onerous and difficult for specialist ships, such as warships. In some cases, strict adherence can be mutually incompatible with a warship's *raison d'être*, to fight and win at sea (and survive).

WARSHIP COMPLAINCE WITH MARINE ENVIORNMENT PROTECTION MEASURES

Background

IMO marine environment protection measures cover a wide and diverse range of polluting and potentially polluting activities and substances, and span the entire ship life from initial design through to disposal. Issues addressed by the IMO include: emissions and the disposal of oil and oily wastes, chemicals, sewage, garbage and atmospheric pollutants; biocides in antifouling paints; the transfer of pathogens and invasive marine species in ballast water; the design of oil tanks to limit the likelihood of loss in the event of collision or grounding; and the design and fit of ships to improve the recovery of recyclable materials and limit the amount of harmful substances requiring disposal at the time of dismantling. The IMO is currently in the process of expanding its mandate to include ship energy efficiency and greenhouse gas (GHG) emissions, the transfer of invasive species via biofouling, and ship strikes with cetaceans. In the future, it may be expected that in addition to a general tightening of existing regulations, the IMO will also address other issues such as greywater.

The principal IMO instrument for marine environment protection is the *International Convention for the Prevention of Pollution from Ships*, as modified by the Protocol of 1978 relating thereto (otherwise referred to as MARPOL 73/78). Article 3(2) of MARPOL 73/78 contains a specific exemption for warships, which states: *This Convention shall not apply to any warships, naval auxiliary, or other ships owned or operated by a Party and used, for the time being, only on government non-commercial service. However, each Party shall ensure, by the adoption of appropriate measures not impairing operations or operational capabilities of such ships owned or operated by it, that such ships act in a manner consistent, so far as is reasonable and practicable, with this Convention.* This warship exemption is mirrored in all other IMO marine environment protection conventions.

Some navies may, by way of implementation policy, choose to interpret and apply Article 3(2) as a 'concession', rather than as an exemption, but this nuance can be pivotal in circumstances where compliance represents a significant challenge, and particularly where it achieves minimal actual environment protection. Just as navies have had to learn, and constantly re-learn, how to manage the technical and operational challenges of acquiring, operating and sustaining warships, there needs to be a parallel shift in thinking concerning IMO compliance. 'Motherhood' statements that imply full environmental compliance for warships are representative of fuzzy, Utopian thinking, devoid of realistic appraisal of what is required to achieve and maintain such compliance, and imply tacit acceptance of inevitable, unavoidable compromise in other aspects of warship design and operation.

The Benefits of Warship Compliance

Apart from the benefits to marine environment protection and public reputation, warship compliance often makes good sense within the domains of maintenance and through-life support costs, operational flexibility and combat capability. For example, minimising garbage generation by reducing wastage can extend ship endurance at sea and reduce costs expended in the acquisition of stores, particularly in the case of victuals. Similarly, inappropriate disposal of garbage to sea can have adverse operational implications, such as floating garbage having to be treated as possible mines, or providing useful information on ship movements to an adversary. Other examples include the flexibility afforded by avoiding limitations on movements in certain sea areas if a ship was otherwise compelled to leave that area or enter port to dispose of waste, the obvious advantages for ship range/endurance from improving fuel efficiency, and the benefits to propulsion efficiency, speed and range and reduced ship noise gained from limiting biofouling on hulls and propellers. In other cases, non-compliance by warships is effectively not an option, one example being the global ban on tri-butyl tin (TBT) paints, where even if desired for application to a particular warship, sourcing such a paint would be difficult due to the absence of supply. These cases demonstrate situations where compliance is a relatively simple proposition for warships, and where it can be achieved without adversely affecting a ship's combat role, if not enhancing it.

Other situations where warship compliance makes good sense and can usually be achieved readily is the case of ships not specifically designed as warships, or those that have analogous civil roles and designs. This cohort includes ships such as tankers, survey ships and other auxiliaries, which are often obtained from commercial sources or based on commercial designs adapted for naval purposes.

The need for ships, including warships, to accommodate marine environment protection requirements into their design and operation is a fairly new development, first initiated in the 1970s and only gaining real substance in the period from the late 1980s onwards. The need for warships to address new requirements is not unique to marine environment protection, as warship design has evolved to adjust to new opportunities and obligations dictated by changing technologies, regulations and community standards. The warship of 60 or 70 years ago had no need to make room in its design and upkeep for bunks or air conditioners, nor did it need to accommodate guided missile launchers, but all of these needs are rightly and routinely incorporated in contemporary designs.

Disadvantages and Incompatibilities of Warship Compliance

When considering the environmental compliance framework for warships, it is germane to note that, by way of its charter, the IMO is focused upon merchant shipping and, with regards to marine environment protection, the particular environmental risk characteristics and operational discharges characteristic of that sector. Where the conundrum arises for warship compliance is in relation to their design complexity, unique operational requirements (e.g. the need to be able to accept damage and continue to float, move and fight), severe constraints in weight and space availability, longevity, and in particular, the characteristically protracted duration of the design and build processes. These factors are particularly acute for surface combatants (e.g. cruisers, destroyers, frigates) and submarines, but also have varying degrees of relevance to most other types of warship. Many of the distinct facets of warship design, operation and purpose are fundamentally different to the norms which underpin, frame and guide IMO marine environment protection actions and ambitions.

Hence the development and focus of IMO marine environment protection initiatives has three pivotal implications for the planning and implementation of environmental compliance for warships which are dichotomous to those pertaining to the IMO's merchant ship reference model. These three salient, inter-related issues are:

1. **Design and operational factors**: This embraces aspects such as typical hull shape and profile of merchant ships; crewing and crew training; watchkeeping arrangements; ship control and manoeuvrability; damage control capabilities; amount and type of fuel/s carried; excess weight and space reserves available for ancillary equipment; and usual ship activities and operating cycles.

2. **Environmental risk profile**: This stems from the environmental risks associated with merchant shipping, predicated upon the way that merchant ships are designed, built, maintained and operated and the materials they may carry. For example, with regard to aspects such as oil spill vulnerability in the event of grounding or collision, the environmental risks are evaluated on the hull form, type/s of fuel used, quantities which may be carried in any single tank, and ship crew and watchkeeping arrangements.

3. **Time horizon for implementation of new or modified marine environment protection rules**: The life span of a typical merchant ship design is characteristically short compared with that of warships, with merchant ship designs constantly evolving through incremental change and in response to individual orders. The normal two to four or five year window permitted by the IMO for planned entry into force of new or modified regulatory requirements for ships, or 'grandfathering' arrangements for ships in service, represents a substantial proportion of a merchant ship's life span, and is based upon making reasonable allowance for ship designers, builders, owners and operators to adopt any required new or revised measures as a synchronous component of the ship design, acquisition and obsolescence cycle.

This last point is a particularly pertinent one. Merchant ships, even the largest ones, can be ordered and delivered within periods typically measured in no more than a few years, and often less. The period from design completion to keel laying to entry into service of a merchant ship can often be measured in months and rarely exceeds more than a few years for any particular class of sisterships. For example, the design of the complex, one-off, 150,000 dwt liner *Queen Mary 2* was completed in mid-1998, and even given an apparent delay in placing the build contract, the ship was ordered in November 2000 and delivered in December 2003. This compares with simpler to design and build, serially constructed merchant ships such as tankers, bulk carriers and container carriers, where it is not uncommon for ships to be delivered within 12 months or less of an order being placed.

Contrast this with a case of a warship design, such as the 10 ship *Anzac* class frigate. The selected design was announced in 1989, with the lead ship commissioning in 1996 and the last of class in 2006. Another example is that of the US *Arleigh Burke* (DDG-51) class destroyers: the design was conceived in the early 1980s, with the lead ship entering service in 1991, and ship 62 expected to commission around 2012. Despite some design modifications through the ship acquisition program, the underlying platform design will be approaching 30 years of age by the time the last of class commissions, and will likely be of the order of 60 to 70 years old by the time the last ship decommissions.

Although the typical individual warship lifespan of 20 to 30 years is consistent with that of merchant ships, the key difference with respect to maintaining currency with marine pollution prevention initiatives is in the ***design age*** of ships. Aside from operational and design limitations, any navy's aspirational goal of 100% warship environmental compliance at time of commissioning, let alone throughout ship life, would require continual evolution and design change almost up to the time of actual commissioning (NB: whereas some IMO requirements base application to a specific ship to 'stage of construction', others relate to 'date of installation' of equipment into a ship, and thus could apply to a ship at any time in its life). A goal of 100% compliance would be difficult in any single, individual ship, but would be especially burdensome and expensive to realise and sustain in an extended acquisition program encompassing many ships of the same class. Such an outcome would also exacerbate through-life support requirements, as would be expected when trying to support ships of the same class, albeit varying ages, with disparate configurations of marine environment protection equipment.

Strict compliance is further exacerbated throughout life-of-type as observation of IMO design and equipment rules can mandate replacement or enhancement of existing, or installation of entirely new, pollution control systems at different junctures in a ship's life. In accordance with many IMO rules, such requirements may be simply as a result of ship age, or be triggered as a consequence of other, potentially even unrelated, refit work. It is also important to consider what maintenance of compliance through life-of-type means in terms of the technical management framework that would need to be in place to undertake the necessary design change process. Developing and implementing design and equipment configuration changes to operational systems requires thorough, properly supported engineering processes implemented by trained and experienced staff, and the associated allocation of adequate funding from a finite financial resources base.

In the case of the *Anzac* class frigates, even if the lead ship had been fully compliant with IMO requirements extant at the time of commissioning, without design change or anticipation of future requirements the final ship would have exhibited deficiencies in several areas including oily water, sewage treatment and atmospheric emissions. Even so, these ships will still have many years of life remaining in 2016 but by that time have the prospect of being non-compliant with a host of IMO requirements, such as ballast water treatment regulations.

The IMO is characteristically a pragmatic forum, restricting itself to promulgating regulations which can be reasonably met with existing or emergent technology. The advent of new and refined technologies permits the IMO to review and tighten regulations; recent revisions to existing discharge regulations for oily water, sewage and diesel exhaust emissions to make them more stringent are but three examples testament to this. New environmental issues, and awareness of developing marine environment risks and stresses, will also generate IMO action, sometimes in entirely new domains. In the last 10 years or so, the IMO has extended the scope of regulations to include ship recycling/scrapping, air emissions and biofouling. It is highly speculative to consider what else may ensue over the next 10 years, let alone the next 50, yet these are the time horizons within which warship designers and builders need to think.

It may be argued that IMO marine environment protection enterprises are a lag indicator of wider community and regulatory concerns in non-maritime fields, and that on this premise, likely IMO requirements five to 10 years into the future can be predicted by ship designers. While there is some appeal and basis to this line of expostulation, the reality is that such forecasting provides minimal valid indication to the ship designer of future IMO requirements. Although the general domain of future ship requirements may be discerned, there is scant material guidance as to specific regulations. Thus, the ship designer needs to be mindful of the risk of misappropriating resources and compromising other aspects of ship design on a gamble on possible future IMO regulations, which may not eventuate or else may do so in such way that the speculative ship design does not satisfy.

For example, a prescient ship designer commencing the ship development process in 1992 for a ship to be delivered in 2002 and remain in service until 2022, would have been aware of various unilateral national and early IMO initiatives regarding invasive marine species and ballast water, but would have had no definitive guidance on actual regulations to be adopted by the IMO until mid-2004. This is despite the fact that the IMO ballast water treatment regulations will apply to the subject ship from no later than 2016 and possibly from as early as 2009. Hypothetically, the prescient ship designer may have fitted the ship with some form of generic ballast water treatment system (assuming suitable technologies existed within that timeframe, which they generally did not, and notwithstanding that acquisition would still be subject to whatever the vendors may have decided to make available to the market), but this system would not have been able to comply with the Regulation D-2 standard of the future BWM Convention. Consequently the ship operator would be confronted with the need to strip this system from the ship and replace it with a compliant unit. It could be suggested that this scenario draws on a complex example, and thus has limited merit in providing a valid argument in more mundane situations. Nevertheless, the core holds true, as even within the relatively simple, long-standing arena of ship sewage treatment requirements, a ship designer of 2005 would have had minimal objective basis to know with any certainty the specific IMO sewage treatment standards applicable to treatment systems intended to be installed in ships from 2010 onwards.

Contending with Weight and Space Limitations

Ship refit and reconfiguration in order to comply with new or revised IMO requirements is an additional challenge. In merchant ships, size is generally dictated by the desired cargo carrying capacity, with the ship hull dimensions dictated by size and layout of payload carrying compartments. Engine rooms and machinery spaces are characteristically spacious and physically separate from the cargo areas. In the case of warships, the design paradigm is to literally 'squeeze' as many sensors, weapons, and other required machinery, personnel and personnel support features (e.g. galley, recreation spaces) into the available space. This has to be done in a manner consistent with ship capability, survivability, habitability and other similar needs. Unlike merchant ships, warships display a more heterogeneous layout, with machinery, operational and personnel spaces interspersed and intermingled with each other.

This premium on weight and space availability makes retrofit to existing warships of new or updated pollution control systems, as may be required for example, for sewage treatment plants or diesel generator exhaust systems, problematic in its own right. Fitting of a larger or additional sewage plant, for instance, in many merchants ships can be done with minimal ship re-design and reconfiguration effort, but the same does not hold for most warships. The compliance equation becomes more acute for warship designers and operators when any available weight and space reserves may be consumed by updated marine pollution control equipment, at the expense of modernised or enhanced warfighting capabilities.

When Warship Compliance Can Make Minimal Sense

Aspects of some new IMO marine environment protection regulations make little sense when applied to warships, resulting in design compromises which can seriously degrade capability and survivability, for little or no beneficial environmental outcome. Two pertinent examples concern protected fuel tanks and ship materials.

In the case of the former, new IMO design rules require any ship designed to carry 600 cubic metres or more of oil fuel to have 'protected tanks', typically achieved by the IMO-prescribed solution of double-hulling around the tanks. Such double-hulling can be relatively easily accomplished in simple, invariably slab-sided merchant ships, which due to their design and operational characteristics have historically shown to experience loss of fuel oil in the event of collision or grounding on a relatively regular basis, and hence acted as the catalyst for IMO action on this issue.

In contrast, warships have features of design, propulsion, steering and crewing which minimise the likelihood of collision or grounding in the first instance, and lessen the risk that such casualty will result in the loss of fuel at all, or the amount of any such loss (i.e. warships typically have a number of small fuel tanks dispersed around the ship, whereas merchant ships usually have a small number of very large tanks). Furthermore, it may be thought that the typical hull projections found on warships (e.g. sonar dome, stabilisers, propeller/s, rudder/s and the rounded hull itself) provide the form of a 'virtual double-hull' compared with the flat bottom and slab sides of merchant ships; this contention is borne out in the record of warships collisions and groundings, where the projection is usually the point or area of contact.

Fitting 1 m of hull separation around a fuel tank in a merchant ship with a beam of 20 m and a relatively square cross-section is an entirely different proposition to a warship, 14 m in the beam and with a rounded hull form. Space taken up by double-hulling around fuel tanks in a merchant ship may result in some loss of cargo capacity, but this can be easily compensated by designing larger holds or increasing the unit cost of cargo carriage to maintain revenue margins. In the case of a 'space-poor' warship, any loss of space results in some linked reduction in weapon or sensor fit, range/endurance, and/or crew habitability, none of which can be readily compensated. This renders the compliance with an IMO direction on reducing oil spill vulnerability, never intended for application to warships, to result in a direct impediment to warship capability and survivability. Clearly, the emasculation of warship potency and survivability which compliance with this rule could entail is counter to the whole purpose and reason for being of a warship. Perversely, such a design measure would likely increase the chance of oil loss in the event that the ship is damaged or sunk in combat owing to any reduced combat capability as a result of having protected the tanks.

It is also important to consider that unlike control of sewage, garbage, air emissions or similar, which are 'active' agents of environmental harm, the protected tank rule deals exclusively with a 'latent' environmental risk. Fuel tank protection will only be of relevance in the infrequent event of a warship experiencing a severe collision or grounding. The protected tanks rule is a particular example where precipitate, dogmatic compliance with IMO prescripts could result in a warship severely compromised in its ability to fulfil its fundamental purpose, ironically without any actual environmental benefit.

The adoption of other recent and embryonic IMO developments will likely compromise warship design and capability, with the result that compliance by warships with IMO requirements will become more divergent from the caveat of "not impairing the operations or operational capabilities of such ships". For example, with current technologies, attainment of IMO Tier III diesel emission standards will require substantially large exhaust gas after treatment systems, and some form of intermediary consumable such as urea. The machinery aspects of Tier III compliance will place further premium on warship weight and space margins, and the need for urea or similar will add another materiel stream to the logistic support requirements for warships alongside and at sea. Full compliance could result in the spectre of underway replenishment tankers needing to carry and transfer urea, with associated replenishment-at-sea (RAS) receiving arrangements for urea in the client combatant.

Another case is the new IMO ship recycling convention. This requires ship designers and operators to select materials for the design and fit-out of ships which are optimised for recovery for recycling or which exhibit minimal toxicity, easing ship dismantling requirements. This is a positive development, and one which should also bring OH&S benefits to ship's crews and dockyard workers throughout a ship's life, by reducing the potential for personnel exposure to hazardous materials. Application to a warship however, needs to be mindful of a material's intended service purpose. Any replacement material which may trade-off its suitability for warship application in-terms of aspects of ability to sustain battle damage, or improved reliability in mission-critical systems, is arguably a poor substitute for the unique and challenging performance needs of warships.

AN ANALOGY: MILITARY AND COMMERCIAL AIRCRAFT

A situation analogous to that confronting warships exists with respect to the application of civil aviation requirements to military aviation. There are numerous parallels between the international management and regulatory framework pertaining to aviation and that which applies to shipping. Similar to the IMO for shipping, a UN based multilateral body, the International Civil Aviation Organization (ICAO), establishes internationally agreed standards and regulations for aviation.

The Convention underpinning the ICAO is the *Chicago Convention 1944*. Similar to Article 3 of MARPOL 73/78, Article 3 of the *Chicago Convention 1944* contains an exemption for 'state' aircraft (i.e. military, customs and police aircraft), although *when issuing regulations for their state aircraft, (Contracting parties) will have due regard for the safety of navigation of civil aircraft.*

The environmental performance of aircraft is one of the focus areas of the ICAO. The ICAO has established the Committee on Aviation Environmental Protection (CAEP). The two foci of the CAEP are aircraft noise and aircraft engine emissions.

Reduction at source is one element of the ICAO strategy to reduce aircraft noise emissions. This is fostered by the requirement for aircraft to meet noise certification standards adopted by the Council of ICAO.

Controls upon aircraft engine emissions, the other main area of CAEP activity, relate to aircraft engines as a source of a range of pollutants and GHGs. CAEP activities are aligned with the aims of the Montreal Protocol on Substances that Deplete the Ozone Layer and the Framework Convention on Climate Change. The CAEP has implemented a range of measures to limit aircraft emissions, including the development of emissions standards and certification for aircraft engines.

Although the ICAO has no direct relevance to military aviation its environmental policies and initiatives may nevertheless be used as a benchmark against which to evaluate military aviation activities and systems. Indirect linkages exist as measures developed under ICAO tutelage to improve the environmental performance of commercial aviation migrate into the military sphere, such as occurs with aircraft and aircraft engines which have both civil and military applications.

For example, the Convention on International Civil Aviation requires aircraft to meet the engine certification standards adopted by the Council of ICAO. These regulations are acknowledged as focusing upon commercial aircraft, and hence are not directly applicable to the engine types or modes of operation of military aircraft. The Regulations do have applicability, however, to the operation of some military aircraft, notably those which use engines developed primarily for civil applications. Examples include VIP aircraft, some transports and tankers and airborne early warning (AEW) platforms. Research undertaken in the preparation of this paper can find no evidence of the ICAO or any national regulator or aircraft or engine manufacturer attempting to apply the ICAO noise and engine emissions requirements to 'combat' aircraft (i.e. those designed, built and operated specifically for military purposes) particularly high-performance aircraft, as opposed to those with dual applications (e.g. B737s used for both military and commercial roles). Nevertheless, aircraft engine manufacturers often seek to reduce engine noise and fuel consumption (and hence emission) rates in new generation military equipment, although combat capability and operational requirements remain the primary focus. Apart from emission reductions, minimising fuel use offers operational benefits to military aerospace platforms in the form of longer range, extended flight endurance and/or reduced weights.

No air force is known by this author to have any specific policy on the application of ICAO environmental standards to military aircraft. Nevertheless, in many instances a default policy effectively exists, where essentially commercial aviation engines in military service often maintain their civil certifications.

It may be surmised that responsible air forces and builders of military aircraft and engines strive to achieve environmental best practice, consistent and sometimes linked with the reason for being of combat aircraft - delivering combat capability. The key differentiator in comparison with the situation for warships is that air forces are not necessarily compelled and do not seek to apply civil environmental protection regulations to combat aircraft in other than a collateral sense.

CONCLUSION

Aside from the benefits to marine environment protection and public reputation, warship environmental compliance often makes good sense within the domains of maintenance and through-life support costs, operational flexibility, tactical applications and combat capability. Warship compliance with IMO marine environment protection requirements is entering into a new paradigm, however, where adherence to some elements of IMO regulations is becoming more onerous and can detract from other fundamental tenets of warship design and operation. This is particularly so for combatant warships - those designed, built and operated to optimise their warfighting capabilities and survivability. Compliance for naval auxiliaries, particularly those which have a close affinity with merchant ship equivalents, is not as difficult. By extension, compliance by combatant warships with IMO requirements will become more divergent from the IMO convention caveat of "not impairing the operations or operational capabilities of such ships". In many respects, the period of straightforward warship compliance with IMO marine environment protection requirements has ended. Navies are at the threshold of an era when strict compliance with many IMO requirements will neither be practicable nor achievable, at least not without severe constraints and limitations on warship operational capability and effectiveness. Naval technical managers and their regulators need to recognise this and plan proactively for realistic IMO compliance, noting that aspirational targets of 100% compliance cannot be considered to produce the best operational or technical outcomes for navies, nor the best investment of public monies.

The unrelenting challenge of remaining up-to-date with perpetually changing IMO marine pollution developments for warships needs also to be addressed. This could be accomplished by declaring a compliance baseline date for each class of ship. Thus the constantly changing 'round hole' could be viewed as static, and the 'square peg' of the subject warship class could be adapted as necessary to best fit that hole.

In this more pragmatic archetype, a class of ships would be designed to achieve the best practicable compliance with the environment protection objectives of applicable IMO requirements extant at the nominated baseline date, balanced with other operational and design requirements and constraints. Navies would be spared the burden and cost of constantly revisiting ship designs and fit-outs through a class's build cycle and service life in order to remain temporarily, at best, abreast of dynamic IMO requirements. The corollary to this is acceptance by navies, their governments and regulators, and the national populations that they serve and which pay for the ships, that the ships of any particular class are unlikely to be fully compliant with IMO requirements at the time of their commissioning, and the extent of any such variance will increase as a build program extends through time and as the ships serve through their commissions.

Practicable application of IMO marine environment protection goals should remain a primary policy endeavour for navies. It is axiomatic, however, that the application of some IMO marine environment protection rules is incompatible with, if not contradictory to, the primary purposes and requirements of warships, particularly combatants. Actual application of IMO rules will need to be approached on a more balanced, objective, judicious and less rigid basis if warships are to optimise their operational capabilities. The mechanisms of warship marine environment protection need to be transformed to incorporate a focus upon risk assessed needs and the adoption of an 'outcomes based' approach, as opposed to the prevailing orthodox construct of strict adherence to prescriptive IMO rules.

Development of the Vulnerability Assessment Tool CVAM to provide Realistic Damage Control Training Scenarios

M Skeen and M Buckland
Defence Science and Technology Organisation, Maritime Platforms Division, Melbourne, Victoria, Australia
michael.skeen@dsto.defence.gov.au

ABSTRACT

Vulnerability assessment tools are used by military institutions world wide as a method of determining the consequences of weapons damage to military platforms. Generally, the vulnerability tools utilise analytical and empirical methods to assess the effect of battle damage. These methods have been developed to allow the rapid assessment of weapons effects with low computational overheads.

Over the last two decades the Defence Science and Technology Organisation (DSTO), Maritime Platforms Division has been developing, validating and using a suite of vulnerability assessment tools. The tools are used to evaluate the effect of blast, fragment and UNDEX damage using established damage algorithms. By including a fault tree of ships systems, it is possible to determine which capabilities will be lost due to a damage event. CVAM V2.1 is the latest version of this tool to be developed by DSTO. The aim of this development program has been to increase the speed at which analysis can be carried out and ensure that realistic assessments of the damage suffered on the ship can be relayed back to the user.

The CVAM V2.1 software is being developed with the aim of it being used to develop realistic damage control scenarios. When used, it will assist training exercises to evaluate casualties, damage to the ship and associated loss of capability due to weapon damage.

This paper describes the CVAM V2.1 suite of software, its development history and the application of this code in the development of realistic damage control training scenarios for use on-board naval vessels.

BACKGROUND

Vulnerability Assessment of naval vessels has traditionally been used to improve the design of vessels prior to construction, to determine the capability of a vessel to survive in a particular threat environment once it has been built and to assess the lethality of a particular weapon against a particular ship design.

Vulnerability Assessment packages generally model the effects of conventional weapons i.e. blast, shock and fragmentation damage to a vessel. The damage to the structure of the ship is modelled, as well as to individual components of the ships systems. When a particular component on a ship is damaged to the point that it is ineffective, the effectiveness of components reliant on the operation of that damaged component are also degraded. The damage to components will result in some functionality of the ship being unavailable, and the amount of loss of capability can be assessed using fault tree analysis. For instance, where a switchboard is damaged or destroyed, the systems that are connected to this switchboard will no longer be operational if an alternative (or redundant) power source is not available.

In reality when a naval vessel does sustain damage during an operation, damage control crews are on hand to contain and repair damage to the ship. For instance, holes below the water line may need to be shored or fires may need to be put out. The crew responsible for this damage control undergo constant training aboard a naval vessel, but for this training to be effective it must incorporate realistic scenarios so that the crew is fully prepared to deal with real world events.

In this paper, the use of realistic damage control scenarios generated by the vulnerability package CVAM V2.1 is described and future improvements to the application of this program are discussed.

CVAM V2.1 AND DAMAGE CONTROL TRAINING

CVAM V2.1 is a vulnerability suite that has been developed at the Defence Science and Technology Organisation, Maritime Platforms Division. The code can be used for the vulnerability assessment of ship designs and also for conducting lethality studies. Another important use of the tool has been identified as being able to be used as an on-board training aide, in the development of realistic damage scenarios resulting from weapons damage. It has therefore been considered important to ensure that the operation of this software be simple to use and realistic enough that it can be used on-board a ship for training purposes.

The CVAM V2.1 suite consists of a number of independent but inter-related modules that are used to prepare the vulnerability model and carry out the various aspects of the analysis. The interrelationship of the programs included in the CVAM V2.1 suite is shown in Figure 1.

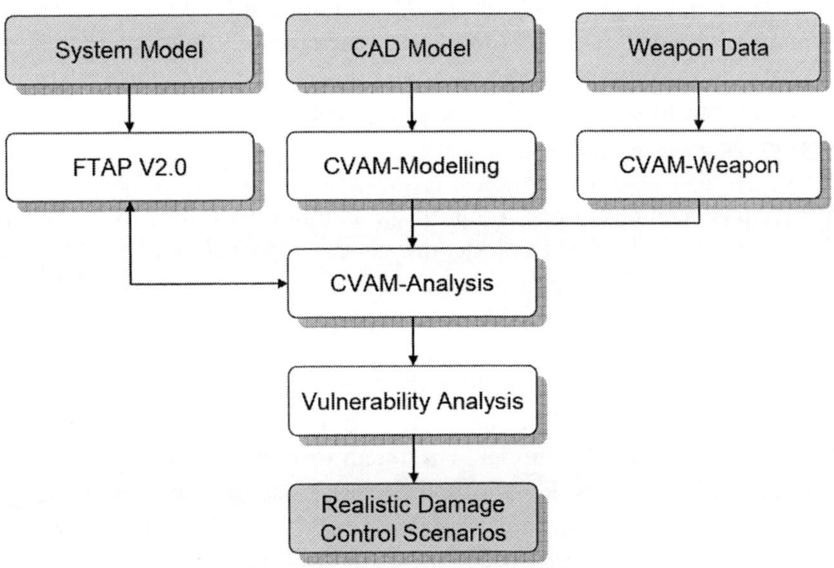

Figure 1 - CVAM V2.1 Flowchart

A vital part of developing a vulnerability model for CVAM V2.1 is to develop a Computer Aided Design (CAD) model of the ship and then apply the appropriate structural details and vital component details which are important for the assessment.

CVAM V2.1 is based around the Dassault Systemes products CATIA [1] and VIRTOOLS [2]. These tools were chosen because of the in-built capability of these packages, when used together, to import many different CAD formats.

As part of the CVAM V2.1 suite, CVAM-Modelling has been created to interact with CATIA allowing custom vulnerability properties to be applied to the CATIA model. The CVAM-Modelling interface is shown in Figure 2 interacting with a CATIA based model. CVAM-Modelling has been designed to communicate with CATIA using a Component Object Model (COM) interface. Using this application, the CVAM V2.1 model and properties can be exported into the CVAM-Analysis package which has been created using the VIRTOOLS gaming environment.

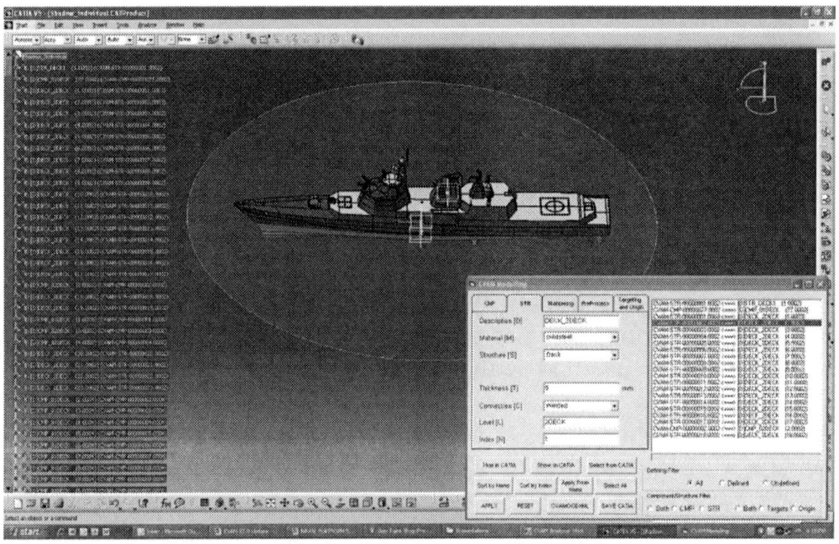

Figure 2 - CATIA model of a ship showing CVAM-Modelling interface

While the geometry of the ship in CAD is important, the inter-relationship between systems and components is just as important when modelling damage to ship systems. For example, damage to a computer system may render a vital weapon system inoperable. It is therefore important to build a system model of the ship which considers the overall effect of damage to the ships systems.

CVAM V2.1 has been developed to be fully integrated with the DSTO developed program FTAP V2.0. This code is used to model the systems of a ship and report on the capabilities that have been affected by weapons damage. This is achieved by developing a fault tree of the ship. An example fault tree is shown in Figure 3.

The fault tree is developed such that there is a node for each component which is considered important to the function of the ship. These nodes are then associated with geometry in the CVAM V2.1 model. The integration of the damage module and the fault tree analysis module allows for the damage assessed in the vulnerability model to be communicated directly to the fault tree analysis, providing a rapid answer of the total extent of damage resulting from a weapon.

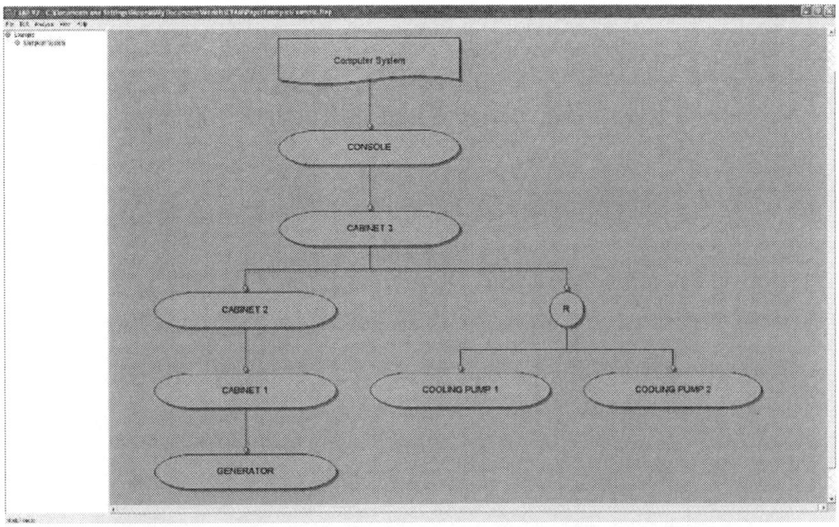

Figure 3 - An example fault tree developed using FTAP V2.0.

The vulnerability assessment is undertaken via a program called CVAM-Analysis. Figure 4 shows an example model in CVAM-Analysis. The model has been made semi-transparent to illustrate the internal details of the ship's systems.

CVAM-Analysis has been developed so that it will be easy to use for both a full vulnerability analysis and by damage control trainers using this software on-board a ship. It is envisaged that only the CVAM-Analysis program be made available to the damage control trainers, with the specialised job of generating the weapon and ship models to be left to the expert vulnerability analysts. A library of ships and weapons will be provided for damage control training purposes.

Figure 4 – CVAM-Analysis V2.1 Program with model loaded

CVAM-Analysis has been written using a gaming engine called VIRTOOLS [2] and Microsoft Visual Studio C++ [3]. The use of a gaming engine allows for advanced display of results. For example, Figure 5 shows how it is possible to "explode" the deck levels of the ship so that details of each deck level are visible. This is particularly useful for viewing the internal damage to the ship and systems.

Figure 5 - Deck levels exploded to show compartment details

The process of loading a model, running the analysis and viewing the results has been incorporated into the CVAM-Analysis toolbar as shown in Figure 6. CVAM-Analysis has been designed to make the analysis process easy for a non-specialist.

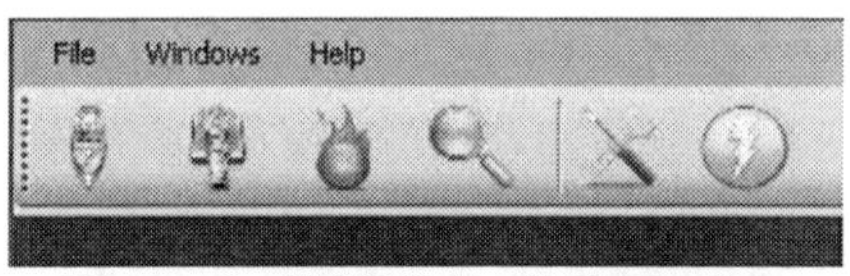

Figure 6 - CVAM-Analysis toolbar

To run an analysis for a ship, the first step is to load the CVAM V2.1 model of this ship. This is achieved by simply clicking on the ship icon which will open a window that allows the user to select from a pre-developed library of ship models.

A weapon is selected for the assessment by clicking on the gun icon. This opens another window which allows the user to select from a list of weapons for use in the analysis. These weapon files are generated by a program called CVAM-Weapon. The weapon files include details of the fragmentation pattern generated by the weapon and the mass and type of explosive used. It is intended that a pre-developed library of weapon files will be included with the software so that the user on-board a ship would simply select the weapon that they would like to use for their training exercise.

The next step in the assessment process is to setup the attack details. This is achieved by clicking on the fireball icon which brings up a window to select the how the weapon will be delivered for this training exercise. CVAM-Analysis has been developed to include a large number of analysis options and includes some advanced features for multiple detonation point analysis; obviously these features are not of interest for damage control training exercises, where single damage events will be considered. CVAM-Analysis has therefore been designed to hide the advanced attack options from a user, leaving only a simple interface that allows the user to select the position and orientation of the weapon.

There are two types of assessments that can be performed in CVAM-Analysis. The most simple is to define the location and orientation of the weapon at which the detonation will occur. This type of analysis is useful for defining training exercises involving damage to specific areas of the ship, such as the bridge or engine room.

Alternatively, the assessment can be setup to calculate the detonation location based on how far the weapon will penetrate into the ship based on a fuse delay and the initial position and direction of the weapon. Using this type of analysis it is even possible to predict the damage caused by the weapon as it penetrates the ship. This type of analysis is useful for larger scale training exercises involving other ships where the initial location and direction of the weapon will be defined by the relative locations of the ships.

Running the analysis is a simple one button click operation. After the results are calculated they are automatically displayed by the program. Damaged structure and components are displayed in various colours depicting the level of damage received.

A more detailed analysis of the results can be examined by clicking on the magnifying glass icon. This opens up the Results window which provides greater control over how the results are displayed, such as "exploding" the deck level, hiding entire decks and controlling the level of transparency of undamaged components.

The Results window also provides access to details of the individual components and structures that have been damaged. This would allow the trainers to identify which systems need to be shutdown or marked as unavailable in order to realistically simulate the effect of the damage.

Another powerful feature of CVAM V2.1 is the integration of this program with the fault tree analysis program FTAP V2.0. At the conclusion of an analysis, FTAP V2.0 is used to evaluate the overall effect of the damage to the systems on a ship.

An example model is shown in Figure 7 where one compartment is full of computer equipment. This computer equipment is powered by a generator in another compartment. The fault tree for this system is shown in Figure 8.

For this example a detonation has been set off next to the generator. Figure 9 shows that the generator is damaged by the detonation, as well as the surrounding structure. The fragmentation pattern for the warhead being considered is also shown.

As the generator has been damaged, the computer systems are no longer operational. This is shown in Figure 10, where the fault tree data has been loaded into the model from FTAP V2.0. CVAM V2.1 and FTAP V2.0 have been integrated such that the fault tree results are automatically loaded into the CVAM V2.1 program. Damage Control training instructors are able to use this information to shutdown parts of the ship to simulate the damage that would have occurred if a real detonation had taken place. The crew would then have to react to the predicted damage accounting for the degraded ships capability due to the damage. This provides a realistic damage control exercise.

If the crew were incorporated as part of the vulnerability model, injured crew could be identified. This would limit the number of crew that are available to deal with the situation and would also provide training for medical staff.

Figure 7 - Example Model

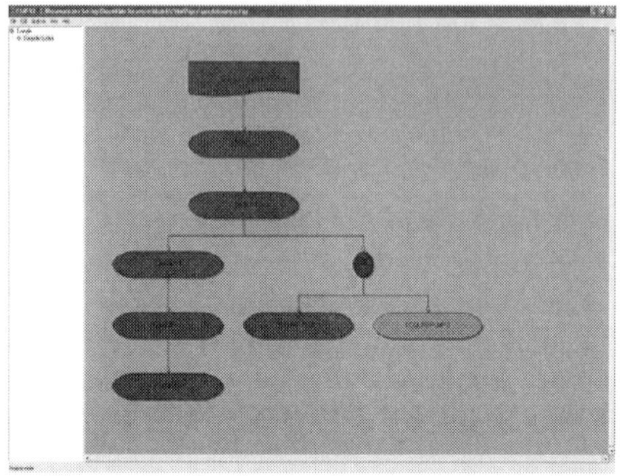

Figure 8 - Fault tree for example model

Figure 9 - Example model showing physical damage

Figure 10 - Example model showing logical damage

ADVANTAGES OF REALISTIC DAMAGE CONTROL SCENARIOS

While Damage Control trainers currently generate scenarios based on weapons damage to the ship [4], the methods used to predict the amount of damage are not as realistic as that provided by vulnerability analysis software. If the CVAM V2.1 software is used each time to generate a unique training scenario for the crew, this would alleviate the monotony associated with repeated training exercises and truly test the ability of the crew to react to unique situations. This way the crew will be better prepared to deal with a real world event.

FUTURE IMPROVEMENTS

There is much scope for developing this realistic damage control training software further to improve both the ease with which this technology can be implemented and also to enhance the training for the crew.

For example, it would be fairly straight forward to modify the software to generate a report of the damage sustained and the crew that have been injured by a particular weapons event.

This could also include details such as safe shutdown procedures for damaged equipment. These measures would improve the ease with which training exercises could be carried out. The realism of the damage prediction could also be improved by including details of the damage to doors, corridors and access hatches. By blocking access routes as part of the training exercise, the training experience would be enhanced by forcing crew to use alternative routes. This could also be used to predict and simulate trapped crew that may require rescue.

Details of any flooding or fire damage could be incorporated into the model, which may add a time-based damage mechanism, as well as also adding other mechanisms where access routes may be blocked.

If this technology was fully embraced, wireless devices (like those used in restaurants to indicate that a meal is ready) could be used to indicate crew members that had been injured and who would be either out of the training exercise or who would act as patients for the training medical team. This technology could also be used to automatically set off devices to simulate fire and smoke associated with the detonation. This type of technology would allow for more realistic training scenarios. The time to reset a training scenario could also be decreased, allowing for more training to be carried out. The exercises would also be able to be carried out with less damage control training staff to administer the training.

CONCLUSION

This paper has described the vulnerability assessment tool CVAM V2.1. This software has been designed to be easy to use, while providing realistic outputs. The use of this software has been described with application to its use as a damage control training tool on-board naval vessels.

This paper has also outlined improvements that could be made to CVAM V2.1 to enhance its capability as a damage control training tool.

REFERENCES

[1] http://www.3ds.com/products/catia/welcome/ (accessed 11 November 2009)

[2] www.virtools.com (accessed 11 November 2009)

[3] http://msdn.microsoft.com/en-us/vstudio/default.aspx (accessed 11 November 2009)

[4] *Damage Controlman*, NAVEDTRA 14057, Naval Education and Training Professional Development and Technology Centre, United States Navy, 2003

The Integration of Flexible Payload Systems into Future Submarine Platforms

Max Ratcliffe and Jon Lynn
Babcock Integrated Technology Australia
max.ratcliffe@babcock.com.au, jon.lynn@babcock.com.au

ABSTRACT

Future submarines will be designed to be more capable than the existing classes. However, diminishing defence budgets dictate that they will also need to be more flexible and be able to accommodate enhancements throughout their service lives.

In order for a submarine to be able to satisfy a broad range of mission requirements and be easily upgraded it is essential that payloads (and payload integration in particular) are considered during the earliest phases of conceptual design.

Payload integration is therefore an essential – and challenging – aspect of the design of any modern submarine, and is likely to become more and more critical in the future, particularly as many payload technologies have experienced recent rapid development and are now either mature and proven or expected to be so in the near future.

Flexible payload systems represent an attractive means of engineering adaptability into a submarine. This paper discusses the current and future flexible payload integration strategies, as well as decision making tools that to provide the most effective systems.

1. INTRODUCTION

The Future Submarine will need to meet an ever-increasing range of mission requirements. This, coupled with continual development of payload capabilities and forms, results in a need for Payload Deployment Systems to be highly flexible and easily upgradable.

Given the duration between the Future Submarine's inception through to the end of its service life, several payload upgrades will be required to maintain state-of-the-art submarine capabilities. A flexible modular payload approach, in conjunction with suitably adaptable launch systems, will provide the RAN – and any potential international customers – with a flexible submarine platform that can successfully complete the range of present and future mission profiles.

This paper begins by briefly discussing the off-board Payload Deployment Systems used on the Collins Class submarine – and elsewhere – to provide an understanding of their capabilities. It then presents decision-making tools used for down-selecting and assessing payload integration options, before discussing in more detail a range of flexible modular payload concepts as well as providing a number of application examples.

2. BACKGROUND

Submarine offensive, defensive and data gathering technology is continually evolving, along with operational scenarios, placing additional demand on submarines to provide a platform from which to deploy a wide range of payloads.

There is a large range of potential off-board payloads available, including weapons, Unmanned Underwater Vehicles (UUVs), Unmanned Aerial Vehicles (UAVs), countermeasures and deployable communications systems. These are displayed schematically in Figure 1.

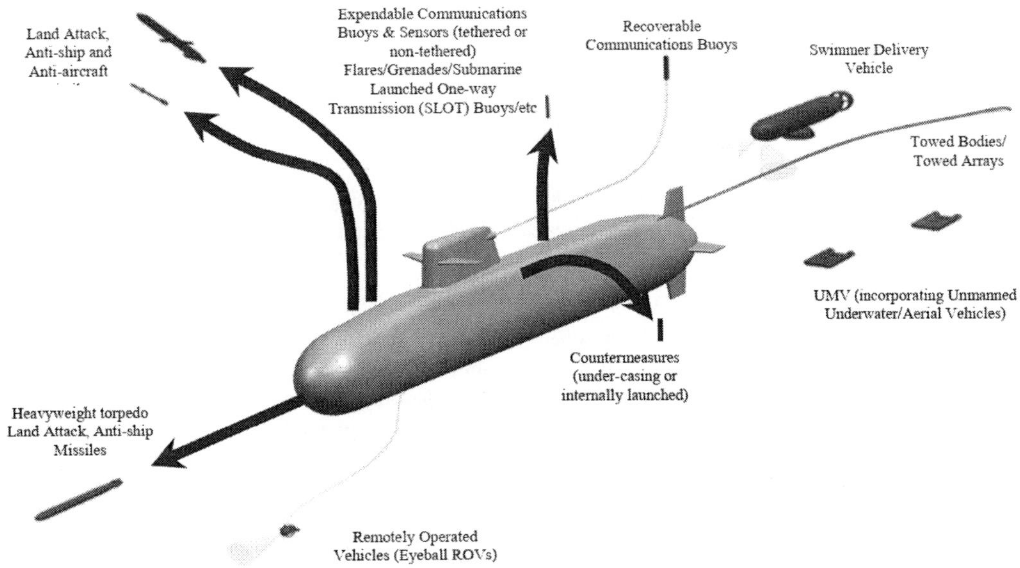

Figure 1 – Example of Off-board Payloads

Current deployment systems have limitations due to significant dimensional constraints. These can lead to a compromise in payload performance (in the form of operational durability, sensor detection or warhead effectiveness) or an inability to integrate into the submarine.

When integrating payloads into conventional (SSK) submarines, the inherent size constraints present difficult challenges for designers and project stakeholders. Flexible payload solutions provide an attractive alternative that will enable deployment of a range of payloads, and provide a means of affordable capability growth when future payloads become available.

2.1 Current Payload Deployment Capabilities

Babcock have been involved in six first-of-class submarine programmes over the past 30 years (including three ocean-going SSKs), which is more than any other single organisation worldwide, and positions us at the forefront of Payload Deployment System delivery.

The Collins Class deployment systems include the forward facing Weapon Discharge Equipment (WDE) consisting of six horizontal 21" torpedo tubes and the forward and aft

Submerged Signal Ejectors (SSEs). These deployment systems are used to launch a variety of payloads including:

- Torpedoes;
- Mines;
- Missiles;
- Signal flares; and
- Submarine Launched One-way Transmission (SLOT) buoys.

The WDE supplied by Babcock for a number of submarine classes uses a highly successful Air Turbine Pump (ATP) system to achieve a low signature and positive discharge launch of a range of payloads. Babcock has recently developed an improved version of the ATP which benefits from greater efficiency and lower power usage.

While the currently-employed WDE – with its standard 21" tubes – is not configured for modular approaches, the ATP system is capable of integrating with flexible payload systems for the Future Submarine, thanks to its Programmable Firing Valve (PFV), enabling customization of launch characteristics and profiles for specific payload types and launch tube configurations.

The SSEs provide a safety critical function on many submarine classes, which is likely to be carried over to the design of the Future Submarine. The current systems supplied by Babcock employ positive discharge and use a prime mover in the form of a water ram, and allow for stores up to around 4" diameter and around 40" length.

Although initially intended only to launch safety critical devices such as emergency flares, the SSEs have increasingly been used to launch additional payloads such as countermeasures, communication buoys and sensors. The current SSE system is well proven and will successfully supplement a flexible payload integration strategy.

2.2 Current Payload Stowage Capability

Payload stowage is conventionally achieved with an internal Weapon Stowage Compartment (WSC) at the fore end of the platform. Although there are a number of different WSC configurations, each one provides payloads with a dry, shock-protected environment and the ability to load the payload into the Torpedo Launch Tube (TLT).

3. ASSESSMENT & SELECTION OF MODULAR PAYLOADS IN CURRENT AND THE FUTURE SUBMARINE

Babcock have developed a structured methodology to assist in the decision making process for Future Submarine systems. The methodology utilises a number of tools and is applicable to situations where a number of interacting subsystems are considered, such as in the assessment of flexible payload types and deployment systems.

Typically as the selection process progresses, the number of concepts considered reduces and the detail of assessment increases, allowing a more detailed assessment to be limited to the most suitable systems. The tools provide methods that assist in making impartial decisions, by breaking each user input into a simple and manageable decision. They also retain full traceability of the decision making process providing justification of decisions.

3.1 Understanding the Requirements

To work through the payload selection process effectively, it is essential that the customer and project requirements of the submarine platform are well understood. Quality Function Deployment (QFD) is a structured approach used to capture the true customer needs for a project. QFD achieves this by listing customer statements for a specific system and identifies it as a "solution", "measure", "target" or "need"; some typical customer statements are shown and classified in Table 1. Where a customer statement is judged to be a "solution", "measure" or "target", the underlying need is derived, and listed along with the "needs" already established which identifies the true customer needs.

	Customer Statement	Solution	Measure	Target	Need
1	System costs should be minimised			X	
2	System should be adaptable to various payloads	X			
3	System shall be able to load, stow and launch countermeasures				X

Table 1 – Customer Statements

An important aspect of understanding high level requirements of a Payload Deployment System is understanding the roles and missions that Future Submarines may be required to undertake, which may include:

- Anti-submarine warfare;
- Anti-surface warfare;
- Strategic strike;
- Mine laying;
- Mine detection;
- Intelligence and data gathering; and
- Special Forces operations.

To support these roles and missions a range of payloads are required. A clear understanding of the payload types required to achieve certain roles and missions is needed. For example, off-board payloads required to achieve successful support of anti-submarine warfare include heavyweight torpedoes and counter-measures.

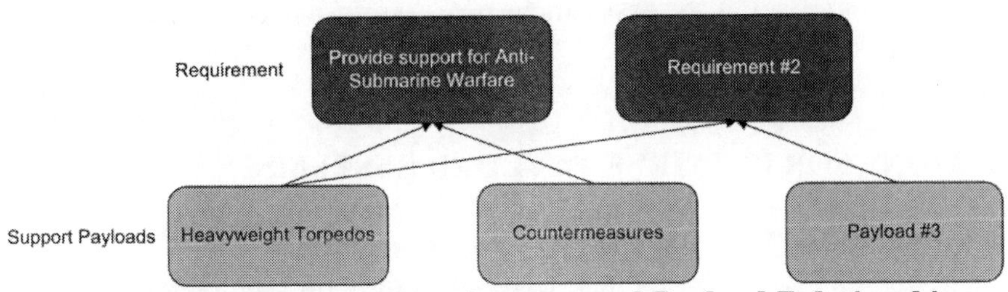

Figure 2 – Platform Requirement and Payload Relationship

These types of high level payload and payload subsystem requirements filter down into more detailed payload deployment requirements that are used in the assessment of the whole Payload Deployment System.

3.2 Morphological Analysis and Cross Consistency Analysis

Morphological analysis lists the available subsystem options and provides a method of developing the possible combinations of those subsystems that could make up the final system. Another tool used to reduce the number of system options is cross-consistency analysis, which establishes which subsystem solutions are compatible. An example cross-consistency analysis is shown in Figure 3. A ⊕ symbol indicates that the two subsystems are consistent and compatible. Those that are not compatible are disregarded and do not need to be assessed further.

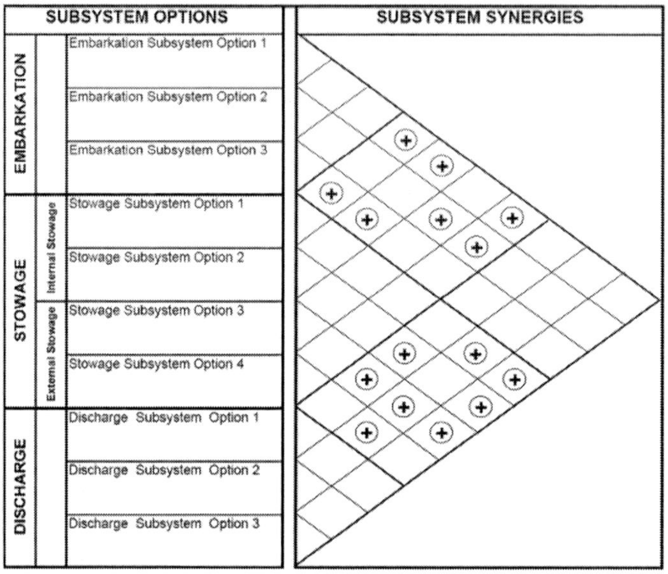

Figure 3 – Cross Consistency Analysis

3.3 System Metrics

Babcock has developed a method of metrics for the assessment of complex systems which can be applied to any number of submarine systems. It allows a complex system to be broken down into manageable system attributes that are given an importance rating and scored. The output of the assessment is an overall system score which is used for comparison against competing systems. By using this form of metric, the assessment remains fully traceable, and can be used to support and record the decision making process. With the appropriate project cross-sectional team (i.e. appropriate personnel from a range of departments) involved in the assessment process, it involves all key stake-holders in the decision making process.

4. OPTIONS FOR FLEXIBLE PAYLOAD SYSTEMS

4.1 Modular and Reconfigurable Payload Systems

A modular payload system is one that uses common main architecture in conjunction with a payload-specific module to allow embarkation and stowage and to launch the payload. This type of payload system has the advantage of providing rapid payload change-over for replenishment or a change in payload type.

A modular approach will provide the ability to configure the submarine platform with the most effective payload combination for any given mission. Mission types can vary from strategic strikes to information gathering, and a modular payload approach will provide benefits throughout the life of the submarine as its operational envelope evolves.

4.2 Internal and External Payload Systems

Payload systems may be characterized as either predominantly internal or predominantly external systems. Where a deployment system is located internal to the pressure hull, the system can often be accessed from within the submarine, and configured with a variety of mission-specific payloads on demand. However an internal payload system has the drawback of introducing additional hull penetrations, thus increasing the system cost and complexity.

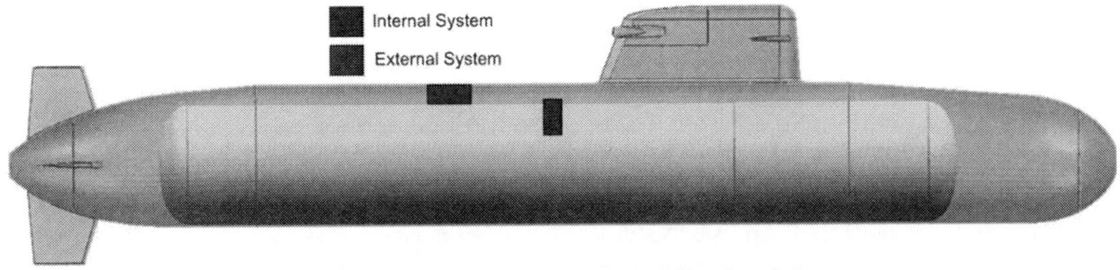

Figure 4 – Internal and External Payload System

For external systems – which are located in the free-flood space – the payload needs to be selected, loaded and configured while the submarine is alongside. This provides a broad flexibility of submarine capability; however, it requires the submarine to return to a port to re-configure the payload should the mission change while deployed. An external payload system has the advantage of not introducing complex hull penetrations.

External modular payloads present additional challenges of serviceability, environmental considerations and shock resistance. An external system must be robust since servicing cannot be conducted while at sea. The payload module must also provide the appropriate shock attenuation as the payload will need to be stored within the launch system, and cannot be located inboard on shock mounted stowage areas.

4.3 Large Diameter Configurable Payload System

One method of providing a modular payload system is to use a large diameter launch tube that interfaces with a wide range of payload-specific modules. This results in a highly flexible system and allows for easy integration of new payloads, providing the submarine with significant potential for capability growth. If the payload is able to fit within the defining parameters of the tube, it is probable that it can be integrated with the deployment system.

The large diameter tube shown in Figure 5 consists of a proposed standard launch tube and hatch, which accommodates a payload module that interfaces with the launch tube and the payload/s.

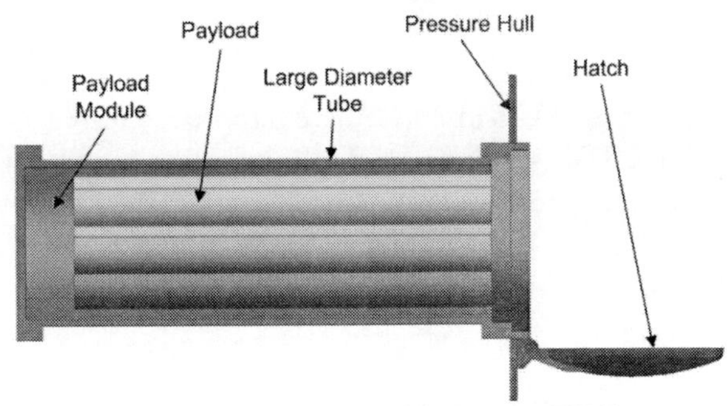

Figure 5 – Large Diameter PDS

4.4 Launch System Options

When developing a flexible modular payload system, it is essential to have an understanding of the launch methods available. Based on the energy source used to achieve launch, the three fundamental principles are:

- Platform Motion;
- Payload Potential; and
- Positive Discharge.

The first two options – with the exception of payload self-propulsion (where the payload motor starts within the tube) – have a reduced launch signature, and do not require additional energy from the platform. Positive discharge requires input energy from a prime mover included as part of the platform. The ATP system discussed in Section 2.1 is an example of a positive discharge system.

Launch System Options	Concepts	Comments
Platform Motion	Drogue Launch	This method has been extensively studied by Babcock Integrated Technology.
	Hydrodynamic Lift	Uses the force developed by the payload form in conjunction with its forward motion. This method is an insufficient means of launch on its own.
Payload Potential	Gravity Launch	Studied for horizontal attitude, this technique has previously been employed for mine laying.
	Buoyancy Launch	Developed by Babcock Integrated Technology and used for experimental devices. Provides a good launch envelope but requires a more sophisticated payload device.
	Self-Propulsion	Swim-out torpedoes/tubes are in service with a number of navies worldwide.
Positive Discharge	Tube Dedicated	Examples include water ram systems included in the SSE, compressed air systems and pyrotechnic gas generators.
	Multi Tube	A large number of systems have been developed which include single acting reciprocating pumps, rotodynamic pumps and accumulator based systems. This technique has largely been employed for torpedoes and other weapons in multi-tube configurations.

Table 2 - Typical Launch System Options

A flexible system must be capable of deploying a range of payloads with varying deployment characteristics. A torpedo for example requires rapid launch to achieve military effect, while a UUV is able to be deployed more slowly. To accommodate the full range of launch characteristics, positive discharge will certainly be required. For this reason a flexible positive discharge system – such as the ATP system – is preferred as it can provide rapid launch, while also maintaining the ability to effectively conduct low energy launches.

5. FLEXIBLE PAYLOAD APPLICATIONS

5.1 Developing Payloads

Moving forward, payload advancements are expected to drive the development of deployment systems. Payloads currently receiving high levels of attention include UUVs, UAVs, short range missiles, vertically-launched land attack missiles and integration of Special Forces equipment. Babcock has and continues to conduct studies on the integration of these payloads into SSK submarines. These technologies will enhance the capabilities of submarine operations, and will increasingly become requirements on future submarines.

5.2 Flexible Payload Concepts

Flexible payload concepts strive to provide maximum utilisation of submarine real estate and equipment. These often have configurable components that allow operators to change the payload system and enable alternative payloads to be embarked, stowed and launched. These principles are illustrated in Figure 6 with an under-casing payload modules concept and a configurable launch tube that uses insert sleeves to configure the launch tube to accommodate a range of payloads.

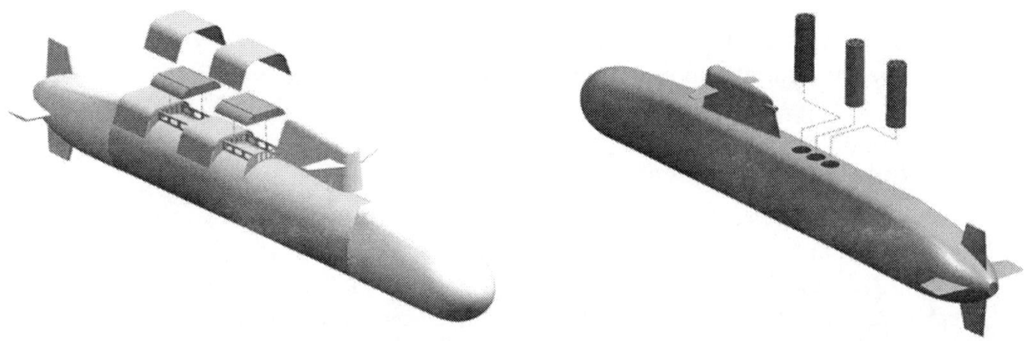

Figure 6 – Flexible Under-casing Payload Stowages and Vertical Payload Modules

5.3 Flexible External Launch System

Babcock have developed, constructed and proven a flexible launch system, in conjunction with our international partners, and this technology is currently being developed for a new torpedo countermeasure system.

It includes a soft kill system that uses a 4" diameter device and a hard kill system that uses an 8" diameter device. Modules of launchers, along with the tube dedicated launch systems are installed externally to the pressure hull, offering high levels of flexibility, rapid launch of a number of countermeasures and rapid payload changeover and replenishment. The launcher and its integration with the platform are shown in Figure 7

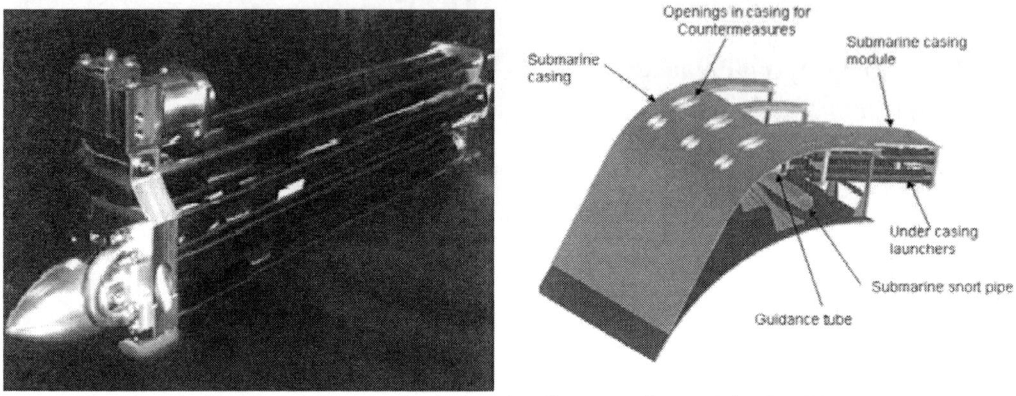

Figure 7 - External Launcher and its Integration with the Platform

5.4 Flexible WSC

Current WSCs are inherently flexible to some degree, capable of stowing a variety of payloads including heavy and lightweight torpedoes, mines, missiles and UUVs. The WSC can be further enhanced by designing the WSC architecture to be more simply configured for multiple stowage scenarios. This allows for a wider range of payloads to be stowed but also benefits specific roles and missions such as for Special Forces operations, where the WSC can be configured to house Special Forces personnel and equipment.

5.5 Industry developments

The Australian submarine community has benefited from a very close level of Australia-US collaboration in undersea warfare capability which is expected to continue. It is interesting to note that the US has adopted a large diameter configurable and modular vertical tube system for their SSGN Virginia Class Block III [1]. Due to its size, this arrangement (proven on in-service SSGN submarines) has the potential to integrate an extremely wide range of payloads, including land attack missiles.

Babcock is investigating the implications of integrating similar large diameter flexible tubes into inherently smaller SSK submarines. There are significant technical challenges involved to successfully integrate the two – even in today's larger SSKs – which require further development work to prove their compatibility. The costs and benefits of adopting such a system also require careful assessment to determine their suitability to meet the Future Submarine's requirements.

Possible benefits of adopting large diameter configurable tubes include the potential use of future payloads developed using the increased volume of a large diameter tube, along with the capability that the payload provides. This may not be an option with current and small configurable payload systems.

6. CONCLUDING REMARKS

For the Future Submarine to retain state-of-the-art capabilities throughout its service life, submarine systems must provide scope for continual and cost effective improvements. For payload capability, this can be accomplished by incorporating maximum flexibility into the payload systems.

When assessing the payload systems it is important that assessment and selection uses a structured methodology. This – in conjunction with an in-depth understanding of the current and Future Submarine capability requirements – will result in the best composition of submarine payload and payload systems.

It is essential that new technology is well understood, and de-risked with technical demonstrators where required. Further studies and development work are required on integration of flexible payload systems into future SSK submarines, which will inform the decision makers of the benefits and costs of available options.

7. REFERENCES

1. Richard R. Burgess, 2009, "On to the Next Block", Seapower, Volume 52 - Number 3, pages 28-30

REFEREED PAPER

Hydroelastic Response of a Segmented Frigate Model

Brett Morris, Defence Science and Technology Organisation (DSTO), Australia. brett.morris@dsto.defence.gov.au , Chris Hutchison, Australian Maritime College (AMC), c.hutchison@amc.edu.au, Giles Thomas, AMC, giles@amc.edu.au, Bernie Phelps, DSTO, Bernard.Phelps@dsto.defence.gov.au & Tim Lilienthal, AMC, timl@amc.edu.au

ABSTRACT

Previous work by DSTO has highlighted potential limits of applicability for numerical seakeeping and wave-induced global loading computer codes based on rigid body equations of motion. In collaboration with the AMC, a hydroelastic frigate model with six segments has been designed, constructed and tested in a variety of regular wave conditions. The results of these experiments can be compared to the results of a planned numerical analysis using rigid body computer codes to investigate their bounds of application. This paper covers the hydroelastic model experiments and initial stages of the numerical replication of the experiments.

INTRODUCTION

Naval architects have used different methods of predicting the wave-induced global loading to which a vessel will be subjected. Early methods utilised approximations that involved balancing a ship on a regular (sinusoidal) wave with a length (λ) equal to that of the ship length (L) and height equal to L/20 [1]. These methods have been shown to over predict the actual wave-induced bending moments due to neglecting the dynamic and hydrodynamic loadings that arise from wave pressures and ship motions [2]. Later methods of predicting the wave-induced global loads have involved their direct computation utilising deterministic approaches to include the dynamic and hydrodynamic effects [2]. Two such methods of incorporating the inertial and ship motion aspects into the calculation of these loads are strip theory and the three-dimensional (3D) panel method.

A previous study undertaken by the Defence Science and Technology Organisation (DSTO) investigated the structural changes that may be required if a frigate's displacement and operating time were increased. This study involved the use of a 3D panel method seakeeping code [3] to evaluate the shear force and bending moment distributions along the hull for a range of speeds, wave heights, wavelengths and relative wave headings. The results were then combined statistically to determine the long-term loads for global strength and fatigue analyses. An outcome of the DSTO study was that it highlighted an issue with the seakeeping code relating to the distribution of vertical bending moments along the hull. This applied to a limited number of load cases, however, the cause was not able to be determined at the time. This prompted the present investigation into the operational conditions that may limit the reliable use of these codes.

The work carried out in this present study has been performed with the aim of investigating the reliable bounds of global loading predictions given using strip theory and 3D panel methods [4]. A segmented model of a frigate hullform was designed, constructed and tested at various speeds and waveheights, with measurements of motions and wave-induced global loads taken. The results of a numerical model analysis using computer codes based on linear strip theory and 3D panel methods and rigid body equations of motion to replicate the hydroelastic model tests will then be used to investigate the reliable boundaries of these methods.

This paper covers the initial stages of the study, describing the design and construction of the model, along with some early results of measurements of motions and loads in response to regular waves.

MODEL DESIGN AND CONSTRUCTION

Segmentation of a model to determine the hydroelastic behaviour of a full scale ship has become a valuable procedure for the measurement of global wave-induced loads [5]. By scaling consistently the natural bending frequencies of the hydroelastic model with the full scale vessel, the whipping and related phenomena responses may be measured [6]. The model flexibility is provided by flexible links at the cuts between segments. The segments themselves are designed and constructed to be of high stiffness so the flexibility in the model is primarily introduced at the elastic links.

The primary consideration when selecting the scale of the hydroelastic frigate model was the size, in particular the width, of the test facilities at the Australian Maritime College (AMC). A study was performed based on wall interference due to incident wave reflection and an effective speed difference created by blockage effects. The most appropriate model size was determined as being 1:44 scale and the model particulars are given in Table 1.

Table 1: Hydroelastic Model Particulars

Length between perpendiculars	2.5m
Displacement	41.23kg
Draft	0.101m
Trim	0 degrees
Longitudinal centre of gravity	1.35m aft of FP

The model was designed with five transverse cuts, giving six segments. These segments are connected by elastic links, machined from a solid aluminium section and bolted into an aluminium section backbone beam. The selection of six segments was made based on the desire for the model to provide bending moment data along its length and achieve correlation with as many vertical bending modes of the full-scale frigate as possible, within the physical constraints of the model size. To avoid a low strain reading towards the stern of the model, the aft segment is 550 mm long and the other segments each 390 mm long.

The elastic links were designed to achieve the required dynamic response of the vessel with respect to the 2-node and 3-node vertical longitudinal natural frequencies. An additional requirement was to obtain reasonable strain values from the strain gauges mounted on the links for the conditions to be investigated in terms of signal to noise ratio. A prototype elastic link was designed to be milled from a 40 x 40 mm length of aluminium to a narrower elastic cross section as shown in Figure 1. The elastic link was fitted to the model using an upper and lower clamping plate with 40 mm recesses and bolts through the clamping plates into the model.

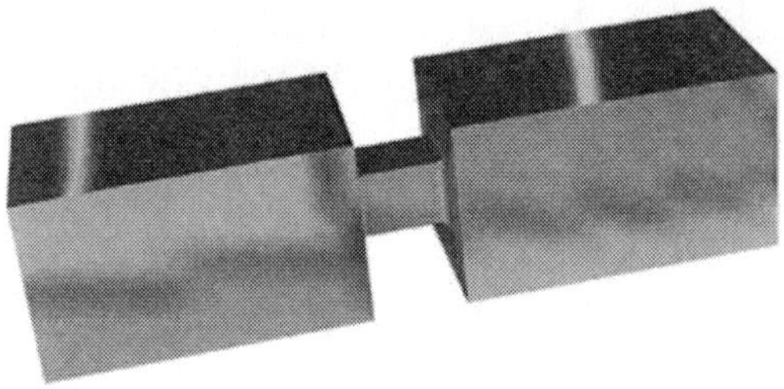

Figure 1: Render of prototype elastic link

To determine the appropriate elastic link size for the hydroelastic model that would obtain as many equivalent full scale ship bending modes as possible, a study was undertaken utilising an existing two segment model of an NPL 6a hullform. The NPL 6a model was fitted with the prototype elastic link and while the model was floating in still water at zero speed, an impulse load was applied to the stern to excite the longitudinal mode of vibration. Accelerations of the NPL 6a model were recorded using an accelerometer that was located just aft of the elastic link joining the forward and aft segments of the model. This acceleration data was analysed using a custom LabVIEW spectral analysis program. The data was high-pass filtered to remove low frequency drift and windowed using a Hanning window to reduce spectral leakage before the power spectrum was calculated. An idealised Finite Element Analysis (FEA) model based on the two segment NPL 6a model was established. Beam elements were utilised for the hull segments and elastic link whilst mass elements were distributed along the model to represent the model mass as well as the added mass. The difference between the predicted and measured modal frequency enabled the effective stiffness of the elastic link to be determined. This stiffness was then utilised in the FEA model to obtain the predicted two node mode for the NPL 6a model.

After verification that this approach was appropriate for the two segment NPL 6a model, an FEA of an idealised six segment frigate model was used to calculate the expected vertical bending modes for various link dimensions. Again, beam elements were utilised for the hull segments and elastic links, whilst mass elements were distributed along the model to represent the model mass as well as the fluid added mass. Following several iterations of link sizes, very good agreement with the full scale ship value for the two node mode was exhibited. The three node mode for the six segment hydroelastic model is slightly higher than the full scale ship, but deemed to lie within acceptable limits.

The model has a variety of materials used in its construction. The hull is a symmetric carbon fibre E-glass laminate, with three layers of 435 g glass as core material. All of the full and partial bulkheads are sandwich construction with a 12 mm balsa end grain core and two layers of carbon fibre either side. The decks are fabricated from two layers of carbon fibre either side of 6 mm plywood. These materials were selected in order to make the model segments extremely stiff, thereby ensuring that the model flexes primarily in the elastic links. An aluminium section backbone was also fixed in each segment longitudinally to further stiffen the segments. The completed model can be seen in Figure 2. The materials utilised, surface finish and dimensional tolerances of the model are all as recommended by international Towing Tank Conference (ITTC) procedures [7].

Figure 2: Six segment hydroelastic frigate model

INSTRUMENTATION

The experiments were conducted in order to investigate the motions and loads of the hydroelastic model frigate in head and following seas. The model was instrumented to measure vertical bending moments (VBM), pitch and heave motions, along with the wave profile. Strain gauges were mounted on the elastic links of the model to record the differential strain on the top and bottom surfaces of the elastic link at each of the connection points along the model. The strain gauges were connected to a Vishay 2100 system signal conditioner as full Wheatstone bridges prior to the signal being recorded. Each strain gauge unit was calibrated to fine tune the complete system and reduce sources of noise from equipment and resistance errors from the cabling. Following calibration of the complete model, the measured strain values were converted to VBM by an in-house custom LabVIEW program.

To measure pitch and heave motions, two Schaevitz 5000 DC-EC Linear Voltage Displacement Transducers (LVDTs) were connected to the centreline of the model, one forward of the longitudinal centre of gravity (LCG) of the model, and one aft of the LCG. These measured the vertical displacement from which the model heave and pitch was determined. The wave profile was measured by a stationary resistance wave probe located close to the wave maker that recorded the wave profile prior to commencing a test run. In addition, a second measurement system, an ultrasonic wave sensor, was attached to the carriage vertically in line with the model's forward perpendicular. This provided details of the encountered wave height and frequency to permit the phase relationships between the wave and the model motions and loads to be determined.

All data was acquired at a sampling rate of 200 Hz by a HP desktop computer via a National Instruments PCI-6036E Multifunction Data Acquisition Card, which was controlled by LabVIEW. All channels were sampled simultaneously so that all measurements were time stamped and could be cross analysed. Each instrument used for data acquisition was calibrated regularly. The linear variable differential transformers (LVDTs), wave probe and force transducer were calibrated on a daily basis during the head sea seakeeping tests. The calibrations included all items of the measurement chain, including signal conditioner, amplifier, filter and analogue to digital converter.

MODEL EXPERIMENTS

The experiments were conducted in the Australian Maritime College's Towing Tank, which is a rectangular tank 100 m long, 3.55 m wide and 1.5 m deep. Waves were produced by a hydraulically driven wave maker positioned at one end of the tank. The computer software that controls the wave maker paddle provides the ability to produce a wide variety of wave forms, including regular waves and irregular wave spectra. A carriage, which has a maximum speed of 4.6 m/s, is used to tow the models.

The model was attached to the carriage using a two post system. The posts can move freely in the vertical direction against mounted linear bearings on the carriage frame. The forward post connection to the model was hinge supported allowing rotation around the lateral axis only and restricting all relative translational modes. The aft post was connected to the model via a ball bearing whose base was allowed to slide longitudinally. Therefore, the model was free to heave and pitch while being constrained in roll, yaw, sway and surge.

Two series of experiments were undertaken in the towing tank in both calm water and regular head seas. The first series of tests were conducted as per the test matrix in Table 2. In the second test series, the model's displacement was increased by approximately 10% and tested over the same range of vessel speeds, wave heights and wave frequencies to investigate the affect of displacement on motions and loads.

Table 2: Test Matrix for Series 1 Hydroelastic Frigate Model Tests

Speed		Wave Height		Wave Frequency
Full Scale (knots)	Model (m/s)	Full Scale (m)	Model (mm)	Model scale (Hz)
12, 18, 24	0.93, 1.4, 1.86	0	0	0
12	0.93	2	45.45	0.4 – 1.4
18	1.40	2	45.45	0.5 – 1.4
24	1.86	2	45.45	0.4 – 1.2
12	0.93	4	90.91	0.5 – 1.2
18	1.40	4	90.91	0.5 – 1.2
24	1.86	4	90.91	0.5 – 1.2
12	0.93	6	136.36	0.5 – 1.2
18	1.40	3	68.18	0.5 – 1.1
24	1.86	3	68.18	0.5 – 1.1

From Table 2, it can be seen that the model was not tested at the full scale speeds of 18 and 24 knots for the full scale wave height of six metres. This was not possible due to large amounts of water being shipped over the bow of the model during the experiments at the greater wave heights.

NUMERICAL SIMULATION

Although there has been substantial development in hydroelastic computer codes over the last two decades, they are not widely available. The application focus for these codes has been very large floating structures (VLFS) , such as floating airport runways [8]. The focus on these structures could be due to them having natural bending modes that are in the range of frequencies that are close to the exciting wave frequencies. In the case of some large cargo ships where the natural bending modes are close to the ship's wave encounter frequency, whipping phenomena has been experienced [5]. Generally, frigate hullform ships will have significantly higher natural bending frequencies than cargo ships due to their relatively short length. Therefore, most numerical seakeeping analyses of these hullforms have assumed that the rigid body motions are appreciably larger than the elastic distortions and that hydroelasticity can be ignored [9].

DSTO has available both strip theory and 3D panel method frequency domain computer codes [10], [3] that are based on rigid body equations of motion. Numerical models of the hydroelastic frigate model hullform were created, using the same parameters as those of the hydroelastic model, for use in each of these codes. The first part of the numerical analysis focused on the development and refinement of these numerical models. The numerical model hydrostatics were verified against those of the hydroelastic frigate model as a check of similarity, with good agreement achieved. The effect of using different size panels within the numerical mesh was also investigated. These can then be used in motion and global loading analyses in the same operational conditions as used for testing in the towing tank. The results of these numerical analyses can be used to investigate the reliable boundaries of their application.

Motions

Computation of the global loads acting on a ship requires that firstly, the motions of the ship are calculated in order to allow the inertial and fluid loads at each time step to be determined [2]. For both the strip theory and 3D panel method codes used for the analysis, a linearised equation of motion is applied for a rigid vessel travelling with constant forward speed, U, at an arbitrary angle to regular sinusoidal waves. The resultant motions in the six degrees of freedom are then governed by the coupled equations in (1) [11].

$$\sum_{j=1}^{6}\left[-\omega_e^2\left(M_{ij}+A_{ij}\right)-i\,\omega_e B_{ij}+C_{ij}\right]\zeta_j=F_i \qquad i=1\ldots6 \tag{1}$$

Where, M_{ij} are the components of the generalised mass matrix of the ship, A_{ij} are the ship added mass coefficients, B_{ij} are the damping coefficients, C_{ij} are the hydrostatic restoring coefficients and F_i are the complex amplitudes of the exciting (wave) force and moment. The symbol ω_e represents the wave encounter frequency and ζ_j is the complex amplitude of the motions [11]. Equation (1) can be solved for six degrees of freedom, using six simultaneous equations, by first estimating the roll amplitude in order to evaluate the nonlinear roll damping forces. A final solution is obtained after iterating until, for a given set of conditions, the roll amplitude converges [12].

The rigid body strip theory and the 3D panel method codes both apply the same assumption of an inviscid, incompressible and irrotational fluid flow [4], [11]. However, due to the discretisation of the hull surface into sectional strips in strip theory and quadrilateral panels in the panel method, differences arise in the approaches used by each to predict the hydrostatic and hydrodynamic forces acting on the hull. The hydrodynamic forces arise due to the incident waves approaching the ship, along with the diffracted and radiated waves created by the ship moving through the fluid. Previous comparisons with experimental ship motion data indicate that generally, the 3D numerical method will give better agreement than strip theory [4].

Global Loads

Numerical prediction of the global loads acting on a ship requires an accurate description of the ships mass distribution [10]. For the numerical model of the frigate, the input masses were the masses of each section of the hydroelastic model and they were located at the longitudinal centre of each section on the centreline. Locating the masses longitudinally within each segment in such a fashion is the only option available with the numerical codes as the mass is uniformly distributed along each segment [10] [3]. The vertical calculation of the global load at a longitudinal location requires that this location lies between two segments of distributed mass [3]. This led to slight differences with the actual hydroelastic model, however, the differences in longitudinal position were all less than 10% of the segment length, with most less than five percent of the segment length.

Both the strip theory and 3D panel method codes predict the global wave loading at a longitudinal location on a ship using the same approach. The approach uses direct integration of the inertial force due to the mass distribution and the previously mentioned hydrostatic and hydrodynamic forces acting forward of the location where the wave-induced global load is being calculated.

A shortcoming of this linear frequency domain calculation and the resulting linear loads is that the response function is given in terms of amplitude and phase. Therefore, there is no difference between the resulting hogging and sagging bending moments [13]. Measurements on board ships, such as the work of Clarke [1], have shown that in reality, ships will experience larger sagging bending moments than hogging.

RESULTS AND DISCUSSION

For the head seas hydroelastic model experiments, both the heave and pitch motions increase with an increase in forward speed. As the speed increases to 24 knots, a resonant peak for pitch is evident at 0.6 Hz. The linearity of the increase in vessel motions with waveheight can be seen at each speed by comparing the non-dimensional motions for the range of wave heights tested. At a speed of 12 knots the heave and pitch motions increase with waveheight with a relationship that is close to linear. However, as the vessel speed increases, this relationship tends to be non-linear. This is shown for the non-dimensional pitch motions in Figure 3, for a full scale vessel speed of 24 knots, where the pitch, non-dimensionalised by wave slope and height, is lower for the larger waves at most frequencies below 0.9 Hz.

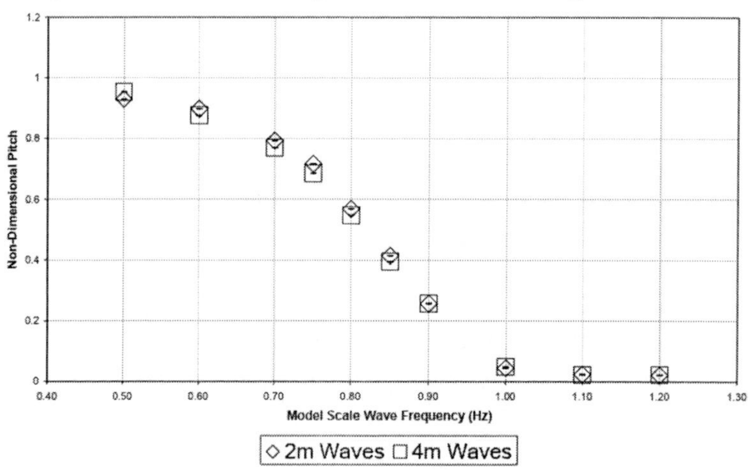

(a)

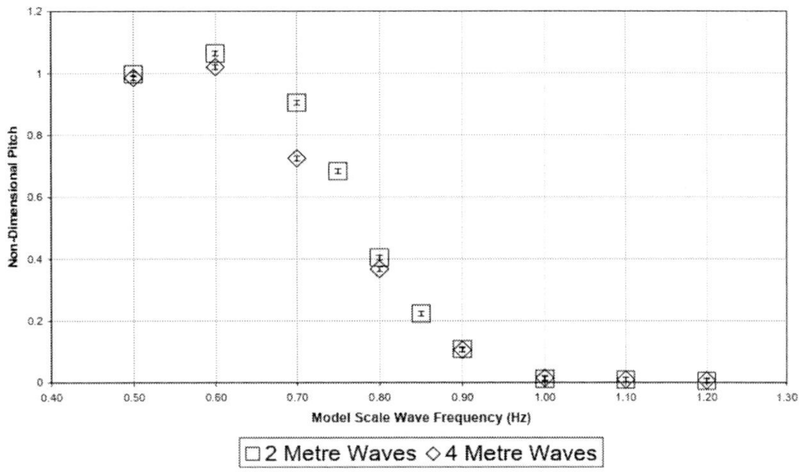

(b)

Figure 3: Experimental non-dimensional pitch motion results for full scale speed of 12 (a) and 24 (b) knots in 2 and 4 metre headseas

From Figure 3 (b) it can be seen in the 24 knots case, that the resonant peak for pitch motion reduces slightly as the equivalent full scale waveheight is increased from two to four metres. At a model scale wave frequency of 0.7 Hz, the reduction in non-dimensional pitch for two metre to four metre waves is more significant. This relative reduction in response at the high speed may be due to the influence of the topside flare reducing motions as it is immersed more deeply with the increased wave height. Figure 3 (a) displays a narrower error range, which were established from repeated runs, than for the higher speed experiments shown in Figure 3 (b).

For the wave-induced global loads measured during the head seas model experiments, the bending moments increased as vessel speed increased. Larger sagging bending moments than hogging bending moments were measured and this difference became larger at the higher speeds, as the sagging moment became more pronounced. The relationship between the increase in measured bending moments and the increase in wave height can be seen at each speed by comparing the non-dimensionalised bending moments. The non-dimensional bending moments (*BM'*) were found using (2).

$$BM' = \frac{BM}{1/2\rho V^2 \zeta_o \nabla^{2/3}} \qquad (2)$$

Where *BM* is the measured bending moment, V is the vessel speed, ζ_o is the incident wave amplitude and ∇ is the vessel volume of displacement. For each of the three speeds investigated during the model testing, the bending moment at midships was non-linear with respect to wave height, as the non-dimensionalised values vary for the different waveheights. This non-linear relationship, particularly for the step between the four and six metres wave heights, can be seen for the measured non-dimensional midships bending moments at a full scale vessel speed of 12 knots, shown in Figure 4, where the negative values are sagging bending moments and the positive values are hogging bending moments.

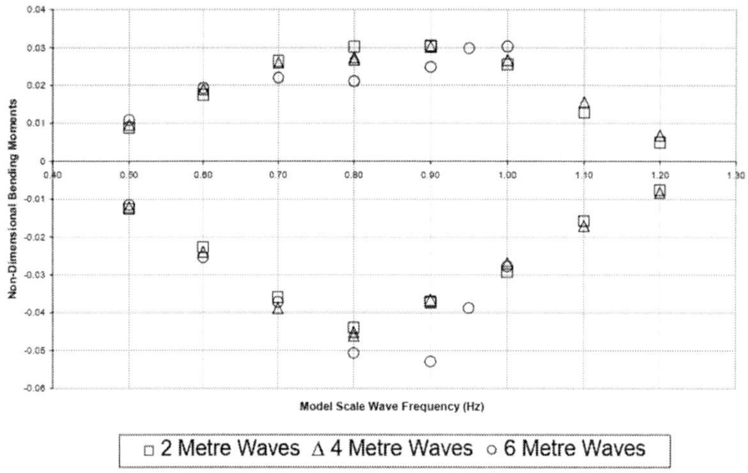

Figure 4: Non-dimensional bending moments for 2, 4 and 6 metre waves at 12 knots full scale vessel speed at midships

Preliminary work on the numerical replication of the hydroelastic model experiments focused on validation of the mesh and refinement of the surface panels to be used in the 3D panel codes. Figure 5 shows a plot of the numerical non-dimensional pitch results, at two mesh refinement levels, for the meshes at full and model scale for a full scale speed of 24 knots in two metre head seas. Initially, the numerical models were constructed at model scale in an attempt to minimise any scaling effects on the results. The hydrostatics of these models agreed well with the values for the hydroelastic model. However, the preliminary numerical results displayed unexpected peaks at the higher frequencies, particularly at around 1 hertz. The software guidelines recommend that the length of the panels is set so that there at least five panels within the encounter wavelength at the highest wave encounter frequency to be investigated [11]. When the numerical calculations are carried out at full scale, the pitch response at 1 hertz reduces significantly even though the panel length to wave length ratio remains the same. Further investigation is required to establish the cause for this irregularity

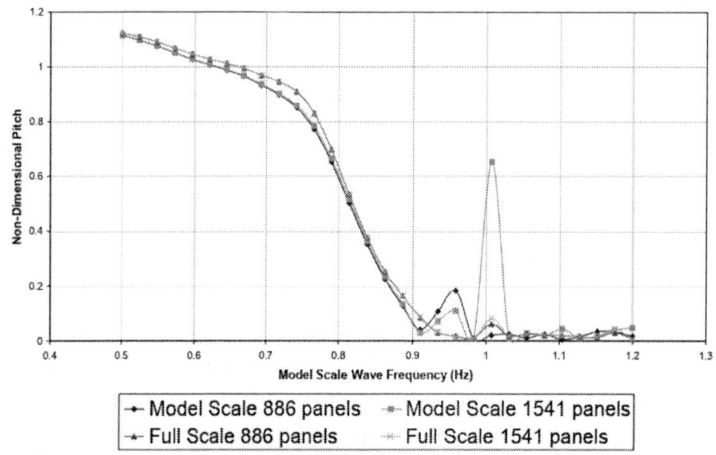

Figure 5: Numerical pitch motions for combinations of scale and number of panels in 2 metre headseas at 24 knots

From Figure 5 it can be seen that the model scale results exhibit irregularities at the upper frequencies and that increasing the number of surface panels at full scale results in very little refinement. Also at the lower frequencies, the pitch response exceeds unity although should trend back to one as frequency approaches zero.

CONCLUSIONS

Experiments have been conducted at the Australian Maritime College to investigate the motions and loads experienced by a frigate in head seas. A specially designed and constructed 1:44 scale hydroelastic model, comprising 6 segments, was tested at speeds corresponding to 12, 18 and 24 knots full-scale for a range of waveheights (2, 3, 4 and 6 metres full-scale). For the conditions tested, the heave and pitch motions do not increase linearly with respect to waveheight. The wave-induced global bending moments that were measured in head seas increased as vessel speed was increased. Testing at different wave heights for the same vessel speed, shows that as the wave height increases, the sag moment becomes more dominant than the hog moment. As with the measured motions, the measured bending moments are non-linear with respect to wave height, thus indicating that a spectral approach for estimating the expected loads in an irregular sea would be unreliable and tank testing would be preferable. The hydrostatics of the numerical models created for use in the strip theory and 3D panel codes correspond well to those of the hydroelastic frigate model. Initial 3D panel method analysis using model-scale numerical models showed that irregular results occurred at the higher encounter frequencies. These irregularities are due to the high encounter frequencies at model-scale operating conditions and led to adopting a numerical analysis using full-scale conditions. The full-scale numerical models will be used in the further numerical analyses to assess the bounds of the strip theory and 3D panel method codes' applicability.

REFERENCES

1. Clarke, J.D. *Wave Loading in Warships*. in *Advances in Marine Structures*. 1986. ARE: Elsevier Science Publishers.
2. Lewis, E.V., ed. *Principles of Naval Architecture*. Second ed. Vol. 1. 1988, SNAME: Jersey City.
3. Martec, L., *FD - WaveLoad 2008 A - User's Guide*. 2009, Martec.
4. Lewis, E.V., ed. *Principles of Naval Architecture*. Second Revision ed. Vol. Volume 3. 1989, SNAME: Jersey City.
5. McTaggart, K., et al. *Motions and loads of a hydroelastic frigate model in severe seas*. in *Transactions - Society of Naval Architects and Marine Engineers*. 1998. Ottawa, Can: Soc of Naval Architects & Marine Engineers.
6. Holloway, D., J. Lavroff, and M. Davis, *Global Structural Design of Hydroelastic Models*, in *HIPER*. 2006: Tasmania.
7. ITTC, *Testing and Extrapolation Methods: Resistance Test*, in *Recommended Procedures and Guidelines*. 2002.
8. Riggs, H.R., et al., *Comparison of hydroelastic computer codes based on the ISSC VLFS benchmark*. Ocean Engineering, 2008. **35**(7): p. 589-597.
9. Clayton, B.R. and R.E.D. Bishop, *Mechanics of Marine Vehicles*. 1982, London: E. & F. N. Spon Ltd.
10. McTaggart, K., *SHIPMO7: An Updated Strip Theory Program for Predicting Ship Motions and Sea Loads in Waves*. 1997, DREA.
11. Veer, A.P.v.t., *PRECAL V6.4 Theory Manual*. 2008, MARIN.
12. Andrewartha, T., et al., *Replenishment at sea: Motions of ships operating side by side in head seas*. Transactions of the Royal Institution of Naval Architects Part A: International Journal of Maritime Engineering, 2007. **149**(3): p. 101.
13. Veer, A.P.v.t., *Internal Loads Upgrade for PRECAL: Discussion on Upgrade Between V4 and V6*. 2003, MARIN.

The Prediction of Wetdeck Slam Loads on High-Speed Catamarans

M R Davis[1] , J Lavroff[1], B French[2], G A Thomas[2] , D S Holloway[1], T R Roberts[3]
[1] School of Engineering, University of Tasmania, Australia
[2] Australian Maritime College, University of Tasmania, Australia
[3] Revolution Design Pty Ltd, Australia
Presenting author email: M.R.Davis@utas.edu.au

1. INTRODUCTION: EVOLUTION OF HIGH SPEED DESIGN

The earliest catamaran ferries (Figures 1, 2 and 3) were essentially conceived for use in very sheltered waters such as river, estuary and bay crossings and for routes protected by offshore reefs where tour crossings were easily postponed in rough weather. They were therefore designed with relatively small wet deck heights (height of the underside of the central deck above the water line) and thus with a relatively low transverse deck structure at the bow. However, the speed capability and length of these vessels gave rise to length based Froude numbers of 1.0 (Figures 2 and 3) and it became apparent that wave encounter induced significant vessel motions. Therefore if such vessels were operated in significant seas there resulted not only increased passenger discomfort but also increased risk of wave slam impact on the bow and wet deck. To overcome these problems the wave piercing and semi-SWATH (small water plane area twin hull) designs were introduced (Figures 4 and 5) so as to reduce the vessel motions by making the bow region relatively soft with regard to vertical wave forces on the structure. This was to a certain extent successful in reducing passenger motions towards the bow of the vessels, but the softness of the bow region exposed such designs to deep bow entry, slamming impact on the central structure and damage in following seas (Figure 5).

Figure 1 Catamaran ferry PS Kangaroo, Tasmanian Government 1855 (6 knots, Froude number =0.25, model in Tasmanian Museum and Art Gallery, Hobart)

Figure 2 Steel hull catamaran ferry Derwent Explorer, INCAT Tasmania Hull 001, 1977 (26 knots, Froude Number =1.0)

Figure 3 Aluminium catamaran ferry, INCAT Hull 004, 1981 (28 knots, Froude Number =1.0). Photo: INCAT Tasmania

Notwithstanding the somewhat reduced Froude numbers for large car ferry designs, and also the application of active motion control systems, ship motions have remained a significant issue and operation in severe sea states can result in heavy bow slamming. This is particularly true for military vessels where severe conditions may be tolerated and where there may be an imperative to maintain speed. However, commercial passenger carrying vessels are generally restricted to operations in seas of around 3 m significant wave height or less, and in some cases forward radars are now fitted so that speed can be reduced to avoid bow damage as in Figure 5. To provide bow region protection against large wave encounter the INCAT Tasmania wave piercing design was developed in which a short centre bow is fitted (Figure 6). The centre bow keel is close to the design water line and has a relatively fine keel entry so that during small wave encounter the centre bow experiences very small forces and encountered waves can break over the outboard demi-hulls thereby reducing wave upward forces around the front of the vessel. However during large wave encounter the centre bow forces are greatly increased and so act to prevent deep bow entry. The design is thus essentially non linear as regards wave response, but is inherently very seaworthy with almost no risk of deck diving or green water on the upper deck surfaces. However during deep water entry the bow structure experiences large upward forces which can become greatly increased if the arches between the outboard hulls and centre bow become filled completely. This leads to a particular type of wet deck slam which can generate substantial short duration forces. This paper sets out to consider factors which determine these slam forces.

Figure 4 Aluminium wave piercing catamaran, INCAT Hull 016, 1985
Photo: INCAT Tasmania

2. SHIP MOTIONS AND SEA KEEPING

Ship motions are the dominant factor in occurrence of wave slamming and so the prediction of ship motions is central to the prediction of wave slamming. A feature of the wave response of slender hulls at high Froude number is the emergence of what might be described as a significant motion resonance, the heave motion often being in excess of twice the wave height. This usually occurs at a dimensionless encounter frequency ($\omega_e^* = \omega_e\sqrt{L/g}$ where ω_e is the angular encounter frequency, g the acceleration due to gravity and L the hull length) of about 4 to 5. The resonant response implies that the motion of the hull as a whole is relatively lightly damped. Thus the prediction of the motion at the resonant maximum becomes sensitive to all contributions to motion damping including wave breaking, friction on the hull surfaces and the formation of separated flows and wakes. Figures 7 and 8 show computed and measured heave response amplitude operators (Davis, [1]) for the NPL round bilge hull form for large and small Froude numbers respectively. The emergence of relatively large heave motions at Fr=0.8 (figure 7) is clearly evident, whilst at the lower Froude number of 0.2 the large peak is absent. These two figures thus demonstrate the problem of increased ship motions at large Froude number which has been associated with the operation of high speed craft.

Figure 5 Wave damage to aluminium semi-SWATH catamaran car ferry HSS1500, 1996 (40 knots, Froude number =0.58)

Figure 6. 112m wave piercing aluminium catamaran, INCAT Hull 066, 2009 (Froude number = 0.61). Photo : LD Lines

Holloway & Davis [3] have developed a time domain method based on the transient Green function solution for spatially fixed sections of the water perpendicular to the direction of motion. The solution for each fixed strip begins as the bow enters the fixed water sections and concludes as the stern leaves the section. The Green function is that given by Wehausen and Laitone [5] as

$$f(z,t)=\frac{Q(t)}{2\pi}\ln(z-c(t))-\frac{Q(t)}{2\pi}\ln(z-\bar{c}(t))-\frac{g}{\pi}\int_0^t Q(\tau)\int_0^\infty \frac{1}{\sqrt{gk}}e^{-ik(z-\bar{c}(\tau))}\sin\left[\sqrt{gk}(t-\tau)\right]dk\,d\tau \qquad (1)$$

where Q is the source strength, k is the wavenumber, c is the source location and $\bar{c}$ its complex conjugate. The complex velocity $W = u - iv$ is obtained by integrating $\frac{\partial f}{\partial z}$ over all sources. This Green function satisfies the linearised free surface boundary condition and therefore it is only necessary to locate sources on the hull surface itself and not over the free surface of the liquid. This simplifies the computation and avoids problems associated with the boundaries of the free surface distant from and at the hull. Application of the hull boundary condition leads to a set of equations for the source strength terms at each time step in terms of the hull motion and the previous time history integrals.

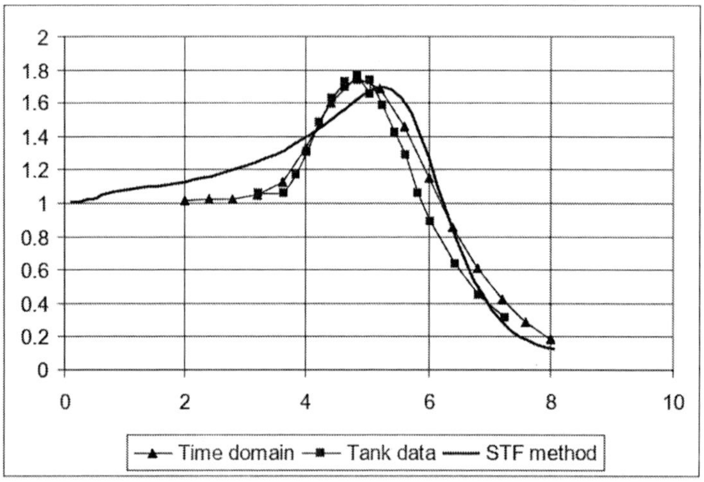

Figure 7 Heave response amplitude operator (ordinate axis, ratio of heave motion to water surface motion) at Fr=0.8 for NPL round bilge hull form. (Tank data of Wellicome et. al.[2]; time domain computation by program BEAMSEA, Holloway and Davis [3]; STF computation using method of Salvesen, Tuck and Faltinsen, [4]).
Abscissa axis: dimensionless encounter frequency ω^*_e.

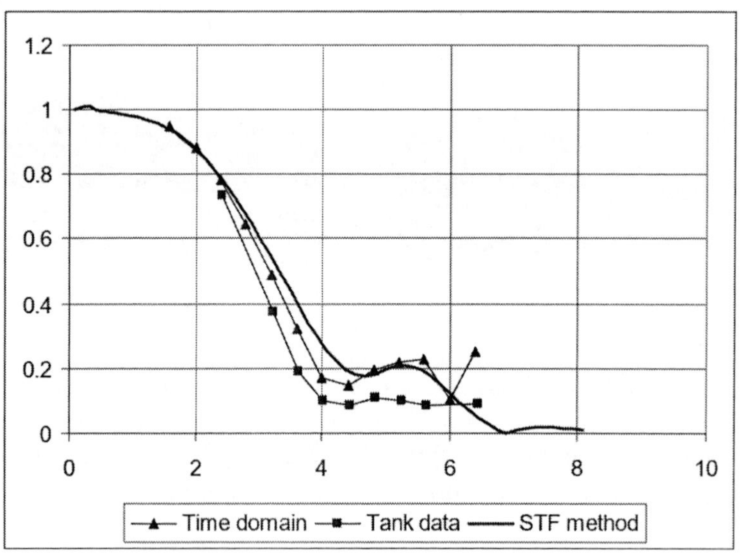

Figure 8 Heave response amplitude operator (as for figure 7, Fr=0.2,)

A variety of other methods for computing high speed ship motions has been developed and Davis et.al. [1] conducted a survey of the outcomes. The survey was carried out from the published results at or near to the frequency of maximum motion, using the same encounter frequency for measured and computed results. Inevitably such a process introduces some inaccuracy as only the published figures were available, but this uncertainty is generally much less than the difference between measured and computed data. The outcomes are shown as the ratio of computed/experimental results for heave in figure 9. Outcomes have been related to the underlying method used, including conventional strip theory, high speed slender ship theory, time periodic 3D Rankine source or Green function and time domain. With regard to heave we see that there is a wide spread of results across the entire range of Froude number with heave predicted by up to 70% more than measured. In general the predicted heave exceeds the measured heave, lending weight to the argument that the neglect of frictional effects in most of these outcomes is at least a significant contributing factor since the data has all been obtained at model scale. However, it is also seen that the time domain method used in the present paper gives quite accurate prediction over the full range of Froude number. With regard to pitch, it appeared that there is a trend for the ratio of computed/measured pitch to increase at larger Froude numbers for most methods. It seems likely that this is due in part to the effects of the separated flow at the transom which will, if not represented well in the computation, in particular affect the outcome of three dimensional methods due to the effect on the three dimensional boundary conditions. This is because any discrepancy near a transom will particularly affect the pitching moment. The time domain method used here again maintains a good prediction across the range of Froude number of interest and is therefore our preferred method for use in the prediction of slamming in terms of relative motion of bow and water surface.

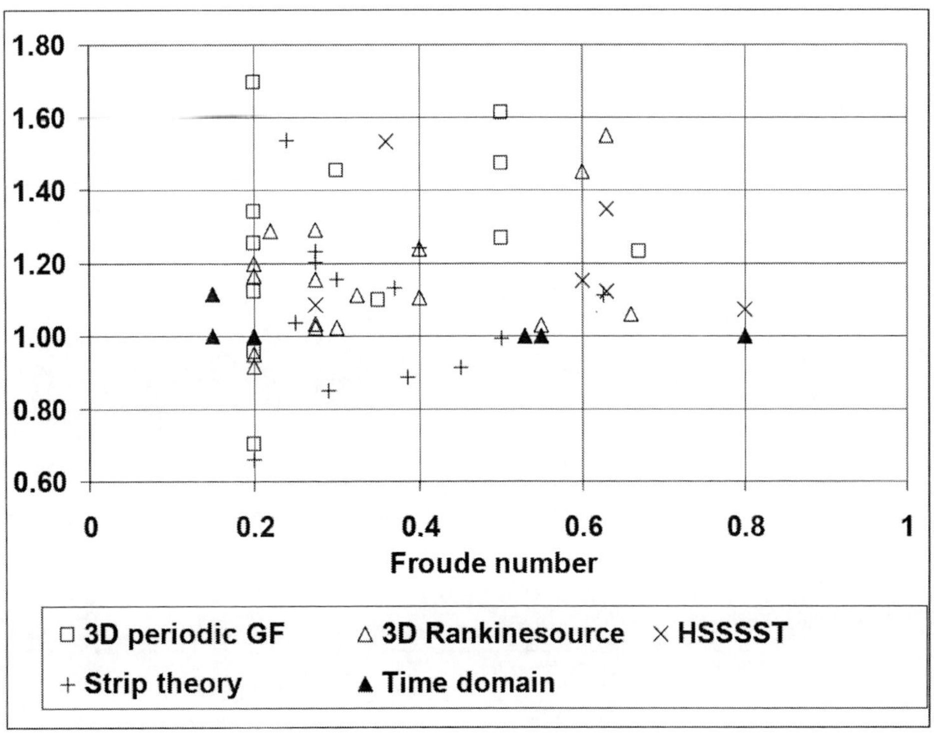

Figure 9 Ratio of computed/measured heave motions (vertical ordinate axis) by a variety of prediction methods (Davis et.al., [1])

3. IDENTIFICATION OF SLAMMING LOADS IN FULL SCALE TRIALS

The hydrodynamic interaction between a moving wave piercer bow and the encountered water surface is clearly a complicated process involving a three dimensional, unsteady transient as the vessel bow moves vertically downwards relative to the water surface [6,7,8]. For this reason it appears that identification of slam loads is best carried out by full scale sea trials and also by model testing since considerable uncertainty attaches to computation of slam events. Slamming is generally followed by whipping, the longitudinal vibration of the hull which occurs at about 3 Hz and has a damping ratio between 0.02 and 0.06 [9,10,11,12].

Sea trials have been carried out on a number of INCAT Tasmania vessels in which the vessel is fitted with a TSK wave radar to measure wave movement, accelerometers to measure vessel motion and strain gauges to measure unsteady stresses in the ship structure. The upper part of Figure 10 shows the time record of large and small slam events recorded on an INCAT 112m vessel whilst the lower part shows the corresponding wavelet transform. The underlying wave loads can also be seen with a period of about 4 seconds. Large slams are fairly easy to identify in the time records but smaller slams are not so easy to identify. It has been found that the continuous wavelet transform provides an effective means of slam identification (figure 10).

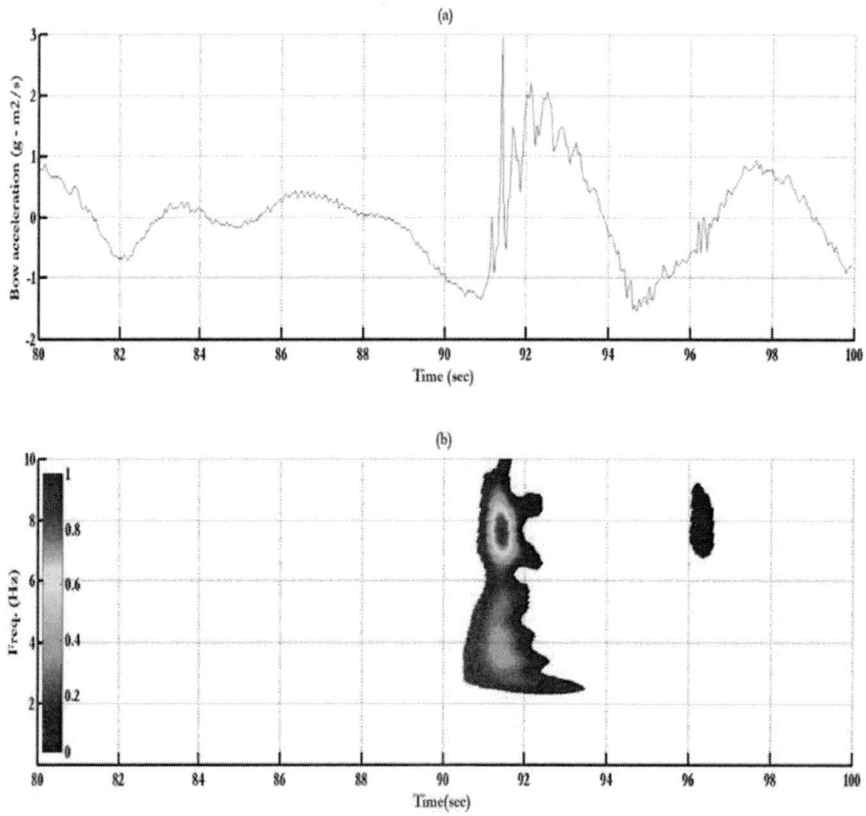

Figure 10 Transient bow vertical acceleration record (upper part) and corresponding wavelet transform (lower part) of large and small slam events for 112m INCAT Wave Piercer Catamaran

The continuous wavelet transform of a time signal $F(t)$ is defined as

$$W(f,\tau)=\int_{-\infty}^{+\infty}F(t)\psi(t-\tau)dt \tag{2}$$

where the Morlet mother wavelet function is applied,

$$\psi(t-\tau)=\sqrt{f/f_c}\psi_m((f/f_c)(t-\tau)) \tag{3}$$

and where

$$\psi_m(t)=\sqrt{1/\pi f_b}e^{2\pi i f_c t}e^{-t^2/f_b} \tag{4}$$

The scale parameter is f/fc (here scale represents frequency) and the time translation parameter is τ. The mother wavelet function is a pre-defined function within a finite time domain characterised by oscillation. Slams are characterised by excitation of the main structural modes of the ship and so values of f_b= 2Hz and f_c= 1Hz have been applied in the computation of the wavelet transform shown in Figure 10. The slam has primarily excited a mode of the ship structure at about 7.5Hz. Finite element analysis of this vessel has shown that the main longitudinal bending mode is at 2.6 Hz whilst there is a higher frequency mode in which most of the bending takes place at 7.7 Hz in the bow region. In the case of the 112m INCAT design it appears from figure 10 that slams mainly excite the localised bow vertical bending mode. This outcome differs from that observed in trials of 98m and smaller INCAT designs in which slams excited the main longitudinal bending mode at a frequency somewhat under 3 Hz. It is thought that the incorporation of a full second tier structural vehicle deck on the 112m has resulted in the main body of the vessel being stiffer relative to the bow area thus resulting in slam energy being imparted mainly to the localised bow mode.

Figure 11 Segmented 2.5m hydroelastic model of INCAT Tasmania 112m vessel

4. IDENTIFICATION OF SLAM LOADS BY MODEL TESTING

Sea trials do not allow a controlled investigation of slam loadings and in particular do not demonstrate what the maximum possible slam loading is. For this reason a program of model testing has been undertaken using a hydroelastically scaled test model as shown in figure 11. The model is constructed from six main longitudinal segments: a rigid centre section, forward and aft demihulls on both sides and a centre bow mounted from the forward demihulls. The segments are connected by elastic links designed to give a first longitudinal bending mode for the three longitudinal segments at 13.8 Hz, this being scaled to represent the main longitudinal bending mode of the full scale vessel at 2.3 Hz. The elastic links connecting the segments are also fitted with strain gauges: the bow mounting links allow determination of the location and magnitude of the impulsive slam force on the bow whilst the links in the demihulls facilitate calculation of the hull bending moments. It was found that model also replicated the damping ratio for whipping of the full scale vessels, this increasing from about 0.02 to 0.03 as speed increased from zero to 38 knots equivalent. Model tests have been carried out for a range of wave heights and encounter frequency and figure 12 shows how the maximum slam force varies with encounter frequency. We see that a maximum slam load of 233 N is observed, which scales to a full scale equivalent of 2129 tonnes (ie 85% of the ship displacement). However the duration of this very large slam force is relatively short, the impulse being 156 Ns (ie 917 tonne seconds at full scale, the effective impulsive duration of the peak load being 0.43 seconds at full scale). A closely similar maximum slam loading of 2252 tonnes has been observed without damage for a 96 m INCAT design during severe weather trials, the slam load being determined from FEA analysis combined with trials strain gauge readings. Moreover the maximum strain energy imparted to whipping in the model tests was 0.84 J (equivalent to 3.47 MJ at full scale), this corresponding closely to the maximum whipping strain energy observed in sea trials of 3.5MJ.

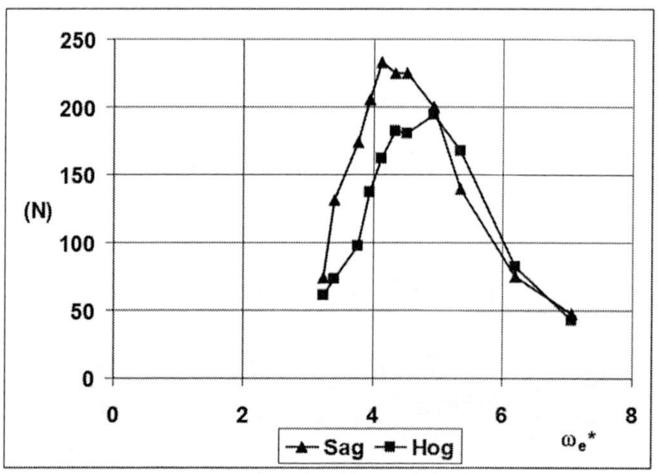

Figure 12 Variation of maximum slam load of 2.5m test model with wave encounter frequency at a wave height of 120 mm (5.4m wave height full scale equivalent)

5. CONCLUSIONS

Computation of ship motions is the first essential component for predicting wave slamming and the time domain seakeeping computation gives good motion predictions across a wide speed range. The wave piercing catamaran with substantial forward buoyancy above the water line in a centre bow is well protected against deck diving damage and green water over the deck is very unlikely to occur. However, slam loads can reach 85 % of the vessel displacement, have a duration of about 0.4 seconds and impart about 3.5 MJ of energy into the whipping which decays with a damping ratio in the range 0.03 to 0.06 at full scale depending on vessel design. Testing of a scaled hydroelastic model has been found to replicate maximum slam loads and imparted energy at full scale quite well. It is found that the wave encounter frequency of maximum slam severity coincides closely with the frequency of maximum vessel motions. This demonstrates that slamming is dominated by vessel motions, a conclusion that is supported by video film records of model tests.

9. REFERENCES

1. Davis M.R. and Holloway D.S. (2003), Effect of sea, ride controls, hull form and spacing on motion and sickness incidence for high-speed catamarans. Proc FAST03, Seventh International Conference on Fast Sea Transportation, Ischia, Section E, pp. 1-10.
2. Wellicome, J.F. Temarel, P. Molland, A.F. and Couser, P.R.(1995), Experimental measurements of the seakeeping characteristics of fast displacement catamarans in long crested head-seas. Ship Science Report 89, Department of Ship Science, University of Southampton.
3. Holloway, D.S. and Davis, M.R. (2002), Green function solutions for the transient motion of water sections. Journal of Ship Research, 46(2), 99-120.
4. Salvesen, N. Tuck, E.O. and Faltinsen, O.M. (1970), Ship motions and sea loads, Transactions of the Society of Naval Architects and Marine Engineers, 78, 250-287.
5. Wehausen, J.V. and Laitone E.V. (1960), Surface waves. Handbuch der Physik' (ed. S. Flügge), Springer, v9, pp.445–814.
6. Davis, M.R. and Whelan, J.R. (2007), Computation of wet deck bow slam loads for catamarans. Ocean Engineering, 34, 2265-2276.
7. Thomas, G.A., Davis, M.R., Holloway, D. and Roberts, T.R. (2002), Extreme Asymmetric Slam Loads on Large High Speed Catamarans. 6th Symposium of High Speed Marine Vehicles, Naples, September, I.15-I.23.
8. Thomas, G.A., Davis, M.R.., Holloway, D.S.. and Roberts, T.R. (2003), Transient Slam Response of Large High Speed Catamarans. Proc. FAST 03, Ischia, Vol. 2, B1.1-B1.8.
9. Thomas, G.A., Davis, M.R., Holloway, D.S and Roberts, T.J. (2008), The vibratory damping of large high speed catamarans. Marine Structures, 21, 1-22.
10. Lavroff, J., Davis, M.R., Holloway, D.S. and Thomas, G.A. (2007), The whipping vibratory response of a hydro-elastic segmented catamaran model, Proceedings of FAST2007, Ninth Int. Conference on Fast Sea Transportation, Shanghai, 600-607.
11. Davis, M.R., Amin, W., Lavroff, J., Holloway, D.S., Thomas, G.A.T. and Matsubara, S. (2009), Global and slam load model testing to support developing HSMV operations in severe sea conditions. Conference on Innovation in High Speed Marine Vehicles, Fremantle, 93-102.
12. Davis, M.R., Thomas, G.A., Holloway, D.S.., Lavroff, J., Amin, W., Matsubara, S. and Roberts, T.R. (2009) Slamming and whipping of wave piercing catamarans. Proceedings of the Fifth International Conference on Hydroelasticity in Marine Technology, University of Southampton, 223-232.

10. ACKNOWLEDGEMENTS

The support of the Australian Research Council and INCAT Tasmania is gratefully acknowledged. The analysis for figure 10 was carried out by Mr Hoai Nam Do.

An approach to improve advanced outfitting practices in ship construction — automatic scheduling system

Y Wei, U Nienhuis
Delft University of Technology, Delft, the Netherlands

1. INTRODUCTION

Building ships involves two basic activities: constructing structural components of the ship, and outfitting the ship by installing various systems and equipment that allow it to operate and perform various missions [1]. Today, the relative importance (at least in terms of cost) of the steel shaping and assembling has decreased substantially. This is true especially for the bigger and ferry vessels where EU shipbuilding still holds a considerable share of the market. The focus of shipyards has been shifted from the pure steel construction towards extensive outfitting work. In a modern passenger cruise vessel for example, the outfitting work can represent, in terms of cost, up to 80% of the total work, the steel work being only 20% [2]. In Germany [3], outfitting processes represent about 38% of the construction volume in building industry while the main construction trade represents only about 30%.

Clearly, any improvement in the advanced outfitting process would have a significant impact on shipbuilding industry. Since this process is characterized by frequent interferences between yard and different subcontractors, disturbances by unexpected material delays, and technological dependencies due to e.g. spatial accessibility and physical connection considerations, scheduling of this process is quite complex. Currently, yards only provide the subcontractors with a general time framework for outfitting and play the role of director. This implies that the outfitting process planning is scattered amongst different subcontractors and each party has its own plan towards its own work, which causes space collision, rework and long-time waiting to happen frequently on the work site. Although both yards and subcontractors believe high levels of advanced outfitting would bring mutual benefit in the terms of decrease in total labor hours, lack of a reasonable and precise scheduling system of outfitting process adversely affects the degree of advanced outfitting they can accomplish.

The fundamental problem is to identify economically and technologically desirable opportunities for advanced outfitting. This requires answering the question related to feasibility, i.e., "Is there sufficient time and resource to have a certain section or block advanced outfitted, and is it technically feasible?"[4]. The main goal of the research described in this paper is to develop an automatic schedule generation system. It is expected to provide to each party involved in advanced outfitting a holistic planning. Hence, the onsite coordination and installation effort maybe reduced and the level of advance outfitting practices maybe increased, which leads to the reduce of cost and lead-time. This paper shows possible approaches to the automatic schedule generation system to generate the advanced outfitting activities and their assembly sequence. In section 2, the characteristics of outfitting process and limitations to higher levels of advanced outfitting will be introduced. The main architecture of the automatic schedule generation system and the activity generation template are explained in section 3. Section 4 describes the sequence generation model in more detail. The test case result is presented and its advanced outfitting plan is generated. Section 5 provides conclusions of the paper.

2. CHARACTERISTICS OF OUTFITTING PROCESS

2.1 The Definition of Advance Outfitting and Outfitting

In the ship construction processes, outfitting usually refers to two distinctive but closely linked phases, i.e. advanced outfitting and outfitting. "Advanced outfitting", or pre-outfitting, generally means installing piping, electrical components, HVAC components, and other system components (engines, winches, blowers, switchboards, etc) in the panels, sections and blocks making up the hull structure. This work, generally, is performed inside the specialized and well-equipped workshops with ideal lighting and access [5]. "Outfitting" means to continue the activities, which have not yet been finished in the previous advanced outfitting stage, or to embark on new activities that can only be done during the hull erection in the slipway or after launching alongside the quay, like cable pulling and further connection work.

Working conditions in the outfitting stage are not ideal because of factors such as difficult access, limited space, and difficult work positions (e.g., overhead welding). The workplace is typically congested, with high material flow costs, and often hazardous conditions. It is now widely accepted that many of these problems can be relieved to some degree by installing more components earlier in advanced outfitting stage. However, so far there is no clear boundary to identify which specific activities should be done in pre-outfitting or which in outfitting. Briefly, the choice of the best stage to perform the advanced outfitting depends upon:

- Production schedule
- Space accessibility
- Lifting capability
- Material availability
- Structural peculiarities

2.2 Limitations to Higher Levels of Advanced Outfitting

From just steel erection, yards in the Netherlands have increasingly become a total systems integrator of a complete engineering, fabricating and outfitting process. However, even when they have close collaboration with a pool of specialist subcontractors, the lack of sufficient design information and unavailability of integral planning become potential constraints to greater levels of advanced outfitting.

Presently many shipbuilding CAD systems such as the Tribon, IntelliShip, Nupas-Cadmatic, Foran, etc. are available. They supply sufficient information for manufacturing the larger outfitting components at the initial design stage. But indeed, it is customary to exclude smaller or light-weight components from the CAD-models, for example, pipes less than 12.50 mm in diameter. They are of course specified in the Bill Of Materials (BOM) that can be acquired from procurement but locations of these smaller components are normally not available. This causes the risk of a collision with other components in the installation.

Meanwhile, the lack of integral outfitting plan renders yards unable to give subcontractors reasonable time for advanced outfitting. For the working principles of yards the slipway work is the most important and estimated durations for section-building are kept as leading. After being built and advanced outfitted, a section will be transported to a conservation hall to be painted.

After consideration according to their previous experience, the yard may only give 8 weeks for the construction of a certain section including one or two weeks for subcontractors to do early outfitting work, which is shown in Figure1. Sometimes, if the section construction is delayed, there is even no time for advanced outfitting because no matter what happens, the hull block erection schedule is fixed. Subcontractors respond to this problem either by working overtime or perform the mounting later in the outfitting stage, which takes more effort and costs. Last but not least, the scattered outfitting plan is the main reason for space collision, which leads to rework and long waiting times. This, to a great extent, increases negotiation on the work site and makes precious advanced outfitting time less efficient.

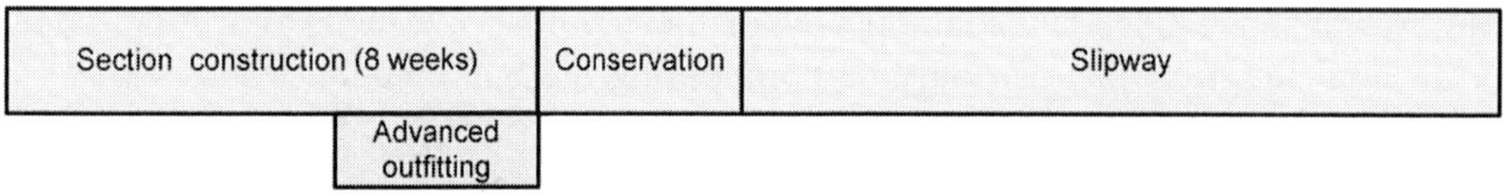

Figure1. Advanced outfitting in the section building stage

3. AUTOMATIC SCHEDULE GENERATION SYSTEM

Automatic schedule generation system is developed as a computer aided technique of generating a reasonable and holistic plan for advanced outfitting within one section. This section covers a brief description of the system [6]. In Figure 2, the planning of outfitting process on the operational level has been divided into three interrelated steps: activity generation, sequence determination and schedule determination, trying to answer questions "What has to be accomplished?", "In what order will it be accomplished?", and "When will it be accomplished?" respectively. The first two steps comprise process planning and the last is scheduling.

Process planning is the bridge between the product design and manufacturing, and determines what tasks to perform and in what sequence. It emphasizes the technological and organizational requirements of a task. Scheduling receives process plans as the input and imposes time ordering and resources allocation constraints while satisfying the precedence relations given in process plans [7]. The objective of scheduling is to balance work with available and required production facilities. A good schedule must reflect a satisfactory compromise among these competing influences [8]. Since scheduling has received considerable attention and many methods and tools have been developed in manufacturing environment similar to shipbuilding, this paper pays more attention to process planning.

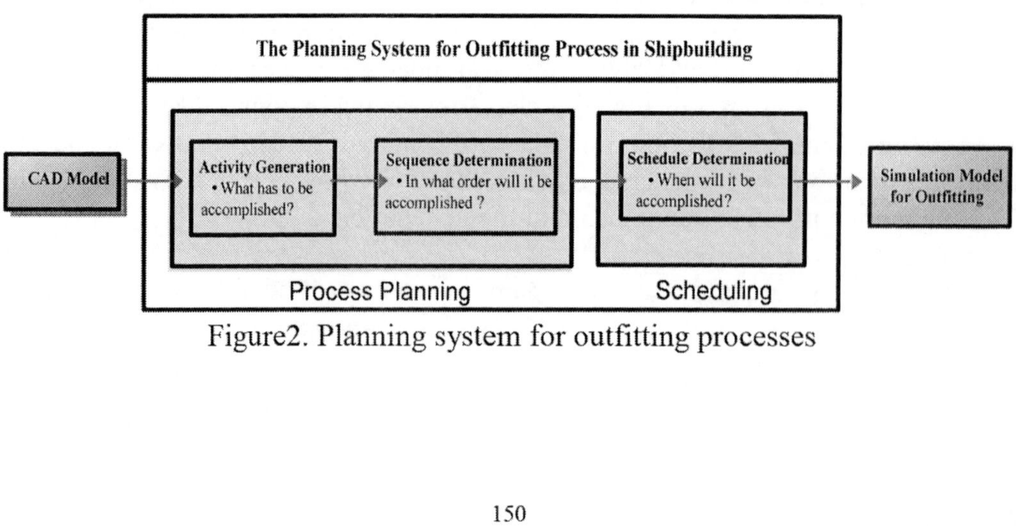

Figure2. Planning system for outfitting processes

Currently, a complete outfitting activity list is not adopted by the shipyard or subcontractors. After getting the information from foremen, like 2D drawings and specification sketches, workers start their installation work according to their experience. Meanwhile, few researches have been focused on how to generate the activity list automatically. Ambiguity in this area has notably remained. The goal of the activity generation method is to find what has to be made, to be done and to be prepared by automatically translating CAD model information into the activity list

For this an activity template, representing the typical installation work, is adopted to generate the activity list. The purpose is to compensate for the missing information in 3D models. Observations necessary for the creation of an activity template were done on the work site in one of the leading shipyards in the Netherlands. In order to record the actual work conditions, the observations were performed for a long period of time.

In fact, not only the types of the component but also its weight, material and position altogether determine the installation activities. For example, for lighter pipes, supports are installed after putting pipes on their positions; for heavier ones, which imply the activity of using a crane, pipes are often installed after the supports' fixation. Supports for pipes cannot be welded directly on the steel parts and the activities of welding angle bar on the frame as a foundation of supports are always needed; however, supports are allowed to be welded on the deck if the pipe is on the top tank which is made of thicker steel plates. All these differences will be included in the activity template which will be registered later as a main part into the activity pool. According to the attributes of a component derived automatically from the 3D model, the activity pool will first search its required activities, then combine them in a sequential order and finally output the activity list. The whole process can be shown in Figure3.

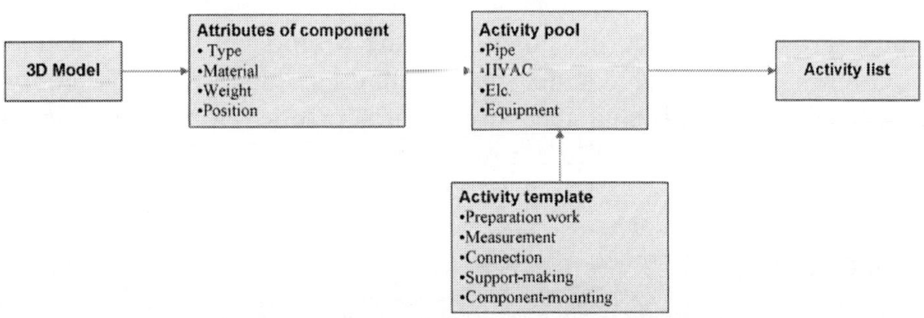

Figure 3: The activity generation process

The function of the activity template is not only to help illustrate the tacit knowledge explicitly in terms of computer programs, i.e. the ability to recognize and extract the rules from expertise and years of experience, but also through deconstructing the outfitting process into activities, to precisely estimate the duration time. This also may help to take the finite capacity of resources into consideration as early as possible.

4. SEQUENCE GENERATION

4.1 Precedence Constraints

In order to decide the assembly sequences of advanced outfitting components, conditions or restrictions of variables are employed as precedence constraints. The more constraints are specified, the more the solution space is restricted and consequently, the more the multitude of possible assembly sequences is reduced. It is in fact a highly constrained problem, with two types of constraints, hard (or strong) and soft (or weak) [9]. Hard constraints define stringent conditions in the assembly process. They must be fulfilled before a work can be started. Soft constraints are set to evaluate assembly difficulties due to spatial accessibility, construction strategy, etc. The hard constraints cannot be violated, while the violation of the soft ones is possible but not preferable, because they might be helpful if no solution is detectable. A feasible assembly sequence is generated fulfilling all hard constraints. The preferred assembly sequence violates the soft constraints as little as possible [3]. An overview of the defined precedence constraints is given in table 1.

#	Constraint	Constraint Type
1	Position Constraint	Hard Constraint
2	Material Constraint	Hard Constraint
3	Weight Constraint	Hard Constraint
4	Size Constraint	Hard Constraint
5	Productivity	Soft Constraint
6	Strategy	Soft Constraint

Table 1: Precedence constraints in outfitting assembly

4.1 (a) Position Constraint

It would be practical to install components closest to the compartment boundaries, like walls, floors or ceilings. From this aspect, the position of the component within a compartment, when taking it to extremes, completely decides its installation order in the assembly process. For example, when considering the spatial accessibility, it is practical to install a pipe which is much closer to the ceiling earlier than installing a duct which is crossing below this pipe and far way from the ceiling.

4.1 (b) Material Constraint

The possibility for an installed component to be damaged by later assembly operations is directly related to what material this component is made of, which implies the material aspect also plays a role in the determination of assembly sequence and should be given considerable attention. For instance, the welding of brackets for a steel pipe may cause the damage to a sheet metal duct that is adjacent to this pipe and which has already been perfectly positioned and supported. Therefore, rework occurs and cost increases.

4.1 (c) Weight Constraints

Weight implies the potential to use the logistic resources, such as cranes, trucks and lift trucks, to transport certain components from storehouse to the appointed assembly area. Since in the early assembly stage the materials and equipment are closer at hand, the heavier outfitting components are, the earlier they should be assembled. Otherwise, the obstruction from structural components would hinder the accessibility to hoist and lower the heavier components or equipment to the right position, which adds to the indirect time required to perform a structural assembly or erection task.

4.1 (d) Size Constraint

Here, the 'size' means the outside diameter for pipes or width for rectangular ducts and cable trays. The bigger components are, the larger area will they occupy in confined spaces and the larger the openings that are needed to put them in. Thus, this type of constraint is primarily concerned with the spatial accessibility. For example, a piece of duct 700 mm in width, 1500 mm in length and 250 mm in height may be put inside the section in a much easier way when the compartment is still open allowing access from above.

4.1 (e) Productivity Constraint

The productivity constraint is a functional benchmark. It will be defined as a soft constraint. The complex coherence between free working space and productivity of workers needs to be considered. For example, a pipe fitter only achieves the scheduled productivity if his required work space is provided. Productiveness will rapidly fall if this cannot be provided, for example, due to overcrowding or inaccessibility.

4.1 (f) Strategies Constraint

Strategies vary from subcontractors and are defined as soft constraint. For example, pipe installation could preferably be done according to the systems that they are part of (exhaust system, fresh water cooling system, hydraulic system, etc), e.g. if different skills are preferred for the associated work. This affects the assembly sequence to a great extent.

4.2 Representation of Precedence Constraints

The goal of the preceding presentation of constraints is to set up a method able to encode the necessary information and generate a feasible assembly sequence. In this paper, an alternative method for the representation of assembly constraints is introduced. This novel method combines the advantages of both explicit and implicit representations, by using an interference matrix as explicit representation to illustrate the height constraint and setting a set of rules as implicit representations to explain the material, weight and size constraints.

4.2.2 Interference Matrix

The interference matrix is used to express the position relationship among the n components in a section or compartment. The method is based on the assumption that the disassembly sequence is the reverse of a feasible assembly sequence. The square matrix A_k of order n is used to detect all the collisions until the component is completely disassembled from the compartment/ section along the direction +k [10]. It is defined as:

$$A_k = \begin{array}{c|ccccc|} & e_1 & e_2 & \cdots & \cdots & e_n \\ \hline e_1 & a_{1,1} & a_{1,2} & \cdots & \cdots & a_{1,n} \\ e_2 & a_{2,1} & a_{2,2} & \cdots & \cdots & a_{2,n} \\ \cdots & \cdots & \cdots & \cdots & \cdots & \cdots \\ e_n & a_{n,1} & a_{n,2} & \cdots & \cdots & a_{n,n} \end{array}$$

$e_1, e_2, \ldots, e_n$, are n non-structural outfitting components in a compartment/ section. Here $a_{i,j}=1$ if the element e_i interferes with the element e_j during the translation (by whatever distance) along the direction +k, otherwise $a_{i,j}=0$. As a convention $a_{i,i}$ is always equal to zero.

In the matrix A_k, there must be at least one component e_i ($1 \leq i \leq n$) satisfying the following condition: $\sum_{j=1}^{n} a_{i,j} = 0$, which implies none of the other components will collide with component e_i when it is translated along the direction +k. Therefore, the component e_i can be disassembled first without any collision with other components. The result shows that several components may be dissembled simultaneously, which means other constraints are required to further differentiate their disassembly orders.

4.2.3 The Establishment Conditions Representations

The basis for the establishment conditions is the expertise obtained from human experts. These experts use both theoretical knowledge about optimality and much empirical knowledge about their specific environment. By extracting the implicit expertise gained through years of experience, information in the form of assembly rules is represented, such as IF-THEN. The rule can then be used to perform operations on data that interfere in order to reach an appropriate conclusion. In the assembly process, the way the material, weight and size of a component decides its own installing sequence, is described qualitatively first with the help of a set of rules. Then the result of each rule is represented as a rank quantitatively, reflecting the absolute order to perform the assembly. So far, with the combination of explicit and implicit representation each constraint of a component is expressed with a value.

4.4 Weighting Coefficients

After representing the hard constraints quantitatively, the contribution of each hard constraint to the determination of assembly sequence can be analyzed in terms of weighting coefficients. Determining a weighting coefficient actually is to decide the relative importance of the hard constraints. In making the comparisons, the decision makers can use concrete data about the constraints, or they can use their judgments about the constraints' relative meaning and importance. A priority vector method can be used to convert these subjective evaluations to numerical values that can be processed and compared over the entire range of the problem [11]. Thus, a numerical weight or priority is derived for each constraint, allowing diverse constraints to be compared to one another in a rational and consistent way.

4.5 Sequence Generation Model

Based on the representation of constraints and derivation of weighting coefficients, a sequence generation model is developed and the basic operating screen from this system is shown in figure5. This model will generate the sequence through the comparison of the final scores.

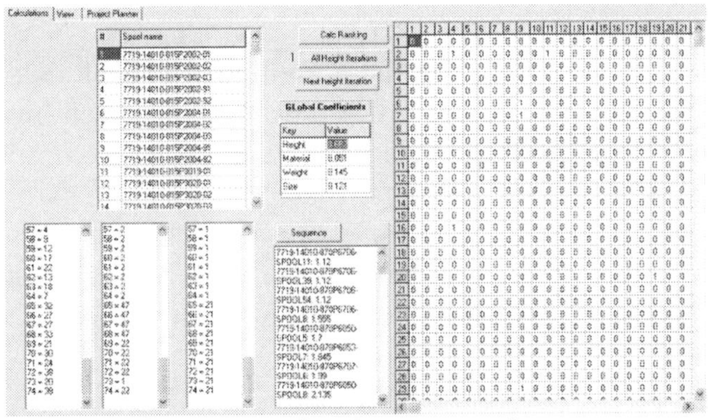

Figure 4 The basic operating screen for the automatic sequence generation system

To verify the model, part of a deckhouse of a multi-purpose offshore support vessel is chosen as the test case because the arrangement of its components is not as complicated as of the engine room but fuller than of the accommodation room. Figure 5 shows the 3D model of section 7719-14010 where the deckhouse is.

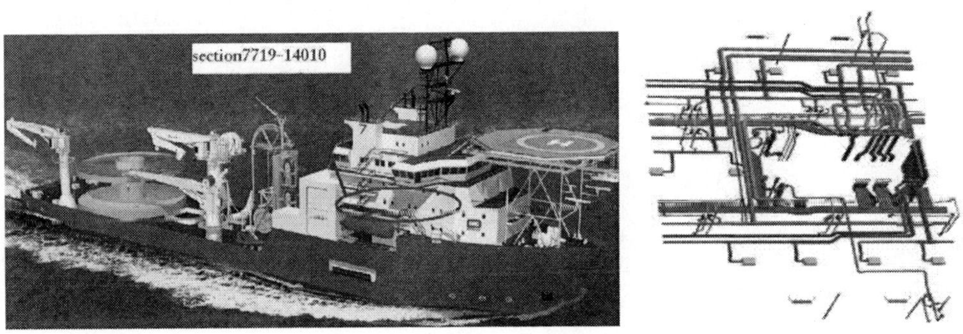

Figure 5 3D model of section 7719-14010

There are 25 components with 74 spools chosen to be installed in the advanced outfitting stage. After the generation of an assembly sequence, the animation of the assembly procedure on a computer display is available. Figure 6 shows the following points in the assembly procedure on a display screen: when 25% components, 50% components and all 74 spools are installed.

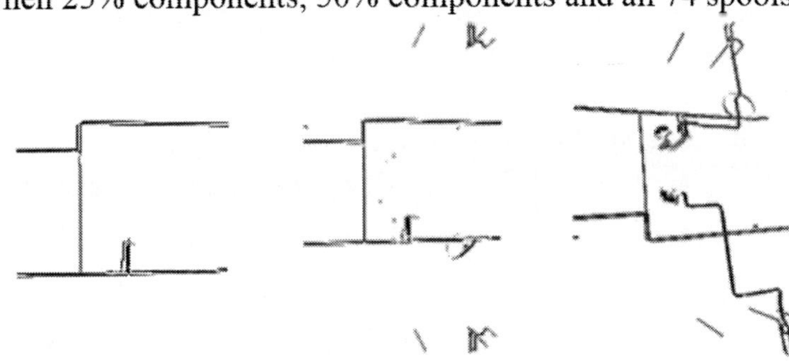

Figure 6 Assembly procedure of section 7719-14010 in advanced outfitting

4.6 Discussion

As a first test case the method can be used together with some assumptions about the shipbuilding process:
A1: Assume the only relationship between the activities is one of sequence or precedence. An activity, 'A', precedes another activity, 'B', if A must be completed before 'B' can be initiated. Thus the mounting order is in sequential other than in parallel.
A2: The various resources, such as drawings, material, equipments and workers with certain skills, etc. required to perform the activities are always available.
A3: The duration for activities is only associated with the properties of outfitting components themselves. Thus, all outside factors which may cause the discrepancy of mounting time are not taken into account.

Together with the method described in this paper, these three assumptions result in the graphical representation of the relationship between activities and mounting time for each activity. Figure 7 shows the plan for advanced outfitting the section 7719-14010, including man-hour calculation.

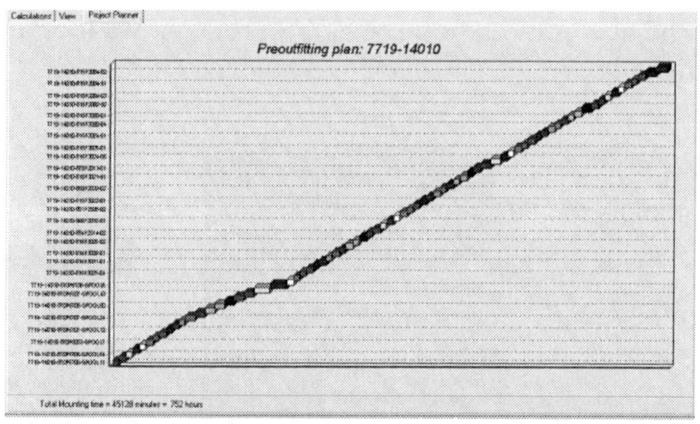

Figure 7 The advanced outfitting plans for section 7719-14010

It must be recognized from the outset that the process thus modeled is only hypothetical and does not represent the existing process. Next versions of the method will eliminate the need for the above assumptions.

At the present time, very few shipyards in the Netherlands employ the integral advanced outfitting planning in production, thus, the proposed model constitutes a significant departure from current standard practice. On the other hand, it is also apparent that interest in this type of methodology is growing in many of the Dutch shipyards, we therefore believe that the proposed model is in line with long term trends in the industry and is based upon some prominent elements currently recognized in the shipbuilding industry.

5. CONCLUSIONS

In this paper, an automatic sequence generation system has been developed and proposed. It provides a systematic framework for determining the technical and economic feasibility of plans of advanced outfitting process. The technological dependencies which specify definite sequences between advanced outfitting tasks are illustrated. The mounting time for each activity is estimated. The foremost benefit of the system is more reasonable integral planning and control of advanced outfitting. It also helps shipyards to estimate the man-hour at the early design stage so that labor resource utilization maybe maximized. The secondary benefit of the model is that it would provide prerequisites for further simulating advance outfitting process. This includes not only indispensible rough data but also knowledge of dependencies [12]. This is predicted to help to accelerate the realization of simulating the whole outfitting process. And in this way, it would prove to be very valuable to optimize the ship construction process.

Further ongoing development will concentrate on the improvement of this system, including: calculating the duration time for each outfitting activity with the application of statistic method, identifying pre-conditions for each activity within the (advanced) outfitting phase of ship construction. Current work is under way to find the solution of mounting spools in parallel instead of in sequential.

REFERENCES

1. SCHANK, J. F., 2005, "Outsourcing and Outfitting Practices: Implications for the Ministry of Defense Shipbuilding Programs". RAND Corporation, 41-60

2. ANDRITSOS, F. and PEREZ-PRAT, J., 2000, "The Automation and Integration of Production Processes in Shipbuilding". European Commission, Joint Research Center, Institute for Systems, Informatics and Safety, June

3. KONIG, M., BEIßERT, U., STEINHAUER, D. and BARGSTADT, H.J., 2007, "Constraint-based Simulation of Outfitting Processes in Shipbuilding and Civil Engineering"

4. GRAVES, R. J., 1982, "The Outfitting Planning Problem: Production Planning in Shipbuilding", Naval Research Logistics Quarterly, Vol. 29, No. 2, 357-384

5. MAFFIOLI, P., DAIDOLA, J. C., OLIVIER, J. and COLTON, T., 2001, "Competitive shipbuilding production practices", Discussion. Author's closure, Transactions- Society of Naval Architects and Marine Engineers 109, 309-339

6. WEI,Y. and NIENHUIS, U., 2009, "An Automatic Schedule Generation System of Outfitting Process in Shipbuilding", International Conference on Computer Applications in Shipbuilding 2009, volume III, 199-207

7. BERRY, P. M., 1990, "Satisfying conflicting objectives in factory scheduling", Artificial Intelligence Applications, Sixth Conference, 101-107

8. LIU, B., 1993, "Problem Acquisition in Scheduling Domains", Expert Systems with Applications, 257-257

9. SEBBAALY, M.F. and FUJIMOTO, H., 1996, "A Genetic Planner for Assembly Automation", Proceedings of the IEEE Conference on Evolutionary Computation, 401-406

10. DINI, G. and SANTOCHI, M., 1992, "Automated Sequencing and Subassembly Detection in Assembly Planning", Annals of the CIRP, Vol. 41, no.1, 1-4

11. SATTY, Thomas L and VARGAS, Luis Gonzales, 2001, 'Models, Methods, Concepts & Applications of the Analytic Hierarchy Process', ISBN 0792372670, 9780792372677, Published by Springer

12. KAARSEMAKER, J. A. J. and U. NIENHUIS, 2006, "Simulation of a maritime pre-fabrication process", The 5th International Conference on Computer Applications and Information Technology in the Maritime Industries

REFEREED PAPER

Control of a System of UAV by Undersea Wireless LAN

Steve Chan[1]
Reynaldo Castillo[2]
King Fai Lai[3]
Mansoor Seyd[4]

Abstract

In PAC08 [1], we presented a paper dealing with aspects of using a system of communicating unmanned undersea vehicles and the realization of a networked sensor. The present paper attempts to follow up with further discussion on the transmission channel. Firstly, we propose that the mother ship is equipped with a mageto-hydrodynamic (MHD) transducer for generating info carrier wave. Such a device consists of a rectangular controlled volume inside a wave-guide with apertures at both ends. There is a static B field and a harmonic E field perpendicular to it. The volume is filled with an electrolyte. The Lorentz force acting on the ions generates a MHD acoustic wave and the heating effect on the ions produces a thermal acoustic wave. The MHD acoustic wave and the thermal acoustic wave together generate a dipole and a monopole acoustic radiation. Through the aperture of the wave-guides the acoustic field is extended into the deep sea. The pressure wave associated with this acoustic radiation is then our info carrier. Secondly, we show that the radiation may be digitally modulated and the transmission realizes the 4-layer model of SNMP. By solving the Poisson pressure equation at the unmanned autonomous vehicle (UAV), we see that the transmitted info is received at the double layer potential term of the pressure distribution. We estimate the impact of globally regular solutions to the transport equations on the sound pressure spectrum at the UAV and indicate how to generalize the proposed scheme to include duplex info transmission.

---

1. stevechan@internode.on.net
2. r.Castillo@uws.edu.au
3. kinglaihonkon@gmail.com
4. mansoor.syed@defence.gov.au

SECTION 1: INTRODUCTION

The present paper attempts to follow up with the discussion of the transmission channel and presents a couple more concrete suggestions on the architecture of such a channel. Firstly, we propose that the mother ship is equipped with a mageto-hydrodynamic (MHD) transducer for generating info carrier wave. In the simplest design, such a device consists of a rectangular controlled volume inside a wave-guide with an aperture. There is a static B field and a harmonic E field perpendicular to it. The volume is filled with an electrolyte such as NaCl or $NaNO_3$. This furnishes us with a simplest model of MHD. The Lorentz force acting on the ions generates a MHD acoustic wave and the heating effect on the ions produces a thermal acoustic wave. To see that these waves are actually generated, we only have to set up the basic equations governing the MHD model and look for plane waves solutions in accordance with the rules in plasma waves. Among the wealth of waves generated, we look for one that is longitudinal, the wave vector and the ion velocity are parallel.

The MHD acoustic wave and the thermal acoustic wave together generate a dipole and a monopole acoustic radiation. Through the aperture of the wave-guides the acoustic field is extended into the deep sea. The pressure wave associated with this acoustic radiation is then our info carrier.

Inside the control volume, the dynamics of the MHD is adequately described by a linear wave equation of the pressure wave. By solving this equation, it is easily seen that the pressure field is a traveling wave with amplitude proportional to the applied E field. We point out that this pressure wave can easily be digitally modulated using technique such as PCM. The hydrodynamics of the deep sea is described by either the Navier-Stokes equation (in the event of sea water is treated as being viscous) or the Euler equation (if viscosity is zero). There is also a continuity equation to express conservation of mass or incompressibility. Successful data transmission requires that there should be a longitudinal wave for us to do digital modulation. Mathematically speaking, the availability of such a carrier wave requires that the basic equation to possess solutions, which are regular all the time or at least over the period in which transmission from and to the mother ship takes place. Once the solution ceases to be regular, turbulence sets in and transmission is no longer possible.

It turns out that this requirement is very difficult and one may not be able to satisfy it without paying down some constraints on the initial conditions of the envisaged MHD transmitter. The quest for a set of general conditions allowing globally regular solutions (in time) to exist is an open question and is intensively pursued by engineers, physicists and mathematicians. Equally difficult is to prove the non-existence of such solutions [2].

It is much beyond the scope of this paper to look for the general set if initial conditions that would allow data transmission to take place all the time. To do that is just the same as to attempt to solve the Clay millennium problem. Our modest goal is to follow up on the suggestions of mathematicians that when certain set of initial conditions are available, the Navier-Stokes equation does possess solutions that are globally regular in time and therefore data transmission can take place [3] [4] [5].

It is not the intention of this work to present a comprehensive review how one can make use of the results of the above references to ensure that data transmission to and from the ship over an extended period is possible. In Section 6 of the paper, we will do a brief review.

When the carrier is determined, we propose an underwater version of SNMP. To actuate the control, we can use a conventional ceramic transducer on the UAV to receive the SNMP-like commands from the mother ship.

Generation of info carrier proceeds as follows. There is a control volume V with a bounding surface S. Inside V, there is an electrolyte such as NaCl of a certain concentration and at a certain temperature T. There is an axial applied static B field and an AC E field perpendicular to it. The control volume V is placed inside a wave-guide with apertures so that the electrolyte MHD system can radiate into the outer space. In this work, we assume that the UAV is never a few kilometers away from the mother ship and ray approximation of the MHD wave equations is applicable. Then by considering energy and momentum balance of in and out of S, we end up with the following ray approximation

$$(\nabla^2 - \frac{1}{c_0^2}\partial_t^2)p = (-\frac{\beta}{\sigma C_p})\partial_t J^2 + div(J \times B)(H(x-a) - H(x-b)) \tag{1.1}$$

Here c_o is the isotropic sound speed defined, as a constant for the electrolyte at constant entropy and p is the pressure fluctuation about a quiescent value. The other symbols on the right hand side of (1) have the following meaning, (a) β is the coefficient of thermal expansion, (b) σ is the ionic fluid conductivity, (c) C_p is the specific heat and constant pressure, (d) J is the current density and (e) H is the step function.

Equation (1) predicts pressure fluctuations give rise to spherically spreading acoustic radiation. This shall be the info carrier between the mother ship and the UAV. This equation is the result of considering mass and momentum conservation inside the wave-guide together with a process of *linearization*. To save space, we skip outlining the derivation steps and just mention that we have used the small amplitude approximation and the Boussinesq approximation [6].

We have now a spherically-spreading pressure variation source and need to examined how is it transmitted to the receiving side.

Out there in the ocean, the governing equations are either the incompressible Navier-Stokes equations or the Euler equations. Let us use the Euler equations.

$$\partial_t u + u \cdot \nabla u + \nabla p = 0$$

$$\nabla \cdot u = 0$$

Boundary conditions i) no slip condition $n.u|_\Gamma = 0 \quad u|_\Gamma = 0, \quad ii) \frac{\partial p}{\partial n}|_\Gamma = 0$ *as a consequence.*

(1.2)

Here Γ represents the walls of the wave-guide. We see that the pressure field at the UAV is obtained from solving the Poisson equation subjected to appropriate boundary conditions in (1.2).

$$\Delta p + \nabla \cdot (u \cdot \nabla u) = 0 \tag{1.3}$$

If density variation is included, this equation has to be modified to include additional terms. See (4.5).

SECTION 2: SOLUTION OF THE PRESSURE WAVE EQUATION

Equation (1) is of the form

$$\nabla^2 p - \frac{1}{c_0^2}\partial_t^2 p = -4\pi f(x,t) \tag{2.1}$$

The right hand member represents the source terms on the right hand side of (1.1). Fourier transform both sides of (2.1) in time, we have

$$\begin{aligned}&(\nabla^2 + k^2)\overline{p}(x,\omega) = -4\pi \overline{f}(x,\omega)\\ &k = \frac{\omega}{c_0}\end{aligned} \tag{2.2}$$

Here, $\overline{p}, \overline{f}$ are Fourier amplitudes. From (1.1), we have

$$f(x,t) = \frac{1}{4\pi}\{(\frac{\beta}{\sigma C_p})\partial_{t'}J^2 - \nabla_{x'}\cdot(J \times B)\}(H(x'-a) - H(x'-b)) \tag{2.3}$$

Here $H(x'-a)$ is the step function. Further Fourier transform in x, we have

$$p(x,y,z,t) = \int d\omega\, e^{i\omega t} p(x,y,z,\omega) \tag{2.4}$$

$$\begin{aligned}J(x,y,z,t) &= \int d\omega\, e^{i\omega t} J(y,z,\omega)(H(x-a) - H(x-b))\\ &= \int d\omega\, e^{i\omega t} J(y,z,\omega)(\int d\lambda\, e^{i\lambda(x-a)}\frac{1}{\lambda} - \int d\lambda\, e^{i\lambda(x-b)}\frac{1}{\lambda})\end{aligned} \tag{2.5}$$

There is a similar representation for the thermo-acoustic term with J^2. Hence (2.2) becomes

$$(\partial_y^2 + \partial_z^2 + k^2 - \lambda^2)p(y,z,\omega,\lambda) = -\{(\frac{\beta}{\sigma C_p}(2i\omega)J^2(y,z,\omega) - J(y,z,\omega)B_0)(e^{-i\lambda a}/\lambda - e^{-i\lambda b}/\lambda))\}$$

(2.6)

This is just the wave equation for an open rectangular wave-guide and is solved by means of the Green's function of the Helmholtz equation. The derivation is given in chapter 3 of [7]. The result is

$$G(y,z;y',z') = \frac{\sum_{m,n} \cos(\frac{2n\pi y}{a})\cos(\frac{2n\pi y'}{a})\cos(\frac{2m\pi z}{b})\cos(\frac{2m\pi z'}{b})}{(\gamma_{nm}^2 - k^2 + \lambda^2)} \quad (2.7)$$

$$\gamma_{nm}^2 = \pi^2((\frac{2n}{a})^2 + (\frac{2m}{b})^2)$$

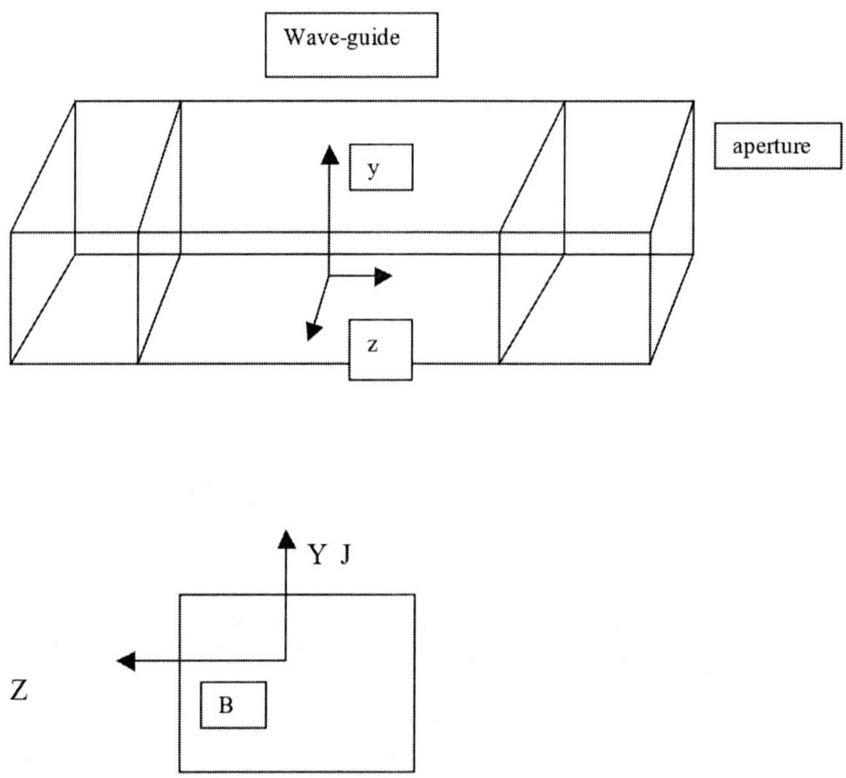

Figure 1 the coordinate system

In writing down the Green's function in (2.6), we have assumed that the correct boundary condition is the vanishing of normal derivative at y = (-a/2, a/2), and z = (-b/2, b/2). This is because the velocity of the fluid has to be zero at the boundary of the wave-guide.

If the inverse Fourier transform in λ, $\frac{1}{2\pi}\int d\lambda\, e^{-i\lambda x} p(y,z,\omega,\lambda)$, is performed, one would find traveling waves of the form $e^{-i\lambda_{\pm}x}\cdot(e^{-i\lambda_{\pm}a} - e^{-i\lambda_{\pm}b})$. Here $\lambda_{+} = \sqrt{(\gamma_{nm}{}^{2} - k^{2})} = -\lambda_{-}$. This shows that we have the following results

1. In order to have pressure waves propagating out to the UAV, we must have high modes n, m excitation.
2. The operating frequency of the applied electric field should be low.

SECTION 3: DATA TRANSMISSION ISSUES

In this section, we shall discuss how to use the pressure wave to carry information. We shall first discuss how to apply PCM to p (x, y, z, ω). We shall see that the outgoing information is encoded in the Fourier components of the applied electric field. Out in the ocean, wave amplitudes cannot be taken to be small. Therefore, the nonlinear nature of the Euler equation (or Navier-Stokes equation) has to be addressed. As a result, the pressure distribution at the UAV is much more complex compared with what we see in the previous section. At the same time, we need to enable a non-trivial decoding algorithm.

We can now write the expression for $p(x,y,z,\omega)$.

$$p(x,y,z,\omega) = \frac{1}{8\pi^{3}}\int d\lambda\, e^{-i\lambda x}\frac{1}{\lambda}(e^{-i\lambda a} - e^{-i\lambda b})\int dy'\, dz'\, G(y,z;y',z';\lambda)\{-\frac{\beta}{\sigma C_{p}}(2i\omega)J^{2}(y',z',\omega)$$
$$+J(y',z,'\omega)\}$$

(3.1)

The first term is the thermo-acoustic contribution while the second term is the MHD contribution. To do the digital modulation of $p\,(x,y,z,\omega)$, we use a variant of the well-known pulse code modulation method since what we have now is a 2D pulse. So let us re-capture the details of the conventional PCM through an example.

Suppose we are given a waveform

$z = \cos(x) - 0.3\cos(3x) + 0.5\cos(5x) - 0.7\cos(7x)$

We also know that the length of the wave-guide along y is 6 in some system of units. Suppose further that the sample "rate" is 0.01 and that we are going to radiate 16-bit signals out to the ocean. There are 65536 (2**16) signal levels and there are 600 sampling points and the same number of sampled values. The simple program that assigns the quantization level runs as follows.

```
# include <iostream>
# include <math.h>
# include <iomanip.h>
# include <vcl.h>
# pragma hdrstop
using namespace std;

# pragma argsused
int main(int argc, char* argv[])
{
        float x;
        char c;
        int level;
        float z[65536], delta, deltax, upper, lower;
        delta = 0.000045776;
        //This is the total variation divided by the total number of signal levels
        deltax = 0.01
        //This defines the spatial sampling frequency.
        x = 0.0;
        int m, n;
        n = 0;
        while (x <=6.0)
        {
                z[m]=cos (x)-0.3*cos(3*x)+0.5*cos(5*x)-0.7*cos(7*x);
                while(n <=65536){
                        if ((-1.5+n*delta) < z[m] < (-1.5+(n+1)*delta))
                        {level = n;
                        cout << "signal level is " << level << endl;
                        lower = -1.5+n*delta;
                        upper = -1.5+(n+1)*delta;
                        cout << "lower is " << lower <<setw(10) << endl;
                        cout << "upper is " << upper << setw(10) << endl;
                        cout << "sampled value is " << z[m] <<setw(10) << endl;
                        break;
                        }
                        else
                        n = n + 1;
                        }
                n = 0;
                x = x = deltax;
                m = m + 1;
    }
        return 0;
    }
```

Segment of the output is shown.

Sampled value	Lower	Upper	Level in decimal	Level in binary
0.5	0.499999	0.500045	43692	1010101010101100
0,501174	0.501144	0.501189	43717	1010101011000101
0.504691	0.504668	0.504714	43794	1010101100010010
0.510528	0.510528	0.510573	43922	1010101110010010
0.518652	0.51863	0.518676	44099	1010110001000011
0.529014	0.528975	0.529021	44325	1010110100100101
0.541553	0.541518	0.541564	44599	1010111000110111
0.556195	0.556166	0.556212	44919	1010111101110111
0.57852	0.572829	0.572875	45283	1011000011100011
0.591426	0.591414	0.59146	45689	1011001001111001

Table 1: Signal Level Quantization

Suppose that the mother ship and the UAV are exchanging SNMP-type of messages. How is this handled using our MHD acoustic radiator? Without going into the details of the SNMP protocol [8], we use the following command as example.

snmptrap -v 1 -c public .1.3.6.1.4.1.2789.2005 127.0.0.1 6\2476317 " .1.3.6.1.4.1.2789.2005.1 s "WWW Server Has Been Restarted′

In this example, an object with identity (OID) .1.3.6.4.1.2789.2005 at address 127.0.0.1 sends a notification to a network monitoring station (NMS) about conditions it should know. The generic id of the message is 6 and the type of the message is 2476317. The OID of the destination is .1.3.6.1.4.1.2789.2005.1. The data type sent is a string, s. The content of the string is "WWW Server Has Been Restarted". On top of the string, the sender adds information that is not relevant to our discussion. Here is the binary representation of the string.

a. The hex dump from an ethereal trace is

736e6d70747261702 02d762031202d63207075626c6963202e312e332e362e312e342
e312e323738392e32303035203132372e30302e3120365c32343736333137
202727202e312e332e362e312e342e312e323738392e323030352e312073
20225757572053657276657220486173204265656e2052657374617274 5
6422

We want to determine what should be the appropriate applied E field if we have to transmit this SNMP command. To do this, we first translate the hex dump into binary. Here is a segment of the result.

b. Binary Representation

Signal Level in Binary	Signal Level in Decimal	Lower	Upper
0111001101101110	13166		
0110110101110000	28016		
0111010001110010	29810		
0110000101110000	24944		
0111011000110001	30257		
0010110101100011	11619		
0111000001110101	28789		
0110001001101100	25196		
0110100101100011	26979		
0010111000110001	11825		

Table 2: Reverse PCM Scheme

Now we have to choose a quantization scheme. Once that is done, we will, for each signal level, fill out the "Upper" and the "Lower" values. We also have to determine the spatial sampling frequency. The upshot of these activities is a series of flat top polygons.

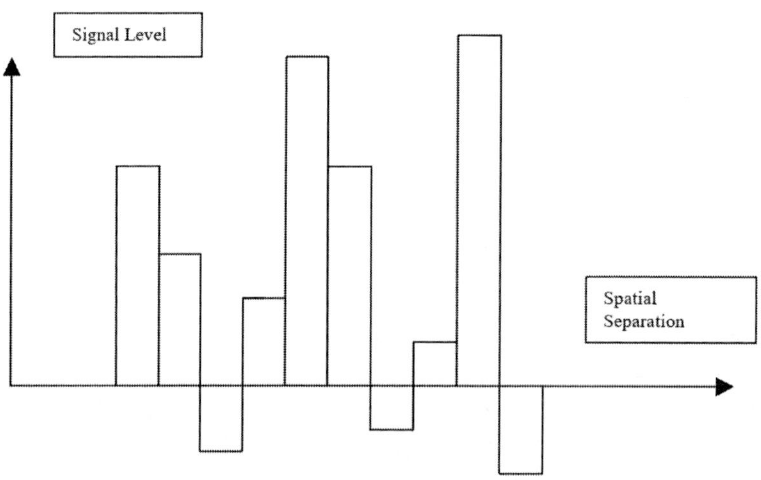

Table 3: Interpolation in Reverse PCM Scheme

Using an interpolation process, one can easily arrive at the profile of the required E field for transmitting the SNMP command "*snmptrap*".

So far, we have only explained how does the dipole term contributes to the transmission of the SNMP command. The working of the monopole term is more complicated but is similar. Because of the lack of time and space, we leave the details to the reader.

Now we turn our attention to the realization of the layered model of communication of SNMP.

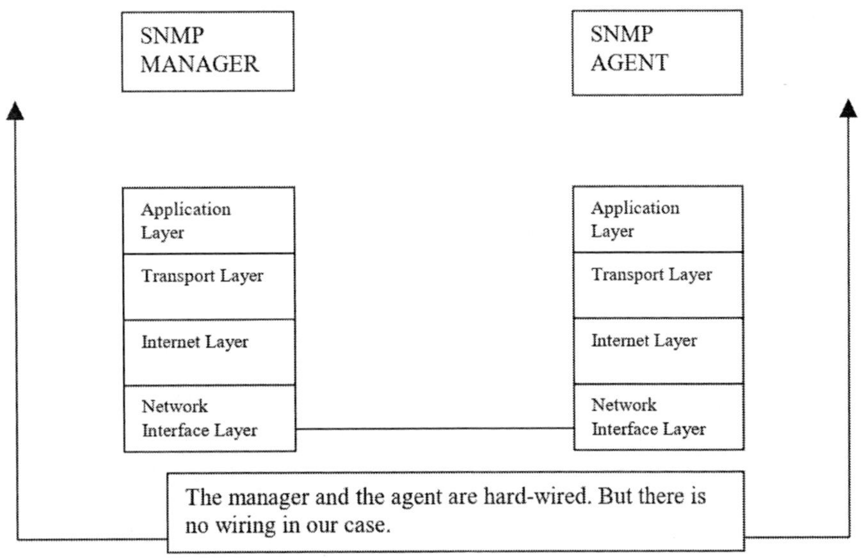

Figure 2: SNMP Layered Model of Communication

To understand how MHD scheme would constitute a wireless implementation of the layered model, we first go through a simple SNMP operation. Suppose the Manager wants to know the Agent's System Name is and prepares a GET message for the appropriate OID. It then passes the message to the UDP layer. The UDP layer adds a data block that identifies the manager port to which the response packet should be sent and the port on which it expects the SNMP agent to be listening for messages. The packet thus formed is passed to the IP layer. Here a data block containing the Media Access addresses (MAC address) of the manager and the agent is added before the entire assembled packet gets passed to the Network Interface layer. The Network Interface layer verifies the media access and availability and places the packet on the media for transport. In the conventional setup, the transport media is usually twisted pair copper, RG58 co-axial or fiber. Also SNMP resides in the Application layer, UDP resides in the Transport layer and IP reside in the Internet layer. The fourth layer is the Network Interface layer.

So far in our discussion, we assume that only the mother ship has the capability to send. This is changed if we have a more symmetrical arrangement, i.e. we allow the UAV to be equipped also with a MHD transducer so that it can also carry out data message generation just as the mother ship. In that case, a wireless-version of the layered model is possible in the following sense

- The aim of the present investigation is to show that it is possible to generate digital message of the SNMP type MHD transducer.
- The first step towards the transmission is to solve the linear sound pressure wave equation with a dipole driving term (due to Lorentz force) and a monopole term (due to Ohmic heating.)
- In solving the wave equation, we impose a "no-slip" boundary condition, so that the message is registered in the tangential derivative of the pressure at the boundary of the wave-guide.
- Out in the deep sea, where the UAV is situated, we have the Euler equations (or Navier-Stokes equations) to worry about. In the next section, we shall see that by taking the divergence of the Euler equation and by invoking the condition that the fluid is incompressible, the UAV arrives at what is known as the Poisson Pressure Equation. This equation has to be solved if the pressure distribution is to be determined. It turns out that one can set up a solution of the equation in terms of the sum of a volume potential and a single layer potential. The density function of the single layer potential is determined by the normal derivative of the pressure at the boundary of the wave-guide. So, by looking at the single layer potential term, the UAV knows what has been transmitted. So in this sense, data has been transmitted from the mother ship without using cables.
- If the UAV is also equipped with a MHD transducer, all the activities happening in the application layer, transport layer and Internet layer can be implemented (or simulated) by a clever program.
- Our MHD transducer implements the Network Interface layer.
- We have now realized the 4-layer communication model of SNMP in a wireless fashion.

Of course, there are a lot of factors impacting on the viability of this scheme. These include convective instability due to density stratification, onset of turbulence due to the uncertainty of whether it is always possible to have globally (in time) regular solution from the Euler equation so that we can do this kind of transmission. Under certain conditions, the answer is yes. This is actually the subject of an open millennium problem posed by the Clay Institute. The scope of this paper is limited so that resolutions of these problems are not available from here. But in Section 6, we shall devote more time to discuss them.

SECTION 4: PRESSURE DISTRIBUTION AT THE UAV

We now have a MHD transducer in place and there is a pressure wave emanating from it. In this section, we want to work out the received pressure distribution at the UAV. To do this, we have to solve the Poisson pressure equation (1.3) /(4.5) outside the rectangle defining the profile of the wave-guide.

Solving Poisson equation is a familiar process in engineering. Unfortunately, the geometry of the domain over which the equation is to be solved is not standard. We have to perform a series of conformal transformation so that we map the original domain $Z = y + iz$ to the upper half of a $W = u + iv$ plane. Once this is done, the solution in the W-plane is immediate.

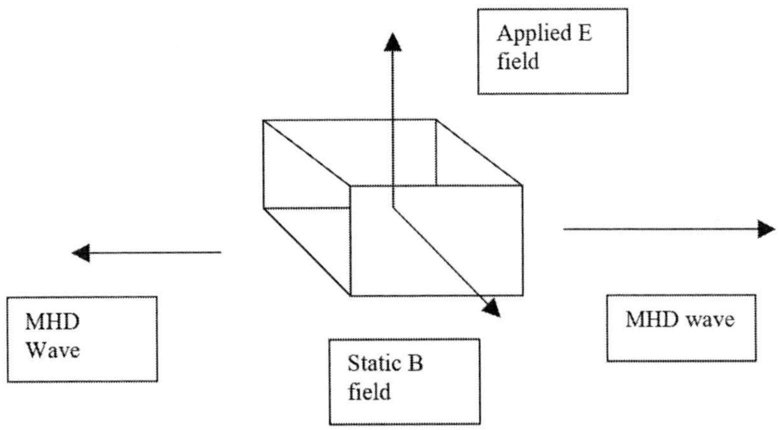

Figure 3: The MHD transducer

In the ocean, the mathematical model is provided by the incompressible Euler equation (or the Navier-Stokes equation if the viscosity is not zero).

$$\frac{Du}{Dt} = -\frac{1}{\rho_b}\nabla p - \frac{g}{\rho_b}\rho\, e_1 \tag{4.1}$$

$$\frac{D\rho}{Dt} = 0 \tag{4.2}$$

$$\nabla \cdot u = 0 \tag{4.3}$$

Here $\frac{D}{Dt} = \partial_t + u \cdot \nabla, u$ is the velocity field and p is the pressure field and

$$\rho\,(x,y,z,t) = \rho_b + \overline{\rho}(x) + \rho'(x,y,z,t) \tag{4.4}$$

is the density expressed as the sum of a constant reference density ρ_b, a linear ambient density $\overline{\rho}(x)$ and a perturbation density ρ'(x, y, z, t) and e_1 is a unit vector in the positive direction of the x-axis opposite to gravity. Also we will use the Boussinesq approximation [6].

The first thing we have to do is to derive an equation satisfied by the pressure. To this end, we simply take the divergence of both sides of (4.1) and use the incompressible condition in (4.3) to arrive at

$$-\Delta p = \rho_b \nabla \cdot (u \cdot \nabla) u + g \frac{\partial \rho}{\partial x} \tag{4.5}$$

SOLUTION OF THE PRESSURE POISSON EQUATION

The ultimate aim of this section is to determine the received (at the UAV) pressure distribution. We can do this by solving the pressure Poisson equation over the domain

$$\text{D:} \quad \begin{aligned} &-\infty < x < \infty \\ &-\infty < y \leq -a/2 \quad a/2 \leq y < \infty \\ &-\infty < z \leq -b/2 \quad b/2 \leq z < \infty \end{aligned}$$

As the forgoing discussion shows, once the x-dimension has been Fourier transformed out, we are effectively determining Green's functions and solving Helmholtz equation in the (y, z) plane outside the rectangle (a/2, b/2), (-a/2, b/2), (-a/2, -b/2) and (a/2, -b/2). Define the Z plane by $Z = y + iz$ and the W plane by $W = u + iv$. Then the intersection of the rectangle with the first quadrant of the Z-plane is the rectangle (a/2, 0), (a/2, b/2), (0, b/2), (0,0). If we use Ω to denote that region of the first quadrant of the Z-plane that is outside the quarter rectangle, then it can be shown that Ω is the image of the following mapping function

$$Z(W) = i(b/2) + \int_0^W \sqrt{\cos\zeta - \cos\alpha} \quad d\zeta \tag{4.6}$$

The angle α is determined by

$$a/2 = \int_0^\alpha \sqrt{\cos\xi - \cos\alpha}\, d\xi \tag{4.6'}$$

$$b/2 = \int_\alpha^\pi \sqrt{\cos\alpha - \cos\xi}\, d\xi \tag{4.6''}$$

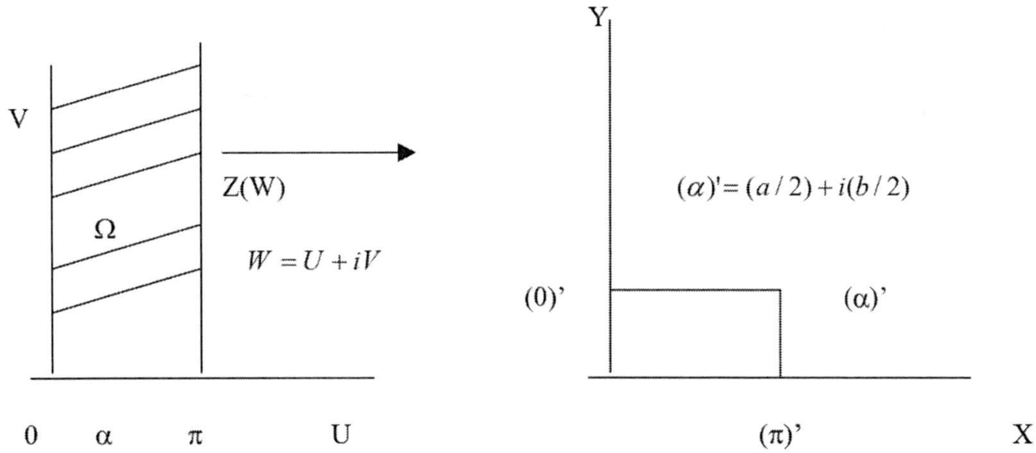

Figure 4: Regions Defining Z (W)

By Schwarz's reflection principle [9], the function F (W) defined by

$$
\begin{aligned}
F(W) &= Z(W) \quad in\ \Omega \cup [0,\pi] \\
&\quad \overline{Z(\overline{W})} \quad in\ \Omega^{*}
\end{aligned}
\tag{4.7}
$$

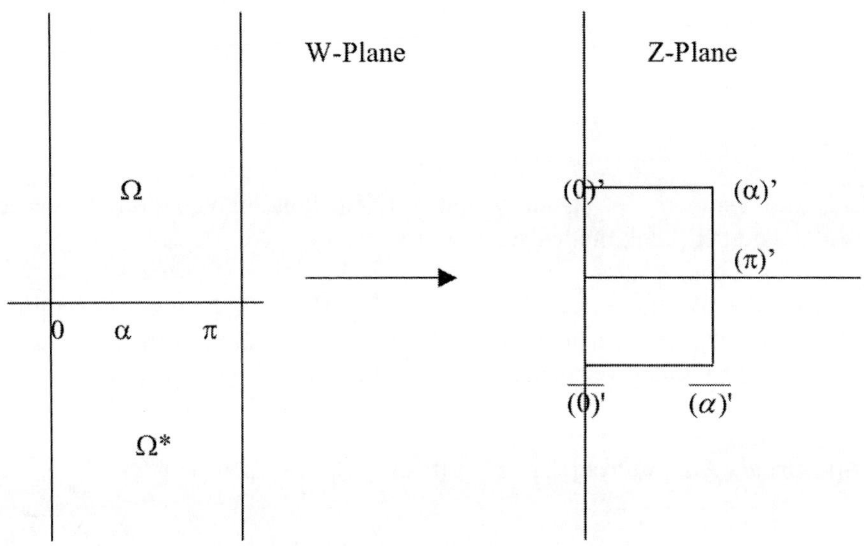

Figure 5: Schwarz Reflection

Now apply the reflection principle in its general form [10]. We arrive at the final form of the mapping.

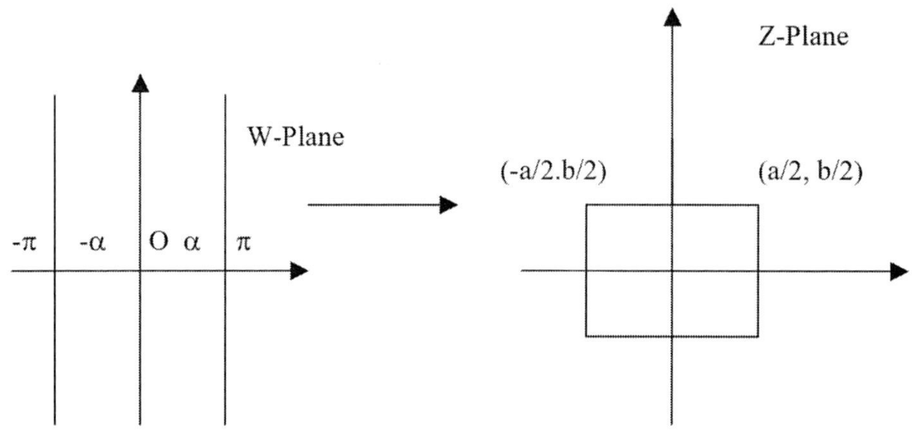

Figure 6: Conformal Mapping of the exterior of a rectangle

Finally, we need one more mapping to stretch out the segment [-π, π] on which the boundary conditions (the normal derivative of the pressure on the wall of the wave-guide) are prescribed.

The transformation that maps ± π to ± ∞ and leaves the origin (of the W-plane fixed is given by

$$W' = \frac{W}{(W - \pi)(W + \pi)} \qquad (4.8)$$

On W'-plane, the geometry is standard and solving Poisson equation can proceed according to standard text-book methods.

Let us represent the W'-plane by $W' = u' + iv'$. The Green's function is conformal invariant. This is also true of the boundary condition of vanishing normal derivative of the pressure [12]. Hence on the W'-plane, the explicit solution is [13]

$$p(u', v') = -[\text{integral of the source}] - \int_{-\infty}^{\infty} du''\, X(u'', v'')\,|_{v''=0}\, \frac{\partial}{\partial v''} G(u', v', u'', v'')\,|_{v''=0} \qquad (4.9)$$

Here $X(u'',v'')$ is the tangential component of the pressure written on the wall of the wave-guide by the magneto-hydrodynamic wave. Hence, through the double-layer potential of the solution of the Poisson pressure equation, the UAV receives the transmitted message.

SECTION 5: MANAGEMENT OF THE UAVS

In this section, we require the mother ship to perform the role of a SNMP agent to manage the system of UAV. In order to have the same type of interactions that exist between an SNMP manager and an agent (residing on the managed device, UAV), we need to equip the UAV also with a MHD transducer. In this way, it can send as well as receive messages. Not only is the hardware difficult to implement, but also the software and the algorithmic architecture are also intricate. If this is done, then

- The UAV can implement via software transports analogous to TCP, UDP and UNIX Domain Sockets. It can receive and transmit SNMP messages through these transports.
- The message processor will be the software that reads the density function of the double layer potential term of the pressure distribution to see what the sender is talking about.
- In SNMP, the main function of the USM module is to decrypt the incoming messages. Since what we have in undersea communication between the mother ship and the UAV, there is no need for encryption. Hence, USM module can be skipped.
- In SNMP, the dispatcher is responsible for routing messages to appropriate destinations. In this set up, there is no routing. The MHD transducer sends sound pressure wave in a broadcast fashion. Each UAV is assigned an OID and is responsible to find out whether a particular message is relevant.
- The VACM module controls the access to the UAV infrastructure and is too intricate to be included in the architecture at this stage.
- The Repository contains the configuration information of the UAV. At this stage, it can be implemented by an embedded database such as SqlLite or embedded MySQL. This is also true of the OID registration handler.
- There is no need to have an AgentX.

So, provided we can enable a duplex data transmission between the mother ship and the UAV, the system may be regarded as a wireless LAN with SNMP messages going around. The mother ship performs the role of manager executing SNMP system management on the UAV.

SECTION 6: IMPACTS FROM POSSIBLE ONSET OF TURBULENCE

Basically, what we have been doing is as follows. We first solve linear MHD wave equation inside the wave-guide subject to the "no slip" boundary condition. To the UAV in the deep sea, this is equivalent to prescribing the boundary value of the normal derivative of the pressure. In order to get this boundary value, the UAV sets up the Poisson pressure equation and reads the density function of the double layer potential.

Crucial to the success of this scheme is that the Euler equation should possess smooth solution for the whole duration of transmission. So do we have what we need? This problem is extremely difficult and is known to be still open [2].

In this section, we will first review the fundamental work of Beale et al [11], which gives a precise statement of the so-called "finite time singularity". Then we will also review a paper by Giga et al, which points to the possibility that if initial data of a certain nature is used, there is no "finite time" singularity. This should then be the basis of our data transmission.

It should be pointed out existence of "finite time" singularity has not been proven. Its non-existence has also not been proven. What has been done is the result of extensive numerical investigations by mathematicians and engineers indicate that it might be present.

For two-dimensional flows, solutions of the Euler equations with smooth initial data remain smooth for all times. However, in three dimensions numerical investigations [Chlorin] suggest very different results. In particular, these computations show that solutions of Euler equations, which at first represent smooth flows, may develop singularities signifying the onset of turbulent behavior. Qualitatively, the formation of singularities is related to the concentration of vorticity on successively smaller sets. In terms of what we have been doing, this will mean that even if the initial velocity at the beginning of a period of time is smooth, transmission is only possible before the singularity develops. Once it is in place, Euler equations no longer govern the motion of the fluid and there will be no Poisson pressure equation for us to use. What makes it worse is we do not know when is this "finite time" singularity going to come.

The contribution of Beale et al is their result that if a solution is initially smooth and loses its regularity at some later time, then the maximum vorticity necessarily without bound as the critical time approaches. Hence what is desirable to see is to ensure that the vorticity remains bounded.

Giga et al looks for further conditions on the initial data so that the Navier-Stokes equation should have a globally regular solution. This is an effort to have additional constraints laid on the initial conditions so that the solution would come out to have nice behavior. The statements of the their results are rather technical and it would be out of place for the present paper to go into the details. Let us state their result in approximate for. If u_0 is the prescribed initial velocity and

- u_0 is divergence free
- u_0 is a function of the 3-D space and has complex valued components (to allows for Fourier transform)

- FM denotes the space in which the Fourier transform of u_0 lives
- The Fourier transform of u_0 is supported in sum-closed frequency sets with no zero frequency components. This requirement covers the important special cases of almost periodic and highly oscillating initial data. These are the cases that are of interest to us.
- The Fourier transform of the initial data has a finite bound, denoted by $\| u_0 \|_{FM}$.
- Then the Navier-Stokes (Euler) equation has a globally regular solution $u(t)$ satisfying

$$\| u(t) \| \leq 2e^{-\lambda t} \| u_0 \|_{FM} \quad for\ t > 0$$

Here λ is a constant. Since vorticity $\omega = \nabla \times u$, therefore if the above inequality is satisfied, there would not be any growth of vorticity and hence "finite time" singularity would not exist.

In our work, we have not specified what kind of initial conditions we need. We have only solved the boundary value problem twice, once at the wave-guide and once at the UAV. In our future works on this subject where initial conditions prescription is necessary, we will tacitly assume that the choice by Giga et al will be adhered to.

SECTION 7: CONCLUSION

In this paper, we have outlined the possibility that by equipping the mother ship with a MHD transducer inside a rectangular wave-guide, it is possible to generate acoustic pressure wave propagating into the deep sea. The source of the wave consists of a monopole term and a dipole term. The monopole term comes from the Ohmic heating on the particles in the electrolyte and is size if proportional to J^2, where J is the current density. This contribution may be termed "thermo-acoustic". The second term is the result of the Lorentz force acting on the electrolyte particle. This term is proportional to J × B, where B is the applied magnetic field. This is the MHD term.

Using the MHD term, we show that it is possible to generate and transmit SNMP-like commands from the mother ship. The generation part comes from using PCM on the applied E field. As a matter of fact, using both PCM and reverse PCM, we have a one-to-one correspondence between the command to be transmitted and the E field to be applied.

At a particular time, the UAV may solve the Poisson pressure equation and express the solution and expresses the solution by means of the Green's formula, i.e. a volume potential term plus a double layer potential term. The message sent by the mother ship is contained in the double layer potential.

Whether such a scheme would be meaningful or not depends entirely on whether we have a solution of the Euler equations that has global regularity in time. The status of this subject is still under intensive research by mathematicians and engineers. However, there is one team [3], who shows that existence of global regularity in time depends on the initial data. Among other more general conditions, they show that if we have bounded almost periodic initial velocity, then globally regular solution exists.

The discussion so far sets up a simplex type communication between the mother ship and the UAV. The usefulness of such as scheme is limited. This is because the UAV is supposed to serve as a sensor and has to upload info to the mother ship. This means that the UAV should also be able to send. As a matter of fact, we want to push the design to cover a system of UAV each is capable of carrying out duplex communication with the mother ship. This is illustrated as follows.

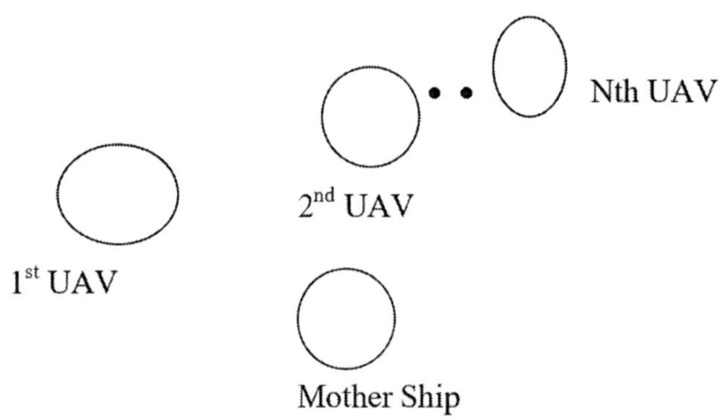

Figure 7: A System of UAV Communicating with Mother Ship

In this diagram, each circle represents a combination of a sending and a receiving device. The solution of the Poisson pressure equation is made up of a sum of a single volume potential terms plus a number of surface terms each containing the encoded message sent out by the particular device. There is also a circle at infinity, which has been omitted.

To demonstrate that such a scheme is actually feasible in the practical sense requires a lot of work, as the problem is complex both in engineering and mathematics. However, it is obvious that a successful scheme in this direction will be enormously useful. We hope to return to return to some of these problems.

REFERENCES

1. S. Chan, K. F. Lai, A Combinatorial Algorithm for Motion Planning in an Autonomous Unmanned Undersea Vehicle Equipped with Forward-Looking Sonar, Proceedings of IMCPAC08, 2008.
2. C. L. Fefferman, Existence and Smoothness of the Navier-Stokes Equation, http://www.claymath.org/millenium/Navier-Stokes_Equations/navierstokes.pdf.
3. Y. Giga, K. Inui, A. Mahalov, J. Saal, Global Solvability of the Navier-Stokes equation in spaces based on sum-closed frequency set, Annales Scientifigue Superieure, Vol. 39, Issue 4, 679-698, 2006.
4. A. Mahalov, B. Nicolaenko, C. Bardos, F. Golse, Regularity of Euler equation for a class of three-dimensional initial data, Methods and Application of Analysis, Vol. 11, No 4, 605-634.
5. Y. Giga, T. Miyakawa, Navier-Stokes Flow in R^3 with measures as initial vorticity and Morrey spaces, Communications in Partial Differential Equations, Vol. 14, Issue 5, 577-618, 1989
6. M. Lelong, J. Riley, Internal Wave-Vortical Mode Interaction in Strongly Stratified Flows, J. Fluid Mechanics, Vol. 232 pp 1-19. http://en.wikipedia.org/wiki/Boussinesq_approximation
7. J. Jackson, Classical Electrodynamics, John Wiley & Sons, Inc. 1967
8. D. Mauro, K. Schmidt, Essential SNMP, Second Edition, O'Reily, 2005
9. R. Schinizinger, P. Laura, Conformal Mapping: Methods and Applications, Elsevier Science Publisher, 1991
10. B. Palka, An Introduction to Complex Function Theory, Springer, 1991
11. T. Beale, T. Kato, A. Majda, Remarks on the Breakdown of Smooth Solutions for the 3-D Euler Equation, Communications in Math. Phys. 94, 61-66, 1984
12. M. Schiffer, Fredholm Eigen-values of Multiply Connected Domains, Pacific J. Math. Vol. 9, Number 1 (1959) 211-269

Use Of Modern Navigation Systems Within Complex Maritime Environments

M. Vierow
L-3 Nautronix, Australia
M. Christou
L-3 Nautronix, Australia

OVERVIEW

The recent transformation of navigation systems and their integration with modern sensors to increase situational awareness has been unquestionably significant; these systems have and continue to transform the way watchkeepers navigate. However, the digitisation of navigation and the extreme advancement of navigational sensors, systems and architectures have together achieved much more than just a virtual host for navigators and command teams. They represent a new capability which is set to rival traditional methods of decision making and achieving situational awareness – principal to navigating safely within a complex maritime environment. This paper will examine the interaction between these "information age" technologies and ships, shore stations and users followed by proposing and evaluating recent configurations and procedures. This discussion is intended to contribute towards the future definition of a pragmatic roadmap to effectively employing modern navigation systems within complex maritime environments.

DEFINING A COMPLEX MARITIME ENVIRONMENT

The significant breadth of industry experience that senior mariners possess creates the potential for application of different interpretations of what "complexity" is within the maritime environment. Traditionally, mariners may describe sea pressure points such as the Dover and Malacca Straits as high traffic and/or demanding shipping areas complicated by the presence of more than 400 transiting vessels daily[1] as well as the unorthodox or asymmetric crossings of fishing vessels, ferries and other non-conforming traffic activity. Modern mariners are more likely to describe complex maritime environments as those which they encounter during transits through or operations in ungoverned bodies of water, usually complicated by the lack of marine safety equipment and/or fixtures aboard vessels, and in certain cases, the presence of insurgent maritime operations. The nature of an insurgent maritime environment consists of many intangible elements but it has been suggested that they include any radical movement that increases political, economic or social control of a region[2] and ranges from piracy to people smuggling, illegal fishing or pollution.

This paper will utilise a broad definition of complex maritime environments. This will facilitate the establishment of a thorough requirements set in order to form a realistic and relevant needs proposition. This paper proposes that complex maritime environments are those which consist of one or more of the following characteristics:

1. The presence of large numbers of static or dynamic vessels whether bound by a regional traffic system (such as a Traffic Separation Scheme) or otherwise, as well as the potential limitation of safe and/or navigable waters commonly encountered in littoral or port-approach areas;
2. An ungoverned, or poorly regulated, body of water whereby vessels operating within the area may not be fitted with standard or regulatory safety equipment, such as navigational lights or VHF equipment, critical to mutually safe navigation;

3. Unstable or inadequately governed sea regions which may include areas where insurgent operations exist such as piracy or other illegal operations, indeed any such area where the safe or innocent passage of vessels is placed at risk.

A BASIS FOR REDEFINING COMPLEX MARITIME REQUIREMENTS

It could be argued that our proposed definition of complex maritime environments has existed since the revolution of purposeful navigation, i.e. vessels navigating from A to B for economic or other trade related purpose. Therefore one could reasonably substantiate that the requirements for navigation within complex environments have and will remain extant. The authors of this paper take the view that there are several factors which collectively form justification for a redefinition of the needs for navigating in a complex environment. The primary factors are as follows:

1. Requirements for navigating in complex environments must be matched, and mature, against advancements in technology. Whilst high level requirements for navigating in complex environments, such as "keeping a good lookout", may not have drastically changed, the "information-age" has delivered an incredibly advanced set of tools which are now available to the mariner. In short, the art of bridge watchkeeping has significantly evolved since the digitisation of equipment; this must be acknowledged and reflected accordingly to ensure the relevance and utilisation/adoption of new toolsets and technologies.
2. Increased levels of international trading, easier access to seaworthy vessels and other similar technological advancements have reduced the barriers to entry for qualified and unqualified mariners to put to sea. Consequently sea lanes and coastal areas are busier, bottlenecks are increasingly pressurised and unorthodox marine behaviour is more prevalent. Complex environments are becoming even more complex.
3. Globalisation and modern connectivity now enables more coordinated forms of illegal or insurgent activities at sea. This goes hand in hand with the ease of access to digital media which spreads the potential windfall to be gained through illegal activities, such as the payment of ransoms. These are likely contributory factors in a 200% increase in armed piracy during 2009, compared to 2008[3] in specific sea regions. Unstable or security-adverse shipping areas are becoming increasingly dangerous year-on-year.
4. A cost-benefit argument for reducing the impact of complex maritime environments on shipping is gaining momentum, making the once unjustifiable costs of a requirements "revolution" far more realistic. This is evidenced through one subset characteristic of complex environments, piracy, which is estimated to be costing more than 16 billion US dollars per year[4].

MODERN INITIATIVES REDEFINING NAVIGATION AT SEA

In 2005 the e-Navigation concept was developed[5] with further clarity of purpose issued by IALA in 2007[6] specifying that:

> *"e-Navigation is the harmonised collection, integration, exchange, presentation and analysis of maritime information onboard and shore by electronic means to enhance berth to berth navigation and related services, for safety and security at sea and the protection of the environment."*

The "e" in "e-Navigation" was at the time undefined, it either represents the mentioning of "electronic" or "enhanced" within the definition. The authors of this paper take the view that "electronic" tools "enhance" navigation and therefore enhanced navigation is an outcome of the capability that "electronic navigation" (hereby referred to as e-Navigation) delivers.

A comparison of the 2007 e-navigation definition with this paper's basis for defining new requirements for complex maritime environments provides close linkage. The interpretation of the title itself, representing electronic or digital-age navigation, reflects that the requirements are intended to be matched with advancements in technology. There is specific reference to the safety of berth-to-berth navigation, a "cradle-to-grave" assessment of the changing complexities at sea in port approaches, coastal waters and the high seas. There is important clarification that e-Navigation is also intended to address security and the environment, which is matched with our assessment of the level of complexity found within ungoverned waters. It is assessed by the authors that purposeful and considered adoption of e-Navigation provides a suitable framework for a modern approach to navigation within complex environments.

PROPOSED REQUIREMENTS FOR NAVIGATION IN COMPLEX MARITIME ENVIRONMENTS

In the four years since the development of e-Navigation there has been conservative progress towards the key themes of the concept. It is the authors' assessment that only rudimentary progress has, to this point, been made towards an effective harmonisation, integration, exchange, presentation and analysis of maritime information for the purposes of safety and security at sea. e-Navigation currently exists in stand-alone formats such as Automatic Identification Systems (AIS), Electronic Chart Display and Information Systems (ECDIS), Vessel Traffic Systems (VTS) with a limited amount of intra-data exchange occurring between platforms and ashore.

These systems do not present harmonisation of data with often varying situational pictures being presented from one ship or shore installation to another. Small amounts of data exchange, such as vessel ID information via AIS, currently exist but not to the point where the exchange of information could justify the description "integrated". Alternatively, "integrated" could be defined as the increasing prevalence of "integrated" bridge systems which do go a long way in demonstrating an integrated capability. Modern "turn-key" solutions offer Integrated Ship Management Systems which provide networked functionality of Platform Management Services, Bridge Sensors and shipboard administration. However, this exists in systems and information for the watchkeeper, i.e. at a singular or unit level, and is not reflective of a more collaborative marine environment.

In examining how information is presented to marine users, the authors assess that information has been sufficiently digitised, or digitally ported, to reflect the vision of e-Navigation. However in saying that, the migration of navigational information from the chart to the monitor has been predominantly conventional. Initially, the scanned charts (Raster chart type) were straightforward scans of traditional navigation presentation into the digital age. In other words, the transfer was aimed at maintaining the navigational "status quo" using digital tools. This strategy during technological revolutions is not without precedence; the migration of newspapers to the internet, at their outset, appeared very similar to their printed genesis. More recently, this strategy has matured with the introduction of Vector charting, otherwise known as "smart charts", which are scalable and have the potential to provide enhanced features via the same data format and operator interface.

This paper intends to propose, examine and evaluate new technologies or technology paths which have the potential to provide true e-Navigation suitable for use in complex maritime environments. The "gap" is assessed in the following comparative table.

e-Navigation Characteristic	Current	Future
Harmonised	No harmonisation of navigational data with situational picture variance from platform to platform.	An agreed situational picture between platforms which allows units to gain a full appreciation and awareness of their surroundings.
Integrated	Minimal integration of systems exists between platforms, E.g. AIS. Integration of navigation systems and bridge within onboard platforms is widely available.	A more embraced approach to integration of systems between ships and shore units. Continued advancement of integrated bridge systems, within platforms.
Exchanging	Minimal amounts of bandwidth and data exchanged between platforms.	Increased bandwidth and high volumes of exchanged information between platforms.
Presentation	Digitised navigational information and situation, scalable and able to display enhanced information in complex environments.	Continued digitised presentation of information and situational information, advantageously using efficiencies in future user interface developments.
Analysis	Little automated analysis, mostly basic analysis conducted by "seaman's eye".	Intelligent analysis of navigational information presented with minimal impact to the user.

Table 1 – Comparison of current and future e-Navigation characteristics

THE DECISION MAKING PROCESS

This paper intends to propose a summary of the systematic decision making processes that occur on a regular (if not constantly) basis, , during operations or transits in complex maritime environments. The purpose of this summary is to describe how delivery of a true e-navigation concept will support a more effective decision making process, rather than serve as a replacement.

The suggested decision making process is a hybrid of decisional and situational theories sourced from dynamic system environments[7] and leadership best practice[8].

Step 1 – Gathering Information; provision of both background and task related information using optimum modes or means to the user or system which is gained from objective, calibrated or from other trusted sources.
Step 2 – Assessment/Analysis of Information; to efficiently gain an appreciation of a situation through the use of a combination of experience, intelligence or automation. Assessments and analyses are used to reach conclusions, appreciate inherent risk and/or gain knowledge.
Step 3 – Evidence Based Action; where a developed structure of information is used as evidence to justify the execution of action, assisted by ease of decision dissemination and timely responses.

PROPOSAL AND EVALUATION

This section will propose, evaluate and summarise recent contributions to research or the product market for use as an architecture, component or element of a future e-Navigation offering. The proposals will be evaluated against the groundwork laid so far in proposing requirements for navigation in complex environments or against the summarised decision making process in complex environments. The intended outcome is an objective and thought provoking contribution towards laying the roadmap for effective employment of modern navigation systems to cope with increasingly complex maritime environments.

Communication Infrastructure: A facilitator of exchanging and gathering information

Aspects of e-Navigation characteristics such as "harmonised", "integrated" and "exchanging" require a communications infrastructure with bandwidth beyond what is readily available in today's licensed marine channels. This has been evidenced through research conducted into the bandwidth requirements for future e-Navigation services which concluded that "AIS frequencies cannot realistically support any significant increase of e-Navigation services"[9]. This is not to say that AIS provides a redundant service. In fact it has demonstrated immense success in providing a "first step" in reliable and automated VHF mode of exchanging ship identification information. However, future reliance on the AIS infrastructure is not deemed to be an "optimum mode" for gathering information and providing this to the user – a more capable infrastructure is required.

A joint initiative between the Infocomm Development Authority of Singapore (IDA) and the Maritime and Port Authority of Singapore (MPA)[10] has pioneered the development of a marine communications infrastructure through the installation of a wireless broadband network infrastructure. This network is available within 15 kilometres (8 nautical miles) of the southern Singapore coastline[11]. This capability offers a high level of connectivity and bandwidth in the complex approaches to the port of Singapore. It has been received with strong interest and take-up and is currently used for various data and internet applications including video conferencing, where the business benefit is realised through the cost savings of removing high cost satellite communications[11].

There are other ground based alternatives to establishing a capable communications infrastructure; the predominant competitor to long range wireless networking implemented by IDA/MPA is the next generation cellular network, commonly referred to as "4G". However, access to the 4G technology, like its predecessor 3G, is likely to be rolled out by commercial carriers with proprietary equipment access. This is in direct contrast to long range wireless networking which will be a ratified and open standard. Consequently, significant cost savings are forecast between the long range wireless and 4G cellular options, these are likely to be in the order of 50%[12] or greater. Furthermore, long range wireless technology provides the potential for a high capacity network using a universal standard, backed with common protocols, which is identified as a critical element of establishing situational awareness[13].

Whilst the "long range" description used within the new wireless technology may reflect a reasonably long distance in metropolitan areas, the scales of the vast oceans are significantly different. The range of eight nautical miles is a significant limitation of this technology as this is merely the beginning of sea coverage and is still considered the "littoral" or "brown water" area. However, in addressing the requirement for a communications infrastructure which provides bandwidth to complex maritime areas in the approaches to ports and navigationally constrained areas, such as the Dover and Malacca Straits, it provides a respectable solution.

Long range wireless installations are not limited to the shoreline and can be boosted from the offshore position of aids to navigation, beacons or buoys, and used for ship-to-ship data exchanges. The security of this technology is worthy of due regard, however it is concluded that adequate levels of encryption and current uses in other "secure" operations widely exist to avoid any risk of discounting the proposal based on security concerns alone. Overall, it is our assessment that long range wireless technology be considered as a viable solution for the purposes of establishing new marine communications infrastructure for optimum mode of data exchange between ship-to-ship and ship-to-shore.

Harmonising the Navigational Picture: Common situation and task related information

An agreed situational and common picture between platforms and shore facilities will become critical to the inter-management of increasingly congested sea lanes and coastal areas. In the "information-age", access to a common picture, which includes background and task related information, is perceived by the authors as an essential toolset which should be available to the modern mariner. This technology should form the framework for a more collaborative appraisal of the situation and risk of collision[14].

Recent Norwegian research has developed a prototype which specifically addresses the need for harmonisation and "common" direction of attention, with the specific aim of reducing the risk of collision. The risk reduction tool, named a "3-dimensional attention zone" provides a dynamic and scalable zone around any marine object of interest for presentation on an ECDIS screen and is distributable via TCP/IP message format across a network[15]. The capability of this concept provides the benefits of a modern solution effectively gathering task related information for vessels passing through high density areas. It offers mariners or Vessel Traffic Services (VTS) the option to increase the risk profile (termed risk multiplication factor) of own-ships, other vessels or other known objects / navigation threats in the area. This "multiplication factor" can be based on changing conditions, a limited ability of the platform or any other condition which justifies increased awareness of that vessel or object to other mariners in the area. It is proposed as an automated function with layered ECDIS presentation available to all platforms in the geographical region[15].

Whilst this is an excellent example of how units and shore installations, such as VTS, can harmonise and create a common picture with common appreciation for vessels or objects possessing high risk categories, there are multiple assumptions pertaining to the roll-out of such an initiative. Like most e-Navigation concepts, the technology assumes the use of electronic charting systems which are integrated across a communications framework. It assumes that only well trained operators use the system otherwise the quality of information would be diluted. It also assumes that optimisation in the presentation of this information can be achieved through the setting of specific "cross-over" rules, whereby the user is only notified on encroaching a particular "attention zone"[15], otherwise this could lead to cluttered ECDIS screens. To counter these assumptions, the risk assessment of vessels and objects could be controlled using a permissions structure controlled by the applicable port or coastguard authority.

Harmonising an Insurgent Navigational Picture

Besides a harmonised picture with a greater fidelity of collision risk, the basis for definition of future requirements included consideration of navigation in insurgent areas. In addressing the needs gap of harmonised insurgent operations summary, accessible by non-Naval users, private enterprise now offers historic and predictive intelligence plotted on an electronic chart interface[16]. This product is intended to support the greater complexity of decision making by providing information for the assessment and mitigation of risk. The value proposition of this service is stated as "Intelligence = Profit"[17]. This value offering links with the cost benefit justification of using modern technology, in this case harmonised and distributed insurgent summaries, to combat the inflicted costs of navigating within complex environments.

As piracy operations evolve, with the longest range attack recently occurring in excess of 1,000 nautical miles offshore[18], advancements must provide better modes of distribution than VHF and navigation telex (NAVTEX). Special attention zones and predictive insurgent summaries are offerings that demonstrate the nature and benefits of situational assessments that can be harmonised across multiple platforms with better technology. They enhance the ability to gain situational awareness and therefore enable more informed decision making in complex maritime environments. Although both offerings remain proprietary products, there are now multiple examples of modularised software development for bridge systems and the Nautical Institute continues to champion S-Mode[19] as a standard bridge system interface. Both modular software development and S-Mode provide technology pathways for common interfaces to provide harmonised data to varying types of bridge or shore system designs.

Integrated Systems: Collaborative situational assessment complex environments
The amounts of information collectively gathered and processed by VTS, ships, and installations, such as aids to navigation, are vast. They are readily available to play a greater collaborative role in assessing the dynamic components of complex environments. Maritime Safety Queensland (MSQ) and the Australian Maritime Safety Authority (AMSA) jointly operate a VTS system along the Great Barrier Reef[20]. The system, "REEFVTS", provides a direct and automated interaction with transiting vessels feeding data to the VTS from a variety of sensors around the Queensland coast and onboard vessels, including radar, AIS and Inmarsat-C. The information is used to create a full and real-time picture of all vessels within the Great Barrier Reef area, and can predict the position of vessels based on intended route, planned/actual speed and traffic forecasts across an area of approximately 1,200 nautical miles of coastline out to a distance of 50 nautical miles, through the use of offshore aids to navigation[20].

This system currently provides alarms and warnings to the system operators when vessels are likely to enter shallow water areas or when they have diverted from designated route corridors. Future enhancements will potentially provide a more tailored warning system based on integrated under keel clearance management, which will relate vessels' specific draft to the available navigable water. There currently exists a vessel advisory service, for transiting and non-transiting vessels, which could also be enhanced by combining static or dynamic "attention zones" creating a far more automated and intelligent service. Importantly, the VTS system has demonstrated its usefulness in providing navigational assistance to vessels in the event of failure of onboard navigation systems; this has been used on multiple occasions to prevent potential groundings. The system also has the capability to monitor and report unidentified or "contacts of interest" to relevant authorities and demonstrates the tools to provide the type of sea governance that is lacking in other regions. In a more complex environment, this capability could mutually function alongside a predictive intelligence product which re-routes vessels based on security assessments.

Similar to other "information age" VTS systems, "REEFVTS" is based on a highly modular and configurable design that allows additional functionality and data sources to be added easily to the system. Colloquially referred to as "future proofing", the system is an open architecture, and allows for future distribution or remote operations from several installations or vessels over a higher bandwidth communications infrastructure or network, such as long range wireless technology. It is expected that long range coordination services such as "REEFVTS" combined with close quarter, or port authority, services such as "attention zones" are "information age" technologies that could be effectively combined to offer an enhanced "berth to berth" navigation service as set out in the IALA e-Navigation definition[6].

Due to the flexibility of these "e" systems, they could collaboratively provide the competence required to meet the ever changing nature of complex environments found in port approaches, traffic schemes and areas where unqualified seamanship exist – and be presented in a scalable format.

Analysing information: developing the automated "seaman's eye"

By no means is developing the "seaman's eye" using automation to be interpreted as a suggestion to irresponsibly handover the duties of safe navigation to "information age" technology. In considering the nature of increasing traffic and insurgency levels and the effect of these on leaner bridge crews, this suggestion refers to using automation in an advisory or assistant capacity. Automation is capable of efficiently processing and analysing large amounts of data to monitor and detect variances to normal seamanlike conduct. It is accepted that "variances" may be noticed by a more labour intensive radar and visual lookout operation, such as the multi-tiered teams employed on warships. Nevertheless, both grey and other hull colours operate within hostile regions and this paper proposes the use of automation to assist in these increasingly complex environments.

Recent research into such a capability has highlighted the level of automation which can be created using algorithms which are established to monitor "seamanlike" rules. The concept has been created to detect possibly threatening activities at sea[21]. It uses algorithms which are established to analyse the behaviour of contacts at sea and has a multi-level alarm system, with adjustable sensitivity, to report the detection of threatening or dangerous behaviour including higher level alarms based on behaviour which directly threatens another object at sea[21]. Whilst guard zones or collision avoidance alarms are a current capability of modern systems, this capability presents the first step in creating total automated water-space monitoring. Initially developed to detect threatening contact behaviour, the future extensions of this system are stated to include automated comparison of AIS, dynamic characteristics and ship database information as a form of "identity control" at sea. Intelligent analysis of threatening or asymmetric behaviour at sea, particularly in poorly governed areas, would be a significant leap forward in the ability of relevant authorities to receive reports or detections to directly combat the increasing prevalence of illegal activities at sea. The ability to embed automated analysis into existing VTS and bridge systems would also rapidly grow the coverage of an automated "seaman's eye".

Integrated Bridge Systems: Presenting, enabling and disseminating timely decisions

Modern integrated bridge systems continue to evolve and are beginning to truly reflect "information age" operability. Recent developments include offerings such as a web-styled navigation interface to switch between Radar, ECDIS, Machinery and Utility pages[22]. In the future, providing a design which more fully acknowledges and understands the multitude and complexity of modern day navigation requirements will become increasingly critical to enabling X and Y generations to slickly and efficiently make readily informed decisions and disseminate these, e.g. through machinery control or rudder movement, in a highly efficient manner. However, this assessment is made with the fundamental assumption that the capabilities of modern Bridge equipment does not lead to hasty or ill informed decision making or excessive trust / reliance on the automated system. Indeed recent shipping accidents, such as the fatal Ouzo collision[23], have highlighted the dangers of over-reliance on user defined system information. The proposed decision making cycle concludes with the vital importance of evidence based action using a developed structure of information.

Although modern systems are able to more competently sense and display information, it is critically important to ensure that the structure of evidence is displayed to a user. In short, this means ensuring that the user is made aware of all the sensors and processes that are active, discounted or disabled so that a full appreciation of the evidence applicable to the decision is both maintained and documented. Electronic charting is assessed as an aspect of bridge systems whereby evidence relating to a navigational decision can often be placed at risk through misinterpretation, a lack of pertinent task related information or incorrect user settings. Customised Vector Nautical Charts (CVNC) are an Australian development which provide the mariner with greater levels of accuracy, survey fidelity and information which is specific to tasks such as port approaches, constrained navigation and berthing or unberthing[24]. Small to medium scale CVNCs can be designed for coastal VTS areas or long range identification and tracking with the potential to display dynamic routing information, attention or maritime zones, radar & AIS coverage boundaries, military boundaries and ranges, or fishery operations[24].

CVNCs intend to offer a vehicle for the presentation of standardised, intelligent and updateable base data layers, such as dynamic models, to the mariner. It is expected that this positive development will contribute to reducing the risks of "assumptions" when operating in complex maritime areas. It will achieve this by using various sources of information to feed a greater amount of evidence into the decision making process. It is assessed that a standardising feature such as S-Mode[19] would be a critical requirement of CVNCs to remove the risk of user defined settings excluding information. Effective employment of customised charts would greatly increase the ability of the mariner to understand the area and the strengths or weaknesses of the evidence being used to make key navigational decisions.

CONCLUSIONS AND RECOMMENDATIONS

This paper intended to contribute towards the future definition of a pragmatic roadmap to effectively employing modern navigation systems within complex maritime environments. It defined the complex maritime environment, provided justification for the shift in "goal posts" and proposed a new set of requirements for improved navigation in these environments. Prior to proposing and assessing recent contributions to the navigation community, this paper proposed a decision making process carried out by the mariner on a regular if not constant cycle. The following recommendations are the result of this paper's analysis and evaluation.

1. There are unavoidable limitations with current e-Navigation communication infrastructures. For background and task related information to be gathered effectively, and in real time, a high bandwidth and reliable communications infrastructure is required in order to exchange data for harmonisation and integration across platforms. The authors assess that long range wireless network technology is a proven and in-service infrastructure that warrants serious consideration for further trials in landscapes larger than the approaches to Singapore. This is a key element of building an integrated marine environment.
2. In order to gather task related information, it must be harmonised and agreed between in-area platforms. Distributed "attention zones"[15] and the broadcast of predictive sea intelligence are excellent examples of the types of "agreed risks" that can be harmonised between assets. Both technologies are proprietary and a suitable commercial framework is required for implementation. However these niche products represent a significant step forward in the ability to assess and mitigate for known and unknown seaborne risks.
3. Basic collaborative assessment of the navigational picture is already being offered within various VTS, such as "REEFVTS". Integration of this information across platforms

allows both ship and shore operators to appreciate inherent risk. However, this is currently unintegrated between larger "high seas" areas, covered by VTS, and port approaches. e-Navigation specifies berth-to-berth navigation and the approach to integrating complex maritime environment information requires to be scaled accordingly.

4. Automation in the "information age" is widely used in manufacturing and operations to reliably substitute or improve manpower efficiency. Automation can be used, with cautious implementation, to assist mariners in the analysis of significant amounts of information to reach conclusions which are escalated to the watchkeeper in a tiered manner. As environments become more complex, and in parallel bridge resources become leaner, it will be more important to explore the benefits of automated analysis systems, such as behaviour analysis, to ease the operational strain on watchkeepers.
5. Bridge systems and ECDIS should enable efficient execution of watchkeeper decisions. They should support this through the provision of gathered information and analysis using a harmonised local picture, presented in an unambiguous structure of evidence to assess risk to platform safety and security. As proposed by a navigation expert; "Navigation displays should be designed to clearly indicate risk and to optimise support for decision making"[19]. The evolution and evaluation of CVNCs suggests that this technology could be readily applied as a vehicle to optimise the use of other risk assessment technologies discussed in this paper.

---

[1] Channel Navigation Information Service (CNIS), 2007, "The Dover Strait".
[2] Moore, S., 2007, "The Basics of Counterinsurgency", p.3.
[3] International Chamber of Commerce, Commercial Crime Services, 2009, "Unprecedented increase in Somali pirate activity".
[4] Chalk, P., 2009, "Maritime Piracy; reasons, dangers and solutions".
[5] IMO MSC, 2005, "Development of an e-Navigation strategy", Agenda item 23.
[6] International Association of Marine Aids to Navigation and Lighthouse Authority, 2007, "Development of an e-Navigation strategy", Agenda item 13.
[7] Endsley, M. R., 1995, "Toward a theory of situation awareness in dynamic systems", page 36
[8] MindTools, 2009, "Information Gathering".
[9] Rodsth, O. & Kvamstad, B., 2009, "Digital Communication Bandwidth Requirements for Future e-Navigation Services", European Journal of Navigation, Volume 7, pages 10-23.
[10] Infocomm Development Authority of Singapore, 2007, "Infocomm @ SeaPort".
[11] Infocomm Development Authority of Singapore, 2008, "First In The World: Wireless Mobile Wimax Access In Singapore's Seaport Now A Reality".
[12] Higginbotham, S., 2008, "LTE vs. WiMAX: A little 4G sibling rivalry".
[13] Australian Department of Defence, 2009, "Defending Australia in the Asia Pacific Century: Force 2030 White Paper".
[14] International Maritime Organisation, 1972, "Convention on the International Regulations for Preventing Collisions at Sea".
[15] Mestl, T., Anderssen, T., Zachrisen, M., Hjollo, B., 2009, "Towards the Collaborative Sea", European Journal of Navigation, Volume 6, pages 14-16.
[16] Dryad Maritime Intelligence Services, 2009, "Decision Support Intelligence".
[17] Dryad Maritime Intelligence Services, 2009, Home.
[18] Reuters, 2009, "Somali Pirates Attack Tanker", 12th November 2009.
[19] Patraiko, D., 2008, "e-Navigation", Nautical Institute, June 2008.
[20] Australian Maritime Safety Authority, 2005, "Shipping Safety: REEFVTS".
[21] Harre, I., 2009, "A Novel Approach for Early Detection of Possibly Threatening Activities as Sea", European Journal of Navigation, Volume 7, pages 4-9.
[22] SAM Electronics, 2009, "Platinum Bridge Product Guide".
[23] Marine Accident Investigation Branch, 2007, "Investigation Report:Ouzo".
[24] L-3 Nautronix, 2009, "Customised Vector Nautical Charts Product Specification Sheet".

MULTI-ENGINE SUBMARINE POWER SUPPLIES: THE OPERATING CASE

John Buckingham CEng MIMechE

Jacob Mann AMIMechE

BMT Defence Services Ltd, Bath, United Kingdom

INTRODUCTION

For many years both conventional and nuclear submarines have used one, two or three diesel gensets to provide electrical power to recharge the submarine main batteries. However, many current gensets are based on older engine designs and more modern designs with their attendant advantages do not exist as products for submarine use. This paper seeks to explore the issues surrounding the use of up to four gensets with a modern diesel engine design to identify the benefits and shortfalls which may exist. The issues of ship impact, system reliability and product development will be considered. The performance assessments will be conducted with the BMT Vidar-36 submarine concept.

CONTEXT

The diesel electric (SSK) submarine has become established as the leading capital platform for many of the world's navies. In modern times they are increasingly asked to conduct a wider range of operations of longer and more distant duration: as a highly valuable asset they need to be able to spend the maximum time on patrol. To achieve this, the machinery needs to have a high availability and low upkeep burden.

The machinery in a submarine comprises a number of key elements: energy consumption; power generation; energy storage; propulsion drives and these have been addressed in a previous BMT paper in 2008 [Ref. 1]. This paper considers the issues relating to the use of diesel engines for power generation in more detail.

The most complex and hard worked equipment item onboard an SSK is the diesel engine generating set (genset) whose continuing good health and reliability is fundamental to the ability to transit and then conduct continuous operations once in the mission area. The current models which serve the world's western navies are well-established designs which have been developed and improved through extensive usage.

The Role of Diesel Engine Generators

Although diesel engine generators are a mission critical feature of a submarine on patrol, they are notorious for being amongst the least reliable of onboard machinery. [Ref 2].

Although the development of the battery energy store and the electric propulsion elements have continued to be greatly developed through modern materials and solid-static electronics, respectively, the engine has an inherent vulnerability due to its arduous operation as a reciprocating engine when snorting. Although the engine technology continues to develop, such testing operating conditions are seldom the focus of mainstream development.

Current SSK Arrangements

The typical SSK has two or three gensets to allow for a high battery charge rate and to allow for a degree of redundancy. However often the increased number of units makes for a cramped arrangement which limits access for upkeep leading to the use of horizontal crankshaft removal in the Hedemora engines in the Australian Collins class [Ref 3]. The Australian Collins class is one of the few SSK with three gensets: this is largely driven by her extended periods away from her base port and the enduring need to ensure that there is always one genset available to allow for the ability to steam home.

Current Engines

Engine Range

However although the current range of world SSK gensets in western navies is confined the following range of genset:

Engine Make & Model	**Power kW (bhP)**	**Speed rpm**
MTU 396: Surfaced : Submerged	600 to 1350 600 to 1250	1800
Caterpillar 3516B	1492 to 1864	1600
Pielstick PA4	590 to 2215 (800 to 3000)	1500
Paxman Valenta	1300	1800
Hedemora HV V18b/15Ub (VB210)	1400 (1877)	1400
A210SM	1020 to 3400kW	1500
ASR1	1420	920

Table 1 - Submarine Snort Engines

Many of these engine models have also been used in fast ferry propulsion whose demand for fast acting engines which can respond reliably to sudden load demands makes for a robust design.

MTU 396

MTU have produced many diesel engines for submarines both with and without super-charging [Ref. 4]. Their current offering is the 396 which has been available with a turbo-charger design since the 1980's after MTU had a public test-bed demonstration of its performance in 1982. Over 210 MTU 396 are in use in SSK applications. They come from a broad pedigree which was introduced to submarines in 1982 and which has also been developed for a range of specialist applications in including low-magnetic applications.

Caterpillar

The Caterpillar 3516B is a well proven engine which is employed in the USN Virginia class of submarines. Following marine tests, in 1998 a more powerful version of the 3516B was introduced which had a longer stroke with 1864kW at 1600 rev/min.

Pielstick PA4

The 12 cylinder Pielstick PA4 engine is used on French submarines [Ref. 5] and those exported to other navies. The best known models are the PA4-185 and the 200-VG models.

The PA4 engines are fitted with a variable geometry combustion chamber and are now supported by MAN.

Hedemora

The Hedemora engines in the Collins class run at 1400 rpm [Ref. 6]. Slower speeds are usually associated with larger engines for the same power and this also usually means a larger engine as a whole which leads to volume constraints and issues with access and the removal of main parts.

Paxman Valenta

Paxman engines have been on UK submarines since the U and V class of World War II. The last design to enter service before Paxman was merged with MAN was the Valenta 16SZ which has a mechanically driven super-charger on the HMS Upholder [Ref. 7 and 8] , now HMCS Victoria and the remainder of that class. The design is reviewed in some detail by Sessions et al [Ref. 9]. His paper shows how supercharging both increases the triples the rated power of a naturally aspirated engine and lowers the specific fuel consumption (Sfc) by over 20%.

Supercharging v Turbocharging

One of the challenges with a supercharged engine is the high exhaust temperatures experienced. When the Valenta was in development temperatures of over 650°C were experienced. [Ref. 9]. Such temperatures make for expensive exhaust pipework materials and also make for larger cross-sectional areas due to the large volumetric flow. A supercharger design is also less fuel efficient than a turbocharger design.

In a turbocharger design, the addition of a turbine in the exhaust stream removes useful energy as work to drive the compressor. This is a more effective means of driving the compressor than in a supercharger design and it also reduces the exhaust gas temperatures with the benefits of smaller volumetric flow, smaller trunking and less need for exhaust gas spray insertion to reduce the exhaust gas infra-red signature.

Alternatives

Many of the current set of long-serving submarine engines run at speeds between 1600 and 1800 rev/min and are thus considered to be high speed engines but for today's market there are many examples of even faster engines which may be more compact. It is those engines which are considered in due course.

Potential diesel engine replacements such as air-breathing Solid Oxide Fuel Cells in hybrid arrangements are some way away they do not offer the compact power density of the diesel engine (they are 50 times worse).

Current diesel engines have shown that they can be operated to achieve such requirements for most of the time. A modern replacement will need to be proven to be substantially better.

Engine Trends

Emissions Led

Future surface ship applications will need to meet IMO Tier III emissions measures and for many this will be achieved with secondary measures such as exhaust gas scrubbers to remove NOx and SOx. This will leave the engine designer more able to design for reduced fuel consumption and hence carbon dioxide emissions. Increasingly such improvements to fuel consumption are being achieved through double or sequential turbo-charging.

Although the greater trend towards the use of double turbo-charged diesel engines will allow for better fuel consumption of engines as a whole, this trend will provide more engines of a comparable design to the market leads of MTU and Pielstick who have led this technology in the past and then applied them to submarine operations.

Hence it is possible that there will be a greater source of submarine capable gensets in the future due to this very trend.

Speed Led

The ability to run diesel engines at speeds above 2,000 rpm has progressed greatly in the past 20 years. As an example, Caterpillar now offers engines which can operate in this speed range. Such engines have a high compression ratio of 13 or over and if their Time Between Overhauls (TBO) is to be used as a measure of reliability, their MTBF has also increased significantly over this period.

General Trends in Engine Failures At Sea

Using the wealth of operating data in the Japanese commercial fleet, in 2001 Kiriya [Ref. 2] showed how there is a discernable set of trends when comparing data for the period 1982 to 1989 to that recorded for the period 1990 to 1997. When all failure sources are considered, diesel engines continue to dominate the burden of unplanned upkeep onboard when at sea.

The general MTBF of diesel engines has increased by almost 100%: increasing from 1000 hours to 2000 hours in the past 30 years. However after fail, the trend for the time to repair was often becoming longer at 30 mins with combustion chamber related issues being the worst going from 40 to 50 minutes. Such an observation is probably due to the increasingly complex nature of the modern engine with its greater number of sensors, interfaces and electrical devices all of which lead to a longer Mean Time To Repair (MTTR). The effort required to rectify the failure is likewise increasing from 10 to 14 man-hours per failure.

Lessons For Submarines

MTTR

Whilst there is not a direct read-across from commercial engine usage to submarine engine usage, the lessons from the trends do indicate that the MTTR is a key aspect of Power and Propulsion (P&P) System availability.

If the complexity of modern engines may mean that the MTTR will be longer it may also mean that some engine failures may not be repairable at sea.

In both cases, it will be important for safety and operational reasons for a high proportion of generating capacity to remain whilst the damaged engine is not available.

Margin of Capacity

The provision of power generation onboard a submarine is usually such that the necessary power to the batteries whilst also supplying ship's services and propulsion when snorting can be met by the gensets with one assumed to be inoperable. There is therefore a margin of capacity which represents a surfeit that can be fully used at once due to the current limitations on the ship's lead acid batteries. This margin is provided purely because of the critical nature of the role of the engine and the perception of its inherent unreliability.

MORE POWER SOURCES

The use of smaller and more modern engines onboard a submarine would enable the current benefits of the emerging engines to be implemented to the advantage of the submarine P&P system as a whole. A set of four engines whose total power equals the maximum power that can be absorbed by the energy storage whilst still supplying hotel and propulsion loads would in effect have no excess capacity.

Graceful Degradation

A multiple genset design would represent a design which has a better ability to tolerate item failure as there would be a genuine scope for "graceful degradation" whereby a loss of one modern genset would allow a remaining, adequate, 75% capacity. The concept of genuine use of a redundant system of elements has been thoroughly examined by Salter [Ref. 10]. Salter shows that systems where there are multiple items whose total capacity meets the top-line requirement, can be capable of accepting individual item losses with the overall capability continuing to be wholly adequate. Salter's message is clear: it is important to consider a system wide graceful degradation as well as the necessary attention to individual item reliability. The battery energy store on a submarine is a good example of how a critical P&P system function is designed, by necessity, to have multiple elements any one of which can be dropped out if not fit for service.

Derating

Submarine models are typically rated at 50 to 60 % of their open atmosphere rating when operating in snort conditions. As an example the MTU 12V 396 is rated at 1200kWe for the TE54 designation [Ref. 11, pg 52] but when employed in the submarine version it is derated to 1,000kW on the surface and 900kW when snorting [Ref.4].

Although the current submarine engine models continue to operate reliably there now appears to be a situation due to modern technology whereby new entrants may enter this small but important market.

PERFORMANCE

The diesel engine is designed to supply the maximum electrical load budget at its maximum continuous rating (MCR) output or below with a margin, usually of the order of 10%.

The new replacements for the currents set of engines will need to be able to withstand some exacting loads: they are likely to be thermally challenged by the testing requirements for a submarine power source. Such requirements are to be achieved with a high degree of reliability. Additionally such engines also need to be able to meet ever exacting emissions standards when operating in open atmosphere conditions and need to acquire an Engine International Air Pollution Prevention (EAIPP) certificate.

A submarine genset needs to be able to operate in a wide range of operating conditions:

- High back-pressure and low inlet pressures;
- Quick load acquisition;
- Good at accepting sharp air supply and load changes;
- Quick shut-down and secure arrangements;
- Open atmosphere and submerged conditions
- Fast shut-down from full load.

Modern engines tend to have electronic control systems, advanced turbo-chargers and intercooler to allow for a better air charge. These features also permit the technical challenges to be addressed with more flexibility.

High Back-Pressure

The high exhaust gas back pressure due to snorting with the exhaust located 5m or so below the sea surface is an obvious challenge to the engine's operations as are the reduced input air pressures. BMT have modelled the behaviour of diesel engines in these cases to identify the impact on the rated power and the specific fuel consumption (Sfc) (See Figures 1 and 2). Of greater concern is the dynamic influence of the varying back-pressure and the dynamic instability this can cause to the turbo-charger behaviour.

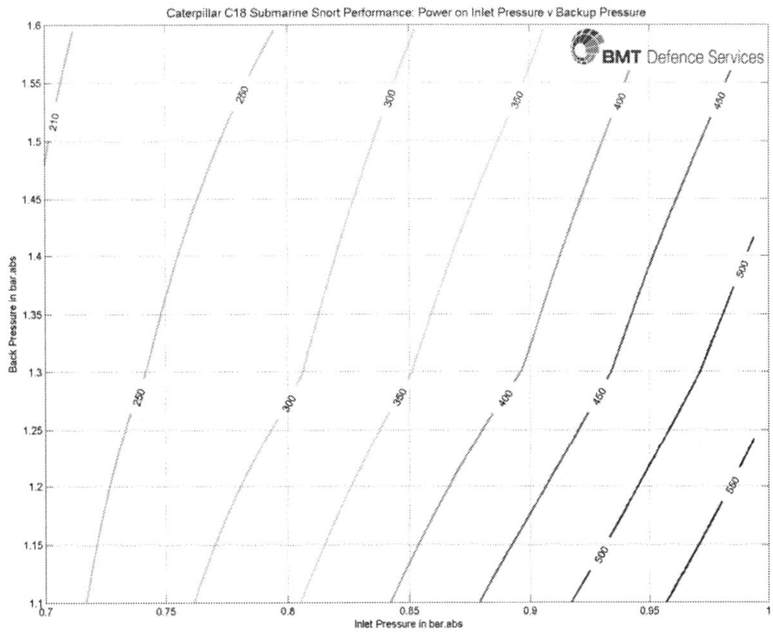

Figure 1 - C18 Rated Power Variations

Figure 1 shows there is a significant loss of rated power with decreasing air supply pressure whilst an increase in the back-pressure has a lower impact.

Low Inlet Pressures

BMT studies indicate that the reduced air inlet pressure to the engine is a more significant influence than the back-pressure as this leads to a reduced air charge to the combustion chamber and has a direct reduction on the rated power for reduced inlet pressures. In essence the impact of the lower inlet pressure is to negate the benefit of the turbocharger such that the power rating tends towards the power of a non-turbo-charged engine.

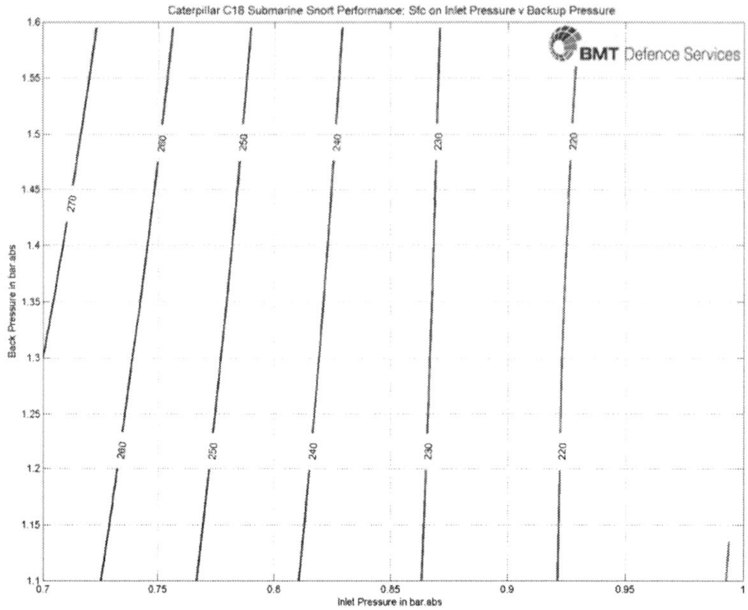

Figure 2 - C18 Specific Fuel Consumption Variations

Figure 2 shows how the Sfc is most sensitive to the reduction in inlet air pressure with a low variation due to the changes to the exhaust gas back-pressure. The studies have found that the back-pressure can be tolerated up to a point when the turbo-charger effectively stalls.

Quick Load Acquisition

The submarine genset is expected to acquire full load within two minutes from a cold start. Although electrical pre-warming of LT, HT and LO lines will facilitate this the engine itself has to be thermally tough and modern engines are designed with greater attention to thermal resilience through more advanced design methods.

Good at accepting sharp air supply and load changes

The engine can experience random and sustained fluctuations in air supply and exhaust transients due to the action of the sea state when snorting.

A modern electrical control system will allow the regulation of fuel to ensure a quality of supply from the generator to meet the STANAG 1008 requirements. Such regulation is required due to the transients seen in the exhaust and air inlet lines due to wave interaction with the air inlet and diesel exhaust. Higher speed engines running at 1800 rev/min have a lower inertia than slower designs and although they would tend to react quicker to an external influence, they too are more responsive to the electronic controls.

Quick shut-down and secure arrangements

Due to the need to avoid enemy detection, it is necessary for an engine to be shut down promptly at any time. The design of the engine needs to be able to handle the thermal soak that follows such a shut-down. The lighter inertia allows for the shaft rotation to be quickly reduced and the turbo-charger also acts to decelerate the motion of the pistons.

Open atmosphere and submerged conditions

The submarine engine needs to be able to cope with a wide range of external conditions: chiefly the need to be able to operate safely and reliably both in open and snorting conditions.

Summary

Modern engine developments such as electronic control and advanced turbo-charging techniques have made it easier to consider the performance merits of modern diesel engines for submarine genset applications.

SAFETY

In the past the supply of a safe genset was a matter for the procurement authority who would impose its own quality system and means for acceptance. Today the classification societies are now able to offer a suitable means of third party assessment for submarine equipment. In particular Germanischer Lloyd have a full set of submarine standards which include the requirements for submarine diesel generator sets. Such a trend has made it easier to adapt a standard genset for submarine use.

Due to electronic controls, modern diesel engines are capable of producing a steady power supply over a wider range of ambient intake and exhaust conditions than was possible with older models.

The engines can be "dead ship started" in the event of complete loss of electrical power. Battery power to the on-engine ancillaries and especially the governor is a key design feature. Hydraulic or other actuator back-ups will also be required for the operation of valves such as those for the exhaust hull valves.

MODERN SUBMARINE ENGINES

The study has identified a modern engine which would appear to suit the necessary conditions for operation a submarine genset. A wide range of high-speed engines were reviewed and one of the best candidates was the Caterpillar C18 [Ref. 12]. The expected performance of four such engines was compared with the performance expected from two MTU 12V396SE84 gensets using information from Ref. 4.

Feature	MTU 12V396SE84	Caterpillar C18	Comments
Number onboard	2	4	
Unit Weight	9.62 tonnes	4.66 tonnes	
Unit Footprint	7m2	4m2	
Unit Volume	18.65 cubes	6.23 cubes	
Power density	53.61 kW/cube	88.25 kW/cube	63% benefit
Standard Sfc: surface (Open)	214 g/kWh	207 g/kWh	4.5% benefit
Standard Sfc: snorting	280 g/kWh [1]	223 g/kWh	3.5% benefit
Unit Rated Power - Open	1,000 kW	550kW	

[1] BMT estimate

Feature	MTU 12V396SE84	Caterpillar C18	Comments
Total installed power - Open	2000 kW	2,200kW	
Total installed power - Snort	2 x 900kW = 1800kW	4 x 420kW = 1680kW	7% penalty
Onboard Total Genset			
Weight	19 tonnes	19 tonnes	Same
Footprint	14m2	16m2	14% penalty
Volume	37 cubes	25 cubes	32% benefit
Fuel for 20 days snort-transit	157 cubes	140 cubes	4.5% penalty
Fuel & Gensets			
Total volume:	194 cubes	190 cubes	2% benefit
Total weight	153 tonnes	159 tonnes	4% penalty
Estimated Time Between Overhauls (TBO)	7,247 hours [1]	4,651 hours	Estimated
Estimated MTBF	14,494 hours [1]	9,301 hours	MTBF is assumed to be twice the TBO
Calculated probability of all gensets failing in service	1.06e-03	6.4e-6	165 times less likely with the Cat C18

Table 1 - Engine Comparisons

Table 1 indicates how the C18 may offer benefits in terms of volume (ie due to power density) and the probability of power supply with small penalties in terms of a reduced installed snort power

C18 Performance

BMT studies sought to identify the likely performance of the C18 when operated as a snort engine in an SSK environment (Figures 1 and 2). Without much of the detailed information necessary to portray an accurate prediction, the general trends were estimated to identify the impact on the rated power and the change to the Sfc at the top rated power for changes in inlet air supply and gas exhaust back-pressures. This engine performance data was then used to assess the impact on platform operations

PLATFORM

The Vidar-36 submarine design from BMT [Ref. 13] was considered as the platform for this assessment. BMT Defence Services developed the Vidar-36 through its Inspira ™ process and its serves as a virtual technology test-bed for many methods and design ideas. The baseline Vidar design is a conventional SSK which comprises the following machinery features

- A displacement of 3,600 tonnes to offer ocean-going performance;
- A brushless DC (active stator) propulsion motor rated at 5,300kW;
- Hydrogen-fuelled PEMFC for submerged endurance;
- Liquid oxygen for submerged fuel combustion;

- Two 1200kW diesel engines installation for snort power generation.

Range

The key requirement is believed to be range, not top speed: a good range will permit operations far from base-port, as well as extended duration in theatre once the submarine has arrived. The transit to the theatre will use snorting and batteries to avoid depleting the AIP fuels and oxidant. Such technologies have been subject to study and development for many years.

Self-noise and indiscretion on such occasions is still critical but to arrive in theatre in a realistic timescale, transit speeds of over 10 knots are therefore to be expected. A reasonable future transit requirement would be over 3200nm range at over 12 knots for a mission of 70 days.

Vidar Mission Details

Once in theatre the submarine travels at four to five knots. The hotel load is assumed to be 150kW with a snort electrical load of 260kW. The snort transit range requirement is for 3,400nm (ie 6,800nm there and back). The endurance when snorting is a key performance criterion for a platform which may have to travel long distances to theatre. The operation of the Vidar with four Caterpillar C18 gensets in snort operations for battery recharge between submerged transits is presented below.

Figure 3 shows the anticipated range and patrol length performance on Submerged Transit Speed (STS) (y-axis) and snort speed (x-axis) for such an arrangement.

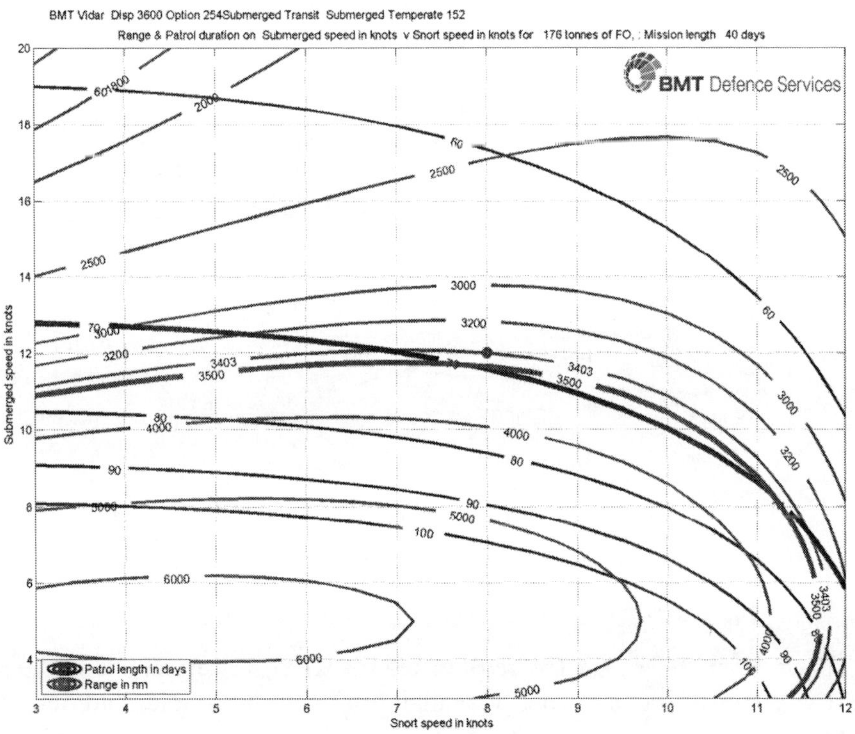

Figure 3: Range and Patrol Length Performance

Figure 3 shows how the provision of 176 tonnes fuel for a mission length of 40days permits a range of 3,400 nm to be achieved if a 12 knot submerged transit speed is combined with a snort speed of 8 knots (red circle). The total patrol length is 68 days (blue lines).

TRANSIT SPEED V MISSION DURATION

Requirements

To be able to operate at ranges far from the home port, and to stay on mission for the longest possible time is a standard question addressed by those in the operational analysis world [Ref. 14]. However a submarine build project requires such issues to be addressed in the early stages of design if the vessel is to be suitable for a range of operations and indeed if the appropriate number of vessels is to be built with the right set of properties as identified above.

The study has identified the set of patrols and mission durations and ranges for a range of submerged transit speeds. The snort speed has been fixed at 8 knots and snort periods in transit are not limited due to the lack of an active threat outside the mission theatre.

Range and Patrol Duration

For the total amount of available fuel burn, the range of the submarine can be deduced from the mission duration and the STS. The variation of range with these parameters is shown in Figure 4.

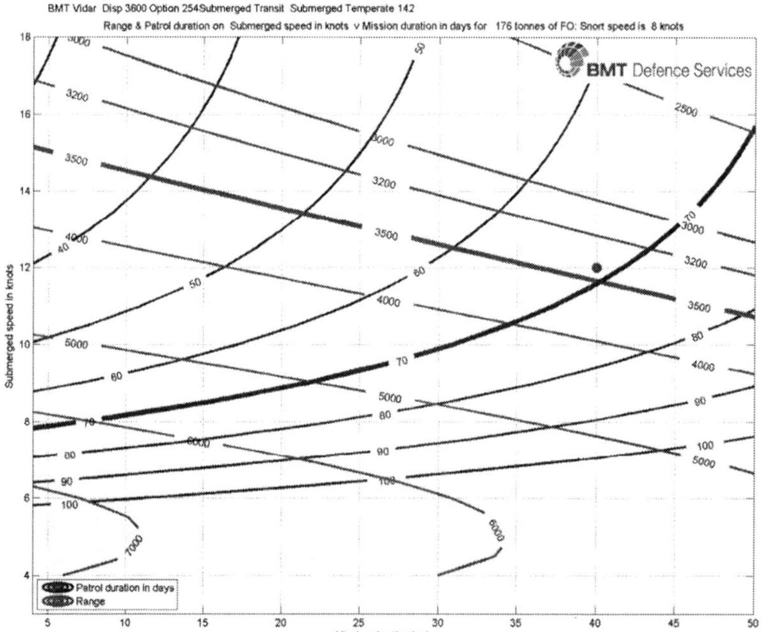

Figure 4: Range & Patrol Duration v Mission Duration

Figure 4 shows how the range (in red) increases with slower STS and shorter mission durations. The effect of reduced energy availability in the lead acid batteries is seen in the rapid falling away of range for the higher STS. Figure 4 also shows how the patrol lengths (in blue) decrease with increasing submerged transit speed.

Conclusions

The study has explored the scope for the use of more but smaller diesel engine generating sets to provide the power for submarine snorting and battery recharging.

The type of solution analysis presented here allows for an assessment of:

- The mix of engines and their sizes;
- The impact of the available fuel;
- The limitations of lead-acid batteries at higher submerged transit speeds;
- The mix of snort and submerged transit speeds in a patrol.
- The impact of mission profiles on maintenance period durations.

The information provided can allow decision makers to review the basis of operations and it can be used to form procurement and operational policy decisions. The right answer in terms of submarine design, fuel allowances and the number and mix of engines is to be derived from the operational and engineering analysis to realise a solution which meets national requirements.

SUMMARY

This paper has explored the scope for the use of smaller gensets in an SSK. The considerations for mission and patrol times as well as operational range have been reviewed and an example of the kind of analysis that could be employed has been demonstrated.

The BMT Vidar-36 submarine design has been used to demonstrate some of the impact on performance due to engine changes.

ACKNOWLDGEMENTS

The opinions expressed in this paper are those of the authors and not those of BMT Defence Services Limited, other companies, the UK MoD or any of its agencies. The authors wish to acknowledge the able assistance of Christopher Hodge, Ian Parker and Matthew Edgar in the engine modelling work associated with this study.

REFERENCES

1 J. Buckingham et al. (2008) "Submarine Power and Propulsion - Application of Technology to Deliver Customer Benefit" UDT 2008 Glasgow, Scotland.

2 Nobuo Kiriya. "Statistical Study on Reliability of Ship Equipment and Safety Management –Reliability Estimation for Failures on Main Engine System by Ship Reliability Database System" Translated from the Journal of MESJ vol. 35, No. (Manuscript Received Aug. 22, 2000)

3 P Yule et al "The Collins Class Submarine Story". 2008. Cambridge University Press.

4 MTU document . "Diesel engines in submarines" Feb-2006.

5 J-L Magnet. "New Concept for Submarine Diesel Engines"

6 MDI. "Hedemora Engines in Australian Type 471 Collins-Class Subs" Martime Defence International Vol 15, Issue 6, Pg 203-204. 1-Jun-90.

7 Cdr J P Bullard RN et al. "The Type 2400 'Upholder' Class submarine". JNE Volume. 28, No. 3.

8 Wrobel P G "Design of the Type 2400 Patrol Class Submarine". RINA. 1984.

9 Lt Cdr Sessions et al. "The mechanically supercharged Valenta diesel engine". JNE Volume 29. No 3.

10 B G Salter et al.(2008) "Facing up to Failure - the defence of graceful degradation" INEC 2008. Hamburg, Germany.

11 MTU Marine Sales Program. Edition 1/09.

12 "Caterpillar Marine Engine Selection Guide". Caterpillar C18. page 58. "Sep 2009.

13 S Binns (2008) "The High Capability Conventional Submarine Solution Space.". BMT Defence Services, RINA Warships 2008: Naval submarines. June 2008. Glasgow. Scotland.

14 H D Torresan and R A Jarvis. (2008) "Modelling of Conventional Submarine Power and Propulsion Systems for Future Capability Assessment".. DSTO. SIA 2008. Sydney 2008.

Liquid Oxygen Recharge System for Submarines

Vincent Grande, © ASC Pty Ltd, vincent.grande@asc.com.au
Tommy Andreasson, ASC Pty Ltd, tommy.andreasson@deepbluetech.com.au
Dr Carl Howard, The University of Adelaide, carl.howard@mecheng.adelaide.edu.au

ABSTRACT
A common feature in modern conventional submarines is the Air Independent Propulsion (AIP) system that can increase the submerged endurance from days to weeks, and hence remain on patrol significantly longer. All non-nuclear AIP systems available today employ Liquid Oxygen (LOX) instead of atmospheric air as an oxidizer in their energy converters.

The amount of LOX carried onboard a submarine in cryogenic tanks is limited by the available storage. For short range submarines returning to port from the patrol area to recharge the LOX storage tank may only take a couple of days, however for long range submarines it could take weeks, and would therefore reduce operational availability. The ability to recharge a LOX storage tank at sea would enable long range conventional submarines with AIP to remain deployed in the patrol area without the need to transit long distances to return to base. A submarine with a LOX Recharge System (RS) would only require a short transit to and from a safe zone where the submarine can snort and recharge LOX storage tanks.

This paper contains a review of current technology for the LOX production and the feasibility of implementing a LOX RS for future long range conventional submarines. A conceptual design has been developed addressing key requirements such as power, volume and safety.

BACKGROUND
Conventional submarines have diesel engines connected to generators to produce electrical energy for propulsion and charging the main batteries. The main batteries provide electrical energy when the submarine is submerged. The power required for submerged operation is limited by the finite capacity of the batteries. All conventional submarines regularly approach the surface to run the diesel engines and generators to recharge the batteries. Surface charging operations leave the submarine in a vulnerable position as it can be visually detected and generates relatively high noise levels compared to submerged operation.

A common feature of modern conventional submarines is the AIP system. The AIP system provides power for submarine propulsion and hotel loads when submerged, hence increasing submerged endurance. Current AIP systems typically increase the submerged endurance from a couple of days to a couple of weeks. [1]

An AIP system converts energy from a fuel source to electrical power without a supply of atmospheric air. The types of AIP systems that are currently installed on conventional submarines or are under development are:

- Polymer Electrolyte Membrane Fuel Cell, [2]
- Closed Cycle Diesel Generator (CCDG), [3]
- Stirling Engine, [4]
- MESMA Steam Turbine, [5]

All non-nuclear AIP systems available today employ Liquid Oxygen (LOX) instead of atmospheric air as an oxidizer in their energy converters. LOX storage offers the optimum solution on a mass and volumetric basis compared to other oxygen storage methods. [6]

The specific oxygen consumption of the AIP system and the submerged endurance defines the amount of LOX stored within the submarine. LOX stored in cryogenic tanks has a significant effect on the AIP arrangement within a submarine. The length of the AIP section within the submarine is predominantly determined by the size of the LOX tank.

The amount of LOX carried onboard a submarine in cryogenic tanks is limited by the available storage capacity. For short range submarines returning to port from the patrol area to recharge the LOX storage tank may only take a couple of days, however for long range submarines it could take weeks, and would therefore reduce operational availability.

The ability to recharge a LOX storage tank at sea would enable long range conventional submarines with AIP to remain deployed in the patrol area without the need to transit long distances to return to base. A submarine with a LOX RS would only require a short transit to and from a safe zone where the submarine can snort and recharge LOX storage tanks.

LOX RS CONCEPT OF OPERATION

The operating profile of a submarine depends on the specific mission scenario. Figure 1 shows an example of mission profiles for a submarine with AIP and a LOX RS.

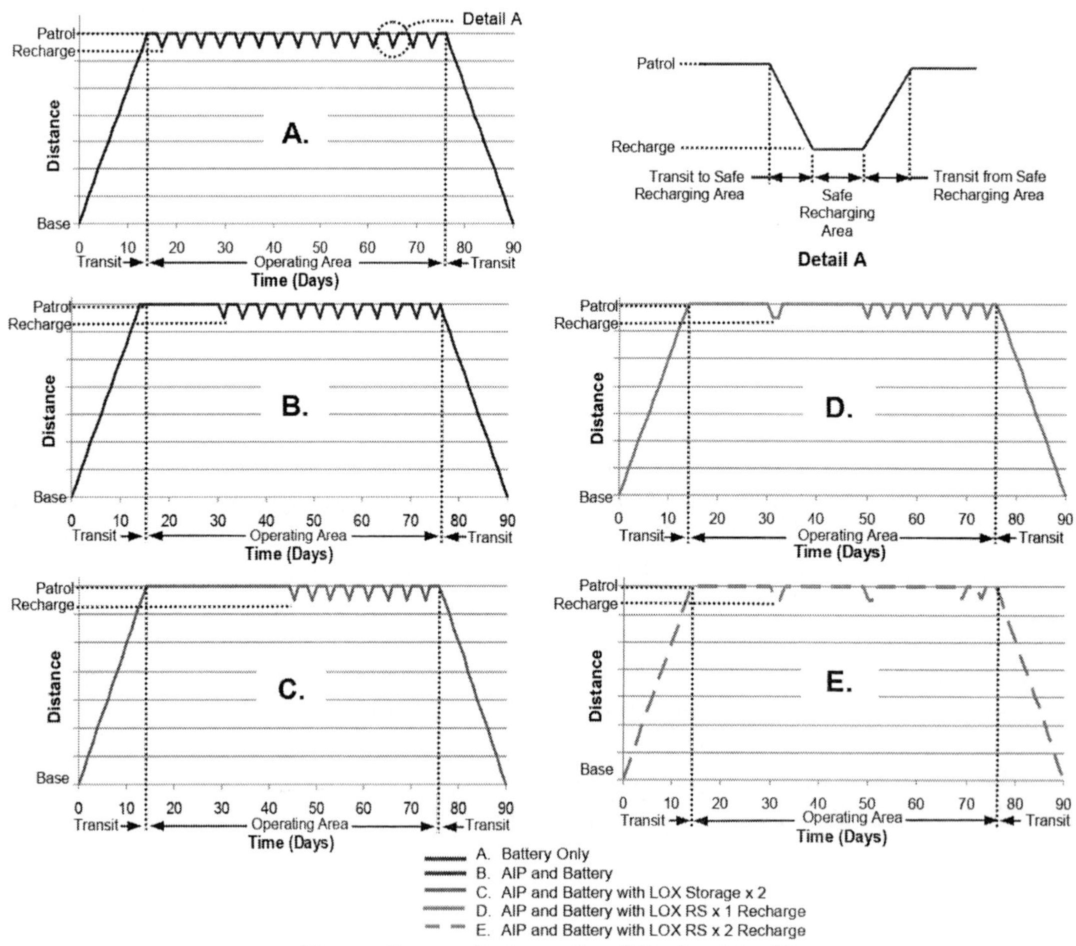

Figure 1: Submarine Mission Profiles

The operating range and the submerged endurance of a conventional submarine with AIP is a function of the quantity of LOX stored and the submarines speed of advance. Figure 2 shows the benefit of having an AIP and with a LOX RS. The ability to recharge the LOX storage tank enables the submarine to increase its submerged endurance. The number of times the LOX storage tank is recharged depends on the submarines mission and how much fuel and other stores that can be carried.

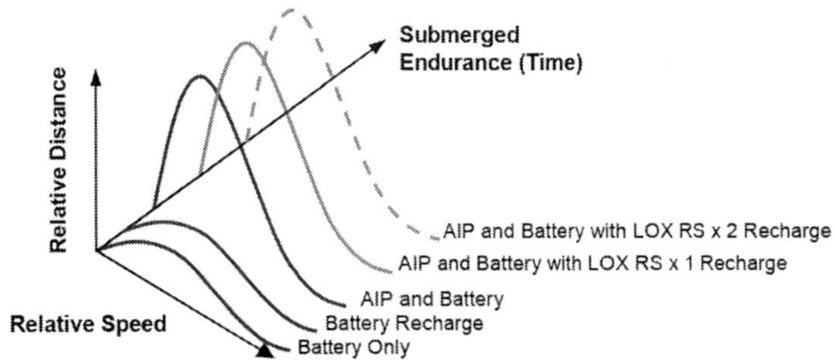

Figure 2: Benefit of LOX RS on Submerged Endurance

Table 1 shows the time on patrol and maximum submerged duration for a long range conventional submarine with varying configurations, for a 90 day mission and 28 day transit duration. The submarine with multiple LOX storage tank recharges from the onboard LOX RS system will spend more time on patrol in the operating area when compared to other conventional submarine configurations.

Table 1: Summary of Submarine Mission Profiles

Submarine Configuration	Time in Operating Area		Ratio of Time On Patrol to Time in Operating Area (%)	Maximum Submerged Duration
	Total Time On Patrol (Days)	Total Time Recharging (Days)		
Battery Only	47	15	76	3 Days x 16
AIP and Battery	50	12	81	17 Days x 1
AIP and Battery with LOX storage x 2	54	8	87	31 Days x 1
AIP and Battery with LOX RS x 1 Recharge	53	9	85	17 Days x 2
AIP and Battery with LOX RS x 2 Recharge	56	6	90	17 Days x 3

SYSTEM REQUIREMENTS

The following system requirements have been developed for the concept design of a LOX RS. The LOX RS shall:

- produce LOX at sea to recharge a LOX Storage tank.
- be effectively integrated within the submarine volume and shape.
- recharge the LOX storage tanks as quickly as possible (24-48 hours preferable).
- use installed power generation capacity from within the submarine.
- have low detectable signatures during operation.

The required amount of LOX to be stored onboard a submarine is governed by the submerged endurance, propulsive loads, hotel load and the specific power consumption of the AIP system. Table 2 shows an estimate of the quantity of LOX required for various AIP systems considered in this paper. Therefore the proposed LOX production rate of 55 MTPD (1600 m^3/hr) with a purity at least 95% was selected as a design goal for the concept design.

Table 2: Quantity of LOX Stored

AIP Technology	LOX Mass(kg)
Fuel Cell	40,000
CCDG	75,000
Stirling	95,000
MESMA	110,000

LIMITATIONS

In this study it is assumed that the maximum power available in the concept submarine is 4.2MW. The snorting propulsive load and the submarines' hotel load is assumed to be 300kW. Therefore the maximum available power for the LOX RS is 3.9MW.

The maximum size of the LOX RS components is limited by the hull diameter of the concept submarine, in this study a hull diameter of 8m was assumed. Modern submarine construction techniques will be utilised and the LOX RS will be assembled into a submarine section that will be integrated into a submarine. Components larger than the maximum hatch diameter will be suitable for the life of the submarine without replacement, typically 30 years.

AIR SEPARATION

A review of current and developing air separation technology identified both cryogenic and non-cryogenic processes to produce oxygen, as shown in Figure 3. Cryogenic processes exploit the differences in boiling points of the constituents of air, whereas non-cryogenic processes exploit the differences in molecular weights, molecular size and other properties for air separation.

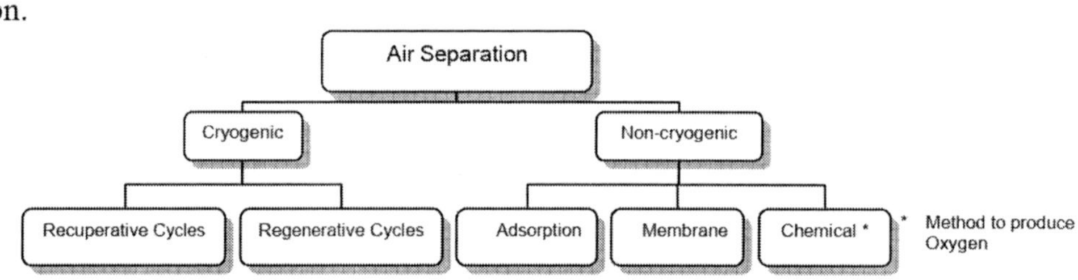

Figure 3: Types of Air Separation

Typically, cryogenic systems have the capability to produce large volume and high purity products in gaseous and/or liquid form. Non-cryogenic systems are used when lower volume and purity products in gaseous form are required. Table 3 contains a summary of LOX production systems.

Table 3: LOX Production System Summary

System Type	Product and State	LOX Production (m^3/hr)	Purity (%)	Power Consumption (kW)	SPC (kW-hr/ m^3)	Approximate Plant Size L x W x H (m)
Cryogenic	GOX, LOX, GAN, LIN	1770	90-99.6	1770	1.01	21 x 23.5 x 27.1
		2230	90-99.6	2140	0.96	21 x 23.5 x 27.1
Vacuum Swing Adsorption (VSA) with Liquefaction	GOX, LOX, GAN	1748	90-95	2218	1.27	15 x 9 x 7
		2331	90-95	2960	1.27	21 x 18 x 7
Membrane (ITM)	GOX, LOX[1], GAN	1457	25-95	NP	NP	NP
Chemical (LPE)	GOX, LOX[1]	6.37	100	100	15.7	1.52 x 1.22 x 2.03

Comment: 1 Only with integrated liquefaction system

LOX PRODUCTION SYSTEM SELECTION

LOX production systems have been compared against the key system attributes such as the quantity, physical state (liquid or gas), purity of products, power consumption and size. Each system has been rated against the system attributes as good, fair or poor and a numerical value of 5, 3 and 1 assigned to these ratings, respectively. The results are presented in Table 4 together with an overall rating of each system, obtained by summation of individual ratings. Each criterion is assumed to be of equal importance and therefore given equal weighting.

Table 4: LOX Production System Comparison

Criteria	Cryogenic	VSA with Liquefaction	Membrane	Chemical
Liquid Products	Good (5)	Fair (3)	Fair (3)[1]	Fair (3)[1]
Quantity	Good (5)	Good (5)	Fair (3)	Poor (1)
Purity	Good (5)	Fair (3)	Good (5)	Good (5)
Specific Power Consumption	Good (5)	Fair (3)	Poor (1)	Poor (1)
Size	Fair (3)	Good (5)	Poor (1)	Poor (1)
Rating	23	19	13	11

Comment: 1 Liquefaction system required

Based on the review of available LOX Production Systems, the cryogenic distillation system was identified as the optimum for development into a LOX RS concept design.

LOX RS CONCEPT

A concept of a LOX RS utilising a cryogenic system has been developed. The system is capable of producing 1770 m^3/hr and 99.6% purity with a specific power consumption of 1.01 kW-hr/m^3. Figure 4 shows the block diagram of the major components of the cryogenic LOX RS.

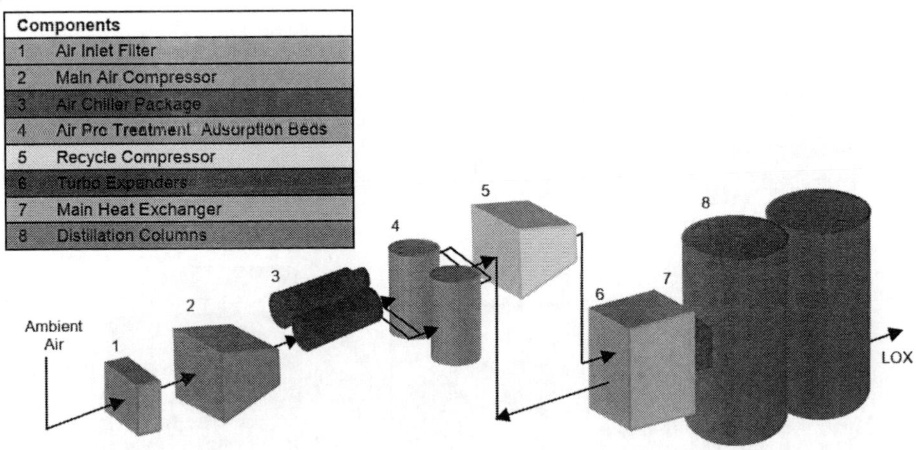

Figure 4: Cryogenic LOX RS Block Diagram

Submarine Integration

Integration of a cryogenic air separation plant on an offshore marine platform is not a new idea, with its origins more than 50 years ago, where major navies operated air separation plants on aircraft carriers [7, 8]. These plants, whilst small, had to address some of the challenges for implementing larger cryogenic air separation plant on an offshore marine vessel. Integration challenges applicable to the LOX RS on a submarine include:

- Generation of a compact design within the confines of the submarine.
- Quick start up.
- Overcoming the effects of marine motion.

Compact Design

To produce a compact design the LOX RS concept has been developed to be installed within its own submarine section. Figure 5 shows the LOX RS concept submarine interfacing systems.

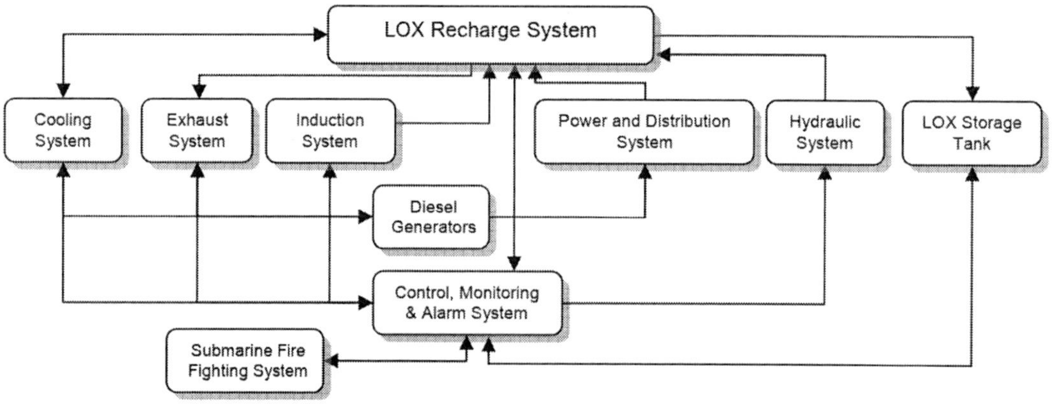

Figure 5: LOX RS Interfaces

The cryogenic equipment would be mounted to shock isolated platforms within the LOX RS section. The use of vertical stacking of components and split oxygen and nitrogen distillation columns produces a compact arrangement that minimises the LOS RS section length. Figure 6 shows a general arrangement sketch of Cryogenic LOX RS submarine section. The overall section length for this LOX RS concept is 5m.

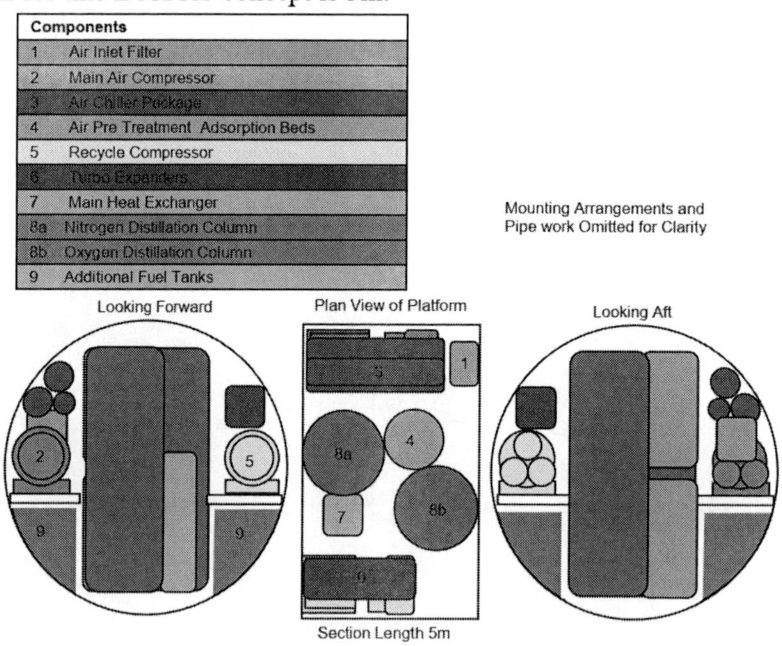

Figure 6: Cryogenic LOX RS Submarine Arrangement

Quick Start up

For commercial cryogenic air separation plants the ability to transition between operating conditions was of little importance because the operating conditions change infrequently. However, for a LOX RS, rapid start up is critical to minimise the operating time to recharge the LOX storage tanks. Research has shown that the start up time can be reduced significantly by reintroducing some of the stored liquid into the air separation system [7]. It is expected that the start up time can be reduced from hours to minutes using this method.

Overcoming the Effects of Marine Motion
Marine motion impacts the performance of the distillation column, reducing the quantity and purity of the liquid products. To overcome the effects of marine motion, the small units developed for operation on aircraft carriers limited the column diameter and use multiple, independent parallel bubble cap units for each separation tray. The use of structured packing as the contacting medium, combined with baffles provides a superior solution to counter the effects of marine motion [8]. Dynamic analysis of the distillation columns performance has shown the product LOX sump at the bottom of the oxygen column averages out purity fluctuations allowing greater tolerance for packing malfunction [8]. These studies show that a periodic loss in distillation performance (100-70% over 20 seconds) caused by marine motion would produce a gradual loss of product purity of 1% over 2 hours. A 1% drop is not a significant decrease in purity and can be corrected by a small increase in feed air flow to the process.

Further Integration
A cryogenic LOX RS produces liquid and gaseous nitrogen and gaseous oxygen as waste products. The gaseous oxygen can be returned to the system for liquefaction and the liquid and gaseous nitrogen waste products can be can be utilised for additional cooling system and an AIP purge system.

LOX RS Safety
A system safety program consistent with the safety case for the submarine will have to be conducted for the LOX RS if the project progresses to system design. Hazards associated with LOX onboard submarines are not new and have already been addressed on submarines with AIP. The primary hazards that need to be addressed with a cryogenic LOX RS are:

- Rapid oxidation
- Pressure excursions due to vaporising liquids
- Oxygen enriched or deficient atmosphere
- Extreme cold temperatures

LOX RS Concept Discussion
Cryogenic air separation was selected as the preferred LOX RS. The Cryogenic LOX RS has a high LOX production rate of 1770 Nm^3/hr with a purity up to 99.6% and a low specific power consumption of 1.01 kW-hr/m^3. Limitations on the layout and height of the ASU make lower purities desirable (95%). The power required to operate this system is within the total power available during snorting.

The length of the LOX RS section is 5m, in order to accommodate the distillation column. The success of the LOX RS is reliant upon the development of a customised split distillation column for submarine use. Techniques that have already been proven on aircraft carriers and other mobile platforms can be utilised to minimise the risk of development.

The mass of the LOX RS will depend on the final components selected, however it is expected that the overall mass of the section will be less than the extra buoyancy created by the submarine section.

Additional fuel for LOX RS operation and other consumables for extended submerged endurance would be required and needs to be accommodated with in the submarine design.

Selection Charts

A series of selection charts have been developed to demonstrate the benefits of the LOX RS, as shown in Figures 7-9. The selection charts can be used to determine the amount of LOX required for the concept submarine and the resultant submarine section length for various AIP systems, LOX storage tank configuration and the LOX RS operating duration.

The LOX RS for the concept submarine in this study has the potential to at least double the submarines submerged endurance, with a shorter submarine section length than the equivalent LOX storage tank section. The overall benefit of LOX RS to a conventional submarine with AIP depends on the submarines mission profile, submerged endurance, submarine propulsive and hotel loads, the type of AIP system and LOX storage tank configuration.

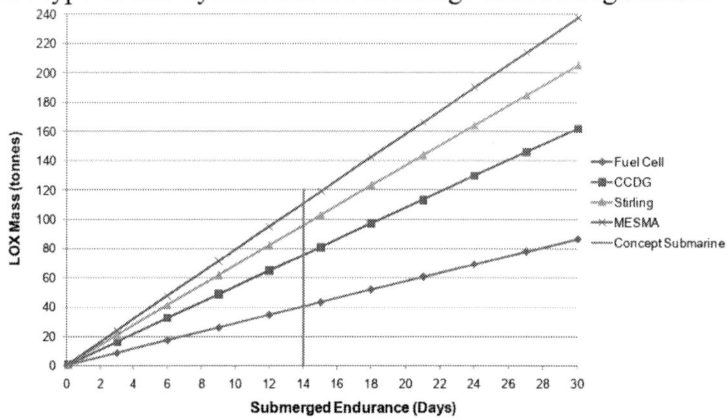

Figure 7: Selection Chart 1 - Submarine Endurance Versus LOX Mass

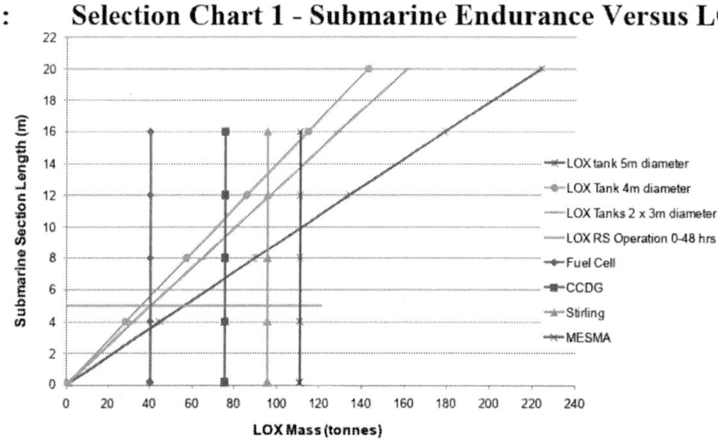

Figure 8: Selection Chart 2 - LOX Mass Versus Submarine Section Length

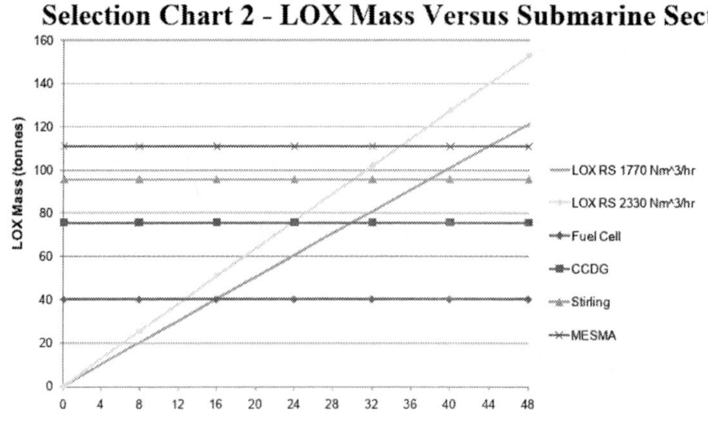

Figure 9: Selection Chart 3 - LOX RS Operating Duration

CONCLUSIONS

The literature review identified that there are cryogenic and non-cryogenic systems that can produce LOX. A comparison of the alternative systems and justification for selection of the preferred LOX production method has been undertaken based on the quantity, physical state (liquid or gas), purity of products, power consumption and physical size of components. Cryogenic and Vacuum Swing Adsorption with Liquefaction systems were identified as viable options to produce LOX and have been developed into LOX RS concepts.

The LOX RS conceptual design addressed key requirements such as power, volume and safety. The optimum system identified was the Cryogenic LOX RS that has a high LOX production rate of 1770 m^3/hr with a purity up to 99.6% and a low specific power consumption of 1.01 kW-hr/m^3. The power required to operate this systems is within the total power available during snorting.

This study determined that a 5m LOX RS section length is required in order to accommodate the distillation column. The development of a customised split distillation column for submarine use would be critical to the success of such a system. Techniques that have already been proven on aircraft carriers and other mobile platforms can be utilised to minimise the risk of development. There is also the ability to integrate the LOX RS with other submarine systems. The waste products, such as liquid and gaseous nitrogen, can be utilised for additional cooling and AIP system purge.

A series of selection charts have been developed to determine the amount of LOX required for the concept submarine in this study, the resultant submarine section length for various AIP systems, LOX storage tank configuration and the LOX RS operating duration.

The LOX RS has the potential to at least double the submerged endurance of the concept submarine with a shorter submarine section length than the equivalent LOX storage tank section. The overall benefit of LOX RS to a conventional submarine with AIP depends on the mission profile, submerged endurance, propulsive and hotel loads, the type of AIP system and LOX storage tank configuration.

REFERENCES

[1] Williams, D, Andreasson, T, 2008, The Australian AIP Dilemma, Deep Blue Tech Pty Ltd

[2] Siemens, 2008, PEM Fuel Cell for submarines

[3] Hammerschmiidt A, Krummrich S, 2007, Fuel Cell Propulsion for Submarines, UDT Europe.

[4] Kockums, 2008, Stirling AIP on board Swedish Navy Submarine data sheet

[5] DCNS, 2008, MESMA AIP modules for SSK

[6] Lakeman, JB, and Browing, DJ, 2003, The Role of Fuel Cells in the supply of Silent Power For Operations in Littoral Waters, Novel Vehicle Concepts and Emerging Vehicle Technologies, RTO AVT Symposium

[7] Miller, J, et al. 2008b, Improving Agility of Cryogenic Air Separation Plants, American Chemical Society.

[8] Goldstone, PG, Tijm, PJA and Miller D, 1998, The safe and Efficient provision of Oxygen for Remote, Offshore Conversion of Natural Gas, Air Products and Chemicals Inc. presented at the 'Remote and Stranded Gas Reserves Conference.

ACRONYMS

Acronym	Description
AIP	Air Independent Propulsion
CCDG	Closed Cycle Diesel Generator
GAN	Gaseous Nitrogen
GOX	Gaseous Oxygen
ITM	Ion Transfer Membrane
LIN	Liquid Nitrogen
LOX	Liquid Oxygen
LPE	Low Pressure Electrolysis
MESMA	Module d'Energie Sous-Marin Autonome
MTPD	Metric Tonne per day
NP	Not Published
PEM	Polymer Electrolyte Membrane or Proton Exchange Membrane
RS	Recharge System
SPC	Specific Power Consumption
VSA	Vacuum Swing Adsorption

Aerodynamic modeling for the design of multi spot helicopter landing operations

Yong-Jin Kim[1], Gordon MacDonald[2], & Tim Rafferty[3]
[1]BMT WBM, Melbourne, VIC, Australia, [2]BMT D & T, Melbourne, VIC, Australia,
[3]Defence Materiel Organization (DMO), Canberra, ACT, Australia
ykim@wbmpl.com.au

ABSTRACT

Availability of aerodynamic information is important for any aviation capable ship design, particularly for Multi Spot Helicopter Platforms (MSHP) to maximize the operating envelope and to ensure the safety of helicopter landing operations. This includes the mean wind speed distribution, turbulence, pressure, and temperature. The exhaust gas from the turbine exhaust stack and idling helicopters is also critical as it may influence helicopter landing and take-off operations and impact adversely on measurement sensors.

This paper presents aerodynamic modeling techniques for analysis of the interface between platform and flight operations using Computational Fluid Dynamics (CFD) for a MSHP in order to predict the aerodynamic environment around the ship. The unsteady RANSE model has been applied to predict the flow and thermal characteristics. To account for the effects of the exhaust gas from the exhaust stack and idling helicopters, as well as the rotor motion of the idling helicopters, a number of different scenarios were investigated throughout the study for a range of cross deck wind conditions.

By performing comprehensive aerodynamic simulations of the MSHP, it is possible to predict the aerodynamic environment by CFD for various operating and environmental conditions. Various methodologies have been developed to assist optimization of ship design and identification and mitigation of potential risks of the helicopter landing operations by characterizing the acquired aerodynamic information.

INTRODUCTION

Investigations of airflow around the flight deck of ships has been an area of interest for a number of years extending as far back as the 1940s [1]. This research continues today but focuses mainly on prediction and modeling of ship flow fields and complex air-wakes to allow safer launch and recovery of aircraft and helicopters to and from these ships. Shipboard launch and recovery is one of the most challenging environments that can be faced by a pilot [2, 3]. Often pilots must endure high winds, low visibility, including landing at night and unpredictable airflow patterns around the ship structure before landing on a moving deck that pitches and yaws along with the ship as it moves across the water [3].

Airflow hazards are very difficult to detect as they are generally invisible, so a pilot can easily stray into an area where hazardous wind conditions exist without knowing it and the situation can quickly progress to the stage where helicopter, crew and the ship may all be in danger. Also, pilots operating helicopters from ships must work in a confined area, just meters from the large superstructure of the ship or island [2, 3]. In these situations the margin for error is very small which only increases the risk to pilot, crew and shipboard personnel who are on the landing deck. As the helicopter operates so close to these large features, the effect of the flow induced by the helicopter itself on the airflow around the ship must also be taken into account [4], a topic which has only received a limited amount of investigation.

CFD is being increasingly used to model the airflow around ships and in particular, ship air-wakes to garner a greater understanding of the airflow conditions encountered by pilots attempting to land on these ships. As computational methods have increased over the past decade, models of increased accuracy have been developed and used to great effect in testing the dynamic interface between the helicopter and the ship [5]. CFD has already been used extensively to expose hazardous flow patterns around ships and helicopter landing platforms. This work has shown these areas to be dominated by large regions of separated flow, shedding vortices from edges on the ship as well as a large unstructured wake region behind the superstructure on flat decks of carrier type ships [4, 6, 7, 8, 9].

A number of turbulence models have been used in past CFD research including κ-ε, MILES, RNG methods, SST turbulence modeling and Detached Eddy Simulation (DES). Also, after comparing steady air-wakes and time-varying air-wakes it was found that a time-varying air-wake has a significant impact on the aircraft response, especially when the helicopter is in the hover position close to the deck [3] and therefore should be used when accurate analysis of ship air-wakes is required.

Some results obtained from past CFD analyses have contributed to the creation of flight simulators that can give a realistic representation of the air-wake and local conditions around a ship which when fully developed will be an invaluable tool for the training of pilots. It is also hoped that this technology can be used to develop Ship Helicopter Operating Limits (SHOL) which would reduce the time and resource cost required in long ship trials currently required to determine SHOL [2]. This ability to predict the flight envelope of a helicopter operating off a ship via the use of CFD would be a useful capability that would save time, money and reduce the risk exposed to pilots when determining the extremities of the safe flight envelope.

However, until this technology is fully developed the goal of any investigation using CFD can be only to create some sort of tool that combines this information so that it can be used by operators to ensure safer and more efficient use of the helicopter/ship combination and not to overload users with great amounts of data that has limited practical use. The objective of the present study is to create a realistic computational model including wind over deck (WOD) of varying directions, influence of induced flow due to the helicopter rotor and exhaust gases from both the helicopter and ship gas turbines to uncover the resultant flow patterns from these real world wind conditions.

COMPUTATIONAL MODEL

The MSHP considered in the study was 223m long, 32m wide, and 46m high to the turbine exhaust stack above the water level. Great geometric detail was considered in the CFD model including the major features of the two islands, the deck and hull shape and the detailed geometry of helicopters represented by a S-70B Seahawk often used in maritime operations. The CFD model was developed to represent the worst case for the helicopter operation and contains an operationally ready helicopter at each of the six landing spots and a further two helicopters on the flight deck. The CFD model geometry of the MSHP is illustrated in Figure 1. The CFD model was constructed to include main rotor blade motion as well as the exhaust gas from idling helicopters and turbine exhaust stacks. The computational domain extended sufficiently far away in each direction to avoid any boundary influence on the flow solution and consisted of 6.3 million tetrahedral elements initially and was converted to 2.18 million polyhedral elements for the simulations.

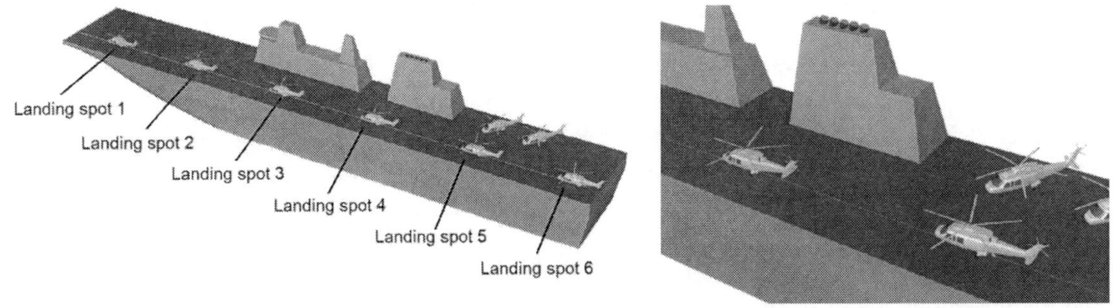

Figure 1. CFD Model geometry of MSHP (left) and zoom-in view of the helicopters and turbine exhaust stacks (right).

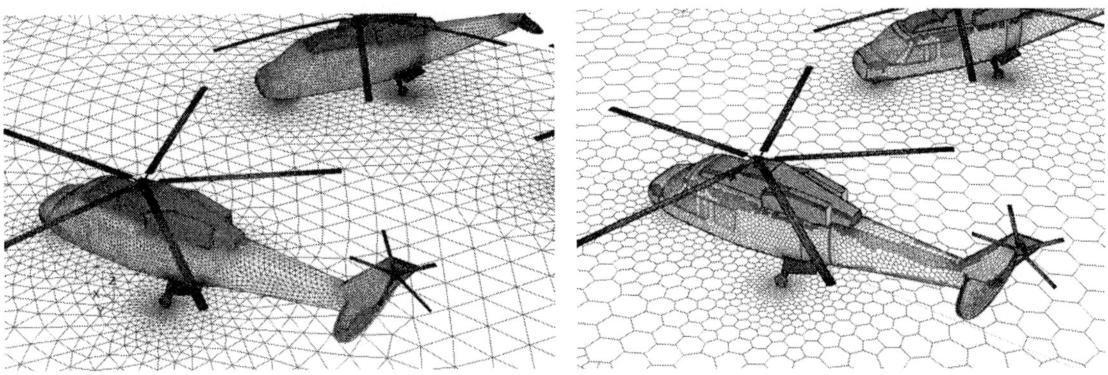

Figure 2. Tetrahedral element (left) and polyhedral element (right) on helicopter surfaces.

The commercial CFD package Fluent was used for this study which is one of the most widely used CFD software packages and provides state-of-the-art geometry creation, mesh generation technologies and a number of advanced turbulence models. In the present study the simulations were performed by solving the unsteady Reynolds-averaged Navier-Stokes (RANS) equations with the realizable two-equation k-ε model. Detached Eddy Simulation (DES) was run for WOD of 0° condition (wind over the bow) and the result was compared with RANS model. The helicopter main rotor blade motion was modeled by Multiple Reference Frame (MRF) technique in Fluent by using the detailed and exact main rotor blade geometries. The motion of stator was not included in the modeling.

MODELLING APPROACH

Base Model

The first part of modeling was aimed at investigating the flow characteristics over the deck of MSHP with helicopters in operation-ready position. The flow over the deck changes substantially as the wind angle changes, so wind from various directions are considered in the Base Model. In total eight wind directions (in 45° increments from the bow of the ship and moving towards starboard) are used for the wind boundary conditions and a uniform 5m/s wind speed was applied at the wind boundary. As the Base Model is aimed at providing fluid-thermodynamic information of the interface between the ship and helicopter operations which is valuable to the ship designers and helicopter operators, the Base Model was constructed not only with the ship but also with helicopters on the deck. In the CFD model one helicopter is positioned at each of the six landing spots and two more helicopters are included on the inactive area of the flight deck behind the ship superstructure. All the relevant details of the helicopters were included in the model but the rotors were kept static.

Model 2, MSHP Model considering idling helicopters and Turbine exhaust

This second part of modeling was aimed at identifying the flow characteristics over the deck including velocity, turbulence and temperature changes by the inclusion of the rotor blade motion, exhaust gas from helicopters and gas turbine. The same wind boundary conditions used in the Base Model were used for Model 2. The main rotor blade motion and exhaust gas ingress to the ship space from idling helicopters at the same six landing spots were included in Model 2 and the same two static helicopters remain in the inactive area of the flight deck. The main rotor idling speed of the modeled helicopters was set to 258 rpm which represents the rotor speed immediately before take off. The engine data used in the helicopter model is from the GE-T700 engine specification which is widely used in maritime and naval operations with two engines used per helicopter. The ship exhaust was modeled as exhaust from a GE LM2500 engine which is commonly used in naval operations and the ship was modeled as having two engines and the exhaust exited from three of the five stacks modeled. Data for both of these engines is shown below in Table 1.

	GE T700 (Helicopter Engine)	GE LM2500 (Ship Engine)
Power (kW)	2,820	24,050
Mass Flow Rate (Kg/s)	4.53	70.5
Exit Temperature (°C)	334	566
Number of Engines	2	2

Table 1: Helicopter and Ship Engine Data

The investigation of the mean velocity above the helicopter landing spots uses a velocity ratio (VR). The VR is a ratio of local wind velocity magnitude above the helicopter to the ambient wind speed. The flow characteristics do not change significantly when varying wind speed so VR is assumed as a good indicator of velocity over the deck flow characterization [6]. In the investigation of temperature and turbulence effects on helicopter operations, the assessment criteria specified in CAP437 was referenced to the study [11]. This civil aviation standard regulates that the temperature in the vicinity of the flight path and over the landing area should not exceed 2 °C above the ambient temperature (averaged over a three second time interval) for safe helicopter operations. The turbulence is defined as the standard deviation of the vertical airflow velocity above the helicopter and shall not exceed 1.75m/s. The temperature difference and standard deviation of vertical airflow velocity is included in the present study to assess the impact of exhaust gases and helicopter rotor blade motion on the flow characteristics over the deck.

These fluid thermodynamic result including local velocity ratios, temperature, and vertical velocity standard deviation is valuable information to identify and mitigate the helicopter operational risks in combining with the existing Ship Helicopter Operational Limits (SHOL) which defines the wind speed and environmental limitations for helicopters operating from MSHPs.

Unsteady RANS and DES Model Comparison

It is generally known that advanced turbulent modeling techniques including Large Eddy Simulation (LES) and Detached Eddy Simulation (DES) predict more accurate turbulence details particularly when investigating complex ship air wake flow field which involves flow separation, reattachment, and vortex shedding. However these turbulent models require significant computer resources and run time which makes CFD modeling prohibitive as part of a design assessment tool for aircraft carrier ship design and helicopter operational envelop studies. As the present study is intended to find and use the most practical but accurate turbulent model, DES and unsteady RANS turbulence models were considered for comparison. These two turbulence models were compared for the Base Model with WOD at 5 m/s and 0°.

The vorticity magnitude iso-surfaces at a value of 1.2 for two models are shown in Figure 3. DES predicted early vortex generation from the bow deck edge and vortex propagation into the landing deck well especially near the surface boundary. The vortices predicted by DES model are not much different in size and location compared with the result of the unsteady RANS model except for the near surface vortices predicted by DES model on the bow deck edge.

Figure 3 Iso surfaces of vorticity comparing two turbulence models (left: Unsteady RANS, right: DES)

The mean flow velocity distribution on the deck is crucial in determining the flow characteristics over the deck. Local VR for the six landing spots are compared for the two turbulent models and it is shown in Figure 4 that the trend of the velocity changes along the landing deck match well for both models. As the current study is targeted to set up the real world modeling approaches to be used for the ship design and helicopter operational risk assessment process, it was decided to apply unsteady RANS model throughout the study.

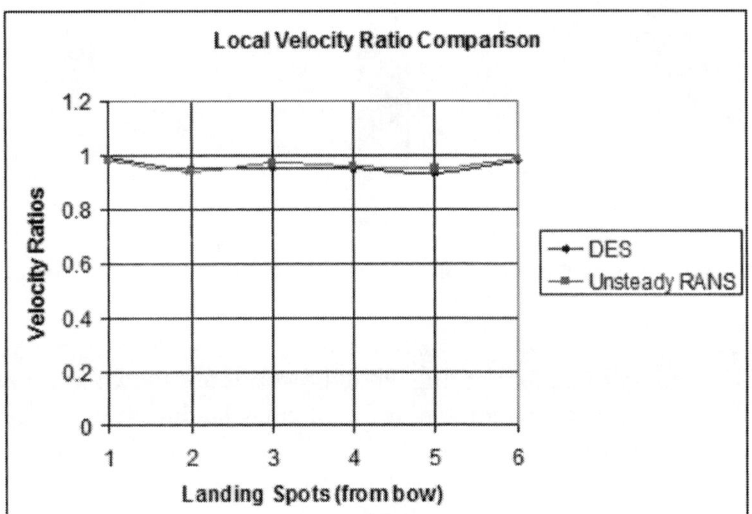

Figure 4 Local VR comparison between two turbulent models

RESULT

Base Model

Steady state simulations for the Base Model were carried out first and transient simulations were commenced using the steady state result as the initial condition. The transient result was processed at 400 seconds of the flow time after examining the consistency and representation of the flow structure throughout the flow time. The changes in the flow characteristics for different wind directions are demonstrated in Figure 5 by using vorticity magnitude iso-surfaces. It is obvious that the flow characteristics change significantly with changing wind directions.

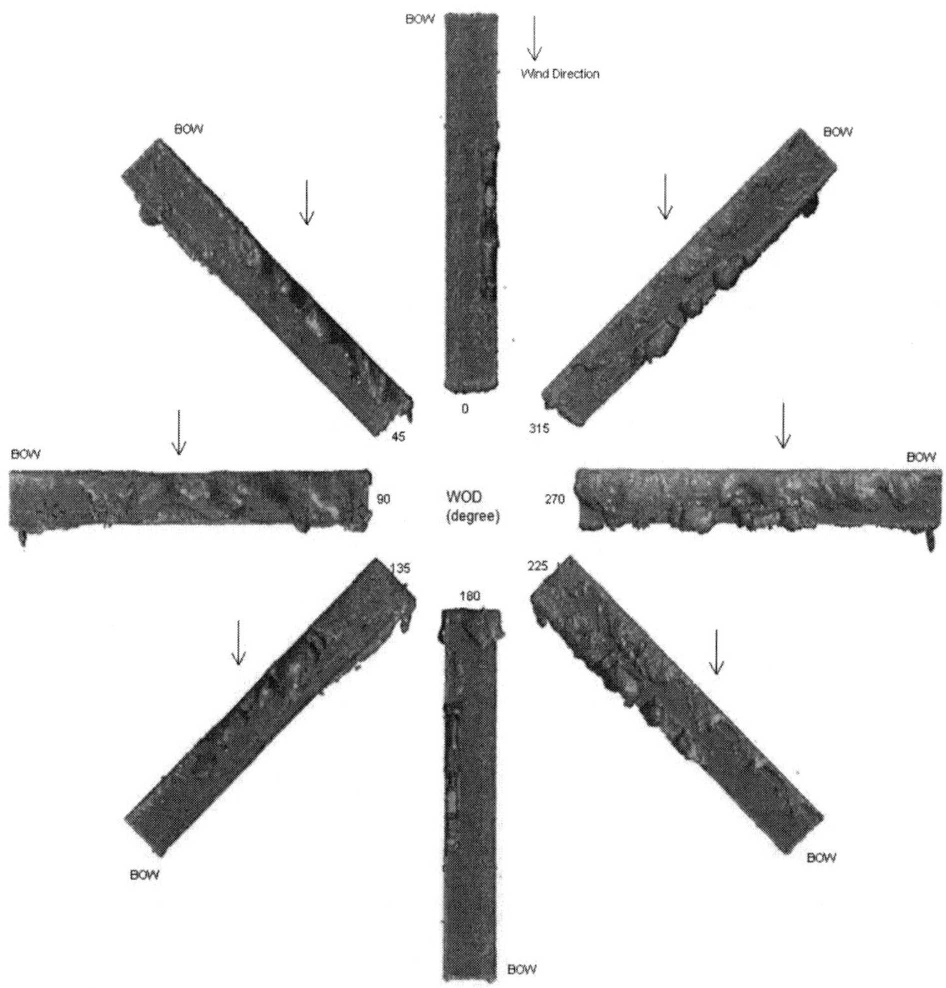

Figure 5 Iso-surfaces of vorticity for wind direction variations (plan view, wind from top of the figure, bow toward outside of figure)

The VR of the six landing spots for varying wind directions are shown in Figure 6. The results indicate that the mean velocity magnitude at a particular landing spot is dependent on wind direction and this value fluctuated for each landing spot. The mean velocity decreased substantially for side wind conditions (wind flow from port side and starboard side) as re-circulation zone is formed on the deck induced by the flow separation started at the port and starboard deck edge. The inboard landing spots, especially landing spot 3 and 4, experienced a lower mean velocity for the side cross wind conditions. A slightly higher velocity than the ambient wind speed is shown at landing spots 1 and 6 for skewed wind directions. This is due to local flow acceleration.

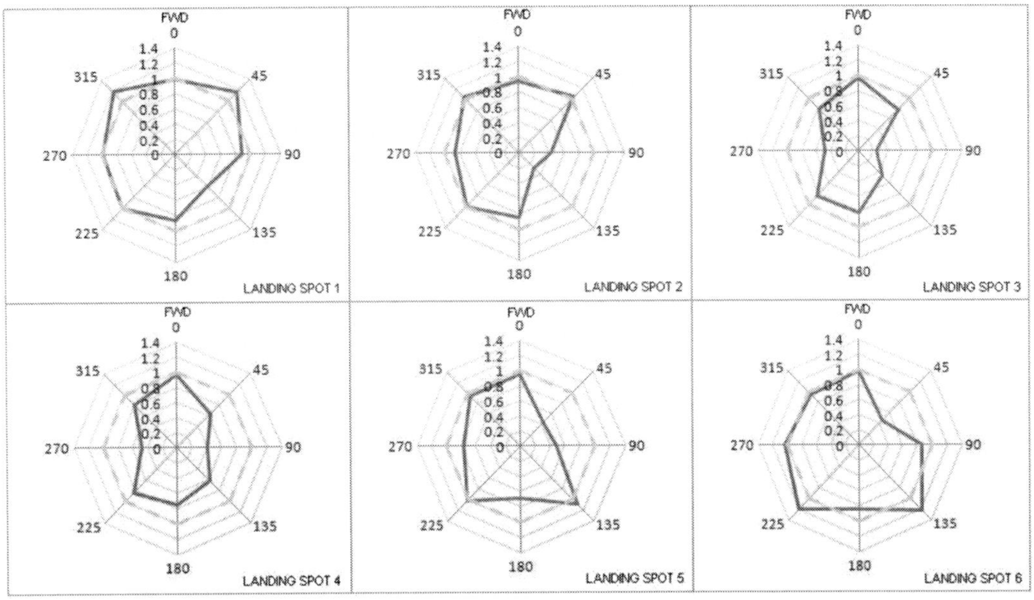

Figure 6 VR for six landing spots for varying wind directions (dark blue solid line: local VR and pale peach dotted line: VR of 1, scale on the outer diameter is WOD angle)

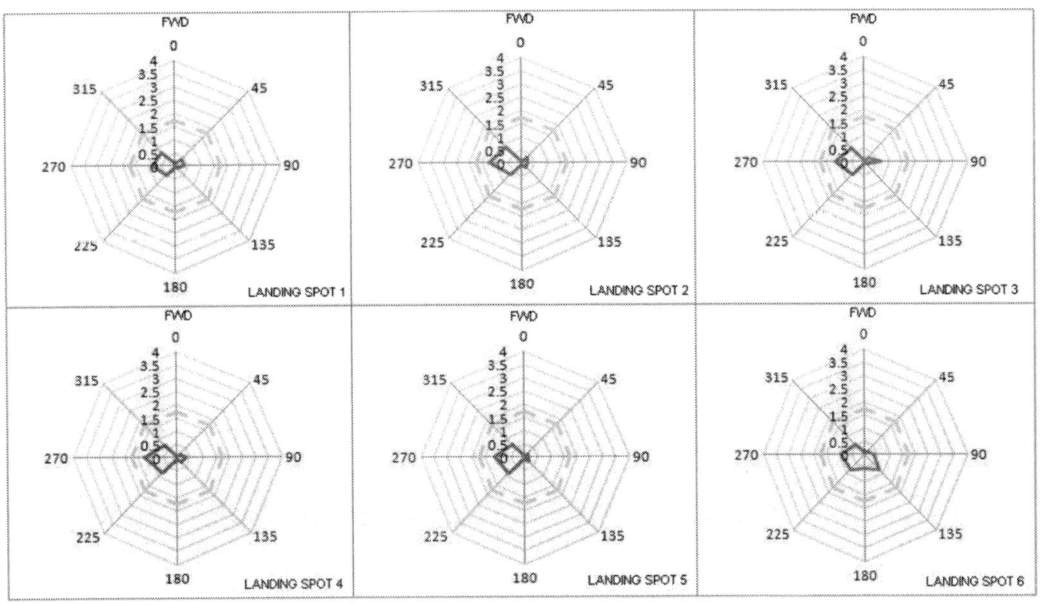

Figure 7 Vertical wind velocity standard deviation for six landing spots for varying wind directions (dark blue solid line: local vertical velocity standard deviation and pale peach dotted line: safety criteria in CAP 437, scale on the outer diameter is WOD angle)

The standard deviation of vertical velocity above the six landing spots for varying wind directions are shown in Figure 7. For all six landing spots, higher standard deviation of vertical velocity were found in the case of wind from port side (WOD of 215°, 270°, and 315°). As the ambient wind speed used is set at a moderate speed of 5m/s, the absolute value of vertical gust does not exceed the limit of the assessment criteria specified in CAP437. But if linear scaling is used, it can be seen that even a modest increase in wind speed could push this result beyond the allowable safety criteria.

Model 2, MSHP Model considering idling helicopters and Turbine exhaust

The flow characteristics over the deck for WOD of 0° are demonstrated in Figure 8. The result indicates that the wind is blown down by the downward momentum created by the helicopter rotor motion in the vicinity of the idling helicopters and exhaust gas from helicopter is dispersed downstream along with the wind over the deck. A large re-circulation created at the stern of the ship impacted the flow over the deck significantly. The landing spot 6 is engulfed in the large unstable re-circulation zone in addition to highly unstable flow and high temperatures. The turbine exhaust gas was dispersed vertically and horizontally for the given wind boundary condition and mixed with the exhaust gas from the helicopters at the rear of the ship.

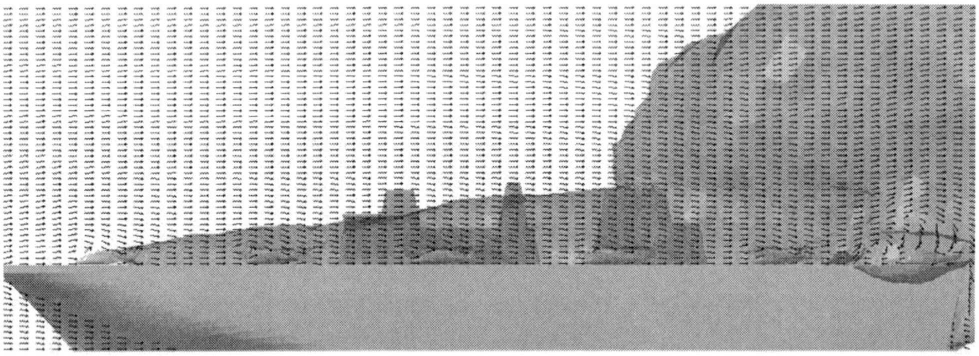

Figure 8 Velocity vectors at the horizontal cross section through the center of the landing spots (on uniform grids) overlaid with thermal plume (red is temperature iso surfaces exceed 2° C above the ambient temperature, WOD of 0° boundary condition)

The thermal plume dispersion from the helicopter exhaust and ship turbine exhaust stacks are shown for varying wind directions in Figure 9. It can be seen that the thermal plume traverses into the helicopter approach and departure path for all landing spots in all wind directions. As expected the thermal plume covers most of the ship space if wind blows from stern (WOD of 180°).

The local VR of the six landing spots for varying wind directions are shown in Figure 10. The result was compared with the Base Model to identify the mean flow changes over the deck. In most of the cases the mean velocity magnitude at the landing spots was increased by the addition of the rotor motion and is significantly increased for the landing spot 6 as the rotor motion enhanced the re-circulation at the trailing edge.

The standard deviation of vertical velocity above six landing spots for varying wind directions are shown in Figure 11. The amplitude of vertical gust was considerably increased from the Base Model result. It is understood that helicopter rotor blade motion and thermal plume outflow from the exhaust gas outlet combine to create higher vertical velocity instability over the deck. The result indicates that the landing spot 6 is unsafe to operate helicopters according to the CAP 437 limit.

The local temperature difference from the ambient temperature above the six landing spots for varying wind directions are shown in Figure 12. The CAP437 safety criteria are referenced in pale peach dotted lines. In most of the cases the temperature variations at the landing spots exceed the limit of the safety criteria specified in CAP437. As expected the temperature above the landing spot 6 was the highest as heat builds up inside the stagnant flow zone induced by the large re circulation created at the stern edge of the ship.

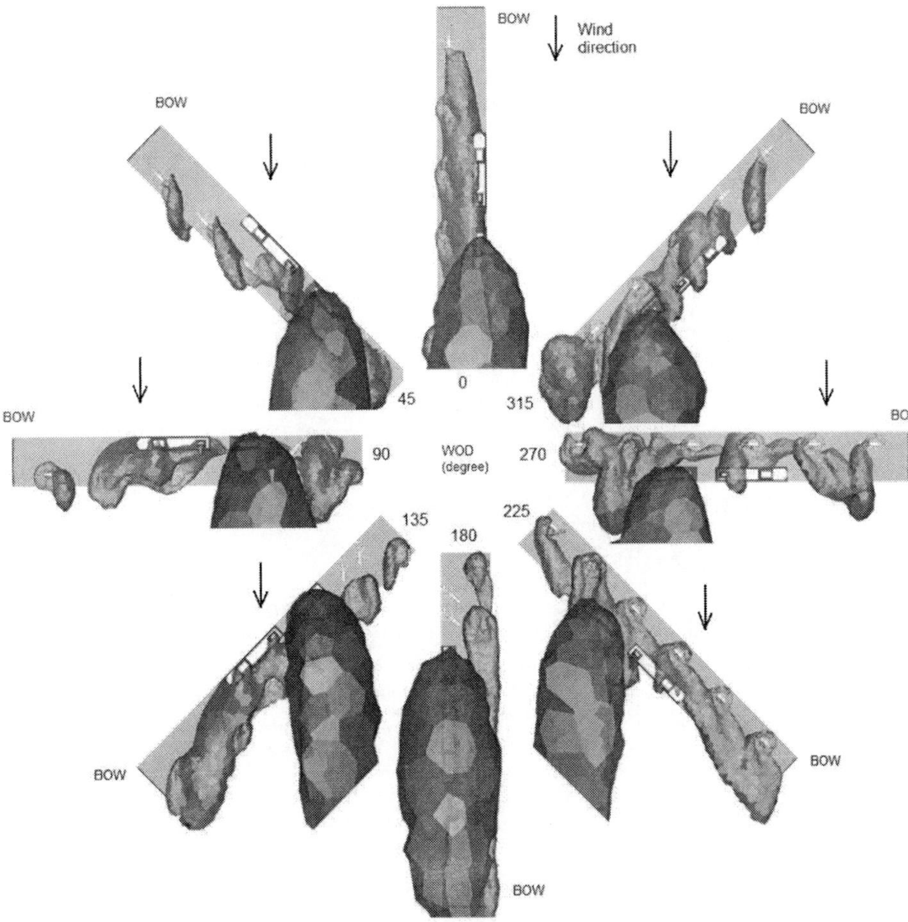

Figure 9 Thermal plume dispersion from exhaust gas outlets (temperature iso-surfaces exceed 2° C above the ambient temperature) induced by the exhaust gas flow

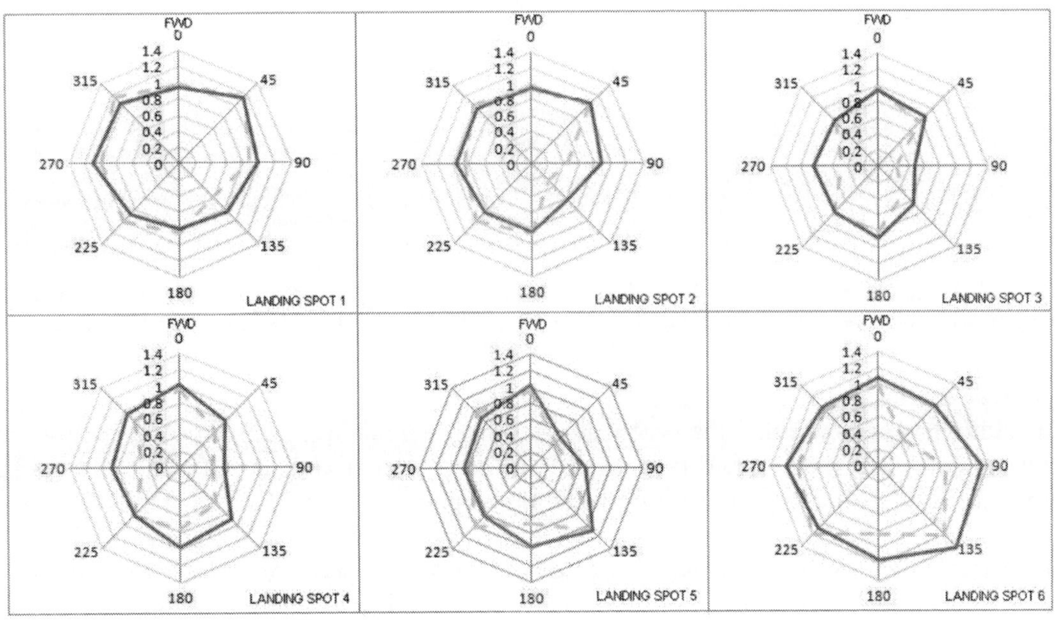

Figure 10 VR of six landing spots for varying wind directions (dark blue solid line: local VR and pale peach dotted line: Base Model result, scale on the outer diameter is WOD angle)

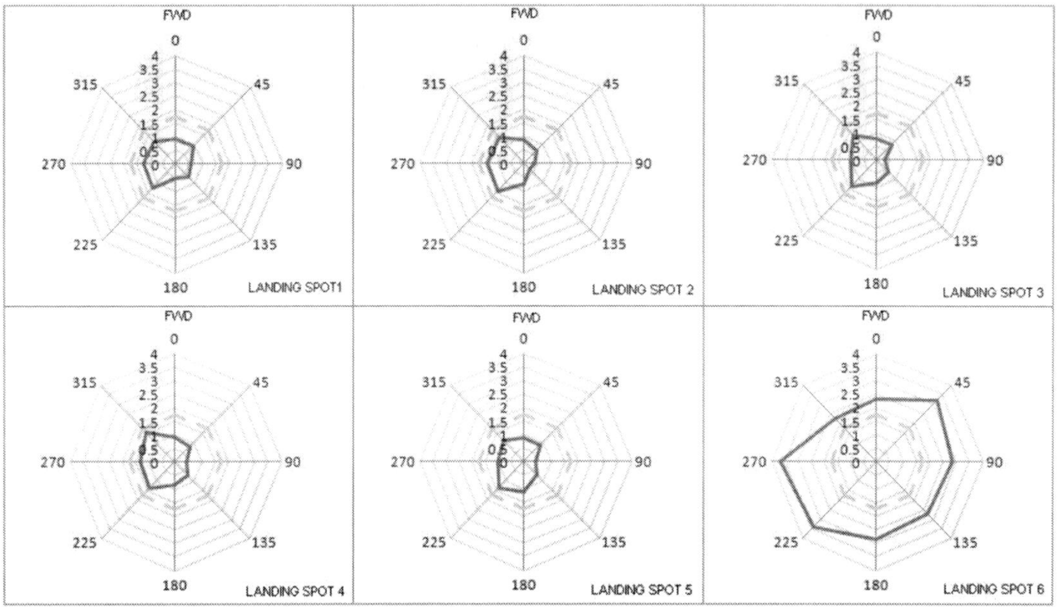

Figure 11 Vertical wind velocity standard deviation for six landing spots for varying wind directions (dark blue solid line: local Vertical velocity standard deviation and pale peach dotted line: safety criteria in CAP 437, scale on the outer diameter is WOD angle)

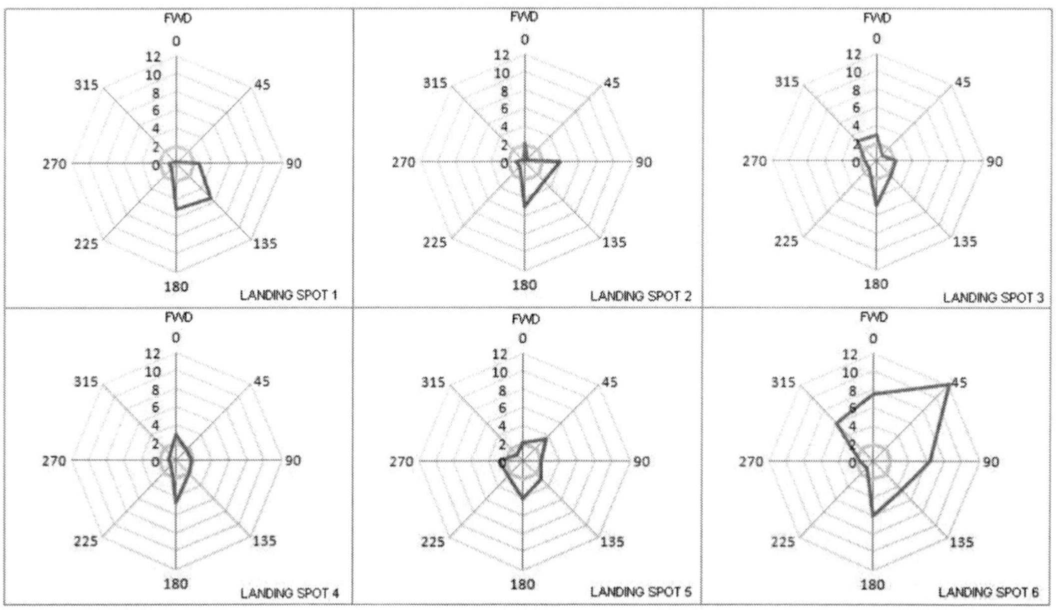

Figure 12 Temperature exceeds 2° above the ambient temperature for varying wind directions (dark blue solid line: local temperature different and pale peach dotted line: safety criteria in CAP 437, scale on the outer diameter is WOD angle)

CONCLUSION

This study has demonstrated the practical use of CFD modeling to predict the aerodynamic environment which is valuable to optimize the ship design and identify and mitigate the potential risk of the helicopter take off and landing operations.

The unsteady RANS turbulent model provides a reasonable prediction as the DES model on the flow characteristics which are intended to be investigated in the present study. It was found that the flow characteristics over the deck are changed significantly by the wind direction. Furthermore, it was found that the helicopter rotor motion and exhaust gases from helicopter and turbine stacks change the flow characteristics over the deck significantly. It was shown that the achieved aerodynamic information can be characterized to assess the risk of the helicopter operations induced by environmental conditions. It was identified that the risk of helicopter operations is significantly increased if the induced flow due to the helicopter rotor and exhaust gases from both the helicopter and ship gas turbines impact to the flow characteristics of the ship space.

CFD modeling and data characterization with realistic environmental boundary conditions has been shown to be a great design tool for detail ship design as well as risk assessment tool for helicopter operations. Future work will include testing of advanced turbulent models, varying wind speeds, and varying ship and helicopter designs as well as increasing the number of WOD directions.

REFERENCES

1. rmi, P Singh. S Seshadri. V, 2007, "Parametric Studies of Exhaust Smoke-Superstructure Interaction on a Naval Ship using CFD" Journal of Computers and Fluids, 36,794-816
2. Aragon C.R, Long K.R, 2005, "Airflow Hazard Visualization for Helicopter Pilots: Flight Simulation Study Results", American Helicopter Society 61st Annual Forum, Grapevine TX
3. Lee.D, Nilay.S, Horn. J, Long, L, 2003, "Simulation of Helicopter Shipboard Launch and Recovery with Time-Accurate Airwakes", American Helicopter Society 59th Annual Forum, Phoenix, Arizona
4. Tattersall. P Albone. C Soliman M, 1998, "Prediction of Ship Air Wakes Over Flight Decks Using CFD" RTO AVT Symposium on Fluid Dynamics Problems of Vehicles Operating near or in the Air-Sea Interface" Amsterdam, The Netherlands
5. Zan. S, 2005, "On Aerodynamic Modelling and Simulations of the Dynamic Interface" Journal of Aerospace Engineering, Volume 219
6. Polsky. S Brunder. C, 2001, "A Computational Study of Unsteady Ship Airwake", RTO AVT Symposium on Advanced Flow Management: Part A – Vorted Flows and High Angle of Attack for Militarty Vehicles", Loen, Norway
7. Reddy.K Roffoletto.R Jones.K, 1999, "Numerical Simulation of Ship Airwake", Journal of Computers and Fluids
8. Syms.G, 2007, "Simulation of Simplified-Frigate Airwakes using a Lattice-Boltzmann Method", Aerodynamics Laboratory, Instidtute for Aerospace Research, Nathional Research Council Canada, Ottawa, Canada
9. Maslov. L Valuev. N Zharinov. A, 1998, "The experience of aerodynamic disturbances research behind an aircraft carrier ship with elements for safe operation of ship based aircraft", RTO AVT Symposium on "Fluid Dynamics Problems of Vehicles Operating near or in the Air-Sea Interface" Amsterdam, The Netherlands
10. Civil Aviation Authority Afety Regulation Group, 2008, "CAP437, Offshore Helicopter Landing Areas – Guidance on Standards".

Using Potential Energy for Static Stability Analysis

Richard Dunworth
Department of Defence, Australia
richard.dunworth@defence.gov.au

The views expressed in this paper are those of the author and are not necessarily endorsed by the Department of Defence, Australia.

INTRODUCTION

This paper covers the development of a method of assessing static stability by considering the potential energy of the floating body system at different inclinations. The original intention was to provide a method of analysis for floating intact and damaged bodies with no well defined axis of stability. The concept was proposed over 150 years ago but application to a complete stability analysis has only become practicable with the advent of modern computing power.

The method has proved useful in analyses at large angle stability - as could most likely be encountered as a result of damage. In order to generate the data for such analyses, hull definition could no longer be made by a series of transverse sections and, as a suitable alternative, panel analysis was adopted.

BACKGROUND

In the mid 1990s, the author was involved in the analysis of the intact and damage stability of a number of Concrete Gravity Structures such as the "West Tuna" shown in Figure 1. These structures were found to have multiple axes of stability and a technique was required for omni-directional analysis. A surface of potential energy was developed over a range of heels and trims. Equilibria and paths of stability could be identified on the surface and a GZ curve produced for each path.

When preparing this paper, a literature search revealed that M. Guyou of the French Navy published a paper (circa. 1879)[1] in which he developed a 'surface of buoyancy' of a floating body with respect to its centre of gravity and went on to develop a number of techniques for explaining the behaviour of the body by analysis of this surface. Local maxima were identified as points of unstable equilibrium, while local minima were shown to be points of stable equilibrium. The vertical distance through which the centres of gravity and of buoyancy become separated by inclination was shown to represent the work done in inclining the body and thus be a measure of dynamic stability. Although the concept of GZ had been introduced over eighty years earlier, M. Guyou appears to have made no connection between his surface of buoyancy and GZ.

Figure 1. West Tuna CGS

The method of producing and analysing a stability surface by potential energy was developed by the author independently of, and without prior knowledge of, the work of M. Guyou.

POTENTIAL ENERGY OF A VESSEL

As a floating vessel inclines, the underwater body shape changes and the vertical shift of centre of buoyancy (B), measured normal to the horizontal, is proportional to the work done in depressing the displaced water. Similarly, the centre of gravity (G) of the solid body moves and the vertical change in height of G is proportional to the work done in raising the mass of the vessel. The change in separation of B and G is proportional to the change in potential energy of the system when the vessel is inclined from upright.

The analysis assumes that the vessel is static at each inclination: kinetic energy and inertia are not involved. It is assumed that the energy component of stability, equivalent to conventional still water stability analysis, is derived solely from potential energy.

RELATIONSHIP BETWEEN POTENTIAL ENERGY AND GZ

At Small Angles

The potential energy (PE) of the system at any free-floating state can be expressed as:

$$PE = WG \cdot \Delta \cdot g + WB \cdot \Delta \cdot g$$

where:

WG is the height of G above the waterplane;
WB is the height of B below the waterplane;
Δ is the displacement of the vessel;
g is the gravitational constant.

Therefore, at any inclination:

$$PE = \Delta \cdot g\,(WG + WB)$$

(WG + WB) is the vertical separation of G and B, so the force Δ g multiplied by the change in vertical separation of G and B is the amount of work done in changing the inclination of the vessel and represents the change in potential energy of the system.

For small angles of heel θ,

$$GZ = GM \cdot \sin\theta$$

The area under the GZ curve from 0 to θ is given by $\int_0^{\theta} GM \sin\theta$

$$= (-GM \cdot \cos\theta) - (-GM \cdot \cos 0)$$
$$= -GM \cdot \cos\theta + GM$$
$$= GM\,(1 - \cos\theta) \qquad (1.)$$

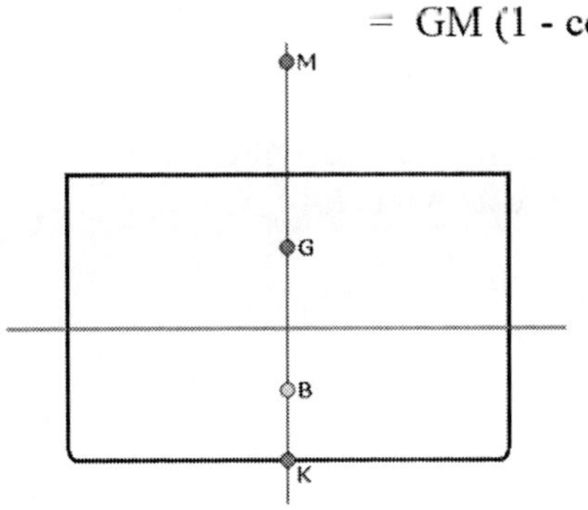

Figure 2. Upright hull

In the upright state, the vertical separation of G and B,

$$BG = KG - KB, \qquad (2.)$$

and

$$BM = KG + GM - KB \qquad (3.)$$

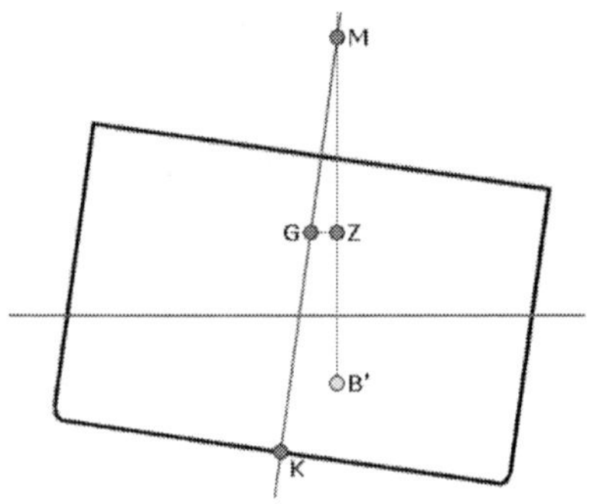

Figure 3. Inclined hull form

In the inclined state, the vertical separation of G and B' = B'Z

$$ZM = GM \cdot \cos\theta \qquad (4.)$$

$$B'Z = B'M - ZM$$

Substituting ZM from equation 4. gives

$$B'Z = B'M - GM \cdot \cos\theta \qquad (5.)$$

The change in separation of B and G

$$= B'Z - BG$$

Substituting B'Z from equation 5. and BG from equation 2. gives the change in separation:

$$= (B'M - GM \cdot \cos\theta) - (KG - KB) \qquad (6.)$$

For small angles of inclination BM may be considered to be constant, so

$$B'M = BM \qquad (7.)$$

Substituting B'M from equation 7. into equation 3. gives:

$$B'M = KG + GM - KB \qquad (8.)$$

Substituting B'M from equation 8. into equation 6. gives the change in separation:

$$= (KG + GM - KB - GM \cdot \cos\theta) - (KG - KB)$$

$$= KG + GM - KB - GM \cdot \cos\theta - KG + KB$$

$$= GM(1 - \cos\theta) \qquad (9.)$$

Equation 9 is the same as equation 1. Therefore, for small angles of heel, the change of vertical separation of B and G (change of potential energy) from upright to θ is equal to the area under the corresponding GZ curve from upright to θ.

Over Large Angles

The relationship between potential energy and GZ for small changes of inclination has been shown above. A series of (n) closely spaced points forming a path across a potential energy surface can be used to derive a corresponding series of (n – 1) GZ values:-

$$GZ_{0.5} = (PE_1 - PE_0) / (\theta_1 - \theta_0)$$

$$GZ_{1.5} = (PE_2 - PE_1) / (\theta_2 - \theta_1)$$

.

.

$$GZ_{n-0.5} = (PE_n - PE_{n-1}) / (\theta_n - \theta_{n-1}) \qquad (10.)$$

The area under the GZ curve can be found using the sum of trapezoids which, at the limit when the spacing between points approaches zero, will give the actual area:-

$$Area_{0..1} = GZ_{0.5} \cdot (\theta_1 - \theta_0)$$

$$Area_{1..2} = GZ_{1.5} \cdot (\theta_2 - \theta_1)$$

.

.

$$Area_{n-1...n} = GZ_{n-0.5} \cdot (\theta_n - \theta_{n-1})$$

$$\text{Total Area} = \text{Area}_{0..1} + \text{Area}_{1..2} + \,..\, + \text{Area}_{n-1..n}$$
$$= GZ_{0.5}.(\theta_1 - \theta_0) + GZ_{1.5}.(\theta_2 - \theta_1) + \,..\, + GZ_{1.5}.(\theta_2 - \theta_1)$$

Substituting GZs from equations 10 gives the total area:

$$= (PE_1 - PE_0) + (PE_2 - PE_1)\,) + \,..\, + (PE_n - PE_{n-1})$$
$$= PE_n - PE_0 \quad (11.)$$

Therefore the area under a GZ curve derived from any path between two points on a potential energy surface is equal to the difference between the potential energies at the end points.

DEPICTION OF THE POTENTIAL ENERGY SURFACE

Calculated values of potential energy (PE) are at discrete values of heels and trims. A full surface may be generated by plotting a grid of tiles in which the calculated values are at the centres of respective tiles.

If PE is calculated for all possible attitudes of a vessel from upright to inverted, by heeling and trimming, the typical PE surface plot for an intact hull will be as shown in Figure 4. Areas of low PE have been shown green and areas of high PE shown red. Upright equilibrium is at the depression in the centre and a valley runs athwartships. A second equilibrium exists at the edge of the valley where the vessel is fully inverted.

Figure 4. Orthogonal surface

In Figures 4, 5 and 6, trim by the bow is towards the top-left, and by the stern towards the bottom-right.

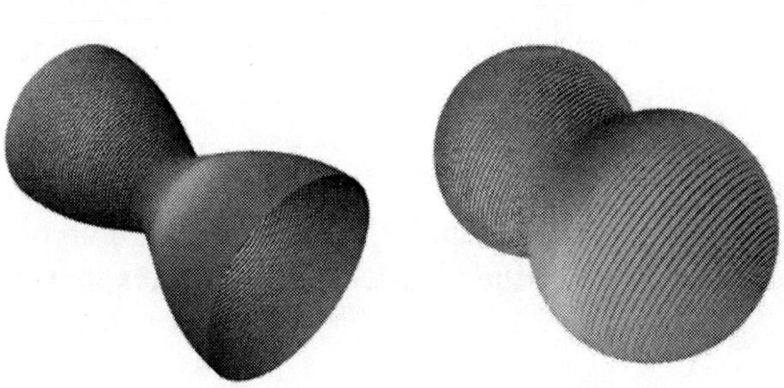

Figure 5. Cylindrical and closed PE surfaces

The edges at $+180^0$ heel and -180^0 heel must be exactly the same, and so it is possible to “roll” the surface into a cylindrical form with a seam at +/- 180^0 heel.

At 90^0 trim, heel is rotation about a vertical axis and the PE at all heels will be the same. It is possible to “pinch” each end of the cylinder into a common point and thereby generate a closed shape.

If the potential energy is described by colour gradient alone, the orthogonal surface will be flat and the corresponding closed shape will be a perfect sphere. However, if a topographical form is shown, the planar surface will exhibit peaks and troughs, while the closed shape will be a deformed sphere with, for conventional hull forms, a lobe at each end (+/- 90^0 pitch) and a neck running around the middle (approx. 0^0 pitch). The topographical surfaces will exhibit a dip at each point of stable equilibrium.

This method allows the effect of catastrophic events to be readily visualised and evaluated. Inundation of an open well causes a sudden decrease in stability with a potential simultaneous large change of trim. The vessel may even be left in a condition of longitudinal loll, with water in the well free to move from end to end with corresponding equilibria at significant trim angles. This will lead to a number of possible paths of stability, all of which will be evident on the potential energy surface.

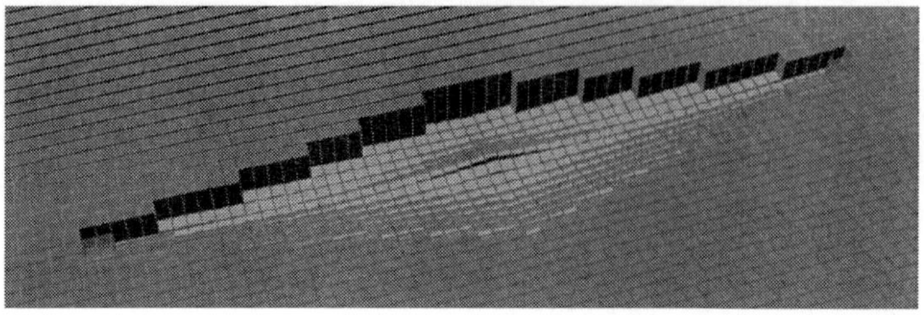

Figure 6. A landing craft with a significant amount of water free to slosh in the cargo well

In Figure 6, although full swamping has not yet occurred, a significant amount of water has been taken aboard and the free water in the cargo well causes an unstable equilibrium close to upright. The green surface applies prior to swamping, the blue surface after swamping and the boundary between the green and blue surfaces corresponds to the attitudes at which the deck edge becomes immersed. In the case shown, swamping via aft trim requires far less energy than via forward trim. After swamping, the green surface will no longer exist and the blue surface will extend over all.

EVALUATION OF THE POTENTIAL ENERGY SURFACE

A point of stable equilibrium is defined as a depression – that is any tile in which all surrounding tiles have greater potential energy. A path of stability is one in which the progressive transition across the surface as the hull inclines involves the least increase (or greatest decrease) in potential energy.

A typical intact ship potential energy surface will have a valley extending to port and starboard, with a low point near upright. Fore and aft, the surface will have steep sides, indicating high longitudinal stability. The path of stability will therefore follow the athwartships valley.

Two nearby depressions indicates a condition of loll; no depression on the upright plot indicates that the vessel is unstable, while no depression on the inverted plot indicates that the vessel is self-righting.

Finding Equilibria

Although the change in potential energy from tile to tile on the surface plot is usually quite small, a tile containing a stable equilibrium is easily identified, being entirely surrounded by higher tiles. The actual equilibrium may lie anywhere within that tile and can be determined by interpolation.

By using a bi-cubic spline fit, a local mathematical surface may be calculated and the equilibrium tile sub-divided by interpolation. Equilibrium is identified, as before, as being on the lowest sub-tile. This can be repeated until the required resolution is obtained and, since no new data has been generated, the original local mathematical surface is used at each repetition.

Finding Paths

The full set of paths can be developed by a number of methods.

One approach, starting at a point of equilibrium, follows three rules:

1. The path is formed by stepping one tile at a time, each step being either orthogonal or diagonal;
2. The centre of the next tile must be further from the starting point than the centre of the current tile;
3. The next tile must be the lowest possible tile which complies with rules 1 and 2.

The smoothness of such a path is dependent on the resolution of the surface grid. This can be overcome by using the method of local bi-cubic spline surface fitting - in a similar manner to that used for determining equilibrium noted above. A series of sub-divided tiles can be generated as the path moves across the surface to give a better resolution.

This will generate a single path from one equilibrium to the next. There may be a number of potential paths from each equilibrium, and the method can be used in each starting direction until all reasonable possibilities have been found.

Another approach is to generate the pattern of 'drainage lines' across the surface as in Figure 7. This is achieved by, assigning a score to each tile, being 1 plus the sum of scores of each adjacent tile which flow towards the current tile. The tiles with the highest scores will be those lying along natural drainage lines. Paths from equilibria are then determined by moving from tile to tile, at each step choosing to move to the adjacent tile with the highest score.

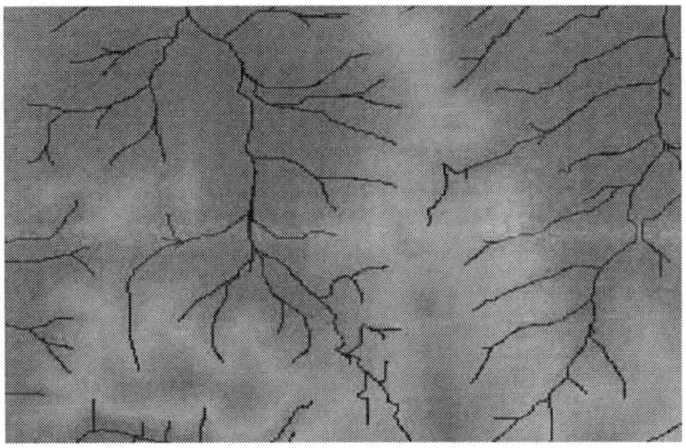

Figure 7. Geographic drainage map

Irrespective of the method chosen, a decision must be made as to which paths the hull is most likely to follow as it inclines. Conventionally, two directions are assumed – to port and starboard of equilibrium – and any method of selection would be expected to identify such paths for a normal-form intact hull.

Selection could be made on the basis of initial "steepness", on total "height" climbed along a path or, in the case of the drainage line method, on the score of the first tile in the path.

The values along the path represent the potential energy at each point and these values can be normalized by subtracting from them the potential energy at a chosen initial equilibrium. The resulting values then represent the energy to incline the vessel from the attitude at equilibrium to the attitude at each point along the path. As previously shown, each value of energy represents the area under a GZ curve from equilibrium to that point and the slope of the path at each point represents the local value of GZ.

Generating GZ Curves from Paths

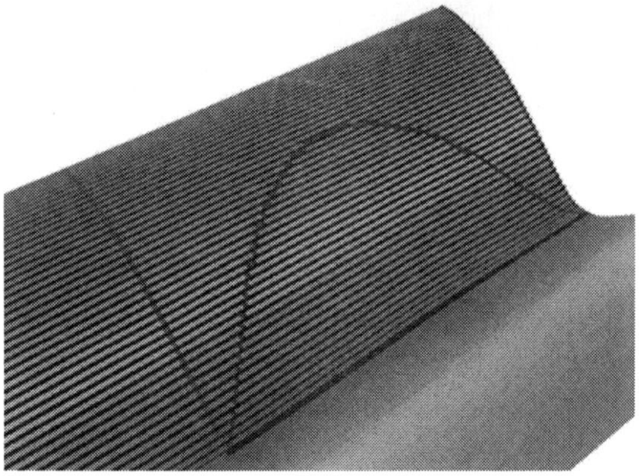

Figure 8. Paths across a PE surface

The potential energy surface shown in Figure 8 is for an intact patrol boat. The following paths have been generated across the surface:-

0°..180° heel; zero trim
0°..180° trim; zero heel
0°..180° heel; trim as a function of heel
0°..180° heel; trim proportional to heel

All the PE paths go from upright to completely inverted and, as would be expected, have very different GZ curves.

These are artificial paths, generated for simplicity of comparison. A realistic path will commence at one point of stable equilibrium and progress across the surface until it reaches another (which, in the case of a self-righting craft, will be the starting point).

GZ curves for each of the four paths are shown in Figure 9. The discontinuity in the Zigzag curve is because of the abrupt change of direction of the path at 90°. The plot of the pitch curve in Figure 8 appears to be discontinuous but this is not the case. At 90° trim all points have the same value, irrespective of heel, so it is permissible for paths to enter and leave the +/-90° trim edges without a discontinuity in the GZ curve.

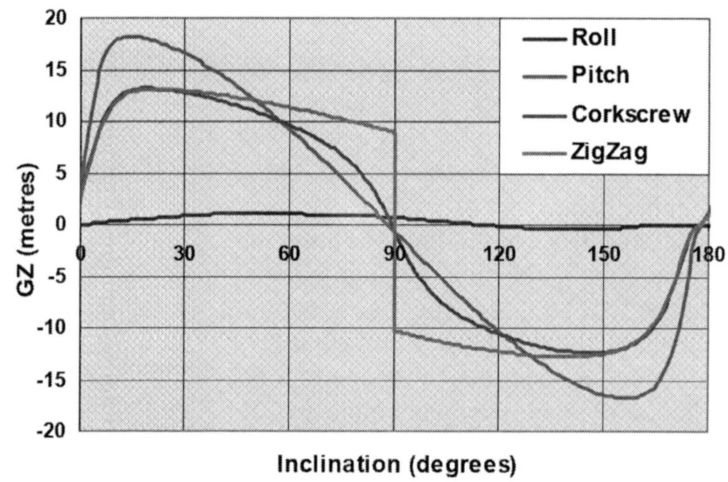

Figure 9. GZ curves for the paths in Figure 8

From	To	Roll	Pitch	C'screw	Zigzag
0	15	3.74	220.83	157.56	149.55
15	30	9.77	263.69	198.95	198.61
30	45	14.65	226.10	171.09	184.55
45	60	15.97	170.73	163.86	187.18
60	75	14.81	103.26	124.56	154.56
75	90	12.69	28.61	49.46	138.78
90	105	7.54	-48.02	-78.87	-167.76
105	120	1.23	-121.17	-140.20	-173.91
120	135	-4.14	-183.97	-178.96	-200.84
135	150	-6.40	-230.70	-174.47	-186.05
150	165	-2.78	-247.69	-174.69	-171.51
165	180	0.77	-113.82	-50.43	-45.32
Total Area		67.85	67.85	67.85	67.85

Table 1. Areas under the GZ curves in Figure 9

Table 1 shows the areas under the four GZ curves depicted in Figure 9. Although the values have differences of, in some cases, more than two orders of magnitude the total area under every curve is the same. This is in accordance with the relationship previously established at equation 11.

EXAMPLE 1 – EVALUATING TIME TO CAPSIZE

A study was undertaken to develop cargo limits before capsize occurred following damage[2].

Figure 10. Landing craft on the point of capsize

The damage was assumed to be a single hole, the size and location of which could be varied as required. For ease of analysis of the water flow through the hole, it was assumed that the hole was circular and sharp edged.

The analysis used added weight to simulate flooding. A detailed model was made of the damaged space to ensure that correct distribution of flood water was modelled at all stages of flooding.

Figure 11. Landing craft PE surfaces: intact, soon after damage and at the point of capsize

In Figures 11, trim by the bow is towards the top-left, and by the stern towards the bottom-right. All surfaces are to the same scale, showing how stability, particularly with trim by the stern, diminishes as flooding progresses. The first two surfaces have two equilibria: one near upright and one inverted. Once capsize has occurred, the upright equilibrium has disappeared and only the one with the craft inverted remains. It is clear that, by the time of capsize, the lobe representing trim by the stern is so diminished that there is no obvious path to inversion – capsize by pitching or by rolling are almost equally likely.

Investigation with a range of heavy loads showed that several different results were possible and that the outcome was very sensitive to the mode of damage (side or bottom damage) and to the location of the load.

Progressive flooding of small increments of water was analysed and equilibrium found at each increment. Concurrently, the flow rate through the damage hole was calculated and, by this means, the time interval between increments determined.

At the time of preparing this paper, a model test program by The Defence Science and Technology Organisation to validate the inflow rates and capsize predictions of the above study had just started.

EXAMPLE 2 – EVALUATING A FLOATING BODY

Using panel definition, a floating human body can easily be modelled and analysed by the potential energy method. This exercise was carried out to demonstrate the suitability of the method for evaluating lifesaving appliances, using typical floating body attitudes and centres of gravity and buoyancy[3]. There was a significant change in flotation between fresh and sea water and, in fresh water, there was a tendency for the body without a lifesaving appliance to sink altogether when exhaled.

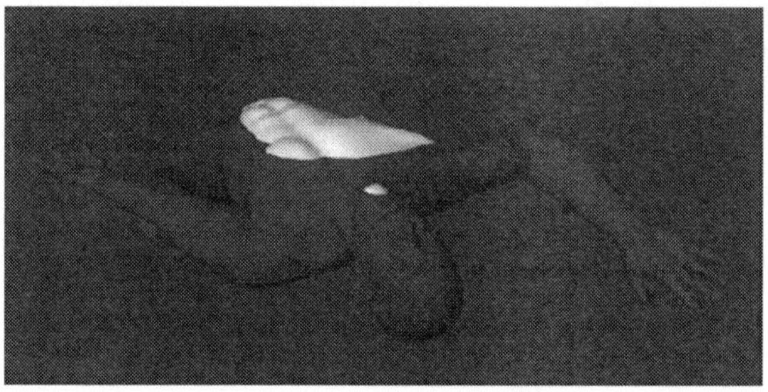

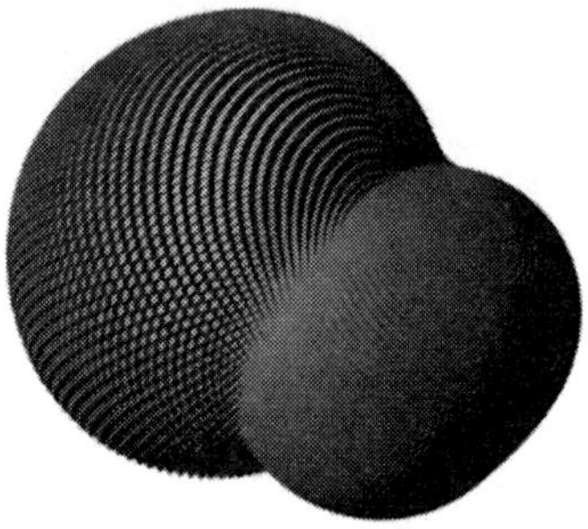

Figure 12. An inert floating body – and the corresponding potential energy surface

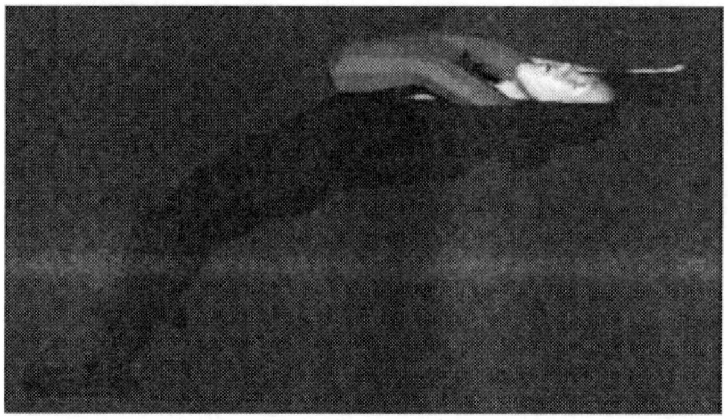

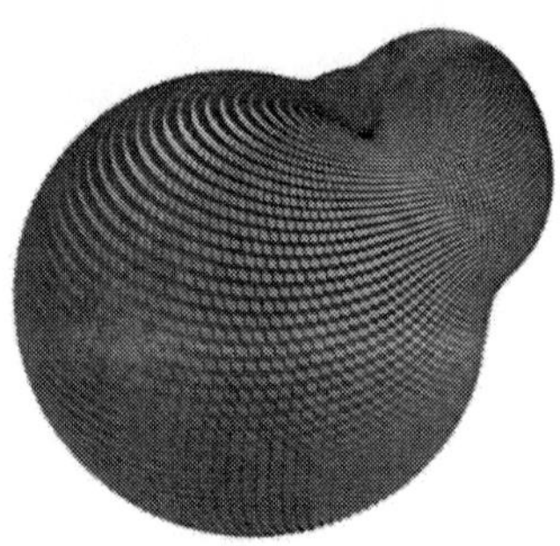

Figure 13. The same figure wearing a lifesaving appliance – and the corresponding surface

The PE surface for the inert body in Figure 12 shows a modest depression in the surface at equilibrium and could easily roll over; that in Figure 13, wearing the lifejacket, has a strong equilibrium and will tend to remain with the face clear of the water.

It is intended to continue this study - for example to investigate the effect of trapped air in immersion suits.

CONCLUSION

The volume of calculation required to generate a potential energy surface has precluded the use of this method as a practical tool until the introduction of modern computers. An analysis which gives a virtually instantaneous answer today would have taken up to 30 years by the calculation methods available to M. Guyou.

Forms of unusual shape or with extensive damage can exhibit abnormal or unexpected stability characteristics which may not be captured by conventional analysis.

Potential energy analysis will find all equilibria, the likely paths of inclination and the GZ curves associated with those paths for any floating form, damage case and loading condition.

REFERENCES

1. Guyou, M., (circa. 1879), *Théorie Nouvelle de la Stabilité de l'Equilibre des Corps Flottans*, Revue Maritime et Coloniale.

2. Hayes, P. & Dunworth, R., (2009), [redacted] *Capsize Study*, Report A017151, Directorate of Navy Platform Systems, Department of Defence, Canberra

3. Kreighbaum, E. & Barthels, K., (1990), *Biomechanics, A Qualitative Approach For Studying Human Movement*, Macmillan Publishing Company, New York, pages 466-474.

Naval Landing Craft Stability Research

Peter Hayes[1], Warren Smith[2], Martin Renilson[3] and Stuart Cannon[4]
[1]DNPS, Department of Defence, peter.hayes5@defence.gov.au
[2]UNSW@ADFA, w.smith@adfa.edu.au
[3]AMC, University of Tasmania, mrenilson@amc.edu.au
[4]DSTO, Department of Defence, stuart.cannon@dsto.defence.gov.au

The views expressed in this paper are those of the authors and not necessarily endorsed by the Department of Defence, Australia.

INTRODUCTION

Landing craft have been operated by various navies around the world for many decades to transport vehicles and personnel, often from ship to shore. Such vessels are usually size limited and the naval architect is typically restricted to producing a hull that resembles a flat bottomed, beamy, monohull vessel.

The RAN assesses the stability of landing craft using contemporary naval stability criteria, perhaps unnecessarily penalising their design and operation. Over the years, multiple landing craft stability issues have arisen, with varying degrees of comfort felt for the solutions developed (adaptations of the RAN stability criteria). The stability criteria that need to be applied to this type of craft are currently being investigated through various research programs by the Defence Science and Technology Organisation (DSTO), the Directorate of Navy Platform Systems (DNPS), the University of New South Wales college at the Australian Defence Force Academy (UNSW@ADFA) and the Australian Maritime College (AMC), with the outcomes now contributing to a collaborative research project between the UK and Australia.

This paper will firstly present a general review of the current Royal Australian Navy stability criteria that are applied to this type of craft and discuss their applicability. The paper will then outline a research program that aims to determine the effects of landing craft parameters on the capsize behaviour of these craft, presenting results to date. The ultimate aim is to derive stability criteria that are based on physical principles rather than quasi-static criteria.

CURRENT STABILITY CRITERIA

Existing stability criteria are based on the still water characteristics of the vessel, incorporating factors to account for operation in severe environments. Some, such as the basic IMO criteria, require nominated characteristics of the righting arm curve, including minimum areas under the GZ curve and minimum GM values. These were based on earlier work by Rahola (1939) who analysed the stability performance of a limited set of vessels that would typically operate near Finland. These types of criteria are strictly only applicable to the data set and the environments used in their derivation. Yet Rahola's work has been extended to many vessel types and sizes not envisaged at the time of his study, and to environments markedly different and more severe than those found around Finland.

Quasi-static criteria have been introduced in more recent decades that attempt to include the effects of wind and waves as overturning forces to be resisted. In these criteria, wave effects are usually introduced to the still water righting arm curve by a 'roll back' angle. Roll back in this discussion can be considered as roll about an equilibrium heel angle which, if non-zero, is usually considered to be caused by the wind. Most contemporary criteria that incorporate wind/wave considerations appear to assume that the vessel will be subjected to extreme waves for its size.

The various stability standards of the world, including from most navies, provide for different categories of service, usually differentiated by design wind speed or pressure and by range from refuge. The more demanding categories typically apply wind/wave criteria, with the only variable being the design wind speed. A common roll back would apply for a given vessel irrespective of service category. Noting that, in severe winds, waves quickly build up and therefore wave height can be considered a function of wind speed, there must be some threshold environment below which less roll back applies. The less demanding categories typically apply simplified criteria but it could be argued that wind/wave criteria, appropriately factored, should apply for all categories of service.

The wind/wave criteria do not address other forms of instability in waves such as may result from a broaching event. There is now considerable research being conducted to include more aspects of dynamic stability. The Cooperative Research Navies (CRN) project, of which Australia is a member, is developing the dynamic stability simulation tool, FREDYN. Use of this type of tool, in conjunction with the results of model and full scale trials as benchmarking tests, will, ultimate allow stability criteria to be derived that are based on physical principles.

Naval Stability Criteria

The naval stability criteria used by the RAN are contained within DEF(AUST)5000 Vol3 Pt2 "Stability of Surface Ships and Boats'. These criteria are not presented in this paper, suffice to note that they are essentially those presented by Sarchin and Goldberg (1962) and later codified by the US Navy in DDS 079-1 (1975), with additional criteria drawn from IMO conventions and other navies. Of most significance is that the naval intact wind/wave criteria are different to the current IMO equivalent criteria, especially with respect to wind effects and roll back angle, and usually the naval criteria are the more onerous. Also, naval damage requirements (extent of damage and required residual stability) are usually far more onerous than commercial damage requirements. This reflects the philosophical differences in operation between commercial and naval practice.

Application of the naval stability criteria to landing craft has highlighted a number of issues:

- The wind/wave criteria apples a roll back angle of 25 degrees. Most naval landing craft are relatively wide and shallow, meaning that the righting arm curve peaks at a low angle and has a limited range. Complying with the wind/wave criteria can be very difficult and may require imposing loading limitations. Noting that the vessels are typically required to operate in modest environments only, in which the 25 deg roll back is unlikely, perhaps there is scope for adjustment. This roll back question is discussed in more detail later.
- Landing craft typically have a large well area formed by the bulwarks and ramp(s). What are the effects if a large green sea were to be taken onboard in rough sea conditions? Could such a green sea event occur in the limited environments usually specified or can an environment be defined that avoids this type of green sea event?
- The usual arrangement for the air supply to the engine room is through baffled intakes situated on the inside of the bulwark. Should the vent sill be used to define the down flood point for intact stability analysis? This is usually the conservative approach. Alternatively, can the air intakes be treated as protected down flood points for the purposes of intact stability (i.e. the down flood point is taken as the top of a full length bulwark)? What is the liklihood of green seas causing progressive flooding through such openings and can

mitigating actions be taken (such as installing sills and float actuated valves)?

- Driving vehicles onto a landing craft will rarely result in a perfect load distribution, especially if the landing craft is still lightly grounded and floating cues are unavailable. How significant are list and trim errors on stability?
- Ramp/ramp transfers have highlighted longitudinal stability and trimming effects as an issue. Though there is likely ample longitudinal stability if the loaded end just submerges the vehicle deck during a transfer, the stability margin rapidly reduces with further submergence. Consequently, the current limit has simply been set at immersion of the margin line. This may be conservative if some temporary immersion of one end of the vehicle deck can be tolerated, especially in the benign conditions during such transfers.
- Longitudinal damage stability is generally assured by limiting sinkage to the margin line, typified by adherence to the floodable length curve. Recent IMO considerations apparently favour removing the floodable length in future standards. Further, some landing craft designers base the margin line on the bulwark deck (presumably, as the bulwarks have enclosed volume, the upper deck is considered the uppermost weather deck, even if narrow). Examination of the implications of this interpretation has revealed that, if the vessel is loaded enough and damaged in the biggest compartment (the ER) to put the (bulwark) margin line on the water, the residual longitudinal stability is considered insufficient to avoid a longitudinal capsize. If immersion of the vehicle deck were to be allowed following damage, what criteria should be applied?

There are any number of other issues yet to be examined including understanding stability in surf. Can surf limits, where the craft is able to withstand any heading, be determined? What are the surf limits where positive control is necessary and achievable?

RESEARCH PROGRAMS

There are a number of studies and research programs aimed at understanding aspects of stability of landing craft being progressed in Australia. These include:

- Landing Craft Damage Stability. A study was completed to develop cargo limits before capsize occurred following damage, Hayes and Dunworth (2009). Due to the size of the damaged space, longitudinal stability was as critical as transverse stability. The analysis was necessarily static in nature and made assumptions as to criteria to apply to the longitudinal case. DSTO has developed a model capsize test program based on the above study. These tests are designed to validate/determine inflow rates, capsize prediction in still water and the influence of waves on capsize. At the time of preparing this paper, the testing program had just started.
- Landing Craft Research. DSTO is orchestrating an extensive research program aimed at understanding the behaviour of landing craft. This includes seakeeping in both deep and shallow water and intact and damage stability. Various undergraduate theses and the research masters described below all contribute to this program. Reports to date include those by Denehy (2009), Rolls (2008) and Rolls et al (2009).
- Landing Craft Stability Research Masters. In 2008, Peter Hayes commenced a research masters into the stability of landing craft, with the co-authors to this paper as supervisors. The main aim of this research masters is to develop practical stability criteria for landing craft with a focus on naval operations. The research will encompass a literature survey, development of any promising concepts, with model testing and dynamic analysis using FREDYN to both develop data and test concepts. To date literature research has

commenced, concepts identified for examination and preliminary model testing conducted. The first experiment series was conducted in February 2009 at the Australian Maritime College (AMC) in Launceston, investigating the effects of freeboard and vertical centre of gravity. A report (Hayes, forthcoming) of this experiment series is in preparation. The remainder of this paper presents material related to this research masters.

Intact Stability Experimental Program

Previous experience at the AMC suggested models less than 1m long could be capsized. Landing craft data for available existing vessels was analysed to determine the range of various ratios that the study series should encompass and from this a basic model representative of a contemporary RAN landing craft was developed. To comply with the above constraints, basic dimensions were set at $Length_{BP}$ = 810mm, Beam = 250mm, with a bow cut-up 162mm long and 45mm high. There was no stern cut-up or any deadrise included to simplify the construction. Maximum draught would be 41mm, giving a loaded displacement of around 7.5 kg. The scale to the intended parent vessel is approximately 1:26. A removable bulwark, 40mm high and 25mm thick, was included.

A series of at least three models was required so that trends could be developed. Freeboard significantly affects stability, especially for barge type vessels, so it was decided that this model series would investigate the effects of freeboard. Hull depths of 50mm, 70mm and 90mm, giving freeboards from very small (9mm) to more than the same as the draught (49mm), were selected. These freeboards are considered to span to either extreme of realistic landing craft. The models are shown in Figure 1.

A stability analysis, utilising typical naval and commercial criteria, was conducted to determine suitable loading conditions (displacement and KG) for the test program. The loading conditions developed are presented in Table 1. The KGs range from easy compliance with contemporary stability criteria through to complete non-compliance. Only the heaviest displacement and lowest KG was systematically tested. A few runs for one model at the middle KG (but at a different radius of gyration) were also conducted.

Table 1: Load conditions matrix.

Parameter		Full Size	Model
Description	Symbol		
Scale Ratio		1	26
Length BP		21.06 m	810 mm
Beam		6.5 m	250 mm
Heavy Displacement (Series 3)			
Draught	T_3	1.092 m	42.000 mm
Displacement	D_3	133.700 t	7.607 kg
KM	KM_3	4.120 m	158.462 mm
Low KG			
KG	KG_{31}	1.976 m	76.000 mm
GM	GM_{31}	2.144 m	82.462 mm
Radius Gyration in Roll	K_{31}	2.600 m	100.000 mm
Natural Roll Period	P_{31}	3.300 sec	0.647 sec
Middle KG			
KG	KG_{32}	2.808 m	108.000 mm
GM	GM_{32}	1.312 m	50.462 mm
Radius Gyration in Roll	K_{32}	2.600 m	100.000 mm
Natural Roll Period	P_{32}	4.500 sec	0.883 sec
High KG			
KG	KG_{33}	3.640 m	140.000 mm
GM	GM_{33}	0.480 m	18.462 mm
Radius Gyration in Roll	K_{33}	2.600 m	100.000 mm
Natural Roll Period	P_{33}	18.000 sec	3.530 sec

Figure 1: Models ready for testing

All three models were partially internally ballasted (the KG was below deck level of the deepest model) and fitted with external weights, some on an adjustable frame, to obtain the required displacement trim, list, KG and radius of gyration. A transverse radius of gyration of around 38% beam was desired as being a typical value. A radius of gyration of 40% beam was achieved.

Figure 2: Models ready for testing

The model testing was conducted in the towing tank at AMC. This tank is 100m long, 3.5m wide and up to 1.5m deep. The wave maker is a bottom hinged, single flap type. Water depth for the model tests was 1.5m. The model was located in the middle of the tank beneath a slave carriage, as shown in Figure 2. It was restrained to remain beam onto the waves using lines across the tank.

Roll was measured by vertical transducers attached with very light line to the outboard ends of the adjusting frame. The slave carriage was moved by hand prior to recording to maintain the vertical alignment of the sensor lines. The motion of the line was observed on several occasions and always observed to be taut, even when roll/heave was most violent.

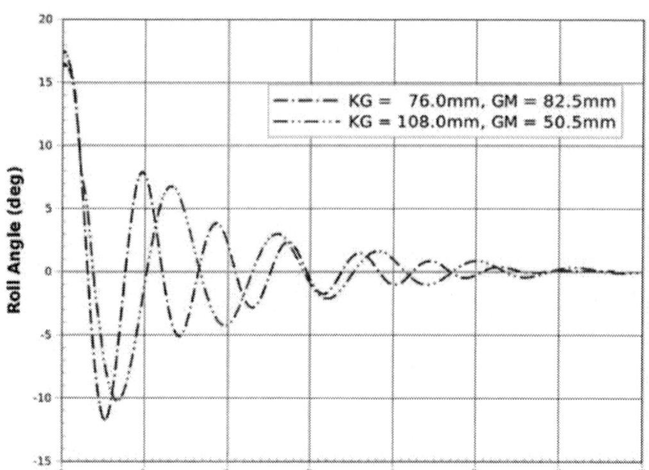

Figure 3: Roll decay tests

Since all models were ballasted to the same displacement, KG and radius of gyration, only the deepest model was subjected to a roll decay test. At a KG = 76mm, the natural roll frequency was 1.14Hz with a period of 0.88 seconds. Very late in the test program, the KG of this model was raised to 108mm and the roll decay test repeated. The natural roll was now a frequency of 0.83Hz and a period of 1.21 seconds. Plots of the roll decay tests are presented in Figure 3.

A range of wavelengths, from near breaking to relatively long, were selected. Table 2 shows the waves used, together with the corresponding waves at the nominal full scale of 1:26. It will be appreciated that only the 100mm and 150mm waves contained periods matching the model. At 200mm, such a period was impossible as steepness was past the breaking limit.

Table 2: Wave parameters

Model						Ship					
Wave Height (mm)	λ/H	Wave Length (mm)	Period (sec)	Freq (Hz)	ω (rad/s)	Wave Height (m)	λ/H	Wave Length (m)	Period (sec)	Freq (Hz)	ω (rad/s)
50	8	400	0.506	1.975	12.411	1.3	8	31.2	4.471	0.224	1.405
	12	600	0.620	1.613	10.134		12	46.8	5.476	0.183	1.147
	16	800	0.716	1.397	8.776		16	62.4	6.323	0.158	0.994
100	8	800	0.716	1.397	8.776	2.6	8	20.8	3.651	0.274	1.721
	12	1200	0.877	1.140	7.165		12	31.2	4.471	0.224	1.405
	16	1600	1.013	0.988	6.205		16	41.6	5.163	0.194	1.217
	20	2000	1.132	0.883	5.550		20	52	5.772	0.173	1.089
150	8	1200	0.877	1.140	7.165	3.9	8	31.2	4.471	0.224	1.405
	12	1800	1.074	0.931	5.851		12	46.8	5.476	0.183	1.147
	16	2400	1.240	0.806	5.067		16	62.4	6.323	0.158	0.994
200	8	1600	1.013	0.988	6.205	5.2	8	41.6	5.163	0.194	1.217
	12	2400	1.240	0.806	5.067		12	62.4	6.323	0.158	0.994
	16	3200	1.432	0.698	4.388		16	83.2	7.301	0.137	0.861

All models experienced significant water across the deck when tested in the steeper waves, as shown in Figure 4. This reduced to no water on deck usually by λ/H = 16 for the higher freeboard models. The lowest freeboard model (50mm depth, no bulwark, 9mm freeboard) experienced water across the deck with most waves and was the only model to capsize – into the wave. When water freely broke over a bulwark into a well, the level stabilized at about a third full and, if not free to drain, caused about 5-6 deg of loll at completion of the run. The GM in this condition was about negative 20mm.

Figure 4: Model 09-10 (90mm depth) being tested

Limited analysis of results has been performed to date. Relationships used in determining the results presented below were:

$$\omega = \sqrt{\left[\frac{2\pi g}{\lambda}\tanh\left(\frac{2\pi d}{\lambda}\right)\right]}$$

$$x'_4 = x_4\left(\frac{H'}{H}\right)$$

$$\bar{x}_4 = \frac{x'_4}{\alpha}$$

$$\alpha = \tan^{-1}\left(\frac{\pi H'}{\lambda}\right)$$

$$\bar{\omega} = \frac{\omega}{\omega_0}$$

where:

d = water depth (m)

g = gravity acceleration = 9.806ms^{-2}

H = wave height as measured (m)

H' = nominal wave height (m)

α = wave slope (rad)

λ = wave length (m)

ω = wave frequency (rad/s)

ω_0 = damped natural roll period of model (rad/s)

$\bar{\omega}$ = non-dimensional wave frequency

x_4 = roll amplitude as measured (rad)

x'_4 = normalised roll amplitude (rad)

$\bar{x}_4$ = non-dimensional roll

All results presented were normalised. Normalising factored the measured roll amplitude by the ratio of measured wave height to required or nominal wave height. The difference was typically less than ±5%.

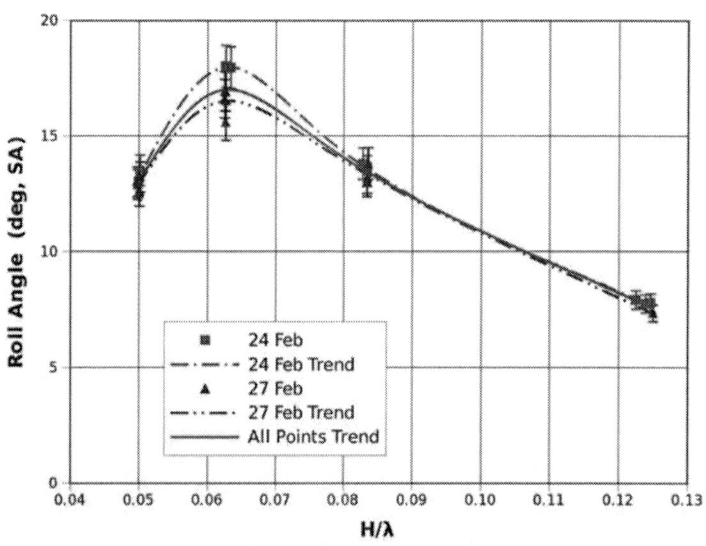

Figure 5: Roll response repeatability

Repeatability was examined by performing a large number of runs on two separate occasions using the 90mm hull. At the time of performing the runs, it was felt that repeatability was proven, though the variability of actual to nominal wave height could have masked an issue. All of the runs with the 90mm hull in 100mm waves, after normalising, were plotted and compared – shown as Figure 5.

Almost all the data was grouped, with the most noticeable variation occurring at the wave height to wavelength ratio of 0.067 where maximum roll occurred. Nonetheless, even for this condition, the spread of data was limited, a maximum of just over 2 degrees roll, indicating that repeatability, over the test period, was good. The average of all the data, shown as the solid line, was used in subsequent analyses. An analysis of the spread of results indicated that error bars of ± 5% on the roll angle appropriately represented the accuracy of the data obtained.

Note: The lines shown in Figure 5, and subsequent figures presenting results, are cubic splines, passing through the average value, both horizontal and vertical axes, of each cloud of data points. The curves do not necessarily show the shape of the underlying motion response but mainly serve to connect a family of data points.

Figure 6 is a plot of non-dimensional roll versus non-dimensional frequency of all model configurations in 100mm waves. The legend has the following meaning:

- 90mm means the 90mm depth hull and so on;
- H means the hull only;
- BW means that the bulwark was fitted;
- Closed/Open/ refers to the state of the freeing ports.

All runs had water on deck to varying degrees except for those in the top two clouds of points (for the purpose of identification, above a non-dimensional roll amplitude of 1.25)

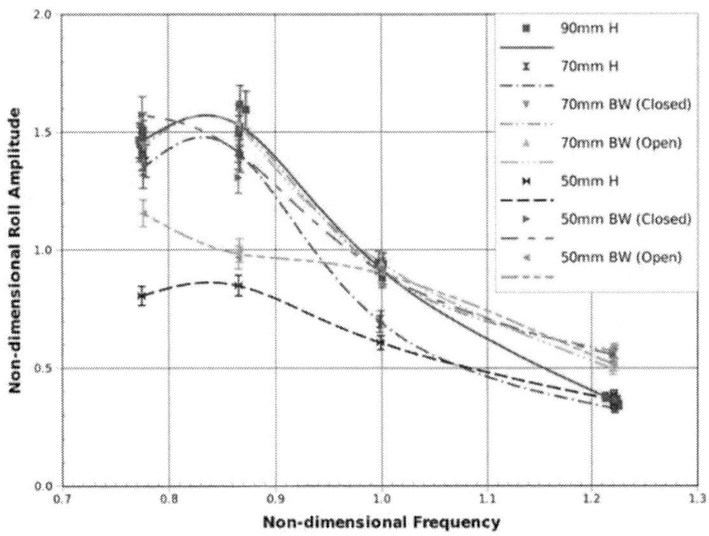

Figure 6: Roll response in 100mm wave

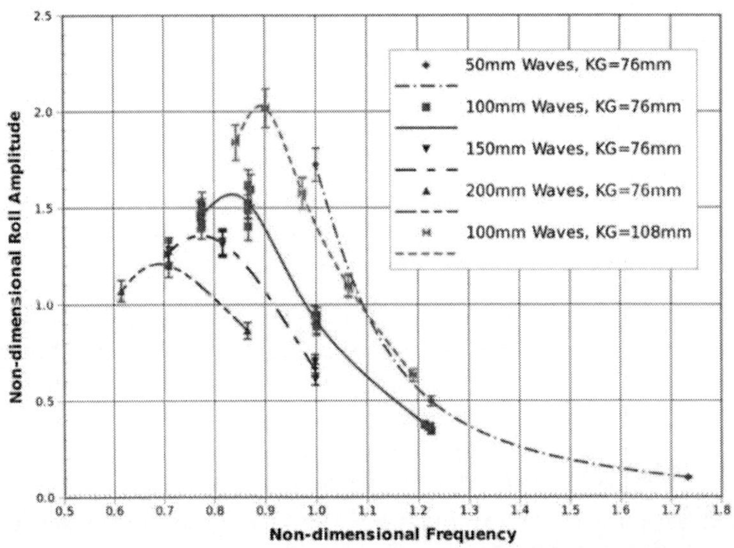

Figure 7: Roll response of Model 09-10 (90mm Depth)

Figure 7 is a plot of non-dimensional roll versus non-dimensional frequency for the 90mm hull only at various wave heights. Included in this plot for comparison is the data from the higher KG runs for this model.

The lowest two frequencies for the 100-200mm waves and the lowest frequency for the 50mm waves did not experience water on deck. There was water on deck to varying degrees for all the higher frequencies.

Figures 6 and 7 present preliminary results. Some observations that can be made include:

- At 76mm KG, all the higher freeboard configurations converged to the same response at a non-dimensional frequency of 1.0. (Note: At this frequency, roll amplitude was about 14 deg, deck/bulwark edge immersion occurred at about 21 deg and bilge emersion occurred at about 18deg.) This similar response then continued for lower frequencies, except for the lowest freeboard model with bulwark and open freeing ports.
- There appears to be a consistent trend to the roll response as wave height increases.
- The 50mm wave response appears to be exhibiting a peak at a non-dimensional frequency of 1.0. More data is required to confirm and further delineate the character of the response.

The maximum roll angle, for 100mm and higher waves occurred at a lower frequency than the roll natural frequency. There was water on deck at the natural frequency, which would have both reduced the effective GM and provided additional damping. Further research is required.

The models were at an equivalent full load condition. This means that there would have been vehicles on deck which would have impeded the water flow and perhaps reduced any roll damping effect. A concentrated load, such as a main battle tank, would have had a minimum effect whereas less dense vehicles, such as armoured personnel carriers, may have a significant effect. This also needs to be researched further.

The data collected from this model test program will be fully analysed. It will be used to validate parallel analysis using FREDYN. The results will be used to help formulate and test possible criteria, some of which will be developed from literature research. An example possible area for development, based on literature and theoretical research is presented below.

Roll Back

The main aim of the research masters is to develop stability criteria focused on naval landing craft. Though dynamic analysis will be used in developing these criteria, it is unlikely that a requirement to perform a dynamic analysis will be included. Rather, it is likely that criteria will be developed that are much like existing stability criteria but tuned properly for naval landing craft and include additional criteria that account for aspects of dynamic stability. One area that has been queried in the past is roll back, especially if a limited environment applies.

The navy wind and wave criteria assume that all vessels roll back 25 deg from the design wind equilibrium angle. It can be appreciated that all vessels, when beam on to critical waves, will roll significantly. It can also be appreciated that, in more severe conditions, the crew are likely to control roll to within acceptable limits by controlling the ship's heading relative to the predominate wave trains. Questioning of a number of naval officers suggests that 25 deg is a little above what most would consider routinely acceptable. It would therefore appear that the 25 deg was a reasonable seamanship limit to roll back in severe waves.

The IMO equivalent wind/wave criteria assume a dead ship in waves. Roll back is determined from simplified formulae that assume a wave matched to the ship's natural roll period on the basis this will cause maximum roll excitation. The effective slope table presents a relationship of wave period to wave slope, which can be used to calculate the nominal wave heights that must apply. These are shown in Figure 8 for the original IMO criteria (labelled IMO 1985), for a modification adopted in 2002 (labelled IMO 2002) and for further exploration of the concept by the author (labelled Proposal). An examination of the original Japanese data presented by Yamagata (1959) and more recently discussions by Francescutto (2004) suggests this wave length ratio assumption to be correct.

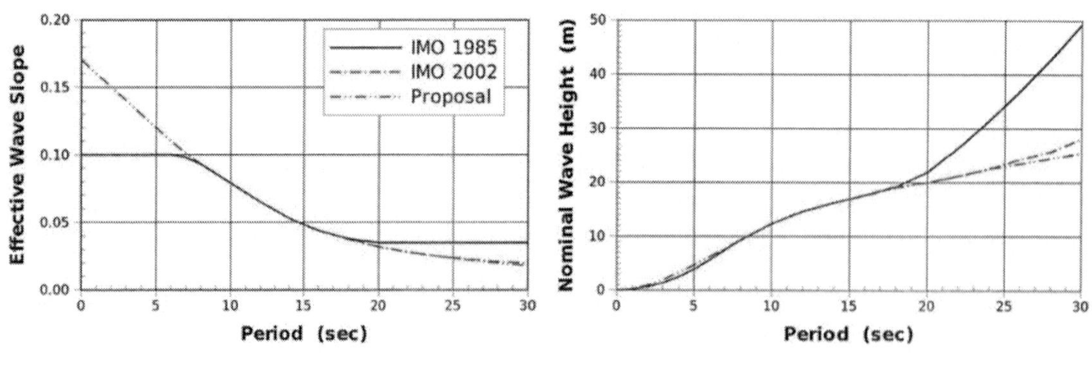

Figure 8: Effective Slope Relationships from IMO Wind/Wave Criteria

The original minimum IMO effective slope was set to 0.035, equating to very large waves at the longer periods. The changes attempted to correct this, giving a much more realistic wave height relationship. It may be reasonable to assign smaller wave effects if the vessel is constrained to operate in lesser environments than implied by the IMO wind/wave criteria. Rollback will be further examined to determine if the IMO approach is valid for landing craft.

CONCLUDING COMMENTS

The Australian Department of Defence, along with ADFA and AMC, is actively pursuing research into landing craft stability. This paper presents the initial findings of a preliminary test program which has highlighted some interesting issues that need to be researched further. The ultimate aim of the program is to significantly advance the science of ship stability in general and of landing craft stability in particular, and to develop more rational stability criteria applicable, in the first instance, to landing craft.

REFERENCES

1. DEF(AUST)5000 ADF Maritime Material Requirements Set, Vol3 Pt 2, *Stability of Surface Ships and Boats*, Issue 2, May 2003
2. Denehy, S. 2009, *Roll Decay Coefficients for a Generic Landing craft at Two Water Depths*, Australian Maritime Model Test Basin, prepared for DSTO (Maritime Platforms Division), Report no. 08/TTMTB/01.
3. Design Data Sheet 079-1 (DDS 079-1) 1975, *Stability and Buoyancy of U.S. Naval Surface Ships*, Naval Sea Systems Command, Washington.
4. Hayes, P. & Dunworth, R. 2009, *Landing Craft Capsize Study*, Report A017151, Directorate of Navy Platform systems, Department of Defence, Canberra.
5. Hayes, P., (forthcoming), *Landing Craft Stability Model Testing Report AMC February 2009*, Report A017159, Directorate of Navy Platform systems, Department of Defence, Canberra.
6. Francescutto, A. 2004, 'Intact Ship Stability: The Way Ahead', *Marine Technology*, vol. 41, no.1, pp. 31-37
7. *FREDYN A Computer Program for the Simulation of a Steered Ship in Extreme Seas and Wind*, Version 9.9, 2008, CRNAV, Maritime Research Institute Netherlands.
8. Rahola, J. 1939, *The Judging of the Stability of Ships and the Determination of the Minimum Amount of Stability*, doctoral thesis, The University of Finland, Helsinki, Finland, May.
9. Rolls, A. 2008, *Preliminary Investigation into Vessel Motions in Deep and Shallow Water*, Australian Maritime Model Test Basin, prepared for DSTO (Maritime platforms Division), Report no. 08/TTMTB/01.
10. Rolls, A., Renilson, M., Macfarlane, G. & Cannon, S. 2009, Effect of Water Depth on Motions in Beam Seas for a Typical Landing Craft, *Transactions of the Royal Intuition of Naval Architects*, vol 151, pt B1, International Journal of Small Craft Technology, Jan-Jun.
11. Sarchin, T.H. & Goldberg, L.L. 1962, 'Stability and Buoyancy Criteria for U.S. Naval Surface Ships', *SNAME Transactions*, vol 70.
12. Yamagata, M. 1959, 'Standard of Stability Adopted in Japan', *Transactions of Institution of Naval Architects*, 101, pp 417-443.

Stability And Configuration Enhancements For The RNZN ANZAC Class Frigates – An Opportunity For Innovation

Brett Longmuir and Peter Goodin
Australian Marine Technologies, Melbourne, VIC, Australia
brett.longmuir@ausmarinetech.com.au

ABSTRACT
The ANZAC Frigates of the Royal New Zealand Navy (RNZN) are undergoing a mid-life upgrade to address obsolescence, to provide some system and supportability enhancements and to facilitate further capability upgrades into the future. The ships are being re-configured to facilitate replacements and enhancements to Integrated Platform Management and HVAC Systems, Propulsion Power Upgrade and to achieve an increase in limiting displacement.

Australian Marine Technologies (AMT) has been contracted to provide the design for the Stability Enhancement and Compartment Changes (SEACC) component of the RNZN ANZAC Ship Platform Systems Upgrade (PSU).

A designer/Navy collaborative approach has provided practical, effective and innovative design solutions to ensure the ships comply with stability criteria at a higher displacement than which they were originally designed. Together with internal changes, quarterdeck changes contribute to buoyancy in critical damage conditions, improve habitability and provide additional "real estate" for other components of the PSU Program. The contracting arrangements employed allow the ship designer to work hand in glove with Navy and Project staff to provide the most cost and time effective implementation solutions whilst resolving issues arising through the application of contemporary standards to in-service ships.

This paper provides an overview of the technical solutions and discusses examples of some of the key areas of designer/customer engagement in obtaining the best overall project results.

PROJECT BACKGROUND

Platform Systems Upgrade

The RNZN ANZACs are undergoing a mid-life upgrade to address issues of obsolescence, maintainability and capability across the ANZAC platform, mission and weapons systems. The upgrade consists of three distinct projects:

- Platform Systems Upgrade (PSU);
- Close-In Weapons System upgrade (CIWS); and
- Self-Defence Upgrade (SDU).

There is a tendency for the displacement of naval ships to increase throughout their life. The original ANZAC ship design included a Service Life Margin to allow for through-life growth with an expected End Of Life Full Load Displacement of 3600 tonnes.

Among other objectives, the PSU creates a significant margin for through-life growth which will enable the SDU and other upgrades to be implemented.

The PSU is broken down into four separate Integrated Work Packages, with Stability Enhancement and Compartment Changes (SEACC) and Power Upgrade (PU) being the first to get started. This paper is focused on the Stability Enhancement and Compartment Changes component of the Platform Systems Upgrade.

Stability Enhancement and Compartment Changes

The Stability Enhancement and Compartment Changes component of the PSU aimed to provide the RNZN ANZAC Class Ships with improved damage stability, a growth margin to enable future capability enhancement and to provide some habitability improvements including a new Gymnasium and improved laundry facilities. Furthermore the opportunity was taken to correct the current inherent list to port and full load trim by the bow.

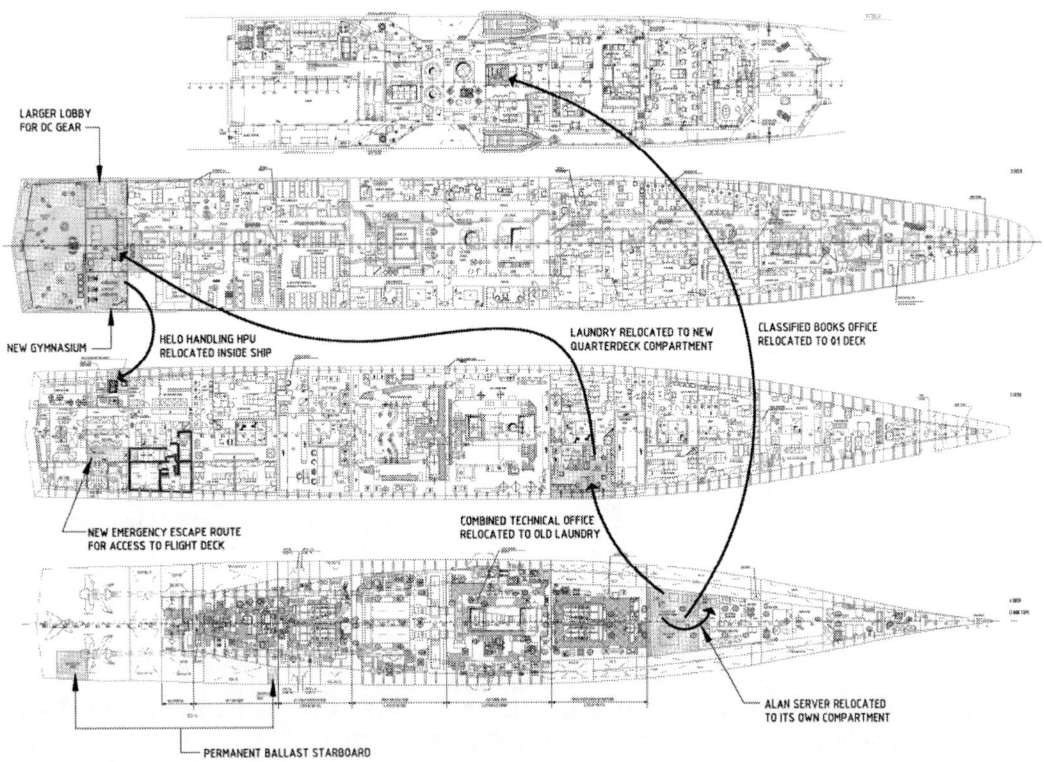

Figure 1 – Summary of compartment changes.

AMT was contracted by the New Zealand Ministry of Defence (hereafter referred to as the Crown) to provide the design solution, engineering analysis and Installation Work Package (IWP) for the modification of both HMNZS TE KAHA and HMNZS TE MANA.

The Installation Work Package was to be installed on the lead ship HNNZS TE KAHA as part of a scheduled maintenance availability commencing in mid 2009. In this period additional work packages relating to the PSU Power Upgrade and other programmed maintenance activities were also occurring. This resulted in a very tight design schedule.

In order to achieve this, AMT had to develop a strong working relationship with the Crown and the shipyard in Auckland (VT Fitzroy). Furthermore, several RNZN personnel were made available to provide expert user advice to assist AMT develop the design solution. Without the co-operative relationship between all stakeholders – it would not have been possible to produce the comprehensive design package needed for the installation in the time available.

Among the many examples of designer /customer engagement to produce the best solution, three (3) particular examples are explored in this paper. They are:

a. Achieving the required stability improvements;
b. The redesign of the Aft Section;
c. Integration of the solid ballast.

INCREASED DISPLACEMENT & STABILITY IMPROVEMENTS

Design Goals

The SEACC project set out to address the following strength and stability-related issues with the RNZN ANZACs:

1. The ships are nearing their strength limiting displacement of 3600 tonnes leaving little margin available for future upgrades and through life growth;
2. Damage stability and reserve buoyancy margins are critical in some damage conditions; and
3. The ships have an inherent heel to port.

Increased Displacement

The Crown advised that increasing the maximum full load displacement to 3720 tonnes would provide sufficient margin for planned upgrades including the SDU, other components of the PSU and an annual allowance for through life growth. AMT utilised global strength data from a finite element analysis previously conducted by Germanischer Lloyd to determine the steel modifications required to meet the increased displacement. The modifications required are relatively minor:

1. Grinding smooth plate edges in the way of several openings in longitudinal bulkheads on 1 Deck and 2 Deck to reduce the notch factor and provide the necessary fatigue resistance; and
2. Additional stiffening of two panels on 2 Deck so that the panels meet buckling requirements under extreme loads.

Damage Stability

One of the key elements of the PSU is the creation of a new watertight compartment in watertight zone 2M. The added buoyancy created by plating in nearly half of the Quarterdeck improves stability when zones J, K and L are damaged. The benefits are two-fold: Reserve buoyancy is increased (i.e. the margin line is less likely to become immersed); and the angle of heel is reduced due to the increased righting moment from the greater waterplane area. Installation of the new watertight bulkhead at Frame 103.2 provides just enough extra buoyancy while still leaving sufficient space for a safe and functional Quarterdeck.

In addition to the margin line aspect of the reserve buoyancy criteria, another requirement is for safe access to enter/exit intact areas without risk. By installing two new watertight hatches in Zone M for access from 3M to the Flight Deck - safe access is provided in JKL damage scenarios.

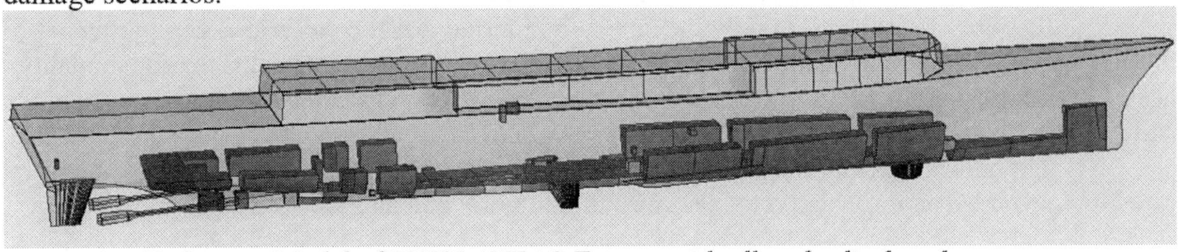

Figure 2 – Hydromax Model of RNZN ANZAC Frigate with all tanks displayed

Using AMT's Hydromax model of the modified RNZN ANZACs, it was calculated the 4L Water Ballast Tanks do not need to be cross connected to achieve compliance with the DEF(AUST) stability criteria.

There are a number of innovative aspects of the design which have helped the Crown achieve project objectives while minimising production impact and hence the cost of implementation. AMT worked together with the Crown to understand their requirements for future growth so that the "best value" solution could be developed. Above 3720 tonnes displacement it may be required to:

1. Conduct more significant strength modifications;
2. Add solid ballast to lower the VCG which would consume some of the margin being created;
3. Cross connect tanks;
4. Plate in a greater part of the Quarterdeck – which could significantly impact mooring operations;
5. Accept a reduction in speed/range.

AFT SECTION RE-DESIGN

Design Goals

The original RNZN ANZAC Ship Quarterdeck arrangement was an open weatherdeck on 2 Deck below the Flight Deck. It extended over the whole watertight section and incorporated a multitude of systems and outfit to support various operations and functions. The configuration is depicted in Figure 3.

Figure 3 – Original Quarterdeck Arrangement

The installation of a watertight bulkhead at Frame 103.2 to facilitate the reserve buoyancy improvements necessitated a major redesign of this area. A key objective was to utilize the internal space created by the new zone for additional or relocated compartments. This was a common theme of the design to incorporate improvements in functionality and maintainability whilst meeting the primary technical requirements.

The allocation of existing systems and outfit to internal compartments or external areas provided an opportunity to maximise these improvements. The design goals included:

External Areas

- A functional mooring arrangement in a compressed area that reduced deck clutter, improved towing facilities and provided better ship handling ergonomics;
- Relocation of the Underwater Decoy (Nixie) Winch into a more sheltered area less vulnerable to quartering green seas;
- Reduction of equipment and systems in the open environment including the helicopter traversing winch with associated maintenance reductions.

Internal Compartments

- Incorporation of a Gymnasium, relocated Laundry and expanded Lobby with optimised independent compartment access.
- A functional Laundry arrangement incorporating workflow requirements for both high volume ship service and personal laundering.
- Optimised air conditioning system design for the high heat loads of the Laundry with segregated dryer system to reduce the ship chilled water loads.

These goals were overlaid with the objective to reduce production implementation costs and minimise disruption to impacted compartments and systems.

Issues and Challenges

The spatial arrangement of the new internal area was not ideal to gain independent access to both the Laundry and Gymnasium compartments. In order to provide this functionality, the original watertight bulkhead needed to be penetrated in a location that did not impact complex systems and suited the compartment layout.

The remaining external space for the Quarterdeck is approximately half of the original deck area whilst still retaining the fairleads and bollards with their associated working zones. This was further complicated by the requirement to have direct leads to the stern mooring capstan without unsafe internal 'bights' in the lines where personnel were located. The RNZN variant of the ANZAC ship class originally had a large Towed Array Sonar winch installed on the Quarterdeck which has been subsequently removed. This required a 300mm deep well in the deck to accommodate its height. The arrangement of the Laundry needed to be designed to incorporate this well to substantially reduce the amount of structural work if otherwise built in.

Converting external spaces to internal compartments provided further challenges in system arrangements. The original fresh air vent diffusers were located on the original forward bulkhead of the Quarterdeck. These needed be relocated aft with ducting extended without clashing with the helicopter haul down grid sump. The base of this sump inclined forward to suit the drainage piping. This piping required judicious rearrangement to suit the new compartments.

The onerous shock and vibration requirements for equipment and associated structure remained for the new design. Significant structural analysis would be required to meet the criteria whilst minimizing the impacts on the complex deckhead systems in the compartments below.

Design Solution

The design solutions developed by AMT utilised a number of methods and tools based on experience with naval ship design. Conceptual designs were developed and refined taking into consideration all spatial and access aspects. Human factors engineering played a large role in the arrangement of the Laundry taking into account the workflows and access locations. The deck well developed into a feature of the compartment layout providing separate work zones and optimised heights for machine access.

The mooring arrangement of the revised Quarterdeck was developed in collaboration with RNZN seamanship specialists. A 3D model of the space was prepared for a joint workshop that refined the arrangement based on detailed user knowledge and design constraints. Using this approach, the arrangement was finalized for detailed design within days. Figure 3 depicts the final layout. Key features include:

- A sheltered Nixie Winch location on a raised foundation to reduce seawater ingress;
- A functional mooring arrangement with optimised line handling utilizing the stern capstan with a new pendant controller;
- Substantial reduction in external equipment with associated maintenance benefits including the relocation of the helicopter traversing system HPU internally.

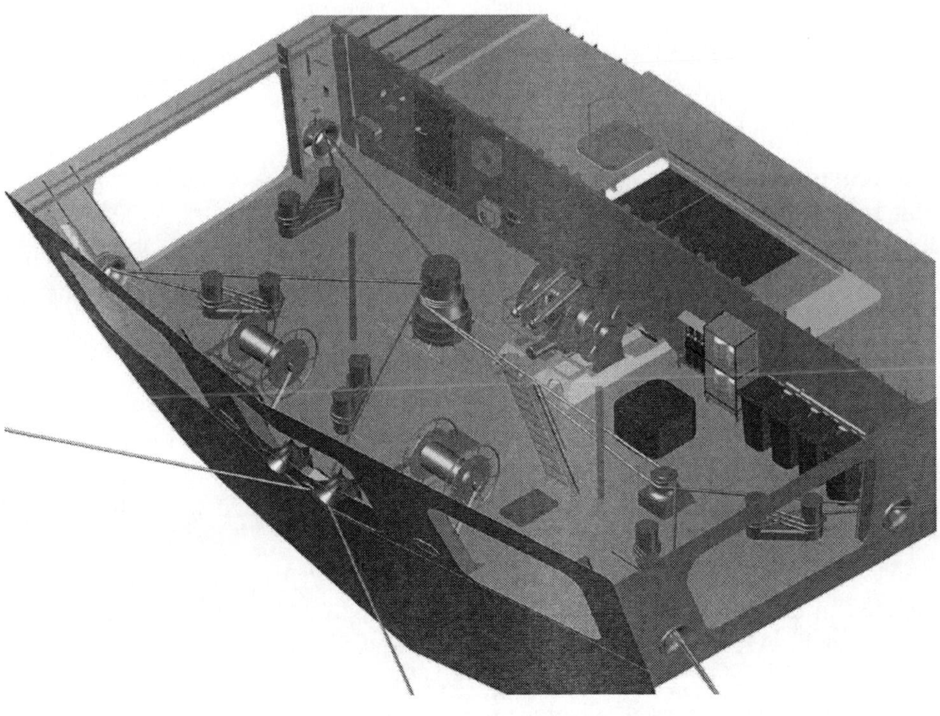

Figure 4 – 3D Model of New Quarterdeck Arrangement

Major equipment foundations were designed using structural FEA software in order to optimise the scantlings and minimise the deck strengthening below given the high shock load requirements. The structural design was iterated to accommodate the systems and equipment constraints in the compartments below.

The redesign for the aft section meets all of the technical requirements and incorporates a number of improvements for maintenance, operations and personnel functionality. AMT has employed a combination of methods in a collaborative approach to develop a functional and cost effective design.

SOLID BALLAST - INSTALLATION

Design Goals

Solid ballast will be installed as part of the Platform Systems Upgrade (PSU) primarily to reduce the inherent heel to port of both ships in the lightship and full load conditions. The Transverse Centre of Gravity (TCG) and associated heel differs between the two ships hence requires different quantities of solid ballast to provide zero heel.
In terms of performance, the design requirements for solid ballast are:

- Provide a transverse moment to bring the ship upright at full load;
- Improve the bow down trim of the vessels in the full load condition;
- Not to degrade the stability characteristics of the vessels by maintaining or preferably lowering the Vertical Centre of Gravity (VCG) of the ship in the post-PSU configuration;
- Providing flexibility in the installed ballast mass to cater for potential variances in the baseline ship condition and PSU weight and moment margins.

In addition, the key technical design requirements and constraints for solid ballast installation are:

- To structurally contain the ballast;
- To provide appropriate air exchange to the space to cater for thermal pressure changes and maintain a low corrosion environment;
- To provide adequate maintenance access and survey facilities to ensure the integrity of the surrounding structure can be monitored;
- Material selection ensuring its properties are benign in nature to ensure that personnel health is not put at risk in the normal state or damage situations such as fire.

From a review of potential ballast material, the Crown selected Perma Ballast. This is a compound of metals mixed with water that flows as a slurry initially then sets as a firm material with a high density. The characteristics of the material are analogous to concrete where the slurry fills the structural space then sets to a firm state that maintains the shape without movement from ship motions.

Whilst utilised in many marine applications such as US Navy vessels including DDG-51 Class Destroyers, T-AKR Strategic Sealift Ships and LSD troop deployment vessels, no evidence has been found of its performance under impulse loading such as under shock conditions. Furthermore, there is no precedent for the installation of Perma Ballast as a ballast material within vessels designed to Federal German Navy standards.

The FGN F123 Class has recently incorporated solid ballast within its aft keel structure. The solid ballast (which sets as a true solid within a bounded tank structure) was integrated to meet the requirements and load cases of the structural standard BV 104. The RAN ANZACs will have a multitude of lead ingots installed directly on the shell encased in a tanktop structure filled with a bitumen compound. The RN (Type 45, RFA Argus and others) are installing Ballast-Crete.

Structural Requirements

Noting the practice used in ballast integration for other navies the structural integration of the ballast material was against the structural standard BV 104. This resulted in the need for some configuration changes to local scantlings. However, noting the doubt on material behaviour, the Crown sought some specific reassurance on the structural performance during a shock event. Consequently, an assessment has been performed to provide the assurance necessary to the Crown.

Given no coverage under naval rules the assessment analysed the worst case solid ballast installation with respect to loads and structural configuration. The analysis covered two primary failure modes:

- Local Panel Strength: This determined the structural limits of the ballast space boundary panels ie. shell, deck and bulkhead and analysed the pressures on a comparative basis with the original design;
- Hull Frame Response: This identified the response from iterating external pressures on a representative frame structure to determine the failure mode on a frame segment of the hull primary structure and its relationship with solid ballast mass.

Local Panel Strength

The analysis method for the stiffened panel structures was performed according to the following:

a. Perform a limit state analysis of the shell, deck and longitudinal bulkhead panels. This identified the failure mode [plate deformation, stiffener bending/tripping or plastic hinge] and quantified the limiting pressure;
b. Determine the transmission factor of pressure for the contents of the ballast space and typical tanks;
c. Apply the relative transmission factors to the deck and longitudinal bulkhead structures and compare with the limiting pressures.

A key aspect in the analysis is the characteristics of the ballast material and typical tank contents (ie fuel, water) and their behaviour in transmission of pressure from an external source. Fluids such as fuel and water are considered with respect to the comparison of the original design and shock event responses.

The characteristics of the ballast material are more difficult to define. Product information describes that in the set state, the composite material contains up to 50% water and the resultant mass is 'generally completely non-shifting'. However given the water content and the expected entrained air of approximately 1-2% by volume, the material is still malleable to the extent that it would deform to a new deflected structure shape under high external load. Hence the ballast material is considered to behave in a similar manner to fluids with respect to the distribution and transfer of pressure on panels.

The transmission of external shock pulses through ship tanks is generally not well covered in engineering literature. AMT has utilised results of an evaluation of this behaviour on an experimental basis performed by the Naval University of Engineering, Hubei China. The results indicated that the transmission of pressures from outer shell to internal tank boundaries within a tank full of fluid is low, down to as low as 5% based on the experimental results.

Due to the low transmission factor of pressure to the internal deck and bulkhead panels, there are significant safety factors for these structures under maximum external loading. These panels would also be subject to the static load pressure due to the ballast mass however the structure is well within the limiting pressures.

Hull Frame Response

A transverse frame segment of the hull girder has been analysed using FEA to determine the response of external loads with the ballast mass included. The segment extends 600mm either side of the transverse frame including plating and longitudinal stiffeners. The transverse frame extends vertically from the keel to 1 Deck.

To understand the load response of the transverse frame, the ballast is treated as a 'solid block'. It is modelled as a series of rigid links connected to transverse frame in the ballast space region. This provides a conservative result as it does not account for stress dissipation of a distributed load.

The analysis method for the hull frame response is as follows:

a. Apply an external pressure load equal to the shell limiting pressure determined through the prior analysis and confirm the frame structure is within the yield strength range;

b. Progressively increase the external pressure until a structural failure of the frame member local to the ballast space boundary is identified;

c. Compare the frame failure equivalent external pressure with that determined for panel limit state.

This result confirms that if an external pressure is applied equal to the hull structure limiting pressure, the frame structure will remain within yield strength by nearly a factor of two. On the basis that the hull structure meets the design shock loads, this confirms the frame structure is adequate.

Figure 5 – Frame Stress Plot – External Pressure

The external pressure was increased until the frame structure reached its Ultimate Tensile Strength. This again is considered conservative due to the simplicity of the model and degradation of the structure that could occur whilst containing the ballast material. The external pressure reached a level of twice that required to reach yield strength.

The results conclude that an external pressure of approximately four times the limiting hull structural pressure would be required for the hull frame to start to fail. This was then equated to an equivalent acceleration of the applied ballast mass and related to the typical accelerations derived from the BV 043(85) shock standard.

With the Perma Ballast material not actually solid, significant deformation can be expected upon subjection to dynamic loading. Consequently, the worst case frame response can be considered to confirm the adequacy of the structural configuration proposed for the integration of the solid ballast material Perma Ballast for the ANZAC PSU.

CONCLUSIONS

This paper provides an overview of some of the challenges faced throughout the design.

In these areas and many others it was found that the engagement with a motivated and interested customer was not only able to assist in overcoming challenges but also lead to an optimised result matching customer expectations but was in many cases able to reduce production cost or schedule. In any naval project with tight timeline or budget engagement with the ultimate customer is a major aid in achieving success.

DISCLAIMER

The opinions expressed in this paper are those of the author and not those of the New Zealand Ministry of Defence, the RNZN, ThyssenKrupp Marine Systems Australia or VT Fitzroy.

REFEREED PAPER

Delamination Charaterisation of Curved Composites Using Acoustic Emission and Fibre Bragg Gratings

Raju[1], **Prusty B.G.**[1], **Kelly D.W.**[1], **Lyons D.**[2], **Peng G.D.**[1]
[1]School of Mechanical Engineering, University of New South Wales, Sydney, Australia
[2]EMP Composites, Sydney, Australia
Email contact: rajuaero@yahoo.com

INTRODUCTION:

Marine structures generally use E-glass and Vinylester composites for the added benefits over metal structures. These composites generally may contain curved shapes such as T-Joints, top hat stiffeners, L-bends, angle bracket, a co-cored web, etc., to accommodate the design and aesthetic features [1, 4, 6]. The damage mechanism of the curved composites is complex and is necessary to analyse the life of the structure. Studies conducted in the past focused on the unidirectional fibres or cross-plied fibres. But in marine applications, short fibre composites play an important role. In marine applications, Chopped Strand Mat (CSM) is used commonly in conjunction with cross ply and unidirectional layers. This paper focuses on the delamination behaviour in laminated composites with CSM

Premature failures of curved composites have occurred in the past, mainly due to the lack of appreciation for the interlaminar tensile stresses, which may be less than 3% of the in-plane strength of a composite laminate. Generally, interlaminar tensile and compressive stresses are developed when curved composite laminates are subjected to flexural loading in the plane of curvature result in induced interlaminar tensile failures [5]. Hence it is vital to understand the interlaminar tensile stress distribution more accurately when designing or analyzing a curved composite structure. The interlaminar tensile strength is a function of the basic laminate constituents, type of reinforcement, stacking sequence, manufacturing process, fabrication quality, aging, type of loading [7, 8, 9]. Many methods have been suggested to arrest the interlaminar failure such as using tougher matrix polymers, using interleaf layers, through thickness reinforcements, improved fibre/matrix strength, optimisation of fabrication, [10, 11].

Of the various failures of composite materials, delamination is one of the predominant modes of failure of curved laminates. Initial failure through delamination may go undetectable as they are frequently embedded within the composite structure. In some cases, the delamination effect on the structure may include a dramatic loss of residual strength of up to 60% [12]. The problem of interlaminar tensile or shear stresses was highlighted at the bonded joints and attachments of marine composite structures [13]. Wisnom [2, 3] was one of the primary investigators to study the delamination failure through experimental, analytical and numerical approaches for interlaminar failure and flexural strength of composite laminate. The Classical Plate Theory (CPT) assumes a state of plane stress in the constitutive relations and almost completely neglects the interlaminar stresses. Higher order theories tend to follow CPT and are in-efficient in calculating the interlaminar tensile stresses. Lekhnitskii's equations are applicable only for pure bending or edge loading. But they cannot sustain the circumferential force without radial restraints. Hence Shenoi & Wang [1] developed equations based on elasticity theory for delamination and flexural strength of curved composites. Also the effects of key variables such as stacking sequence, radius of curvature and thickness of skin of sandwich beam on stress distribution within a curved layered beam and sandwich beam were studied.

For a loaded structure, elastic energy is produced and stored in the structure. If the elastic energy exceeds a critical value, a crack or defect is produced and rapid release of elastic energy is observed. This rapid release of energy is generally termed as acoustic emission [14]. The examination of AE is a very successful tool for the sensitive detection and location of active damages in polymer blends, particle-filled and FRP composites [15, 16]. AE measurement is based on the detection of the surface movements caused by stress waves of the fracture process in a microscopic scale [17]. Two approaches of AE study are possible: Parametric AE analysis and Transient AE analysis. In the current study, parametric AE analysis is used based on the extraction of a number of parameters (such as arrival time, peak amplitude, rise-time, signal duration, counts, Energy release, RMS, etc) from individual AE signals [18]. One of the initial studies [25] showed only two levels of failure, matrix cracking and fibre fracture. Over the years, the advancement in the data acquisition technology, other failure mode detection is possible. A Number of investigators have attempted to use the AE technique for detection and evaluation of damage mechanisms in composites [17, 19, 20, 21, 24].

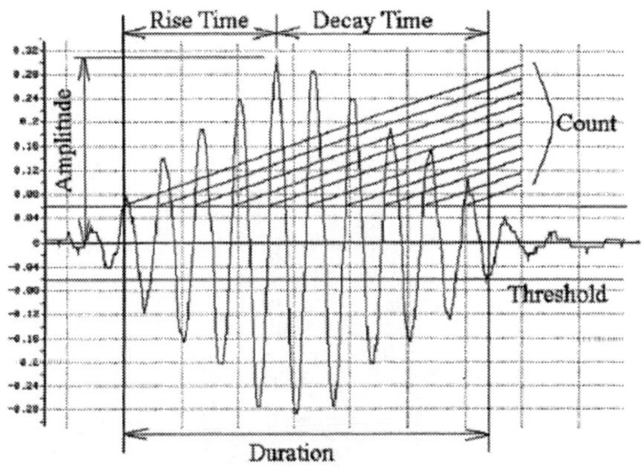

Fig.1 A typical AE Signal [24]

Huget et al [22] studied the damage mode and the corresponding amplitude output of GFRP composites. They proposed that the matrix cracking had amplitude of 40-55dB, debonding had 60-65dB, 65-85dB to fibre pull out and 85-95dB accounts to fibre fracture. The study of Ageorges et al. [23] showed that the interface debonding had amplitude of 40-60dB, fibre breakage of 75-90dB. Failure mechanism based on even duration can be classified as Debonding ~ 10ms, Fibre breakage ~15ms and Delamination – 25ms.

Various studies have been conducted to assess the curved composite delamination due to interlaminar tensile failure. Martin et al. [4] studied failure of curved composites with unidirectional and cross ply mats. Most of loading parameters used were either pure bending or pure edge loading based on Lekhnitskii's solutions. The L-bends laminates were fixed at the base and a pure moment was applied on the top arm until failure and the failures were analysed. But the structures in general experience various types of loading, not just the pure bending and pure edge loading. Boat builders tend to use uni-directional fibres to increase the strength. An attempt to check the effectiveness of addition uni-directional fibres is made.

The aims of this paper are three-fold, to study:

- the through thickness stress distribution of the curved composite laminate with different material layup – Chopped Strand Mat (CSM), Double bias (DB) and Uni-directional fibres (UD) and the combination of all the three fibre mats.
- effect of using uni-directional layer in the layup
- the failure behaviour of curved composite subjected to bending load using Acoustic Emission (AE) and Fibre Bragg Gratings (FBG)

EXPERIMENTAL INVESTIGATION:

Composite laminates were manufactured using E-glass fibre and Vinylester resin (FGI – SPV 6036) by hand layup and cured under room temperature for 24hrs. The specimens were 25mm wide with an arm and leg length of 100mm. The radius of the bend was 19mm as shown in Fig.2. The lamina properties of the fibre and resin are presented in Table 1. The specimens were made up in three layups having 25mm width, and total thickness of 4.62mm with each lamina thickness being 0.66mm. Bending load was applied using two hinges arrangement using Instron test machine shown in Fig.3. The loading was 15mm/min for four minutes and later 5mm/min until 100mm total deflection. The specimens were clamped at the base of the leg and bending load was applied using two hinge arrangement. A hole was drilled in the specimen and connected to the hinge by a bolt. The loading mechanism was designed such that bending load was applied and the structure failed through delamination by interlaminar tensile failure. Special care was exercised while installing the specimens within the grips to ensure proper alignment.

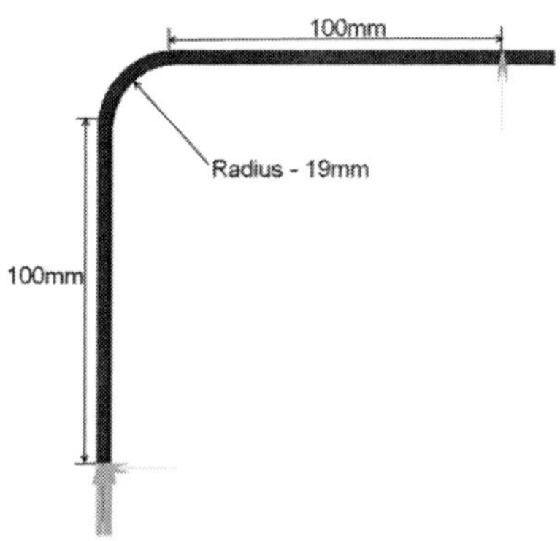

Fig. 2. Specimen Configuration

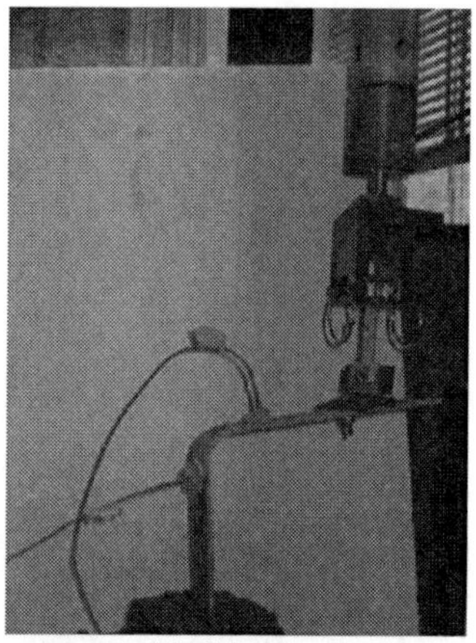

Fig 3. Experimental setup with hinges

Laminate Configuration			
	Layup 1 – 4.62mm	Layup2 – 4.64mm	Layup3 – 4.62 mm
Layer 1	CSM 450gsm	CSM 450gsm	DB 611gsm
Layer 2	CSM 450gsm	DB 611gsm	CSM 450gsm
Layer 3	CSM 450gsm	CSM 450gsm	UD 451gsm
Layer 4	CSM 450gsm	DB 611gsm	DB 611gsm
Layer 5	CSM 450gsm	CSM 450gsm	CSM 450gsm
Layer 6	CSM 450gsm	DB 611gsm	UD 451gsm
Layer 7	CSM 450gsm	CSM 450gsm	DB 611gsm
Weight	100%	115%	110%

Table 1- Laminate Configuration

Mechanical Properties			
	CSM	DB	UD
E_{11} (MPa)*	9600	7850	23500
E_{22} (MPa)+	6062	6673	6560
E_{33} (MPa)+	9500	7500	6560
v_{12} (MPa)+	0.108	0.133	0.086
v_{12} (MPa)+	0.139	0.146	0.249
v_{12} (MPa)+	0.347	0.62	0.369
G_{12} (MPa)+	1847	2050	2265
G_{23} (MPa)+	1847	2050	1847
G_{31} (MPa)+	2602	7157	2265
ILTS (MPa)+	9.5	10.5	10.5
ILSS (MPa)+	28	30	28

Mechanical Properties (contd.)			
	CSM	DB	UD
S_{11T} (MPa)*	176	74	440
S_{11C} (MPa)+	240	150	475
S_{22T} (MPa)+	176	74	58
S_{22C} (MPa)+	240	240	145
S_{33T} (MPa)+	60	80	58
S_{33C} (MPa)+	300	300	283
S_{12} (MPa)+	87	110	59
S_{23} (MPa)+	50	60	50
S_{23} (MPa)+	50	60	50
Flex Modulus (MPa)*	1693	1158	4196
Flex Strength (MPa)*	214	76	433

* Experimental testing
+ Obtained through 'Component Design & Analysis (CoDA)" software using resin and fibre properties

Table 2. Material Properties

A four-channel Acoustic Emission equipment of Physical Acoustic Corporation was used to acquire the stress waves generated by curved composite. Two methods of analyzing the AE data are the parametric analysis and transient analysis. Parametric analysis focuses on the parameters such as amplitude, energy released, even duration, rise time, counts, etc. Transient analysis focuses on the wave form analysis of the AE signals. In the current study parametric analysis performed. Two sensors were used, and AE events were acquired by the sensor as analog signals. They were pre-amplified and converted into digital signals by an A/D converter. These signals were recorded in a parametric AE file. A 40dB system gain and a 45dB threshold were used for the AE acquisition. The AE data acquisition was synchronized with the mechanical loading. The acoustic emission information collected was stored in a parametric AE file. This allowed correlation of the AE parameters with the load-deflection at the time the AE signal was produced.

EXPERIMENTAL RESULTS:

The curved laminates were loaded such that the radius of curvature increased and interlaminar stresses were induced in the curved laminate. Theses interlaminar tensile stress was maximum within 50% of the thickness and zero at the edges [26]. The tangential stresses were compressive at the outer edge, tensile the inner edge and minimum at the point of maximum interlaminar tensile stress. The maximum interlaminar stress occurs at the mid width of the specimen as the loading point is at the mid-width of the specimen. Five specimens each of the three different layups were tested. Figs. 5, 7 & 9 shows the typical load-displacement plots for all the five specimens of each layup. The tests were terminated when the displacement reached 100mm.

For CSM specimens, the initial delamination was observed at 162.92N at a displacement of 66.27mm (Specimen No. A05). This is shown in the load-displacement plot Fig.5 in the form of a small kink. A crack was developed between the second and third layers as shown in Fig. 4a. Upon continuing application of load, the specimen began to straighten up, creating excessive interlaminar tensile stresses at the bend. Outer layers were subjected to interlaminar compressive stresses and the fibres sheared off in compression, shown in Fig.4d. No delaminations were observed between other plies. During the final stages of loading, a small delamination occurred between the 6th and 7th layers. This caused further delamination in the specimen and ultimate collapse occurred at 616N with 85.58mm displacement.

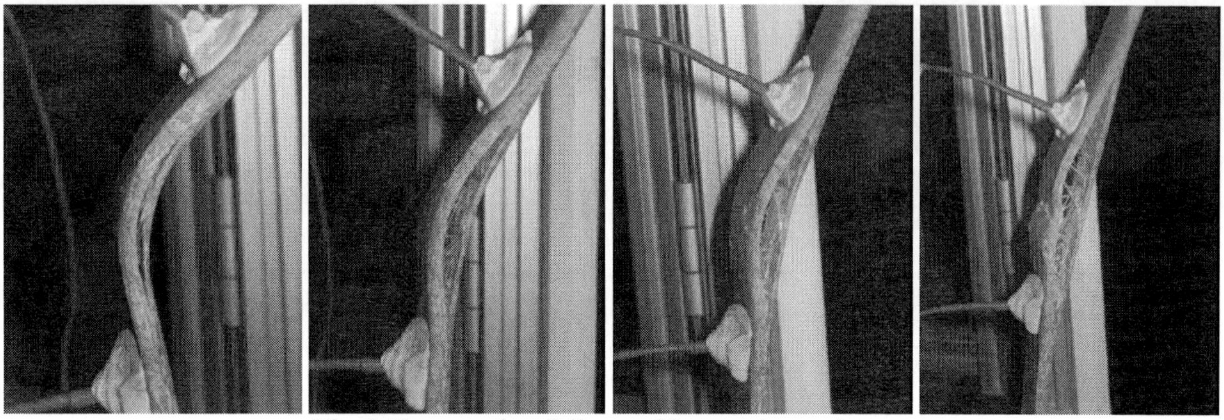

Fig 4. Progressive failure of Layup 1 specimens

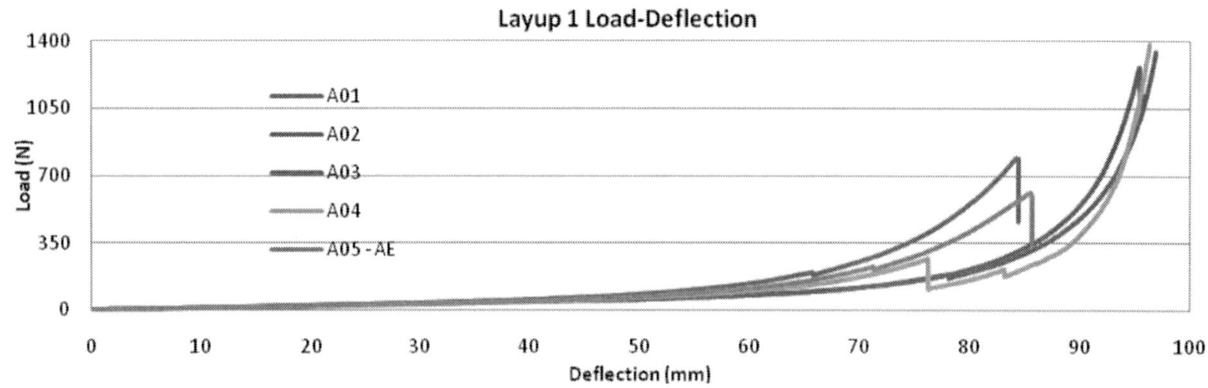

Fig 5. Load-deflection plot for Layup 1 specimens

The second layup consisting of CSM+DB followed the same trend, failing at 573.93N with 93.03mm displacement. The load deflection plot is shown in Fig.7. Initial crack was generated between the second and third layers at a displacement of 61.04mm with a load of 57N. The crack initiation and progression is shown in Fig 6. Further loading caused the crack to propagate in the longitudinal direction. Later another major crack began between third and fourth layers. At the final stages of loading, outer layers were subjected to compression and the fibres in those layers sheared at the mid-length of the specimen. Finally the specimen with embedded FBG failed at 573.6N with a displacement of 93.03mm.

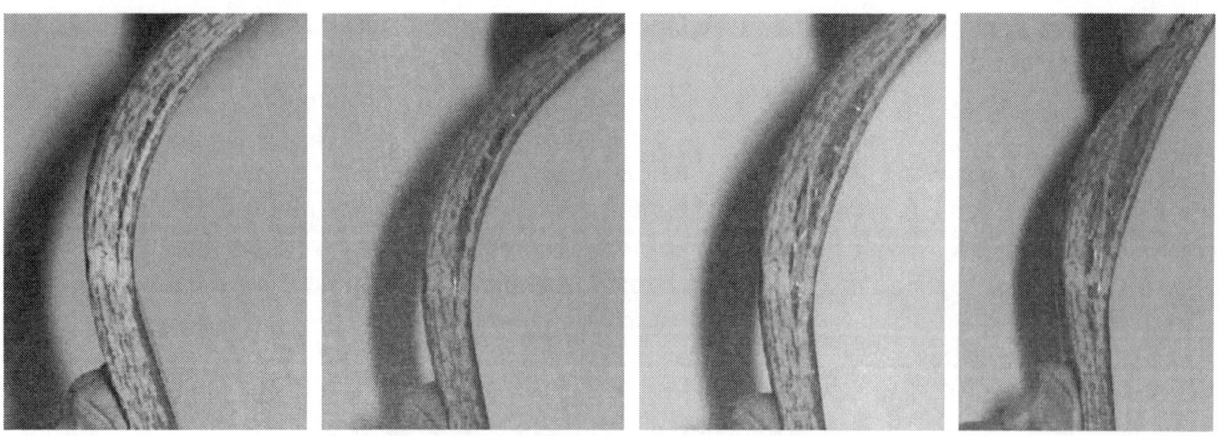

Fig 6. Progressive failure of Layup 2 specimens

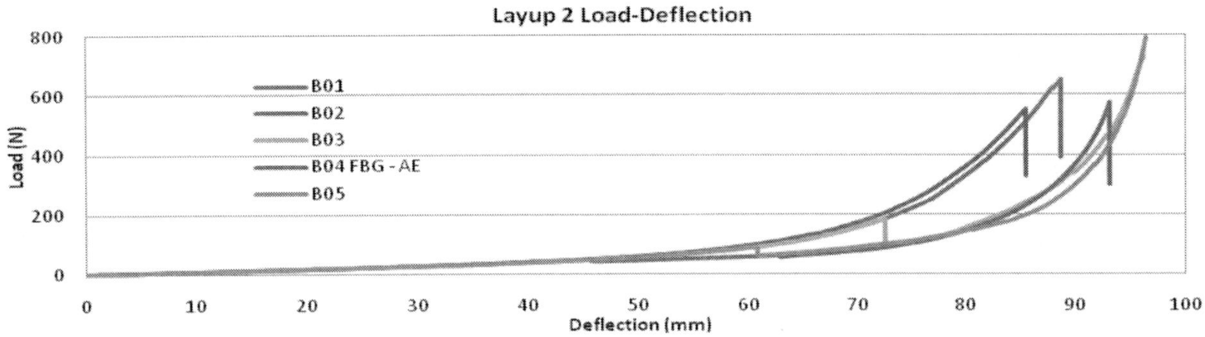

Fig 7. Load-deflection plot for Layup 2 specimens

Interestingly the third layup consisting of CSM+DB+UD had high strength but low stiffness until the first ply failure. However the post-failure behaviour observed had increased residual strength as shown in Fig. 9. Further application of load past the first failure point, the CSM fibres sheared away in tension shown as kinks in the load-displacement plot. The DB layers were more flexible, and remained intact. Upon further loading, the bend started to vanish and the specimen began to straighten up, thus behaving like a tensile specimen, as shown in Fig. 8. Only the uni-directional fibres carried the entire load until final collapse. Hence addition of uni-directional fibres in marine composite curved panels proved beneficial.

Fig 8. Progressive failure of Layup 3 specimens

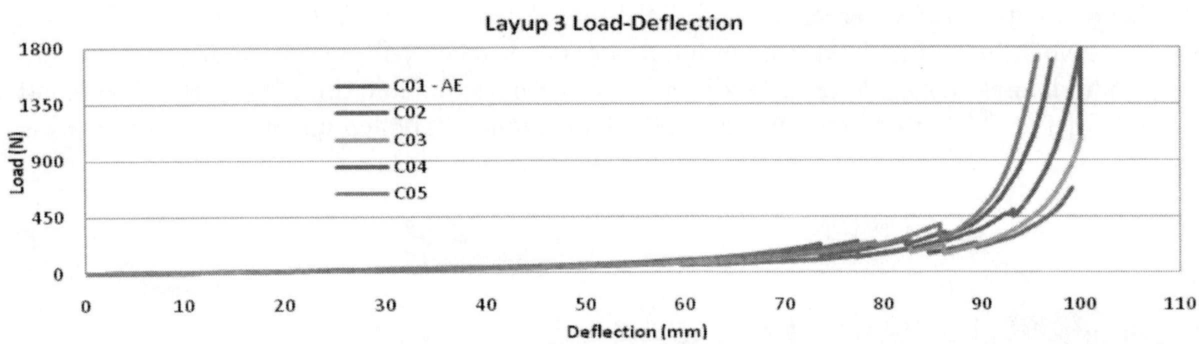

Fig 9. Load-deflection plot for Layup 3 specimens

The stiffness of the first layup (all CSM) was higher than the other two layups. The second layup (CSM + DB) experienced reduced stiffness compared to the other two layups. The addition of the double bias (+-45^0) layers added more flexibility to the structure. Odd numbers of layers were chosen to minimise the exact location of neutral axis for better failure analysis. Interestingly, the initial delamination was observed between the second and third layer irrespective of the layup.

FINITE ELEMENT VERIFICATION:

Finite element analysis is used to predict the out of plane displacement and failure prediction for comparison with the experimental results. MSC MARC MENTAT 2008r1 finite element system was used. The first layer consists of five elements and the rest of the layers consists of three elements in the through thickness direction for uniform distribution of inter-laminar tensile stress (ILTS). The element was eight-node, isoparametric, arbitrary hexahedral brick element and the stiffness of this element was formed using eight-point Gaussian integration, with a global edge length of 1mm around the bend. Two boundary conditions modeled were the actual representation of the experimental setup. The nodes at the base of the leg are fixed while the nodes on the arm are loaded in y-direction until failure. Three layups, CSM, CSM+DB, CSM+DB+UD were used with 7 layers in each laminate. The layup is presented in Table 1(a) and the basic mechanical properties obtained from the coupon tests are presented in table 1(b). Delamination calculations were performed using cohesive elements which combine the aspects of strength based analysis to predict the onset of damage at the interface and fracture mechanics to predict the propagation of a delamination. These are designed to represent the separation at the zero-thickness interface between the layers of 3D elements during delamination and are sufficiently stiff in compression to prevent interpenetration of the delaminated surfaces. Each layer was modeled as separate lamina material and the delamination was defined as the splitting of mesh between two materials. The maximum stress criterion was applied allowing the mesh to split between the materials as shown in equation (1). The stresses were extrapolated to the nodes and transformed to the material interface.

$$\left(\frac{\sigma_n}{S_n}\right)^m + \left(\frac{\sigma_t}{S_t}\right)^n > 1 \qquad (1)$$

Where σ_n is the normal stress, σ_t is the tangential stress and S_n, S_t, are the critical values of normal and tangential stresses, m and n are define the model as either bilinear or exponential.

As the material was same throughout the laminate in layup 1, the interlaminar stress curve was linear until the mid-thickness and drops to near zero on the surface as shown in Fig. 11. The theoretical interlaminar tensile strength on the inside surface is zero, but due to elemental complex behaviour, this cannot be represented in a 3D numerical modeling. The critical interlaminar tensile and shear stresses of all the three layups are presented in Fig.10 & Fig.12 and also in Table 3. In the first layup, the initial failure occurred at a displacement of 64.08mm purely due to interlaminar tensile stress at 9.65 MPa, between second and third layer, mid-width and at an angle of 18.26^0. Further loading caused the laminate to straighten up, creating tensile stress in the inner surface and compressive stress on the outer surface.

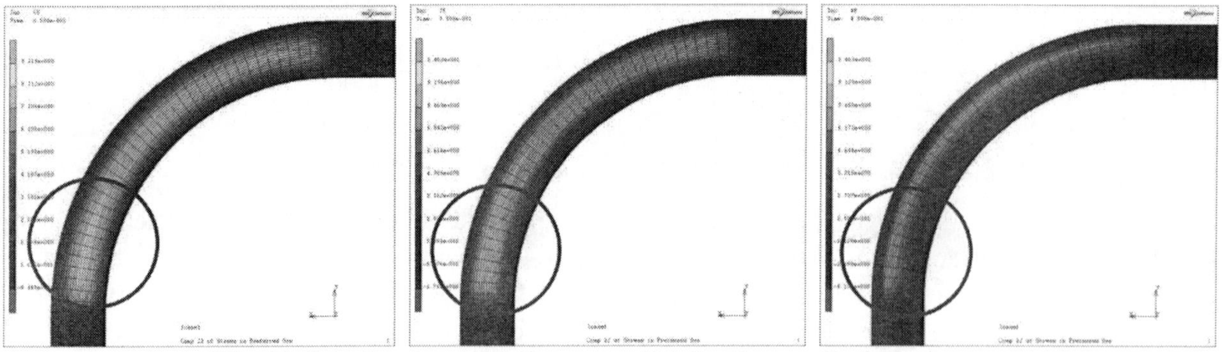

Fig 10. Interlaminar tensile stress distribution – all three layups

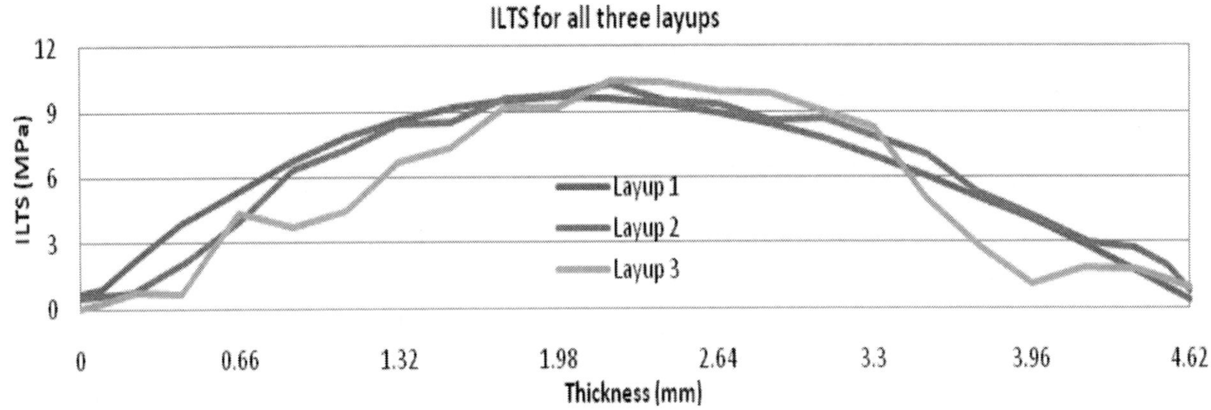

Fig. 11. Interlaminar Tensile Stress distribution just before first failure – all three layups

The second layup consists of four layers CSM and 3 layers DB. The failure occurs between the second (DB) and third (CSM) layer. Delamination occurred at an angle 21.89^0 due to peak interlaminar tensile stress of 10.03 MPa. Delamination progressed until all the layers failed and the structure collapsed. Addition of DB layers increased the load carrying capacity as well as the deflection up to 66.80mm compared to all CSM of 40.80mm. The shear stress distributions for all the three layups are shown in Figs. 12 & 13.

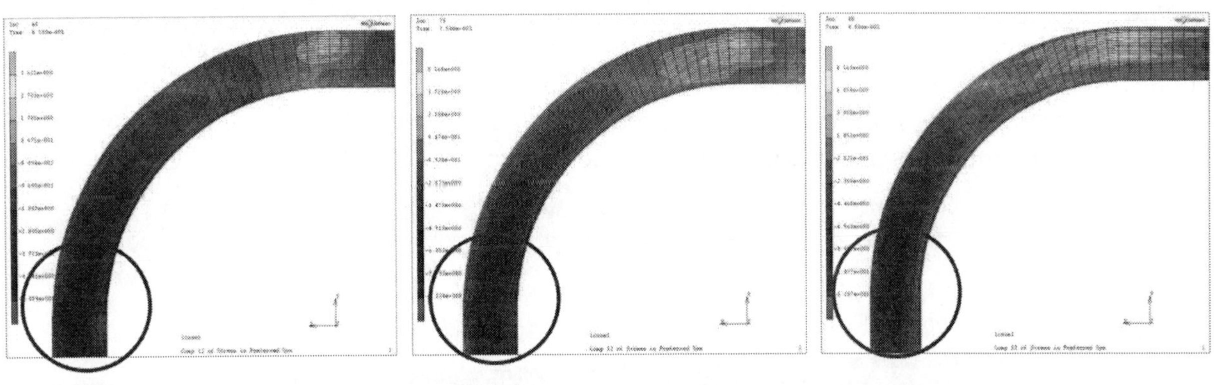

Fig 12. Interlaminar shear stress distribution- all three layups

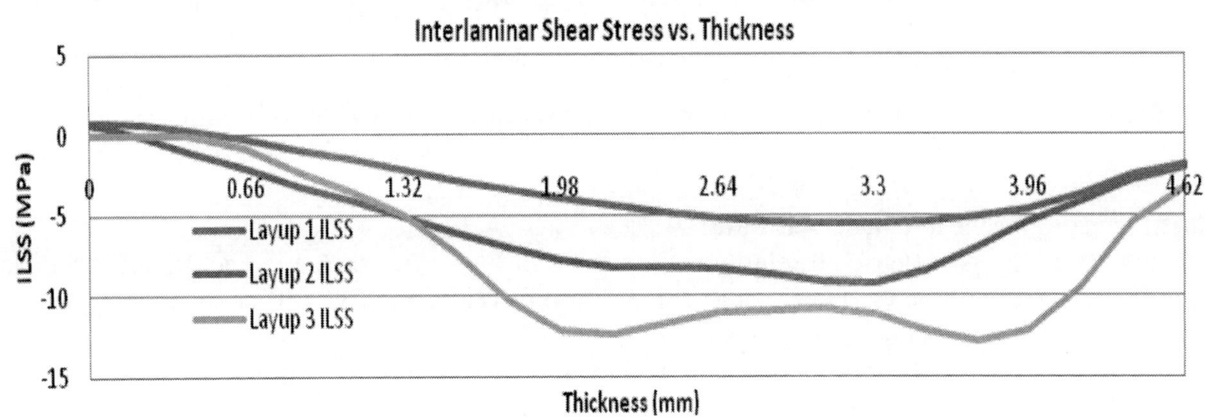

Fig. 13. Interlaminar Shear Stress distribution just before first failure – all three layups

With the third layup, delamination occurred between third and fourth layers, uni-directional and Double bias respectively. Addition of unidirectional fibres increased the stiffness, failed 79.40mm. The maximum interlaminar tensile stress was at an angle of 14.60^0 and maximum shear stress of 10.22MPa at the lower end of the laminate curvature.

	1st failure deflection	Max. ILTS (MPa)	Max. ILSS (MPa)
Layup 1	64.08 mm	9.65	5.56
Layup 2	66.80 mm	10.03	7.52
Layup 3	79.40 mm	10.39	10.22

Table 3. FEA results - Location of initial delamination and peak ILTS

As the specimen was not loaded by pure moment, the failure does not occur at 45^0. FEA results of layups 1 and 2 were comparable to the experimental data. FEA prediction for layup 3 was not comparable with the experimental results as the residual strength could not be modeled. The maximum shear stress for all the three layups occurred at the lower end of the laminate bend. The failure criterion for the finite element analysis was a function of both the interlaminar tensile and shear stresses. Hence failure initiated near the bottom curvature as shown in Fig.10 & 12.

ACOUSTIC EMISSION – PARAMETRIC ANALYSIS:

The history of AE events represents the damage initiation and progression in the material. The non-linear behaviour of the specimens, the AE counts, was studied as a function of time as shown in Fig. 14. When the laminated composite starts to generate more AE events, it indicates that the damage has been initiated in the material. The CSM+DB layup showed signs of saturation with the accumulation rate being stagnant at failure. The other two layups showed increased AE counts, demonstrating the damage progression. The accumulation rate of AE increased even though a further decrease in load was observed as shown in Figs 15, 16 & 17. Load drop at the end of the curve indicates the extensive damage development in the material.

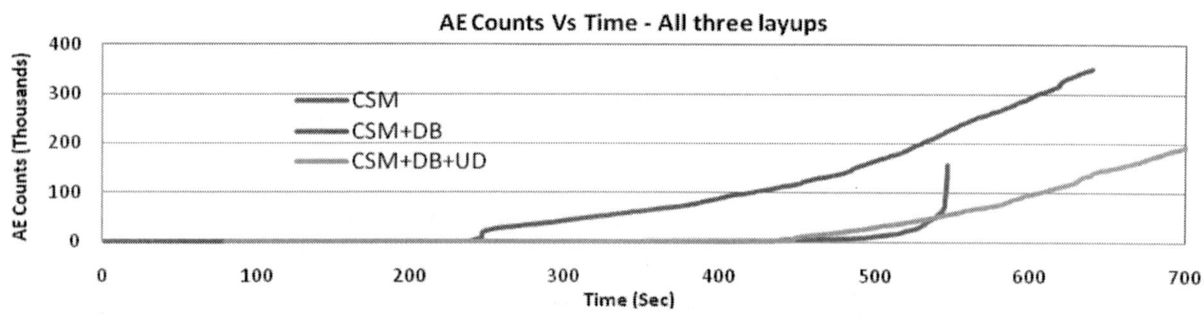

Fig. 14. AE- Cumulative counts vs. time for all three layups

All the three different layups displayed more extensive damage due to non-homogenous stress fields between the layers composed of strongly anisotropic plies with different fibre orientations.

Parametric Correlations with AE data:

The results of the parametric correlations are shown in Figs. 15, 16 & 17. Each graph contains four plots, (1) Amplitude vs. Time, (2) Event duration vs. Time, (3) Event Rise time vs. Time and (4) Energy release vs. Time. In case of a fracture, elastic energy is released and can be measured by acoustic emission. In this experimental program the progressive failure behaviour of the specimens having different composite layups were analysed by comparing the four acoustic emission parametric variables along with the load vs. time plot.

Layup 1 AE Correlation: All layers were chopped strand mat; hence there was no difference in the elastic modulus between the individual layers. Four distinguishable damages were analysed by AE as shown in Fig.15. The initial scatter up to 375 sec accounts for matrix cracking. The first major damage was observed at 375 sec between second and third layers at 30% of ultimate load. The primary sign initial delamination is shown as the energy release of 563 mJ, amplitude of 78 dB, event rise time of 510μs and event duration of 7139 μs. The small load drop in the load-time plot complements the failure analysed by AE. Further damage up to 438 sec was the matrix cracking and delamination growth. Next major delamination occurred at 438 sec between third and fourth layer, with energy release of 521mJ, amplitude of 92dB, event rise time of 546 μs and event duration of 4327 μs. Later the length of delamination extended between which is shown in Fig.15. The third major delamination occurred between fourth and fifth layer at 499 sec with energy release of 490mJ, amplitude of 99dB, event rise time of 726 μs and event duration of 1819 μs. No considerable reduction in load was observed between initial delamination at 375sec and final collapse at 546sec. The final collapse at 546 sec was a combination of matrix cracking, fibre failure and delamination with energy release of 15609mJ, amplitude of 100dB, event rise time of 464 μs and duration of 162365 μs.

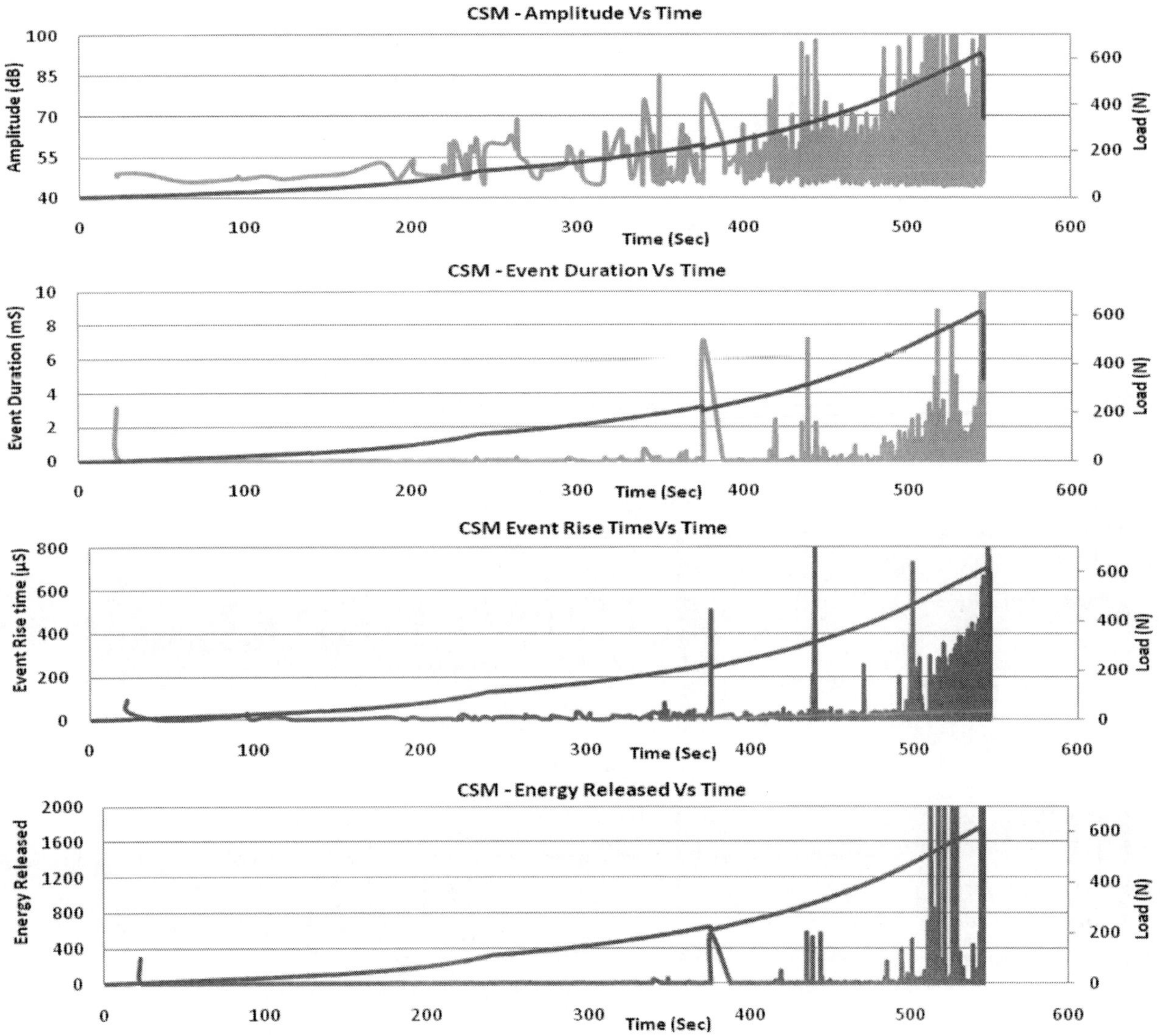

Fig. 15. Parameter Correlation for Layup 1 specimens

Layup 2 AE Correlation: The parametric distribution of AE events for CSM+DB layup is shown in Fig. 16. The flexure modulus and strength of double bias layer were very low, 1158 MPa & 78MPa respectively and the laminate was very flexible. The first damage occurred at 10% of ultimate load at 247sec, delamination between second and third layers. The parametric values at 247sec were Amplitude – 100dB, Energy – 2680mJ, Rise time -708μs and duration - 53027μs. Further damage was characterized by matrix cracking and the growth of delamination. The amplitude of the signal was widely distributed and steady from the initial failure till final collapse; hence progressive damage was assessed with other parameters mainly with energy released. Delamination between third and fourth layer occurred at 416 sec with a duration of 5521μs energy of 1091mJ and an amplitude of 91dB. The third highest peak in the duration plot occurred at 485 sec with a duration of 8602μs, energy of 1475mJ and an amplitude of 95dB. Further AE data was widely scattered accounting for continuous matrix cracking and delamination growth. Final collapse of the structure occurred at 636sec with an amplitude of 92dB, duration of 2331μs at 573.6 N.

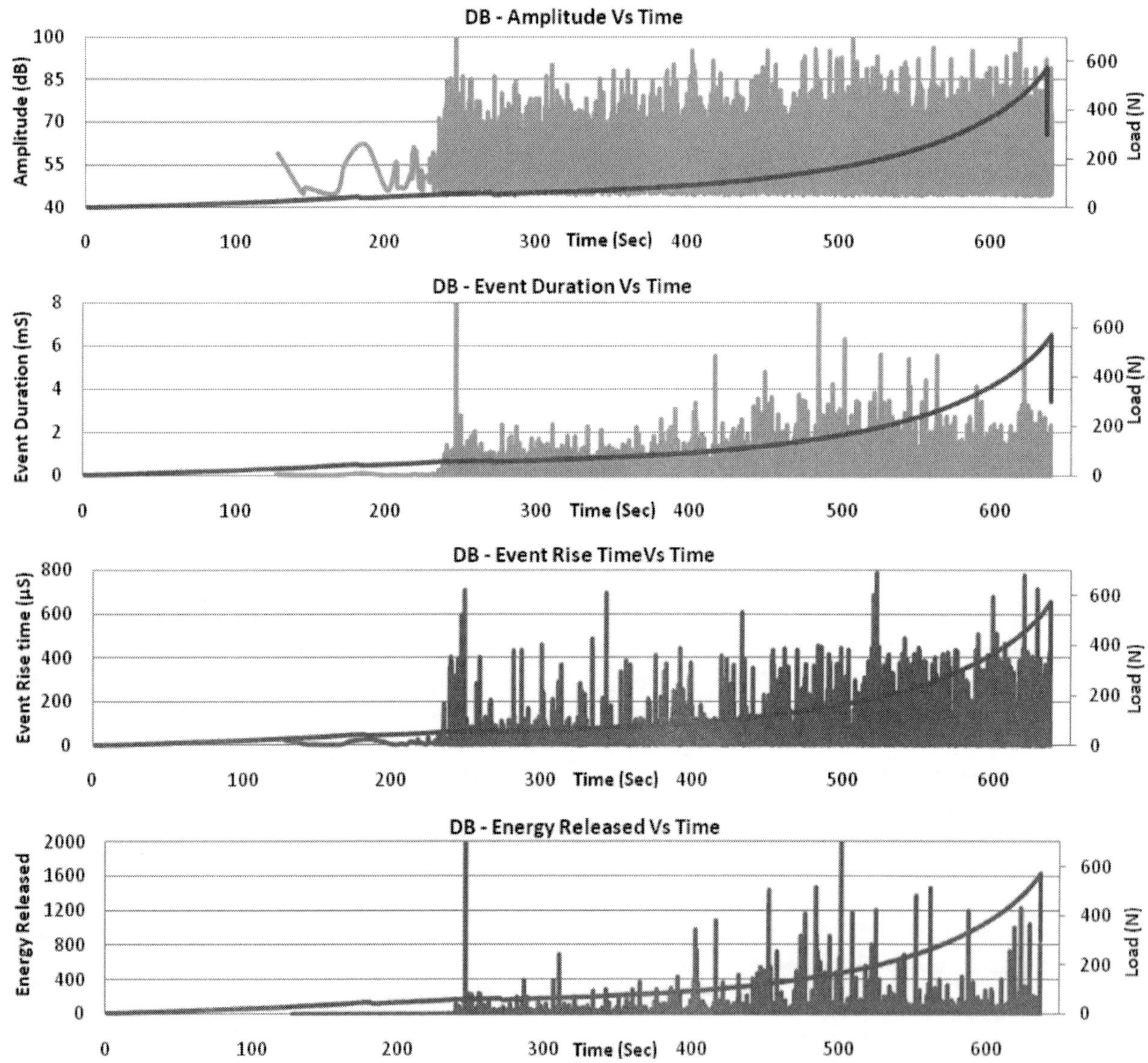

Fig. 16. Parameter Correlation for Layup 2 specimens

Layup 3 AE Correlation: This has two CSM, two UD and three DB layers. The initial stage of damage was observed at 9% of the ultimate load at 370 sec shown in Fig.17. Though no reduction of load was observed due to the presence of UD fibres which carried most of the load, other AE parameters showed fibre extensive matrix cracking and fibre breakage – energy of 2103mJ, amplitude of 95dB and duration of 4076μs. Six peaks in the energy and even duration plots demonstrate progressive delamination until failure. The first major damage was observed at 438 sec at 14% of ultimate load, delamination between second and third layers. The energy released was 6496mJ, amplitude of 100dB, rise time of 469μs and duration of 20912μs. Further delaminations were observed at 550sec (14% of ultimate load), 587sec (17%), 639sec (26%) and final collapse at 720 sec (1800N). Up to 26% of ultimate load, the load deflection plot was linear. At this point, all the CSM and DB layers were completely damaged and were incapable to carry further loads. Unidirectional fibres carried the entire load until final collapse. The residual strength of the unidirectional fibres is demonstrated by, carrying the majority of the load even after the failure of other layers. This is the primary motive of using unidirectional fibres in marine structures, to hold the structure in position beyond failure.

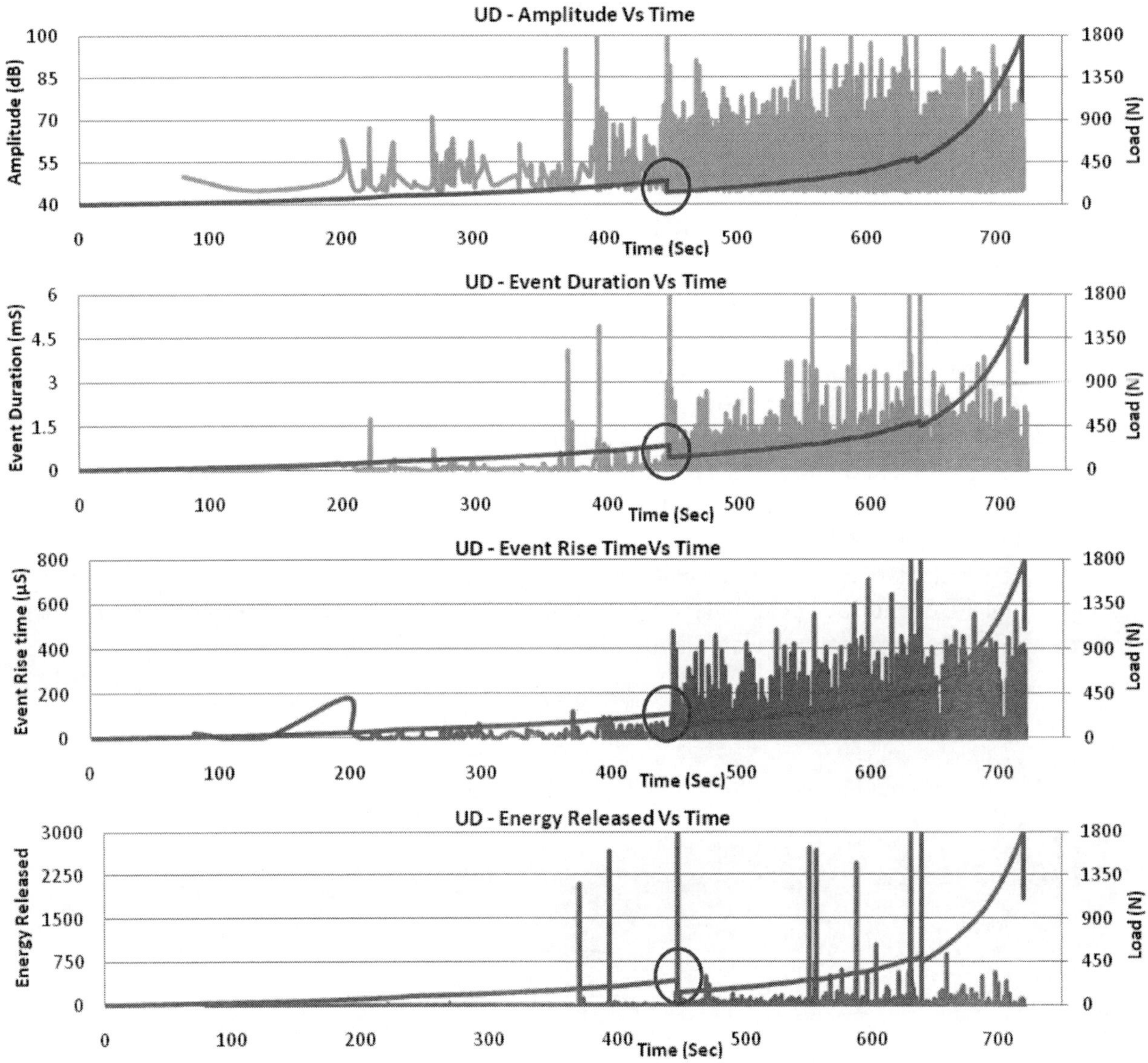

Fig. 17. Parameter Correlation for layup 3 specimens

CASE STUDY – Investigation of failure behaviour of laminated composites by embedded Fibre Bragg Gratings (FBG):

Optical fibres can transmit light over a long distance with minimal loss of light and the properties of the light inside the fibre are affected by physical parameters outside the fibre. This meant that the fibre itself could be used as both the sensing element and the communication path for the signal between the sensor and the optical interrogator. An FBG is created in the optical fibre core by illuminating the side of the fibre with alternating high/low stripes of UV light that induces the refractive index change in the fibre.

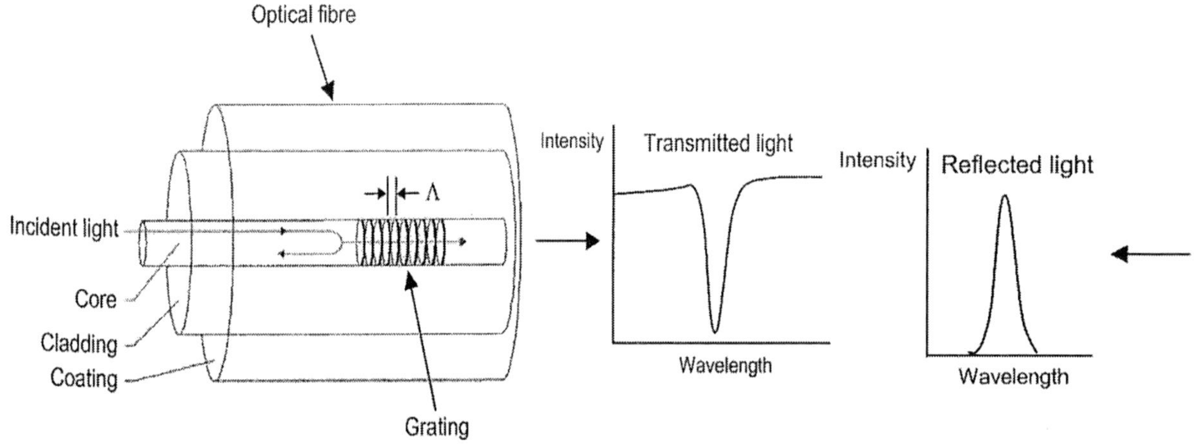

Fig18. The Fibre Bragg Grating Sensor

In the current study Fibre Bragg Grating (FBG) is manufactured using a photosensitive single mode GF1 optical fibre was exposed to a 150mW UV source provided by an argon ion laser. The optical fibre used in this laminate was a single mode fibre with an acrylate coating. The fibre diameter was 250μm with the coating. The length of the FBG was 9mm and was placed at the centre of the bend. FBG reflects the wavelength of light that is dependent on the pitch of the stripes in the grating, shown in Fig. 18. AS the FBG is strained, the grating pitch changes so does the wavelength of the reflection. Monitoring the wavelength reflected from an FBG therefore provides an absolute method of detecting point strains at the location of the grating [27]. The FBG is very sensitive to temperature variations. In the current study, the manufacturing and testing of the FBG and the specimens were maintained at almost constant temperature (22^{0}C) so as to minimise the temperature related anomalies.

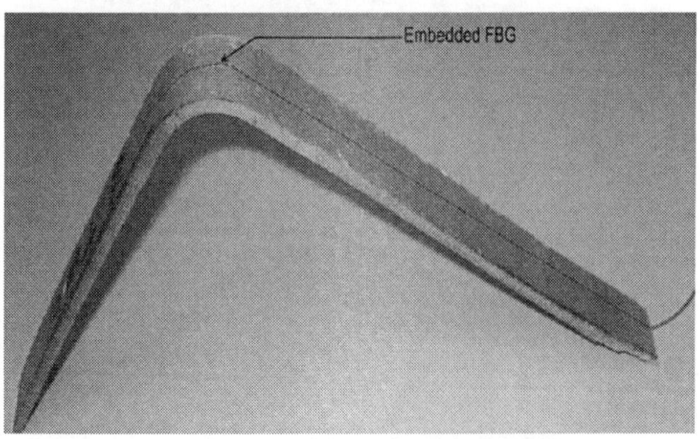

Fig 19. Specimen with embedded FBG

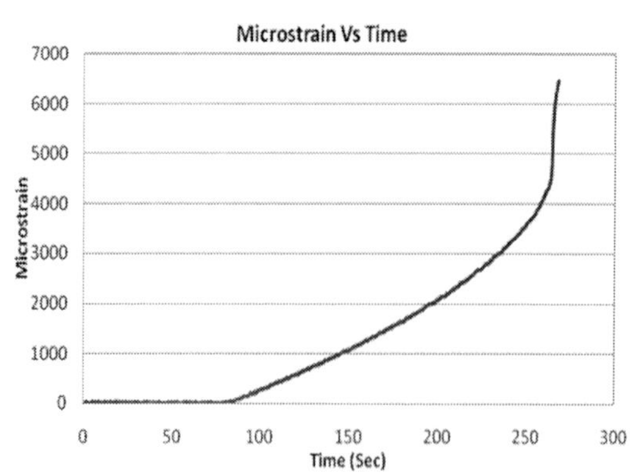

Fig.20. FBG response during the test

The optical fibre was spliced with the second optical fibre with another FBG which was unstrained and used as the reference FBG. The specimen with embedded optical fibre is shown in Fig.19 and the FBG response is shown in Fig.20 as micro-strain. The FBG reflectivity wavelength was 1556.1 nm before embedding and 1545.635 nm after embedding. The reference FBG wavelength was 1556.451 nm. The basic material of the optical fibre was silica, and the fibre thickness was just 250 μm. Hence the cohesion between the optical fibre and the fibre mat was assumed to be satisfactory with minimum local stress concentration.

Optical fibre was embedded between first and second layers during manufacturing of laminate of CSM+DB. On application of the mechanical loading, the first three layers subjected to tension, hence the optical fibre was subjected to tensile stress. The real time strain measurement by embedded FBG provided an understanding of the strain in the laminate. At 266sec, a crack occurred between second and third layer. Just before the crack, huge tensile stress was experienced and this is clearly shown in Fig.20 as the rapid increase in the strain. There was a sudden increase in the amplitude at the same time with a reading of 100 dB with the AE. The rise time was 760μs and the duration was around 52250μs. The longer duration AE signals accounts for macro damage such as delamination. Finite element model showed that the critical ILTS was 12.71 MPa at 66.80mm which correlated with the AE and the strain measured by the embedded FBG. The structure delaminated at 266 sec and within 2 sec, and the optical fibre failed in tension, as the critical tensile strain limit of the fibre was exceeded. The embedded FBG was successful in capturing the structural behaviour until initial failure. This study was a pilot study for the future work of top hat stiffeners for keel structures. Hence only one optical fibre subjected to tensile stress was tested. Future experiments will focus on embedding two optical fibres, one in tension and one in compression so as to compare the strain sensitivity of FBG.

CONCLUSION

The behaviour of curved laminated composites subjected to bending load was studied. Specimens with laminates were tested until failure. Load carrying capacity, failure initiation and progression were studied for each laminate. The failure was completely dominated by the matrix failure in the through thickness direction. Each layer has different Young's modulus and Poisson's ratio in longitudinal direction. Hence each layer deforms at different pace, e.g., small deformation in the Uni-directional fibres. The bending load in this study generates more of interlaminar tensile stress compared to interlaminar shear stress. Also, the difference in the Young's Modulus gives rise to high interlaminar tensile stresses at the bend, and this may cause premature failure (delamination) in the structure for the second and third layup. Hence Finite element modeling was used to predict the failure load and the interlaminar stress distribution around the bend.

The Damage was assessed with two different techniques, Acoustic Emission and Fibre Bragg Gratings. Considering that the fibre composite systems used are fibre mats, and the specimens were subjected to an out-of-plane bending load, these failure mechanisms are reasonable. Embedded FBG can be vital in obtaining critical strain just before delamination. Further investigation with more FBG embedded laminates could confirm the need to be used in marine structures for efficient structural design. The residual strength of the structure was identified so as to assist the boat builders in designing the curved structures. Addition of unidirectional layers had higher residual strength compared to other two layups. The improved knowledge of the low ratio of delamination strength to ultimate tensile strength for the laminates tested (particularly the mixed laminates) is of benefit for designers.

REFERENCES

1. Shenoi R.A., Wang W. (2001), Through-thickness stresses in curved composite laminates and sandwich beams, Composites Science and Technology, 61, 1501-1512.
2. Wisnom M.R. (1996), 3-D Finite element analysis of curved beams in bending, Journal of Composite Material, 30(11), 1178– 1190.
3. Kaczmarek K, Wisnom M.R, Jones M.I. (1998), Edge delamination in curved (04/_456)s glass-fibre/epoxy beams loaded in bending, Composite Science and Technology, 58, 155–161.
4. Martin R.H, Jackson W.C., (1993), Damage prediction in cross plied curved composite laminates, Composite materials: Fatigue and Fracture, Fourth Volume ASTM STP 1156, 105-106.
5. Kedward K.T., Wilson R.S., Mclean SK, (1989), Flexure of simply curved composite shapes, Composites, 20 (6).
6. Wisnom M.R., Petrossian Z.J., Jones M.I. (1996), Interlaminar failure of unidirectional glass/epoxy due to combined through thickness shear and tension, Composites Part A, 27A, 921-929.
7. Heinz. D., Ritcher B., Weber S. (2000), Application of advanced materials for ship construction – Experience and problems, Materials and Corrosion, 51, 407-412.
8. Davies P., Petton D. (1999), An experimental study of scale effects in Marine composites, Composites Part A, 30, 267-275.
9. Kim J.K., Sham M.L. (2000), Impact and delamination failure of woven-fabric composites, Composite science and technology, 60, 745-761.
10. Mouritz A.P., Leong K.M., Herszberg I. (1997), A review of the effect of stitching on the in-plane mechanical properties of fibre-reinforced polymer composites, Composites Part A, 28, 979-991.
11. Mouritz A.P., Gallagher J., Goodwin A.A. (1997), Flexural strength and interlaminar shear strength of stitched GRP laminates following repeated impact, Composite science and technology, 87, 509-522.
12. Antonio M., Miguel A.J. (2002), Application of finite element method to prediction of onset of delamination growth, Applied mechanical review, 55(2).
13. Dodkins A.R., Shenoi R.A., Hawkins G.L. (1994), Design of joints and attachments in FRP ship structures, Marine Structures, 7, 364-398.
14. Liptai R.G., Harris D.O. (1972), Acoustic emission – An introductory review, Materials research and standards, 11(3), 8-10.
15. CARP Aerospace/Advanced composites subcommittee, (1993), Guidance for development of AE applications on composites, Journal of acoustic emission, 11(3), C1-C24.
16. Bohse J., Kroh G. (1992), Micromechanics and acoustic emission analysis of the failure process of thermoplastic composites, Journal of Material Science, 27, 298-306.
17. Bohse J. (2000), Acoustic emission characteristics of micro-failure process in polymer blends and composites, Composites science and technology, 60, 1213-1226.
18. Hartmut V. (2002), AE testing fundamentals, equipment, applications, www.NDT.net, 7(9).
19. Bhat M.R., Majeet M.A., Murthy C.R.L. (1994), Charaterisation of fatigue damage in unidirectional GFRP composites through acoustic emission signal analysis, NDT & E international, 21(1).
20. Leszek G, Kanji O. (1995), Acoustic emission analysis of laminate failure mechanisms with reference to failure criteria, AECM-5: Fifth international symposium, 5.
21. Benmedakhene S., Kenane M., Benzeggagh L. (1999), Initiation and growth of delamination in glass/epoxy composites subjected to static and dynamic loading by AE monitoring, Composites science and technology, 59, 201-208.
22. Huguet S., Godin N., Gaertner R., Salmon L., Villard D. (2002), Use of acoustic emission to identify damage modes in glass fibre reinforced polyester. Composite Science and Technology, 62, 1433-1444.
23. Ageorges C., Friedrich K., Schuller T., Lauke B., (1999), Single-fibre Broutman test: fibre-matrix interface transverse debonding. Composites: Part A; 30: 1423-1434.
24. Yuris A.D., Jie Q., (2001) Analysis of microdamage evolution histories in composites. International Journal of Solids and Structures: 38: 1831-1854
25. Nahas M.N. (1985), Analysis of composite failure mechanisms using acoustic emission and ultrasonic scanning techniques, Composites, 16(2).
26. Li R., Kelly D., Ness R. (2003), Application of first invariant strain criterion for matrix failure in composite materials, Journal of Composite Materials, 37(22).
27. Li H.C.H., Herszberg I., Davis C.E., Mouritz A.P., Galea S.C., (2006), Health Monitoring of Marine composite structural joints using fibre optic sensors, Composite Structures, 75, 321-327.

Marine Corrosion of 5XXX Aluminium Alloys

P.L.Mart, E. Pescott, S. Russo* and L.M. Pedrina
Defence Science Technology Organisation, Australia
* formerly DSTO, now Thales Australia
peter.mart@dsto.defence.gov.au

ABSTRACT
Aluminium 5XXX series alloys are increasingly used in the naval shipbuilding industry and are susceptible to sensitization due to segregation of β-phase at grain boundaries. This gives rise to intergranular corrosion (IGC), exfoliation corrosion, and stress corrosion cracking (SCC) under certain environmental conditions. This paper reports the corrosion behaviour of AA5083, AA5383 and AA5059 aluminium alloys, with significant variation between the alloys with respect to sensitization and associated corrosion. AA5059 and AA5383 were found to be most susceptible to intergranular corrosion, and AA5059 significantly more susceptible to exfoliation corrosion.

INTRODUCTION

Marine grade aluminium alloys are increasingly used for high speed vessels, naval ships and patrol boats, due to their reasonable strength, light weight and corrosion resistance. The AA5XXX series of aluminium-magnesium alloys are non-heat treatable and so are strain hardened. Alloy AA5083, originally developed for land-based and armour applications, has been used for several decades by shipbuilders, with two tempers H116 (strain hardened for exfoliation corrosion resistance) and H321 (strain hardened and stabilized) used interchangeably. The alloy AA5383 was subsequently developed for marine applications, with both enhanced mechanical properties and superior corrosion resistance. More recently, AA5059 was developed to give mechanical properties superior to both AA5083 and AA5383, and corrosion resistance at the same level or better than AA5083.

A problem with high magnesium AA5XXX alloys, when Mg>3%, is that magnesium–rich Mg_2Al_3 or β-phase can form at the grain boundaries after a sensitization heat treatment (1). The β-phase is anodic to the alloy, and this renders the alloy susceptible to IGC, exfoliation and SCC. The nitric acid mass loss ASTM G 67 (NAMLT) Test (2) is used to quantitatively determine an alloy's susceptibility to IGC, and the ASTM G 66 (ASSET) Test (3) is used to provide a visual assessment of an alloy's exfoliation corrosion susceptibility. Following recent North American experiences (4) of significant cracking of AA5083-H321 in marine applications, a new ASTM standard B 928/B928 M was developed (5) that requires AA5XXX alloys to pass both tests. The sensitization of the North American AA5XXX alloys arose from a change in mill processing (4). However, sensitization can also result from in-service conditions. ASTM B928/B928M-07 notes that alloys 5059, 5083, 5086, 5383 and 5456 should not be used for service which provides prolonged exposure to temperatures exceeding 65°C (whether continuous or discontinuous exposure) because of the risk of sensitization and the resultant susceptibility to IGC and SCC.

SCC requires three conditions to be fulfilled: the alloy must be susceptible, there must be a specific corrosive environment, and there must be a component of tensile stress perpendicular to the grain boundaries to open them up. There have been numerous SCC studies of AA5XXX alloys (6).

As vessels may operate in a tropical environment, where deck temperatures can easily exceed 65°C, there is a need to determine the risk of sensitization and associated corrosion in the range of AA5XXX alloys and tempers that may be used. This is receiving considerable attention from US Office of Naval Research (7), as the US Navy accelerates use and acquisition of high speed aluminium vessels. Australia has two world-leading designers and builders of such vessels, which have undergone extensive US operations. Austal USA has won US Navy orders to design and build two Littoral Combat Ships, and also to design and build the US Department of Defense's first Joint High Speed Vessel.

The RAN's Austal-built Armidale Class Patrol Boats (ACPB), constructed from marine grade AA5083-H321 hull plate, are based and operate in tropical waters. Landing Craft Mechanized (LCM 2000), also constructed from AA5083-H321 alloy, similarly operate extensively in a tropical environment. The RAN has experienced IGC, exfoliation corrosion and SCC (in conjunction with corrosion fatigue) in AA5083-H321/H116 deck plating on Landing Platform Amphibious ships.

DSTO has commenced generic investigations into the corrosion performance of marine grade AA5XXX alloys, including alloys in current usage and as well those that may be used in future acquisitions. The aim is to inform the through life maintenance support of existing vessels constructed from these alloys and tempers, and the selection of appropriate materials in future acquisitions, from a corrosion perspective. This paper reports some results from DSTO investigations which have been reported in more detail elsewhere (8). The focus of this paper is to raise awareness in the maritime community of the potential for sensitization and associated corrosion in marine grade aluminium alloys, and the need for further studies.

EXPERIMENTAL

Testing was conducted on three AA5XXX alloys, and involved sensitization heat treatments, followed by optical microscopy, NAMLT and ASSET testing, and SCC testing.

Preparation of Samples

Details of the sheet aluminium alloys used are shown in Table 1, and composition limits (5) are shown in Table II.

Alloy	Temper	Thickness (mm)	Comments
AA5083	H116	3.0	Complies with ASTM B928, DNV approved
	H321	5.0	DNV approved
AA5383	H116	3.0	Alcan Sealium Marine Grade, DNV approved
AA5059	H321	3.0	Corus ALUSTAR™ H321

Table I. Details of aluminium alloys used

Alloy	Si	Fe	Cu	Mn	Mg	Cr	Zn	Ti	Zr	Aluminium
AA5083 limits	0.40 max	0.40 max	0.10 max	0.40-1.0	4.0-4.9	0.05-0.25	0.25 max	0.15 max	-	remainder
AA5383 limits	0.25 max	0.25 max	0.20 max	0.70-1.0	4.0-5.2	0.25 max	0.40 max	0.15 max	-	remainder
AA5059 limits	0.50 max	0.50 max	0.40 max	0.60-1.2	5.0-6.0	0.30 max	0.4-1.5	0.20 max	0.05-0.25	remainder

Table II. Composition limits of the alloys used (5).

For the NAMLT tests, specimens were prepared as per ASTM G 67-04 by bandsawing and machining to final dimensions 50 mm by 6 mm by product thickness, with the 50 mm dimension parallel to the rolling direction. For the ASSET tests, specimens were initially prepared as per ASTM G 66-99 (2005) by bandsawing and machining to final dimensions 20 mm by 50 mm with the 20mm direction parallel to the rolling direction. Samples were also machined to remove 10% of the thickness so as to retain one as-manufactured surface and one machined surface, required by ASTM B928.

DSTO used the ASCOR (Automated Stress Corrosion Ring) test (9) to determine the SCC behaviour of AA5XXX alloys. The ASCOR test is a constant strain test, and uses direct stressing of a pin-loaded flat tension test specimen in a loading ring instrumented with strain gauges to measure the load. Dogbone test specimens of the required geometry (9), with the length of the specimens in the long transverse (LT) direction of the sheet, were manufactured from 3 mm sheet by milling to a final thickness of 2.0 mm after removing 0.5 mm from each face. Specimens of all the alloys were prepared for exposure in ASCOR tests with the exception of AA5083-H321.

Sensitization of Samples and Corrosion Testing

As well as testing as-received machined samples, samples of each machined alloy were heated in an oven for 100 hours at three temperatures (150°C, 100°C and 65°C respectively) and at 65°C for different times (100, 336, 672 and 1344 hours respectively). The highest temperature was chosen to promote a significant amount of sensitization in a reasonable time, while the lowest temperature was selected in accordance with the lower limit for onset of sensitization in ASTM B928/B928M-07. As-received and sensitized samples were prepared for metallographic examination by polishing and then etching in 40% phosphoric acid at 35°C for 5 minutes.

The NAMLT testing procedure was conducted as per ASTM G 67-04, and mass loss per unit area calculated. The ASSET testing procedure was conducted as per ASTM G66-99, except that ASSET samples were one quarter of the recommended surface area, in order to reduce the volume of test solution required.

Stress Corrosion Cracking Tests

The samples were secured in the ASCOR loading device as shown in Figure I, and loaded to 90% of the yield stress. All rings were pre-calibrated using a strain-gauged cylindrical specimen. The ASCOR tests were conducted using the specified 0.6M sodium chloride solution initially, but when this was found to be insufficiently corrosive, later tests were conducted by exposing samples to a more corrosive exfoliation corrosion (EXCO) solution according to ASTM G34. The composition of the EXCO solution was NaCl 234 g/litre, KNO_3 50 g/litre, HNO_3 (68%) 6.8 mg/litre. Solution was dripped onto the specimen at a constant rate of 0.4 ml/hour, using a peristaltic pump. The load was measured at intervals of 300 seconds and logged by a computer.

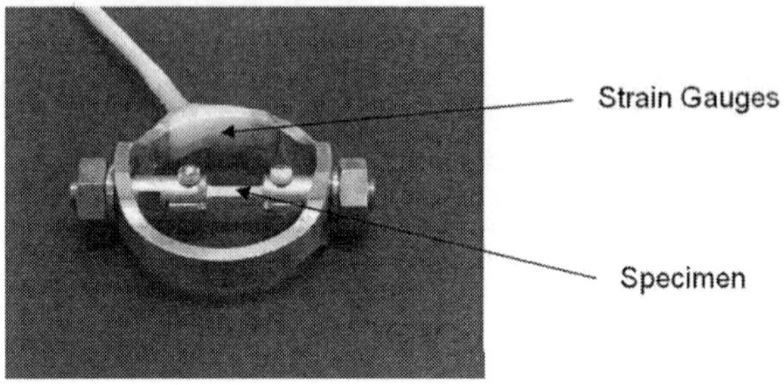

Figure I. ASCOR Loading device and specimen

RESULTS

Sensitization

Figures II and III show optical micrographs of two of the etched alloys, as-received and aged for 100 hours. No significant grain boundary attack was found for either of the AA5083 tempers or the AA5383 alloy in the as-received condition or after heat treatment at 65°C. However, grain boundary attack was observed for these alloys after heat treatment at both 100°C and 150°C. The AA5383 sample heat treated at 150°C suffered widespread general corrosion, as did the AA5059 sample. The AA5059 showed a similar extent of grain boundary attack compared with AA5083 and AA5383 alloys after heat treatments at 100°C and 150°C, but showed a greater degree of grain boundary attack in both the as-received condition and after heat treatment at 65°C.

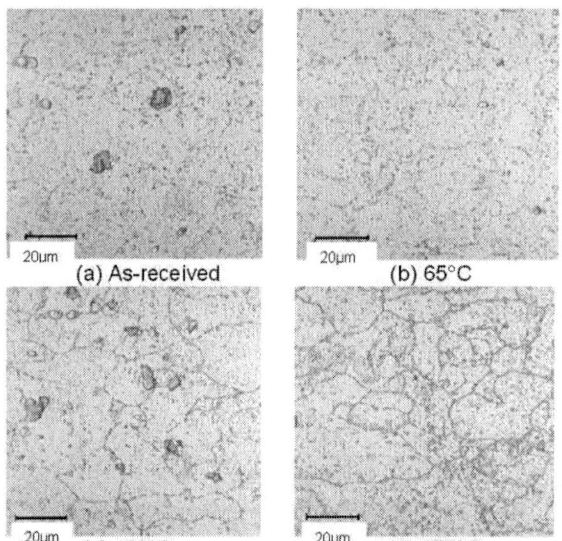

Figure II. Microstructures of the top surface of AA5083-H116 sheet
(a) as-received and after 100 hour heat treatments at (b) 65°C (c) 100°C and (d) 150°C.

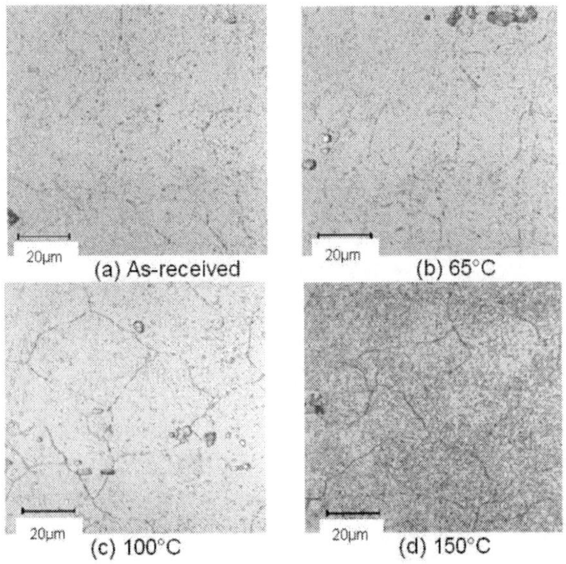

Figure III. Microstructures of the top surface of AA5383-H116 sheet
(a) as-received and after 100 hour heat treatments at (b) 65°C (c) 100°C and (d) 150°C.

NAMLT Testing

The NAMLT values for samples aged for 100 hours at various temperatures are shown in Figure IV. The values for the as-received samples are shown at a nominal 25°C. The NAMLT values for samples heat treated for times up to 1350 hours at 65°C are shown in Figure V. Both figures show the NAMLT values and the regions where ASTM B 928/928M-07 designates samples to be susceptible (>25 mg/cm^2) or resistant (<15 mg/cm^2) to IGC,

All alloys showed no susceptibility to IGC in the as-received condition and after sensitization at 65°C for 100 hours (Figure IV). AA5059 and AA5383 both showed susceptibility to IGC after sensitization at both 100°C and 150°C while both the AA5083 tempers only showed susceptibility after sensitization at 150°C. In the extended time test, at 65°C, the AA5059 and AA5383 alloys failed after approximately 670 hours sensitization while neither of the AA5083 tempers failed during this test.

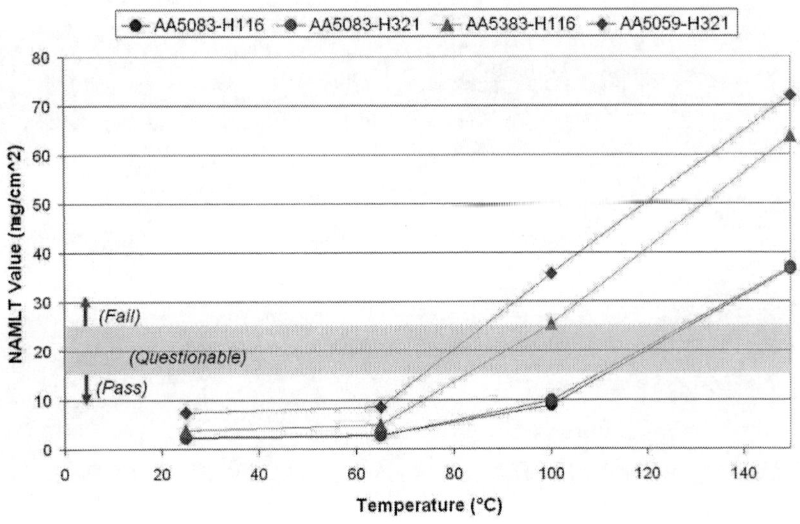

Figure IV. NAMLT values for as-received (25°C) alloys and for alloys heat treated for 100 hours at 65°C, 100°C and 150°C. The pass/fail criteria are with reference to ASTM B 928.

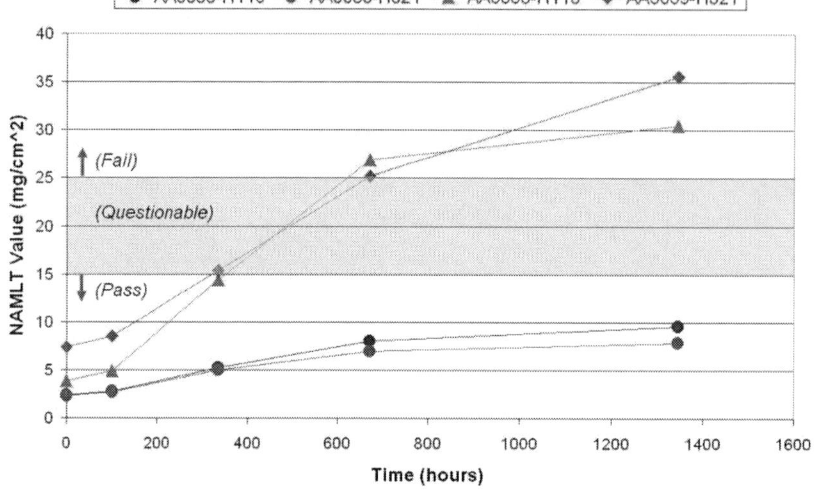

Figure V. NAMLT values versus heat treatment time at 65°C for 5XXX alloys. The pass/fail criteria are with reference to ASTM B 928.

ASSET Testing

In Figure VI, samples of the as-received alloys and samples heat treated for 100 hours at 150°C are compared after ASSET testing. The upper surface shown for each sample is the machined face. Samples of both AA5083 tempers heat treated at 150°C showed etching of both the machined and non-machined obverse surfaces, and very few edge pits. The AA5383-H116 alloy showed moderate pitting over both machined and non-machined surfaces, with a few signs of exfoliation around some pits, and no edge attack. The AA5059-H321 alloy showed high exfoliation of both machined and non-machined surfaces, but little edge attack. No appreciable surface attack was seen for three of the alloys in the as-received condition, or after heat treatment at 65°C, (both AA5083 tempers & AA5383), but slight pitting was seen for the AA5059-H321 alloy in both cases. A similar gradation of attack was seen for the alloys heat treated at 100°C, where only the AA5059-H321 showed moderate pitting all over the surface, and extensive exfoliation.

Figure VI. Samples heated at 150°C for 100 hours, machined face, after ASSET test.
(a) AA5083-H116 (b) AA5083-H321 (c) AA5383-H116 (d) AA5059-H321

The ASSET ratings of the samples, judged against the photographs and captions in ASTM G66-99 (2005), are summarized in Table III in the specified format. ASTM B928/B928M-07 requires AA5XXX alloys to exhibit no evidence of exfoliation corrosion and a pitting rating of PB or better. While all of the as-received alloys meet this criterion, the AA5059-H321 alloy fails to do so after sensitization heat treatments at both 100°C and 150°C for 100 hours.

Sample	AA5083 - H116	AA5083 - H321	AA5383-H116	AA5059-H321
As-received	N	N	N	PA
Heated 65°C for 100 hours	N	N	N	PA
Heated 100°C for 100 hours	N	N	N	EC
Heated 150°C for 100 hours	N	N	PA	ED

Table III. ASSET test rating of as-received and heated 5XXX samples.
N No appreciable attack, P Pitting, E Exfoliation, grading from A (low) to C (high) for pitting, and A (low) to D (high) for exfoliation.

ASCOR Testing

The load-time records for the alloys heated at 65°C, shown in Figure VII, displayed considerable curvature. This corresponded to the observed general corrosion of the specimens, due to the high corrosivity of the EXCO solution. Alloys heated at 100°C for 100 hours and loaded at 75% yield stress in EXCO solution also underwent extensive general corrosion. Figure VII also shows steps in the load-time graphs, corresponding to multiple SCC events prior to eventual complete sample failure.

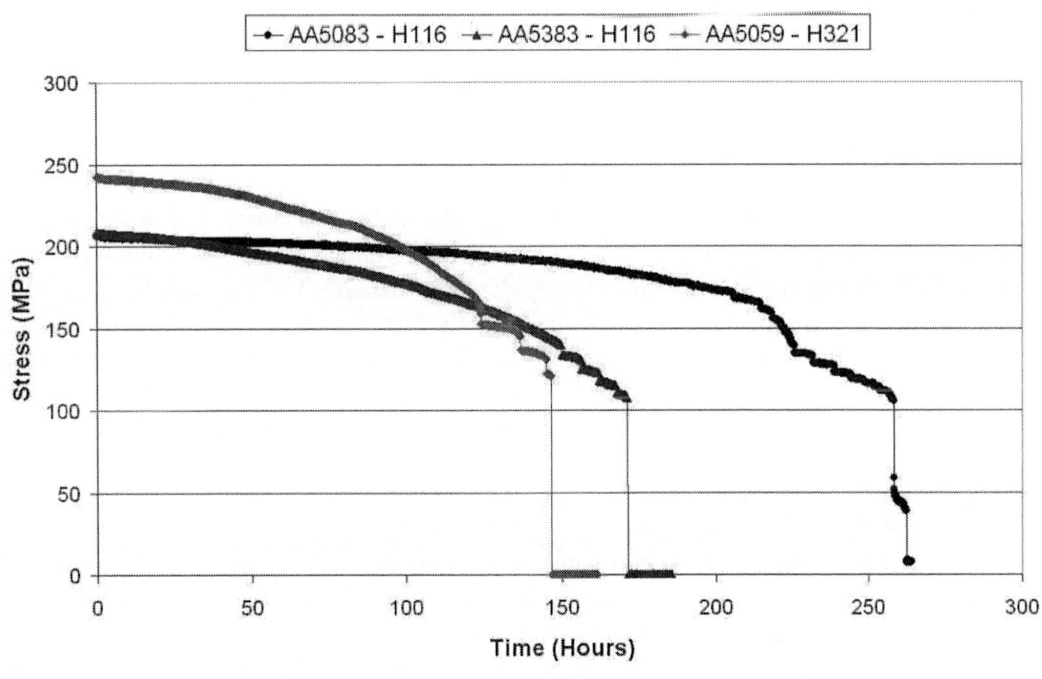

Figure VII. ASCOR Load-Time results for AA5XXX alloys heated at 65°C for 100 hours, then exposed to EXCO solution at 90% yield stress.

The samples heat treated at 150°C for 100 hours (not shown) each displayed no curvature in the load-time graphs, and a single SCC event leading to sample failure.

The SCC behaviour for the alloys exposed to EXCO solution is summarized in Table IV, where the time to sample failure is given rather than the initiation of SCC, due to the curvature of many of the load-time graphs. It is evident that AA5083-H116 was considerably more resistant to SCC than the other AA5XXX alloys examined.

Sample Heated	Load % Yield	Solution	AA5083-H116	AA5383-H116	AA5059-H321
65°C for 100 hrs	90	EXCO	270*	170*	140*
100°C for 100 hrs	75	EXCO	455*	65*	132*
150°C for 100hrs	90	EXCO	110	2.5	20

* General corrosion of sample prior to failure

Table IV. ASCOR Stress Corrosion Cracking of 5XXX alloys: Times to Failure (Hours)

DISCUSSION

The optical micrographs and NAMLT tests showed that none of the alloys showed sensitization in the as-received condition, or when heat treated at 65°C for 100 hours. When heated for 100 hours at 100°C, all the alloys were sensitized, but only AA5059 was appreciably susceptible to IGC, with AA5383 marginally susceptible, and both AA5083 tempers not susceptible. All four alloys were sensitized when heated for 100 hours at 150°C, and were susceptible to IGC, with AA5059 most susceptible, followed by AA5383, while both AA5083 tempers were considerably less susceptible.

The NAMLT tests for extended times at 65°C show sensitization of all the alloys. AA5059 and AA5383 were susceptible to IGC after 700 hours, but both AA5083 tempers were still resistant to IGC after 1350 hours. Although the AA5083 curve in Figure V is levelling off with time, it is possible that over far longer periods the NAMLT values may move into or beyond the intermediate region, resulting in susceptibility to IGC at comparatively low environmental temperatures. Longer term trials are required to confirm this.

The ASSET results show a similar gradation to the NAMLT results. The combined results show a high degree of sensitization, IGC and exfoliation corrosion in the AA5059 alloy, a smaller degree of sensitization and IGC in the AA5383 alloy, and little evidence of sensitization and IGC in the AA5083 alloys. All alloys in the as-received condition passed both ASTM G66 and ASTM G67 as stipulated in ASTM B 928/B 928 M. The NAMLT and ASSET results differ from literature results (10) where, after ageing at temperatures between 80°C and 140°C for times between 7 and 200 days, the corrosion resistance of AA5383-H116 was reported as better than that of AA5083-H116 for IGC and exfoliation corrosion. However the present results are consistent with a recent study (11) which reported that AA5059 when heated for up to 100 days at temperatures from 50-125°C was found to subsequently corrode substantially by exfoliation, whereas AA5083 tended to pit.

The results from the ASCOR stress corrosion cracking tests are less conclusive, due to the corrosivity of the EXCO solution. This caused general corrosion of the samples, and prevented a clear determination of the initiation of SCC in all except the samples heat treated at 150°C. For these highly sensitized samples, the AA5083 alloy had the greatest resistance to the initiation of SCC, but there was a less clear distinction between the SCC response of the AA5383 and AA5059 alloys.

As sensitization is a diffusion controlled process, heat treatments involving extended time at low temperature should achieve a similar result as heat treatments involving a shorter time at higher temperature. Some studies (12, 13) have however suggested that there is a change in sensitization behaviour between samples heat treated at 175°C compared to 120°C and 100°C. This may relate either to a critical degree of ß-phase precipitate continuity at the grain boundaries or to a critical sensitization temperature (12).

A recent study (14) of the effect of cyclic and constant thermal exposure on sensitization and IGC of AA5083-H321, AA5083-H116, AA5083-O (fully annealed) and AA5456-H116 alloys has shown that the alloys have greater susceptibility when exposed to thermal cycling between room temperature and 70°C (4 hours at the elevated temperature on each cycle) than when held at a constant 70°C for an equal number of hours. This was attributed to thermal stress facilitating β-phase nucleation. The effect was most pronounced for 5456-H116 alloy, less for 5083-H321, and not discernable for 5083-H116 or 5083-O alloys. The greater increase in corrosion rate (as measured by NAMLT test) with number of temperature cycles for the 5456 alloy was consistent with its higher magnesium concentration compared to the 5083 alloys. Thermal cycling at 43°C produced a smaller but still significant corrosion rate, largely independent of number of cycles, and in the order 5456-H116 (Fail) > 5083-H321 (Questionable) >> 5083-H116 = 5083-O (Pass). It must be noted that no such distinction between the IGC susceptibility of 5083-H321 and 5083-H116 tempers has been observed in the present work, as shown in Figure V, for constant temperature exposure at 65°C for 1400 hours. This suggests further research should be conducted into the comparison of the IGC susceptibility of AA5083 alloys of various tempers and from various manufacturers, and performing multiple NAMLT tests on each sample, to determine whether such differences are significant.

CONCLUSIONS

From a practical viewpoint, sensitization of AA5XXX series marine grade aluminium alloys is a concern where vessels will operate for extended periods in tropical waters, where deck temperatures can easily exceed 65°C, above which sensitization is known to occur.

As sensitization is linked to Intergranular Corrosion, Exfoliation Corrosion and Stress Corrosion Cracking, it is important to assess the susceptibility of AA5XXX alloys that may be used in shipbuilding, as all these modes of corrosion degradation have potentially serious consequences for ship structural integrity and through-life maintenance.

This study has shown that AA5059-H321 and AA5383-H116 are more susceptible to intergranular corrosion than AA5083-H116 or AA5083-H321 at both elevated temperature (150°C) for 100 hours and at lower temperature (65°C) for extended periods of time. The ASSET tests showed that AA5059-H321 and AA5383-H116 are more susceptible to pitting and exfoliation corrosion after sensitization, with AA5059-H321 also susceptible in the as-received state. The ASCOR tests indicate that AA5083-H116 is less susceptible to stress corrosion cracking than AA5383-H116 and AA5059-H321, for sensitization at temperatures from 65°C to 150°C.

DSTO plans further research to assess the risk to current and future RAN vessels, which incorporate marine grade AA5XXX alloys. Research is required to assess whether the Heat

Affected Zone in welded structures is more or less susceptible to sensitization than as-received alloy (6). As SCC requires a sensitized alloy, a corrosive environment, and a component of tensile stress perpendicular to the grain boundaries to open them up, alloy susceptibility to SCC must be correlated with residual weld stresses and operational stresses that the alloys experience in actual ship structures, to assess the likelihood of SCC in service.

This research will help inform possible mitigation strategies. While it is premature at this stage to speculate upon such strategies, a prudent approach might be use of Low Solar Absorbance coatings to reduce deck temperatures and hence delay the onset of potential sensitization and the risk of associated corrosion, as well as offering improved habitability and reduced air conditioning load for vessels operating in tropical environments.

REFERENCES

(1) Dix, E.H, Anderson, W.A. and Shumaker, M.B., (1959), "Influence of Service Temperature on the Resistance of Wrought Aluminum-Magnesium Alloys to Corrosion", Corrosion, 15(2), 55t-62t. Discussion of paper 15(6), 337t-338t

(2) ASTM G67-04, (2004), "Standard Test Method for Determining the Susceptibility to Intergranular Corrosion of 5XXX Series Aluminium Alloys by Mass Loss After Exposure to Nitric Acid (NAMLT Test)," ASTM International, West Conshohocken, PA

(3) ASTM G66-99, Reapproved (2005), "Standard Test Method for Visual Assessment of Exfoliation Corrosion Susceptibility of 5XXX Series Aluminium Alloys (ASSET Test)," ASTM International, West Conshohocken, PA

(4) Bushfield, H., Cruder, M., Farley, R. and Towers, J., (2006), "Marine Aluminum Plate – ASTM Standard Specification B 928 and the Events Leading to its Adoption", World Maritime Technology Conference, London, 1-18

(5) ASTM B 928/B 928M-07, (2007), "Standard Specification for High Magnesium Aluminium-alloy Sheet and Plate for Marine Service and Similar Environments," ASTM International, West Conshohocken, PA

(6) Braun, R., (2007), "Environmentally assisted cracking of aluminium alloys", Matt.-wiss. u. Werkstofftech., 38(9), 674-688

(7) Sielski, R.A., (2007), "Research Needs in Aluminum Structure", 10th International Symposium on Practical Design of Ships and Other Floating Structures, American Bureau of Shipping, Houston, TX

(8) Mart, P.L., Pescott, E., Russo, S., Pedrina, L.M. and Gage, N., (2008), "Corrosion of Marine Grade Aluminium Alloys", Proceedings Corrosion and Protection 2008 Conference, Australasian Corrosion Association, Wellington, NZ, Paper 067

(9) Schra, L. and Groep, F.F., (1993), "The ASCOR Test: A Simple Automated Method for Stress Corrosion Testing of Aluminum Alloys", Journal of Testing and Evaluation, 21(1), 44-50

(10) Lequeu , Ph., Raynaud, G.M., Cottignies, L., Dif, R. and Ehrström, J.C., (1998), "Fatigue and Corrosion Resistance of High Strength 5383 Aluminium Structures", Ausmarine'98, Fremantle, Australia, 174-178

(11) Field, A., (2007), "Sensitization of 5000 Series Al Alloys", Mega Rust Conference, San Diego, CA

(12) Bovard, F., (2004), "Sensitization and Environmental Cracking of 5XXX Aluminium Marine Sheet and Plate Alloys", Proceedings of the Joint International Meeting of the Electrochemical Society, Honolulu, Oct. 3-8

(13) Oguocha, I.N.A., Adigun, O.J. and Yannacopoulos, S. J., (2008), "Effect of sensitization heat treatment on properties of Al-Mg alloy AA5083-H116", J. Mater. Sci., 43 , 4208-4214

(14) Kar, S., Rodchanarowwan, A. and Free, M. L., (2009), "The Effect of Cyclic and Constant Thermal Exposure on AA5083 and Related Alloy Susceptibility to Intergranular Corrosion", NACE Corrosion Conference, paper 09190

Resistance Prediction for Round-bilge and Hard-chine Catamarans

Phillip J. Helmore
University of New South Wales, Australia
p.helmore@unsw.edu.au

Felix W. Scott
High Modulus, New Zealand
felix.scott@highmodulus.co.nz

Daniel I.H. Wong
Fulsail Shipyard, Malaysia
danielwongih84@hotmail.com

ABSTRACT

A numerical investigation of the resistance in calm water of high-speed displacement and semi-displacement catamarans with symmetric demihulls in calm water has been conducted. The project used the linearised potential-flow program, *Hydros*, as a numerical towing tank to determine the residuary resistance of a series of round-bilge and hard-chine catamaran hullforms, having a wide range of length/beam ratios, hull separation/length ratios and slenderness coefficients. These results were checked against individual towing-tank results and found good agreement. The whole set of results was then run through a regression analysis to determine expressions for the residuary resistance in terms of the main hull parameters. The results are presented in the form of equations with numerical coefficients for ease of use in the design office.

1. INTRODUCTION

The genesis of this investigation was a group of four final-year student design projects on the design of a 35 m 30 kn catamaran, for which the resistance predictions using available methods were coming up with installed powers which were about fifty percent higher than being installed in current vessels. The hullforms had been modelled in *Maxsurf*, but this was before the advent of the slender-body method in *Hullspeed*, so the offsets were exported via *Hydrolink* and massaged into a form for input to *Hydros* (Doctors 1993, 2004). The resulting predictions of resistance and power tied in closely with powers being installed in current vessels, and solved the immediate problem.

However, there is a dearth of methods available for the prediction of the resistance of catamaran hullforms in preliminary design. *Hullspeed* now provides good predictions using the slender-body method, but relies on the whole hull shape and not just the principal parameters of the hullform, so that the hull must first be modelled in *Maxsurf*. CFD methods are improving all the time, but are not yet at the stage where they can be relied on directly for a resistance prediction.

A project was therefore proposed to analyse the resistance of a series of catamaran hullforms, both round bilge and hard chine, with a wide range of principal parameters, and at speeds up into the semi-planing regime, using *Hydros*. The resistance results would then be run through regression analyses to derive equations for the prediction of resistance based on the principal hullform parameters for preliminary design purposes.

2. PREVIOUS WORK

The earliest known work on catamaran resistance prediction was presented by Clement (1962) in a series of graphs for predicting the resistance of hard-chine, planing catamarans. The graphs showed the lift coefficient, centre of pressure ratio, and resistance-per-unit-weight for a range of hull trim angles and aspect ratios for deadrise angles of 0, 5 and 10 degrees. Interference effects between the demihulls were not included in the analysis.

Turner and Taplin (1968) tested symmetric and asymmetric catamaran models with hullforms based on Todd's Series 60 merchant ship hullforms. Catamaran interference resistance was given, together with an interference factor. Graphs were produced showing effective horsepower versus speed for a range of drafts, and the interference factor versus demihull spacing for a range of speed/length ratios. An interesting conclusion was that the interference factor can be negative in a narrow range of cases of specific speed/length ratios (around 1.1) and demihull spacings.

Insel and Molland (1992) looked at the calm water resistance components of symmetrical demihull, round-bilge, high-speed, displacement catamarans, looking primarily at viscous interference. Model testing was carried out at the University of Southampton, and they tested a Wigley hull and three hulls of NPL form for Froude numbers from 0.1 to 1.0, with demihull/length spacings of: s/L 0.2, 0.3, 0.4 and 0.5. They came up with an interference factor, and concluded that the form factor is independent of speed, and therefore should be kept constant over the speed range.

Following Insel's initial work, Molland extended the range of testing in Molland, Wellicome and Couser (1995), looking at the effect of length/displacement ratio and breadth/draft ratio on catamaran wave resistance; Molland and Lee (1997), investigating the effect of prismatic coefficient and hull spacing on residuary, wave and frictional resistance, and Utama and Molland (2001), attempting to quantify the viscous resistance interference factor by testing a pair of ellipsoids in a wind tunnel.

Millward (1992) investigated the effect of hull separation on wave resistance. His results showed that, at high Froude numbers, all hull spacings performed better than the demihull in isolation. It was also found that there was a peak in resistance at a Froude number of 0.5, exacerbated by closer hull spacings.

Hanhirova et al. (1995) performed a regression analysis on a range of vessels using existing resistance data to provide a correction to the thin ship theory of Michell (1898) for resistance prediction. Results included a comparison of the regression correction to the actual difference between the thin ship theory and model tests. All forms of catamaran were included in the group, including round bilge, hard chine and swath forms.

Zips (1995) and Müller-Graf et al. (2002) both performed regression analyses on the VWS Hard Chine Catamaran Hull Series '89 hull forms which were developed by Müller-Graf (1993). Both came up with expressions for the residuary resistance in terms of remarkably few parameters: the length/beam ratio, demihull deadrise angle amidships, and the transom wedge angle (if fitted).

Of the available methods, the latter two are the only ones easily programmed. However, a limiting factor is that the series had a fixed demihull spacing/length ratio, and the resulting resistance formulation does not take account of a number of parameters which affect the resistance, such as varying displacement. There is therefore still a need for a method for prediction of catamaran resistance which relies only on the principal hullform parameters.

3. PARENT HULLFORMS

3.1 Round Bilge

The parent needed to be indicative of modern, high-speed catamarans. The hullform chosen was based on the Series 64 high-speed displacement forms developed by Yeh (1965). These hulls are slender, with a round-bilge and transom-stern shape, but were slightly modified from the original series by removing the tumblehome above the waterline and moving the maximum section location to midships. The Australian Maritime College has a pair of these hulls in their inventory, and so they could be used in catamaran configuration for verification testing. The principal particulars of one demihull model are given in Table 3.1, and the body plan of the demihull is shown in Figure 3.1.

Table 3.1 Parent Round-bilge Demihull Particulars

Item	Symbol	Value
Length overall	L_{OA}	1.531 m
Length waterline	L_{WL}	1.498 m
Beam	B	0.112 m
Draft to DWL	T	0.046 m
Displacement	Δ	3.745 kg
Block coefficient	C_B	0.488
Prismatic coefficient	C_P	0.669
Maximum-section coefficient	C_M	0.729
Demihull slenderness ratio	$L/\nabla^{1/3}$	9.646

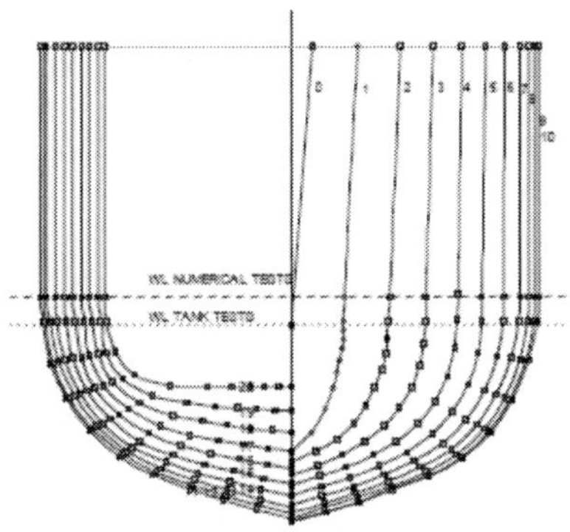

Figure 3.1 Modified Series 64 Parent Body Plan

3.2 Hard Chine

A hard-chine parent hullform was obtained from industry as representative of good practice. A model of this hullform had already been tested in the towing tank at the Australian Maritime College, so the data was available for verification purposes. The principal particulars of one demihull at model scale are given in Table 3.2, and the body plan of the demihull is shown in Figure 3.2.

Table 3.2 Parent Hard-chine Demihull Particulars

Item	Symbol	Value
Length overall	L_{OA}	1.702 m
Length waterline	L_{WL}	1.622 m
Beam	B	0.104 m
Draft to DWL	T	0.080 m
Displacement	Δ	6.591 kg
Block coefficient	C_B	0.478
Prismatic coefficient	C_P	0.736
Maximum-section coefficient	C_M	0.723
Demihull slenderness ratio	$L/\nabla^{1/3}$	8.724

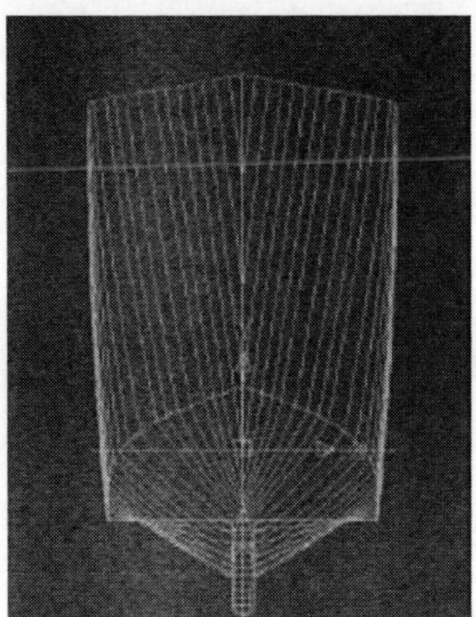

Figure 3.2 Hard-chine Parent Body Plan

4. VERIFICATION

For this project the round-bilge parent hullform was tested in catamaran configuration over a wide range of demihull spacing/length ratios and at speeds up to a Froude number of 1.0 in the towing tank at the Australian Maritime College.

The demihulls and tank testing in progress are shown in Figure 3.3.

(a) Parent Hulls (b) Test in Progress at $s/L = 0.2$, Fn = 1.0

Figure 3.3 Tank testing

Hydros is a linearised potential-flow program which uses the Michell integral to calculate the wave-making resistance R_W, takes account of the flow separation behind a transom stern and the consequent loss of hydrostatic pressure there, R_H, which was first noted by Armstrong (2000), and can be set to use any one of a plethora of friction formulations.

The residuary resistance has been taken to be the sum of R_W and R_H calculated by *Hydros*. Armstrong (2000) reported that values of $(1+k)$ significantly closer to 1.0 than found by previous researchers resulted if this loss of hydrostatic pressure was taken into account. However, not having a reliable numerical predictor of values of $(1+k)$ for a wide range of hullform parameters, we have opted to use $(1+k) = 1.0$ throughout.

The configurations which had been tank tested were then run through *Gencat*, the resistance-prediction subroutine of *Hydros*, with the friction set to use the ITTC 1957 formulation with $(1+k) = 1.0$. The results for the round-bilge parent hullform at model scale and various spacing/length ratios are shown in Figure 3.4. The results for the hard-chine parent hullform at full size at the actual spacing/length ratio and using $C_A = 0$ are shown in Figure 3.5.

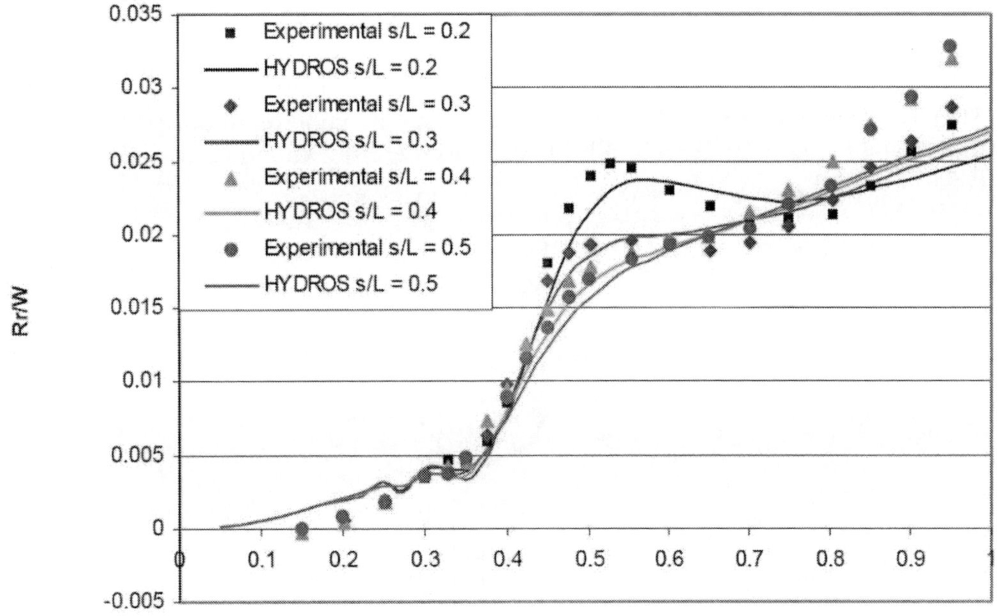

Figure 3.4 Experimental and Numerical Predictions for Round-bilge Parent

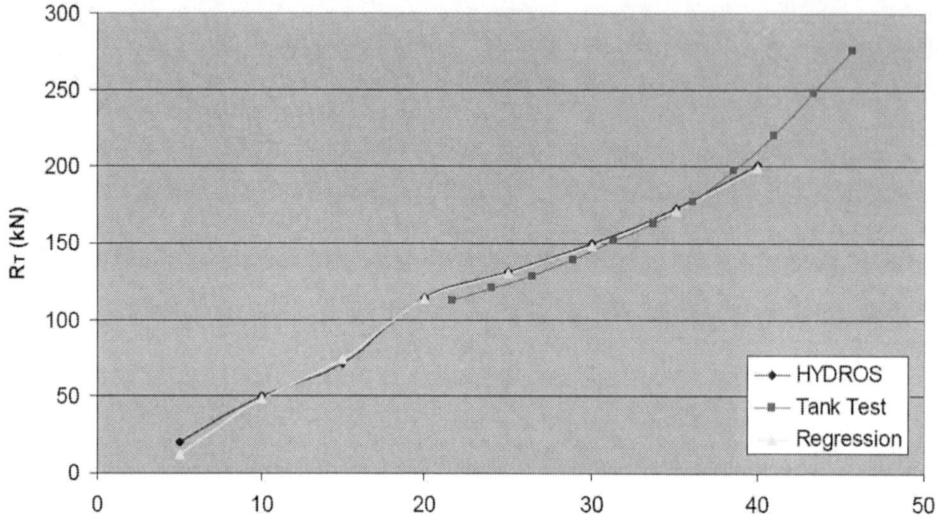

Figure 3.5 Experimental and Numerical Predictions for Hard-chine Parent

It may be seen from the figures that *Hydros* provides a good prediction of the resistance and, in particular, picks up and accounts for the effect of the demihull spacing/length ratio for the round-bilge parent model, and satisfyingly close to the experimental results for the hard-chine parent at full size.

5. VARIATION OF HULLFORMS AND RESISTANCE PREDICTION

The series of round-bilge and hard-chine hullforms for the project were created using the *Merger* subroutine of *Hydros*. This subroutine takes any number of parent hulls, and uses specified proportions of each for the longitudinal, transverse and vertical offsets to create a new hullform. Fine and full variations of the parents were therefore created for the mergers. The deckline, profile and body plans were varied to create extreme but realistic hullforms for each.

Details of the fine and full hullform variations are given in Scott (2004) and Wong (2007), and the body plans for the round-bilge variations are shown in Figure 5.1.

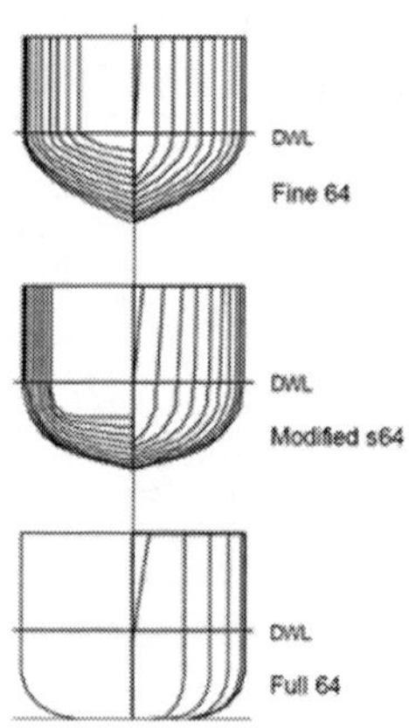

Figure 5.1 Body Plans of Fine and Full Hullforms

Merger was then used to create variations of the hullforms, taking various percentages of the fine and parent hullforms, or various percentages of the full and parent hullforms, to vary the block (and other) coefficient(s). An example combination would be 20% of the fine hullform and 80% of the parent. Steps of 20% between fine and parent and full and parent were used, to create 11 variations (including the fine, full and parent) of slenderness ratio, $L/\nabla^{1/3}$.

Length/beam ratio was then varied by directly varying the beam of each variation. For the round-bilge variations, eight L/B ratios were used, combined with eight demihull spacings and taking the total configurations to 704. These were combined with eighteen Froude numbers, giving a total of 12672 resistance data points for the regression analysis. Using experience gained in the analysis of the round-bilge hullforms, fewer combinations were used for the analysis of the hard-chine hullforms. The details of the variations are shown in Table 5.1. It is considered that the extremes easily encompass the principal parameters of modern catamaran hullforms.

Table 5.1

Parameter	Round Bilge		Hard Chine	
	No. of Variations	Values	No. of Variations	Values
s/L	8	0.15, 0.2, 0.25, 0.3, 0.35, 0.4, 0.45, 0.5	5	0.2, 0.25, 0.3, 0.4, 0.5
L/B	8	8, 11, 13, 16, 21, 26, 31, 36	5	8, 12, 16 24, 32
$L/\nabla^{1/3}$	11	6.4–13.9	11	6.4–11.8
Fn	18	0.1–1.0	8	0.1–1.1
Data pts	12672		2200	

These hullform variations were then run through *Gencat* in *Hydros* to obtain the resistance components for each.

6. REGRESSION ANALYSIS

It was decided at the outset that the Froude number should be an independent variable. Other variables which could influence the resistance were considered to include the slenderness ratio $L/\nabla^{1/3}$, length/beam ratio *L/B*, beam/draft ratio *B/T*, demihull spacing/length ratio *s/L*, block coefficient C_B, prismatic coefficient C_P, maximum immersed section coefficient C_X and the immersed transom area/maximum immersed area ratio A_T/A_X.

It was found, after running the hulls through *Merger*, that many of the hull parameters which were initially considered were closely correlated. Because of this, it was decided that C_B, C_P, C_X, and A_T/A_X could be discarded in the regression analysis. Also, if the draft is held constant whilst beam is altered, then both *L/B* and *B/T* are closely correlated, so *B/T* was also discarded as a variable. The final decision was to have an equation for dimensionless residuary resistance of the form $R_R = f(s/L, L/B, L/\nabla^{1/3})$. This combination was, coincidentally, the one recommended by Armstrong (2000).

Many polynomial combinations of these and the Froude number were tried, together with sin and cos terms in an attempt to replicate the humps and hollows of the residuary resistance, all without much success.

van Oortmerssen (1971) published the results of a regression analysis of the residuary resistance of a set of 93 unrelated models of tugs and trawlers which had been tank tested at the Maritime Research Institute Netherlands (then the Netherlands Ship Model Basin). He obtained a parametric expression for the residuary resistance in terms of the principal hull parameters, and this method generally reproduces resistance humps and hollows well.

van Oossanen (1979) says that some extrapolation of van Oortmerssen's equation to higher speeds is permissible because of the theoretical nature of the basic expression. van Oortmerssen based his theory on that of Havelock for the waves produced by a travelling two-dimensional pressure disturbance, but simplified so that it could be formulated in terms of the vessel's parameters.

It was therefore decided to try a van Oortmerssen-type equation to see if better results could be obtained. Initial result were encouraging, and so this method was explored with many combinations of parameters, with and without cross-product terms.

The final form of the equation for prediction of residuary resistance is given by

$$R_R / W = c_1 e^{-m/9F_n^2} + c_2 e^{-m/F_n^2} + c_3 e^{-m/F_n^2} \sin(1/F_n^2) + c_4 e^{-m/F_n^2} \cos(1/F_n^2) \tag{1}$$

where W is the weight of the vessel, F_n is the Froude number, $m = -00036(L/\nabla^{1/3})^2$ and the c_i coefficents are given by:

$$\begin{aligned} c_i = {} & d_{i,0} + d_{i,1} s/L + d_{i,2}(s/L)^2 + d_{i,3} L/B + d_{i,4}(L/B)^2 + d_{i,5}(L/\nabla^{1/3}) + d_{i,6}(L/\nabla^{1/3})^2 \\ & + d_{i,7}(L/B)(L/\nabla^{1/3}) + d_{i,8}(L/\nabla^{1/3}) s/L + d_{i,9} s/L(L/B) \end{aligned} \tag{2}$$

Fn, *S/L*, *L/B* and $L/\nabla^{1/3}$ are all based on the length on the waterline, and ∇ is the displacement volume of one demihull. The $d_{i,j}$ coefficients of the regression analysis are shown in Tables 6.1 and 6.2.

One measure of the "goodness of fit" is the correlation coefficient (or coefficient of determination):

$$R^2 = 1 - \sum_{i=0}^{N} (y_i - \hat{y}_i)^2 / \sum_{i=0}^{N} (y_i - \bar{y}_i)^2 \tag{3}$$

where $\bar{y}_i$ is the mean of the N values of y_i and N the number of data sets in the sample. If $R^2 = 1$, then the equation fits the data exactly whereas, if $R^2 = 0$ then the equation does not fit the data at all well. In ship resistance work, correlation coefficients greater than 0.99 are considered an excellent fit, and above 0.95 acceptable. The R^2 values obtained for the final prediction equation were 0.9712 for the round bilge data, and 0.9875 for the hard-chine data.

Table 6.1 $d_{i,j}$ **Coefficients for Round-bilge Hullforms**

i =	1	2	3	4
$d_{i,0}$	0.0797685	0.259148	0.0946618	−0.137417
$d_{i,1}$	−0.0129356	0.144608	−0.331370	0.131802
$d_{i,2}$	0.00925812	0.00233958	0.0920495	−0.0830109
$d_{i,3}$	0.00379405	0.0123867	−0.00560952	−0.00386564
$d_{i,4}$	−0.0000399500	0.000338495	0.000150753	−0.00000208829
$d_{i,5}$	−0.0168401	−0.0770948	0.00547200	0.0270261
$d_{i,6}$	0.000484087	0.0080109	−0.00301451	−0.00144753
$d_{i,7}$	−0.0000333417	−0.00318107	0.00158550	0.000338047
$d_{i,8}$	0.000972250	−0.0279309	0.0442693	−0.00605957
$d_{i,9}$	−0.000293306	0.00572622	−0.00773861	0.000497995

Table 6.2 $d_{i,j}$ **Coefficients for Hard-chine Hullforms**

i =	1	2	3	4
$d_{i,0}$	0.02615	0.36362	0.14427	-0.21284
$d_{i,1}$	−0.018577	0.060224	−0.22251	0.28815
$d_{i,2}$	0.040777	−0.073227	0.15499	−0.17626
$d_{i,3}$	−0.0014833	0.0093802	0.0039113	−0.0048556
$d_{i,4}$	-0.000036925	0.000055282	0.00004442	−0.000027469
$d_{i,5}$	0.0080266	−0.079438	−0.035817	0.038129
$d_{i,6}$	−0.0010771	0.0045556	0.0023099	−0.0019458
$d_{i,7}$	0.00038315	−0.0010804	−0.00057170	0.00050910
$d_{i,8}$	−0.00074602	−0.0078552	0.021075	−0.016417
$d_{i,9}$	0.0000030890	0.0016163	−0.0031581	0.0011812

The results from the regression equation are compared to the original tank-test data for the round-bilge parent hullform in Figure 6.1, and against the *Hydros* results for the hard-chine parent hullform in Figure 6.2.

It may be seen from the figures that the regression equation successfully accounts for the the humps and hollows in the prediction of residuary resistance, and for the effect of demihull spacing.

As an example of the application of the regression equations, the resistance of a round-bilge hullform is calculated in the Appendix, and the resistance is compared with that of a hard-chine hullform having the same principal particulars, shown in Table 6.3. The results are compared in Figure 6.3.

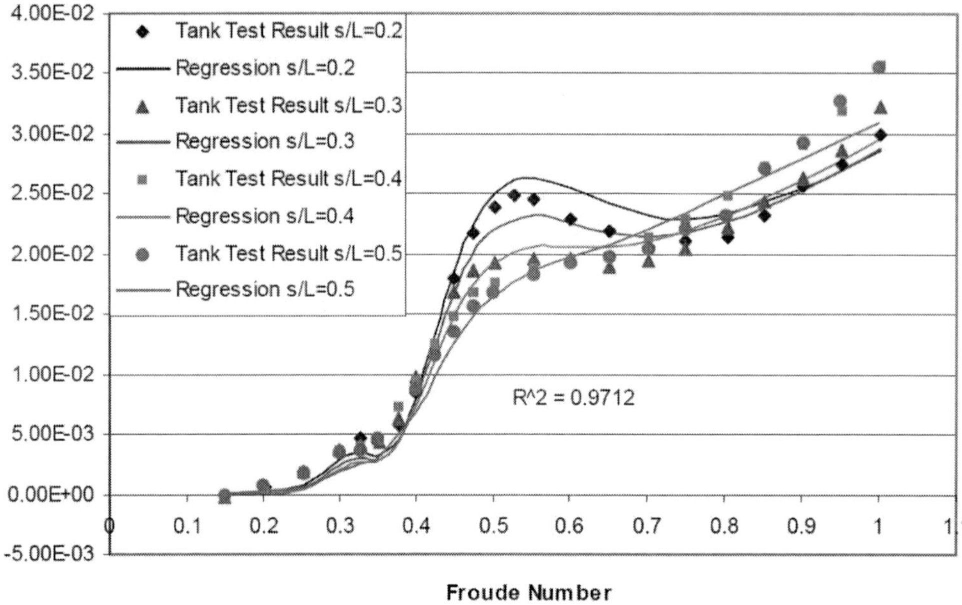

Figure 6.1 Regression Results vs Experiment for Round-bilge Hullforms

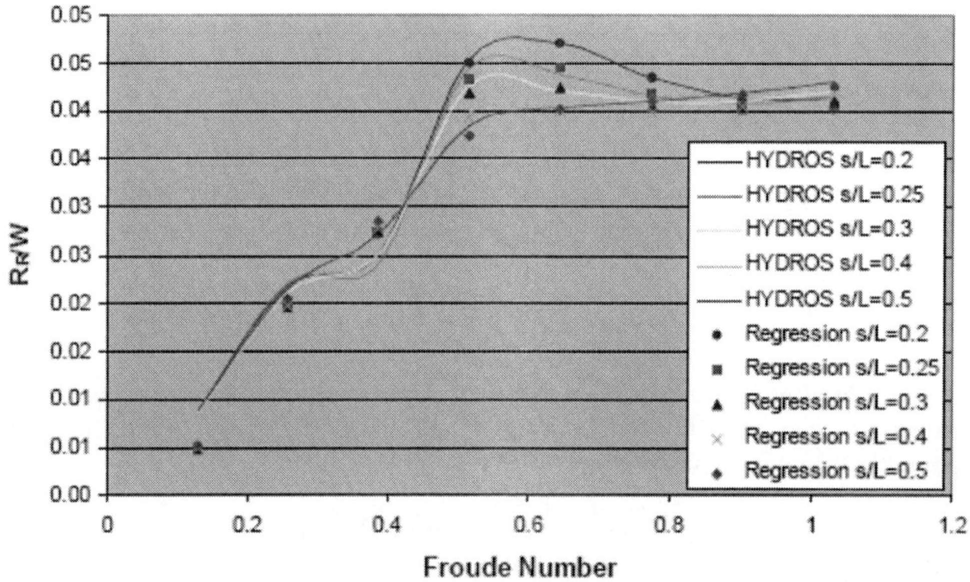

Figure 6.2 Regression Results vs *Hydros* for Hard-chine Hullforms

Table 6.3 Round-bilge and Hard-chine Catamaran Particulars

Item	Symbol	Value
Length waterline	L_{WL}	34.30 m
Beam demihull	B	2.85 m
Spacing of demihills	s	10.29 m
Displacement	Δ	131.2 t
Wetted surface area	S	295 m^2
Demihull slenderness ratio	$L/\nabla^{1/3}$	8.575

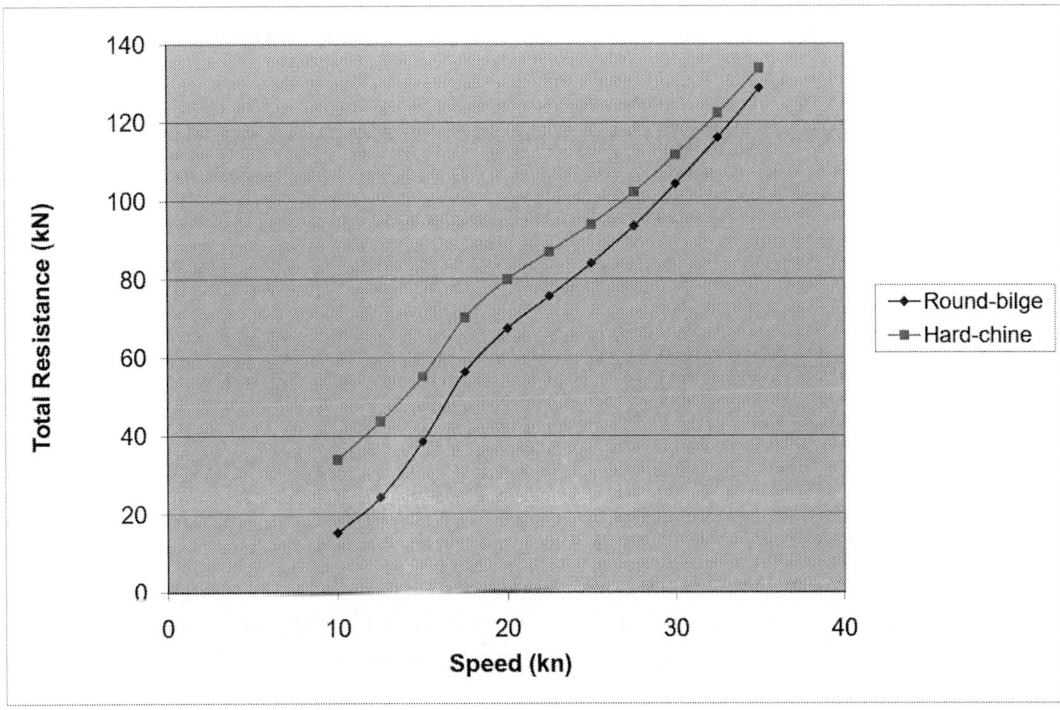

Figure 6.3 Round-bilge and Hard-chine Catamaran Resistance

7. CONCLUSION

A numerical investigation of the resistance of high-speed catamarans has been conducted. The resistance was determined using the linearised potential-flow program, *Hydros*, to determine the resistance of a series of round-bilge and hard-chine catamaran hullforms, having a wide range of length/beam ratios, hull separation/length ratios and slenderness coefficients. These results were then used in a regression analysis and an expression found which represents residuary resistance well in terms of the main hull parameters for use in preliminary design.

8. REFERENCES

Armstrong, N.A. (2000), *On the Viscous Resistance and Form Factor of High-speed Catamaran-ferry Hullforms*, PhD Thesis, University of New South Wales, Sydney.

Clement, E.P. (1962), Graphs for Predicting the Ideal High-Speed Resistance for Planing Catamarans, *International Shipbuilding Progress*, v. 9 n.99, pp. 464–477, November.

Doctors, L.J.(1993), Integrated Software for the analysis of the Hydrostatics and Hydrodynamics of Marine Vehicles, *Proc. 10th International Maritime and Shipping Symposium*, University of New South Wales, 1993.

Doctors, L.J. (2004), *The* Hydros *Programs: Description and Documentation*, University of New South Wales, Sydney.

Hanhirova, K., Rintala, S. and Karppinen, T. (1995), Preliminary Resistance Prediction Method for Fast Mono- and Multihull Vessels. *Proc. High Speed Vessels for Transport and Defence*, Royal Society of Arts, London.

Michell, J.H. (1898), The Wave resistance of a Ship, *Philosophical Magazine*, London, Series 5, v.45, pp.106–123.

Millward, A. (1992) The Effect of Hull Separation and Restricted Water Depth on Catamaran Resistance, *Trans. Royal Institution of Naval Architects*, v.134, pp.341–346.

Molland, A.F. and Lee, A.R. (1997), An Investigation into the Effect of Prismatic Coefficient on Catamaran Resistance, *Trans. Royal Institution of Naval Architects*, v.139, pp.157–162.

Molland, A.F., Wellicome, J.F. and Couser, P.R. (1998), Experimental Measurement of Sideforce and Induced Drag on Catamaran Demihulls, *International Shipbuilding Progress*, v.45 n.443, pp.225–235.

Müller-Graf, B.(1993), SUS-A The Scope of the VWS Hard Chine Catamaran Hull Series '89, *Proc. Second International Conference on Fast Sea Transportation (FAST'93)*, Yokohama, Japan, v.1, pp. 223–238.

Müller-Graf, B., Radojčić, D. and Simić, A. (2002), Resistance and Propulsion Characteristics of the VWS Hard Chine Catamaran Hull Series '89, *Trans. Society of Naval Architects and Marine Engineers*, v.110.

Scott, F.W. (2004), Regression Analysis of Round-bilge Catamarans, *Bachelor of Engineering Thesis*, University of NSW, Sydney.

Turner, H. and Taplin, A. (1968), The Resistance of Large Powered Catamarans, *Trans. Society of Naval Architects and Marine Engineers*, v.76, pp.180–213.

Utama, I.K.A. and Molland, A.F. (2001), Experimental and Numerical Investigations into Catamaran Viscous Resistance, *Trans. Royal Institution of Naval Architects*, v.143, pp.295–306.

van Oortmerssen, G. (1971), A Power Prediction Method and its Application to Small Ships, *International Shipbuilding Progress*, v.18 n.217.

van Oossanen, P. (1979), Resistance Prediction if Small High-speed Displacement Vessels: State of the Art, *Proc. Australian Symposium on Ship Technology*, RINA/IMarE, The University of New South Wales, Sydney.

Wong, D.I.H. (2007), Prediction of Resistance for Hard-chine Catamarans, *Bachelor of Engineering Thesis*, University of NSW, Sydney.

Yeh, H.Y.H. (1965), Series 64 Resistance Experiments on High-Speed Displacement Forms, *Marine Technology*, v.2, pp.248–272.

Zips, J.M. (1995), Numerical Resistance Prediction based on the Results of the VWS Hard Chine Catamaran Hull Series '89, *Proc. Third International Conference on Fast Sea Transportation (FAST'95)*, Lübeck-Travemünde, Germany, v.1, pp. 67–74.

APPENDIX

Round-bilge Catamaran Resistance ver 1.1

Length WL	34.3 m	
Beam WL demihull	2.85 m	
Displacement volume total	128 m^3	
Spacing of CLs demihull	10.29 m	
Wetted surface total	295 m^2	
Water density	1.025 t/m^3	1.0250 at 19 deg C
Water viscosity	1.07854 x 10^{-6} m^2/s	1.07854 x 10^{-6} at 19 deg C
Correlation allowance	0.0001	

∇ demihull	64 m^3	
s/L	0.3000	0.15-0.5 Range
L_{wl}/B	12.035	8-36
$L_{wl}/\nabla_{dh}^{1/3}$	8.575	6.4-13.9
10^3C_1	5.789	
10^3C_2	49.337	
10^3C_3	-10.980	
10^3C_4	-5.766	
m	0.2647	

Speed kn	Fn on L_{wl}	Cf+Ca ITTC'57	Rf kN	Rr/W	Rr kN	Rt kN
	0.1 min					
10	0.2805	0.002042	8.17	0.005438	7.00	15.17
12.5	0.3506	0.001983	12.40	0.009246	11.90	24.29
15	0.4207	0.001937	17.44	0.016381	21.08	38.51
17.5	0.4909	0.001899	23.27	0.025691	33.05	56.32
20	0.5610	0.001867	29.88	0.029202	37.57	67.46
22.5	0.6311	0.001840	37.26	0.029822	38.37	75.63
25	0.7012	0.001816	45.41	0.030019	38.62	84.03
27.5	0.7714	0.001794	54.30	0.030540	39.29	93.59
30	0.8415	0.001775	63.93	0.031414	40.42	104.35
32.5	0.9116	0.001758	74.30	0.032508	41.83	116.12
35	0.9817	0.001742	85.40	0.033691	43.35	128.74
	1.0 max					

Design and Build of Composite Patrol Boats

D Firth, V Corniani
Gurit Australia, Mona Vale, NSW, Australia

INTRODUCTION

This paper sets out to work through and discuss the design process and options a structural engineer should consider to complete a successful design of a composite patrol boat. The topics covered will range from determining the basic performance criteria for the boat through to the final design of the structural elements.
Included will be information to help support the design choices made to use composite materials and provide additional ideas for improvements and developments over existing composite design solutions. Developments will be based on actual experience of designing and engineering composite patrol boats to improve performance criteria.
Topics covered will include operation requirements, through life costs, arrangement implications, manufacturing technology, materials and efficient structures. There will be information covering the recent developments in the fire safety for composite materials and additional data that can be used to influence the selection of a composite patrol boat.

OPERATION AND PERFORMANCE REQUIREMENTS

The design of composite patrol boats starts with a definition of the performance criteria required of the vessel. The difficulty for the structural engineer is determining the effect and priority of each different criterion on the structural arrangement. For small patrol boats the driving performance criteria is generally high speed which will result in a requirement for light weight. High speed will mean high design accelerations which results in higher slamming loads for the hull structure which can lead to an increase in structural weight.
The selection of the correct design acceleration is crucial as a starting point for the structural design however the value needs to be realistic to match the expected operational conditions. The design acceleration is a relationship between the vessel speed, displacement and expected sea state. Many patrol boats will have a max speed of 50knots however not all of them will do this speed in high sea states, as crew safety concerns will limit high accelerations. The structural engineer needs to have guidance from the naval architect and operator to correctly determine the right sea state and speed; there may have to be several iterations to determine which case develops the highest pressure loads on the structure.
The final design can be a compromise where an accelerometer is positioned at the helm station with the validity of the structure dependant on a recorded maximum acceleration and not sea state or speed, relying on a properly trained crew to control the craft to suit.
Performance criteria is not just limited to high speed, the boat may have a requirement for high payload capacity. A light structure will, for a given hull form; allow for greater payload capacity, this can lead to multi role functions for a single vessel design. If a craft has a requirement for long service range building a lighter structure will permit a greater range at similar speeds for the same power and fuel, this argument is also valid for reduced fuel consumption.

MANUFACTURING TECHNOLOGY AND MATERIALS

After determining the performance criteria of the design the structural engineer now needs to review what materials and construction processes will be used to construct the final designed vessel.

An efficient structural design will involve open dialogue between the naval architect/client, structural engineer and builder to determine the most appropriate manufacturing technology and materials. The first question to ask generally is whether the design is going to be a series production or a one-off build as some construction process may not be best suited to one off construction or series build. Generally most patrol boats will be built as a series to replace an existing fleet. To save on the cost of painting and fairing a hull and deck the builder will construct female moulds to match the technology level to be used.

Hand Lamination

Hand lamination was the first technology used when composites were introduced in the marine industry. This technology is still used for cheap and low specification constructions. Fibres are wetted with resin and hand laminated over the mould. Materials used are usually glass fibres and polyester resins as these are the lowest cost resin and fibre types. Sandwich panels using a structural core are usually limited to panels above the hull bottom. This type of construction is the lowest technology level requiring the simplest mould, however to get good laminate quality the skill of all the laminators needs to be high.

Vacuum Consolidation

The composite laminate is still hand laminated with resin and dry reinforcement however a vacuum bag is used to cover and consolidate the laminate, this uniform pressure over the laminate helps to remove voids and improve laminate quality. Vacuum consolidation is also used for bonding of the core to ensure an accurate fit against the first skin. Using a vacuum bag will allow excess resin to be drawn off reducing the weight of the laminate. This level of lamination will require a better quality mould to ensure it is air tight; the skill of the laminators will still have to be high to ensure good quality laminates are created before the vacuum bag is applied.

Resin Infusion

With resin infusion, developed more than 15 years ago, all dry materials including the core of a sandwich panel are placed in a mould then vacuum consolidated. Once the dry laminate has been properly consolidated and inspected for air leaks, resin is introduced into the bag directly on to the laminate and pulled through the dry material by vacuum lines on the edges of the mould. This process wets out the fabric and core surfaces producing a high quality lightweight laminate. It is critical for this process that the mould is high quality and air tight as the vacuum will pull any air leaks into the laminate rather than the resin. The skill level required for infusion is lower as the laminate is assembled rather than laminated.

Pre-Preg

This method initially developed for the aerospace industry to produce the lightest and highest quality laminates. Dry fibres are pre-impregnated with a temperature dependant catalysed resin before they are laid into the mould. This process gives the laminate the ideal fibre/resin ratio to achieve the optimum laminate properties. The fibres combined with the resin are laminated into a mould and vacuum consolidated before the mould is placed in an oven which has to be heated at the required temperature to start the cure process of the resin, typically 70°C to 100°C for marine applications. This process requires a high quality thermally stable mould, an oven and a freezer (to store the material). The skill level is the same as vacuum consolidation to ensure minimal voids before the vacuum consolidation.

Sprint

This is a new fabric type developed by SP Systems (now Gurit) which combines the best aspects of resin infusion with the precise laminate quality of Pre-Preg. A temperature dependant catalysed resin film is sandwiched between dry fabrics layers. The combined material is place in a mould and then vacuum consolidated, as the fibres are still dry, these act as a pathway to the vacuum for evacuation of any trapped air between layers. Once the material is consolidated the mould is placed in an oven which has to be heated at the required temperature to start the cure process of the resin, typically 70°C to 100°C. This process requires a high quality thermally stable mould, an oven and a freezer (to store the material). The skill level required however is lower than pre-preg as layers are easier to lay down and less care has to be paid to ensure no air is trapped between layers (i.e.: no need for de-bulking).

The average weight savings with each of the above construction methods are listed below in Table 1. The similarity between vacuum consolidation and resin infusion is driven by the additional weight of the resin infused in the core which doesn't happen with vacuum consolidation.
Cost comparisons between different construction methods are dependent on too many variables and harder to define in one table.

Construction Method	Overall Structural Weight %
Hand Laminated Single Skin	100
Hand Laminated Hull Bottom Cored	85-90
Vacuum Consolidated	65-80
Resin Infusion	65-80
Pre-Preg or SPRINT	55-65

Table 1, Construction method weights

OH&S Requirements

In recent years the health and safety requirements of the composite boat building industry have been increased significantly. This has lead to developments in process technology and changes in the materials used. The main concern is exposure to resin and emissions of styrene from curing of polyester and vinylester resin. Epoxy resins do not have this problem and with careful selection of the correct hardeners epoxy provides a low OH&S risk for wet resin laminating. The concerns about styrene emissions have lead to the widespread adoption of vacuum infusion in the composite boat building industry for vinylester construction (polyester based resins are generally not ideally suitable for infusion). The infusion process is a closed mould technology so the emission of styrene is greatly reduced as there is little exposure to the air. SPRINT and Pre-Preg both give the additional OH&S advantage of not requiring the user to mix resins.

Toughness

When designing a composite patrol boat toughness is often one of the driving factors. Impact toughness can be measured quite easily. A research conducted by VT Halmatic in UK (ref 1) has shown that the impact toughness of modern advanced composites is on par with existing low tech thick heavy lamination and aluminium plate. The study also showed that Kevlar is not the best composite material to provide good impact toughness with T700 carbon and glass testing very well. The best performing laminate was a T700 carbon skinned sandwich panel.

Composites also have the advantages of natural damping and insulation to protect against shock pressure loads, an advantage which has been well proven in the use of composites in mine hunter hulls. The US Navy are currently trialling a full composite high speed patrol boat which has been designed in composite specifically to take advantage of the damping to reduce the effect of the high impact slamming forces on the crew onboard the boat.

Moreover, structural damage to composite ships resulting from minor collisions with concrete wharves and even minor reef groundings has generally been shown to be confined to the gelcoat and easily repairable.

Hull Bottom Core Material

Traditionally the use of core materials in the hull bottom has been frowned upon; this is more to do with poor choice of materials. The cores traditionally chosen for hull bottom slamming regions have proven not suitable for dynamic slamming loads.

A core material to support the high slamming loads experienced in a fast patrol boats hull bottom must have a combination of high shear strength and shear elongation.

The cheaper structural core like balsa and the strong cross linked PVC foams that have traditionally been used have the required high shear strength but their shear elongation is low which results in core shear failures from high dynamic slamming loads.

Linear PVC and SAN foam cores have the combined benefits of adequate shear strength and excellent shear elongation. This allows the cored panels to support dynamic slamming loads without core shear failures.

Fire Protection

Whilst it is well known that composites have poor fire performance, to reject the use of composites on this basis is as kin to rejecting steel and aluminium as boat building materials because they sink.

The fire safety and protection requirements for patrol boats are not govern by a specific rule, whilst passenger craft are typically governed by the IMO codes. However the fire requirements of the IMO HSC code are commonly adopted for patrol boats.

The HSC code has two types of designations *fire resisting divisions* and *fire restricting materials*. However to be a fire-resisting division the division must be constructed from fire restricting materials.

In this area of fire restricting materials the approach for composite materials has been to either coat a panel in an intumescent paint or coating, use fire retarded versions of existing resin systems or use materials with inherent fire restricting properties.

The latter two options whilst providing the fire protection required have draw backs of reduced mechanical properties or increased cost. Recent developments have been focused on producing resins and coating systems which will provide the required fire restricting properties.

For fire resisting divisions a light weight structural fire protection is also required for all composite materials to achieve the requirements of the HSC code, this is similar to a boat built of aluminium.

STRUCTURAL DESIGN

With the performance criteria for the design and the manufacturing technology determined, the structural engineer needs to focus on the structural arrangement of the vessel and the main structural components.

Global Stiffness

Additional research performed by Gurit (ref 2) has looked into the effect of common arrangement complications affecting patrol boats. This study demonstrates the impact of several arrangements on the longitudinal stiffness of a vessel.

As an example, the following findings can be taken from the research.

Patrol boats often have small deckhouses to give plenty of working deck space however it can be shown that whilst the deckhouse is small its contribution to overall longitudinal stiffness is significant and cannot be discounted as a classification society may do.

Larger patrol boats will commonly have a dropped aft main deck for easier deployment and recovery of personnel. Whilst an initial inspection could discount the longitudinal effect of both the upper and lower main decks as providing no contribution to the vessels longitudinal strength, it can be shown that both decks provide a greater than expected positive effect on the hull longitudinal strength. The same effect can also be shown with discontinuous lower decks.

Production Friendly Structures

Where patrol boats are designed for production or series builds the structural engineer should provide solutions to reduce build time. When building a series of composite boats the hull and deck shells will be produced in a female mould, to speed up the production time of the series it is important to reduce in-mould time. The structures should be designed to be made out of the boat at the same time as the hull. Parts made out the boat can be made on separate moulds or on flat tables where the use of vacuum consolidation or infusion can be utilised to reduce weight and improve laminate quality.

The layout of the systems installation should be created at the same time as the structure is designed to remove any unnecessary clashes and incorporate penetrations into the initial build. Having the system penetrations already designed and built will save time when building the structure and during the installation of the systems.

The choice of hull bottom laminate needs to be considered not just on the cost difference between the materials used in the basic laminate of a sandwich hull bottom or a single skin hull bottom. Typically a single skin hull will require more transverse frames between bulkheads than a cored hull bottom. The additional labour and material costs required to build the additional transverse frames of a single skin laminate has to be included when comparing hull bottom laminate costs. See figures 1 and 2 for images of typical hull bottom supporting structure for a 9m RIB as example of the additional material and labour that would be required.

Figure 1 Example of internal structure of single skin hull bottom

Figure 2 Example of internal structure of sandwich hull bottom

Pre-Made Hull Beams

An efficient way of saving time and weight in the build of a composite boat is to build the longitudinal and transverse supporting beams out the boat on a flat table. A vacuum consolidated beam will be lighter and have higher laminate quality than a foam top hat beam laminated in the boat. The pre-made beam will use less core materials saving cost as well. A simple wooden former mould is required on the table however this is a simple detail as good surface finish is not required. See figure 3 for typical beam cross sections

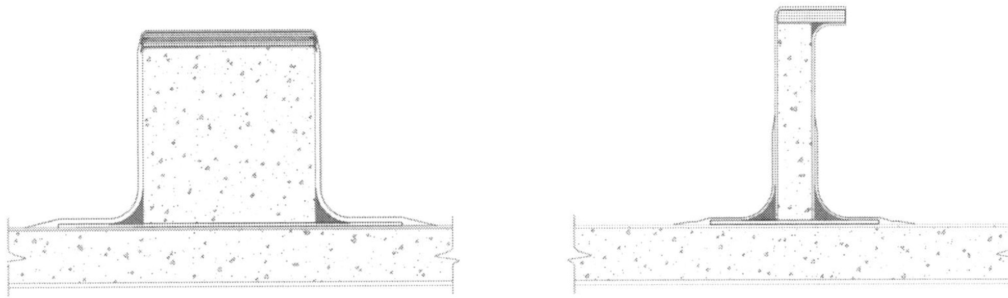

Typical top hat beam laminated in the boat

Typical pre-made beam laminated on a table

Fig.3 Typical Beam Sections

One Piece Frames and Bulkheads

If a total approach is taken to the design of the hull bottom, topsides and deck supporting structure to ensure continuity between the different elements the complexity of the structure can be simplified. This will lead to savings in material and labour cost. Combining the structural elements into ring frames or partial bulkheads allows these elements to be built as a single moulding on a flat table using either vacuum consolidation or infusion. With the structural elements combined there is a reduction in the total build time as less parts have to be laminated in to the boat.

THROUGH LIFE COSTS

Maintenance

A composites structure offers the clear advantages of having no susceptibility to corrosion or fatigue cracking, combined with relative ease of localised repair.

In-service experience with composite naval ships has demonstrated that the elimination of the need for structural maintenance serves to minimise dry-dock downtime and additionally frees-up crew from traditional day to day painting for corrosion protection.

The benefit to the navy is greater operational availability of both the crew and the vessel.

Hull Longevity

Steel and aluminium ships structures may experience fairly constant maintenance costs over a 15 year service life, thereafter maintenance costs begin to rapidly soar with the onset of steel corrosion and aluminium fatigue cracking. For small naval vessels with relatively light hull scantlings, 20% of steel hull plating typically needs replacing at the fifteen year mark. By comparison, today's composites technology has eliminated osmosis, so composite vessels retain their aesthetics, exhibit no structural deterioration, and experience no escalation in maintenance costs over and beyond a 20-year lifecycle. The comparison is represented in Figure 4.

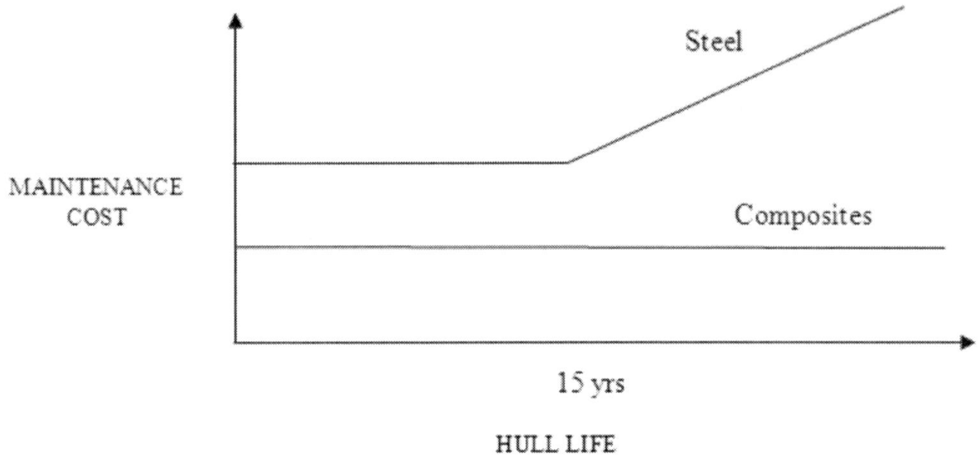

Figure 4: Structural Maintenance Costs over Lifecycle

Moreover, the additional downtime required for maintenance of both steel and aluminium hulled patrol boat fleets beyond year 15, would reduce further their relative operational availability.

The useful operational life of a composite vessel is not limited by its structural integrity or any escalating maintenance in the final years of operation. However, extended life could be limited by the aging and obsolescence of the equipment outfit. These issues can be separately addressed by designers, in particular in the selection of the installed configuration of the propulsion system so as to optimise its serviceability and extend the system service life.

Operational Costs

A composite patrol boat structure can be engineered to be significantly lighter and as a consequence require less installed power for equivalent performance. This enables the selection of smaller, lighter engines, creating better access for machinery maintenance which can help to minimise vessel downtime and maximising operational availability. Smaller engines will have reduced fuel and consumables consumptions which provide significant cost savings per year

The careful selection of composites hull material and associated installed power could achieve the maximum fleet availability and minimum support costs over the longest term. The higher boat availability achievable through adopting composites vessels means that mission requirements could be achieved from a smaller fleet of boats – effectively delivering direct savings in fleet acquisition and support costs, in addition to the projected operational fuel savings.

MILITARY REQUIREMENTS

Composite materials have several properties that are of advantage when considering the implications of military requirements of a patrol boat.

Radar/Stealth Signature

Being a non metallic material composites have a very low radar signature and combined with the ability to maintain very flat panels gives composite patrol boats a low radar signature. Further reductions can be included with the use of radar absorbing materials into the laminate. The thermal insulation properties of a sandwich composite panel combined with the low thermal conductivity of composite laminates will help to reduce the infra red signature of the vessel.

The combination of low thermal properties and low radar signature give composite boats the possibility of excellent stealth characteristics without the need for additional components.

Magnetic Signature

The suitability of composites for use in boats operating in mine fields has been well documented and proven with the use of composite materials in modern day mine counter measure vessels.

Ballistic Protection

There are currently developments in the field of ballistic protection incorporated into composites however these technologies are not yet commercially available. The use of aramid in bullet proof vests is well documented however the use of aramid in a structural composite laminate does not give the same protection. The main reason for this is not only that the type of aramid used in bullet proof vest is a different grade from the one used for structural laminates, but also that, when laminated, the aramid is locked in the layup by the resin, not allowing membrane deformation but forcing the fibres to work in shear.

EXAMPLE: A 15M PATROL BOAT

Gurit was recently contracted to re-engineer an existing design with the design brief to reduce weight and simplify the construction. After several design loops exploring the various possible structural arrangements we defined the optimal solution for our client.

The initial structure was designed with a single skin hull bottom and cored topsides and deck. The hull and deck panels were supported by numerous foam top hat beams. The materials chosen for this structure were pre-preg carbon with Nomex cores for the main mouldings and wet laminated carbon and foam for the supporting structure.

Gurit's approach was to switch to a foam cored hull bottom and change the technology level to infusion to reduce the costs. The main mouldings are laminated with carbon fibre and Corecell (SAN) foam vacuum infused with epoxy resin, the supporting structure is, where possible, laminated on vacuum tables with carbon fibre epoxy resin. The amount of transverse hull structure has been reduced significantly; the remaining transverse structure has been combined to incorporate beams into complete frames or partial bulkheads.

The weight and cost savings made are summarised in tables 2 & 3 below, the summary has been broken down in to sections to show where the main savings have been made.

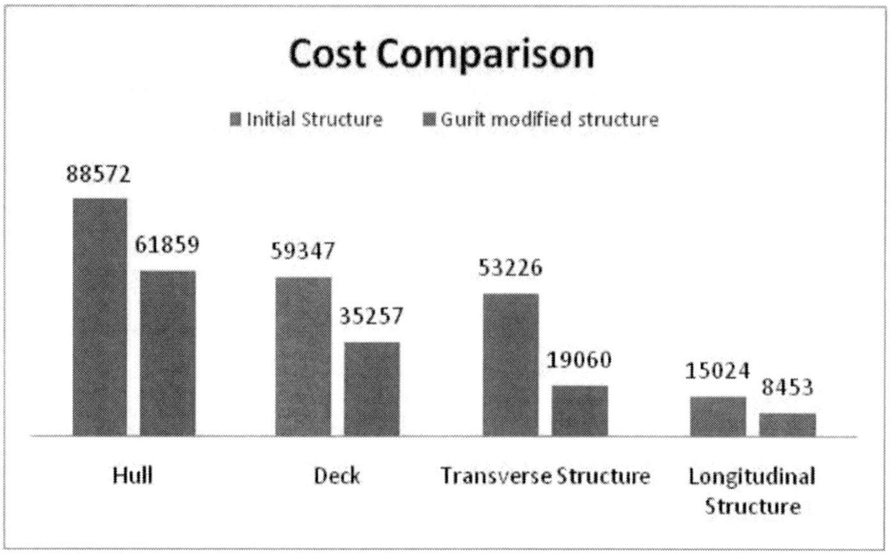

Table 2 Cost Comparison

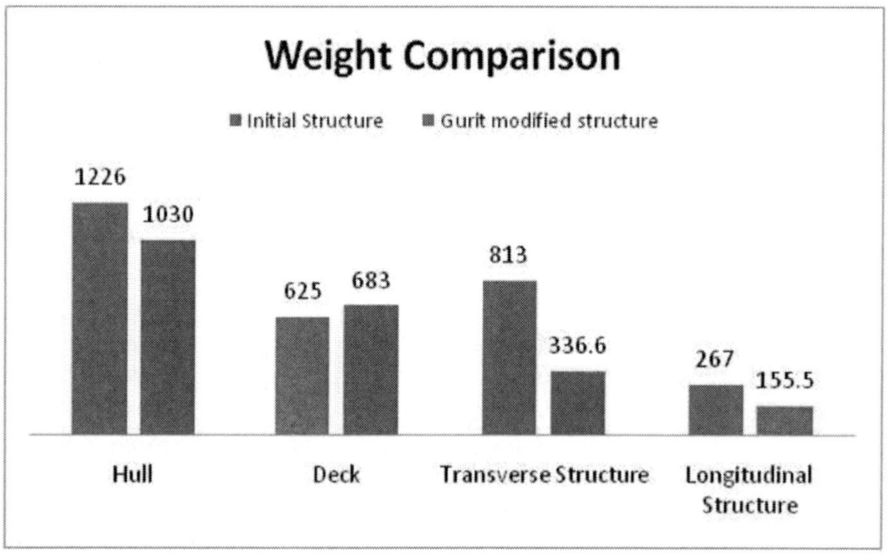

Table 3 Weight comparison

The above tables show that whilst a light weight high quality laminate option has been chosen, without including a lightweight structural arrangement further weight and cost savings would have not been realised.

REFERENCES

(1) Lee, Steven; Findon Ed; Impact testing of High Technology Laminates and Structures, METS 2005 Presentation

(2) Schrum, Molly; Wadia, Mal; Hull Girder Strength of a Composite Patrol Boat, SNAME 2009 Annual Meeting & Expo

Uniform Operating Limitations for High-Speed Craft

Rob Gehling
Consultant – Technical, Ship Safety, Maritime Standards Division
Australian Maritime Safety Authority, Canberra
rob.gehling@amsa.gov.au

ABSTRACT

The author has been deeply involved with matters relating to the International Code of Safety for High-Speed Craft (HSC Code) since its draft stages over 15 years ago.

During development of the 2006 amendments to the 2000 HSC Code, a need was identified for establishment of uniform procedures for determining the operating limitations covered by the craft's Permit to Operate. In particular, the need was foreseen to ensure that the operating limitations take account of the craft's stability and seakeeping characteristics in addition to the hull strength related sea-state limitations that have generally been imposed by the classification society.

The paper provides documentation for the resulting guidelines on uniform operating limitations that have now been promulgated by IMO.

INTRODUCTION

Australian industry was an early user of the international *Code of Safety for Dynamically Supported Craft (DSC Code)*[1], particularly in applying it to high-speed planing craft and catamarans.

Through this experience, particularly in using the Code as the statutory standard for the first 74m Incat catamaran, *Hoverspeed Great Britain*, it became apparent that its recommendatory and voluntary status relative to the mandatory instruments of SOLAS and the 1966 Load Lines Convention presented these craft with significant problems in being accepted for international operations.

The author had the honour of chairing the IMO's Ship Design and Equipment Sub-Committee's working group that developed the *International Code of Safety for High-Speed Craft (1994 HSC Code)*[2] as a replacement for the DSC Code. This Code subsequently became mandatory as an alternative to normal SOLAS certification through SOLAS Chapter X which entered force on 1 January 1996.

In accordance with its provision stating that it should be subjected to regular review, the 1994 HSC Code was revised in 2000 resulting in the *2000 HSC Code*[3], which was itself reviewed and amended in 2006[4] with effect from 1 July 2008. IMO subsequently published a 2008 consolidated edition of the 2000 HSC Code[5] to incorporate the 2006 amendments, which also incorporated into the text of the 2000 HSC Code matters that had previously been covered by interpretations promulgated as a circular[6].

One of the features that distinguishes the 1994 HSC Code and its successor documents from traditional ship safety standards such as SOLAS[7] is that the HSC Code does not provide for "go anywhere" certification but has a two-part certification system. The construction and equipment of the craft is covered by the High-Speed Craft Safety Certificate, while an additional certificate, the Permit to Operate High-Speed Craft is issued in consultation with port States who are required to provide details of "any operational conditions associated with the operation of the craft".

During the 2006 review of the 2000 HSC Code, it was pointed out to IMO's Ship Design and Equipment (DE) Sub-Committee[8] that imposition of operational conditions or limitations was subject to non-uniform interpretation. For example, some Administrations only impose operating limitations relating to loading on the craft's structure in accordance with classification society requirements, whereas others may consider the totality of the vessel's safe operations including considerations pertaining to:

- intact stability;
- damage stability;
- manoeuvring/seakeeping;
- ability to safely evacuate in an emergency; and
- environment.

Examples of this final point are the imposition by the United Kingdom and New Zealand of operational limitations to protect the environment and persons in the vicinity from wake wash waves that may be generated by high-speed craft in certain circumstances.

The need for the HSC Codes to provide for the active involvement of the port State(s) in setting operational limitations is obvious when it is considered that high-speed craft certificated under the Codes are restricted to operating on specified routes or geographic areas and that those routes/areas may lie entirely outside the waters of the flag State.

As a result of the variations in matters considered by different Administrations, RINA 2005[(8)] provides several examples of discrepancies such as operational limitations on significant wave height varying from 3.5 m to 5 m that have been imposed on "one family of closely similar 96 m to 98 m catamarans" by different Administrations.

IMO's Maritime Safety Committee subsequently agreed in May of 2006[9] that the DE Sub-Committee should develop guidelines for uniform operating limitations for high-speed craft.

DEVELOPMENT OF THE GUIDELINES

The task was first considered by the DE Sub-Committee at its 50th meeting in March 2007, but the main work on it was delegated to a correspondence group coordinated by the author. The group was instructed to develop draft guidelines taking into account submissions on the subject to the Sub-Committee's current and preceding sessions (RINA 2005, etc.[10],[11]) and comments that may be made by the other relevant Sub-Committees.

It is not intended for this paper to provide a detailed description of the workings of the correspondence group. However, it is instructive to elaborate on some of the issues considered by the group.

The proposal by China (China 2006) for standard formulae of waveheight/speed limitation curves did not attract any support and so was discarded. Based largely on RINA 2005a, an initial draft of the guidelines was prepared for consideration by the group. It provided for the operating limitations, drawing on the parent provisions of the amended 2000 HSC Code, to take into account:

- Maximum distance from place of refuge
- Available rescue and operational support resources
- Physical operational environment (wind force, minimum air temperature, visibility and depth of water)
- Sea state limitations – significant wave height
- Trials demonstrating compliance in relation to operating limitations
- Navigational matters
- Display of operating limitations.

Much of the resulting text was uncontentious and remained virtually unamended through to finalisation of the guidelines. However, a number of issues were the subject of vigorous discussion and some are likely to remain "on the boil" even beyond the finalisation of the guidelines. These are outlined below.

Evacuation considerations

As part of the operational philosophy of high-speed craft is to evacuate into survival craft in the event of significant emergency, it has been a prime concern since AMSA's first involvement with high-speed craft under its jurisdiction that these craft should not operate in sea conditions where safe evacuation has not been demonstrated. This was of particular concern in early high-speed craft where passengers were required to climb down the side of the craft and into a life-raft without the benefit of a marine evacuation system (MES).

During the development of the Guidelines, a leading Australian liferaft and MES manufacturer submitted to AMSA that the requirement of the life-saving test procedures[12] for heavy-weather sea trials of MES in seas of significant wave-height of 3.0 m should not restrict craft on which the MES is fitted to operations up to that sea state. The submission was along the lines that, although the MES could no doubt be safely used in worse conditions, the personal safety hazards associated with more severe sea conditions would negate any safety benefits of further testing those conditions. On this basis, the satisfactory completion of MES type approval should not restrict the craft using those MES to the sea conditions in which they were tested.

As other Administrations took a similar view to this manufacturer, AMSA agreed to accept this outcome and change its policy approach accordingly. This removed one factor from those that may define operational limitations.

Understandably in light of the *Sleipner* tragedy[13], Norway was not happy with accepting evacuation arrangements involving evacuation directly into the survival craft without the use of MES. Accordingly, a clause was inserted into the guidelines providing the Administration in such cases with the right to require satisfactory evacuation trials in weather and sea conditions up to the worst intended conditions stipulated on the Permit to Operate.

Extrapolation of sea trial results

Prior to the 2006 amendments, the HSC Code placed no restriction on the extent to which linear extrapolation could be applied to sea trial results when setting operational limitations. An amendment to paragraph 3.2 of Annex 9 of the Code reduced this permissible extrapolation to 150% of the trial sea conditions.

In this regard it should be noted that weather and sea conditions closely approximating the worst intended conditions may not be readily available for a high-speed craft builder to prove the safe operations of his product. Nor is it practicable to require a craft that is to enter commercial service to be tested in full scale to the worst intended conditions, where the safety of the craft (eg. in relation to bow diving of a catamaran or skirt tuck-in of a hovercraft when manoeuvring) may be marginal. On the other hand, technological developments since the 1994 HSC Code was adopted are such that non-linear seakeeping calculations can be conducted using software that is freely available from commercial sources.

However, for the two years that the guidelines were under development by the correspondence group, the United Kingdom insisted that they were unhappy with accepting any substantial extrapolation, but could if necessary accept a small extrapolation. This position received no support. But in recognition of the United Kingdom's concerns, a limit of 130% (ie. 30% extrapolation) was finally included in the guidelines. Nonetheless, when the guidelines were finally adopted[14], this concession did not prevent the United Kingdom from again protesting at the 130% value when the Guidelines were being finally approved by IMO's Maritime Safety Committee.

The discrepancy between the 150% value in the Code and the 130% value in the Guidelines presents an operational problem for those high-speed craft builders who cannot readily access sea trial conditions requiring less than 30% extrapolation to worst intended conditions. It is expected that this discrepancy will be remedied when the 2000 HSC Code is next amended.

Available Rescue and Operational Support Resources

This factor, which allows for retention of overall safety standards while facilitating weight reduction of high-speed craft relative to conventional vessels, can be readily overlooked if the design of a high-speed craft service is undertaken with a concentration on hardware rather than a total safety perspective.

It has therefore been given appropriate emphasis in the guidelines.

Environmental Matters

Mention has previously been made of wake wash waves. Interests centred in the United Kingdom were enthusiastic about having the United Kingdom's requirements on the subject incorporated into the guidelines. The group noted, however, that other Administrations have their own ways of dealing with this phenomenon if and when it becomes a problem, so the final decision was to deal with it in broad terms only rather than introduce specifics.

Minimum safe water depth with regard to bottom scouring and adverse effects on seabed flora and fauna are among the other environmental aspects mentioned on the final guidelines.

Measurement of Significant Wave Height

Since the significant wave-height is generally a significant if not predominant part of the operating limitations placed upon a high-speed craft, the means of measuring such a wave-height in an operational environment is an important part of the implementation and effectiveness of the Guidelines in an operational context. While it is recognised that many craft have systems for directly measuring the sea state or accelerations to which structure is subjected in service, the Guidelines incorporate a low-tech method by which operational wave-height can be estimated.

Other Safety Matters

During the process of development of the 1994 HSC Code, mention was made of the brittle failure of a main engine tail-shaft on a hovercraft operating in the Canadian Arctic that had been built in tropical Australia. One would normally expect that any environmental (eg. temperature) limitations on the operations of a craft would be reflected in restrictions imposed by the classification society when classing the craft, but the HSC Codes provide for the imposition of such limitations on the Permit to Operate, whether in concert with or in addition to class restrictions.

To cover such occasional cases, especially where the operational routes/areas are predominantly or entirely outside the waters of the flag State, is there is an obvious need for port State involvement to ensure that the operating limitations are comprehensive and consistent relative to similar craft operating under other flag(s) in adjacent waters.

FINALISATION OF THE GUIDELINES

The Guidelines (IMO 2009) were finalised by the DE Sub-Committee at is session in March 2009, without amendment to the draft finalised by the correspondence group, and subsequently promulgated by the Maritime Safety Committee at its 86th session in June 2009.

CONCLUSION

Development and promulgation of the Guidelines fills an important gap in the standards framework associated with the 2000 HSC Code, particularly in relation to achieving international uniformity in the operational limitations imposed under Permits to Operate issued under the Code.

In particular, the Guidelines draw the attention of Administrations, when establishing operational limitations for the Permit to Operate for individual craft, to consider much more than the sea state limitations that may have been imposed by the classification society from a structural perspective

ACKNOWLEDGEMENTS

The author would like to express his appreciation firstly to the Australian Maritime Safety Authority for its support for Australian participation in the work outlined in and for the writing of this paper, and secondly to the Australian Shipbuilders Association for the input provided by its members through its Technical Coordinator, Dr Tony Armstrong, that has permitted this work to be drawn to a successful conclusion.

*****

REFERENCES

[1] International Maritime Organization (IMO) 1977, Res. A.373X) *Code of Safety for Dynamically Supported Craft*, adopted 14 November 1977, IMO, London

[2] IMO 1994, Res. MSC.36(63) *International Code of Safety for High-Speed Craft*, adopted 20 May 1994, IMO, London

[3] IMO 2000, Res. MSC.97(73) *International Code of Safety for High-Speed Craft, 2000*, adopted 5 December 2000, IMO, London

[4] IMO 2006, Res. MSC.222(82) *Adoption of amendments to the International Code of Safety for High-Speed Craft, 2000*, adopted 8 December 2006

[5] IMO 2008, *International Code of Safety for High-Speed Craft, 2000, 2008 Edition*, IMO, London, 2008

[6] IMO 2005, *MSC/Circ.1177 Unified interpretation of the 2000 HSC Code*, IMO, London, May 2005

[7] IMO 2004, *Consolidated text of the International Convention for the Safety of Life at Sea, 1974, and its Protocol of 1988: articles, annexes and certificates – incorporating all amendments in effect from 1 July 2004*, IMO, London

[8] RINA 2005, *DE 49/5/3 Consistency in determination of operating limitations*, IMO, London, December 2005

[9] IMO 2006, MSC 81/25 *Report of the Maritime Safety Committee*, paragraph 23.45, IMO, London, May 2006

[10] RINA 2005a, *DE 49/INF.5 Consistency in determination of operating limitations*, IMO, London, December 2005

[11] China 2006, *DE 50/18 Proposal on speed limitations of high-speed craft*, IMO, London, December 2006

[12] IMO 1998, Res. MSC.81(80) *Revised recommendation on testing of life-saving appliances*, IMO, London, December 1998

[13] *The High-Speed Craft MS SLEIPNER Disaster 26 November 1999*, Official Norwegian Reports 2000:31, Government Administration Services – Information Management, Oslo, 2000

[14] IMO 2009, *MSC/Circ.1329 Guidelines for uniform operating limitations for high-speed craft*, IMO, London 2009

The Implications of High Speed Transits on Personnel

Paul Elischer, Darren Sanford, Darren Wiese.

Defence Science and Technology Organisation
Platform and Human Systems Laboratory
Maritime Platforms Division

Paul.elischer@dsto.defence.gov.au

ABSTRACT

This paper describes the approach and outcomes of a study undertaken on two high speed craft (HSC) of the type used by "thrill ride" operators, rescue organisations and the military to assess the implications of high speed transits on the health of crew and passengers. There are significant occupational health and safety issues for personnel undertaking fast boat transits and these can be attributed to both continuous whole-body vibrations and intermittent hydrodynamic loadings (accelerations) due to the HSC impacting with the waves. The study assessed the discomfort levels and the potential risk of injury to personnel during a range of high speed transits.

The level of discomfort was quantified by analysing the average of the 1/10 highest accelerations measured on the deck of each HSC and at the lumbar back of the operators. The potential risk of injury was quantified using a biomechanical model (International Standard, ISO 2631 Part 5) to determine the static compressive stress (Spine Stress Dose (Sed8)) value and an adverse health effects factor "R". The Sed8 value relates the calculated spinal accelerations, determined from measured acceleration data, to the likelihood of personal injury. This is achieved by comparing the Sed8 values against proposed injury tolerance thresholds. The adverse health effects factor "R" provides an indication of the potential risk of personnel sustaining long term injury due to occupational exposure.

1. INTRODUCTION

Whilst the use of HSC allow for faster response in rescue operations and for a fast and exciting experience for the general population there are human factor issues associated with whole body vibrations and intermittent shock exposures due to the HSC impacting with waves. During fast transits in heavy seas, personnel can be subjected to significantly high accelerations for considerable periods of time. Prolonged exposure to these loadings can have a debilitating effect on the physical and cognitive performance of the personnel involved in HSC operations. This can have both short and long term occupational health and safety (OH&S) ramifications for all personnel.

The loadings resulting from wave impacts are transferred directly to the back, neck, knees and supporting muscles causing discomfort, lower back problems and acute and chronic pain. In the more extreme cases fracture of the neck vertebrae can also occur. The ultimate effect can include prolonged pain and discomfort, the increased costs associated with medical care and potential compensation [1].

This paper presents an assessment of the implications of high speed transits, on personnel, in terms of human discomfort and adverse shock loadings. It relates these characteristics to the potential for injury that personnel may experience or sustain during high speed transits. The analysis used is the average of the 1/10 highest accelerations, the static compressive stress value experienced by the lumbar region of the back (Spine Stress Dose (Sed8)) and the adverse health effects factor "R" which provides an indication of potential long term health effects. The Sed8 and "R" values are determined from the algorithms described in the international standard, ISO 2631, Part 5 [2]. This standard adopts a biomechanical approach to define a method of quantifying the effects of whole-body vibration containing multiple shocks, such as those experienced during fast boat transits, to human health and identifies the effects on the lumbar spine as being the dominating health risk.

2. METHODOLOGY

2.1 Trial Procedure

Two different HSC designs were assessed during a sea trial which was conducted in sea conditions ranging from smooth seas (sea state 0) to moderate seas (approx. sea state 2, 1.5 m waves). The sea conditions remained essentially the same throughout the trial period and therefore there was no opportunity to obtain acceleration data in higher sea states. The manner in which HSC respond to the prevailing sea conditions and the subsequent effect on personnel was characterised by measuring the accelerations at specified locations on the HSC and on personnel at their lower back (lumbar spine).

Each HSC was manned by experienced operators (crew A and crew B) and two passengers. Two different data runs were undertaken. A "controlled" run during which the HSC maintained a constant but specified speed irrespective of the heading taken and, an "operational" run during which the coxswain adopted a driving strategy that reflected the procedures they would use during operations given the same prevailing conditions.

The data runs, summarised in Table 1, were conducted at five headings relative to the prevailing sea (Figure 1); head, following, starboard beam, quarter bow and quarter stern. Usually eight headings are undertaken [3, 4] for a complete data run however, as the HSC have symmetrical shapes any accelerations measured during the equivalent starboard or port headings would be the same. To provide an adequate statistical sample of the wave impacts, five minutes of acceleration data were collected for each heading, providing a total of 25 minutes of data for a completed data run. The HSC were not loaded to their maximum displacement and therefore the trial was conducted with both HSC being in a part loaded condition.

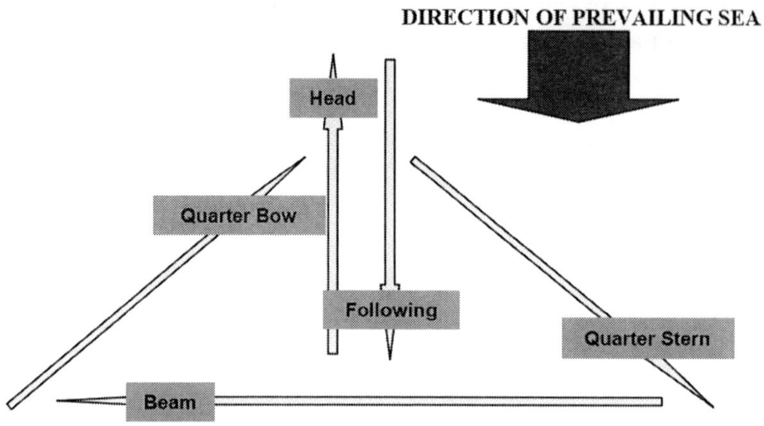

Figure 1: Transit headings with respect to the prevailing sea

Table 1: Summary of the data runs undertaken during the trial.

Run	HSC	HSC Speed (knots)	Sea State	Wave height (m)	Crew	Driving Mode
4	1	22	2	1-1.5	A	Controlled
4	2	22	2	1-1.5	B	Controlled
5	1	31	2	1-1.5	A	Operational
5	2	31	2	1-1.5	B	Operational
6	1	20	2	1-1.5	B	Controlled
6	2	20	2	1-1.5	A	Controlled
7	1	32	2	1-1.5	B	Operational
7	2	32	2	1-1.5	A	Operational

2.2 Instrumentation

The configuration of the sensors and acquisition units (Figure 2) were based on the methodology adopted during previous HSC trials and is in line with the procedures used in similar overseas trials [5]. This approach is consistent with a draft measurement methodology [6] currently being prepared under the auspices of the American, Australian, British, Canadian and Dutch (ABCD) ad hoc Working Group investigating human performance at sea.

Each HSC was equipped with an identical data acquisition unit powered from the onboard power supply. Tri-axial accelerometers were fitted to rigid areas on the deck at the fwd, mid and aft positions along the longitudinal centre line of the HSC. The mid accelerometers were placed as close as possible to the coxswain's workstation to measure the local structural acceleration and the shock loading that the coxswain would be exposed to via the deck. To complement these measurements and to measure the loading at their lower back (lumbar spine), the coxswain on each HSC was required to wear a kidney belt which was also fitted with a tri-axial accelerometer.

The datum for all accelerometer locations was at the aft position of the HSC. The coordinate system for the accelerometers was as follows; "X" represents the forward-aft direction (longitudinal), "Y" the port-starboard direction (transverse) and "Z" the vertical plane relative to the deck (vertical).

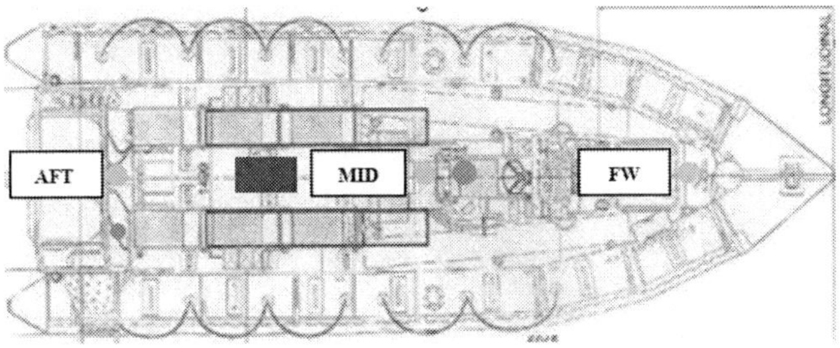

- Tri-axial accelerometer
- GPS
- Coxswain with kidney belt accelerometer
- Data acquisition unit

Figure 2: Sensor and acquisition configuration onboard the HSC

2.3 Data acquisition

The acceleration data was acquired using a LSD Nicolet (Liberty) acquisition unit configured to record 12 channels of data at 4000 samples per second with 16-bit sample resolution. The acquisition units were set to record five minutes of data for each of the five data runs. Acquisition commenced by activating a "push button" external trigger when the coxswain's signaled that the HSC were at the required speed and heading.

The accelerations in the "Y" and "Z" directions were measured using PCB ICP accelerometers; model J353B33 with a measurement frequency range of 1 to 4000Hz. The accelerations in the "X" direction were measured using Dytran model 3041A2 accelerometers which have a frequency range of 1 to 3000Hz. All accelerometers have ± 50 g operating range and were calibrated against National Association of Testing Authorities (NATA) certified standards.

The loadings at the lower back (lumbar spine) were measured using a tri-axial accelerometer which was imbedded within the rubber matrix approximately 200 mm in diameter and 5 mm thick. The accelerometer was secured to the inside of a kidney belt. This allowed the accelerometer to fit firmly and directly against the lumbar spine when the kidney belt was fastened around the waist. These accelerometers have an operating range of ± 10 g and frequency range of 0.5 to 1000 Hz. Factory calibration values were used to calculate the accelerations recorded by the lumbar accelerometers.

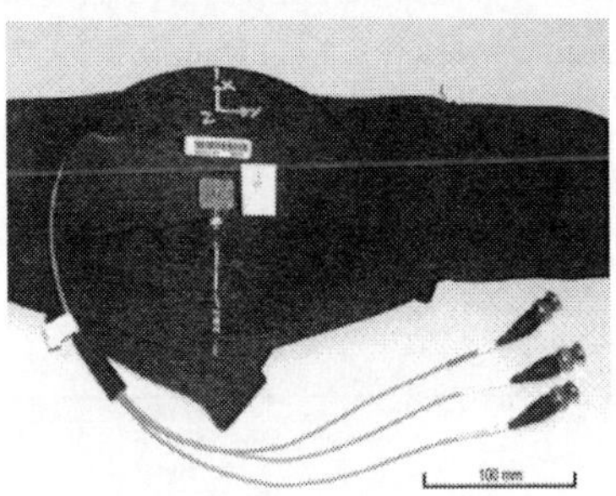

Figure 3: Kidney belt fitted with the lumbar accelerometer and the kidney belt fitted to a coxswain.

3. RESULTS

3.1 Acceleration data

The acceleration data were analysed using the average of the 1/10 highest accelerations to assess crew comfort levels, and in accordance to the guidelines provided by the international standard, ISO 2631 Part 5 standard [2], to quantify the health risks due to whole body vibrations containing multiple shocks. The acceleration data analysed was restricted to that collected during data runs 4 to 7 (Table 1) which represented head sea transits in the rougher sea conditions. Data runs 4 and 6 were conducted under controlled conditions with the speed remaining constant at around 22 knots irrespective of the heading taken. Data runs 5 and 7 were operational runs and typified the driving strategies and speeds that the coxswain would employ under the prevailing conditions.

3.1.1 Crew comfort levels determined from the average of the 1/10 highest accelerations

Accelerations in the vertical "Z" direction, into head seas, are the most severe responses during fast transits and can be better understood by calculating the average of the 1/10 highest accelerations. These accelerations can be correlated to the human discomfort levels suggested by Savaitsky and Kobel [7] shown in Table 2. The analysis to assess the relationship between sea keeping and human performance and the acceleration data is based on this guidance together with the outcome of a USA study [8]. The plots for head sea transits at 22 knots and 32 knots are shown in Figure 4.

All the acceleration data was down sampled to 160 samples per second and then processed using a Matlab® routine prior to calculating the average of the 1/10 highest accelerations. The routine removed accelerometer drift, identified the crests and troughs after each zero crossing and determined the average of the 1/10 highest of the positive and negative accelerations.

Table 2: The effects of vertical accelerations on personnel

g's Above Static	Human Discomfort levels
0.6	Minor discomfort
1.0	Maximum for Military function long term (over 4 hours)
1.5	Maximum for Military function long term (over 1-2 hours)
2.0	
3.0	Extreme discomfort
4.0	Physical Injury
5.0	
6.0	

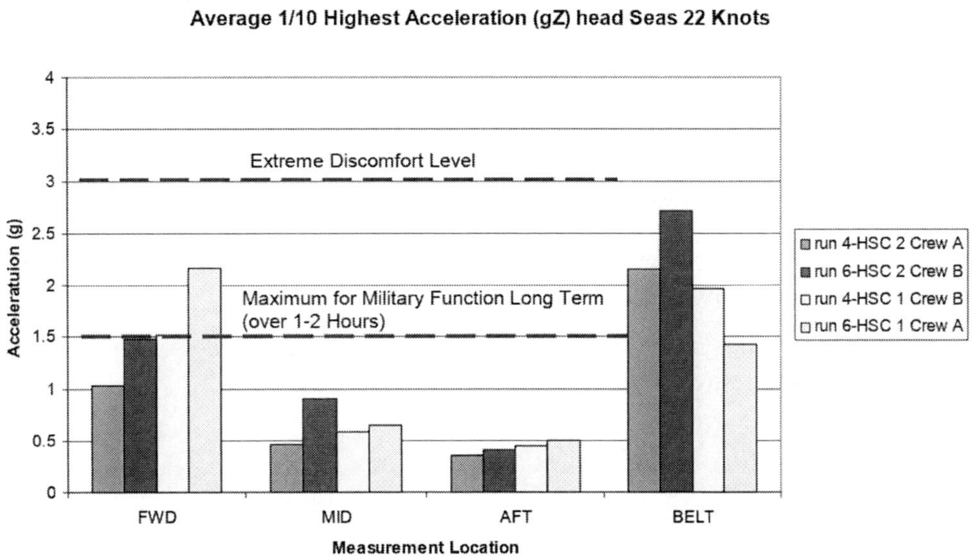

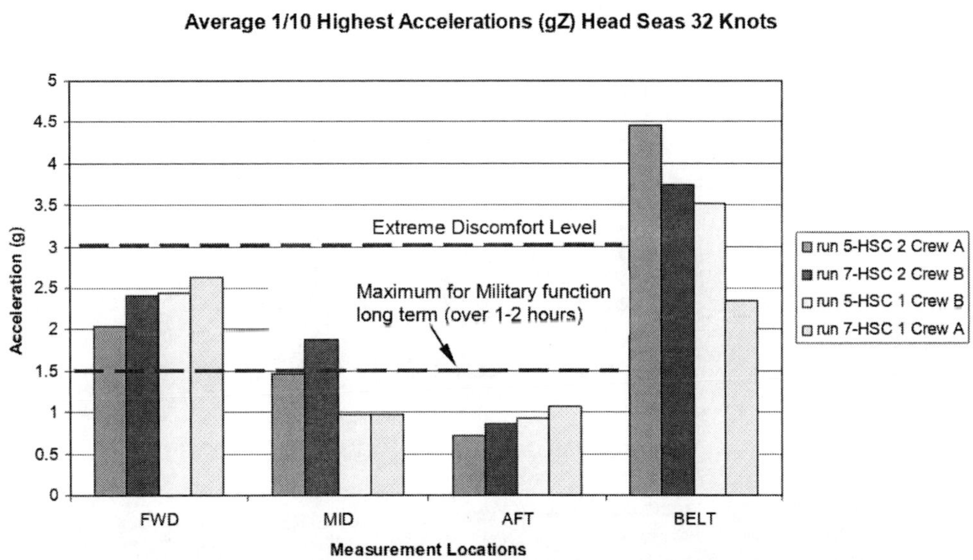

Figure 4: Average of the 1/10 highest "Z" accelerations for transits into head seas at 22 and 32 knots.

Evaluation of crew health/ discomfort using the ISO 2631 Part 5 Guidelines

The international standard ISO 2631 Part 5 is the generally accepted standard [9, 10, 11] for quantifying health risks on personnel exposed to whole body vibrations containing multiple shocks. It defines two procedures; the static compressive stress value (spine stress dose (Sed)) and adverse health effects factor "R".

The standard is a biomechanical model which in its normal application is based on accelerations (X, Y and Z directions) measured at the seat cushion of a seated person. This situation does not necessarily apply to all HSC transits as, depending on the design, seating options may not be available or personnel may adopt their own individual posture for coping with wave impacts. The accepted alternative is to measure acceleration data at the lumbar spine of personnel transiting on HSC [12].

3.1.1.1 Evaluation of Health Effects using the Static Compressive Stress Value (Spine Stress Dose (Sed8))

The standard defines algorithms to evaluate health effects based on acceleration data. It has been shown that there is a linear relationship between the compressive stress that is due to the wave impacts and the peak acceleration response in the spine. These accelerations are used to calculate a Sed value which can be normalised to an average 8 hour work day to produce a Sed8 value. Normalisation of the Sed value provides a means of comparing the results obtained from other trials.

The Sed8 values calculated for the head sea runs at 22 and 32 knots, data runs 4 and 6 and 5 and 7 respectively, are shown in Figure 5.

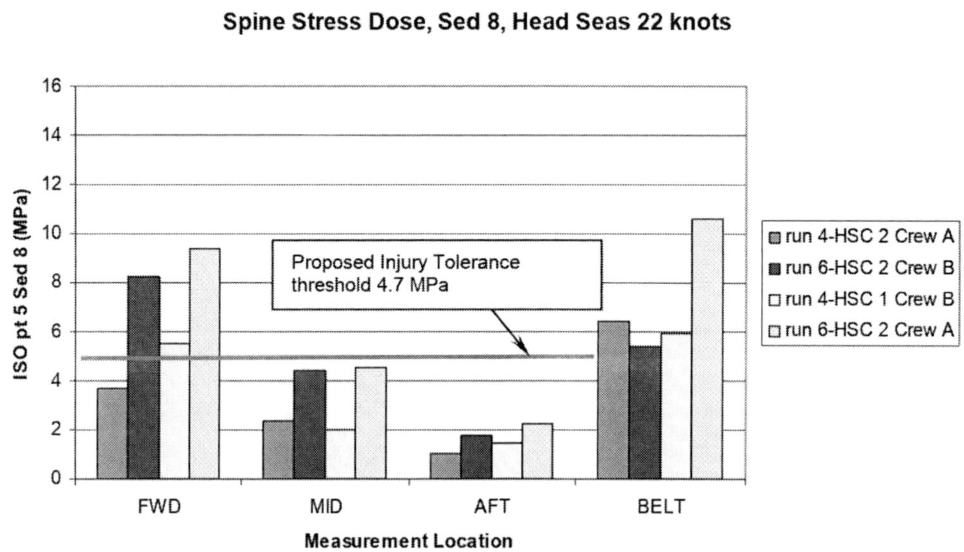

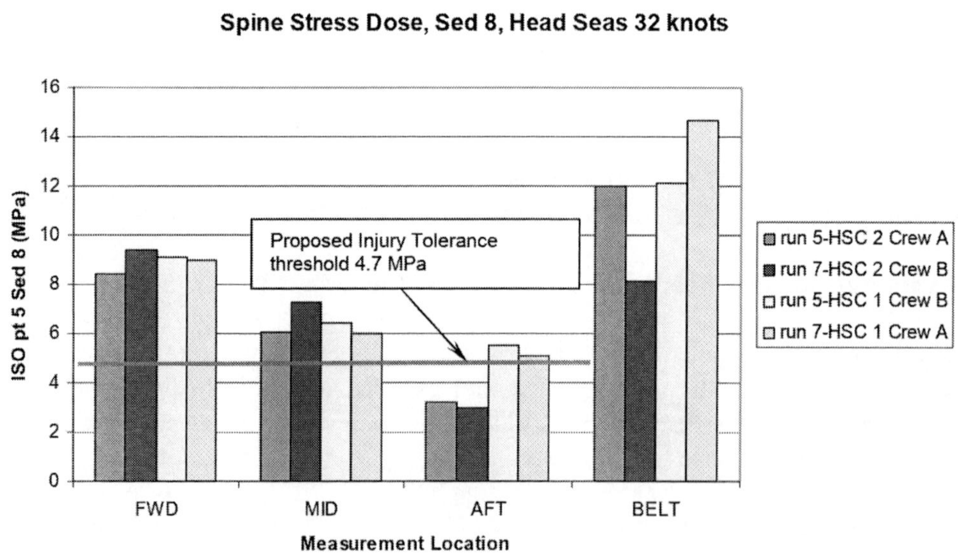

Figure 5: Sed8 values for transits into head seas at 22 and 32 knots.

3.1.1.2 Adverse Health Effects Factor "R"

The ISO 2631 Part 5 standard also describes a method to determine an adverse health effects factor "R", which is defined as being the risk of long term injury to personnel due to occupation exposure to shock and whole body vibration. This is based on the Sed values described above. To calculate "R", additional data such as the age of personnel, their duration and frequency of exposure is required. This information was not available and therefore a hypothetical demographic data set was created. It was assumed that personnel commenced work as a coxswain at the age of 20 years and were exposed to the accelerations measured during data runs 4 to 7 for an average of 4 hours per week for 50 weeks of the year (50 days exposure per year) over a ten year period. The calculated "R" values using this data set are summarised in Table 3.

Table 3: Adverse Health Effects values "R" for the lumbar region (belt) for combined headings for data runs 4 & 6 and 5 & 7.

Crew	Data Run	Speed (knots)	HSC 1 "R"	HSC 2 "R"
Crew A	Controlled run	22	3.9	3.5
Crew B	Controlled run	22	2.8	2.4
Crew A	Operational run	32	5.6	4.9
Crew B	Operational run	32	4.6	3.4

4. DISCUSSION

4.1 Crew comfort levels using the average of the 1/10 highest accelerations

The results, Figure 4, show the expected trend of the vertical acceleration (Z) loading for transits into head seas, decreasing from the fwd to aft locations and the overall increase in accelerations as the speed increases. It also shows a higher loading being recorded at the lower back (lumbar spine) of personnel when compared to the accelerations recorded at the mid locations. At a speed of 22 knots the discomfort levels described in table 2 vary depending on the location on the HSC. Personnel occupying mid and aft positions on the HSC would experience discomfort levels ranging from minor (aft) to the maximum for military function long term, over 4 hours (mid). At a speed of 32 knots personnel occupying the same positions as above would experience discomfort levels ranging from the maximum for military function long term, over 4 hours to around the maximum for military function long term, 1-2 hours (mid). At both speeds the discomfort levels at the forward location ranged from around the maximum for military function long term, 4 hours to discomfort levels approaching the extreme. At this location the implications for personnel is not an issue as they do not occupy any positions that are forward of the mid location.

Applying the discomfort values to the accelerations at the lower back (lumbar spine) should be done with caution as the discomfort levels have been suggested for "Z" accelerations that have been measured on fixed, rigid locations on the platform and not on personnel. Nevertheless these have been included to provide an indication of the response of the lumbar spine (belt) relative to the discomfort levels.

4.2 Risk of injury using Static Compressive Stress Value (Spine Stress Dose (Sed8) values

The Sed8 values, Figure 5, calculated for the fwd, mid, aft locations and the lower back (lumber spine) follow a similar trend to that observed for the average 1/10 highest accelerations shown in Figure 4 and also show the expected increase in Sed8 value with increased speed. Any variation in the Sed8 values calculated at the lower back (lumbar spine) could be attributed to the manner in which the coxswain's responded to the prevailing conditions and their agility of personnel to manoeuvre their bodies commensurate with the motion of the HSC.

At a speed of 22 knots personnel occupying the mid and aft positions would be below the Sed8 injury tolerance threshold of 4.7 MPa as proposed [13, 14] for the international standard ISO 2631 Part 5. At a speed of 32 knots personnel occupying the mid positions would exceed the proposed injury tolerance threshold of 4.7 MPa. Personnel occupying the aft positions would be just or below the proposed threshold.

As yet there are no Sed8 injury tolerance threshold proposed for a standing person and this is an area that is being addressed by overseas research establishments. In collaborative discussions between ABCD members [15] it was indicated that the injury threshold value for a standing person could be higher than the Sed8 range currently cited in the standard.

4.3 Risk of injury using the Adverse Health Effects Factor "R"

The adverse health effects factor "R" results, Table 3, are the predicted long term health effects calculated over a ten year period for the hypothetical data set used. The variation in the results can be attributed to the differences in the design of the two HSC and the manner in which the HSC are driven by the coxswain's implying that the overall performance of the HSC and hence the well being of personnel may be related to the driving strategy adopted by the coxswain.

Although the "R" values are hypothetical and based on a relatively small data set they do provide an indication of the potential for long term health implications for personnel transiting on HSC.

5. CONCLUSION

It is known that high speed transits can have an adverse effect on both short and long term well being of personnel and is the subject of international interest. Assessing the OH&S implications in terms of human discomfort and well being using the approaches described in the paper provides a means for both obtaining a better understanding and for quantifying the problem. The assessment concluded that OH&S issues are both dependant on the design of the vessel and the driving strategy adopted by the coxswain.

The information presented can be used to develop mitigating strategies that would minimise the risk of injury to personnel.

6. REFERENCES

1. Stanton, D. (2006) *Small Craft Incidents, Army Small Craft & Equipment Operation Safety Case Study,* Version 1.2 OHSC. Occupational Health Safety and Compensation Management Information Directorate, Safety Information Management Unit.
2. ISO 2631-5:2004, *Mechanical Vibration and Shock-Evaluation of Human Exposure to Whole Body Vibration Part 5 Method for Evaluation of Vibration Containing Multiple Shocks*. International Standard, First edition 2004-02-15.
3. Haupt, K. (1997) *High-Speed Craft Motions: a Case Study.* Naval Surface Warfare Center, Carderock Division, Combatant Craft Department, Norfolk, Virginia, USA.
4 Peterson, R., Price, B. & Pierce, E. (2005) *Active Shock, INC. (ASI) NSW RIB Semi-Active Suspension Seat*, Naval Surface Warfare Center, Panama City, Florida 32407.
5. Ullman Dynamics, (2001), *Ullman Cockpit Test Report.*
6. Gunston, T. (2007) *Draft Measurement Methodology for Small High Speed Craft-UK View SIG SCP.*
7. Savitsky, D. & Koebel, G. J. (1993), *Sea keeping of Hard Chine Planing Hulls.* Society of Naval Architects and Marine Engineers.
8. Coats, T. & Stark, J. (2008) *High Speed Craft-Sea keeping and human Performance,* Pacific 2008 International Maritime Conference and ABCD Symposium, Sydney, Australia.
9. Dobbins, T., Rowley, I. & Campbell, L. et al (2008) *High Speed Craft Human Factors Engineering Design Guide*, ABCD-TR-08-01 V1.0.
10. Coats, T. & Stark, J. (2008) *High Speed Craft-Sea keeping and human Performance,* Pacific 2008 International Maritime Conference and ABCD Symposium, Sydney, Australia.
11. Alem, N. (2005) *Application of the New IOS 2631-5 to Health Hazard Assessment of Repeated Shocks in U.S. Army Vehicles.* Industrial Health, 43, 403-412.
12. Peterson, R., Price, B. & Pierce, E. (2005) *Active Shock, INC. (ASI) NSW RIB Semi-Active Suspension Seat,* NSWC-PC, test report to USSACOM.
13. Peterson, R. & Pierce, E. (2005) *High Speed Craft Biodynamics & Human Performance. ONR* IFO Human factors Workshop. Poole, UK.
14. Bass, C., Ash, J., Salzar, R., Peterson, R. & Pierce, E. (2008) *Prediction of Injury Risk for Occupants of High Speed Planing Craft.* Pacific 2008, International Maritime Conference and ABCD Symposium, Sydney, Australia.
15. Workshop discussions (2008) ABCD Working Group Human Performance at Sea, Virginia, USA.

Biofouling: Implications for Performance

Lyn Fletcher, Lou Montelli & Andrew Scardino,
Defence Science & Technology Organisation, Melbourne, VIC, Australia;
lyn.fletcher@dsto.defence.gov.au

ABSTRACT

Biofouling is a complex problem that can substantially impact ships' performance if poorly controlled. Biofouling can significantly increase drag which has a deleterious impact on fuel consumption, speed and range and increases engine loads, leading to higher cleaning and maintenance costs and the potential translocation of invasive marine species. Military vessels have a very different operational tempo than commercial vessels and therefore require different antifouling strategies. In addition, Royal Australian Navy ships operate in a range of geographical regions and can spend substantial periods in northern tropical waters where biofouling pressures are severe. This paper examines the problems with biofouling on military vessels and outlines DSTO's S&T program to control and reduce biofouling on RAN Ships including: methods to quantify the penalty from biofouling; investigating current commercial antifouling coatings and emerging novel coatings; and techniques to reduce biofouling settlement on stationary vessels.

INTRODUCTION

Fouling of vessels is a universal problem and the process of fouling begins within minutes of immersion. Australia's marine environment, particularly the tropics, has some of the strongest fouling pressure in the world. Fouling of submerged surfaces is an enormous problem both economically and environmentally (Lewis & Houridis, 2001).

The main method for controlling fouling is the use of effective antifouling coatings. Tributyl tins (TBT) coatings have traditionally been the mainstay of coatings; however, these are now banned due to the environmental risks associated with TBT accumulation (IMO, 2007). Currently the two main classes of antifouling coatings commercially available are copper self polishing coatings (SPC) that contain a variety of booster biocides (Yebra et al. 2004) and the new non-toxic fouling-release coatings (FRC) (Anderson et al. 2003). All biocidal coatings must be registered and approved for use in Australia which can restrict the availability of coatings. In addition, there are continuing issues associated with amount of volatile organic compounds used with the current stable of SPC antifoulants.

Fouling-release coatings are designed to self clean by reducing the adhesion strength of the fouling organisms causing them to release when the vessel reaches operating speed. Due to their design FRCs are not effective when a vessel is stationary for long periods or if the vessel operates at low speeds (< 8 knots). Once certain types of fouling organisms are firmly attached to FRCs they will not be released even at high operating speeds.

At present the majority of the RAN fleet are coated with copper SPC coatings. The Armidale Class Patrol Boats (ACPB) are painted with a FRC due to the incompatibility of aluminium hulls and copper SPC coatings.

FRC coatings are gaining greater market share in the merchant and cruise fleets that have an operational profile with greater consistency in steaming than observed with military vessels. FRC are marketed to have the potential to improve fuel consumption of up to 8% on newly applied coatings for new merchant ships due to the inherent smoothness of FRC coatings. DSTO has a program in place to investigate the efficacy of different coatings for military vessels and a program to investigate improved hull management options.

THE HULL FOULING PENALTY

Fouling can substantially affect ships' performance by increasing the hydrodynamic drag because of increased skin friction. The total resistance of a hull is the sum of skin friction, or frictional resistance, and residuary resistance. As shown in Figure 1, the impact of frictional resistance and the impact of fouling will be most noticeable at moderate cruising speeds (0-20 knots). At higher speeds (25 knots or more) residuary resistance, which is predominately made up of wave resistance, will have the greatest impact on hull resistance. The skin friction is dependent on surface roughness and biofouling contributes directly to surface roughness.

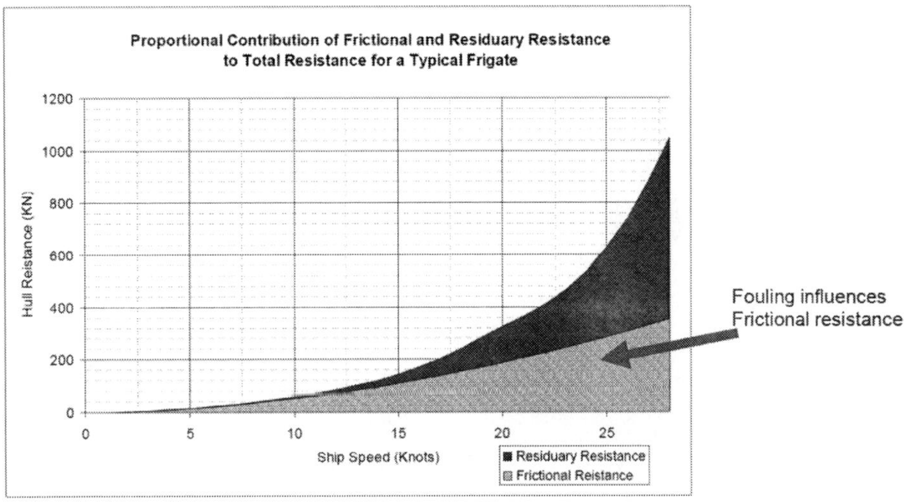

Figure 1: Proportional Contribution of Frictional and Residuary Resistance to Total Resistance for a Typical Frigate (MoD 2007)

In the last couple of years, attempts to better quantify the effects of fouling on ship performance has been the subject of a number of studies (Townsin 2003; Schultz 2007; Walker & Atkins, 2007). Fouling organisms come in many different shapes, sizes and hardness. Most of the initial forms of fouling such as slimes, algae and ascidians are soft bodied. It is extremely difficult to quantify the roughness of this type of fouling as it requires a non-contact method in water to measure it. The more advanced stages of fouling involve hard bodied, calcareous organisms such as barnacles, tubeworms, bryozoans and shellfish. The roughness of these organisms can also be difficult to measure due to their irregular shape and size and often the capability of the roughness measuring tools is exceeded. The numbers presented in Table 1 from the literature (Schultz, 2007) can only be taken as a guide and are most likely to be conservative estimates.

Table 1: The roughness associated with various conditions of fouling. Source: Shultz (2007)

Condition	Roughness (μm)
Hydraulically smooth	0
Newly applied antifouling coating	150
Deteriorated coating or light slime	300
Heavy slime	600
Small calcareous fouling	1000
Medium calcareous fouling	3000
Heavy calcareous fouling	10,000

It is evident from Table 1 that as soon as hard bodied calcareous fouling is established on a surface the roughness increases dramatically. DSTO has measured similar results on current FRCs (Scardino et al. 2009). It is considered that Navy ships operating in the high fouling conditions of northern Australia will accumulate medium to heavy calcareous fouling over extended periods of their docking cycle.

Table 2 outlines some recent studies performed by the Royal Navy and the US Navy on frigate sized ships which compared changes in shaft power and fuel consumption with observed biofouling accumulation on ships coated with SPC coatings.

Table 2 Summary of the impact of biofouling accumulation on shaft power and fuel consumption.

Condition	**FFG-7 Roughness (μm) *(Schultz 2007)***	**FFG-7 ΔSP at 15 knots *(Schultz 2007)***	**> fuel consumption *(Walker & Atkins 2007)***
Newly applied antifouling biocidal coating	150	2%	-
Deteriorated coating or light slime	300	11%	~ 6%
Heavy slime	600	21%	12%
Small calcareous fouling	1000	35%	18-24%
Medium calcareous fouling	3000	54%	61%
Heavy calcareous fouling	10,000	86%	92%

These studies demonstrate that poorly controlled biofouling can increase fuel consumption substantially. The annual consumption of marine fuel by the RAN is approximately 106 million litres (six-year average, 2002-03:2007-08, Joint Fuels and Lubrication Agency). It has been estimated that our ships, operating in tropical northern waters, have biofouling levels in the small to medium calcareous range and this could be imposing an additional fuel consumption penalty of around 20-60%.

For example the ACPBs operate in a waters with high fouling pressure and from observations it is highly likely that they have fouling conditions of at least 'small calcareous fouling' level. This equates to anticipated shaft power increases of approximately 35% (Table 2) and a corresponding increase in fuel consumption of at least 24% (Table 2). As can be seen from

Figure 2 substantial hull fouling occurs on ACPBs. The medium calcareous fouling shown on these vessels have a potential fuel consumption penalty of approximately 60%.

Figure 2: a) HMAS Larrakia – Star board bilge keel (Darwin, December 06); b) HMAS Armidale – Hull surface, aft (Darwin, May 06); c) HMAS Larakia – Hull surface, aft (Darwin, December 06) d) HMAS Childers (Cairns, August 09); e) & f) HMAS Bundaberg (Cairns, Feb 09)

In addition to increased fuel usage, hull fouling can lead to higher maintenance costs. Figure 3 shows a typical graph of changes in engine load with speed for a clean hull, while Figure 4 compares this with the engine load for a fouled hull. Figure 4 indicates that a fouled hull requires a 30% increase in shaft power to achieve the same speed (MoD, 2007). It is considered that hull fouling will cause these propulsion systems to be engaged at lower speeds, which will eventually impact on maintenance budgets. As shown in Figures 3 this is of particular importance with ships that operate gas turbines where maintenance costs can be high.

In addition, as is shown in Figure 4, there is potentially a loss of 1 knot in top speed for a 30% increase in fouling for this Type 23 frigate. In the MoD (2007) study of a Type 23 frigate, the effect of biofouling on fuel efficiency was greatest at 15 knots. Similar trends can be expected for the RAN, although the actual variations will vary from class to class.

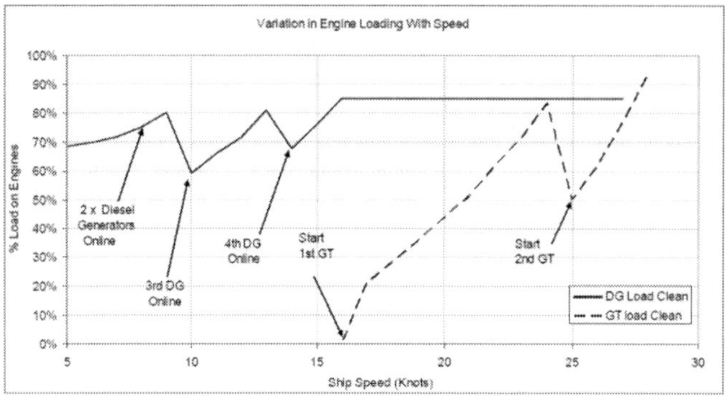

Figure 3: Variation in engine loading with speed for a Type 23 frigate (MoD, 2007)

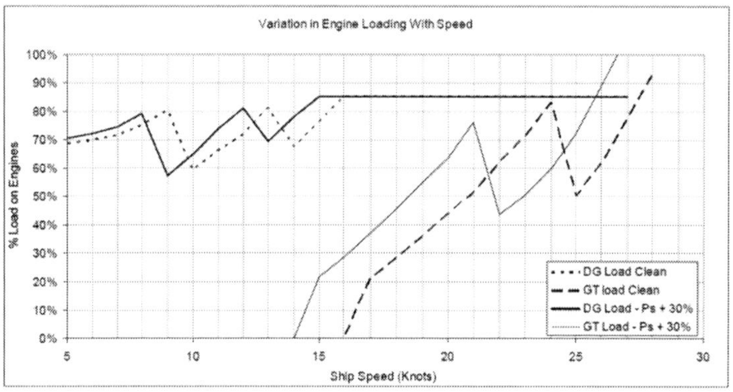

Figure 4: Variation in engine loading with speed including a 30% increase in shaft power for a Type 23 frigate (MoD, 2007)

In Australia we have a tight regulatory regime that must be considered when managing ships biofouling due to risks associated with invasive marine pests being transported in biofouling from ships entering and transiting Australia (Lewis & Houridis, 2001). The National Biofouling Guidelines (2009) and the Australian and New Zealand Environment Conservation Council (ANZECC) Code of practice for Antifouling and In-water Cleaning and Maintenance (1997) prohibit in-water cleaning in Australia when traditional biocide antifouling paints are applied. The guidelines allow for in-water cleaning of ships with FRCs in some circumstances however strict permit conditions are required. This limitation on in-water cleaning severely restricts the ability of the Navy to in-water clean ships' hulls.

ROYAL NAVY SHAFT POWER MONITORING SYSTEM

The RN has developed a system for monitoring the impact of biofouling by monitoring changes in shaft power with the use of a specialised torque meter that measures RPM and torque. Monitoring shaft power under controlled conditions allows assessment of the influence of fouling on the ship. The baseline shaft power is determined as soon as the vessel is out of refit or if it has just been cleaned. Any increases in shaft power from this baseline can be equated to increased fuel consumption and appropriate remediation can be taken.

The RN operates a trigger system whereby shaft power is monitored and cleaning is considered when it reaches a critical threshold. This threshold is set at the point where the cost of cleaning will be recovered by the fuel savings in the following 3 months. Upper and lower threshold limits have also been initiated. The lower trigger limit is set at an 18% increase in shaft power as it coincides with the emergence of hard fouling. The upper trigger limit is set at a 30% increase in shaft power as at this level any further development of fouling could embed into the coating system and cause damage during cleaning.

To measure shaft power each vessel in the RN is required to undertake trials on a monthly basis to manually record torque measurements (Walker & Atkins, 2007). The requirements for the shaft power monitoring trial are:

- Wind speed < 12 knots.
- Sea state 3 or less.
- Stabilisers set to zero incidence.
- Minimum rudder to be used
- Water deep enough to reduce shallow water effects to < 1% of resistance. For ships < 90m long, 20m water depth is required, for 90-160m long vessels 35m is required
- Ship speed set at 13 knots by GPS.
- Shaft power and speed recorded 5 times at 2 minute intervals.
- If the ship has exceeded its trigger value then two further runs are undertaken at 8 and 18 knots. If trigger value is still exceeded then a ship clean is considered.

Actual shaft power increases could be measured on ACPBs if torque meters were installed and could be used to quantify just how much extra fuel each vessel is consuming. In addition, it is important to consider the type of cleaning regime used, as Australia has stringent in-water cleaning regulations and abrasive cleaning of foul-release coatings could easily damage the coating.

It is proposed to trial the RN toque meter on a RAN vessel and to monitor increases in propulsion power in a similar way to the RN method. HMAS Glenelg, an Armidale Class Patrol Boat (ACPB), has been highlighted as the most suitable vessel for the following reasons:

1. HMAS Glenelg already has a suite of complementary sensors and associated data logging system as part of a joint DSTO-Austal aluminium alloy hull performance study.
2. The vessel is subject to extreme fouling pressure as it operates in tropical waters.
3. Evidence to date indicates the silicone fouling-release coatings on the ACPBs have moderate to high levels of fouling.

The torsionmeter trial will provide the following benefits to the RAN:

1. Quantify the increase in fuel consumed by ACPBs as the result of fouling (which could be in excess of 50%).
2. Provide guidelines on hull cleaning regimes to minimise fuel consumption.
3. Provide more consistent vessel performance (speed and range).
4. Reduce maintenance through lower stress on the vessel power train.
5. Provide a framework for the installation of torsionmeters and monitoring of shaft power fleet wide.
6. Ultimately, achieve substantial savings for Defence.

FOULING PREVENTION ON STATIONARY FOULING-RELEASE COATINGS

With the ban on in-water cleaning DSTO has investigated other methods to prevent the build up of fouling on stationary vessels. Early research on air bubbles for antifouling dates back to a 1937 patent (Branner 1937) and Smith (1946) examined air bubbles on glass panels and on a 29-foot vessel in short term exposure trials. The optimal size of perforations, spacing and flow rate are described for plastic tubing and copper pipes.

DSTO has been investigating the air bubble technique using modern sparger materials to deter fouling from settling on stationary surfaces, with emphasis on FRCs. Initial experiments involved delivering compressed air through spargers fitted to v-shaped wings that were angled to mimic the angle of a ships hull. Streams of air-bubbles were released onto test panels and the subsequent fouling was assessed monthly and compared to control test panels that did not have air bubbles. The air spargers were then scaled up to be active over a section of a vessel and its efficacy assessed over the peak fouling season in Victoria, Australia (Scardino et al. 2009).

A prototype test with a scaled up version was conducted to determine if the air bubble curtains could be practically scaled up to be effective on the hull of a vessel. The spargers were changed for this trial to RAUBIOXON PLUS pipe aerators (REHAU, Germany) with a membrane made from silicone elastomer. Three air spargers (1 metre in length) were deployed at HMAS Cerberus, Victoria, Australia (38° 18' 16"S, 145° 11' 1"E) approximately 3 metres below the water line on the 29-11-2007. The spargers were fixed to PVC piping and bolted to the wharf (Figure 5), when the vessel docked the air bubbles were effective in approximately a 1 metre section of the hull towards the forehead starboard side of the vessel. The vessel, a twin diesel powered aluminium workboat (overall length -12.42m, Depth amidships - 2.33m, Displacement 12850kg, Power – 2 x Caterpillar 3160M diesel engines @ 137SHP each, Survey – Class 2C, Speed – Cruising 8.5Kt Maximum speed 10Kt) was coated with a FRC and cleaned the day before the air bubble experiment commenced.

The air was supplied at 8 litres per minute per linear metre using an *Evolution Aqua*® (Airtech AT-40) aerator.

Video footage of the vessel's hull was taken at regular intervals throughout the trial. The trial ended when the vessel was slipped on the 12-03-2008. A full assessment of the fouling cover was then undertaken and the vessel was cleaned with a high pressure hose. The vessel was mostly inactive over the trial period; however, it was used occasionally in rough conditions at a speed of up to 10 knots.

Hull roughness measurements were taken on several sections of the hull at the completion of the trial and compared to hull roughness values for the clean vessel. A BMT Mark 111 hull roughness analyser (BMT SeaTech Ltd, Southampton, U.K.) was used.

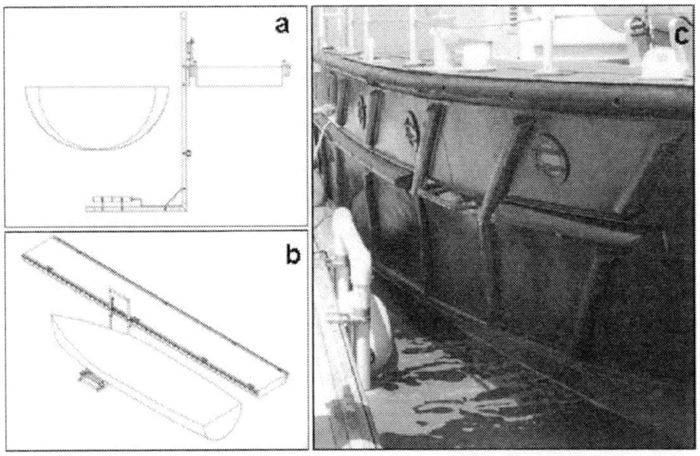

Figure 5 Experimental set up for the vessel patch trial, a) front view of the vessel patch trial set-up, b, layout view of the system on the floating wharf, and c) photograph of bubble curtain emerging form under the test vessel.

The results of the vessel patch air bubble trial were largely successful. The air bubbles were continuously active over the target zone despite large changes in tides, wind and currents. The air bubbles clearly reduced the amount of fouling cover, including reduced thickness of fouling, lower species diversity and generated far less hull roughness than the other areas of the vessel not in the air bubble zone (Figure 6).

Fouling in the air bubble zone of the hull was limited to slime (90% cover) and other soft fouling species such as hydroids and erect bryozoans. There was very little hard calcareous fouling in the bubble zone which comprised a very light scattering of tubeworms and barnacles (< 3%) (Figure 6a). In contrast there was heavy calcareous fouling on other areas of the hull not in contact with air bubbles including Spirorbid tubeworms (28%) and encrusting bryozoans (39%) (Figure 6b).

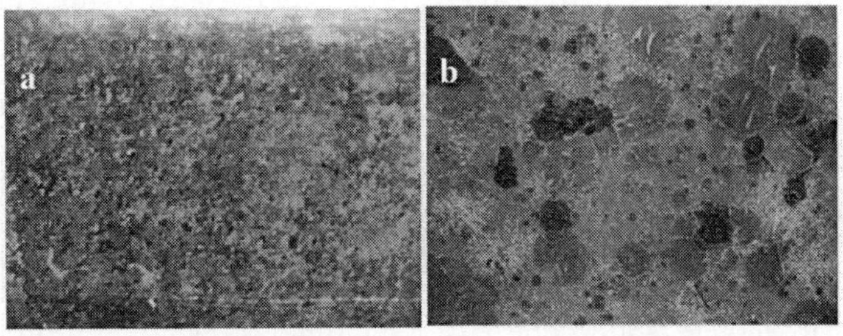

Figure 6 Fouling cover on the vessel patch trial – a) Air Bubble zone b) Outside air bubble zone

Hull roughness measurements were taken on several sections of the hull at the completion of the trial and compared to hull roughness values for the clean vessel. It was found that areas outside that air bubble zone had substantially higher hull roughness (1019 ± 85 μm) than the area of the hull exposed to air bubbles (269 ± 15 μm). Hull roughness in the air bubble zone was only marginally higher than the cleaned vessels' hull roughness (224 ± 16 μm) (Figure 7).

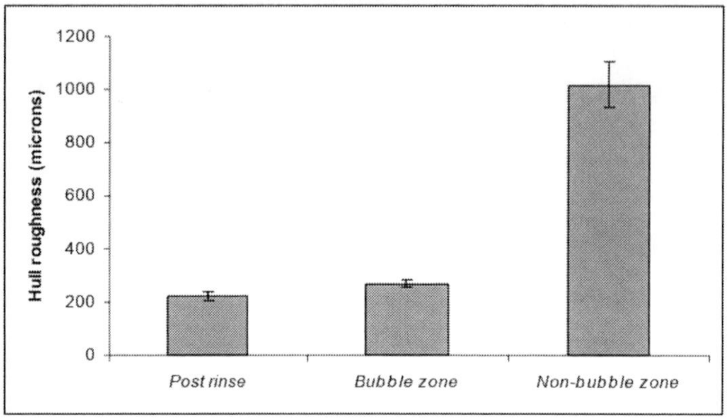

Figure 7 Average hull roughness of patches of the vessel at the conclusion of the field trial. Hull roughness measurements were taken from several sections both within and outside the air bubble zone and after the hull had been rinsed clean.

The air spargers were also placed on the heavily fouled control surfaces at the completion of the field trial. In just three weeks significant reductions in fouling cover were observed, particularly on soft-bodied fouling organisms such as colonial ascidians. The remediation of fouled surfaces by air bubbles indicates that the air spargers do not need to be in continuous operation.

The use of air bubbles to control fouling is a novel, non-toxic technique with potentially wide application.

The impact of fouling reduction due to the use of the bubble curtain technique on fuel consumption can be estimated by examining increases in a vessel's shaft power. For a frigate operating at 15 knots a slight slime covering is expected to increase shaft power by 11% and increase fuel consumption by 6% (Table 2). In contrast a vessel with calcareous fouling will have increases in shaft power of between 35-86% and increases in fuel consumption of 24-92% depending on the severity of the fouling (Table 2). These figures indicate that the air bubble fouling control system could considerably reduce the fuel penalty from ship hull biofouling for ships coated with FRC coatings.

CONCLUSION

Poorly controlled biofouling can result in a substantial increase in fuel consumption, maintenance costs and the risk of transporting potential invasive marine pests. This is of particular interest to vessels that do not have a continuous operating profile, such as military vessels where biofouling can quickly develop when stationary in ports for extended periods.

DSTO is continuing to review a number of techniques to quantify and understand the development of fouling on warships operating in Australian waters and under RAN operational profiles. The aim of this research is to better guide decisions on coatings and hull husbandry options and investigate other novel techniques to reduce the settlement of biofouling. It is anticipated that for effective biofouling control multiple strategies will be required.

REFERENCES

Anderson, C., M. Atlar, M.E. Callow, M. Candries, and R.L. Townsin. (2003), "The development of foul release coatings for seagoing vessels", Journal of Marine Design and Operations, B4, 11-23.

Australian and New Zealand Environment Conservation Council (ANZECC), (1997) Code of Practice For Antifouling and In-water Cleaning and Maintenance.

Branner, F.G. (1937) Means and Method for Protection from Marine Parasites, U.S. Patent, 2,138,831.

International Maritime Organisation, (2007), "Harmful ships' paint systems to be outlawed as international convention meets entry into force criteria", IMO News 4:6.

Lewis, J. A. & Houridis, H., (2001), "Hull fouling as a vector for the translocation of marine organisms: Phase 1 study - Hull fouling research", Ballast Water Research Series: Report 14, AFFA, Canberra.

MoD Hull Fouling Working Group, (2007), "Surface ship hull fouling management." Issue 1. Maritime Acquisition Publication Number 01-026. http://www.nakmo.co.uk/OpenDocument.aspx?Type=library&ID=586

Scardino, A. J., Fletcher, L. E. & Lewis, J. A., (2009), "Fouling control using air bubble curtains: Protection for stationary vessels", Journal of Marine Engineering and Technology, A, 3-10.

Schultz, M.P. (2007), "Effects of coating roughness and biofouling on ship resistance and powering", Biofouling 23(5), 331-341.

Smith, F.G.W., (1946), "Mechanical control of ship-bottom fouling", Quarterly Journal of the Florida Academy of Sciences, 9(3-4), 153-61.

The National System for the Prevention and Management of Marine Pest Incursions, (2009), National Biofouling Guidelines, Commonwealth of Australia.

Townsin RL., (2003), "The ship hull fouling penalty", Biofouling 19, 9–15.

Walker, M. and Atkins, I., (2007), "Surface ship hull and propeller fouling management." In Warship 2007: The Affordable Warship", Royal Institute of Naval Architects.

Yebra DM, Kiil S, and Dam-Johansen K., (2004), "Antifouling technology - past, present and future steps towards efficient and environmentally friendly antifouling coatings". Progress in Organic Coatings, 50(2), 75–104.

Implications of the NATO Naval Ship Code

Bob Simpson, CEng, MBA, FIMarEST
Lloyd's Register

The NATO Naval Ship Code, ANEP 77, provides navies with a Goal Based regulatory framework to demonstrate that the safety of their seafarers is at least as good as that required by statutory legislation. However it also brings with it significant challenges in applying a Goal Based approach and the assumption of compliance being demonstrated by third party verification.

This paper will consider the development of the Naval Ship Code: the principles and concepts adopted in the Code, experience of applying the code in practice; and recent developments of the code which are moving it further towards a purely Goal Based position.

The author is the Lloyd's Register's nominated member of the International Naval Safety Association (INSA), the group responsible for the development, promulgation and promotion of the NSC and has recently completed leading the rewrite of the Machinery and Electrical Chapters. This position gives a unique insight into the challenges that Goal Based safety represents and the solutions being implemented to support this development.

---

1 INTRODUCTION

In December 2003, the then NATO Naval Armament Group formally approved the formation of the NATO NG6 Specialist Team develop a NATO equivalent of SOLAS[1] that would allow navies to demonstrate that they meet the requirements of this international convention on maritime safety [Ref 1]. Since then, work has progressed through various Working Groups established by the Specialist Team to adapt the requirements of SOLAS into the naval environment.

In October 2008 and more recently in October 2009, the representatives of the various navies and Classification Societies met to review the progress to date. Subsequent to the 2008 meeting, the Naval Ship Code was passed to NG/6 (now referred too as Maritime Capability Group 6 (MCG/6)) for their endorsement and publication. The meeting in 2009 took the opportunity to review proposed changes to the requirements from the feedback received and made proposals for the changes to be accepted by NATO.

This paper will now look at the results of these efforts and the impact it is likely to have on the management of safety within navies

2 APPLICATION OF SOLAS TO NAVAL VESSELS

Chapter 1 of SOLAS, Regulation 3, item a) defines that the convention does not apply to "Ships of war and troopships"[2]. This clause has led many to believe that nothing contained within the following sections of the convention has any relevance to a naval vessel and has therefore been ignored.

With the introduction of the principles of Classification to naval ships though, the symbiotic relationship between IMO conventions and Classification Rules have become more clearly

---

[1] International Convention for the Safety of Life at Sea

[2] Ch 1, Reg 3 a) (i)

defined and the need to put in place at least some of the components has been highlighted. In fact when SOLAS and other associated conventions such as MARPOL[3], ILO Convention[4], ColRegs[5] and the Loadline Convention[6], there is much that is of value to consider and can be applied. What has been missing is the mechanism by which the variable operational requirements of naval ships can be allowed for in attempting to satisfy the intention of the conventions.

As an interim measure, some Class Societies had applied specific sections of SOLAS et al to naval ships, but with varying degrees of success. To allow a more flexible approach Lloyd's Register had introduced with their Rules and Regulations for the Classification of Naval Ships January 2005, a set of Classification notations that paraphrased the requirements of SOLAS in the terms of 'Goals' and then demonstrated compliance through meeting specific clauses of the convention or the related codes, subject to it being appropriate within the operational requirements of the vessel [Ref 2].

Where the SOLAS solution was not appropriate, the design was judged against the goal directly using a formal engineering justification process as set out in SOLAS Chapter II-2, Regulation 17, Alternative Design Justification [Ref 3].

As is observed in Ref 2, one of the key differences between the underlying safety philosophies of naval vessels is that they may subjugate crew safety for that of a third party in either conflict or humanitarian situations. However, when not in these situations a vessels' crew should be able to expect that their place of work is as safe as any other seafarer.

3 REQUIREMENT FOR A NAVAL SHIP CODE

This approach using either SOLAS directly, or via a subset of Classification Rules while of value, only provided a limited degree of compliance and therefore had its limitation. Recognising this amongst other factors, navies such as the Royal Navy, the Royal Australian Navy, the Royal Netherlands Navy and others came to the conclusion that a navalised SOLAS would provide a more integrated and holistic solution as it would be based on generic naval practice from the start.

Such an approach would also allow navies to better demonstrate that they did in general comply with statutory requirements as the overarching process would be similar to that applicable to commercial shipping. The commercial model is summarised as shown below, with International Conventions being developed by the International Maritime Organisation, ratified and adopted by Flag States by incorporation into their maritime laws and compliance demonstrated by surveys either carried out by the Flag State or more commonly by Classification Societies [Ref 1].

The naval model using NATO as the equivalent to IMO and Naval Authorities as quasi Flag States. The similarities can be seen in the figure below.

[3] International Convention for the Prevention of Marine Pollution
[4] International Labour Conference Conventions
[5] International Regulations for Preventing Collisions at Sea
[6] International Convention on Load Lines

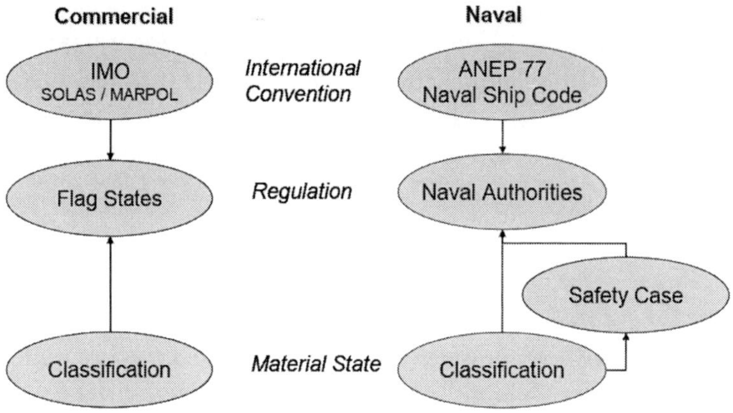

Figure 1 – Comparison of Commercial and Naval Safety Regulation

With the requirement established, the NATO Naval Armaments Group, Maritime Capability Group 6 (NNAG, MCG/6) were approached for their support of the idea and this was established with the setting up of the Specialist Team in 2003.

Apart from the navies involved with the ideas behind the Naval Ship Code, various Classification Societies were invited to be involved. Because of the implications of the Code, it was clear that interfaces with other marine industry bodies would be required and in time, there would need to be some process by which the Code could be maintained and updated. This resulted in the following organisations being involved with the development of the Code and the relationships between them are shown in Figure 2

Organisations include:

- NATO Maritime Capability Group 6 (NATO MCG/6)
- International Naval Safety Association
- Naval Ship Classification Association

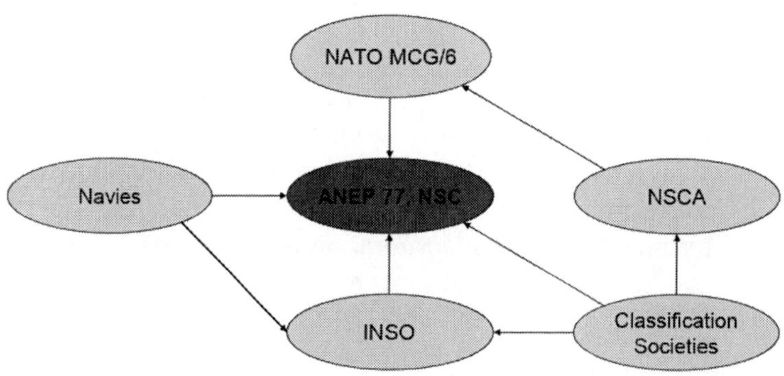

Figure 2 – Organisations involved with the Development of the Naval Ship Code

4 MEETING A GOAL BASED REGULATION

There has been much comment over the value of a 'Goal Based' approach to safety management [Ref 1], but in reality there is little experience in how this works in practice. The explanation in Ref 1, results in the prescriptive verification being assumed to be the required solution and whilst a goal has been identified, it is very easy to miss the need to think outside the box.

When the Working Group for the redrafting of Chapters IV and V was set up, the opportunity was taken to develop the idea of a Goal Based Regulatory environment by ensuring that the requirements had no prescriptive solutions stated in the Code itself, unless it was collectively agreed that this was the ONLY acceptable solution. Instead the guidance provided to the Code would direct designers to consider the selection of an appropriate standard that addressed the goal, but also took account of the operational culture of the navy and the geographical environment in which the vessel would operate.

As this needs a degree of expertise in acceptable alternatives, this creates a need for the Naval Authority to maintain deep knowledge of alternative standards and to be willing to advise at an early stage of the project what may be appropriate. In addition, as verification would need confirmation through an independent organisation, they would need to be selected at an early stage so that they would be familiar with the issues related to integration of the various standards to be adopted.

As noted in Normal Rattenbury's paper on the subject [Ref 2], when selecting commercial standards for naval application, it is imperative that the cultures of naval practice and operational environment must be carefully considered. This is shown in fig 3 below which shows how these impact on the selection of standards when being used to satisfy Naval Ship Code 'Goals'

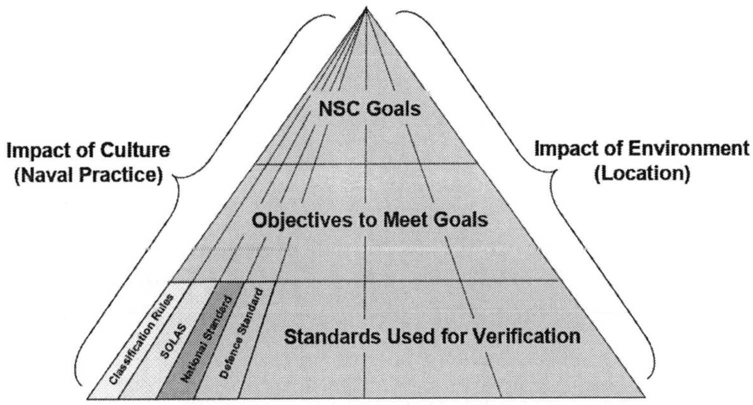

Figure 3 – Effect of Culture and Environment on Selection of Standards within the Naval Ship Code

This approach has been proven to work as it was applied to projects such as the UK's Queen Elizabeth Class Aircraft Carrier; the RAN Canberra Class Landing Helicopter Dock; and the UK MARS Fleet Tanker projects amongst others.

This then also fits in with the concept of the Vee Model of Naval Procurement [Ref 2] as shown in Fig 2, which shows that as one moves from a holistic 'Requirements' position by atomising the components into User Requirements and System Requirements Documents (URD's and SRD') and finally to a Technical Specification, to demonstrate that expectations are being fulfilled, a mechanism for acceptance needs to be evolved. This is frequently referred to as the Verification and Validation (V&V) process. By adopting an independent set of criteria based on technical standards, a project can set in place a process that will be mutually acceptable to both parties in demonstrating technical aspects of the project contract.

This approach can best be understood by the concept of using a library to assemble a set of documents as shown in Figure 4.

Figure 4 – Standards as a Bookshelf

The approach then is to select from the whole library a set of standards that are appropriate, integrated in a suitable regulatory process and supported by various guides and tools.

This development of the philosophy of using a Goal Based approach has exposed one significant learning point in that whilst one can have a Goal Based Regulatory environment this requires some prescriptive technical standard to show achievement. This would suggest that the term 'Goal Based Standard 'is, in reality, a myth as it cannot exist.

5 CURRENT APPLICATION OF ANEP 77

At the Annual General Meeting of the participating organisations in the development of ANEP 77, the NATO Naval Ship Code in 2009, the majority of the navies present confirmed that they had either already instigated the application of the Code, or were putting in place policies that would mandate the Code to their fleets of vessels. This is a very comprehensive ratification of the work done to date and shows the commitment that navies have to be able to demonstrate through independent verification that their safety management is at least as good as required by Statute.

However, in order to deliver this, there is a significant amount of work to do as the understanding of why the Naval Ship Code is both necessary and appropriate, now needs to be promulgated through naval organisations.

In addition, each navy will need to establish a 'Naval Authority' to oversee the requirements and to provide a number of other functions. These can generally be summarised as:

- Issue certification to ships confirming compliance with the Naval Ship Code
- Maintain knowledge of appropriate verification standards to meet the goals of the NSC
- Maintain knowledge of own navy's specific requirements where these are not dealt with by either the NSC or other publicly available standards
- Appoint Recognised Organisations to undertake verification activities on behalf of the Naval Authority
- Monitor activity of the appointed RO's through periodic audit

- Facilitate the feed back of experience from certification activities to the appropriate standard 'owner'

These issues are considered in more detail below:

6 RESPONSIBILITIES OF A NAVAL AUTHORITY

6.1 ISSUE OF CERTIFICATES

Whilst the actual process of writing a certificate can seem relatively straightforward, the necessity to have a fully auditable process behind this activity means that there are a number of facets that need to be implemented. We are all well aware of the number of competent people who exist and are able to carry out inspections of ships, but in addition to this capability, they will need to be able to demonstrate both a level of experience and knowledge to confirm that they are deeply familiar with the standards to which they are working and can make recommendations or take decisions to support their inspection findings.

For a Classification Society, this demonstration that surveyors are Suitable Qualified and Experienced Personnel (SQEP), is dealt with through the Quality Management System (QMS) that they will be certified against by a suitable Accreditation Body (for LR this is our ISO 9000 QMS issued by British Standards). The training and ongoing maintenance of technical knowledge for our surveyors incurs significant overhead and we need to maintain a continuous updating and support process for them. This makes Class Societies highly integrated 'Knowledge Management' organisations that exist for the promulgation of the inherent expertise they have and not a consultancy organisation as some parties see them.

Even the simple requirement to train a person to have a minimum level of expertise can take several years and for a surveyor to be fully authorised to issue certificates without peer review requires a large number of surveys in different situations to have been completed. For a Class Society with the large number of vessels with which we are engaged, this is relatively easy to achieve, but for a navy with a much smaller number of vessels, this may prove a challenge. In addition, naval personnel will be expecting to progress through positions at a fairly brisk turnover rate, so developing this deep expertise may be very challenging.

6.2 KNOWLEDGE OF TECHNICAL STANDARDS

In order to advise projects of which standards they could adopt, or to advise if there are problems with the library of ones chosen, the Naval Authority will need to develop a knowledge across a very large spectrum of options. For one project that LR was involved in this initial count of potential standards to be adopted was 30,000. This was obviously untenable and through a process of rationalisation this was reduced to a workable number.

It must of course be recognised that whilst a designer may select a standard that meets their requirements, this acceptance criteria needs to be flowed down through the procurement chain with great care if the intent is not to be compromised.

In recent projects LR that has been involved in we have found situations where the supplier actually had no knowledge of standards referenced in the specification and was unable to source them, so technically was in default of contract. In another, the manufacturer recognised the difficulty in obtaining materials to a particular standard so requested (and obtained a contract variation) to use equivalent commercially certified materials, saving both them and the client significant costs. Ensuring that the designer was still content with this change needs a good flow of information along the procurement chain and close monitoring to ensure designs are not compromised unintentionally by what may appear sensible commercial decisions on cost.

6.3 MAINTENANCE OF TECHNICAL STANDARDS

As previous commented on in other technical papers, one of the driving forces for the development of naval specific Classification rules, was the understanding in defence organisations that the cost of developing and maintaining unique technical standards was becoming prohibitive.

The current approach where Classification Societies effectively take responsibility for the development and maintenance is now in place, but with the introduction of the Naval Ship Code a very important nuance has to be recognised. This is the issue of how to deal with aspects that are not directly related to Classification Rules. These could either be the:

- intent and content of the Naval Ship Code itself
- other standards such as ISO, IEC or other
- Codes of practice such as the SOLAS Fire Safety Systems, Life Saving Equipment or other national Code of Practice
- Or defence standard where there is no commercials standard that could be applicable

The solution adopted in the UK has been to appoint a Naval Safety Assurance Support Service (NSASS) as described in Ref 4

This joint cooperation between Lloyd's Register and BMT Defence Services provides amongst its other functions a service to consider these issues and to prepare submission to the appropriate bodies for changes to their standards. To date this seems to be functioning very well, but as with all support services, it has a clearly recognisable cost and its long term existence will require the UK MoD to commit to its funding if future projects are to benefit from current lessons learnt [Ref 4].

6.4 APPOINTMENT OF RECOGNISED ORGANISATIONS

Exactly as occurs with Flag States where they appoint suitable competent organisations to act on their behalf the same is allowed for within the Naval Ship Code. This process of appointing a 'Recognised Organisation' to act on the NA's behalf however, is not to be assumed to be a simple way of transferring responsibility or risk, but an inherent obligation of the NA.

Within the commercial regulatory framework, to ensure that there is a consistent approach to this, the IMO provide guidance through Resolution A.739(18) [7] which sets out the requirements and process to be adopted. This has been similarly adopted by the Naval Ship Code such that we have a commonality of approach.

The major difference for navies is that the trust that exists between Flag States and Class Societies has been built up over many years and the closeness that allows for an inherent understanding of the respective roles and responsibilities that each organisation has. This

[7] IMO Resolution A.739(18) - Guidelines for the Authorization of Organizations Acting on Behalf of the Administration

deep understanding is still being evolved between navies and Class Societies, so Naval Authorities are quite rightly approaching this at a measured pace.

Typically, a Class Society will be recognised on a 'Partial' basis to undertake surveys on behalf of the NA who once they have received the survey report, will review the findings before issuing a certificate. Once levels of competence are demonstrated, then the Class Society will be granted Full Authorisation to issue certificates directly on behalf of the RO.

One other model that has been adopted where a Naval Authority is still developing its processes and expertise, is simply to appoint a Class Society as its 'Advisor'. This transfers no responsibilities to the Class Society, but does give the NA access to advice from the Class Society. This has been used in the UK where a NA had been established and needed to issue a certificate for a new vessel, but due to the contract requirements, the Class Society has not been involved and so was not in a position to offer survey reports at the time when they were required. However, LR was able to comment on the findings and submissions provided which justified the level of safety being achieved.

One aspect that this has re-emphasised is that where a Formal Safety Assessment approach using a Safety Case Report is applied, someone needs to take a holistic view of the risks as it has identified and to consider whether any stovepiping of hazards has left any significant open risks. As we all recognise, accidents are very rarely due to a single initiating event, but occur as a result of combinations of failures. It is quite right that we should not require the authors of Safety Case Reports to undertake a multi mode failure approach as the cost would be become far to high, but where items are of concern, then other factors need to be considered. As this is a subjective expertise, it requires a technical review by a competent body. This in turn means that a Safety Case Report needs to be reviewable and not be judged, as sometimes seems at present, on the volume of shelf space it occupies.

6.5 MONITORING A RO'S ACTIVITY

As previously mentioned, there is a significant level of trust between Flag States and Class Societies due to the close relationship they have. However, having this degree of trust is not a legally acceptable defence, so in order to demonstrate that Class Societies are fulfilling their obligations, an extensive auditing programme exists.

This is not just constrained to the mandatory Quality Assurance certification that is required by the IMO Resolution for the appointment of a Recognised Organisation, but also requires the Flag State to personally audit the Class Society against the delegation agreement. The result of this for a major Class Society is that it becomes one of the most audited organisations in existence when the number of Flag States that delegate IMO Conventions to them is considered. For LR, this is well over one hundred Flag States all who have the right to audit any relevant aspect of LR's services, anywhere in the world.

It is reported that this means LR is being audited somewhere in the world virtually every day of the year. This means there is little room to hide any lapses of performance in this environment.

For a Naval Authority, they too will need to set up a suitable audit process, with competent auditors who understand the service obligations that delegation agreements impose. Where a navy has a global existence and the Class Society is acting on the NA's behalf wherever the navies vessels are operating, this is likely to impose some significant logistical and cost implications for them.

This cost should not be seen as a new additional cost of adopting a Naval Ship Code based regulatory safety management system, as the same system would have required the NA itself to undertake surveys anywhere in the world in any case. With the cost of international travel and the concern regarding 'carbon footprints', then delegation would be expected to be highly cost effective if set up carefully.

7 SUMMARY

The gradual adoption by navies of the commercial classification model has exposed the need to implement a parallel safety regulation system. This has brought with it a much deeper and more thorough understanding of the dichotomy between commercial and naval safety objectives.

Now that this is better understood and elucidated by the Naval Ship Code, the underlying philosophical objectives of being able to provide a place of work for our naval crews that is at least as 'safe' as that would be required for a merchant ship, whilst not compromising the imperative of being able subjugate their own safety for that of others is more clearly defined.

With this understanding now comes the responsibility for Naval Safety Management systems as represented by an effective Naval Authority, to take that role on and provide a 'safe' place of work, without limiting operational capabilities. In discharging this responsibility, the tasks that need to be undertaken can now be defined.

The need to have a deep understanding of how technical standards are used in the maritime environment to demonstrate an acceptable level of safety, mean that naval authorities will be required to develop a wide spectrum of support competences, both internally and externally if the objectives are to be fully realised. Some must be developed internally and will require changes to any policy of externalising expertise, but some can certainly be subcontracted.

Where external support is adopted, then the relationships necessary will take time to develop. The mutual trust that exists between Flag States and Class Societies has been created by a very open and honest stance from both parties. This will need to be replicated such that vessel operators do not see either Naval Authority or Class Society as a barrier to be overcome, but as a facilitator for them to be able to provide and demonstrate that they are providing safe places of work.

When this is in place we will then have a demonstrable naval safety management system that is properly 'ship centric' in its delivery.

8 REFERENCES

1. Rudgley/ Boxall/ ter Bekke/ Humphrey, 2005, "Development of a NATO Naval Ship Code", RINA Safety Regulations and Naval Class II
2. Rattenbury, N. 2004, "Selection and use of standards for naval ships", INEC
3. IMO Convention for the Safety of Life at Sea, Chapter II-2, Regulation 17
4. Buckley/ Cuckson/ James/ Marshall, 2008, "Naval Ship Assurance - A New Approach", Pacific International Maritime Conference

9 AUTHORS BIOGRAPHY

Robert M. Simpson, CEng, MBA, FIMarEST. - A former seagoing engineer, presently Business Manager – Naval Ships for Lloyd's Register (LR), responsible for the development of LR's services to naval clients. He is currently involved with all of the UK MoD's major naval projects along with various MoD Naval Authorities as they develop a robust internal regulatory process modelled on international and national statutory legislation. He is also working extensively with a number of other navies as they adopt the Naval Ship Code and Naval Classification as their assurance methodology.

Optimization Procedure as Decision Tool for Ship Design

J.M. Vasconcellos
COPPE/UFRJ – Federal University of Rio de Janeiro, Rio de Janeiro, Brazil

ABSTRACT

The decision tools for ship and offshore design following the computer ability to treat more complex problems have improved. Currently, optimization uses very time consuming approaches. Genetic algorithms and multi objective problems where the mathematical models used are complex and take computer time to evaluate are very common nowadays.
Economic efficiency of shipping investments was and still is the main concern of ship owners but safety, performance and environmental protection have become increasingly important in the ship design decision process.
In the last International Marine Design Conference (IMDC 2009-Trondheim) professor Nowacki made a historical overview and indicated some future trends. He established the necessity to treat economy and safety as the leading motivations for innovation in ship design. But environmental protection (damage survival, subdivision, dangerous cargo) should also be considered. Risk-based design procedure indicates the necessity to improve the design and designers have to test many different conceptions. In this scenario optimization may be a good decision design tool.
This paper highlights the application of optimization procedures for ship design decision making. A brief overview of the classical optimization procedures is indicated and case studies are used to illustrate. Multi objective non linear programming is discussed and the author presents a catamaran preliminary design case study. Finally genetic algorithm (GA) method and Monte Carlo simulation developed to handle uncertainty in the variables and constraints problems are discussed.

Keywords: Ship Design, Optimization, Multi-Criteria Optimization, Catamaran Design

INTRODUCTION

Problems of optimization are problems of maximizing or minimizing a function of one or more variables in a given domain, with the existence or not of restrictions in some variables. The majority of the problems in real life present uncertainties: they are inherent to the majority of physical, chemical, biological, geographical systems, etc. Even in the deterministic context there are systems which are so complex that they do not allow a precise measurement of their parameters. The area of stochastic optimization (also known as optimization under uncertainty) studies models and methods that deal with such situations: they incorporate uncertainties in the modeling through the inclusion of random variables with a known probability distribution (Bortolossi e Pagnonceli, 2006),[1].
The algorithms used to solve these problems can be basically: deterministic and probabilistic or stochastic. The optimization methods based on deterministic algorithms (the majority of the classical methods) generate a deterministic sequence of possible solutions requiring in most cases the use of at least the first derivative of the objective function in relation to the design variables. In such methods the objective function and the restrictions are given as mathematical functions and functional relations. Furthermore, the objective function should

be continuous and distinguishable in the search field (Bastos, 2004),[2]. This type of problem can be represented mathematically in the following way:

$$\text{Max/Min } f(x_1,x_2,\ldots x_n)$$
$$\text{Constraints}$$
$$g_i(x_1,x_2,\ldots x_n) \leq \text{ or } \geq b_1$$
$$.$$
$$..$$
$$g_m(x_1,x_2,\ldots x_n) \geq \text{ or } \leq b_m$$
$$\text{And}$$
$$x_1,x_2,\ldots x_n - \text{design variables}$$
$$f(x_1,x_2,\ldots x_n) - \text{objective function}$$
$$g_1,g_2,\ldots g_n - \text{constraints}$$

Many different methods can be found in literature including excellent text books and papers to solve the deterministic optimization mathematical problem.

On the other hand the optimization methods based on probabilistic algorithms are not so well spread. The methods use only the evaluation of the objective function and they introduce stochastic data and parameters into the optimization process. Because they do not use the derivative of the objective function, they are considered to be zero-order methods (Holtz, 2005),[3].

According to Bastos (2004),[2] some of the advantages of probabilistic algorithms in relation to deterministic algorithms are: the objective function and the restrictions do not necessarily need to have a mathematical representation; the objective function does not need to be continuous or distinguishable; they work satisfactorily both with continuous and discrete parameters, or with a combination of both; they do not need complex formulations or reformulations for the problem; there is no restriction as to the starting point within the search field of the solution; they carry out simultaneous searches in the field of possible solutions through a population of individuals; they optimize a large number of variables, provided that the evaluation of the objective function does not have an excessively high computational cost. The greatest disadvantage in relation to classical methods is the processing time.

Among the many existing stochastic optimization techniques, this article presents the Genetic Algorithm and Monte Carlo Simulation methods.

In the mid-1970s, John H. Holland pushes forward an optimization search technique based on the principles of genetics and natural selection. This technique was popularized by one of his students, David Goldberg, who used the algorithm in his dissertation to solve a difficult optimization problem (Goldberg, 1989),[4].

A genetic algorithm allows a population of possible solutions composed of many individuals to develop, under rules of specified selection, to a state which minimizes the cost function (Haupt and Haupt, 2004),[5].

In relation to the traditional search and optimization functions, Genetic Algorithms (GA) differ principally in four aspects: GAs work with a codification of the set of parameters and not with the parameters themselves; they work with a population and not with a single point with each interaction of the algorithm; they use cost or fitness value information and not derivatives or other auxiliary knowledge, and they use probabilistic and not deterministic rules of transition.

The majority of the real optimization problems bring uncertainties. Even in the deterministic context there are systems that are so complex that they do not allow an exact measurement of their parameters. The stochastic optimization area, also known as optimization under

uncertainty, studies models and methods to deal with such situations: they incorporate uncertainties into the modeling through the inclusion of random variables with a known probability distribution. The objective is then to find solutions which are admissible for all the possible occurrences of the random variables that are part of the modeling (Bortolossi and Pagnoncelli, 2006),[1].
Genetic Algorithms have been used in some research studies that deal with optimization in the offshore area considering uncertainty related to the variables. Subbey et al.(2004),[6], Soleng(1999),[7], Romero and Carter(2001),[8], Saemi et al.(2007),[9], Ballester and Carter(2007),[10] have presented studies concerning the characterization of reservoirs where genetic algorithms appear as the main tool or hybrid systems, where the GAs operate with other techniques such as Neural Networks and Fuzzy Systems.
The present study highlights the Monte Carlo simulation technique to deal with uncertainty concerning variables in optimization projects. This technique was originated during the Second World War at a conference in Los Alamos, USA, where Stanislaw Ulam, an Australian mathematician, presented the development of a method which he had discovered when he was trying to simulate the diffusion of neutrons in nuclear material for the construction of the atomic bomb, and was later implemented by John von Neumann.
The Monte Carlo method is a simulation technique used in solving probabilistic problems in which distributions of probabilities are attributed the entry variables by means of a random process and the distributions of probabilities of the exit variables are obtained as result. The random process used consists in generating random numbers to select the values of each entry variable for each attempt. This process is repeated numerous times, obtaining numerous results from which one builds a distribution of probability of the exit variables. It is a useful method also for resolving deterministic problems that are difficult to solve.
The Monte Carlo Simulation offers a number of advantages, among which the following are of special importance: the distributions of the variables of the model do not need to be close; correlations and other interdependencies can be modeled; the computer performs all the work of generating the random values; the level of precision of the simulation can be improved through a simple increase in the number of interactions calculated; the validity of the Monte Carlo simulation theory is widely recognized, which means that the results are easily accepted; alterations in the model can be made rapidly and the new results can be compared with the previous ones.

CASE STUDY IN SHIP OPTIMIZATION DESIGN

Watson, D.G.M and Gilfillan, A.W., (1977),[11] presented an optimization technique applied to merchant ship design. Afterwards, many others authors developed mathematical models for ship preliminary design. Different optimization techniques were used since and approaches were also changed to achieve better results. Vasconcellos J.M, (1995),[12] developed a semisubmersible platform preliminary design and applied the non linear goal programming to solve the multi-objective problem. Moraes H., Vasconcellos J.M and Almeida M.P.,(2007),[13] used the same optimization technique to solve a high speed catamaran preliminary design. The case study presented herein was extracted from this last reference to follow up and highlight the uncertainty application in preliminary ship designs.

The Amazon area in Brazil (Figure 1) was selected for this case study. Belém-Manaus line, presented in table 1, represents a long line (1646km ~ 890 miles) with many intermediate scales.

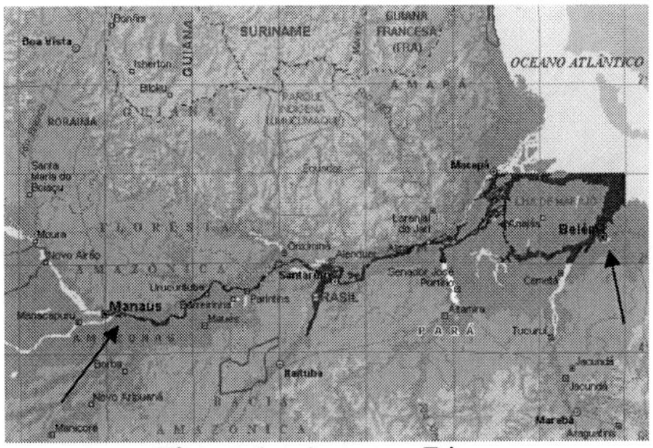

Figure 1 - Amazon River

Table 1 - Belém – Manaus (Route Data)

Distance	889 miles / 1646 Km
Actual Fleet	11
Fleet Age (Average)	15 years
Time	Conventional boat - 100 h
Ticket price (Average)	US$ 40
Passenger Capacity (Average)	316
Stops:	9

Passenger and cargo transportation is performed at a very low speed with many stops (nine in the case study presented herein). Another important route aspect is the current effect (2 knots upstream). Because of a restricted area (at southeast of Marajó Island in Breves) the cruising speed has also to slow down to 16 knots over 167 km. Preliminary design requirements established a 300 passenger vessel to achieve the current demand. Speed was established around 30 knots. The presence of small fishing and passenger boats in the same route requires speed reduction in many places. Belem-Manaus route needs to reduce operation speed in many areas due the presence of objects in the river as tree-trunks and small sand islands.
The mathematical model was applied for preliminary design and goal programming used to find out a compromise solution. The mathematical model included the following aspects:

a) Hull hydrodynamic resistance and power. The method is based on slender body theory for wave resistance and flat plate theory for frictional resistance. Moraes H.B. et al (2004),[14] researched and compared the slender body theory and the 3D theory results. A broad-spectrum analysis was carried out to compare wave resistance for different models. Monohull and catamaran models were tested to determine the interference phenomenon. Even though research was in place, the slender body application theory was applied in preliminary design phase to select a suitable model. The 3D theory models are more complex and time consuming. They use a CFD approach and in the author's point of view should be reserved to advanced design phases.

b) The weight mathematical model includes the structure weight and equipments evaluated by Karayannis et al (1999),[15]. Figure 2 shows power and associated weight for Karayannis et al (1999),[15] model and several main-engines weight (MTU, Carterpilar, Zvezda, and Wartsila) were plotted for comparison purposes. Data was selected from High-Speed Marine Transportation, Jane´s Book (1996-97),[16] and engine catalogue.

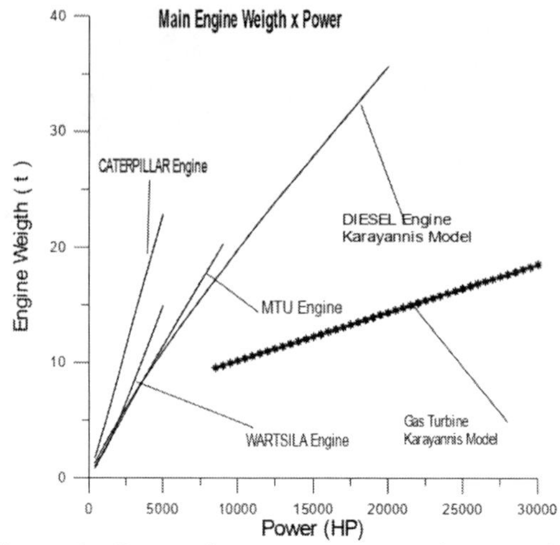

Figure 2 - Power between 1000 and 30000 hp.

c) The operational weight is a function of distance and time. It consist of the sum of crew, luggage, fuel, lube oil, fresh water and food.

d) Cost Evaluation - High speed, smooth hull lines and passenger appreciation are some of the major challenges for catamaran designers. It is common to see high costs in most of high-speed vessel design. New technology is always more expensive and high speed is always associated with higher oil consumption. In high speed catamaran preliminary phase it is necessary to assess some costs such as: investment, operational and infrastructure.

The mathematical model for a preliminary catamaran design can be solved by an optimization procedure. Goal programming proposed by Ignizio, J.P., (1976),[17] was used as an optimization technique. Full mathematical model can be found in Moraes H., Vasconcellos J.M and Almeida M.P.,(2007),[13]. The main results were selected and presented herein.

The optimization procedure using multi-objective non linear programming indicated the following results presented in table 2. The service speed should be reduced to 16 knots during 167 km. The main reason is the restricted area close to Marajó island.

Table 2 – Catamaran Preliminary Design (Multi-Objective non linear programming results)

Length	35,77 m
Twin hull breath	3,56 m
Draft	1,19 m
S/L (S is the distance between twin hulls)	0,23
Power	5017 hp
Speed	30 knots
Reduced speed = 16 knots in 167 km	restricted area

Skoupas S. , Zaraphonitis G. and Papanikolaou A., (2009), [18], use the commercial optimization software modefrontier^{TM} to solve a high-speed RORO passenger preliminary design. The software is well know and has many features useful to optimization application.

The authors applied the some tool to investigate the uncertainty optimization with multi-objective. The catamaran Belem-Manaus case study was used and best procedure selected was the Robust Optimization with GA application Monte Carlo Simulation.

Robust optimization problems arise where there are uncertainties involved in the data or in the model. The presence of uncertainty is due to errors in measuring, or difficulties in sampling, or moreover can depend on events and effects in the future that cannot be known with certainty. The uncertainty makes the traditional deterministic optimization approaches especially insufficient because these methods do not consider the impact such variations. These uncertainties should be included in the optimization procedures, so that prescribed robustness can be achieved in the optimal designs.

The uncertainty consider in the catamaran preliminary design model was DTR (Distance with reduce speed) and VP2 (reduced speed for the DTR distance). The main reasons for speed reduced are: small boats in the area, river margin approximation (wave effect – washing), night navigation and floating objects. The main results are presented in the following graphs shown in figs. 4-10 and table 3.

Figure 3 presents the workflow modeled in the modefrontier™ application. The mathematical model includes seven variables (length, draft, beam, deadrise angle (AD1),region of front hull (ce), VP2 and DTR), three objectives (power, light weight and cost) and four constraints (Cb, B/T, S/L and V/Vol$^{1/3}$). The main equations were written in Fortran and the solver used was the Robust Optimization Method with genetic algorithm and Monte Carlo simulation for the uncertainty included in DTR and VP2 variables.

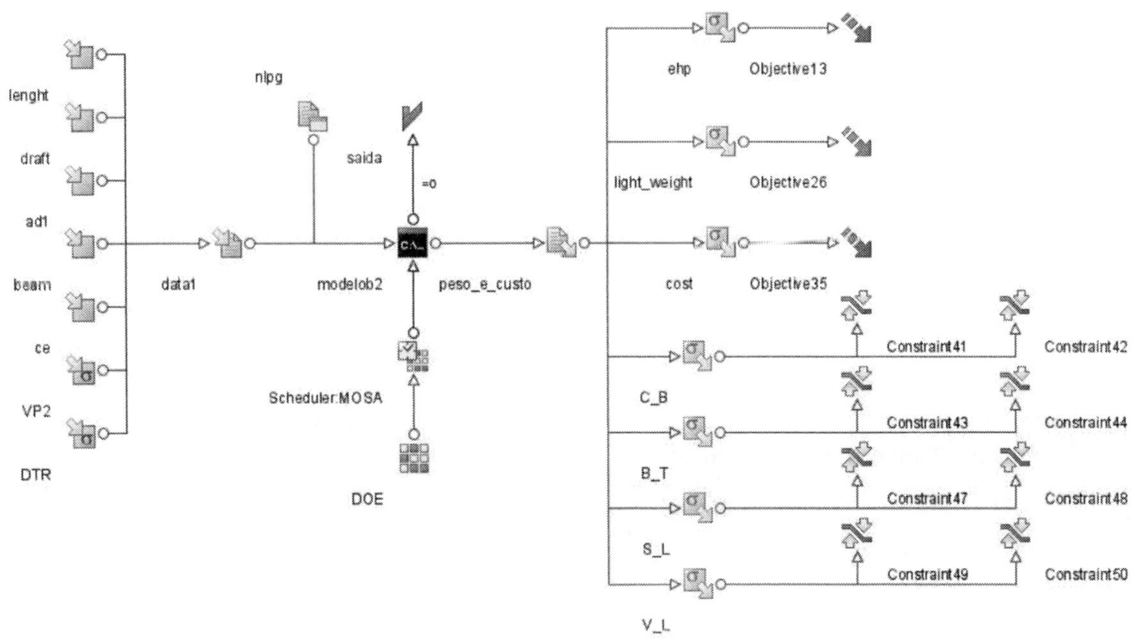

Figure 3 – Optimization Workflow

Figure 4 shows the graph with power (ehp) and light weight feasible designs. The best design considering only power and light weight is indicates by the arrow.

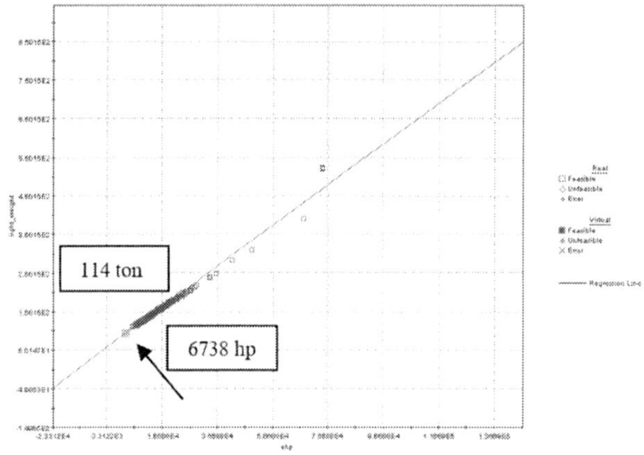

Figure 4 – Light weight (tons) versus Power (ehp)

Figure 5 presents the mean cost and power (ehp). The feasible and Pareto points are shown

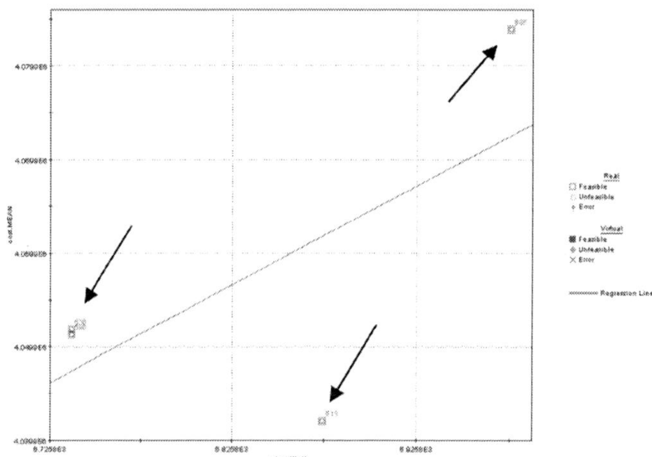

Figure 5 – Mean Cost versus Power (ehp) – Minimum Feasible points

Figure 6 presents the mean cost and the distance with mean reduction speed (DTR). The feasible and Pareto points are indicated.

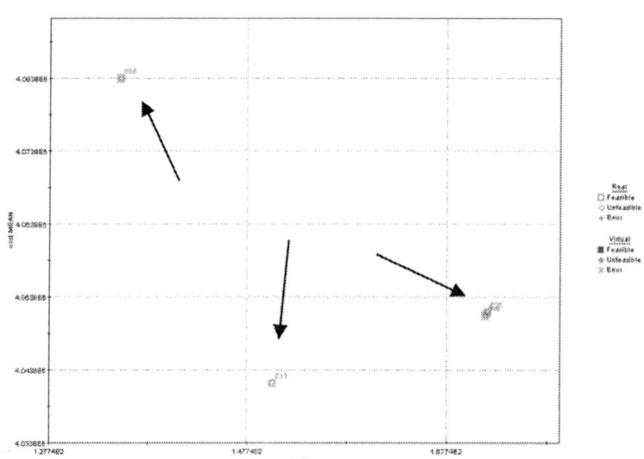

Figure 6 – Mean Cost versus DTR (Distance with Reduced Speed) - Minimum Feasible points

Table 3 indicates the main optimization results. Variables, objectives and constraints

Table 3 – Multi objective optimization under Uncertainty – River Catamaran Design

Results/Design Nº	685	511	636	596
DTR.Max	139	155	175	176
DTR.Mean	135	150	172	172
DTR.Min	131	147	168	168
DTR.STDEV	1.74	1.78	1.65	1.71
VP2.Max	29.5	31.5	18.5	18.4
VP2.Mean	27.2	28.3	15.8	15.8
VP2.Min	26.1	26.4	12.8	12.9
VP2.STDEV	0.8	1.0	1.0	1.1
Length	22.481	27.638	28.553	28.569
S	15.737	12.437	11.992	11.999
S/L	0.7	0.45	0.42	0.42
Beam (B)	2.256	2.299	2.255	2.256
Draft (T)	2.006	2.002	2.001	2.000
B/T	1.12	1.15	1.13	1.13
Cb	0.72	0.75	0.72	0.72
Power (ehp)	6976	6875	6738	6738
Light Weight	114	115	115	115
Mean Cost	4083876	4042039	4051802	4051289
Cost STDEV	3020	3067	10103	10028
Min Power			X	X
Min Weight	X			
Min Cost		X		

Figure 7 shows the Cost Standard Deviation related to the reduced speed (VP2)

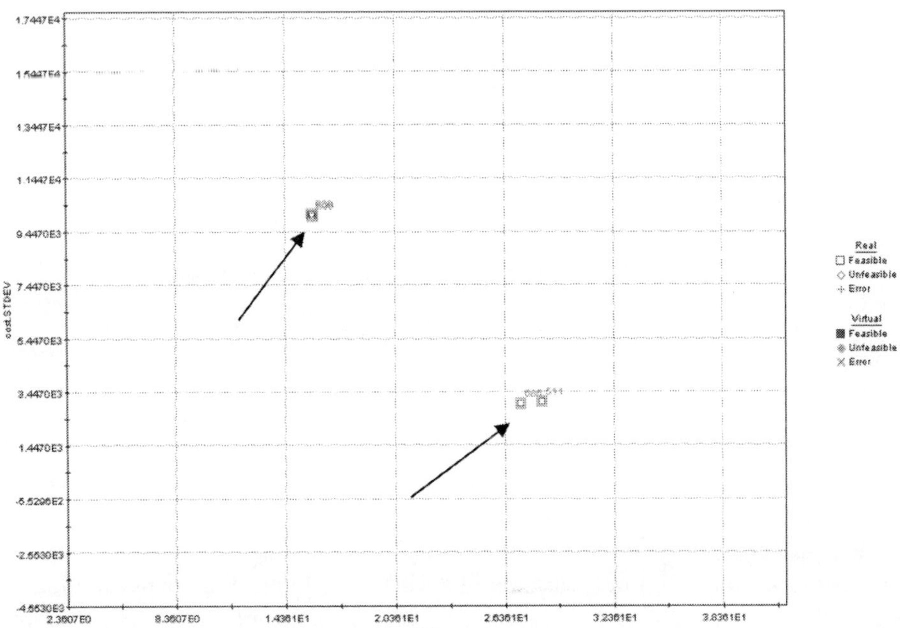

Figure 7 –Cost Standard Deviation versus VP2 (Reduced Speed)
Minimum Feasible points

Figure 8 shows the mean cost and mean distance for reduced speed to four selected designs in Pareto frontier.

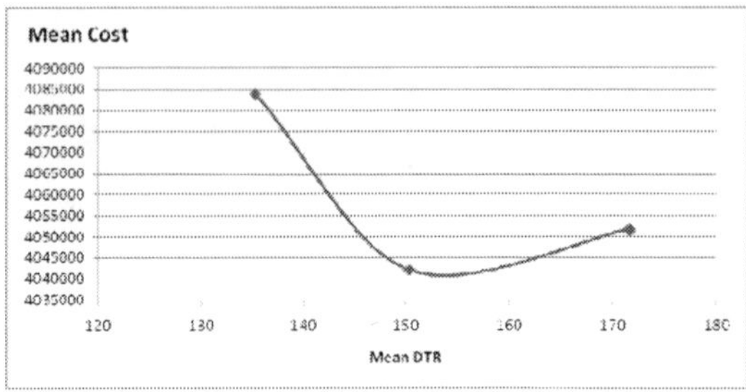

Figure 8 – Mean Cost versus DTR (Distance for Reduced Speed)

Figure 9 presents the relation in length and distance between twin hulls (S). For the last three designs the relation is quite the same.

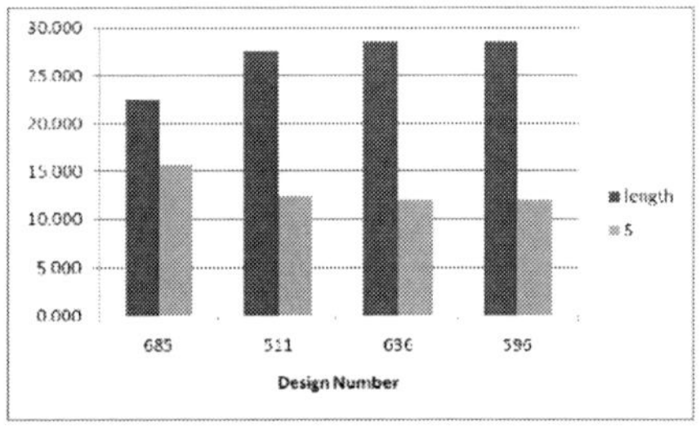

Figure 9 – Length versus Hull distance (S) (four best designs)

Figure 10 shows the beam and draft relation for the four best selected design in the Pareto boundary.

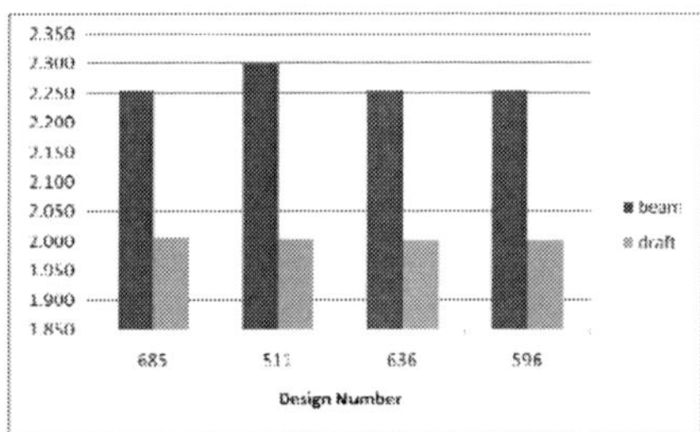

Figure 10 – Beam versus draft (four best designs)

CONCLUSION

Optimization using Robust Method applying Genetic Algorithm and Monte Carlo simulation seems to be an interesting tool to help ship designer in the preliminary variable selection. Uncertainty can be well considered in the mathematical model and results indicate the possibility to include the risk assumed in the initial ship design phase. Although the case highlights the procedure application using a simplified mathematical model, the method can be used including a more complex ship characteristics evaluation as panel methods and finite elements calculation. Computational time increase and a computer cluster is required. A more complex preliminary catamaran design is now under construction to evaluate the robust techniques application.

ACKNOWLEDGEMENTS

The author would like to thank the Brazilian National Scientific and Technical Development Board (*Conselho Nacional de Desenvolvimento Científico e Tecnológico*, CNPq for supporting this study.

REFERENCES

[1] H. J. Bortolossi and B. K. Pagnoncelli, 2006, "Uma introdução à Otimização sob Incerteza", *III Bienal da Sociedade Brasileira de Matemática*, Rio de Janeiro, Brasil.

[2] E. A. Bastos, 2004, "Otimização de seções retangulares de concreto armado submetida à flexo-compressão oblíqua utilizando algoritmos genéticos". *Master Thesis*, Rio de Janeiro, Brasil.

[3] G. C. C. Holtz, L. F. C. R. Martha and L. E. Vaz, 2005, "Envoltória de esforços internos devido a ação de trens-tipo em pontes usando estratégia evolutiva", *Congress of Bridges and Structures*, Rio de Janeiro, 2005

[4] D. E. Goldberg, Genetic Algorithms in Search Optimization and Machine Learning. *Reading, Mass: Addison-Wesley*, 1989.

[5] R. L. Haupt and S. E. Haupt, 2004, "Practical Genetic Algorithms", *2 ed., Wiley.*

[6] S. Subbey, M. Christie and M. Sambridge, 2004, "Prediction under uncertainty in reservoir modelling", *Journal of Petroleum Science and Engineering* 44,143–153, Canberra.

[7] H. H. Soleng, 1999, "Oil Reservoir Production Forecasting with Uncertainty Estimation Using Genetic Algorithms", *Oslo.*

[8] C. E. Romero and J. N. Carter, 2001, "Using genetic algorithms for reservoir characterization", *J PETROL SCI ENC*, 31, 113 - 123.

[9] M. Saemi, M. Ahmadi, A. Y. Varjani, 2007, "Design of neural networks using genetic algorithm for the permeability estimation of the reservoir", *Journal of Petroleum Science and Engineering,* 59, 97-105.

[10] P.J. Ballester and J. N. Carter, 2007, "A parallel real-coded genetic algorithm for history matching and its application to a real petroleum reservoir", *Journal of Petroleum Science and Engineering* 59, 157–168, London.

[11] Watson, D.G.M and Gilfillan, A.W., 1977, "Some Ship Design Methods". Trans of the Royal Institution of Naval Architects, Vol. 119.

[12] Vasconcellos, J.M., 1994, "Semisubmersible Platform Preliminary Design, *D.Sc Thesis, Rio de Janeiro, Brazil.*

[13] Moraes, H.B., Vasconcellos, J.M., Almeida M.P., 2007, "Multiple criteria optimization applied to high speed catamaran preliminary design", Ocean Engineering Journal, Elsevier, 34, 133-147.

[14] Moraes H.B., Vasconcellos J.M. , Latorre R.G., 2004, "Wave resistance for high-speed catamarans", Ocean Engineering, Vol. 31, Issues 17-18, December 2004, pgs 2253-2282.

[15] Karayannis, T., Molland, A. F. and Williams, Y. Sarac, 1999, "Design Data for High-Speed Vessels", FAST-99.

[16] "High-Speed Marine Transportation", 1996-97, Jane's Book, 29° edition.

[17] Ignizio, J.P., 1976, "Goal programming and extensions", Lexington Books. Lexington, MA.

[18] Skoupas S. , Zaraphonitis G. and Papanikolaou A., 2009, "Parametric Design and Optimization of High-speed Twin-Hull RORO Passenger Vessels, IMDC 2009, Trondheim, Norway.

Safety Management of the Introduction of the Tomahawk Land Attack Missile into Royal Naval Service

Charles Brooking, CEng Institute of Mechanical Engineers, c.brooking@fnc.co.uk
Ann Hodgetts, CEng Institute of Mechanical Engineers, a.hodgetts@fnc.co.uk
Dr Paul Hogan, p.hogan@fnc.co.uk

ABSTRACT
This paper describes the development of the Safety Case for the Royal Navy Tomahawk Land Attack Missile (RN TLAM), which has supported the successful certification of the system for use onboard Royal Navy submarines. This paper provides a brief description of the weapon system and then explains how the major challenges surrounding the development of the Safety Case were overcome.

1. INTRODUCTION

The Tomahawk missile is a long range cruise missile used for land attack warfare. Tomahawk missiles are designed to fly at extremely low altitudes at high subsonic speeds, and are piloted by several mission tailored guidance systems. Ultimately, the purpose of the Tomahawk missile is to deliver a high explosive bomb, from long range to a precise location.

The first operational use of Tomahawk was by the US military in Operation Desert Storm 1991, and the missile has since been used in several other conflicts, including recent operations in Afghanistan. In 1995 the governments of the US and UK signed a Foreign Military Sales Agreement for the acquisition of several Tomahawk missiles, and after a launch and live warhead test in November 1998 the UK Ministry of Defence (MoD) declared operational capability. The UK MoD has procured the Tomahawk missiles for introduction into Royal Naval service on Nuclear Powered Fleet Submarines.

In 1995 Frazer-Nash Consultancy Limited (Frazer-Nash) became involved in the development of the RN TLAM Safety Case, which has been key to the acceptance and ongoing in-service use of the missile. The Safety Management Plan (SMP) developed by Frazer-Nash at the start of the programme ensured that the certification and acceptance of the RN TLAM weapon system was achieved to schedule, in order to ensure Royal Navy operational capability by 1998. The first operational use of RN TLAM was during the Kosovo conflict, the first missile being fired on 24th March 1999.

In developing the RN TLAM Safety Case to obtain safety certification and acceptance, Frazer-Nash had to overcome challenges associated with stakeholder management, the certification process, missile integration, safety assessment methods, and information availability. The ways in which Frazer-Nash overcame these challenges to ensure acceptance into service, and support ongoing in-service use, are discussed in this paper.

2. WEAPON SYSTEM

Two variants of TLAM are currently in service with the Royal Navy; Block III and Block IV. Block III was introduced into service in 1998 and Block IV in 2008. Both variants are winged, high-subsonic speed missiles which provide the submarine with an attack capability against a variety of land targets. The missiles include a warhead, solid-propellant rocket motor, cruise engine, and various underwater protection devices. The missiles are contained within steel capsules, which also serve as a smooth bore launching devices when loaded into the submarine torpedo tube, from which the missiles are horizontally launched.

The Block IV missile provides improvements in system capability over the Block III as follows:

1. Increased system flexibility;
2. Reduced system response time;
3. Improved lethality against a broader target set;
4. Significantly lower acquisition and ownership costs.

2.1 Missile Launch and Propulsion

Route planning, initialisation of the missile, downloading of data to, and launch of RN TLAM is controlled by the Tomahawk Weapon Control System (TWCS), which interfaces with the submarine Weapon Handling and Launch System, and the submarine Fire Control Equipment.

At launch RN TLAM is ejected from the steel capsule held within the submarine torpedo tube. When the missile is clear from the submarine, a lanyard initiates the firing circuit for the rocket motor. The guidance system is programmed to pitch the missile nose up using Thrust Vector Control (TVC) on the rocket motor nozzle.

The underwater protection devices consist of two jettisonable wing slot plugs, an engine inlet cover and continuity shrouds. These devices prevent the entry of seawater into the missile during underwater launch. After broach, the missile jettisons the engine inlet cover, wing plugs and the shrouds between the missile and rocket motor allowing the fins to deploy.

Upon reaching a set velocity, the engine inlet and wings are deployed and the shaped trajectory auto-pilot is activated. Once the rocket motor thrust has decayed, the rocket motor is jettisoned, and the cruise engine is initiated. The missile then descends to cruise altitude and is guided to the target by the guidance unit.

2.2 Missile Flight to Target

One of the most prominent characteristics of TLAM is its accuracy. TLAM can fly 1000 miles and hit a target the size of a garage. In addition to this capability, the missiles are also very effective at evading detection due to their low altitude flight path which is beyond the range of most radar systems. The four different systems that help guide TLAM are briefly described in the following paragraphs.

1. Inertial Guidance System (IGS)
The IGS is an acceleration-based system that can keep track (approximately) of where the missile is, based on the accelerations it detects in the missile's motion.

2. Terrain Contour Matching (TERCOM)
TERCOM uses an on-board 3D database of the terrain the missile will be flying over. The TERCOM system compares its stored set of digital maps with the ground elevation readings supplied by the missile's radar altimeter. The TERCOM system is what enables TLAM to sustain a low altitude flight path.

3. Global Positioning System Subsystem (GPSS)
A Global Positioning System Subsystem (GPSS) is used to perform periodic navigational updates to the missile. The GPSS receives signals from GPS satellites, processes the signals, and provides navigational data to the Cruise Missile Guidance Set (CMGS). This ability to receive GPS navigational data provides flexibility and higher reliability in accomplishing mission objectives.

4. Digital Scene Matching Area Correlation (DSMAC)
Once the missile approaches its target, the missile switches to a terminal guidance system to determine the impact point. Stored digitised images or scenes are used to perform terminal updates. A sensor in the DSMAC set takes a visible wavelength image of the ground features, digitises the images and then compares them with the stored digitised images.

3. RN TLAM SAFETY CASE

The RN TLAM Safety Case was developed by Frazer-Nash to provide assurance that the introduction of TLAM into Royal Naval service was safe for all UK military operations. In doing so Frazer-Nash recognised that the following challenges had to be overcome:

- There were a large number of stakeholders associated with the missile and its introduction into service from whom buy-in was required;
- The process that had to be followed to obtain safety certification and acceptance was complex and was modified during the procurement process;
- There was a need to integrate a missile designed for US military equipment into Royal Navy submarines;
- The need for safety assessment expenditure to be proportionate;
- Availability of information required to support the Safety Case and its ongoing validity.

The way in which Frazer-Nash overcame these challenges are discussed in the following sections.

3.1 Multiple Stakeholders

Due to the nature of the procurement, the need to integrate into RN submarines and the certification process surrounding the safety of explosives there are several stakeholders associated with RN TLAM, as shown in Figure 1.

The safety responsibilities of the stakeholders involved with the RN TLAM programme are briefly described in the following paragraphs.

<u>Safety Case Owners</u>
Project Safety Manager (PSM): Responsible for the safe introduction into UK submarine service, and for the continued safe use, of RN TLAM. The PSM is a member of the RN TLAM Project Team.

Project Safety Management Co-ordinator (PSMC): Responsible to the Project Safety Manager for developing the RN TLAM Safety Case. Frazer-Nash fulfils the role of the PSMC.

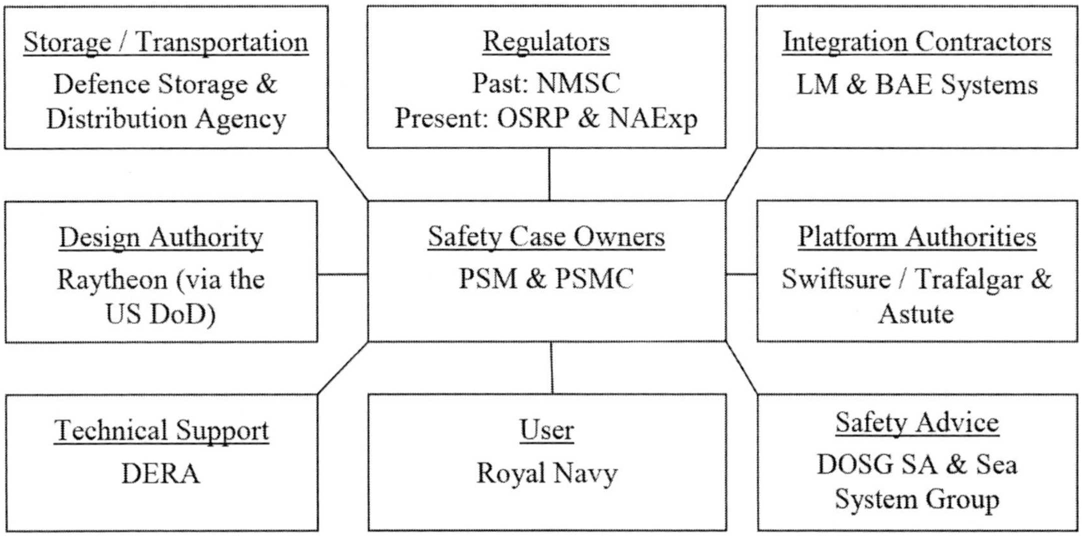

Figure 1: RN TLAM Stakeholders

Regulators

Naval Magazine Safety Committee (NMSC): The NMSC (now replaced by the OSRP and NAExp) was the regulator for RN explosive safety, both from the intrinsic and platform integration perspectives, when Block III was introduced into service.

Ordnance Safety Review Panel (OSRP): The OSRP are primarily concerned with intrinsic explosive system safety, and as such the OSRP periodically provide independent review the RN TLAM Safety Case. Once satisfied that the risks associated with the RN TLAM are being managed, the OSRP issue a time limited Certificate of Safety.

Naval Authority Explosives (NAExp): NAExp is responsible for certifying the safety of integration of approved explosive systems onboard RN vessels.

Integration Contractor

Lockheed Martin (LM) and BAE Systems: Responsible for the integration of RN TLAM hardware and software onboard the launch platforms. This includes integration of TWCS into the submarine Fire Control Equipment (FCE).

Platform Authorities

Swiftsure / Trafalgar & Astute Project Teams: Responsible for the nuclear powered submarines on which RN TLAM is used.

Safety Advice

Defence Ordnance Safety Group Safety Advisor (DOSG SA) and Sea Systems Group (SSG): The DOSG SA and the SSG provide safety advice to the PSM that supports the development of the Safety Case.

User

Royal Navy: Responsible for the in-service use of RN TLAM.

Technical Support

Defence Evaluation Research Agency (DERA): Various departments within DERA were tasked to provide advice and assessments on the missile software, hydrodynamic, shock, spigot intrusion characteristics.

Design Authority

Raytheon (via the US DoD): Technical authority for the design and manufacture of RN TLAM.

Storage / Transportation

Defence Storage & Distribution Agency (DSDA): Responsible for the safe storage and transportation of RN TLAM within the UK.

To surmount the challenge of managing the safety activities of all of these stakeholders with respect to RN TLAM, the RN TLAM Safety Working Group (SWG) was set up. The SWG provides a focal point for addressing all aspects of safety for the RN TLAM project. In particular it provides a forum for discussion and agreement of safety issues, and endorsement of hazard analyses and the Hazard Log. Through ensuring that the SWG comprised representation from all of the stakeholders associated with RN TLAM, it was also possible to obtain early and decisive agreement in areas such as safety requirement identification and safety issue resolution.

The co-ordination of activities and deliverables required by the SWG and the stakeholders involved in the RN TLAM project was achieved through a series of programme plans managed by Frazer-Nash. Regular reviews and revisions were made to the programme plans in order to keep abreast of the developments in the programme, and to monitor deviations from the schedule as they occurred. These reviews included stakeholder review of the plans prior to critical stages of the programme to ensure that interferences between stakeholders were correctly identified and communicated to all relevant parties. In this manner it was ensured that the safety activities were comprehensive, timely and sufficient to achieve the various acceptance stages.

3.2 A Complex and Changing Certification Process

The RN TLAM Safety Case was originally developed to provide assurance that the introduction of Block III TLAM into Royal Naval service was safe for all UK military operations. At this time the NMSC was the regulator for RN explosive safety, both from the intrinsic and platform integration perspectives.

When the US and UK governments subsequently signed a Foreign Military Sales Agreement for the acquisition of the newer Block IV missile, the scope of the Safety Case was increased to provide assurance that the introduction of Block IV into Royal Naval service was also safe. By this time the regulatory structure had changed such that the OSRP and NAExp were individually responsible for intrinsic and platform integration safety respectively.

Both regulatory structures had certification processes that needed management of procedural safety requirements from across the stakeholder set. The challenge here was to ensure that all of the requirements associated with the relevant certification process were fulfilled. Frazer-Nash achieved this by developing and maintaining a Safety Requirements and Certification Matrix to define and monitor the safety requirements. As there were multiple requirements to be fulfilled, and because they came from a range of sources, it was of great importance that the Certification Matrix was agreed by the requirement stakeholders as including all the requirements to secure safety endorsement for the introduction of the RN TLAM weapon system.

To that end the development of the matrix was an iterative process between Frazer-Nash and the requirement stakeholders, utilizing the Safety Working Group as a discussion and review forum, until a consensus was achieved.

Due to the ongoing development of MoD safety requirements and certification processes, the matrix proved a useful tool in tracking and agreeing the way forward for many of the requirements applicable to RN TLAM.

3.3 Integrating an Off-the-Shelf Missile

The need to integrate a missile designed for the US military into Royal Navy submarines presented challenges to the project, as discussed below.

3.3.1 Interface with the Astute Submarine Integration

Integration of TLAM onto RN Submarines has required modification of the interfaces with the submarine Fire Control Equipment (FCE). Engineering of this interface has comprised the introduction of the Tomahawk Weapon Control System (TWCS). TWCS interfaces with the existing submarine software and enables automatic or manual route planning, initialisation of the missile, downloading of all data to and the launch enable of RN TLAM.

Although the TWCS software was developed by the US Design Authority and was provided with a Computer Program Hazard Analysis, this did not include consideration of the interface between TWCS and the remainder of the FCE, and this presented a challenge to the project.

To overcome this Frazer-Nash carried out a risk assessment utilising analysis of the RN TLAM system performed by DERA. The DERA analysis was undertaken to determine system safety vulnerabilities with respect to TWCS. In particular the analysis focused on whether particular signals could occur inadvertently, and what the resulting response of the system would be.

System level behavioural analysis was used to check that the safety of the RN TLAM system was not disproportionately dependent on any one component. The alternative approach would have been to verify each component from the source code, which was not considered to be a feasible task. The system level analyses involved constructing mathematical models of the overall RN TLAM system. The models were then interrogated to determine the system's behaviour in the event of inadvertent signals being sent or received, and under normal conditions.

The specific scenarios identified for investigation were formulated by Frazer-Nash following a detailed review of the Hazard Log. In total 32 relevant potential hazards were identified from the Hazard Log, which were subsequently grouped into 8 Hazard Categories. Investigation areas identified for DERA Malvern included:

1. Inadvertent issue of Intent to Launch;
2. Inadvertent issue of Rocket Motor Arm;
3. Unable to Safe Rocket Motor;
4. Inadvertent launch;
5. Incorrect Rock Motor Status Report;
6. Misidentification of weapon;
7. Inability to abort launch;
8. Misfire.

The system level analyses focused on modelling the safety interlocks within the system. The design is such that the safety interlocks are distributed throughout the system in order to reduce the dependency on individual components.

DERA Malvern provided responses relating to the probability of a particular event occurring. Frazer-Nash subsequently used this information in carrying out a risk based assessment, which was conducted according to the process defined within the RN TLAM Safety Management System. From this assessment Risk Classifications were identified for those system hazards identified in the Hazard Log, and the hazards were subsequently endorsed by the Safety Working Group.

3.3.2 Astute Submarine Integration

Originally RN TLAM was integrated into Swiftsure and Trafalgar Class submarines. As these were existing platforms all of the hazards associated with integration of the missile were managed by the RN TLAM Project Team. More recently however, work has been undertaken to integrate RN TLAM onto the new Astute Class submarine. One of the Astute requirements is to safely store and launch RN TLAM, and so there was a challenge in avoiding the potential for confusion regarding ownership of the integration safety issues, which could have resulted in a lack of safety management.

Frazer-Nash was pro-active in developing a safety management process that facilitates clear identification of responsibility for safety issues across the RN TLAM – Astute interface. This management process recognises that separate RN TLAM and Astute Safety Management Systems exist; it does not aim to replace them, but complements them by ensuring that all integration safety issues are not only being managed, but by being managed by the most appropriate system.

3.4 Safety Expenditure

To ensure that the cost of the development of the RN TLAM Safety Case was proportionate to the overall procurement costs of the missile, Frazer-Nash and the RN TLAM Project Team had to take a pragmatic view with regard to the level of safety assessment undertaken. It was understood that in-depth analysis of all of the hazards associated with RN TLAM would require expenditure that was not commensurate with the missile procurement costs.

The approach taken by Frazer-Nash was therefore to initially carry out a high level risk assessment to identify the most significant risks. This enabled the requirement for more detailed assessment to be prioritised such that only high risks received in-depth analysis. This process of assessing risk was co-ordinated and controlled by a central Hazard Log and the Safety Requirements and Certification Matrix.

In addition a combination of Deterministic Safety Justification and Risk Based Assessment methods were utilised by Frazer-Nash to minimise the cost of any analysis. These assessment techniques are briefly described in the following paragraphs.

a) Deterministic Assessment

Where practicable assessments took a deterministic approach, demonstrating why the hazard will not occur by referring to pre-existing supporting evidence, for instance reference to US DoD test results, procedural documents etc. In doing so the RN TLAM Safety Case acknowledges that RN TLAM is essentially an 'off the shelf' weapon, and as such uses the existing US TLAM Safety Case to underpin the claimed level of safety.

Where a deterministic approach was not appropriate, a risk-based approach was taken.

b) Risk Based Assessment

A risk based assessment, centred on accident frequency and severity assessments made by a suitably qualified and experienced panel, was used when sufficient supporting evidence to support a deterministic assessment was not available. Where detailed analysis techniques were required to support these risk based assessments, these were selected carefully to ensure that costs were proportionate; for example modelling or technical reviews were undertaken where possible to avoid the need for expensive trialling.

3.5 Availability of Supporting Information

To develop the Safety Case information was required from the US DoD to support claims that the RN TLAM is tolerably safe. Ideally this information would have included detailed design data and the full results of trials; however, to protect US military interests, not all of this information was available. Additionally, as the procurement of RN TLAM was via a Foreign Military Sales Agreement, formal communication processes had to be followed, which built in some delay in receiving any information that was available.

To ensure that the sufficient and pertinent information was received in a timely manner, Frazer-Nash and the RN TLAM Project Team spent considerable effort developing good working relationships with the US stakeholders. In particular they began attending the biannual US Tomahawk Weapon System Safety Group, which allowed up to date US DoD safety issues to be understood, and provided a forum for UK MoD safety information requirements to be presented.

Developing relationships ultimately helped to increase the efficiency of data transfer to the UK MoD; however in some cases the required data was ultimately not available. In these instances an alternative approach to resolving the issue in question was developed by Frazer-Nash. Particular examples include:

- Service Life Safety Management;
- Spigot Hazard Assessment;
- Anti-Circular Run Function Analysis;

The alternative approaches developed by Frazer-Nash to take the most benefit from the information that was available in each of these examples are discussed in detail in the following paragraphs.

3.5.1 Service Life Safety Management

Based on the level of data made available, the DOSG SA was only able to qualify certain TLAM components as safe and suitable for very time limited periods. This necessitated that a different method for justifying the ongoing safety of RN TLAM be adopted. The agreed approach was to participate in the US Service Life Assessment Programme (SLAP) that tests components that have been exposed to the missile life cycle to ensure that they still function correctly. In fact one of the provisos on which safety certification has been granted is that the UK MoD continues to participate in the US SLAP programme. Analysis of the SLAP data has enabled the DOSG SA to gain a better understanding of the ongoing safety of TLAM explosive components.

3.5.2 Spigot Hazard Assessment

At an early stage of the programme it was recognised that, because the US DoD explosive safety requirements are not exactly the same as those contained within UK MoD Safety Requirements, the US DoD safety disclosures did not include an assessment of the encapsulated missile dropping on a spigot, or a spigot impacting the encapsulated missile.

The missile is exposed to spigot hazards primarily during the logistics cycle and weapon handling operations. Owing to the nature of the hazard, it was not considered appropriate to develop a deterministic justification in order to demonstrate that the hazard would not occur due to safeguards or procedural controls. Rather a specific Spigot Risk Assessment was conducted.

The purpose of the Spigot Risk Assessment was to demonstrate that the risk associated with the hazard was acceptable to the project. Frazer-Nash achieved this in conjunction with DERA through a process of:

1. Conducting surveys and defining generic spigot hazards.
2. Modelling the effects of an encapsulated missile being dropped onto one of these generic spigot shapes.
3. Conducting the Spigot Risk Assessment.

Surveys were initially conducted of all the sites where it was envisaged that RN TLAM weapon handling operations would be conducted. Photographs and measurements were taken of the specific spigot hazards identified at each site. From the total number of identified hazards, a small group of generic spigot hazards were subsequently defined.

DERA were then tasked to model the consequences of dropping an encapsulated missile onto the defined generic spigot hazards. A variety of drop heights were modelled, and the subsequent consequences assessed. It was conservatively assumed that if the spigot penetrated the capsule, this would result in explosion of the rocket motor or warhead.

Having determined which of the generic spigot hazards would result in penetration of the capsule, the risk assessment continued by revisiting the sites where these hazards had been originally identified order to assess the probability of a missile being dropped from above the safe height, and impacting the identified spigots.

The outcome of the Spigot Risk Assessment was an assessment, which demonstrated that the risk of a spigot hazard to RN TLAM was acceptable. Ultimately a trial was conducted on two live Rocket Motors and an inert filled warhead case, the results of which supported the risk assessment. Corresponding Hazard Log entries were developed and endorsed by the Safety Working Group.

3.5.3 Anti-Circular Run Function Analysis

During the hazard identification process, one of the identified hazards addressed the potential that after launch, during the underwater boost phase, a failure occurs which causes the missile trajectory to become uncontrolled. This is followed by failure of the Anti-Circular Run (ACR) system to function and terminate boost. The hazard scenario is completed by the uncontrolled missile making a circular run, and impacting with the launch platform.

Limited detailed information relating to the ACR function was available during the development of the RN TLAM Safety Case. In addition there was no possibility for introducing a design change for RN TLAM.

Frazer-Nash adopted a risk based approach in order to assess the hazard. Using worst case assumptions it was concluded that single point failure of both the missile guidance control and the ACR function could occur. Using knowledge of the missile boost characteristics and submarine design an argument was developed which demonstrated that the risk of circular run leading to launch platform impact and catastrophic damage was tolerable.

Independently SSG developed an alternative argument that also demonstrated that the risk of circular and launch platform impact and catastrophic damage was tolerable. Based upon these two arguments the Safety Working Group subsequently discussed, accepted and endorsed the risk assessment in accordance with the SMP.

4. CONCLUSIONS

Key factors in the successful development of the RN TLAM Safety Case include Frazer-Nash and the RN TLAM Project Team:

- Effectively planning the development of the Safety Case to ensure that it meets all identified safety requirements;
- Achieving buy-in, and obtaining beneficial input, from a large number of stakeholders associated with the missile;
- Undertaking a proportionate and pragmatic approach to risk assessment that derives the most benefit from the information available.

These factors were significant in enabling the introduction of both Block III and Block IV into Royal Navy service within programme targets and are invaluable in supporting the ongoing service use of RN TLAM.

5. ACKNOWLEDGEMENTS

The RN TLAM Safety Case was successfully developed through the close working relationship between the RN TLAM Project Team, Frazer-Nash and all of the stakeholders associated with the project, i.e. the US DoD; Raytheon; Royal Navy; DERA; DOSG; SSG; OSRP; NAExp; DSTA; Lockheed Martin; BAE Systems; and Swiftsure, Trafalgar and Astute Project Teams. Frazer-Nash would like to acknowledge the support of these stakeholders in the continued maintenance of the RN TLAM Safety Case, and in the development of this paper.

The Development and Application of a Damage Dynamic Stability modeling capability for Naval Vessels.

T. Turner[1], E. Ypma[2], G. Macfarlane[3] & M. R. Renilson[3]

[1]*Defence Science and Technology Organisation, Melbourne, Vic, Australia*
[2]*Maritime Research Institute Netherlands, Wageningen, The Netherlands*
[3]*Australian Maritime College, Launceston, Tasmania, Australia*

ABSTRACT

The Maritime Research Institute Netherlands, MARIN on behalf of The Cooperative Research Navies (CRNav) have been developing a semi empirical and a physics based numerical tool, FREDYN, which can be utilised to study the seakeeping behavior of both intact and damaged naval ships in wind and waves. The Defence Science and Technology Organisation, DSTO, in collaboration with The Australian Maritime College, AMC, have recently undertaken a research program to support MARIN in the validation of the progressive flooding module within FREDYN.

Once fully validated this numerical tool has a variety of applications which are of significant benefit to the Navy. FREDYN can be utilised for the development of operator guidance tools to be used by navies in their decision making processes for damage control procedures following an incident whereby the vessel has sustained damage and consequently flooded. The application of FREDYN will also allow for the development of more appropriate intact and damaged dynamic stability criteria for naval vessels. Another application of this numerical capability is for the through life management of the naval ship's growth margins which will ensure that the highest level of safety of the vessel is maintained at all times.

This paper not only provides an overview of some of the recent experimental studies undertaken for the validation of the progressive flooding capability within FREDYN but also demonstrates several applications that this tool can be utilised for in support of the various stability requirements of naval vessels.

INTRODUCTION

The Cooperative Research Navies (CRNav) Dynamic Stability group was established in 1989 to initiate a research program focussed on increasing the understanding of the dynamic stability of naval vessels from a more physics based approach rather than an empirically derived approach that was developed by Sarchin and Goldberg [1] in the 1960s. The work of CRNav has led to the development of a time domain numerical tool, FREDYN, which simulates the motions of both an intact and a damaged vessel whilst operating in extreme seas. In FREDYN, the dynamic effect of wind, waves and progressive flooding on the vessels motions are all considered when analysing the stability of the vessel.

Despite the Sarchin and Goldberg criteria being currently applied by many navies, what is not fully understood is the applicability of these criteria to modern day naval platforms, and the level of safety assured by compliance to these stability standards. It is also unclear if the quasi-static approach of Sarchin and Goldberg reflects the true dynamic behaviour of ships in extreme conditions. For these reasons the CRNav committee formed the Naval Stability Standards Working Group (NSSWG) to investigate the applicability of the Sarchin and Goldberg criteria and to develop a shared view on the future of naval stability assessment. The intention of the NSSWG is to develop a set of stability guidelines which can be utilised by the participating navies at their discretion. The NSSWG consists of:

- Department of Defence (Australia).
- Department of National Defence (Canada).
- Ministère de la Dèfense (France).
- Ministerie van Defensie (Netherlands).
- Ministry of Defence, DPA (UK).
- United States Coast Guard Naval Engineering (USA).
- Naval Sea Systems Command (USA).

The above members are supported by the following guest members:

- Australian Maritime College (Australia).
- Defence Research and Development Canada (Canada).
- Bassin d'Essais des Carènes (France)
- Maritime Research Institute (Netherlands)
- QinetiQ (UK)
- Naval Surface Warfare Centre (USA).

Until recently most of the focus of NSSWG has been on intact vessels. However a new flooding module has recently been developed that will allow the assessment of the dynamic damaged stability of a damaged vessel. This new capability allows the simulation of progressive flooding mechanisms and the resultant influence this flooding has on the motions of the damaged vessel. The CRNav tasked The Defence Science and Technology Organisation, DSTO, to assist MARIN in the validation of the new flooding module. Recently DSTO, in collaboration with The Australian Maritime College, AMC, undertook an experimental program to collect the data required for validation. This experimental series comprised of recording the motions and progressive flooding of a model scaled destroyer after sustaining damage.

Once the FREDYN flooding module has been fully validated, it can be applied to many aspects relating to the damage stability of a navy vessel. The progressive flooding modelling capability within FREDYN can also be exploited in the development of operator guidance tools to be used by the RAN in their decision making processes for damage control procedures following an incident whereby a vessel has sustained damage. Another application of FREDYN is the identification of the level of safety in applying the existing stability standards to modern day hull forms. The strengths and weakness of these standards can be determined. In the longer term FREDYN can be utilised to assist in both the acquisitions of new naval vessels and the through life management the vessels growth and stability margins.

This paper not only provides an overview of some the experimental studies that were undertaken by AMC but also discusses the application of FREDYN in the development of a rapid assessment tool that assesses the stability of the vessel after sustaining damage. A brief discussion will be presented outlining the potential application of FREDYN during the acquisition process and through life management of the vessel.

VALIDATION STUDIES

A research program has been undertaken to fully validate the progressive flooding capability within FREDYN. A brief overview of the experimental program that supports this validation process is presented in the following section. Results from the numerical studies will be presented in future publications.

Experimental Studies

DSTO was tasked by CRNav to undertake some validation studies of the newly formulated flooding module within FREDYN. This program of work consisted of undertaking a series of model tests at The Australian Maritime College and supporting numerical analysis by MARIN and DSTO. A 3.27 metre long model of a compartmentalised generic destroyer hull form was constructed. The generalised internal geometry consisted of a series of decks comprising of both symmetric and asymmetric compartments. Openings between compartments through which the flooding process could proceed were via a combination of doors and hatches. A taut membrane covering an opening in the starboard hull, see Figure 1 , was punctured resulting in a rapid inflow of flood water and subsequent progressive flooding through out the compartments and decks. The rapid inflow of water made it possible to not only validate the simulation of the flooding process but also the changing dynamic behavior of the vessel.

Figure 1 A photograph showing the membrane covering the damage opening

The vessel was instrumented to measure vessel motions and the levels of water and air pressures in various compartments. An internal video camera was also installed to assist in the understanding of the progressive flooding mechanism. For the calm water tests, the rapid ingress of water resulted in an initial roll to starboard (into the damage), after which as the port side compartments started to flood the vessel began rolling to port. As the progressive flooding continued the vessel gradually rolled back to the starboard side and eventually settled at a final equilibrium heel angle. Figure 2 shows the change in the roll angle of the vessel versus time. Figure 3 shows a photograph of the final equilibrium heel angle of the destroyer model after sustaining damage.

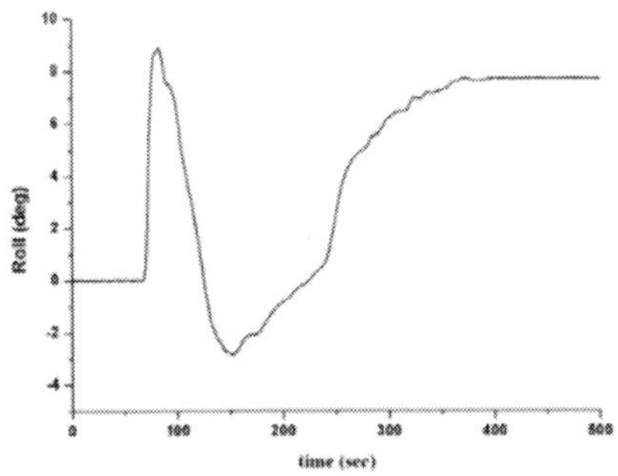

Figure 2 A plot showing the roll of the vessel versus time

Figure 3 A photograph showing the final equilibrium position of the flooded destroyer

Numerical Model

FREDYN [2] is a time domain numerical tool that can be used to predict the motions of a ship in moderate to extreme seaways. These motions incorporate both small and large amplitude motions including broaching and capsize. Recently the capability within FREDYN has been extended to predict the progressive flooding of a damaged ship and the implications that this flooding has on the overall motions and stability of the vessel. Estimates of time to sink or capsize can also be determined.

For any FREDYN simulation of the dynamic stability of a damaged ship, a compartment model is generated which includes details such as: (a) compartment boundaries, (b) openings, (c) liquid densities, (d) initial fluid volumes and (e) compartment permeabilities. A numerical model of the compartmentalised generic destroyer was generated by PARAMARINE [3] and analysis of the model test data undertaken using FREDYN. This analysis is part of the ongoing work program within CRNav to develop a fully validated numerical tool which can be utilised for various applications which are outlined in the following sections.

ONBOARD OPERATOR GUIDANCE DAMAGE CONTROL TOOLS

Initial Assessment of Safety

In recent years there have been several incidents where naval ships have sustained damage either through hostile, eg USS Cole [4] or through accidental events, HMS Nottingham [5]. In both instances it was imperative that operator guidance was given to the crew onboard the vessel as to the best way to not only maintain the safety of the vessel but also undertake the recoverability of the vessel's capability. The Maritime Platform Division within DSTO is currently developing an operator guidance tool, (RapSDAT) [6] that can provide advice to the crew of a warship that has sustained damage. The tool being developed needs to span the spectrum of incidents and provide advice to the crew for all mission objectives. At this stage the tool has the capability to provide advice as to the safety of the vessel in terms of residual structural strength but eventually, a capability to assess the damage dynamic stability of the vessel will be also incorporated. The progressive flooding modeling capability within FREDYN will formulate the basis for the stability aspect of this tool.

RapSDAT currently contains a database of probable damage templates for both accidental and offensive damage mechanisms. These templates are generated from the vulnerability assessment tools that currently exist within DSTO [7]. The user is required to select the damage type from a list of internal blast, external blast, collisions or groundings and then to locate the centre and extent of the damage. The RapSDAT tool will provide a traffic light system to indicate the probability of failure. A green colour is good and suggests that the crew can continue to devote their efforts to damage control. An amber or red light indicates that there is an area of concern. In terms of stability, the operator will then be able to assess whether changing speed, heading angle or counter flooding will improve the situation. The tool will be relatively simple to use and provide advice in seconds. An example of a possible graphical user interface for the stability component is shown in Figure 4.

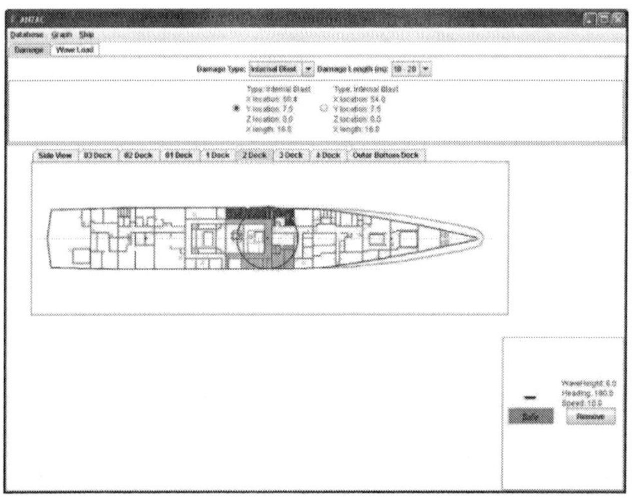

Figure 4 Screen shot from the RapSDAT showing flooded compartments and the green traffic light indicator

Evacuation and Time to Sink Guidance

The progressive flooding capability within FREDYN can also be utilised to give an indication as to the time to sink for a damaged vessel. This information is critical when determining the appropriate time to evacuate and the associated evacuation processes. The polar plots shown in Figure 5, are one example of the time to sink information that could be presented to the operators. This could be a typical display screen within RapSDAT. The plot on the left is an example of a scenario where the ship may be operating in sea state 3 and moving at 5 knots. As long as the vessel maintains a heading within the green zone, and the sea state remains the same, then it will remain afloat for the time indicated on the plot. The polar plot on the right is the same scenario except that the seas have deteriorated to sea state 4. It is evident that the time to remain afloat is significantly reduced and therefore the decision to evacuate would be required to be made at an earlier stage within the damage control process.

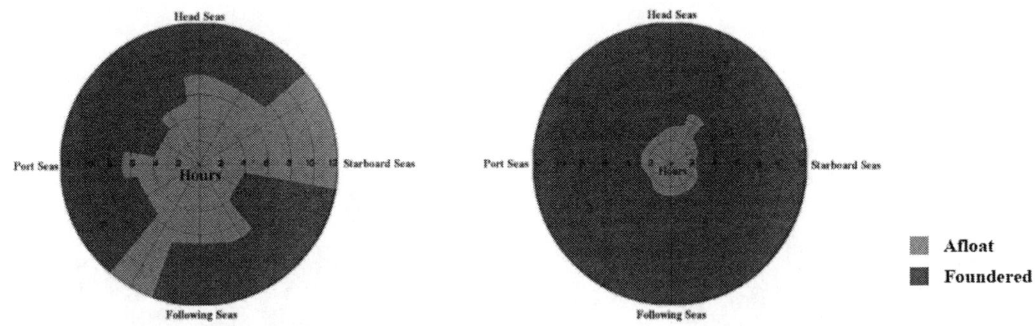

Figure 5 Polar plots showing areas of operability for a damaged ship in two different sea state 3 (L) and sea state 4 (R)

Recoverability Decision Tool

The outputs from FREDYN's progressive flooding modeling capability can also be applied to the decision making processes for the recoverability of the ship's systems.

For an incident where the risk of foundering is sufficiently small, the aim of the damage control process is to recover some of the capabilities of the platform. The overall capability of the platform is now dependant upon which systems are damaged during the event. Therefore it is extremely important that the crew can determine the status of all systems prior to any advice being given. To demonstrate the recoverability tools being developed, it is useful to consider a particular example and follow through the decision making processes.

Let us consider a situation where a surface ship has experienced an offensive strike and the resulting structural damage is fairly limited. The damage event has caused some internal flooding of the hull and a resultant heel of 10 degrees is encountered. The crew have stopped further flooding through the breach in the hull and determined that it may be possible to allow some form of cross flooding which will result in the heel angle reducing from 10 degrees to 0 degrees. Prior to completing this damage control, the crew wishes to determine whether or not this will improve the capability of the platform. In particular they wish to determine whether or not a particular ship system eg weapons system, will be more operable in the given environment. A particular restriction might be that the system can only operate when the maximum roll angle is limited to 18 degrees from the upright condition.

A polar plot representing areas of speed and heading angle combinations which pass these criteria can then be generated to determine operational parameters where the system can be used. Figure 6 shows a typical plot for the vessel in the upright position and then a plot with the vessel heeled over to 10 degrees. The red areas represent combinations of speed and heading where the system will not operate. For the upright condition this represents about 23% in sea state six. Whereas prior to damage control processes, i.e. 10 degree heel, the red areas increases to approximately 58% of the combinations. Also visible from the plot is the asymmetry in the red area due to the damaged heel angle either being towards or away from the encountered waves.

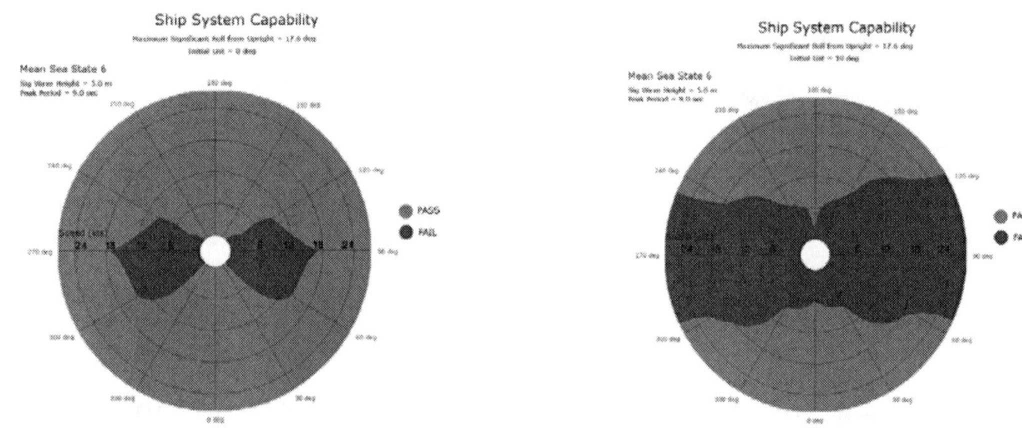

Figure 6 Polar plots showing areas of a ships systems operability for an intact ship (L) and a damaged ship heeled over to 10 deg (R) [6]

If the crew wishes to increase the range of options available, then further damage control is required. Provided there is a range of polar plots available for differing degrees of heel, then the crew can determine whether or not their efforts will be rewarded. These polar plots can be generated by utlising the progressive flooding modeling capability within FREDYN.

OTHER FREDYN APPLICATIONS

FREDYN can also be used to assist Navy in several other stability related areas. These applications range from the upgrading of stability standards, the stability assessments undertaken during the acquisition of a new vessel to the through life (weight) management of the naval platform. The following sections in this paper briefly outline the application of FREDYN in these areas.

Intact and Damage Stability Standards

In addition to the development of onboard operator guidance tools the dynamic stability capability within FREDYN can be utilised to assess the applicability of the Sarchin and Goldberg empirically defined static stability criteria which are applied to modern day naval hull forms. The level of risk associated with applying such criteria is not fully understood and for this reason the

NSSWG is undertaking an extensive investigation into the risks associated with this by applying a more physics based approach, FREDYN, to determine the appropriateness of the Sarchin and Goldberg criteria to modern hull forms. Initially the work of NSSWG is focused on identifying the level of safety of the existing standards and its strengths and weaknesses and the intention is to eventually develop methodologies for assessing stability characteristics and operator guidance for design and through life (weight) management. At the moment the focus has mainly been on intact vessels but another program of work within NSSWG has been initiated with a focus on capturing the dynamic effects of progressive flooding and the implications this has on the damage stability standards.

Acquisitions

Throughout the preliminary stages of the acquisition cycle of naval vessels, a risk assessment is often undertaken for the various options for both intact and damage stability. At this stage these assessments are undertaken comparing the results of static stability tools against the Sarchin and Goldberg based stability criteria. The application of the damage dynamic capability within FREDYN within this acquisition process will result in an increased understanding of the possibility of having to impose any operational limitations on any of the options on offer. The NSSWG is formulating a report outlining the approaches for evaluating the dynamic stability of a vessel in the acquisition process. [8]

Through Life (Weight) Management Process

Another application of FREDYN is in support of the through life management of naval vessels. As part of a naval vessel compliance check with stability standards a limiting KG curve for the vessel is generated. This limiting KG curve encompasses both intact and damage stability. Currently, these curves are generated using calm water static stability analysis techniques. In a damaged, dynamic stability analysis the effect that the progressive flooding and the wave environment has on the vessel's motions will both be considered. These effects may have implications on the limiting KG curve that are not captured through a static stability analysis. Using the static approach the limiting KG value is based upon the final static heel angle. In the experimental program conducted by AMC described earlier in the section on Validation Studies, the maximum heel angle recorded was not always the final equilibrium heel angle. For three out of the four KG values analysed the maximum heel angle was the initial heel angle due to the initial rapid ingress of water. As the vessel continued to flood the vessel rolled across to port and then recovered slowly back to starboard until finally coming to rest at a heel angle smaller than that initially recorded. These results are shown in Figure 7 below. In addition, it should be noted that the results shown in Figure 7 demonstrate the dynamic nature of progressive flooding whilst the vessel is in calm water: in reality the ship would more likely to be in an environment comprising both wind and waves and therefore it is even more important that this dynamic progressive flooding phenomena is accurately determined.

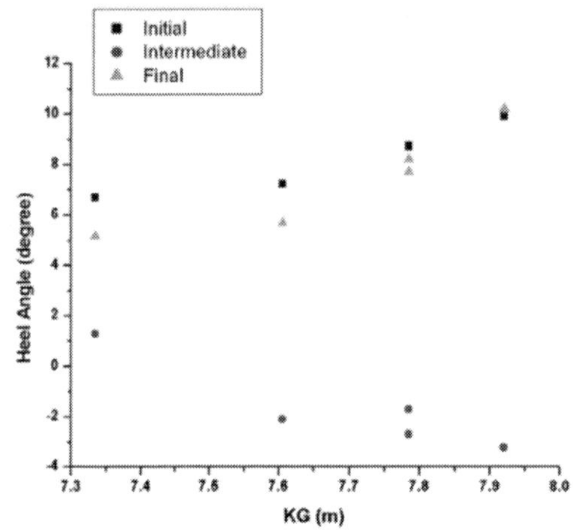

Figure 7 A plot of the resultant heel angle of a damaged ship due to flooding for various initial KG values

The application of FREDYN to the through life (weight) management cycle of the vessel would then allow these dynamic effects to be captured resulting in an increased understanding of the implications that any growth in the vessel's weight margins has on its stability and hence increasing the safety of the vessel during its operations.

CONCLUSION

This paper has demonstrated various potential applications for which the numerical simulation tool FREDYN can be applied to assist Navy in stability related issues. As discussed the progressive flooding capability within FREDYN has undergone some preliminary validation studies and future programs of work are focused on developing and validating this capability even further.

Once fully validated the development of a damage methodology framework is required to enable a complete study to be undertaken of the dynamic stability of a (damaged) warship. This study will also encompass the integration of this capability into a suite of tools where by an overall assessment of a naval warships susceptibility, survivability and recoverability can be undertaken. Once developed, this Integrated Survivability Suite can be applied throughout the life of a vessel ranging from the concept stage right through to the through life management of the platform and its capability.

The Cooperative Research Navies along with the Naval Stability Standards Working Group have ongoing programs of work developing an approach to apply this dynamic stability capability to both the acquisition cycle and the certification process of naval vessels.

REFERENCES

1. Sarchin, T.H. and L.L. Goldberg, *Stability and Buoyancy Criteria for U.S. Naval Surface Ships*. SNAME Transactions, Society of Naval Architects and Marine Engineers, 1962. 70.

2. *FREDYN Version 10.0 A Computer Program for the Simulation of a Steered Ship in Extreme Seas and Wind*. 2009, CRNAV, Maritime Research Institute Netherlands.

3. *PARAMARINE Version 6.1*. 2009, GRC Ltd.

4. *USS Cole Bombing*. Available from: http://en.wikipedia.org/wiki/USS_Cole_bombing.

5. Lt Cdr Ian Groom RN, *HMS Nottingham - a story of damage control*, in *IMarEST Technical Meeting*. 2003: United Kingdom.

6. Turner, T. and S. Cannon, *Onboard Rapid Assessment Tools for Damaged Naval Vessels*, in *Maritime and Systems Technology*. 2007: Genoa Italy.

7. Buckland, M.E., N.M. Burman, and M. Webster, *XVAM - MRL's ship vulnerability assessment Procedure*, in *DSTO-TR-0000*.

8. Naval Stability Standards Working Group, *Approaches for Evaluating Dynamic Stability in the Acquisition Process*, in *(in preparation)*.

An Investigation into the Wavemaking Resistance of a Submarine Travelling Below the Free Surface

S Wilson-Haffenden[1, 2, 3], M Renilson[1], D Ranmuthugala[1], E Dawson[2]
[1] Australian Maritime College (AMC), Launceston, Tasmania, Australia
[2] BMT Design and Technology, Melbourne, VIC Australia.
[3] swhaffenden@gmail.com.au

1. INTRODUCTION

The diesel-electric systems that power conventional submarines require air periodically to run the main engines and replenish the stored battery power. In order to reduce the chance of detection at this stage of operations, submarines usually travel just below the surface taking in air through a snorkel. During this period, wavemaking resistance makes up a significant proportion of the overall resistance experienced by the boat. This resistance either results in a slower surface speed or prolonged battery recharge duration. Future powering options, such as closed cycle engines or fuel cells, may reduce or eliminate the reliance on the atmosphere for power however mission requirements still often involve operating close to the surface.

There is typically a limit to the depth a submarine can snorkel, usually dictated by the backpressure on the snorkel. As the depth of snorkelling increases, the engine must work harder to ingest the air and as a result becomes less efficient (Zimmerman, 2000). Furthermore, snorkels are mechanically complicated and extending them beyond approximately two times the diameter of the vessel is unfeasible (Vine, 1991). The creation of waves due to the submarine travelling close to the free surface is therefore unavoidable. At moderate to high speeds, wavemaking can contribute a large proportion of the total resistance experienced by the boat. This means that relative to a deeply submerged submarine, the amount of additional power required by a submarine to maintain speed when close to the surface is significant. Thus, predicting the increased power requirements due to wavemaking resistance is an important aspect in submarine design.

While the use of Experimental Fluid Dynamics (EFD) on submarines is possible, there are elevated costs, and resource and time requirements commonly attributed to such extensive experimental research, particularly when design optimization is required. In addition, detailed local flow characteristics are unattainable without the use of a Particle Image Velocimeter (PIV) which further adds to the costs and complications of the experiment. These drawbacks to EFD are compounded by the inherent errors of the method; such as those related to scaling.

Through the use of Computational Fluid Dynamics (CFD) methods in partnership with EFD it is proposed that a more cost effective and robust analysis tool can be developed. The tool would support the early stages of development in hull form design, powering options and operational analysis. The current research is the first step in a series of experimental and computational work aimed to provide a platform for further – more detailed – investigations including: the analysis of a fully appended submarine near the free surface or an examination of the effects of sea state on the body and manoeuvring of a submarine close to the surface.

2. EXPERIMENTAL SET-UP

Experimental testing was performed in the towing tank at AMC, which is 100 m long, 3.55 m wide, with a variable water depth of up to 1.6 m. The facility has a powered carriage to tow models up to a maximum speed of approximately 4.6 m/s. The model was a 2.8:1 scale version of the un-appended DARPA SUBOFF geometry (Groves et al, 1989); having a diameter of 181 mm, length of 1556 mm, and surface area of 0.382 m^2. The model was attached to a Horizontal Planar Motion Mechanism (HPMM) located on the carriage via a sting mounting system as shown in Figure 2.1. The depth was varied using an insertion piece between the sting and the carriage. A load cell, located at the forward end of the model, recorded the longitudinal force due to the model/fluid interaction.

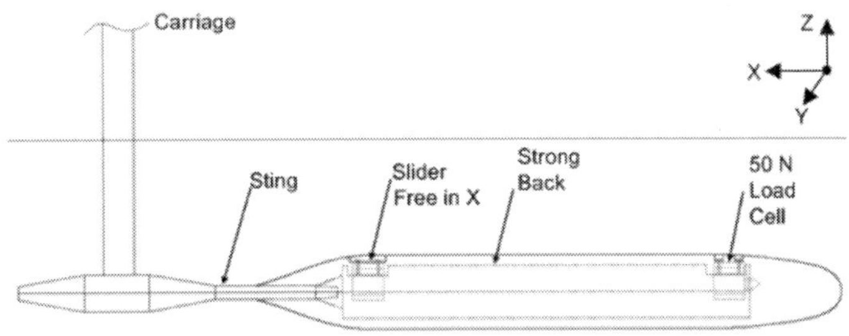

Figure 2.1: Balance and Sting Arrangement with Model Attached

The depths tested experimentally relative to the diameter of the model and the depth of the towing tank are shown in Figure 2.2 where D^* is defined as follows:

$$D^* = \frac{H}{D} \tag{2.1}$$

H is the depth from the free surface to the centre line of the model and D is the diameter of the model.

The speeds tested ranged from 0.5 to 2.5 m/s corresponding to a Froude Number range of 0.128 to 0.640.

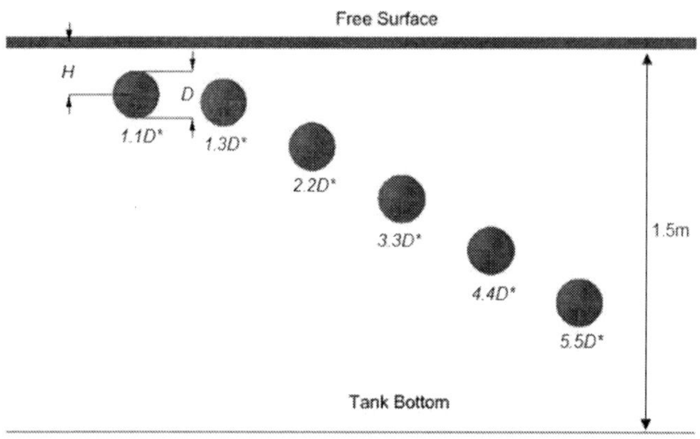

Figure 2.2: Testing Depths

In an effort to ascertain the uncertainty in the experimental results, throughout the testing period a number of runs were repeated when testing each depth. The *1.1D** condition, which was the first condition analysed, was also repeated at the end of the test programme to ascertain the uncertainty between repeat conditions. The uncertainty in an individual run was the same as the uncertainty for the repeated condition and thus the uncertainty due to changing conditions can be assumed to be negligible for the purposes of this paper.

A final check was performed by deliberately misaligning the model approximately five degrees to the direction of flow and measuring the side force and resulting change in axial drag. This angle was clearly visible and thus assumed that any errors due to a misalignment throughout the testing would be smaller. As a result a maximum misalignment of three degrees was assumed and the relationship between the uncertainty and the angle was deemed to be linearly related giving a maximum uncertainty of three percent.

The resulting quantifiable uncertainty was the sum of both the repeatability uncertainty and the alignment uncertainty i.e.:

$$\textit{Total Uncertainty} = \textit{Repeatability Uncertainty} + \textit{Alignment Uncertainty}$$
$$\textit{Total Uncertainty} = \pm(0.049N + 0.03 \times \textit{Force Measured})$$

An additional uncertainty also exists when comparing the experimental results with the numerical results, the largest of which is due to the mounting system. As the sting mounts into the rear of the body some degree of pressure recovery is lost. In addition, the sting arm had its own pressure system, which will include a high pressure region in front of the arm. The model was also fitted with turbulence inducing studs towards the bow whereas the numerical simulations were preformed without these. This uncertainty was not quantified however it is acknowledged as a discrepancy between the EFD and the CFD results.

3. COMPUTATIONAL TECHNIQUES

In order to eliminate any scale effects, model scale was also adopted for the computational (CFD) work. The fluid domain was based on the towing tank in an effort to account for any possible boundary effects. For the purposes of this work a block structured grid using matching cell faces was chosen. The mesh was constructed using ANSYS-ICEM-CFD, utilising an O-C topology for the body and an H topology to refine the free surface. The simulations were carried out using ANSYS-CFX on a computer cluster at AMC. A steady-state model was utilised although transient simulations were also conducted during the validation period. Details of the mesh and the simulation settings, including the time step and free surface options, are described in Wilson-Haffenden (2009).

Due to the symmetrical nature of the flow only half the domain was modelled and a symmetrical boundary was applied to the centre line surface. This reduced the mesh size by half, and thus the computational time. If future work includes non-axial velocity components, it will be necessary to mesh the full domain as the flow will no longer be symmetrical. The final boundary conditions applied are shown in Figure 3.1.

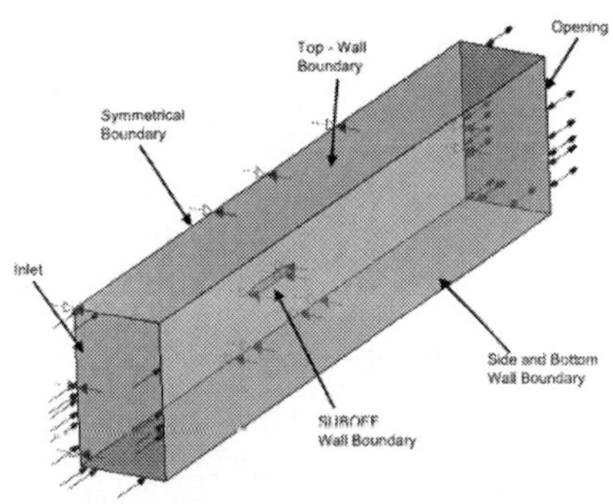

Figure 3.1: Simulation Boundary Conditions

A grid independence study was conducted on the final mesh in order to ascertain the appropriate grid density. The depth of $1.1D^*$ and free stream velocity of 1.5 m/s was chosen for the study as it was considered to be one of the more complex combinations for the flow simulations. The mesh density was incrementally changed between 1.5 and 3.7 million elements. As a means of comparing the results, the drag coefficient (C_D), the pressure coefficient (C_p), and the friction coefficient (C_τ), were monitored. This also allowed for comparisons against available experimental data for the un-appended SUBOFF geometry. Figure 3.1 shows the variation in C_D with respect to the results for the 3.7 million element mesh.

$$C_D = \frac{2R}{\rho L^2 U^2} \qquad (3.1)$$

$$C_p = \frac{2(p - p_0)}{\rho U^2} \qquad (3.2)$$

$$C_\tau = \frac{2\tau_w}{\rho U^2} \quad (3.3)$$

Where, R is the total resistance, ρ is the density, L is the length, U is the velocity, p is the total pressure, p_o is the atmospheric pressure, and τ_w is the wall shear.

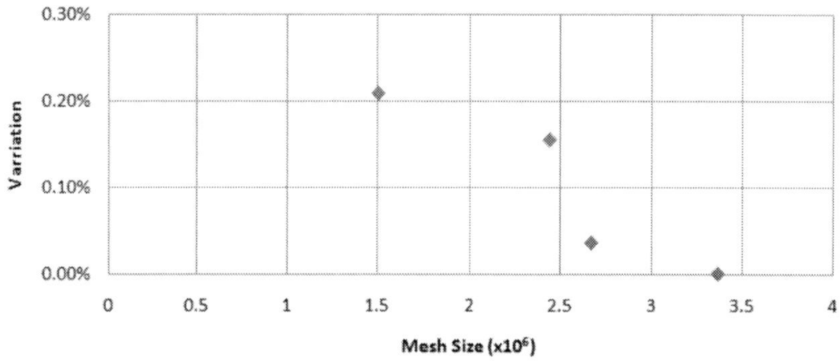

Figure 3.1: Variation of C_D as a Function of Mesh Size

In changing the grid size globally by more than 50%, a variation of less than 1% in both the drag and pressure coefficients was noted, while the friction coefficient (which is typically the most sensitive of the three) remained relatively constant. Thus, the 1.5 million element mesh was assumed to produce grid independent results in line with the suggestion of Sung et al (1996), and was selected for the remaining simulations.

4. RESULTS

The experimentally measured drag coefficients as a function of Froude Number for all depths tested are shown in Figure 4.1.

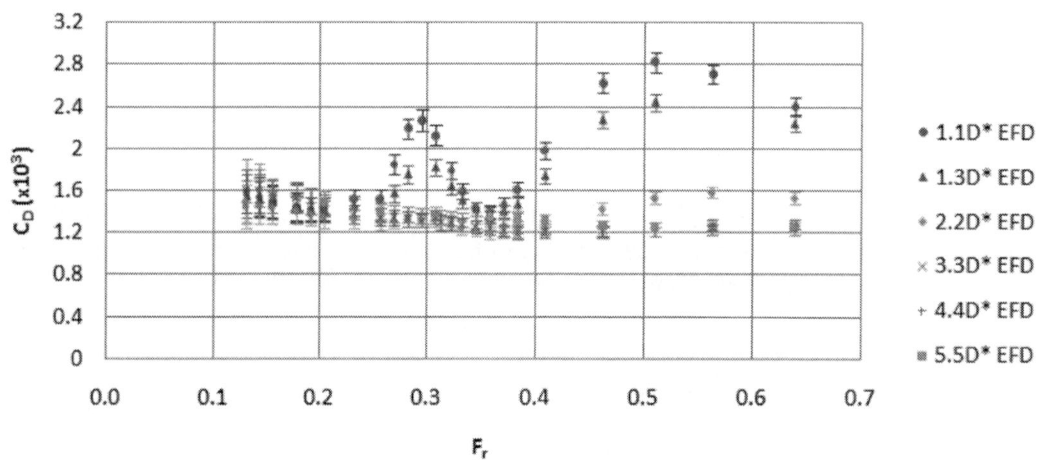

Figure 4.1: C_D as a Function of F_r for all Depths Tested

The drag coefficient, measured both experimentally and numerically as functions of D^*, at Froude Numbers of 0.13, 0.28, 0.36, and 0.64 are shown in Figures 4.2 to 4.5. The maximum variation of numerical results with respect to experimental results was 8.8%, whilst the average variation was 1.2%. Table 4.1 summarises the variation for each depth tested. With the exception of the 2.2D^* condition, the variation clearly decreases with respect to submergence leading to the conclusion that the source of this variation can largely be associated with the simulation of the free surface. The 8.8% variation seen at the 2.2D^* condition occurred at a Froude Number of 0.5, whilst further variations of 8% and 8.1% were seen at Froude Numbers of 0.36 and 0.41 respectively for these conditions. At these Froude Numbers an increase in drag coefficient was also noted in the experimental results when compared to the 4.4D^* case, further supporting the conclusion that the variation can be associated with the free surface.

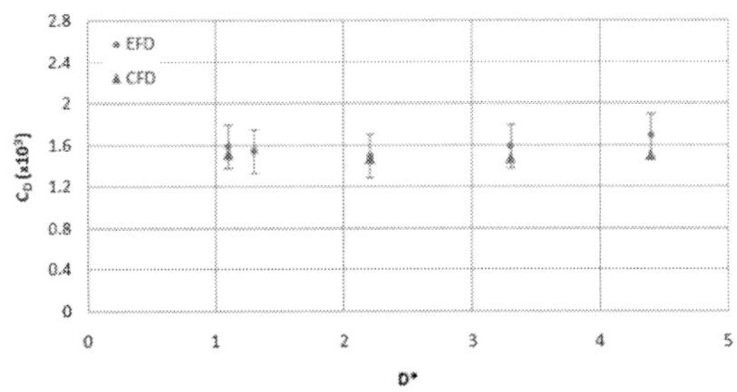

Figure 4.2: C_D as a Function of D* at F_r=0.13

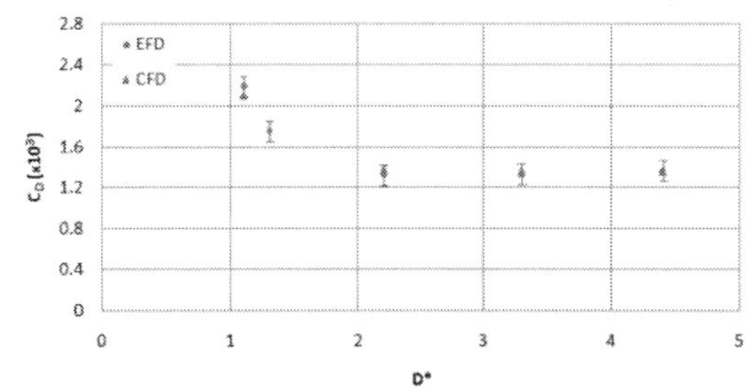

Figure 4.3: C_D as a Function of D* at F_r=0.28

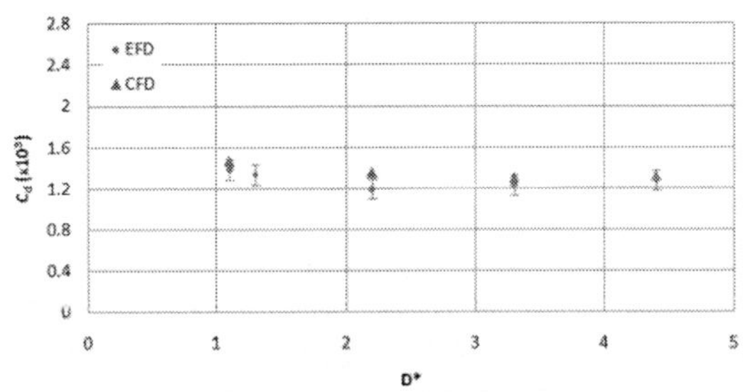

Figure 4.4: C_D as a Function of D* at F_r=0.36

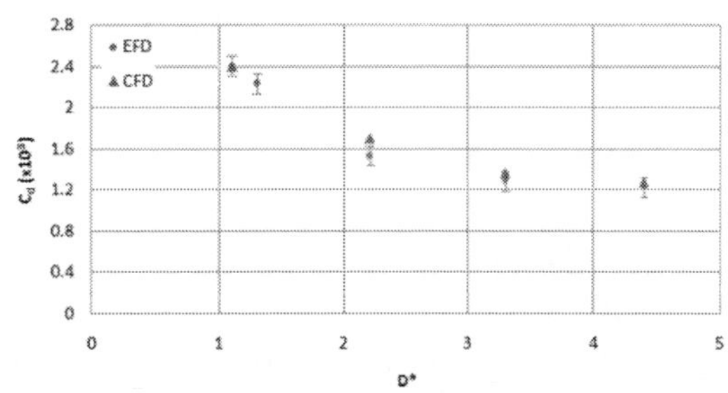

Figure 4.5: C_D as a Function of D* at F_r=0.64

Table 4.1: Variation of CFD with Respect to EFD

Condition	Maximum Variation (%)	Average Variation (%)
1.1D*	3.8	0.7
2.2D*	8.8	4.6
3.3D*	2.2	0.7
4.4D*	0.4	0.1

From Figure 4.1 it is seen that the drag coefficient increases significantly (up to 67 %) at depths of 1.1D^* and 1.3D^* when compared to the 4.4D^* condition. This corresponds to previous studies by Hoerner (1965) and Vine (1991), who both predict the maximum increase at a Froude Number of approximately 0.31. At a Froude Number of approximately 0.36 the increase in drag coefficients is negligible and is assumed to be due to interference effects between the various wave systems acting to almost completely cancel one another out.

It is also seen that very little increase in the drag was noted for the 2.2*D** and 3.3*D** conditions when compared to the 4.4*D** condition; although a slight increase in the drag coefficient is seen between the previously mentioned Froude Numbers of 0.28 and 0.35. Figure 4.6 shows a plot of the drag coefficient for the 2.2*D** condition as a function of Froude Number, with the trend of the 4.4*D** case superimposed as a solid line. An approximate increase of 10 % in the drag coefficient is seen between the 2.2*D** case and the 4.4*D** trend line at a Froude Number of 0.3.

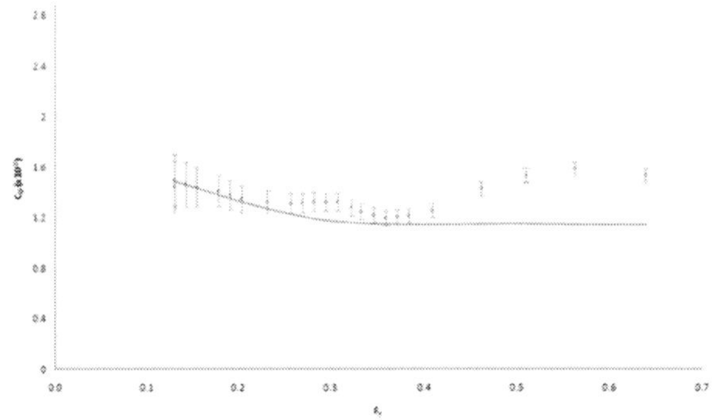

Figure 4.6: C_D as a Function of F_r at *2.2D**
Solid Line Represents *4.4D Condition**

Full Scale Relevance

In order to determine the increase in resistance experienced by a full size submarine, a full scale length equivalent to that of the Australian Collins Class submarine was assumed. Employing the standard ITTC procedure as seen in ITTC Reccomended Procedures (1999), the residual resistance was seperated as,

$$C_{Tm} = C_{Rm} + (1 + k)C_{Fm} \tag{4.2}$$

Where subscript *m* denotes the model scale, C_T is the total drag coefficient, C_R is the residual drag coefficient, C_F is the frictional drag coefficient and *(1+k)* is the form factor.

$$C_T = \frac{2R}{\rho S U^2} \tag{4.3}$$

$$C_R = \frac{2R_R}{\rho S U^2} \tag{4.4}$$

$$C_F = \frac{2R_F}{\rho S U^2} \tag{4.5}$$

Where *S* is the surface area, R_R residual resistance and R_F is the frictional resistance.

The 4.4D* case was then assumed to have no residuary resistance, thereby reducing the equation to,

$$C_{Tm} = (1 + k)C_{Fm} \qquad (4.6)$$

C_{Fm} was then calculated using the ITTC correlation line as,

$$C_{Fm} = \frac{0.075}{(log10R_{nm} - 2)^2} \qquad (4.7)$$

It is acknowledged that the use of this line is not strictly correct as it is a correlation allowance that was developed for surface vessels rather than a friction line, however the purpose of this extrapolation to a Collins sized body is only to compare the approximate percentage that wavemaking resistance contributes to the total resistance. The form factor *(1+k)* was then calculated from equation (4.6).

This process was then repeated for all speeds tested and Figure 4.7 shows the fluctuation of the form factor as a function of Froude Number. The larger fluctuations at the lower Froude Numbers can be attributed to the larger experimental errors associated with this speed range, whilst the clear increase at the higher Froude Numbers may indicate the influence of some free surface effects. An average was then taken resulting in a *(1+k)* of 1.04, which was then used for further calculations.

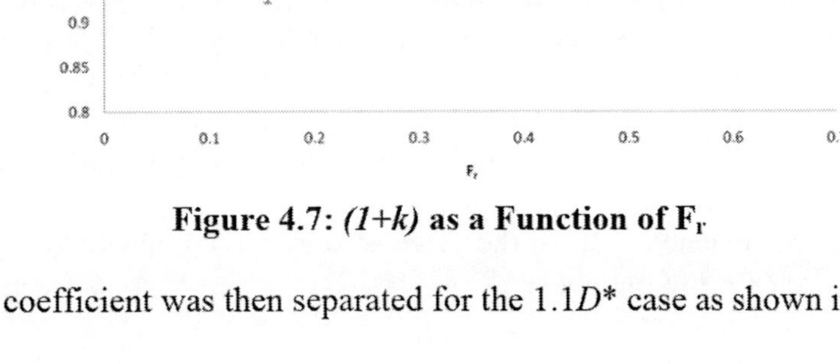

Figure 4.7: *(1+k)* as a Function of F_r

The residuary coefficient was then separated for the 1.1D* case as shown in Figure 4.8.

$$C_r = C_{Tm} - (1 + k)C_{Fm} \qquad (4.8)$$

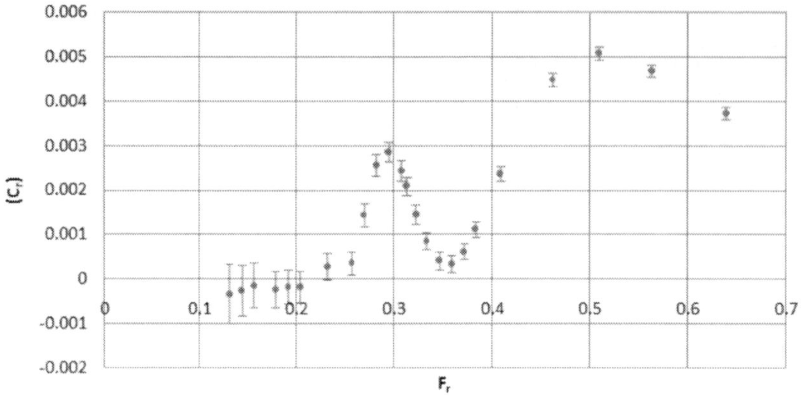

Figure 4.8: C_r as a Function of F_r at 1.1D*

Thus, for a given Froude Number the full scale resistance was calculated assuming that the form factor and residuary resistance coefficient remained constant.

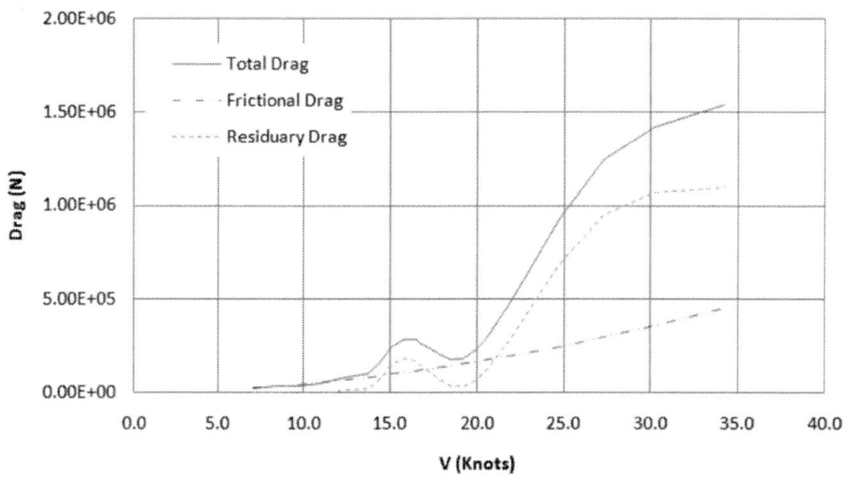

Figure 4.9: Drag Components as a Function of Speed at 1.1D*

Figure 4.9 shows the decomposition of the various components of the resistance for a full scale non-appended submarine travelling at 1.1D*. It is seen that for a given Froude Number the residuary resistance can make up more than 50 % of the total resistance at speeds which can be considered attainable when snorkelling by an SSK, such as the Collins Class. It should be noted that it is, however, unlikely for a typical submarine to be snorkelling so close to the free surface. It is far more likely a submarine will operate in the *2D** to *3D** range where the free surface effects are significantly less. However, the bridge fin may be operating in this region and as the fin is shorter in length, higher Froude Numbers will be reached. It is therefore strongly recommended that this aspect be investigated further in the future.

5. CONCLUSION

It has been shown that the increased drag due to wave making of a submarine travelling just below the free surface is significant for Froude Numbers ranging from 0.27 to 0.35. It has also been shown that the CFD code ANSYS CFX can be used to predict this increase whilst also simulating flow characteristics, such as the pressure distribution and free surface elevation, to an acceptable level. The increased resistance at full scale has also been shown to be significant at attainable speeds for the modern SSK.

Results also show that for the un-appended body the wavemaking resistance below Froude Numbers of 0.35 can be considered negligible at submergence levels at or greater than $2.2D^*$.

7. ACKNOWLEDGEMENTS

The authors would like to thank Mr Gregor Macfarlane and the staff of the Hydrodynamics Centre at AMC and Mr Brendon Anderson and Mr Ronnie Widjaja of DSTO for their contributions and assistance throughout the project.

8. REFERENCES

1. Hoerner, S. (1965). *Fluid - Dynamic Drag.* New Jersey: Author.

2. Groves, N., Huang, T., & Chang, S. (1989). *Geometircal Characteristics of DARPA Suboff Hulls.* Bestheda: David Taylor Research Centre.

3. Sung, C., Griffin, M., & Huang, T. (1996). Validation of Incompressible Flow Computations of Forces and Moments on Axisymetirc Bodies Undergoing Constant Radius Tuirnging. *David Taylor Model Test Basin* .

4. Vine, A. (1991). *An Investigation into the Drag Forces Acting on a Submerged Hull Travelling Close to a Free Surface.* Launceston: Australian Maritme College.

5. Wilson-Haffenden, S. (2009). *An Investigation into the Wave Making Resistance of a Submarine Travelling Below the Free Surafce,.* Launceston: National Centre of Maritime Engineering and Hydrodynamics.

6. Zimmerman, S. (2000). *Submarine Technology for the 21st Century.* Victoria: Trafford Publishing.

CONFIGURATION MANAGEMENT AS A RISK-BASED TOOL IN MANAGING DEPENDABILITY OF COMPLEX SOFTWARE-BASED SYSTEMS

Bernard Twomey & Renny Smith
Lloyd's Register & Lloyd's Register Rail
bernard.twomey@lr.org renny.smith@lrrail.com

SUMMARY

The increasing reliance upon complex systems in marine operations brings a corresponding need to ensure that these systems are demonstrably dependable in use. Furthermore, in order to be truly effective integrated systems they must be developed such that they function as a seamless whole fulfilling operational needs. Ensuring that the integrated systems are traceable throughout the systems lifecycle have been addressed in numerous standards, but the application in the marine sector has been limited, resulting in significant financial loss to the client. This paper will emphasise the importance of shipowners' configuration management (CM) during a systems lifecycle and identifies the benefits of having a CM discipline in place to help manage the risks associated with software maintenance.

Numerous standards have been produced to deal with configuration management and the paper will look at CM with respect to the International Standards ISO 17894 'General Principles for the Development and Use of Programmable Electronic Systems in Marine Applications', and BS ISO 10007:2003, Quality Management Systems -- Guidelines for configuration management.

From the platform of these standards, the paper promotes good systems engineering practice in managing the risks associated with through life software maintenance and seeks to rationalise the ideals with the current practices of the marine sector. In adopting the methodology, it is contended that the goal of effectively managing the automation systems now found in the marine sector can be realistically achieved.

Key CM functions (i.e.,CM planning, CM identification, change control, Impact analysis and CM audit) and their respective roles for newbuild and existing ship implementation are described. In addition, this paper identifies some new opportunities in identifying future risks associated with obsolescence of the programmable electronic systems.

NOMENCLATURE MAJOR HEADING

SCM Software Configuration Management
CI Configuration Item
CR Change Request
SA Status Accounting
IAF Implementation Approval Form

1. INTRODUCTION

Configuration management was first developed by the United States Department of Defence in the 1950s as a technical management discipline. The concepts have been widely adopted by numerous technical management models, including systems engineering, integrated logistics support, Capability Maturity Model Integration (CMMI), ISO 9000, Information Technology Infrastructure Library (ITIL), product lifecycle management, and application lifecycle management. Many of these models have redefined configuration management from its traditional holistic approach to technical management. Some treat configuration management as being similar to a librarian activity, and break out change control and change management as separate areas of discipline, some break out the traditional elements of revision control and engineering release into separate management disciplines; others treat CM as an overarching management discipline. In many other sectors such as the Rail Industry, CM is a mature discipline and applied rigorously in safety related systems

Within the marine sector there is an expectation that the integrated software-based systems for a ship will be realised by using several manufacturers to provide the design solution that will "hopefully" meet the owners specification.

Each manufacturer will have their own quality management procedures for the development of their software based systems, and this will be part of the Classification approval process [1][2] Where standards such as ISO 17894 [3] are specified, there is a requirement in this standards for configuration management.

Naval and passenger ships have traditionally been at the leading edge of many new developments, though the availability of computer technology and component based development have led to software-based systems appearing on virtually all types of ships. As a result the safety operation of ships is becoming dependent to varying degrees on these systems.

Increasing reliance on complex, highly integrated computer-assisted systems presents a problem for most maritime stakeholders. The combination of complexity, use of computer hardware and software, and practices in the purchase, operation and maintenance of computer-based systems leads to a range of potential defects that cannot be guarded against by current marine procedures.

Few ships' engineering staff have an understanding of the implication of software being present in the equipment, and the interaction between the sub systems to ensure the system functions as an integrated whole. Consequently the modern seafarer is becoming reliant on the correct operation of the system to ensure the safe operation of the vessel.

To ensure systems continue to meet the technical and business needs it is usual for software systems to be modified; this modification is referred to as software maintenance. According to Martin and McClure[4] software maintenance must be performed in order to:

- Correct errors.
- Correct design flaws.
- Interface with other systems.
- Make enhancements.

- Make necessary changes to the system.
- Improve design.

In marine applications we also need to consider obsolescence as a cause for maintenance, which would require the introduction of new technology that may be required to interface to older systems and software platforms.

To ensure the systems continue to meet the operational requirements of the owners, consideration should be given to managing these changes throughout the system lifecycle; failure to do so has resulted in significant financial loss and the potential for physical or environmental damage.

2. CM AS A PRACTICAL RISK REDUCTION TOOL

Very few operators have processes in place to properly control the changes to embedded software in their equipment and systems and reliance is placed on the manufacturers and OEM's to provide this process.

It is not sufficient to simply track the modification request; the software product and any changes made to it must be controlled. This control is established by implementing and enforcing an approved Software Configuration Management Process (SCM). For this process to work, management must support the concept of software configuration management. The SCM process is implemented by developing and following a configuration management plan which includes all system documentation, base line software in the systems and procedures for managing software changes, thereby providing tools for the ships' staff to manage changes to their systems.

A fundamental tool to support SCM is impact analysis, where the effect of change within the system and on interrelated systems is analysed and the risks associated with the change is assessed.

Software Configuration Management process is therefore seen as a risk mitigation tool that will enable the owners to successfully identify and manage their risks throughout the systems life.

SCM is part of the software maintenance process, so as to give the context of SCM, further discussion of maintenance is given below.

2.1 Why is there confusion about Software Maintenance?

Software maintenance is a much maligned and misunderstood area of software engineering. Although, software has been maintained for years, relatively little is written about the topic of maintenance. Funding for research about software maintenance is essentially non existent [4], thus, the academic community publishes very little on the subject. Practitioners publish even less because of the corporate fears of giving away the "competitive edge". There are a few text books on software maintenance but these are outdated. Periodicals address the topic infrequently and this lack of information contributes to the misunderstanding and misconception that people have about software maintenance.

2.2 What is software maintenance?

What then is software maintenance? And where does it fit into the lifecycle? If we consider some generally accepted traditional descriptions of software maintenance like;

(a) "...changes that have been made to computer programs after they have been delivered to the customer or user" [4].

(b) "...the performance of those activities required to keep a software system operational and responsive after it is accepted and placed into service." [6].

(c) "Modification of a software product after delivery to correct faults' to improve performance or attributes, or to adapt the product to a modified environment." [6]

(d) "...software product undergoes modifications to code and associated documentation due to a problem or the need for improvement. This objective is to modify existing software product while preserving its integrity [7].".

The common theme of the above descriptions is that maintenance is an "after the fact" activity. Based on these definitions, no maintenance activities occur during the development process. Maintenance occurs after the product is in operation.

2.3 What is the impact of not managing the risk?

A recent incident took place on an LNG vessel that resulted in significant financial loss to the owners. The report from the OEM highlighted the fact that the owners were reliant on the manufacturer's SCM procedures being applied. The result of this failure was the ship coming off hire at considerable cost to the owner ($Millions).
"The upgrade of the reliquifaction software was requested by the shipowner in conjunction with the reliquifaction process system supplier" ."it is very important to understand that the upgrade was not a standard operational procedure and requires secure high level access to the software based system and detailed planning and documentation" "...prior to import of new software pre system configuration status audit was not undertaken to establish a firm base line. Pre works backups were not made, import procedures were not fully documented and associated technical risks not advised to the ship owner"

Unfortunately, reports of this nature are becoming more frequent due to the increase in software based systems being installed in marine applications.

Upon investigation, the above incidents could have been avoided if the shipowner had a SCM process in place to identify and manage the risks associated with this upgrade.

In land based applications, there are numerous reported failures due to software upgrades

- *In 2004, PayPal based in San Jose, Calif., and owned by eBay Inc., is the major online payment system for eBay auctions as well as smaller e-commerce sites. The company suffered intermittent outages. AlertSite Inc. reported that PayPal was available only 35 percent of the time. Keynote Systems Inc. also was unable to connect to the site from 10 locations nationwide. PayPal is blaming the outages on the software update, and officials did not know how many of its 50 million user accounts were affected*

- *Thanksgiving holiday weekend in 2001, Amazon.com suffered a series of outages, due to a software upgrade which cost the retailer an estimated $25,000 per minute of downtime.*

We have to accept that certain operators don't know what software is in their system and the training requirements of STCW [8] fall far short of providing adequate training for the operators to address the complexity now found on modern vessels. The MAIB report[9] on the MV Prospero, another vessel suffering an incident caused by software modifications, identified that the basic elements of STCW training for engineering officers is unlikely to equip them to deal with modern technology.

2.4 Establishing control

The approach described for managing software change and introducing SCM into an organisation have been trialled and implemented by a number of ship owners. Each organisation has adapted the procedures to take into consideration their own risk identification processes and management structure, the benefit of this approach has been a SCM procedure that is tailored for each organisation and aligns with their existing processes.

2.4.1 Ships' configuration management procedure

This process, based on BS ISO 10007:2003[10], applies to specified programmable electronic systems on board a vessel and require information to be provided by suppliers of systems but the procedures dose not apply to the configuration management used by suppliers.

The Configuration Management of software, firmware and configuration data is achieved by controlling the media upon which they are held. For the type of systems used onboard typical examples are computer files for personal computer based systems, electronic memory devices, such as EPROM, embedded electronic cards and removable magnetic and optical media.

Whilst the procedure can be applied across the whole fleet, each ship has a ship specific Configuration Management Plan, which is used to document the local arrangements for the implementation of the procedure.

For newbuild projects, ideally, the configuration management process for a particular ship's system should be brought into use when one or more of the following criteria has been met:

- the ship's system is first integrated with other ship's systems;
- the ship's system is brought into operational use;
- the approval for the ship's system has been given using the New System Development

During the implementation phase suppliers may make changes to their systems under their own configuration management systems without invoking the ship's configuration management processes provided, the systems are not integrated with other systems and have not been put into operation.

In practice, because shipowners' do not have responsibility for, or formal control over systems on ship until formal hand over, it is not possible to mandate the ideal configuration management processes for newbuild projects.

The actual configuration management arrangements that can be achieved for specific newbuild project will be recorded in a newbuild Configuration Management Plan for the ship. Shipyards will not accept a manufacturer's SCM procedure being imposed on the whole project, so it is important that the shipowner specifies SCM requirements at the contract stage with clearly defined deliverables at the time of delivery of the vessel or system. As a minimum the shipowner should:

- Define the version control information that it requires from system suppliers;
- Define and implement its own activities necessary to have confidence in the configuration information that will be provided when systems are eventually handed over. Activities which should include defining configuration management information required from supplier, its oversight activities, and configuration audit process;
- Seek the version control information previously defined from system suppliers before accepting the handover of the vessel.

When a ship is commissioned into service (handover of the ship) the New Build Configuration Management Plan for the ship will be replaced by the ship's In-Service Configuration Management Plan.

Configuration management is only one of the processes required for the proper management of new and changed systems on a ship. The other processes, such as systems engineering, contract, commercial, and project management processes are normally not addressed by this procedure. It is also assumed that there will be cooperation and coordination of activities, of the different processes and that the outputs from the configuration management process will be used to support some of the other processes. For instance, the output from the impact analysis, which is a key part of the configuration management process, could be used as an input to the risk assessment process that should be performed as part of the best practice processes for the development of programmable electronic systems in marine applications, as specified by ISO 17894 [3].

The procedures are normally written in a style that assumes equipment is installed on ship and the necessary configuration management information is available and within the ship's configuration management system.

For existing ships, which have not implemented this procedure, a separate document that plans and controls the introduction of the procedure will be required.

The basic configuration management model used in this procedure is represented diagrammatically by Figure 1.

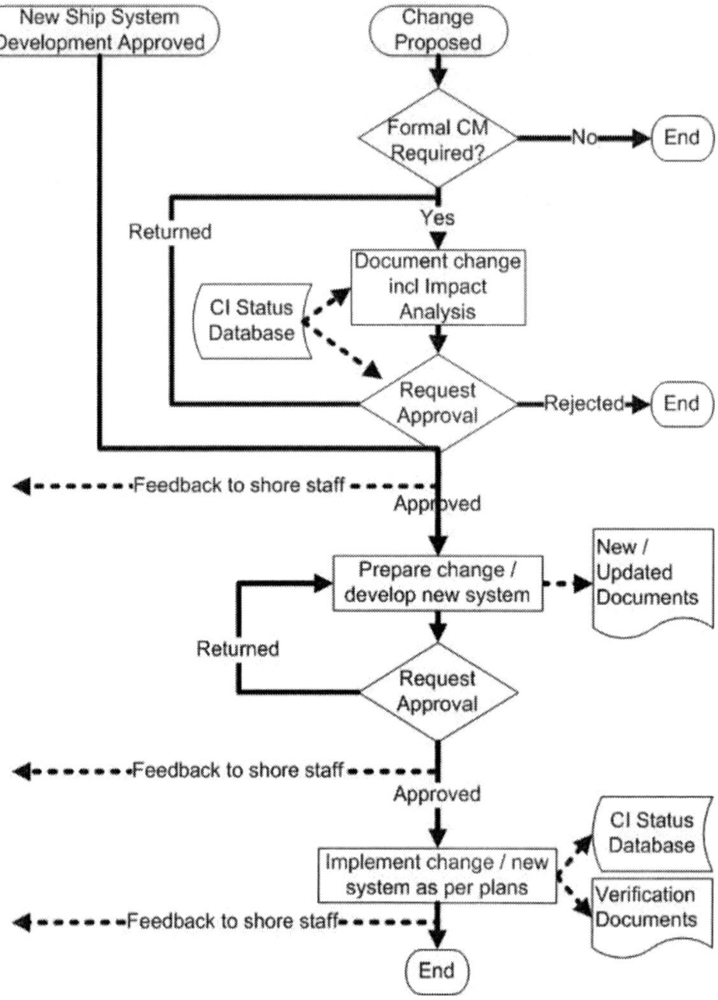

Figure 1 Configuration management model

Items that are required to be held under configuration control are referred to as configuration items (CIs).The model assumes that;

- All CIs are defined and recorded in a database;
- All documents and records related to the configuration items that are in existence and available;
- Staff with the competence and authority to assess and approve requested changes are available.

2.4.2 Documenting Proposed Changes.

The decision to invoke SCM for a particular change needs to be evaluated in accordance with clearly defined and agreed criteria as not all changes require the level of rigour required by these procedures, e.g. tuneable parameters (set points, hysteresis, etc. deliberately provided to be changeable by ships' staff). For these arrangements, alternative mechanisms to the configuration management procedure need to be in place to ensure the initial or default state of the parameter is defined and recorded and that when changes are made, these are recorded and communicated to other staff using the system/equipment.

The limits of allowable change should be defined and staff should be aware of the limits of changes.
Before a change is formally raised as a "proposed change" and subsequently processed in accordance with this procedure:

- a change request shall be produced for all proposed changes;
- initial analysis of the impact of the change shall be performed for all changes.

The scale of and the risk associated with a change shall be taken into account when specifying the proposed list of deliverables on the Change Request; less complicated changes such as changes to application data may be performed by local agreement with ships staff with minimal planning documentation, whereas replacement of equipment to prevent obsolescence, would need stringent and detailed planning.
The model shown in Figure 2 outlines a process for managing software maintenance. There are three key areas for consideration, the data required to identify and control the change, the stakeholders' responsibilities and guidance to ensure the risks associated with software maintenance are successfully managed.

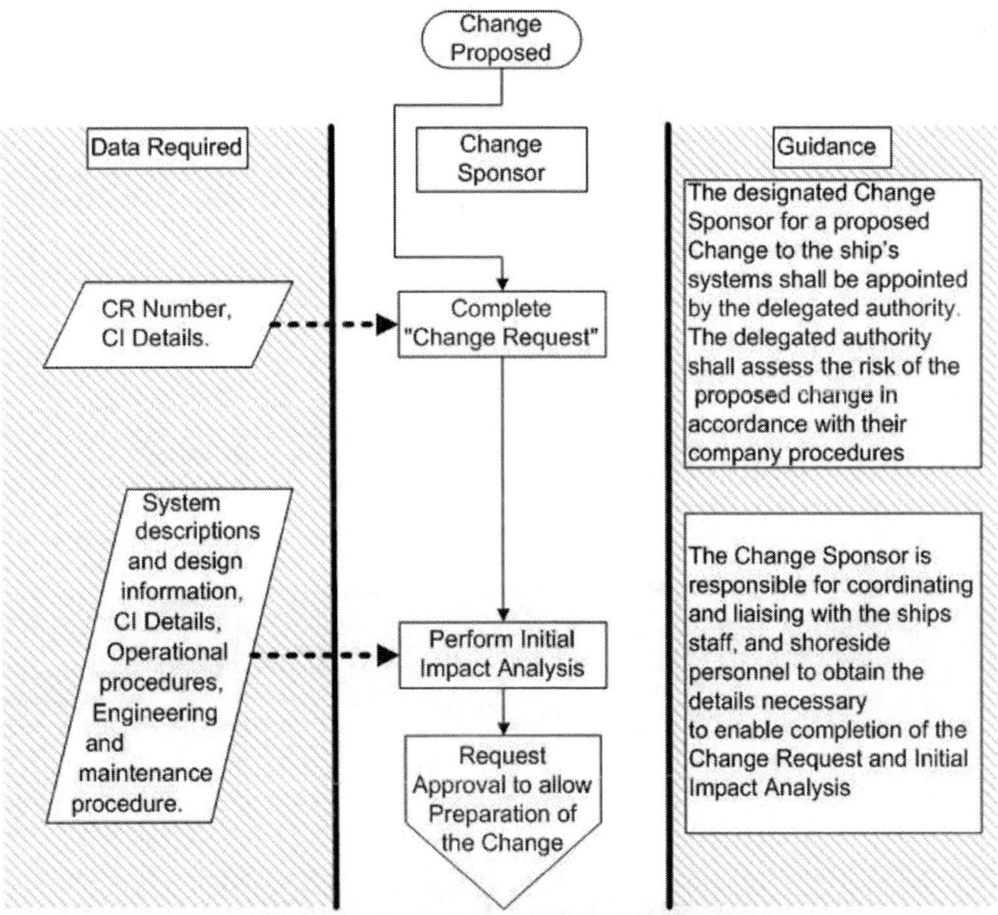

Figure 2 Documenting Proposed Change

2.4.3 Approving Preparation of the Change (see Figure 3)

The next stage in the process is approval of the Change Request (CR). Approval responsibility for a CR shall be assigned to a competent person based on the results of the risk assessment performed at the time of preparing the documentation for the change. In this example, we have chosen the Chief Engineer, but this can be modified to deal with companies specific processes.

The approval process shall be conducted in a way that ensures that the CR and supporting impact analysis is reviewed by persons competent to assess the information being presented.

If the review of a change request raises doubts about the adequacy of the information available relating to the change, consideration should be given to performing a risk assessment of the proposed change before approval.

When a decision has been reached, the approval authority shall record it on a Preparation of Change Approval Form, together with any additional information including, the reasons for rejection, the reasons for returning the Change Request and an explanation of what needs to be done before approval can be given, or any conditions that apply to the approval.

A copy of the CR, impact analysis, supporting information, and Preparation of Change Approval Form, shall be provided to shore staff, based on the results of the risk assessment performed at the time of preparing the documentation for the change

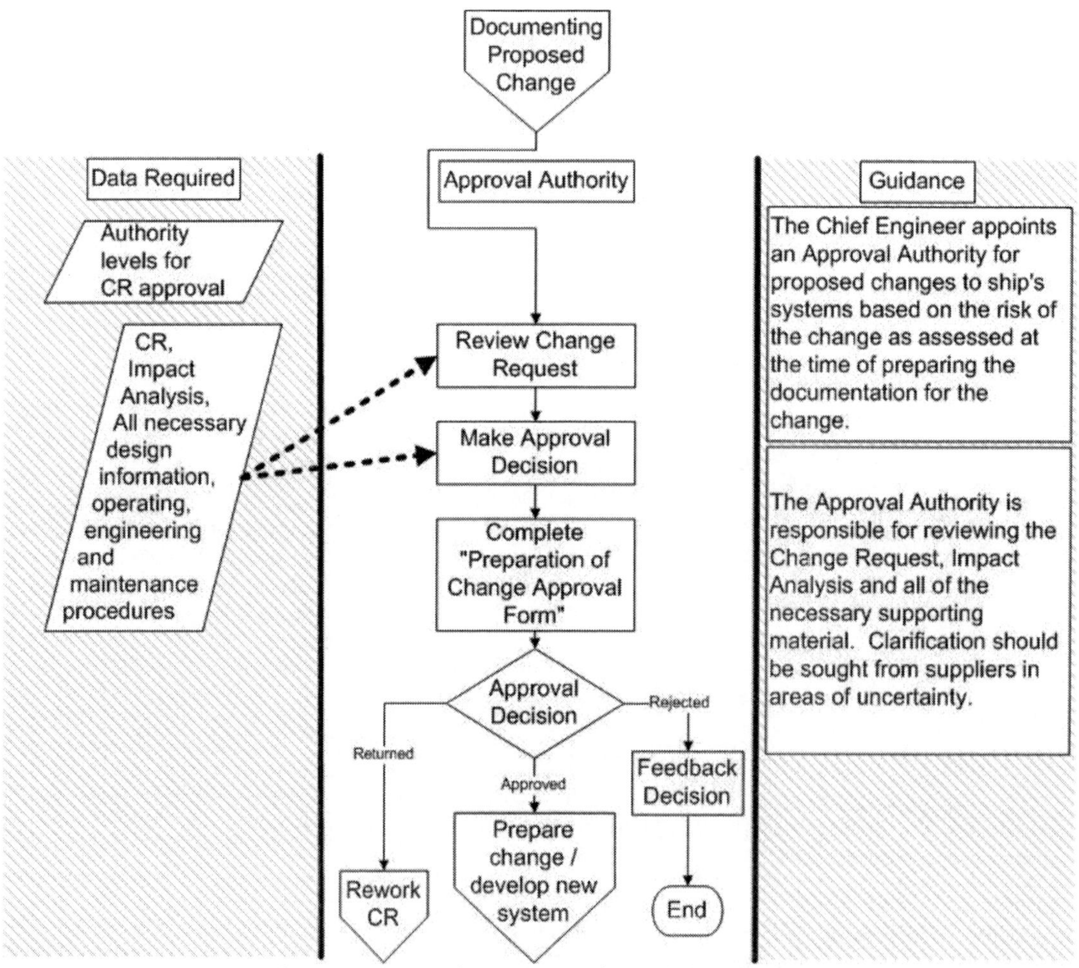

Figure 3 Approving Preparation of the Change

2.4.4 Prepare change/ develop new system (see Figure 4)

Once the approval has been given to develop a change, or a system for a new ship, action shall be taken to prepare all necessary systems, equipment, products, processes, documentation and records in accordance with the details specified in the Change Request/New System Development Approval Form.

The level of management and oversight activities to be performed by the provider of the system and the ship owner's staff, or their agents, shall be agreed, as appropriate to the scale of the change/development and the level of risk involved. The arrangements shall be recorded as appropriate, typically in one of the deliverables proposed in the Change Request/New System Development Approval Form, such as a project or quality plan.

This part of the process introduces requirements for the supplier to develop the approved change/new system and all agreed deliverables in accordance with the approved CR.

During their production, the implementation stage documents, such as Installation, Testing and Commissioning Plans, shall be agreed with ship's staff to ensure that:

- Adequate provision is made for the operation of the ship during the implementation stage;
- Fallback and transition strategies have been prepared and are adequate;
- The testing provides sufficient coverage to demonstrate the success of the changes/development.

On completion of the development activities, but prior to installation, the Chief Engineer shall inform the approval authority of the status of the development and arrange for the approval authority to assess the readiness of the change/development for implementation.

Problems that arise during this phase which may require a change to be made to the functional, performance, or contract requirements of the system being changed or developed, shall be treated as a change and shall be managed using this procedure.

For newbuild, where the shipowner has no formal responsibility or control over suppliers' attempts should be made to influence them to adopt or cooperate with the shipowners' process on a voluntary basis.

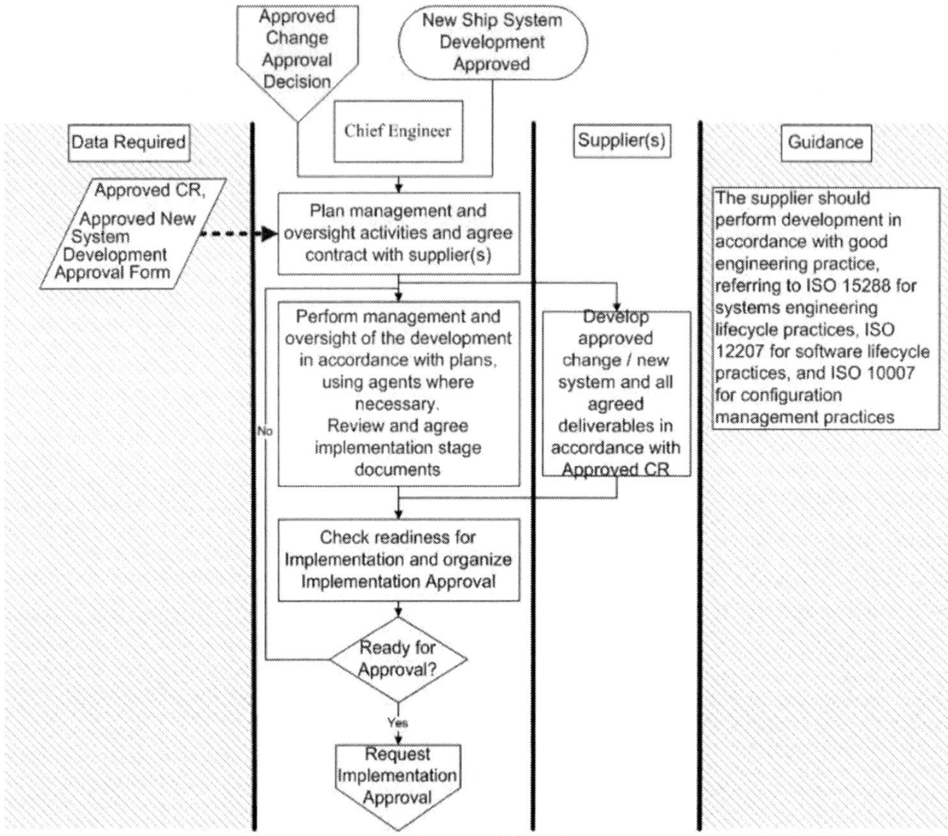

Figure 4 Prepare Change / develop New System

2.4.5 Approval for Implementation (see Figure 5)

On notification that the development activities are finished, the approval authority shall make arrangements to assess the completeness and readiness of the change/development for implementation on ship.

When a decision has been reached, the approval authority shall record it on an Implementation Approval Form (IAF) (to be developed by the shipowner) together with any additional information, including, the reasons for the return (non approval) of the IAF and an explanation of what needs to be done before approval can be given, or any conditions that apply to the approval. Clarification should be sought from the suppliers in areas of uncertainty to ensure there is full understanding of the risks. If necessary, after clarification, re-approval may become necessary before implementation takes place.

If it is considered necessary for a configuration audit to be performed this shall be specified on the IAF.

For newbuild, where the shipowner has no formal responsibility or control over suppliers, the process defined in this section should be applied, as far as practicable, given the limitation of available data and documentation.

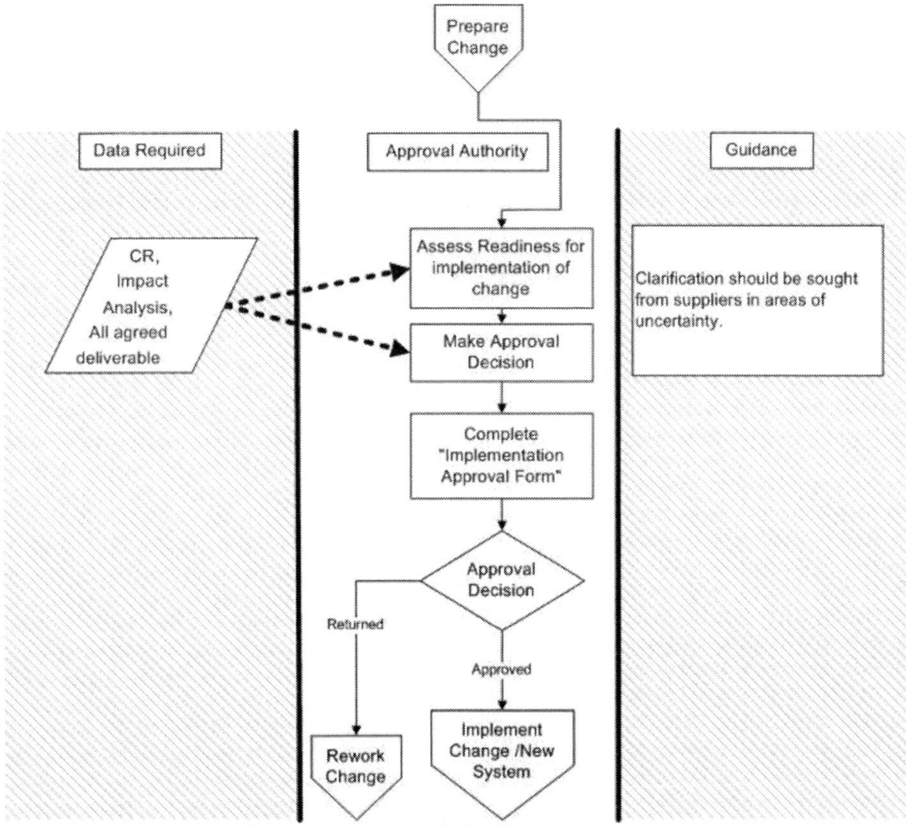

Figure 5 Approval for Implementation

2.4.6 Implementing changes

The software implementation shall be installed, tested and commissioned in accordance with the plans produced during the development phase.

Installation, testing, and commission requires a high degree of communication and co-ordination between ship's staff, shore-based ship's team staff, potential multiple suppliers and installation organisations, and other agencies such as Classification Societies. The activities should be closely overseen and managed by the ships staff to ensure that configuration management good practice is maintained, particularly in respect of on-ship changes.

Temporary changes outside formal configuration management process should be exceptions and only allowed with approval from the Chief Engineer

The supplier/installer shall co-ordinate its activities with ship's staff to ensure that the revision status of all systems, equipment, and products is known to the ship's staff and suppliers of other interfacing systems that are implementing their systems at the same time. This should allow the ship's staff to use the operational and maintenance procedures appropriate to the current status and minimize incompatibility of integrated systems.

During the implementation phase suppliers may make changes to their systems under their own configuration management systems without invoking the ship's configuration management processes provided the systems are not integrated with other systems and have not been put into operation.

All systems shall be put under the ship's configuration management procedure before being brought into operational use or integrated other systems.

Typically as it is difficult to maintain good configuration control during this phase, it is therefore recommended that a member of the ship's staff is assigned to monitor what changes are made and to promote good configuration management practice (this role need not be an exclusive roles and could be one of a number of roles held by the staff member.

Following each discrete stage of implementation (if there is more than one) the CI Status Database shall be updated to ensure that the integrity of the database, with the installed equipment, is maintained throughout the implementation activities.

If a configuration audit has been specified in the IAF, the performance of the audit shall be co-coordinated with the implementation activities to ensure that the audit records show the final status of the systems after changes have been finalized.

On successful completion of the implementation, and any remedial action, copies of all changed or new software, firmware and configuration data should be made and held as archive versions on ship and by the relevant manufacturer or supplier. The Chief Engineer shall be responsible for the management and documenting of archiving and backup.

When the implementation has been completed to the satisfaction of the installer and the ship's staff, the Chief Engineer shall be informed.

Problems that arise during this phase which may require a change to be made to the functional, performance, or contract requirements of the system being changed or developed shall be treated as a change and shall be managed using this procedure.

For new build, where the shipowner has no formal responsibility or control over suppliers', attempts should be made to influence the suppliers to adopt or cooperate with the process defined in this section on a voluntary basis. As a minimum, version control information for all systems should be sought before accepting handover of the vessel.

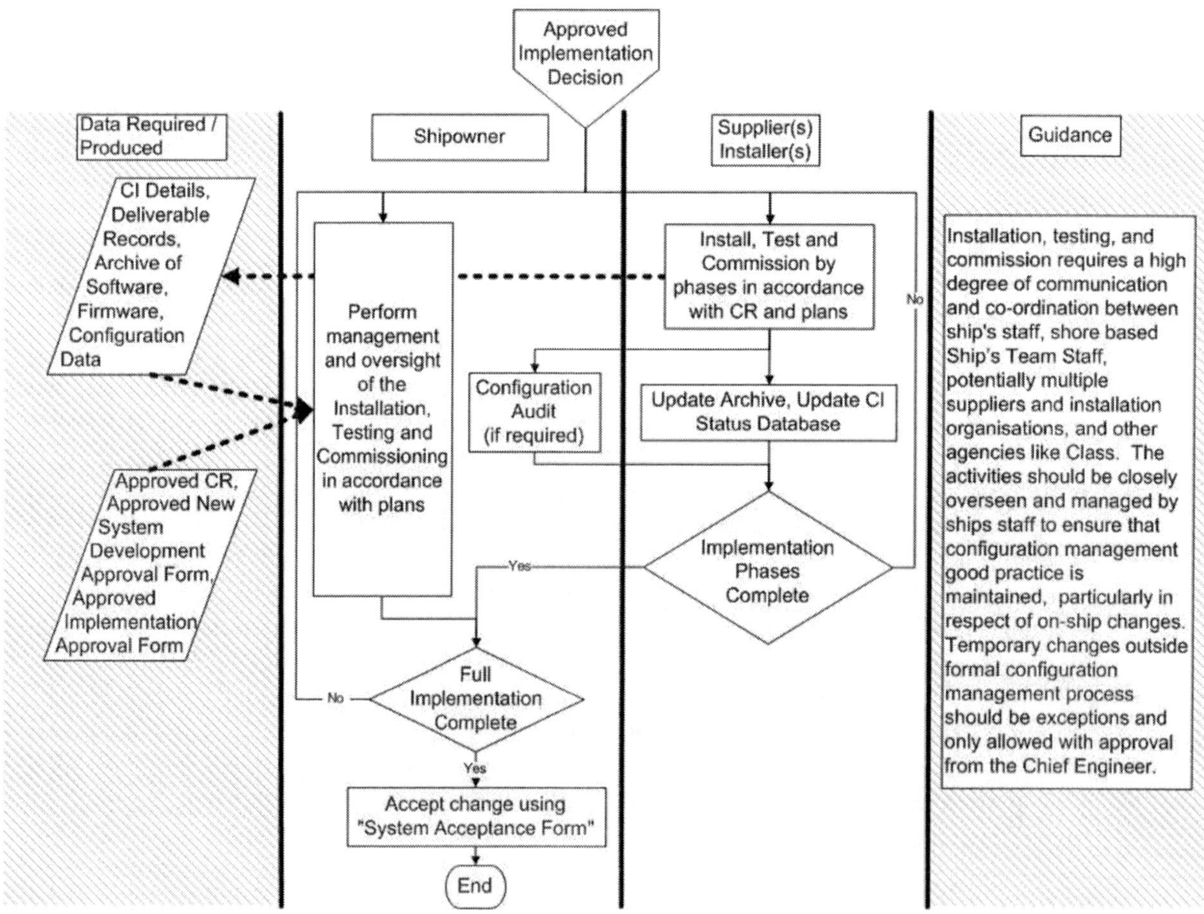

Figure 6 Implementing Change

2.5 Impact Analysis

Impact analysis is the analysis of the effects of a proposed change on existing ship's systems. The term ship's systems are used in the widest sense to include equipment, processes and people[1].

---

[1] systems, which may include such things as:

- electrical systems;
- propulsion systems;
- monitoring, alarm and control systems;
- hotel equipment and services.

These physical systems will include electrical, electronic, mechanical, hydraulic, programmable and data configurable subsystems, and components.
In order to use the equipment effectively and correctly, processes are developed and implemented. These are typically recorded in a suite of operating procedures and maintenance procedures. At the highest levels of the operating procedures, such as hotel services, the link with the physical equipment might not be explicit. However, failure of the hotel services infrastructure, such as the hotel computer systems, public address, or fire detection systems, will all have an effect on the operational procedures used to manage the guests.
The people element of a system concerns the competence of the staff to operate and maintain equipment and implement processes effectively and correctly. Competent staff are established by:

- the selection of people with adequate education and the correct underlying skills, knowledge and personal characteristics for the functions they are required to perform;
- training staff in the processes and procedures to be used for the functions they perform;
- monitoring performance of staff and providing additional training, education and experience as required.

physical equipment might not be explicit. However, failure of the hotel services infrastructure, such as the hotel computer systems, public address, or fire detection systems, will all have an effect on the operational procedures used to manage the guests.
The people element of a system concerns the competence of the staff to operate and maintain equipment and implement processes effectively and correctly. Competent staff are established by:

For any software change, it is usual, and expected, that the supplier analyses the effect of the proposed change on the complete software to ensure that there will be no unintended effects of the change. The analysis guides the way the changes are implemented and tested.

Similarly, at a system level the effects of proposed changes, whether equipment, process or people related, should be considered on the other parts of the system.
The responsibility, and skills necessary, to perform the impact analysis on the internals of a software system lie with the supplier and is not included in this paper.
The analyses of the effects of software changes on the ship's systems is the subject of this paper. The change sponsor has been given responsibility to ensure that the analyses are performed, but to do that the Change Sponsor might need the support of people with knowledge and experience from the ship's staff, shipowners' shore based staff, and suppliers of other systems.
Impact analysis involves considering the effect of the change proposed in detail on all elements of ship's systems. Ultimately, it is the approval authority for the change that decides whether sufficient analyses has been performed. His decision should be based on the criticality to the safe, effective and profitable operation of the ship, of the system being changed and the other systems affected by the change. The analyses should continue until it is considered that the risks of not having identified problems associated with the change are tolerable.

- the selection of people with adequate education and the correct underlying skills, knowledge and personal characteristics for the functions they are required to perform;
- training staff in the processes and procedures to be used for the functions they perform;

monitoring performance of staff and providing additional training, education and experience as required.

2.6 Benefits

The key theme of this paper has been the mitigation of risk associated with software maintenance by implementing SCM. The most obvious benefit of mitigating software maintenance risk is the correct and continued operation of the systems containing software. While the operation of systems and equipment might be the primary focus of the ship's engineering staff, management sees other benefits from implementing SCM.

As discussed earlier in this paper, the failure of software to operate correctly can have very expensive consequences when the operation of a vessel is affected. The mitigation of risk through SCM, with the resulting improvement in the success of software maintenance, is therefore an important contribution to minimising cost of ownership.

Generally implementing SCM returns control of software based system to shipowners', specifically removing the consequences of unilateral and poorly considered changes by suppliers. By taking up the control provided by SCM, shipowners' will be better able to understand the implications of software maintenance being carried out on their behalf by suppliers and contractors.

SCM also provides a demonstration of due diligence in respect of control over software maintenance which will address questions that are starting to be raised by the national authorities, such as the MCA, USCG, accident investigation boards such as the MAIB and NTSB and legal companies who are questioning the shipowners' SCM processes in the event of an incident.

A number of shipowners' have linked the SCM procedures into their planned maintenance system and through this process are tracking obsolescence in their installed systems. This approach provides a significant benefit by enabling the shipowners' to form and implement an appropriate obsolescence strategy.

3 CONCLUSIONS

Not managing the maintenance of software based systems properly can have serious consequences both in terms of operational control of ships and consequential financial losses.

SCM is an important tool that can be used to mitigate some of the risks associated with software maintenance. The resulting benefits include optimising the availability of software systems, minimising the cost of ownership, returning control of change to the shipowner, and demonstrating due diligence to National Authorities.

The procedures described in this paper are expected to be tailored to the specific needs of shipowners' taking into consideration their current organizational structure, existing risk mitigation practices and the competence of the ship's staff.

The implementation of SCM procedures will ensure that shipowners' are able to fully understand the implications of the software maintenance being carried out and that they have control over their systems.

6. REFERENCES

1. Lloyd's Register Rules and Regulations for the Classification of Ships, Part 6, Chapter 1
2. Lloyd's Register Rules and Regulations for the Classification of Naval Ships, Vol 2, Part 9.
3. ISO 17894 General Principles for the Development and Use of Programmable Electronic Systems in Marine Applications, 2005
4. J Martin and C McClure. Software Maintenance: The Problem and its Solution. Englewood Cliffs, NJ: Prentice-Hall
5. TM Pigoski, Practical Software Maintenance, ISBN 0471170011
6. IEEE 1219: Standard for Software Maintenance: 1998
7. ISO/IEC 12207 Systems and software engineering -- Software life cycle processes, 2008
8. International Convention on Standards of Training, Certification and Watchkeeping for Seafarers STCW, 1995 Amendments
9. MAIB Report no 24/2007. Published 21 December 2007
10. BS ISO 10007:2003, Quality management systems -- Guidelines for configuration management
11. ISO/IEC 90003, Software engineering – *Guidelines for the application of ISO 9001:2000 to computer software*
12. RTCA/DO-178B, "Software Considerations in Airborne Systems and Equipment Certification," December 1, 1992.
13. Boeing Co, ' The impact of configuration management during the software product's lifecycle', Digital Avionics Systems Conference, 1999. Proceedings.
14. DOD-STD-2167A, "Military Standard Defense System Software Development," AMSC No. N4327, February 29, 1988.
15. AC Messer, System Integration and Complexity: managing dependability. Lloyd's Register.
16. MIL-HDBK-61A(SE) Configuration management Guidance, 7 February 2001
17. ANSI/IEEE Std 1042-1987 IEEE Guide to Software Configuration Management, IEEE Standards Board, September 10, 1987
18. IEEE Std 828-1998, IEEE Standard for Software Configuration Management Plans, 25 June 1998
19. California power outages suspended--for now, by Rachel Conrad, *CNET news.com*, January 18, 2001 http://news.com.com/2100-1017-251167.html

Perils in the Pacific: Solutions for Interisland Shipping in the Pacific

J Knox, Lightning Naval Architecture Pty Ltd, Sydney, Australia
navlight@bigpond.com

THE ISSUES

There are over 20,000 islands scattered throughout the Pacific Ocean. The populations on many of these islands are geographically isolated. They rely on sea transport for access to the outside world: for transport of goods and people and for economic development. Sources of economic development are limited: mainly trade in copra and other agricultural products, garment manufacture (Fiji) and tourism. A number of the economies depend on remittances from citizens in the US New Zealand and Australia

Due to the Pacific Ocean atoll geology, access to some islands is limited to using small boats; therefore stevedoring is undertaken in hazardous open-sea conditions. Where ports are accessible they are often basic and have limited facilities. The economic status of some of these nations means that low budget solutions have been implemented in a piecemeal way over time: the majority of the vessels in use were not originally designed for their current conditions. Passengers are sometimes transported on barges with fuel and other goods. Many interisland ferries were built to be operated on short trips in protected waters rather than voyages of more than one day in the open seas.

There are therefore many issues of safety and comfort, reliability and speed, as well as the need for cost effective solutions. It is not possible to speak of hardware solutions in isolation as any boat or ship is only part of a transport service. All components of the system need to be considered including provision of berthing infrastructure or alternative landing facilities, safe manning of vessels, sound maintenance programs and strong regulatory control. The diversity of arrangements in different island nations makes it impossible to cover all the various situations in this text. The paper will instead examine some examples of current and proposed operations. Firstly it will consider what is not working in one island nation and then it will consider an innovative transport scheme owned and operated by rural islanders.

A MARITIME DISASTER WAITING TO HAPPEN: THE LOSS OF PRINCESS ASHIKA

On August 5 2009 the fifty-metre (690GT) open-deck Ro-Ro passenger ferry Princess Ashika capsized and sank in seas of less than one metre on a night-time passage between Nuku'alofa and Ha'afeva with the loss of 75 lives. Many of those who died were women and children who were trapped inside the accommodation block on the vessel. The vessel was 86 km north east of Nuku'alofa more than five hours into its passage when the tragedy occurred. There were 54 survivors all of whom were men and only two bodies were recovered. The owner and operator, Shipping Corporation of Polynesia Ltd had stated less than two months before the disaster that the ship was in "good" condition and that it had been well maintained. Shipping Corporation of Polynesia is a state-owned corporation.

The Kingdom of Tonga has around 102,000 people living on 36 islands (of total of 169). The Princess Ashika had arrived in Tonga on 1 July 2009 from Fiji and commenced operations on 4 July when she carried 360 passengers from Nuku'alofa to Ha'apai and Vava'u. The voyages in Tonga were of many hours duration, often overnight and in open ocean conditions. From evidence presented to the current Royal commission of enquiry we can see that the vessel's operation in Fiji islands had been greatly restricted by Fiji Maritime Safety Administration. That evidence included surveys that had been carried out by the Fiji Islands Maritime Safety Administration, none of which were sighted prior to purchase. In the months before it was purchased, the vessel had only legally been allowed to travel in Fiji on a route of less than one hour in duration in sheltered waters and it could only sail with "reduced cargo". It was also only to travel at a "reduced speed" and carry a maximum of 160 passengers.

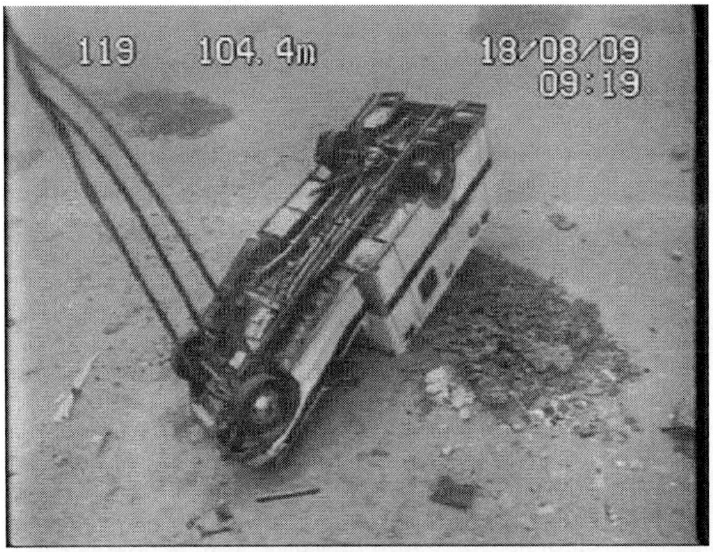

Further evidence indicates that at least one senior marine engineer with the marine and ports division of the Tongan ministry of transport who surveyed the vessel on July 2 i.e. once it had arrived in Tonga attempted to have the vessel's permit to sail withdrawn however his requests were not successful.

Defects in the vessel included:

1. Cargo deck (main deck) and surrounding shell plating was heavily corroded with holes visible at many places;
2. Bow and stern ramps were heavily corroded, not watertight, poorly secured with unsatisfactory improvised locking arrangements. On the stern ramp there were no locking devices; there were dozens of holes in the outer plating of stern ramp; on the bow ramp rubber pads had been fixed to the outer surface to stop leaks;
3. Weathertight doors on the cargo deck leading to spaces below corroded and with no sealing surfaces;
4. Hatch on cargo deck to void space below with rusted off hinges and no securing mechanism;
5. Holes in watertight bulkheads below cargo deck;

6. Scuppers on cargo deck missing non-return valves, blocked with debris or with size reduced by plating over;
7. Corroded hand rails, missing gates in hand rails in way of evacuation station;
8. No visible load line or draft marks
9. No evident safety management system onboard;

Full details of the shocking state of this vessel and of the chilling impact of that condition are available in the transcripts of the royal commission which includes survivors' testimony.[1] Briefly an accumulation of water on deck, which entered though the bow door, and subsequent cargo shifting caused the vessel to capsize trapping the majority of passengers in the accommodation block from which they were unable to escape.

Who will take responsibility for these preventable deaths? Will the ship owner's management admit that one cause of the loss of life was lack of due diligence in examining the vessel and her documentation prior to purchase? Will the company's board admit that one cause of the loss of life was their lack of oversight in approving the purchase? Will the head of the Marine division of the Tongan Ministry of transport admit that one cause of the loss of life was their lack of action in withdrawing the vessel's permit to sail? Will the master admit that one cause of the loss of life was his lack of emergency training so that he did not take action until the situation reached catastrophic proportions? Will the master explain why he had not issued lifejackets to all passengers? (The master has yet to give evidence to the commission of enquiry.) What part all these elements played in the avoidable deaths will hopefully be judged by the commission of enquiry and strong action taken to prevent such a terrible situation again arising.

The Minister of Transport resigned his ministry following the sinking. The CEO of Shipping Corporation of Polynesia was suspended from his position following around one week of hearings of the Royal commission of enquiry. The enquiry is expected to report around January 2010.

A Japanese government aid program is delivering a new built ferry from a Japanese shipyard to Tonga sometime in 2011.

SOLOMON ISLANDS: PELENA EXPRESS

Solomon Islands has a population of around 507,000 living on 330 inhabited islands; sea transport is critical to society and economy.

The Pelena Express is a recent innovative concept for a transport service in Solomon Islands. Pelena Express Limited was registered in Solomon Islands in April 2006. The company is owned by Solomon Islanders committed to living in the rural areas and has a licence from the Western Province Assembly to operate a passenger transport service. Initially, the Pelena Express vessel is only operating in the Western Province.

Common forms of transport in Solomon Islands

Private village-based shipping

[1] Royal commission of enquiry into the sinking of the MV Princess Ashika http://www.rcimvprincessashika.to/transcripts.html

- Dugout Canoe
    - Locally manufactured using materials & skills at the village level
    - Very limited use for trade
    - Slow speed
    - Manually propelled
    - Typically used in vicinity of home village only.
- Fibreglass or aluminium canoe with outboard motor (OBM)
    - High capital cost
    - High running cost
    - Motors require ever high priced, high-quality fuel, not locally available
    - Limited capacity – in cargo weight and size
    - High speed when unloaded
    - Good access to villages and accepted design
    - Draft (with 25HP OBM) at least 0.8 metres with more than 3 persons
- Good access to where people reside, but poor cargo capacity & unable to provide fast inter-island service without very high fuel costs

Commercial Shipping

- Landing barge
    - Reasonable access to many beaches so long as deep water is available.
    - Requires at least 1.2 metres of water depth, more when loaded.
    - Slow speed
    - Inefficient hull
    - Easy loading and unloading
- Inter-islander passenger & freight
    - Mostly slow speed vessels
    - Limited access to villages due to deep draft (at least 2 metres)
    - Mostly irregular timetables
    - One large high speed vessel is operating (Pelican) but is passenger only.
- International Freight
    - Requires deepwater ports or auxiliary smaller vessels such as landing barges to access seaside.
    - Normally no access to villages except through large scale operation, eg logging.
- No or very poor access to villages – villages where more than 85% of population reside

Pelena Express Concept

- Excellent access to majority of seaside villages, including those sheltered by coral reefs, such as lagoon areas
- Ability to take passengers and cargo in order of 3 tons
- Fast, efficient, reliable, and predictable service.
- Locally managed by reliable and respected team.

At present there is only one vessel in the Pelena Express fleet. However, it is planned to increase this number to develop an interconnected rural transportation network across the country. Access to funds is the limiting factor in the expansion of Pelena Express. Pelena Express plans to be more than a ferry service. The company plans that the vessel will be a "multi-function water transport vessel" to enable sustainable development in rural Solomon

Islands. This is the first step in a planned transport service owned and operated by rural Solomon islanders the hardware for which will be 48 similar vessels.

The Pelena Express business is totally privately funded by the owners taking out a private loan. No government or aid funding has contributed to this essential service. Unfortunately, none of the Solomon Islands financial institutions would lend money to the project. Fortunately, a private lender was found in Australia. This is a commercial arrangement where the loan must be repaid with interest.

Pelena Express (PE) and private businesses: PE offers to assist other private businesses with their rural business ventures. They offer to arrange a regular collection of produce. In addition they suggest that use of the Pelena Express can provide the reliable and affordable transport solutions removing the requirement for other private businesses to maintain marine transport hardware for business use. The Pelena operates a Token system which can allow other businesses to provide transport for their staff and customers without the hassles of operating outboard motors and canoes.

Pelena Express and Aid Projects: The excellent load carrying capacity of the Pelena Express boat combined with its very low draught which allows the boat access to many village beachfronts eliminates the need for multi-transfers of cargo from multiple boats to multiple canoes. Additionally, many aid projects suffer from the need to purchase a piece of machinery, such as an outboard motor. The lack of regular transportation services in rural Solomon Islands, often results in the outboard motor and other capital items being used for private use. The Pelena Token system can allow for clear budget forecasting and project planning, without the hassle of worrying about whether or not a piece of machinery will be available and operational for the project.

The vessel is of aluminium construction with an extensive flat deck area for carrying passengers and/or cargo. The very shallow draught (400 mm when idle, 200 mm when at speed), allows the boat access to many village beachfronts. This eliminates the need for multi-transfers of cargo from multiple boats to multiple canoes. The hull is a proven design used extensively in the open ocean oyster-boat industry where wave piercing, shallow draught and high load carrying capacity are essential design elements.

Passenger and Cargo carrying capacity: The vessel carries 24 adult passengers and 10 children, plus crew and luggage; or alternatively 3 tons of cargo. Future vessels will have a slightly larger capacity. The vessel is powered by a Cummins diesel driving a Doen jet unit which allows access to shallow areas previously inaccessible by any water transport except outboard canoes without damage to the boat or the environment. A screen or trash rack stops logs & coconuts from hitting and damaging the propeller. The enclosed propeller design is also a much safer alternative to open propellers which can easily injure people or animals in the water.

The maintenance program for the vessel is strictly controlled. The equipment that makes up the boat has been chosen for its reliability AND ease of spare parts availability and supply. Preventative maintenance of the vessel and regular checks of all components is routine. The crew members have been chosen because of their experience working with other Pelena products, especially the Pelena Energy micro-hydroelectric turbines. Good people and good equipment and good support results in reliable operation.

Safety on Board the Pelena Express: The most important safety feature is the training of the crew. All crew are rural Solomon Islanders with extensive personal experience in safety at sea. Additionally, all crew members have completed extensive Solomon Islands recognised maritime training in fire-fighting, first aid, survival at sea techniques, occupational health and safety. Additionally, some crew members undertook training in Australia towards the final stages of vessel construction and motored the vessel from Australia to Solomon Islands. The vessel is in survey with Solomon Islands Marine Division and has been approved for use as a passenger vessel for daytime operations. The vessel was constructed to USL codes in Australia for use in waters up to 30 nautical miles from the coastline. A full complement of lifejackets, EPIRB (Emergency Position Indicating Radio Beacon), fresh water, bilge pumps, multiple radios (HF & VHF) and satellite phone are also employed. Below deck, the vessel has sufficient buoyancy foam to keep the vessel afloat should the hull of the boat be ruptured.

Fuel: Pelena Energy of Australia has developed technology for reliably operating diesel engines on coconut oil (CNO). It is understood that the vessel is operating on pure coconut oil. The aim is to eliminate the copra processing in villages and instead produce oil directly at the village level. This will provide income for villagers and minimise the dependence of Solomon Islands on imported oil.

The boat was constructed by Pelena Energy, located in Dorrigo NSW Australia http://www.pelena.com.au and the author acknowledges Pelena Energy for their kind permission to use all information in this paper concerning the current and future plans for operations of Pelena Express[2]. For links to video of operations of the vessel refer to footnote [3]

The following maps show a possible expansion program originating in the Western Province (as detailed in the February 2006 Pelena Express Business Plan) and expanding across the country to potentially 48 vessels. The time scale for such a plan is dependent on the success level of the first vessel, but could possible commence as soon as the end of 2009.

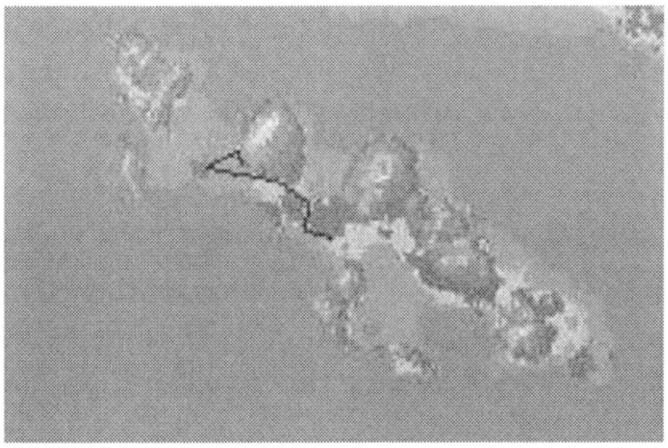

Stage 1 per PE business plan Western Province only 1 vessel only - current situation

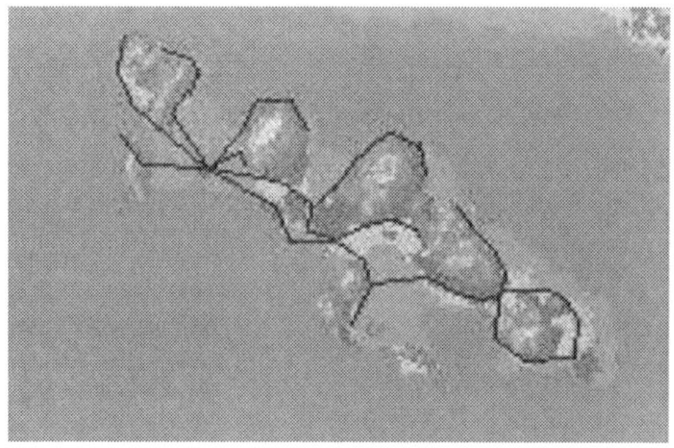

Stage 2 Expansion across Western Province 8 vessels total

Stage 3 Expansion to Choiseul, Isabel, and Guadalcanal 32 vessels total

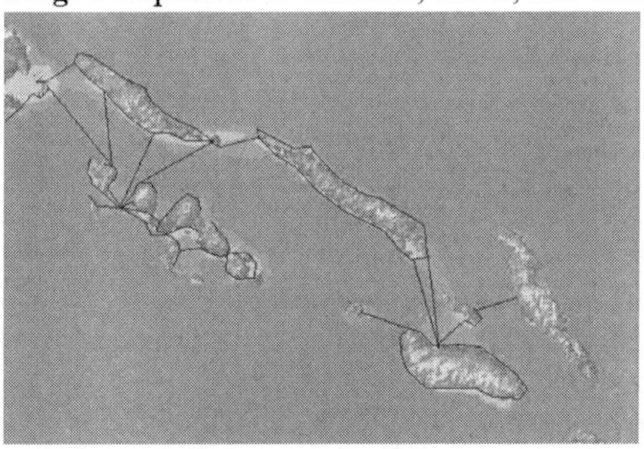

Stage 4 Expansion to Malaita, Central, Makira 42 vessels total

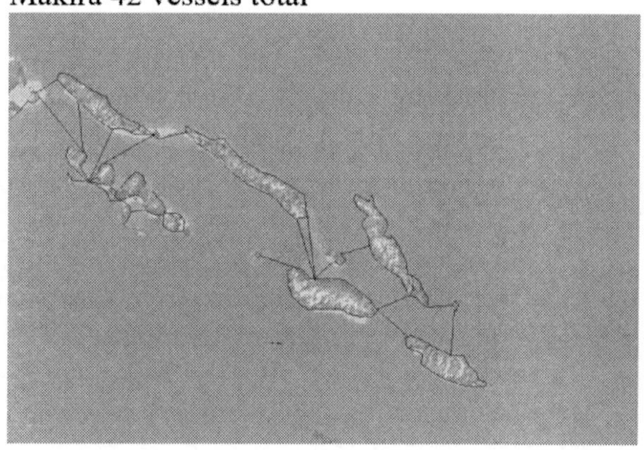

---

[2] Permission is granted to use the material for the paper to be presented at PACIFIC 2010 in Sydney late January 2010 as discussed. All material used to be fully and clearly acknowledged as being sourced from Pelena Energy.

[3] Some links showing the performance of the Pelena Express vessel (actual called the *Namita Rose*): Typical operation. Note the young mothers and their babies. They feel safe with the vessel and crew:
http://www.youtube.com/watch?v=WaPFjmwEoMs
2007 Solomon Tsunami relief effort – Start at 4min 54 sec:
http://www.youtube.com/watch?v=OYd3NkNgcOM

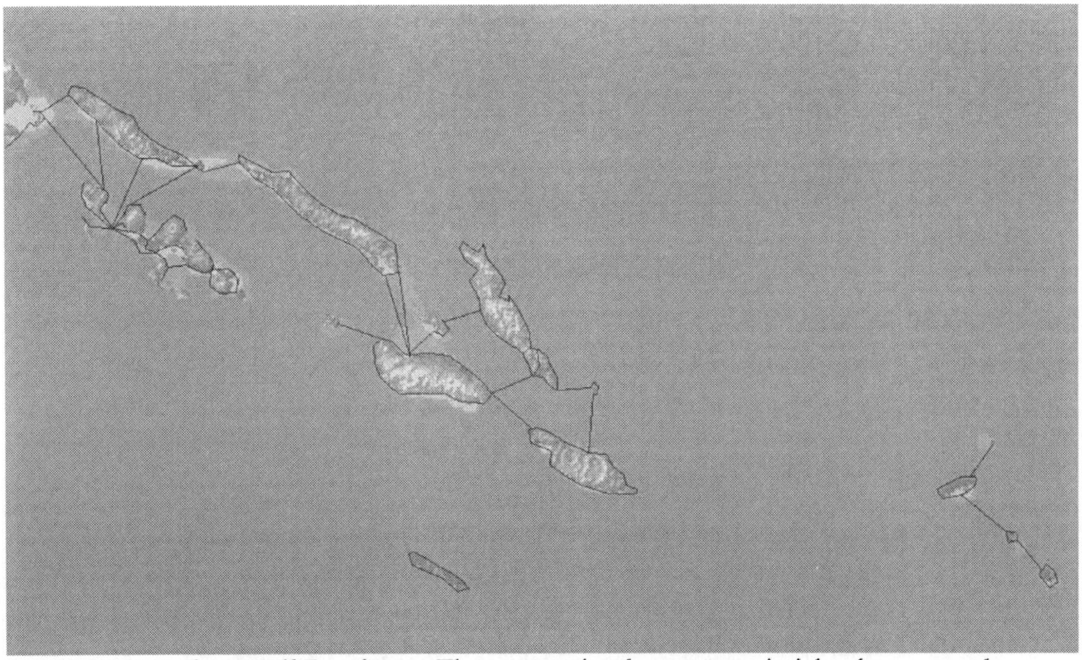

Stage 5 Expansion to all Provinces: The connection between main islands may not be a focus as this would require significant open water travel – not a focus of this service. Maximising rural village inter-connectivity is a focus of the Pelena Express network. Travel to distant Provinces such as Temotu and Rennell-Bellona would be likely better served by larger vessels better suited to extended open sea operation. 48 vessels total.

It is understood that Pelena Express with its safe and innovative transport service, locally operated and providing employment in rural Solomon Islands, may now be seeking Australian aid funding to expand its operations however as it is a private for-profit company there are difficulties in accessing this aid funding.

CONCLUSIONS

The role of international aid

This is a difficult area to assess without access to information not currently held by the author. It seems that in at least two recently-announced cases of ferry new buildings being delivered as foreign aid the aid is tied to the vessels being built in the aiding country. Without more information on these projects it is not possible to comment on whether this arrangement delivers a vessel suitable to the local conditions. These vessels will form only one element of a maritime transport system which without proper training of seafarers and strong regulatory enforcement of safety standards cannot operate safely and effectively. The provision of a budgetary component for vessel maintenance with any hardware delivery would also seem essential. There does seem to be an emphasis on provision of hardware and infrastructure rather than on effective transport services where the development of the human element in operation of services is included in any aid program seeking to improve maritime services.

The role of government providers

Where the government is both regulator and provider of shipping services there will always be the possibility for conflict of interest. The public service requirement of governments to provide maritime transport to a population has at times over-ridden the requirement that that transport should comply with international rules and regulations for safety of life at sea. Separation of responsibility between service provider and regulator must be clearly maintained. The regulator must not be able to be influenced by government shipping company board or management to permit vessels to sail which otherwise would not be so permitted. The development of a ship safety culture must be set as a core goal for shipping providers in both private and government spheres.

The role of private providers

There exists a shortage of suitable second-hand tonnage in the region; one example was found of a private company in Cook Islands paying twice as much to bring suitable tonnage from Norway as they paid for the vessel itself.

An effective maritime transport system is not just hardware

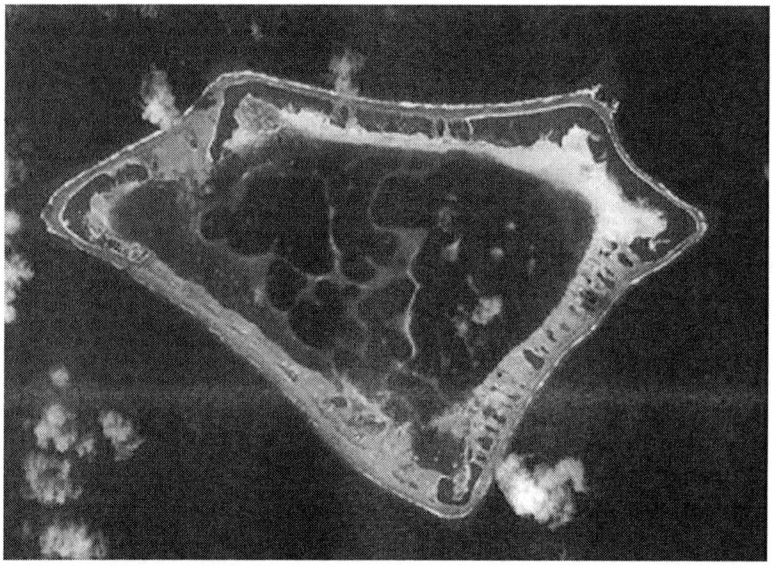

The people of the Pacific Islands have many skilled mariners in their populations. Providing high standards of maritime education and training and then allowing those skilled people to responsibly implement a culture of safety both in shipping companies and in maritime regulators is an obvious path to improving the levels of safety for maritime transport in the Pacific Island nations.

End of text

Control Design for Positioning of Underwater Vehicles with Control Allocation

Tristan Perez[1,2] , Alejandro Donaire[1]

[1]Centre for Complex Dynamic Systems and Control—CDSC, The University of Newcastle, Australia, Email: Tristan.Perez/Alejandro.Donaire@newcastle.edu.au

[2]Centre for Ships and Ocean Structures (CeSOS), Norwegian University of Science and Technology, Trondheim, Norway—NTNU

ABSTRACT

In this paper, we address the control design problem of positioning of over-actuated underwater vehicles. The proposed design is based on a control architecture with combined position and velocity loops and a control tuning method based on the decoupled models. We derive analytical tuning rules based on requirements of closed-loop stability, positioning performance, and the vehicle velocity dynamic characteristics. The vehicle modelling is considered from force to motion with appropriate simplifications related to low-speed manoeuvring hydrodynamics and vehicle symmetry. The control design is considered together with a control allocation mapping. This approach makes the control tuning independent of the characteristics of the force actuators and provides the basis for control reconfiguration in the presence of actuator failure. We propose an anti-wind-up implementation of the controller, which ensures that the constraints related to actuation capacity are not violated. This approach simplifies the control allocation problem since the actuator constraints are mapped into generalised force constraints.

1-INTRODUCTION

Underwater vehicles perform missions that often require different vehicle configurations to accommodate on-board mission-specific equipment. Variations in vehicle configuration can result in changes of the vehicle motion dynamic response due changes in mass distribution, and, in some cases, even changes in hydrodynamic characteristics. If accurate positioning control is required during different vehicle configurations, it is then necessary to re-tune the control system to keep the required control performance.

In this paper, we propose a particular positioning control architecture based on velocity and position loops and develop analytical tuning rules that ensure stability and desired performance. The control system is then considered in conjunction with a control allocation function that maps the desired control forces into appropriate actuator force commands. This implementation allows considering the control tuning in terms of generalised forces associated with the degrees of freedom being controlled, and thus, independent of the characteristics of the force actuators. In addition, we use a particular implementation of the controller based on anti-wind-up proposed in Perez (2009), which takes into account the limiting authority of the actuators by appropriate scaling of the generalised forces. This scaling ensures that the control demands processed by the control allocation function are always feasible. Therefore, this approach results in a simplification of the control allocation problem since it can be considered without constraints.

The performance of the proposed control system is demonstrated based on a simulation scenario of horizontal-plane positioning of a general open-frame vehicle with four actuators.

2-POSITIONING CONTROL AND OPEN-FRAME VEHICLE MODELLING

Positioning control of marine vehicles refer to the use of the propulsion system to regulate the position and orientation despite environmental disturbances. In this paper, we concentrate on the positioning problem in the horizontal plane, that is, surge, sway, and yaw.

For the positioning problem and degrees of freedom being considered, the vehicle dynamics can be represented as

$$\dot{\boldsymbol{\eta}} = \mathbf{R}(\boldsymbol{\eta})\,\boldsymbol{\nu}, \tag{1}$$

$$\mathbf{M}\dot{\boldsymbol{\nu}} + \mathbf{C}(\boldsymbol{\nu},\boldsymbol{\nu}_c) + \mathbf{D}(\boldsymbol{\nu},\boldsymbol{\nu}_c) = \boldsymbol{\tau}. \tag{2}$$

The *generalised position, velocity and force vectors* are given by

$$\boldsymbol{\eta} \triangleq \begin{bmatrix} n \\ e \\ \psi \end{bmatrix}, \quad \boldsymbol{\nu} \triangleq \begin{bmatrix} u \\ v \\ r \end{bmatrix}, \quad \boldsymbol{\tau} \triangleq \begin{bmatrix} X \\ Y \\ N \end{bmatrix},$$

where n and e denote the North and East positions relative to a local geographical frame with a coordinate system convention of North-East-Down, and ψ denotes the heading or yaw angle relative to the North direction. The velocities are expressed in the vehicle's (body-fixed) frame with a forward-starboard-down coordinate-system convention. The surge and sway velocities are denoted by u and v respectively, and the yaw rate is denoted by r. The generalised velocity vector $\mathbf{v}_c$ represents the velocity of the current expressed in body-fixed coordinates. The surge and sway forces along the body-axis are represented by X and Y, and the yaw torque about the z-axis is represented by N. These forces include the propulsion forces and slowly-varying environmental forces due to ocean currents.

Equation (1) is a kinematic transformation that takes the body-fixed generalised velocity vector into the time derivative of the generalised position vector. The matrix $\mathbf{R}(\eta)$ is single axis rotation matrix about the vehicle's vertical axis, that is $\mathbf{R}(\psi)$. Equation (2) is a vector equation of motion. The parameter $\mathbf{M}$ is the total mass and inertia matrix, which includes rigid-body and hydrodynamic added mass components. The second term in (2) represents Coriollis-centripetal forces, which appear as a result of expressing the equations of motion in body-fixed coordinates. This term has a component due to rigid body and another due to hydrodynamics (potential component of the current forces). The third term represents damping forces, which reflect the work done on the fluid due to the motion of the vehicle. We assume that the vehicle is neutrally buoyant, and therefore we do not consider gravity and buoyancy in (2).

For vehicles manoeuvring at very-low speed, a few simplifications can be made to the model (2). First, the Coriollis-centripetal terms can be neglected since these terms have a quadratic dependency on the velocities. Second, the nonlinear damping term in (2) is dominated by a linear damping at low speeds, and therefore we can separate the effect of the current forces and consider them as a slowly varying disturbance. These leads to the following model:

$$\dot{\boldsymbol{\eta}} = \mathbf{R}(\boldsymbol{\eta})\,\boldsymbol{\nu}, \tag{3}$$

$$\mathbf{M}\dot{\boldsymbol{\nu}} + \mathbf{D}\,\boldsymbol{\nu} = \boldsymbol{\tau}_c + \boldsymbol{\tau}_{ctrl}. \tag{4}$$

Third, or open frame vehicles, like the one shown in Figure 1, the hydrodynamic coupling between the yaw and sway can also be neglected (Smallwood and Whitcomb, 2003). The latter assumption leads to diagonal matrices in (4), and thus, to a fully decoupled model.

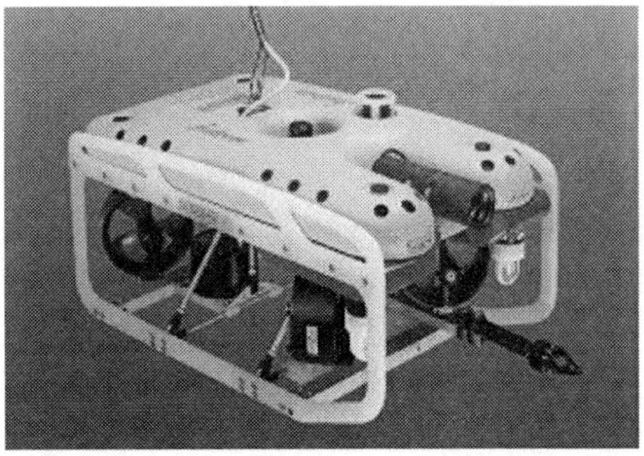

Figure 1. Example of open frame underwater vehicle.

3-POSITIONING CONTROL WITH CONTROL ALLOCATION

3.1 Positioning Controller Architecture

Figure 2 shows a block diagram describing the main functionality of underwater vehicle motion control system. In this diagram, we can distinguish the following subsystems:

- **Navigation**: This system uses various sensors to provide reliable signals related to the vehicle's velocity and position.
- **Controller**: The controller processes information related to desired position and the actual position and velocity generated by the navigation system. This information is used to generate generalised force commands that correct the motion of the vehicle.
- **Allocation**: The allocation system maps the generalised force commands into appropriate force commands for the various actuators on the vehicle.

The separation between generalised force control and actuator force allocation makes the control tuning independent of the characteristics of the force actuators and provides the basis for control reconfiguration in the presence of actuator failure. The vessel actuator configuration is usually designed such that the open frame vehicle is over actuated. Therefore, the control allocation mapping has no unique solution, that is, given a desired generalised force vector computed by the controller, there are infinite combinations of actuator commands that implement such generalised force vector. Therefore, to single out a solution, the allocation is posed as a constrained optimisation problem and solved

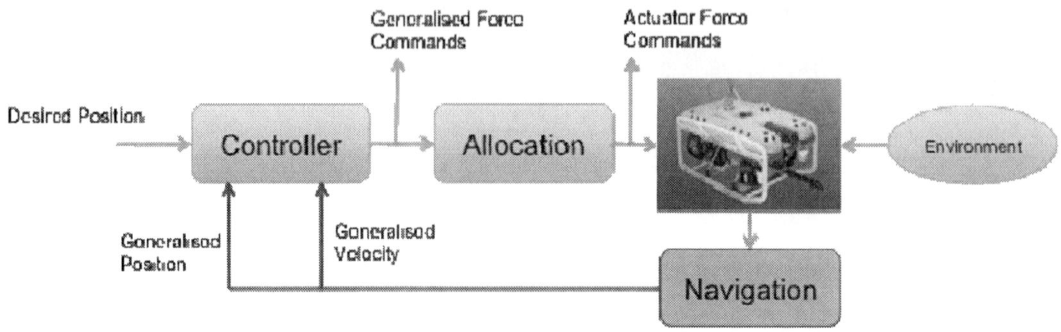

Figure 2 Uninhabited underwater vehicle motion control system with control allocation.

numerically on-line to find the actuator commands that implement the desired generalised forces whilst optimising an index related to the performance of the propulsion system—see, for example, Fossen and Johansen (2006), Peterson and Bodson (2006), Fossen et al. (2008), and references there in.

Figure 3 shows a block diagram of the proposed positioning controller. The position controller is a proportional controller and the velocity controller is a proportional-integral controller. This control architecture provides integral action, that is, it regulates position despite constant force disturbances. In the proposed controller, the measured positions are transformed by the inverse of the kinematic transformation (1). This transformation results in a linearization of the kinematics provided that the vehicle is rotating slowly (Fossen, 2002):

$$\dot{\boldsymbol{\eta}}_p \approx \boldsymbol{\nu} \tag{3}$$

The advantage of using the rotated variables (non-linear controller), is that the control tuning can be addressed within a linear framework. The control tuning is addressed in Section 3.5.

3.2 Unconstrained Control Allocation

The force vectors produced by the actuators can be decomposed into rectangular coordinates (along the longitudinal and transverse direction on the vessel) and combined into a single vector **u**,

$$\mathbf{u} = \begin{bmatrix} T_{X1} & T_{Y1} & T_{X2} & T_{Y2} & \cdots & T_{XN} & T_{YN} \end{bmatrix}^T, \tag{4}$$

where N is the number of actuators. This vector is mapped into the generalised forces via the actuator configuration matrix **B**, which depends only on the location of the actuators on the vehicle:

$$\boldsymbol{\tau} = \mathbf{B}\mathbf{u} \tag{5}$$

Since **u** has more components than $\boldsymbol{\tau}$, there are different vectors **u** satisfying (5) for a given value of $\boldsymbol{\tau}$. In order to limit the number of solutions, one can pose the problem as an optimization problem:

$$\mathbf{u}^{\star} = \arg\min_{\mathbf{u}} \mathbf{u}^{\mathrm{T}}\mathbf{u} \quad (6)$$
$$\text{subject to } \boldsymbol{\tau}_d = \mathbf{B}\mathbf{u}.$$

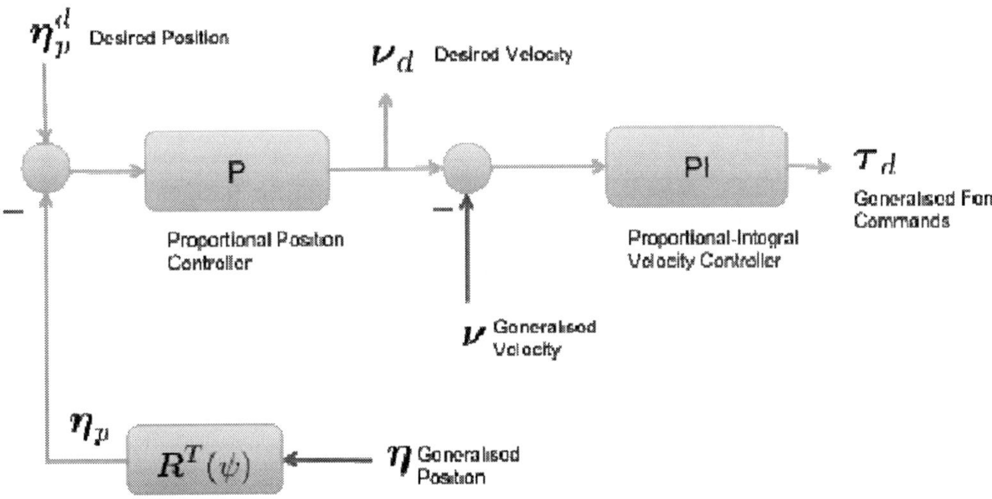

Figure 3. Details of motion controller—proportional (P) position controller and proportional-integral (PI) velocity controller.

The objective function $\mathbf{u}^{\mathrm{T}}\mathbf{u}$ is representative of the total "energy" or control effort. Thus, the control allocation seeks the solution that implements the desired generalized force $\boldsymbol{\tau}_d$ whilst minimizing the control effort. The solution of the above problem is given by

$$\mathbf{u}^{\star} = \mathbf{B}^{\dagger}\boldsymbol{\tau}_d, \qquad \mathbf{B}^{\dagger} = \mathbf{B}^T(\mathbf{B}\mathbf{B}^T)^{-1} \quad (7)$$

Note that since **B** depends only on the location of the actuators on the vehicle, the $\mathbf{B}^{\dagger}$ can be pre-computed.

The optimisation problem (6) does not take into account the fact that the vector **u** must belong to a constraint set due to the maximum force that the various actuators can produce. Adding this constraint to (6) requires on-line numerical optimisation, and the solution (7) is no longer the optimal solution. Constrained numerical optimisation brings out feasibility issues since the optimisation algorithm must always provide a feasible solution within the required sampling period. An alternative to this approach is to constrain the desired generalized force $\boldsymbol{\tau}_d$ such that the constraints on **u** are always satisfied (Perez, 2009). By so doing, the force controller is also informed about reaching constraints, which prevents performance degradation due to the combination actuator saturation and integral action (Goodwin et al, 2001). This control method is described in the next section.

3.4 Constrained Control via Input Scaling

One of the key issues in control design for systems that require integral action and present a potential for actuator saturation is that of integrator wind-up. That is, if the actuators saturate and the integral controller is not informed about the saturation, the integrators continue integrating the error signals but the control action is not seen by the system. This often produces a degradation of performance in terms of undesirable oscillations and even instability. Different anti-wind-up techniques have been discussed in the literature (Bernstein and Michel, 1995).

If a linear controller *C(s)* is minimum phase and bi-proper (as in the case of a PI and PID controller), then anti-wind-up can be achieved using the implementation shown in Figure 4 (Goodwin

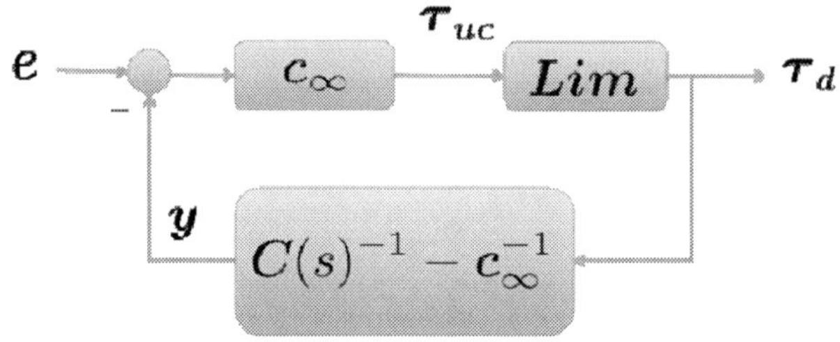

Figure 4. Anti-wind-up implementation of a minimum-phase bi-proper controller.

et al, 2001). In this controller, ***Lim*** represents the saturation (magnitude, rate, or a combination of both), and the gain

$$c_\infty = \lim_{s \to \infty} C(s) \tag{8}$$

Not that if the limitation block were not present in the loop of Figure 4, reduces to the controller:

$$C(s) = [\mathbf{I} + c_\infty(C(s)^{-1} - c_\infty^{-1})]^{-1} c_\infty \tag{9}$$

When the limitation is present, it prevents the control signal τ_d to exceed its limits and the constrained signal drives the states of the controller, which are all on the feedback path. This anti-wind-up strategy is simply a particular implementation of the original controller. That is, it does not require any additional design.

This anti-windup scheme can be applied to the velocity PI controllers of the vehicle positioning control:

$$C_{vel}(s) = \begin{bmatrix} C_u(s) & 0 & 0 \\ 0 & C_v(s) & 0 \\ 0 & 0 & C_r(s) \end{bmatrix} \tag{10}$$

where for $i = u, v, r$,

$$C_i(s) = K_p^i \frac{T_I^i s + 1}{T_I^i s} \tag{11}$$

In order to constrain τ_d we need to construct a set such that

$$\tau_d \in \mathcal{T} \quad \Leftrightarrow \quad u \in \mathcal{U} \tag{12}$$

One way of enforcing these constraint is by computing an unconstrained control τ_{uc} and then scale it down if it is outside the constraint set, that is

$$\tau_d = \begin{cases} \tau_{uc} & \text{if}\, \tau_{uc} \in \mathcal{T} \\ \alpha\,\tau_{uc} & \text{if}\, \tau_{uc} \notin \mathcal{T};\ \alpha < 1 : \alpha\,\tau_{uc} \in \partial\mathcal{T}. \end{cases} \tag{13}$$

By scaling the vector, we preserve its direction. In order to implement (13) we need to compute τ_{uc} and then obtain α. To simplify this computation we can consider a polyedron set to bound the rectangular actuator force components:

$$\mathbf{L} \leq \mathbf{u} \leq \mathbf{U} \tag{14}$$

These constraints can be combined to a single inequality

$$\mathbf{F}\,\mathbf{u} \leq \mathbf{f}, \qquad \mathbf{F} = \begin{bmatrix} \mathbf{I} \\ -\mathbf{I} \end{bmatrix}, \qquad \mathbf{f} = \begin{bmatrix} \mathbf{U} \\ -\mathbf{L} \end{bmatrix}. \tag{15}$$

Using these constraints the following algorithm can be used to determine the scaling factor

1. Compute the unconstrained control, and evaluate $\mathbf{f}_{uc} = \mathbf{F}\,\mathbf{B}^{\dagger}\tau_{uc}$
2. Set $\alpha = 1$,
3. If $f_{uc}(i) > f(i)$ and $f(i)/f_{uc}(i) < \alpha$, then set $\alpha = f(i)/f_{uc}(i)$. Do this for all the components $i = 1,\ldots, 4$ N.

Once the scaling factor is computed the control can be implemented as

$$\tau_d = \alpha\,\tau_{uc} \tag{16}$$

This implementation leads to the complete control scheme depicted in Figure 5 (Perez, 2009).

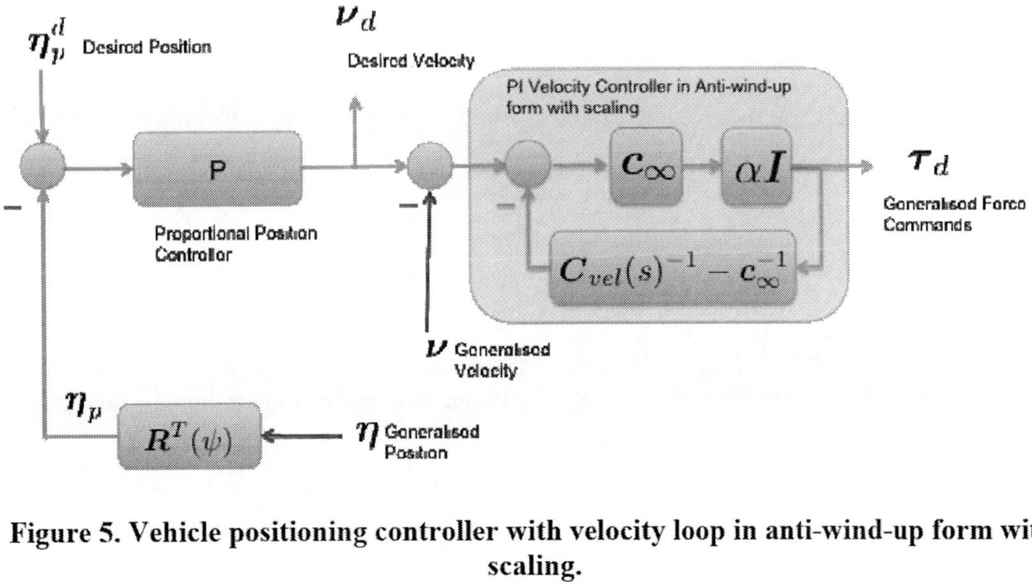

Figure 5. Vehicle positioning controller with velocity loop in anti-wind-up form with scaling.

3.5 Stability and Control Tuning

One key issue is the stability of the closed-loop system. The proposed controller ensures asymptotic stability for position regulation provided that the following conditions hold for the parameters of the proportional-integral velocity controllers (11) and the proportional position controller:

- $K_p^i > 0, \quad T_I^i > 0, \quad \text{and} \quad T_I^i \neq T_v^i$
- $0 < P^i < 1/T_v^i.$

where T_v^i is the open-loop time-constant of the velocity response. The first condition ensures that there is no zero-pole cancelation between the velocity controller and the velocity response of the vehicle. The second condition establishes that the position control loop must be slower than the open-loop velocity response.

The stability proof follows from re-arranging the closed-loop system with control allocation into a Lure system and application of Lyapunov and passivity theories (Khalil, 2000, Brogliato et al., 2007). Details of the stability proof go beyond the scope of this paper.

Based on the above conditions on the controller parameters, practical tuning rules can be obtained using, for example, the root-locus technique. The main idea is to create a faster velocity loop response and then set the proportional gain to obtain a position loop response that is slower than the open loop velocity response. Therefore, we propose the following set of tuning rules.

Given the velocity response models,

$$G_v^i(s) = \frac{K_v^i}{(T_v^i\, s + 1)} \tag{17}$$

Then,

1. Set the zero of the velocity PI at a higher frequency than the pole of the velocity response

$$T_I^i = \beta^i\, T_v^i, \quad \beta^i = 0.2 \tag{17}$$

2. Determine the proportional gain of the velocity PI controller, K_p^i such that the velocity closed loop poles are real and above the zero of the controller. These conditions are satisfied when K_p^i is greater than the critical value K_{pc}^i given by

$$K_{pc}^i = \left[(2\beta^i - 1) + \sqrt{(2\beta^i - 1)^2 - 1}\right] / K_v^i \tag{18}$$

3. Set the closed loop position gain based on the stability constraint. Faster response results when P^i is selected closer to $1/T_v^i$.

$$0 < P^i < 1/T_v^i. \tag{19}$$

The above tuning rules are practical and simple to follow. Note that the only data necessary for the tuning are the parameters of the simplified vehicle open-loop velocity response models (16). These parameters can be obtained from simple zig-zag identification tests performed when the vehicle is launched.

4-SIMULATION CASE STUDY

In this section, we present simulation experiments that both verify and illustrate the performance of the proposed control system. We consider a model of general open frame vehicle with four thrusters in an x-type configuration, namely, bow-starboard, bow-port, stern-port, and stern-starboard. Further details of the vehicle model are given in Appendix I.

Figure 6 shows the results of an experiment in which the vehicle is in regulation under the action of a constant ocean current and then after 20 seconds we set reference steps on surge, sway and yaw positions. This figure shows the demanded and actual surge, sway and yaw positions, velocities and generalized forces. Figure 7 shows the corresponding forces of the four actuators. As we can see from the latter figure, the forces remain within the constraints on the maximum force magnitude. The control system presents a good performance despite the actuators reaching the saturation levels.

Figures 8 and 9 show the results of a second experiment with the same environmental conditions and position set-point change as in the first experiment, but in this case the anti-wind-up is inactive (that is $\alpha = 1$). In this case the controller is unaware of the limitation of the actuators, and this, as expected, produces a severe degradation in the positioning performance. Although the vehicle finally goes to the desired position (in an inadmissible way), the stability of the closed-loop is not theoretically guaranteed.

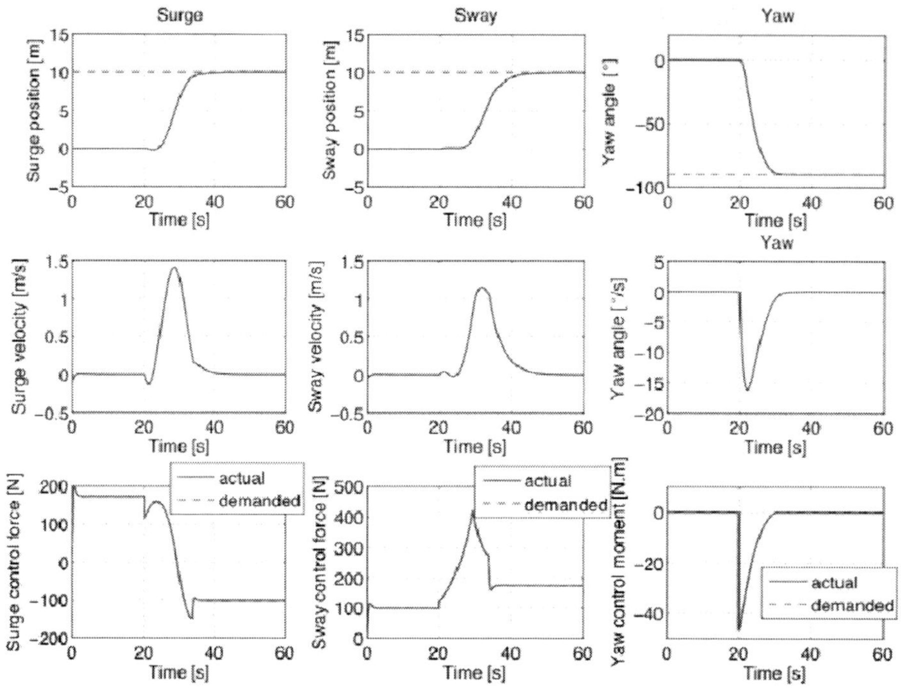

Figure 6. Performance of vehicle position regulation controller in current and position set point change.

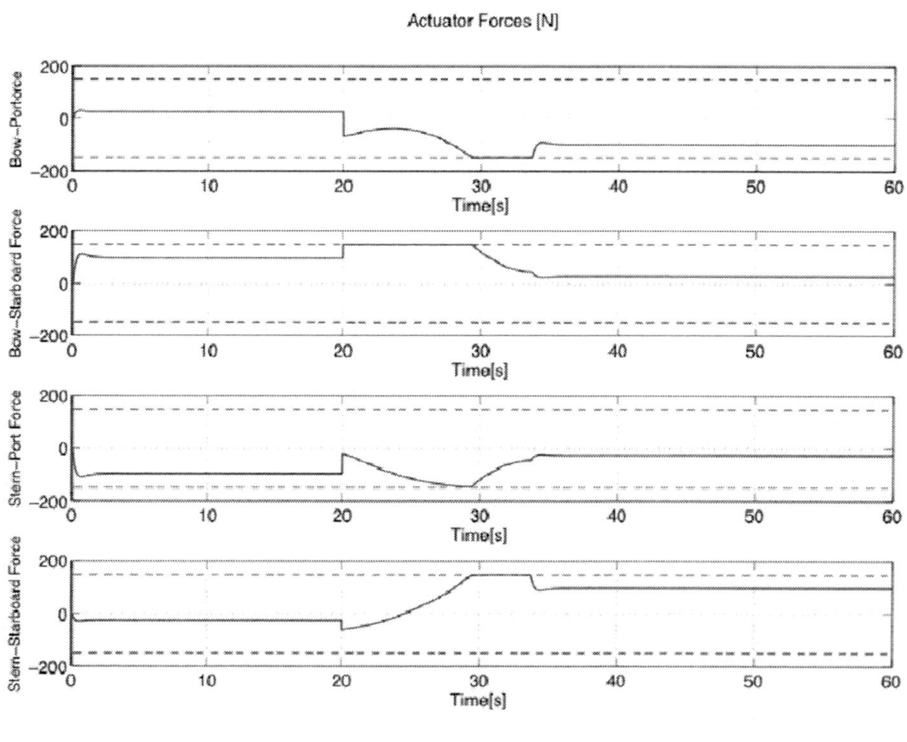

Figure 7. Actuator forces for a vehicle change in position while in current.

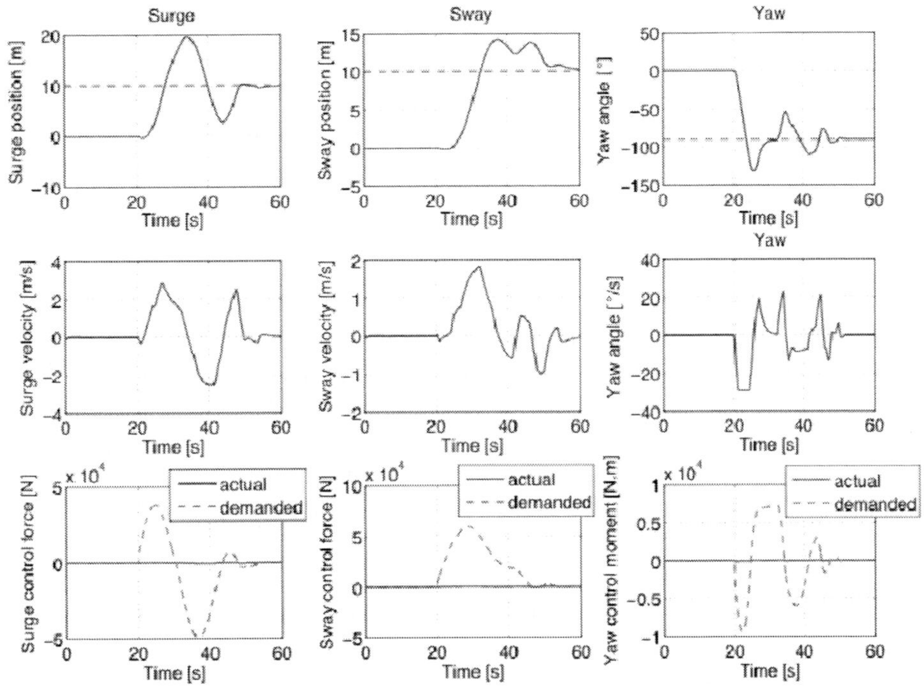

Figure 8. Performance of vehicle position regulation controller in current and position set point change, for the case where the anti-wind-up is not active.

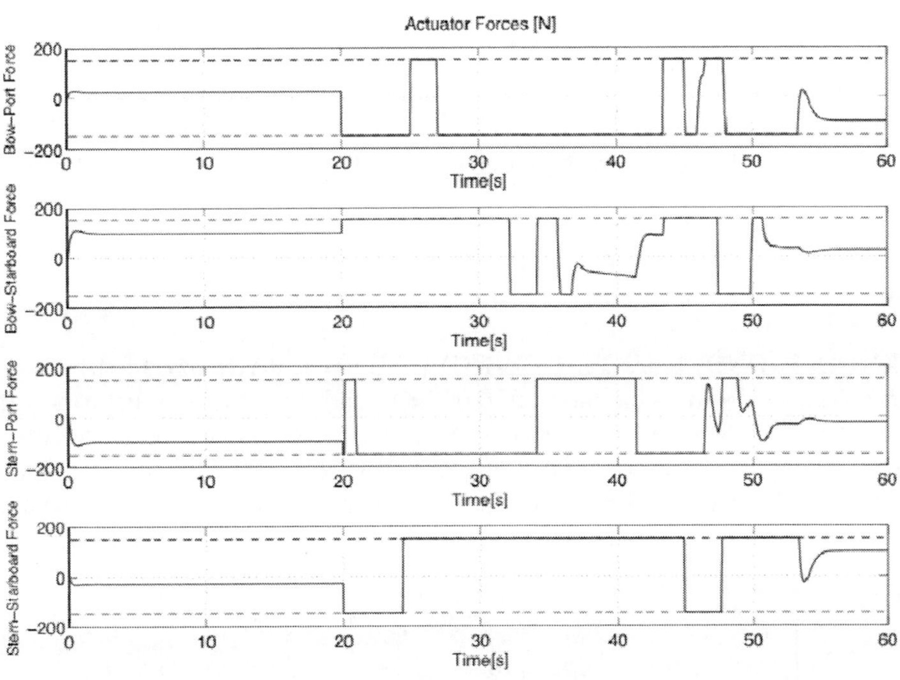

Figure 7. Actuator forces for a vehicle change in position while in current for the case where the anti-wind-up is not active.

5-CONCLUSIONS

In this paper, we have considered the design of a positioning controller in conjunction with control allocation for over-actuated open frame underwater vehicles. The proposed control architecture is based on a velocity loop implemented with a proportional-integral controller and a position loop implemented with a proportional controller. This control architecture provides integral action for position regulation, that is, it rejects constant disturbances due to ocean currents.

The use of control allocation allows the control design to be focussed on the generalised forces related to the degrees of freedom of concern rather than on actuator forces. The use of control allocation also provides the basis for actuator re-configuration in the case of actuator faults, but this goes beyond the scope of this paper. In the proposed control system, we map the constraint set of actuator forces due to the limited authority of the actuators into a constraint set for the generalised forces. This approach results in a constrained control design with a simple, unconstrained, control allocation problem. We provide an algorithm for an on-line implementation of the mapping.

Since the positioning control requires integral action, it is fundamental to consider a constrained control design to avoid performance degradation due to actuator saturation. In the proposed controller, this design is approached using multivariable anti-wind-up. The proposed anti-wind-up scheme scales the demanded generalised forces such that they always belong to their corresponding constraint set.

The tuning of the proposed control system is based on simple practical rules that follow from close loop stability considerations and performance. The key assumptions are that the vehicle's open loop velocity response can be modelled by first order systems and that the hydrodynamic coupling between the different degrees of freedom is negligible. Open-frame underwater vehicles satisfy these assumptions during slow manoeuvres. The tuning rules use information related to the open-loop velocity response. The parameters of the model can be obtained from simple zig-zag test conducted when the vehicle is launched.

The performance of the proposed controller is illustrated based on simulation case scenarios for a generic open frame underwater vehicle with four actuators.

APPENDIX I –UNDERWATER VEHICLE MODEL PARAMETERS

The vehicle considered has a mass of 10,000 Kg, and it is of a similar form as the vehicle shown in Figure 1. The vehicle has four thrusters in an x-type configuration, namely, bow-starboard, bow-port, stern-port, and stern-starboard. The maximum force produced by each thruster is 150N, and for simplicity, we consider a symmetric force characteristic. The thruster configuration matrix is given by

$$\mathbf{B} = \frac{\sqrt{2}}{2}\begin{bmatrix} 1 & 1 & -1 & -1 \\ -1 & 1 & -1 & 1 \\ (l_1^{BP} - l_2^{BP}) & (-l_1^{BS} + l_2^{BS}) & (-l_1^{SP} + l_2^{SP}) & (l_1^{SS} - l_2^{SS}) \end{bmatrix} \quad \text{(A1)}$$

and the distances that determine the horizontal location of the thrusters in the vehicle are given in Table A1. These distances are given with respect to the point at the centre of the box that encloses the vehicle.

The vector equation of motion is formulated at the centre of the box that encloses the vehicle. Table A2 gives the parameters of the velocity response dynamic model and those of the controller.

Table AI. Actuators location

	BP	BS	SP	SS
l_1	0.6 m	0.6 m	0.6 m	0.6 m
l_2	0.35 m	0.35 m	0.35 m	0.35 m

Table A2. Model and Controller Parameters

	Model Parameters		Controller Parameters
Surge	$T_v = 0.0105$ s		$K_p = 1718$
	$K_v = 3.0384$ N.s/m		$T_I = 0.6077$
			$P = 0.3258$
Sway	$T_v = 0.0083$ s		$K_p = 2160$
	$K_v = 3.5383$ N.s/m		$T_I = 0.7077$
			$P = 0.2798$
Yaw	$T_v = 0.0095$ s		$K_p = 1894$
	$K_v = 1.2564$ N.s.m		$T_I = 0.2513$
			$P = 0.7879$

6-REFERENCES

Bernstein, D. and A. Michel (1995). A chronological bibliography on saturating actuators. International Journal of Robust and Nonlinear Control, 5:375–380.

Borgliato, B., R. Lozano, B. Maschke and O. Egeland (2007). *Dissipative Systems Analysis and Control*. Second Ed. Communication and Control Engineerign Serties. Springer-Verlag, London.

Fossen, T.I. (2002) *Marine Control Systems*. Marine Cybernetics.

Fossen, T.I. and T. A. Johansen (2006). A survey of control allocation methods for ships and underwater vehicles. In 14th IEEE Mediterranean Conference on Control and Automation, Ancona, Italy, June.

Fossen, T.I. , T. A. Johansen, and T. Perez (2008). A Survey of Control Allocation Methods for Underwater Vehicles, chapter 7 in Underwater Vehicles, pages 109–128. In-Tech, Vienna, Austria.

Goodwin, G.C., S. Graebe, and M. Salgado (2001). *Control System Design*. Prentice-Hall Inc.

Goodwin, G.C., M.M. Seron, and J.A. DeDona (2005) *Constrained Control and Estimation: An Optimisation Approach*. Communications and Control Engineering.Springer, London.

Khalil, H. (2000) *Nonlinear Systems*. Second Edition Prentice-Hall.

Perez T. (2009). Anti-windup Designs for Ship Dynamic Positioning with Control Allocation. 8th IFAC International Conference on Manoeuvring and Control of Marine Craft. September, Guaruja, Brazil.

Peterson, J. A. M. and M. Bodson (2006). Constrained quadratic programming techniques for control allocation. Control Systems Technology, 14(1): 91–97, January.

Smallwood, D. A. and L. L. Whitcomb (2003). Adaptive Identification of Dynamically Positioned Underwater Vehicles. IEEE Trans. on Cont. Sys. Tech. Vol 11, No 4, pp505-515.

Numerical and Experimental Analysis of a Surfaced Submarine in Beam Seas

B Letter[1], D Ranmuthugala[1], J Duffy[1], and Y Tso[2]
[1] Australian Maritime College (AMC), Launceston, Tasmania, Australia
[2] Defence, Science, and Technology Organisation (DSTO), Melbourne, VIC Australia.

1. INTRODUCTION

Conventional submarines can spend extended periods of time on the surface as part of their operational requirements. It is therefore important to ensure that a surfaced submarine has an adequate level of stability to withstand the effects of hazards such as beam winds and high sea states. Unlike surface vessels and previous generations of submarines, the design of modern submarines is optimised for submerged operations, thus their seakeeping performance on the surface is generally poor. Significant rolling and pitching motions can be experienced in rough seas. An extreme example was the Canadian submarine *HMCS Chicoutimi*, which was immobilised on the surface in sea state eight due to the ingress of seawater resulting in an onboard fire.

The Defence, Science, and Technology Organisation (DSTO) and the Australian Maritime College (AMC) are jointly conducting an investigation into the roll motion of a surfaced submarine; with an emphasis to predict this motion under varying sea and vessel conditions, identify parameters that influence it, and investigate methods to reduce this motion. The work-to-date has concentrated on investigating the motions of surfaced submarines under varying conditions including: roll damping in calm water including the configuration of appendages, varying wave and sea states, vessel intact and damaged conditions, entrained water in the casing and the size of the drainage ports, vessel heading, and free-to-drift conditions. The analysis has mainly used experimental work on a model submarine in AMC's towing tank and model test basin. The investigation also included the use of commercial software to predict the motion of surfaced submarines. Et-Atm (2006) used the newly developed vessel hydrodynamic software package *ShipMo3D* by Defence Research and Development Canada-Atlantic for the first time to ascertain its suitability and validate the code against experimental data.

In the most recent work ANSYS-CFX has been used to develop a six degrees-of-freedom (six-DOF) model within the domain to simulate rigid body dynamics. This is a β-function within CFX V12 and thus the current work is uncharted territory for CFX users, although such simulation tools are available in other commercial CFD packages. The physical scaled model experimental work carried out during this phase was in parallel to the analytical work, providing a range of data for validation purposes for different configurations and conditions.

2. PHYSICAL SCALED MODEL EXPERIMENT SET-UP

A generic scaled submarine model was developed jointly by AMC and DSTO to study the roll behaviour of surfaced submarines. The model has a ballast plate within it (Fig 1), enabling adjustment to the model's vertical, longitudinal, and transverse centres of gravity to ensure appropriate static position. In addition, the ballast plate was orientated such that the model's vertical centre of gravity and moment of inertia were consistent throughout the project. For the damaged condition, a mass was added to the forward section to simulate a flooded section between 10-12% of its length. Figure 7 shows a photograph of the physical scaled model in the damaged condition.

In order to investigate the submarine's roll response due to the orientation of the vessel's stern control surfaces, the stern cone could be rotated. This resulted in either a "+"or "x" configuration, while a blank ("o") stern cone provided a reference setting, (Fig 11). The two different tail cone pieces were ballasted to have an identical mass, which ensured that interchanging the tail would result in minimum variations to the parameters. The openings of the drainage ports of the casing were adjustable, thus taking into account the flow through the casing, including the effects of entrained water.

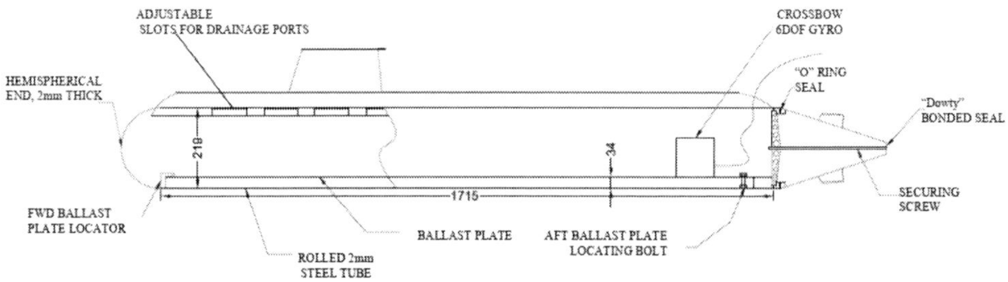

Fig 1 - AMC 05-06 Generic Submarine Model

The investigation into the effect of added appendages on roll required the independent addition of wingtips (winglets) to the stern control surfaces and bilge keels to the parallel body. The bilge keels were sized according to Ikeda and Umeda (1988), with the alignment optimised for flow in a submerged condition resulting in the bilge keels running straight along the hull. The bilge keels were 1/3 of the model length, with a 45 degree angle of entry and exit, and a depth of 20 mm. The model had removable wingtips of an aerofoil shape fitted on to the control fins and optimised for forward motion.

Testing was carried out in a number of phases, *i.e.* investigating the roll motion characteristics of a surfaced submarine in beam seas (Lund, 2005), the same for a damaged submarine (El-Atm, 2006), and roll damping (Davies, 2007 and Letter, 2009). Details of these experiments and results are presented in the above references.

Test Facilities

The experiments were carried out in AMC's hydrodynamic test facilities, *i.e.* the towing tank and the model test basin. The former is 100 m long, 3.55 m wide, with variable water depth up to 1.5 m. The facility has a powered carriage to tow models up to a maximum speed of approximately 4.6 m/s. Waves are produced by a hydraulically driven wave-maker positioned at one end of the tank. The computer software that controls the wave-maker paddle can produce a wide variety of waveforms, including: regular waves, most commonly available wave spectra, and user defined wave spectra or sequences. A large wave energy dissipation device (beach) is fitted to the opposite end of the tank.

The model test basin is 35 m long, 12 m wide, with a variable water depth up to 1 m. The bottom of the basin is flat, providing the ability to conduct experiments in very shallow water depths, if required. The combination of the increased width, solid flat bottom, variable water depth, model towing winch, and wave generation capability makes this facility extremely versatile. Waves are produced by an electrically driven wave-maker positioned on one side of the basin. The computer software that controls the 16 individual wave-maker paddles can produce a wide variety of wave forms, including: regular waves, most commonly available wave spectra, and user defined wave spectra or sequences. All free-to-drift tests were conducted in the model test basin, since its larger breadth allowed a wider degree of un-obstructed motion. Although both facilities are capable of producing a large number of wave height and wave frequency combinations, for model testing purposes it was imperative that the combination resulted in regular wave trains, *i.e.* did not break or spill.

The model was restrained by cables from the bow and the stern, which were connected at the centre of roll of the model, (Figs 2 and 3). These cables were then threaded through pulleys at each side of the tank and restrained by weights, which depended on the wave amplitudes being tested. Data was gathered using a six-DOF gyro located within the model, measuring vessel velocity, from which the relevant motions were obtained by integrating the raw data response curve for each cycle. The wave amplitude and wave frequency were obtained from measurements of the free surface elevation using a static wave probe.

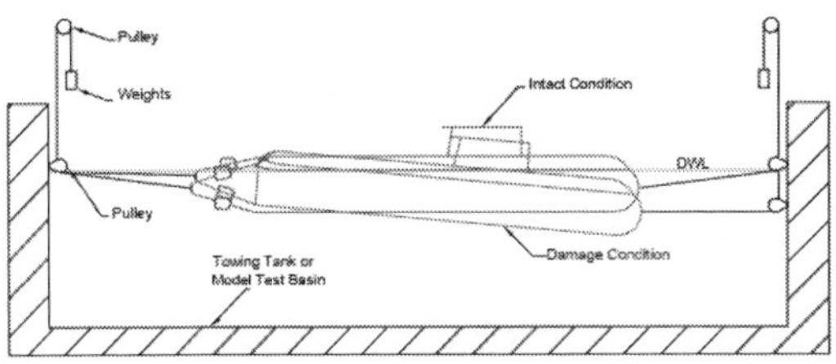

Fig 2 - Model Set-Up

Fig 3 – Intact Model in Tow Tank

3. EXPERIMENTAL RESULTS

Only a brief review of a selection of results is presented here, concentrating on those used to validate the CFD model. Since a detailed summary of the early results were published in Davies *et al* (2008), some conditions such as drift and heading are not reproduced. Lund, (2005), El-Atm (2006), Davies (2007) and Letter (2009) provide detailed analysis of the results.

Effect of Flooding Rates of the Casing in both Intact and Damaged Conditions

The tests for the three different casing slot heights in the damaged and intact cases revealed some distinctive characteristics in the roll response graphs for both conditions (Fig. 4). General characteristics include:

- Lower natural frequencies in the damaged condition for casings 50% and 100% open, and higher natural frequencies for casings 0% open.
- For intact and damaged conditions, with casings 50% open roll response is greater when compared against other casing open conditions.

For casing ports 100% open and variable sea states for the damaged and intact cases (Fig. 5), the following general trends were found:

- Roll RAO's decrease with increasing wave amplitudes.
- Response shifts to lower natural frequencies for intact and damaged conditions as the wave amplitudes increase.
- At lower sea states the damaged condition exhibits smaller roll response compared to the intact condition.
- Higher sea states cause the opposite effect, with the damaged condition having a greater roll response compared to the intact condition.

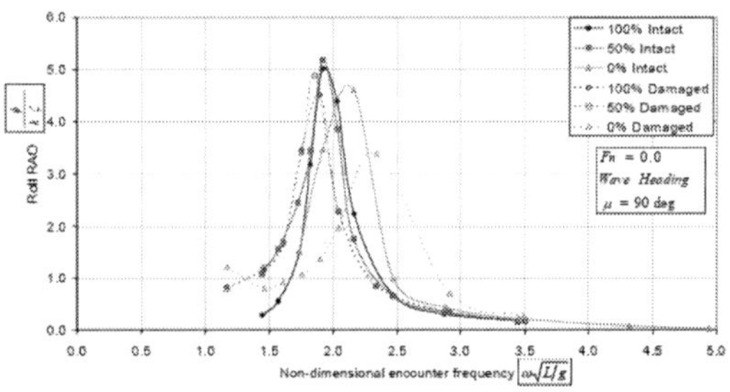

Fig 4 – Varying Casing Opening, Intact vs. Damaged, Sea State: 4, Height: 54 mm

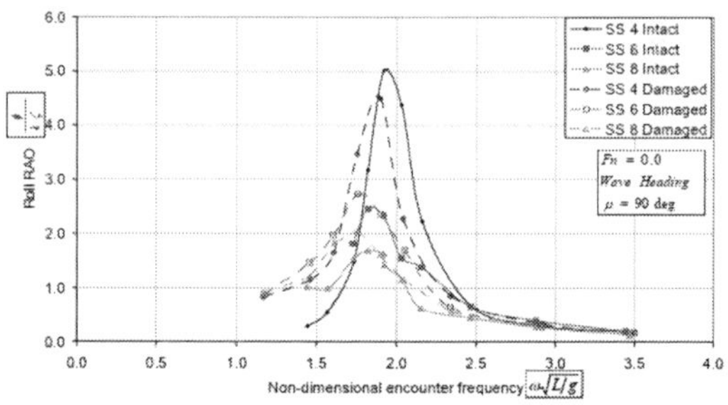

Fig 5 - Casing 100% Open, Intact vs. Damaged

The shift in natural frequency to a lower value is attributed to the increase in added mass of the vessel during rolling motions. Since the metacentric height of a submarine does not change significantly during rolling motions, it can be confidently assumed that the natural frequency is related to wave height and roll radius of gyration. Higher wave amplitudes will have a greater exciting force impacting on the vessel and therefore cause greater velocities and accelerations. In the damaged condition, the vessel has a higher roll radius of gyration due to the coupling effect of yaw. In addition, in a bow down plane, it will have greater added mass than for the intact condition, due to the asymmetrical wetted hull body. These two characteristics, linked with accelerations due to waves, cause a shift in the natural frequency to a lower value. Further, the magnitude of overall shift in the resonant peak is greater in the damaged condition than in the intact condition, due to a higher roll radius of gyration and asymmetric underwater body.

Effect of Control Surface Orientation

Another design feature of a submarine that affects the roll is the orientation of the control surfaces on the stern cone of the vessel. As stated earlier, three conditions were tested, *i.e.* the "+", "x", and "o" stern cone configurations, (see Fig 11).

Upon comparing the damaged and intact test results (Fig 6), there are some distinctive characteristics noticed in the roll response graphs for both conditions. General characteristics of the roll response graph include:

- Roll response of the damaged condition is greater than that of the intact condition.
- Natural frequency is lower for the damaged condition.
- Effect of tail configuration has greater influence on the magnitude of roll motions in the intact condition.
- The "x" tail is most effective in reducing roll motions in damaged condition.

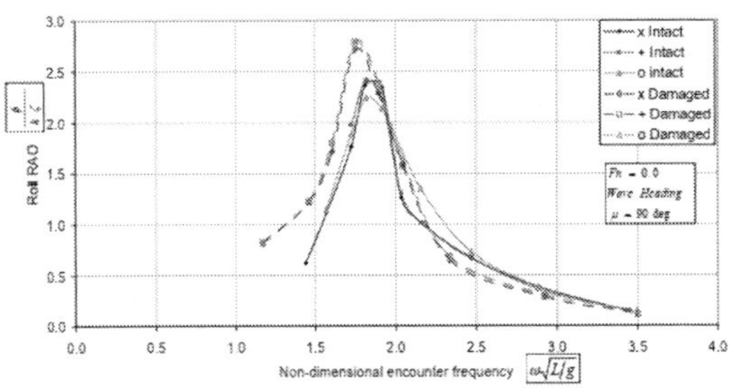

Fig 6 - Tail Cone, Intact vs. Damaged, SS 6, Hw = 143 mm, Casing 100% Open

Fig 7 - Damaged Condition in Tow Tank

The roll response in the damaged condition is larger than in the intact condition and is attributed to the immersion of the control surfaces. At a level trim, the control surfaces are almost completely submerged, however trimming the vessel causes the stern to rise (Fig 7) and reduce the wetted area of the control surfaces. The magnitude of roll damping is attributed to the wetted surface area of appendages (A_{app}), the distance to (the base of) the appendage from the centre of gravity (d_1), and the vessel's form of bilge (*i.e.* hard chine or round bilge). The more roll damping present, the less the vessel will roll. This is also recognised by Ikeda and Umeda, (1988), where damping is given by,

$$damping = \frac{0.55\left[A_{app}\sqrt{W} + 0.0024\,L\,B\sqrt{d_1}\right]d_1^{\,2.5}}{\Delta B^2} \qquad \text{...... [1]}$$

Where W is the width or height of appendage, L is the length of waterline, and B is the breadth of waterline. In the damaged condition, the "x' stern configuration provides greatest roll damping as there are always two control surfaces submerged, while the "+" stern has its horizontal appendages continually entering and re-entering the water when the vessel is rolling. Since the size of control surfaces is small with respect to the vessel's length and displacement, the difference in roll damping is also small. It is concluded that the effect of the tail orientation has only a minimal effect on the roll damping and hence the roll motions in damaged condition for a surfaced submarine.

Roll Decay

In order to ascertain the roll damping coefficients of the model submarine at various roll angles, roll decay tests were performed in calm water in the towing tank. This was achieved by releasing the model from a pre-determined static roll angle and recording the resulting roll decay curve. These tests were used to measure the amount of roll damping present during large amplitude roll motion of the surfaced submarine. A submarine in beam seas will generally experience large roll amplitudes at its natural frequency.

The tests were carried out for three different stern configurations, *i.e.* "o", "+", and "x" tails, with the starting roll angle changed to values between 10 and 50 degrees to determine the differences between the roll decay at large and small initial angles. The model was released from both the port and starboard sides to test if the response was uniform for both sides. The values of the damping coefficient were taken as the average of the first five rolls to port and five rolls to starboard. It should be noted that the first roll to port and starboard were not used in this analysis as the model's motions may have been irregular due to the initial excitation.

The results for the three tail configurations are shown in Figs 8, 9, and 10 respectively, with the tail configurations shown in Fig 11. It is seen that:

- The larger variations of the error bars at the higher angles are mainly due to the submersion of the casing at these angles. The damping effect of the submarine casing also increases the damping coefficient at the larger initial angles.
- As expected, the bare tail role damping coefficient at initial starting roll angles below 30 degrees are consistently lower than the roll damping coefficients of the "+" and "x" tail configurations, which is due to the added damping of the appendages. Beyond this angle, the damping coefficients for the three configurations show no significant difference, which may be due to the damping effect of the casing.
- The damping coefficient for the "x" tail is generally higher than that for the "+" tail, which is possibly due to two control surfaces always being submerged, (Eq. [1]).

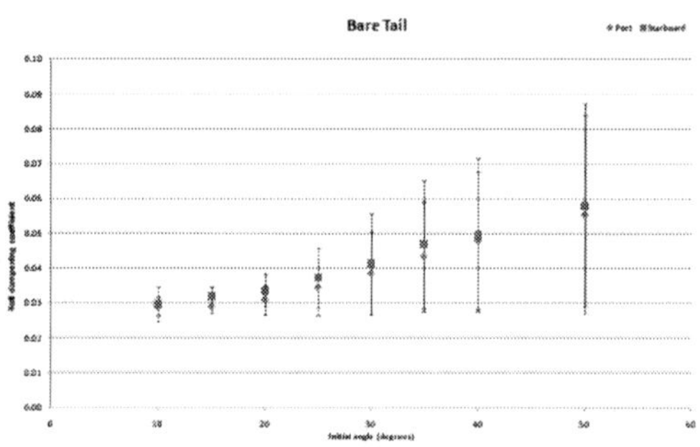

Fig 8 – Damping Coefficients for Bare "o" Tail at Differing Starting Angles

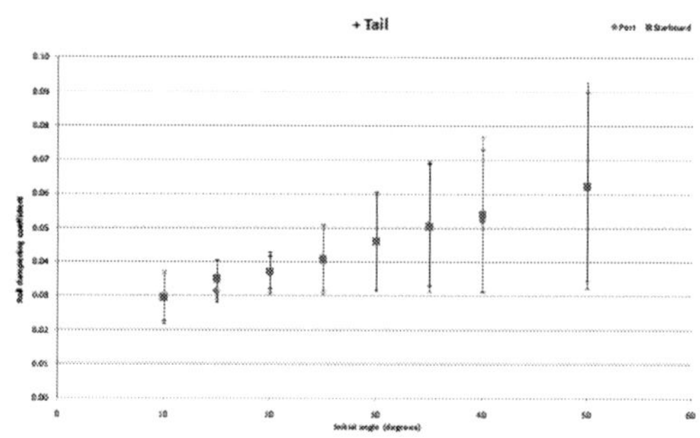

Fig 9 - Damping Coefficients for "+" tail at Differing Starting Angles

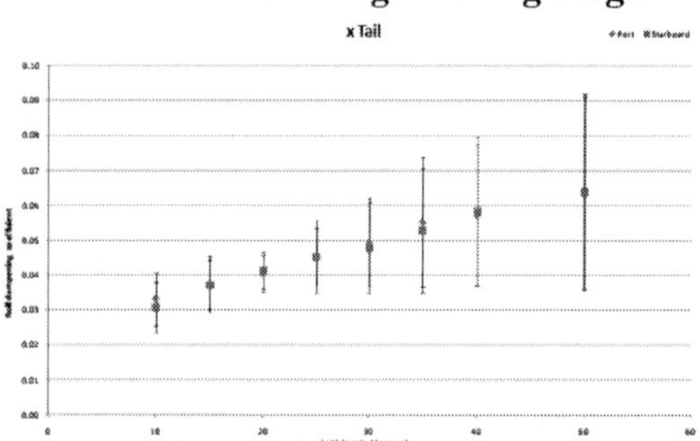

Fig 10 – Damping Coefficients For "x" tail at Differing Starting Angles

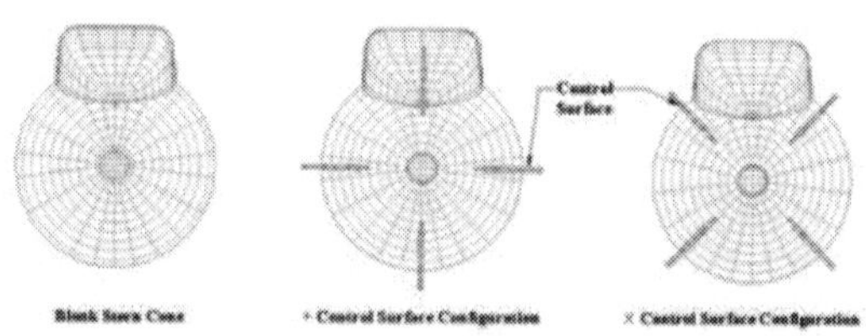

Fig 11 – Stern Cone Configurations

Davies (2007) conducted similar roll decay tests to evaluate the affects of adding wing tips and bilge keels to the model in beam seas. At low sea states (Fig 12), it is seen that the wing tips in the "x" stern configuration have a minor effect in reducing the peak of the RAO, while the bilge keels have a substantial effect in reducing this peak. From Fig 13, comparing the roll damping coefficients shows the improvement due to the installation of wing tips, and the larger improvement with the fitting of bilge keels. This is expected in this lower sea state, since the addition of any roll damping device to a submarine hull will be an improvement, as the submarine hull offers very little in the way of roll damping.

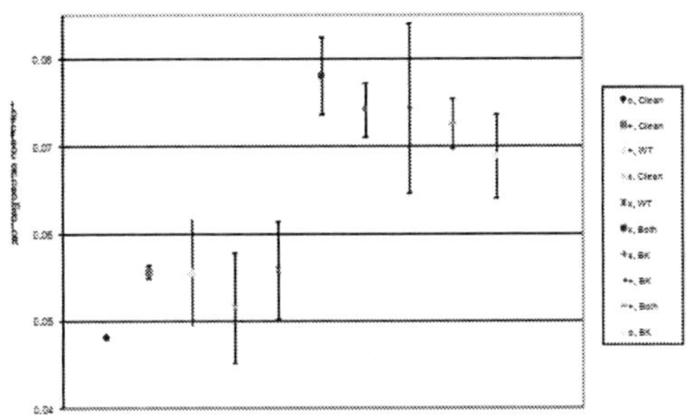

Fig 13 - Roll Damping Coefficients for Tail and Appendage Configurations

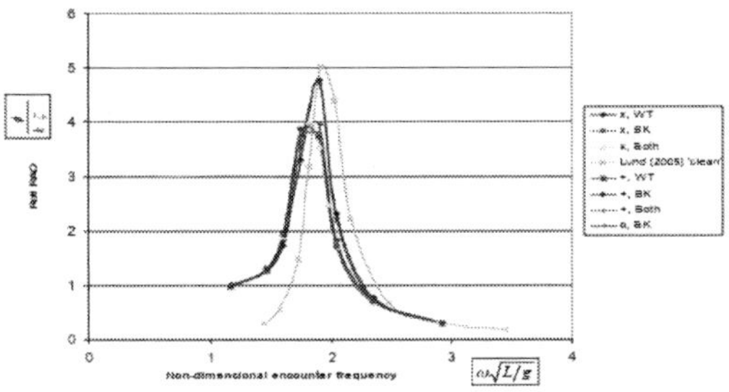

Fig 12 – Sea State 4, Roll RAO

At the largest sea state tested, the bilge keels appear to have an adverse affect on the roll motions, increasing the peak height of the RAO. The reasons for this increase in roll motion is attributed to the lack of reserve buoyancy and the lack of roll damping of the submarine shape, thus the bilge keels increase the excitation effects of the wave motions. The wing tips in this case provide better damping than the bilge keels in lower frequency waves, thus suggesting that to minimise roll motions in larger sea states the installation of wing tips is beneficial. The added benefit of the wingtips is that they also offer directional stability.

4. CFD MODEL

As mentioned previously, in parallel to the experimental work, analytical methods were also employed to predict the roll motion of the vessel. The first software tested was *ShipMo3D* (El-Atm, 2006), with Davies (2007) using CFX to capture the flow past the submerged control surfaces during roll. Following the introduction of a rigid six-DOF tool to CFX, (although this has been available for sometime in other commercial packages), the software was employed to model the submarine in waves and calm water to predict the roll motion and roll coefficients. Note: Thornhill and Hermanski (2008) conducted a CFX simulation of a 2D submarine, however it utilised CFX to calculate the forces on the body and used external routines written in FORTRAN to obtain the motion.

In order to assist with the validation, the simulation consisted of the scaled model submarine used in the experiments, with the fluid domain representing the towing tank facility, thus reducing the errors due to boundary and scaling effects. However, as the rigid body tool in CFX V12 is a β-function, its implementation was far from straight forward, as CFX is yet to provide detailed information on this feature. A number of parameters, settings, and options were identified and employed through discussions with the supplier, trial and error, and importantly, by comparison with experimental data.

The CFD analysis was carried out in two phases, *i.e.* developing the six-DOF model setup using a 2D simulation, followed by a full 3D simulation to replicate the experiments conducted in the towing tank. Initially the 2D model was developed with the rigid body encapsulated within an elliptical sub-domain (Fig 14), thus preventing excessive mesh deformation around the geometry. Later models excluded the ellipse, thus requiring further refinement to the mesh around the body. Although the former does give some errors, it is a useful tool for initial runs for bodies with complex shapes, providing a robust model that is less likely to fail.

To ensure accurate representation of the actual physical model, it was essential to calculate and enter realistic parameters such as the mass, added mass, inertia, centre of gravity, waterline and thus the centre of buoyancy. Errors in these values may cause inaccuracy and/or instability. It was also important to incorporate appropriate boundaries and relevant conditions to prevent the reflection of waves off the domain boundaries. The time steps and the relaxation factors were also adjusted to ensure stability and accuracy.

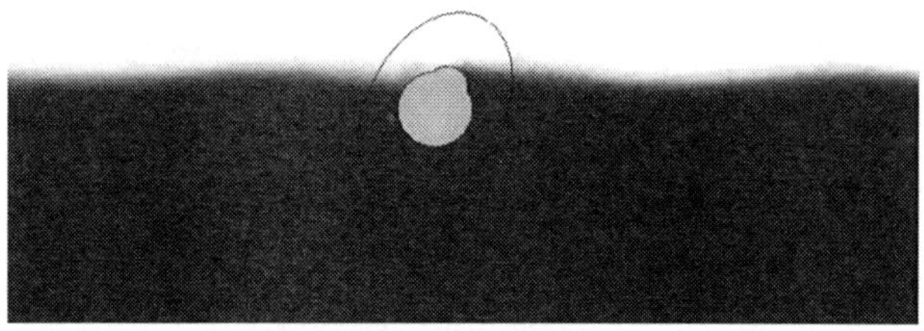

Fig 14 - CFX 2D Model in Waves

Once the 2D model was deemed sufficiently accurate and stable, a 3D model was developed, representing the scaled model submarine in the towing tank, thus including the end effects and providing motion in all six degrees-of-freedom. However, in the initial models the domain length in the direction of flow, *i.e.* along the length of the tank, was limited to keep the mesh size low during the testing phase. This unfortunately prevents proper wake resolution and wave dispersion. It is intended to increase the flow direction domain length in the near future to test the full domain and increase accuracy to enable complete validation against experimental data.

As in the 2D model, single and two domain models were created and compared. The latter proved to be more robust than a single domain with the same number of elements, however, it took considerably longer to solve. In addition, the two domain model produced some errors at the interface, especially around the free surface, which adversely affected the motion. This was more pronounced in the 3D model, possibly due to the 3D flow field. Thus, all 3D simulations were carried out using the single domain model, as the increase robustness was negated by the increase in run time.

As with the 2D model, it was essential to get the correct initial and boundary conditions. Unlike most surface vessels, the vertical centres of gravity and buoyancy (VCG and VCB) of a surfaced submarine are extremely close to each other. Thus, slight errors in the mass, added mass, inertia, waterline, VCG, and surface roughness will cause significant variation in motion amplitudes and frequencies. Therefore, these values must be predicted as accurately as possible.

Fig 15 shows a comparison between the CFD and experimental results for roll decay of the submarine model. Given that the simulation was carried out on a shorter than required domain (Fig 16), the results are seen to be very close. However, the CFD model tends to over-predict the roll decay, which indicates that the settings need further adjustment to reflect those in the experimental model. Further upgrading and testing of the CFX model is currently being undertaken to improve accuracy.

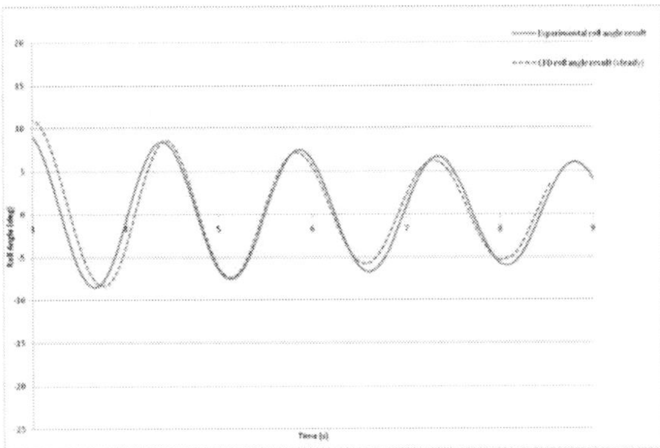

Fig 15 - Comparison of CFX and Experimental Roll Decay

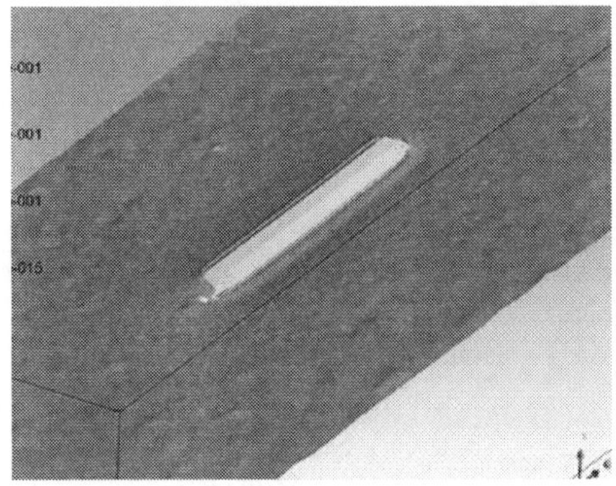

Fig 16 – 3D Roll Decay Simulation

5. CONCLUSION

The motion experienced by a surfaced submarine in varying sea and vessel conditions has been investigated through conducting physical model scale experiments and predictions using computational fluid dynamics.

At smaller wave heights, the damaged condition exhibits smaller roll amplitudes than the intact condition, however this changes as the wave height increases. For both conditions with casing at 50%, roll motions are greatest for all wave heights. The natural frequency reduces with increasing sea state. However, there is an increase in the damage condition when the casing is closed. This is attributed to an increased displacement to keep a bow down trim of five degrees.

The "+" and "x" stern configurations both exhibit higher damping over the bare stern at roll angles below 30 degrees. However, beyond this angle there is no significant difference between the three configurations, possibly due to the influence of the casing. In addition, the "x" tail tends to have the highest roll damping due to the immersion of two control surfaces. The addition of bilge keels helped reduce roll amplitudes at low sea states, but has an adverse effect at large sea states. Although the wing tips do improve damping in all cases, the improvement is marginal. In general, the damping due to the immersed hull body is deemed a greater contributor to damping than the control surfaces.

The 3D CFX simulation of the surface submarine utilising the new six-DOF rigid body dynamic model within a single domain was capable of replicating experimental values, but proved highly sensitive to input data. The model is currently being improved and is undergoing further validation.

6. ACKNOWLEDGEMENTS AND REFERENCES

The authors would like to thank Mr Warren Lund, Mr Jay El-Atm, Mr Andrew Davies, and the staff of the Hydrodynamics Centre at AMC for their contributions and assistance throughout the project.

1. Davies, A., (2007), "The Effect of Appendages on Roll Damping on a Surfaced Submarine", BE Thesis, AMC.
2. Davies, A., Ranmuthugala, D., El-Atm, J., and Tso, Y., (2008), "Reduction of Roll Motion of a Surfaced Submarine in Beam Seas", Pacific 2008, Sydney

3. El-Atm, J., (2006), "Roll Motion of a Damaged Surfaced Submarine in Beam Seas", BE Thesis, AMC
4. Ikeda Y. and Umeda U., (1998), "Application of Slender Body Theory to Rolling Motion of a Moving Ship with Bilge Keels", SNAME, May, Japan
5. Letter, B., (2009), "Numerical and Experimental Roll Response and Decay of a Surfaced Submarine", BE Thesis, AMC
6. Lund, W., (2005), "Parameters that Effect the Roll of a Surfaced Submarine in Beam Seas", BE Thesis, AMC
7. Thornhill, E. and Hermanski, G., (2008), "Numerical and Experimental Analysis of Surfaced Submarine Roll Decay Behaviour". Journal of Ocean Technology, Ocean Sovereignty, Vol. 3, No. 1

ANZAC Ship Margin Management: Challenges and Opportunities

M. WAUGH, ANZAC Systems Program Office, Rockingham, WA, Australia
Z AHMED, ANZAC Systems Program Office, Rockingham, WA, Australia

ABSTRACT

Warship design is an iterative process of dovetailing requirements with technology where trade-off studies are undertaken for size, weapon systems, platform performance and cost resulting in what could, arguably, be referred to as a great compromise of requirements when the general arrangement is first presented. Establishing realistic margins are part of these early design studies. In the art versus science of Naval Architecture the management of margins comes under the science realm while future proofing, the art component, requires estimating what may be demanded such that major integration difficulties are not encountered in service. Challenges will occur as the platform typically has a service life of 25 years whilst most combat system components, with a shorter technology cycle, become obsolete or ineffective before 10 years has typically passed so grasp opportunities in the design phase.

The ANZAC Class Frigate was born as a long-range patrol frigate. By early adolescence the Fitted For But Not With capability had largely been installed and by mid life preparations to install Anti Ship Missile Defence (ASMD) were underway. This evolution has transformed the ANZACs into a warship capable of force protection. However the Service Life Margins have been depleted in the process.

Experiences in margin management of ANZACs for weight, real estate and electrical power are discussed. The phrase coined by Brown of *innovate or die* should propel designers towards future proofing through a combination of art and science. The tools and models developed for ANZACs can be adapted to Future Frigate design.

INTRODUCTION

Warship design is a great compromise of requirements. Margin management is as much customer expectation driven as it is proven naval architecture systems engineering. Feasibility studies and trade-off options lead to better utilisation of real estate and weight margin for warfighting improvement. Sir Rowland Baker [1] was a great innovator of the Royal Corps of Naval Constructors and he often maintained that any new design should be 25% innovation and 75% well-tried practice as *more novelty would make it unreliable and less would make it obsolete in concept.* D K Brown [2] expressed this philosophy more bluntly in coining the phrase *Innovate or Die* therefore a combination of art and science is required for future proofing frigate in the typical 25 year service life of a warship.

Friedman [3] comments that *controlling margins was a way of controlling displacement.* However margins must have been realistic from the early design stage. In the art versus science of Naval Architecture the management of margins comes under the science realm. However margins can be subjective allowances based on risk dependent upon at what design phase the ship is at. Heather [4] suggests future growth margins to be allocated at delivery, to allow for potential upgrade, to be *5% for weight, no propulsion power margin other than that associated with displacement increase, 5% electric power only if generators not to be replaced at half life and no space margin other than specified future fit items to have space allocated.* With these low margins at delivery the detailed design would need to be accurate and the technical risk assessed as low. The UK guidance document for margin policy [5] proposes varying margin percentages depending on design maturity whilst still allowing a Contract Modification Margin (CMM), Capability Upgrade Margin (CUM) plus In-Service Growth Margin (IGM). The ANZAC Frigate was consuming its minimal margins whilst steel was being cut through contract requirement modifications.

Future proofing, the art component, requires estimating what may be demanded such that major integration difficulties are not encountered. Service life margins are minimal and their through-life management, the science component, requires planning models and tools to be developed to optimize warfighting capability insertion to end-of-life.

FRIGATE MARGINS

Margins have a cascading effect on a ship's size: the provision of one margin results on growth of a particular system, which in turn requires another system to grow proportionately, consistent with classical design spiral. Margins are generic and typically broken down into Technical Subject Code such as lightship and real estate or weight and space, electrical power plus heating, ventilation and air conditioning. These margins are not originally presented in capability components such as above water warfare and underwater warfare. Therefore clearly identified capability insertion margins cannot be quarantined or defended if it is not identified in the original requirements documents.

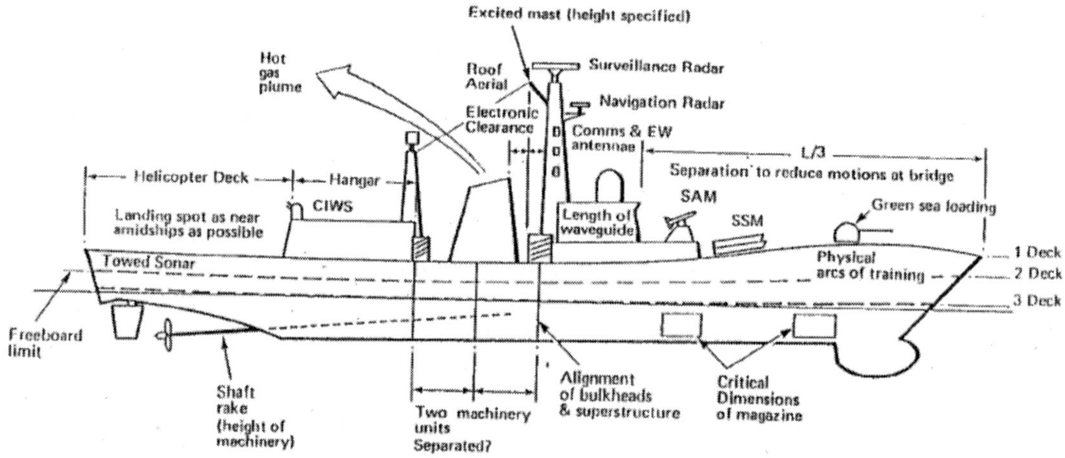

Figure 1: Combat system design drivers for monohull frigates from Brown [7]

The flight deck and aviation facility requirements significantly influence the design of the modern frigate. However, the overall upper deck disposition of the combat system major elements, along with operational features, drives the early design as shown in Figure 1 from Brown [7] for a traditional monohull frigate. This results in there being little room for innovation in the design of a helicopter capable frigate. If there is little room for innovation then there should be low risk in the design decisions. This assumption can be a pitfall should there be little to no CMM in the design and a 3” main gun is changed to a 5” main gun. The perception of low risk has a downside in low margin percentages being allocated at early design for both CUM and IGM which may lead to integration difficulties later in life.

For a 3500 tonne frigate with a 1000 tonne variable load for fuel, ammunition, stores and provisions the lightship would be 2500 tonnes. If the lightship or weight margin is typically five percent [6] then in this example the lightship or weight margin would be 125 tonnes for the service life of the ship. How much should be reserved for each warfighting capability when there are so many consumers vying for this all-important margin?

Real estate or space in a frigate is finite. This margin is often disguised under terms of Spare Compartment or Fitted For But Not With (FFBNW) space which is reserved for future capability. This utopian condition does not match the reality which follows, more or less, a use-it or lose-it philosophy.

Typical service life margin goals in the US Combatants [8] have been:

- 20% electric power reserves
- 20% air conditioning capacity reserves
- 10% full load displacement growth
- 1 ft of full load center of gravity rise

The US designs have very conservative early-design margins which in the case of the *Oliver Hazard Perry* (FFG 7) were ‘eaten up’ during the later stages of design and during construction [9]. FFG-7 underwent an extensive stern redesign to enable the ship to accommodate the LAMPS-MK III helicopter, its hauldown system, and the towed sonar system [10]. Intending to restrain size and cost of these ships the service life margins of the FFG-7 frigate were minimal from the design stage. However this redesign of the stern, predominantly driven by the requirement to modernise the helicopter capability, resulted in an extension of 2.5 metres of the hull to achieve the weight and space required. The minimal margins from the outset required a costly stern extension to carry a more capable helicopter as the ship was not future proofed for its 25 year service life. The lessons learnt from FFG-7 modernisation difficulties need to be captured into corporate knowledge databases.

MARGIN MANAGEMENT

It is prudent to allow varying margin percentages depending on design maturity whilst still allowing a Contract Modification Margin (CMM), Capability Upgrade Margin (CUM) plus In-Service Growth Margin (IGM). All of these specific margin requirements are included in the vessel Functional Performance Specification. Good in theory – but what is the practical experience? The ANZAC Frigate was consuming the minimal margins whilst steel was being cut through introducing contract requirement modifications It can be difficult to determine if the CUM is absorbed by osmosis into the IGM or alternatively the IGM is incorrectly assumed to cater for the CUM in a half-life upgrade program. Clear understanding of the various margins is required to avoid confusion especially when questioned on margin depletion. Figure 2 illustrates the lifecycle approach [6].

Margins are defined. The IGM *allows for unattributable growth to the platform throughout its service life* [6] and is not intended as a supplement to the CUM. It is most important to understand that the intention of the IGM is not to facilitate warfighting capability insertion. It is possible, however, to upgrade marginless designs by extensive substitution of new equipment for old. Examples were illustrated in old Soviets where Surface to Air Missile was upgraded in *Krupny*-class destroyer and bow sonar was modernized in *Kanin* Class [9].

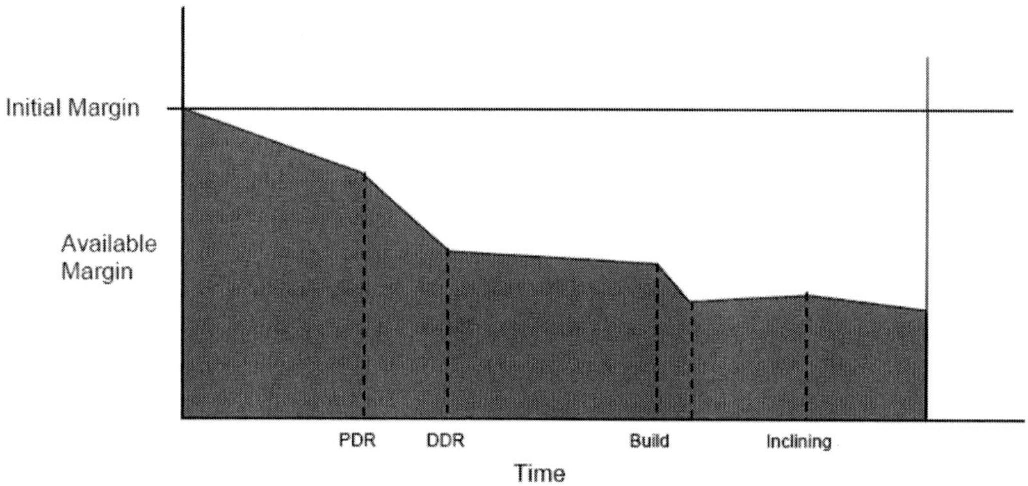

Figure 2: Example Lifecycle Plot (Can be generically applied to any margin)

Service life margins are not reserved for each warfighting capability and organisations almost compete to expend the available margins in as quick a time as possible in populating the FFBNW capability. In addition, these margins may not have been sufficient at time of build if unrealistic expectations were envisaged in the use of the ship. Margins in weight, real estate and electrical power need to be reserved for the aviation capability such that when the time comes to undertake capability insertion there will be far fewer integration difficulties.

Service life margins are called for in build tender documentation with compliance to be demonstrated as part of acceptance into service. Thereafter, the organisation responsible for management of these margins is often not able to wield influence over those organisations requiring capability improvements. Projects are often stovepiped with little regard to accumulative effect or full integration requirements. In addition, there is also more than one customer to manage the expectations of. Upon acceptance into service the CUM plus the IGM should be available to manage the impact of changes and unaccountable growth out to end-of-life.

ANZAC FRIGATE CAPABILITY GROWTH

The acquisition strategy of ANZAC Ship Project sought an *existing design* in the interest of risk reduction and schedule pressure [11]. Whilst there was no endorsed Detailed Operational Requirement (DOR) for the project, the following three technical documents collectively defined the requirements:

- Combat System Tactical Operational Requirement,
- Ship Performance Specification, and
- Basis for Acceptance into Naval Service.

The selected MEKO 200 ANZ was based on the existing MEKO 200 PN design, under construction at that time for the Portuguese Navy but during the design development contract, a number of major engineering changes were incorporated in the MEKO 200 ANZ to suit the requirements of RAN and RNZN. Changes affected the propulsion system, ship systems, communication systems, combat systems and aviation system integration. The scope of these design changes together with the configuration changes incorporated prior to contract award, represented a substantial engineering impact on the existing MEKO 200 PN design. That shifted the emphasis from the original strategy of building an existing design with a minimum of changes, to the maintenance of function and performance. The implication was that the weight margins, in particular, was consumed before the start of production work.

Over a period of 15 years since HMAS ANZAC was launched in 1994, the ANZAC Frigates have undergone number of significant capability upgrades. The displacement grew from 3500 tonnes full load at delivery to currently 3700 tonnes with the introduction of Evolved Sea Sparrow Missile, Mine and Obstacle Avoidance Sonar (MOAS), ASTIS and a Harpoon Missile Launch capability. The Nulka missile decoy and a replacement surface launch torpedo have also been fitted in addition to at least 50 other engineering changes with a weight and moment impact.

The Antiship Missile Defence (ASMD) capability is being fitted on HMAS PERTH as first of Class with a CEA developed Phase Array Radar system and will be rolled in to the remainder of the class in due course. As a consequence, the ships will soon receive a second service life weight and stability margin through stability enhancement modifications. Approximately 100 tonnes will become available for warfighting capability upgrades to be installed to end-of-life. Table 1 depicts high level generic commitment of this attractive 100 tonnes. The weight and location of these changes are shaped in the stability software program under Whole Of Ship Management (WOSM) planning models allowing a strategic outline to be maintained at WOSM level.

Platform system 20 tonnes sub total		Combat system 60 tonnes sub total		Unallocated sub total	
Propulsion upgrade	4 tonnes	Anti Ship Missile Defence	15 tonnes	Obsolescence	10 tonnes
Generator upgrade	2 tonnes	FFBNW 2nd VLS Module	32 tonnes	Uncontrolled	10 tonnes
HVAC upgrade	2 tonnes	Replacement Helicopter	3 tonnes		
Cooling water upgrade	2 tonnes	Future capability	10 tonnes		
Future capability	10 tonnes				

Table 1: Typical 100 tonne Monohull Frigate Service Life Margin allocation

SHIP PERFORMANCE CHARACTERISTICS

ANZAC Frigate Whole of Ship Characteristics is managed holistically through the Technical Management Model (TMM) as illustrated in Fig 3. TMM is a high level plan within ANZAC SPO. Elements include speed, range, strength, weight and stability, manoeuvrability, survivability and shock. Note that separate sections of the TMM address margins and signatures.

Speed is one of the four characteristics (plus stability, range and strength) that contribute to the Contractual Limiting Displacement for the ANZAC ships. When built the ships met the Contractual requirement. Full power sea trial reports provide continuing data from In-

Service procedures for testing full power, in both Gas Turbine and Propulsion Diesel Engine. Increased displacement (above that originally contracted for end-of-life) will result in full power speed reductions. Theoretical calculations plus model testing have been undertaken for ANZAC Frigates at a predicted 3900 tonnes end-of-life displacement. Full power sea trials will be undertaken to validate the research work.

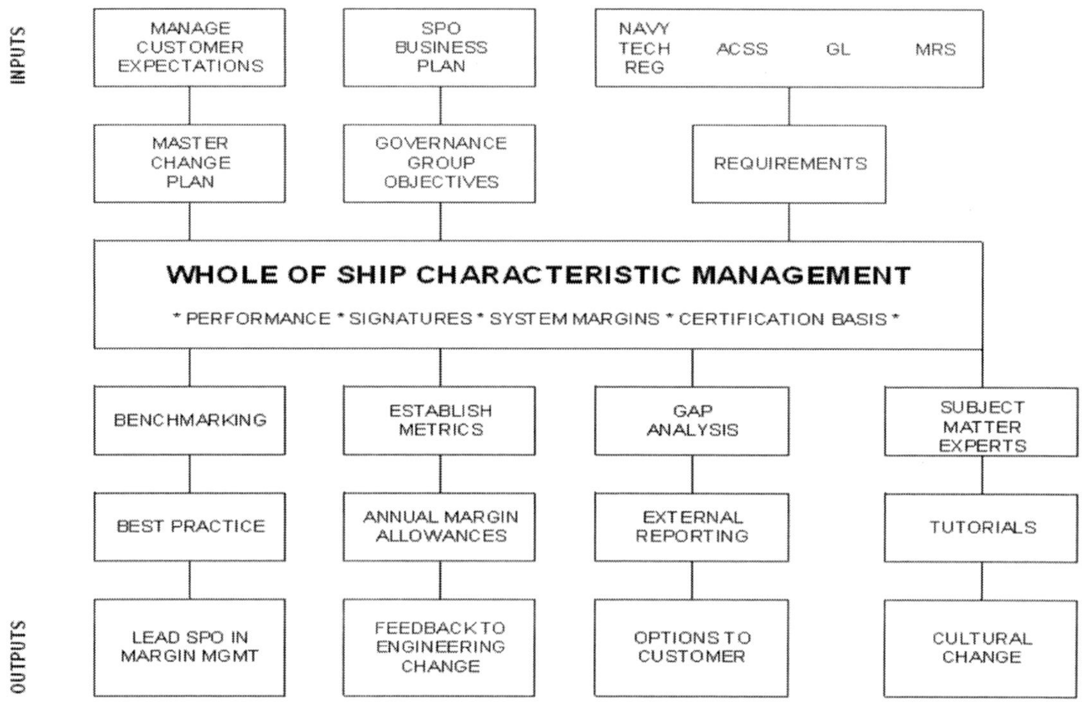

Figure 3: Holistic Approach to Managing Whole Of Ship Characteristics

Increased displacement elevates fuel consumption proportionally with the required speed which in turn reduces the range achieved for a given capacity of fuel. As delivered, ANZAC Frigates exceeded the contracted range requirement. The actual fuel capacity for the contractual range calculation indicates *considerable excess fuel* for the ship to steam the required distance. This excess allows for a potential significantly greater range. Whilst in some load conditions the ship stability condition may require some of the excess fuel to be partitioned and carried as liquid ballast, the available fuel will enable the contracted range requirement to continue to be achieved despite the predicted displacement growth towards end of life.

For a warship, few individual engineering changes are expected to have a significant impact on global ship strength. Whilst many engineering change tasks require investigation of the local strength impact there will not be many that require a re-assessment of global strength. However, collectively the engineering change tasks and additional outfit items will gradually add significant weight to the ship. For ANZAC Frigates, three DSTO studies present the results of their investigation into structural strength and fatigue life at the predicted end of life displacement of 3900t whilst operated at 4800 hours per year at sea for 30 years. The Classification Society, Germanischer Lloyd undertook independent studies with good correlation to those results of DSTO in global ship strength at 3900 tonnes end-of-life displacement.

Stability is one of the four characteristics that contributed to the Contractual Limiting Displacement for ANZAC Frigates. The ships stability is extremely important and it is paramount that it be managed as a whole of ship characteristic to maintain safety and certification. Stability margin management is detailed in the next section.

As the displacement of the ship increases the response to helm will degrade slightly. The ship will typically display greater directional stability however the tactical diameter, or turning circle, will increase. Engineering changes to date plus ASMD and solid ballast will change the displacement and position of the Vertical Centre Of Gravity of the ANZAC Frigates as the displacement increases from 3550 tonnes to 3900 tonnes. Various manoeuvring trials (turning circles at various speeds, "S" turns, forced roll etc) have been proposed post ASMD to determine if the ships manoeuvring has been unacceptably affected by changes since delivery.

STABILITY MARGIN MANAGEMENT

Stability is the only margin that is controlled under a Certification Basis whereby the Approved Limiting VCG Curve forms the certification basis. The ANZAC Frigates are stable and safe to operate as delivered, however, whatever SLM that did exist at delivery has been consumed by the large number of configuration changes that have been accomplished to date and the significant increase in outfit items in support of current operations. ANZAC Class Ships are currently at or approaching the limiting displacement in the Full Load Condition as well as at or approaching the limiting VCG in the Minimum Operation Load Condition. There is a problem to manage!

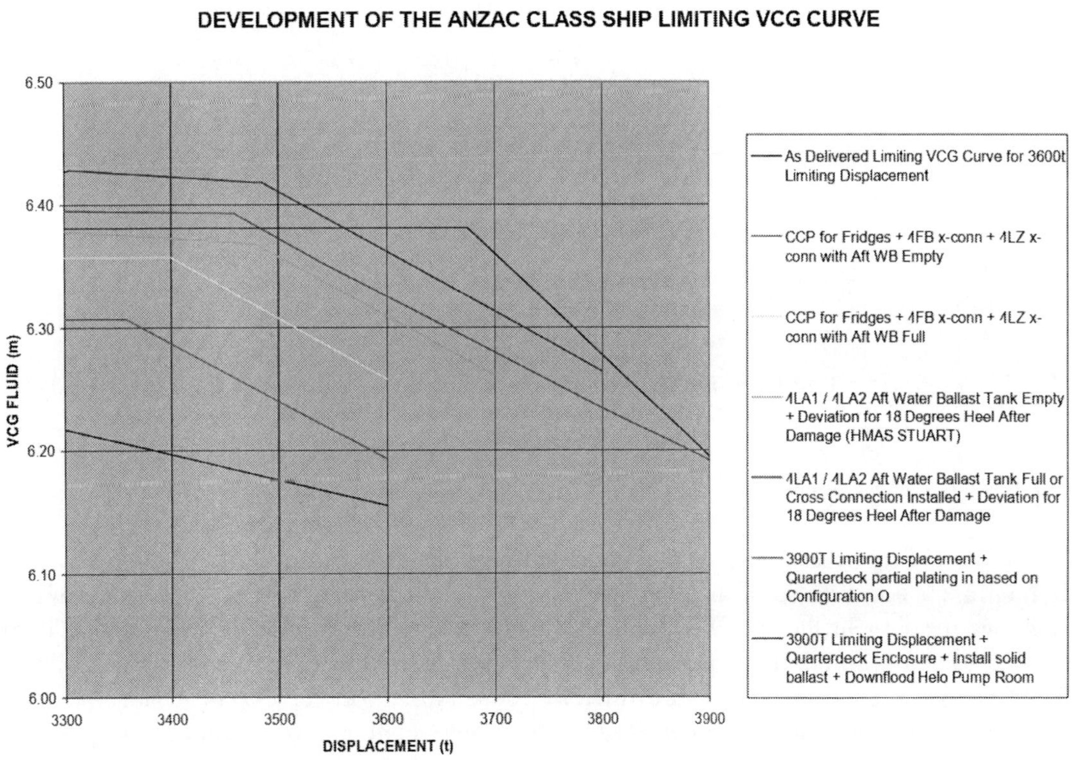

Figure 4: Limiting VCG Increase from Stability Focused Engineering Change

Stability is being reassessed to provide capacity for planned major capability upgrades and give a modest growth allowance to end-of-life. Given the weight growth anticipated it was acknowledged the ships would experience some difficulty meeting the current stability standard. Investigations were undertaken to identify modifications to the ships to restore the stability margin. The stability rectification tasks involve cross-connection of tanks, plating in of the quarterdeck and the installation of a substantial amount of permanent ballast. Figure 4 shows improvements to the Limiting VCG Curve achieved from the original design while Figure 5 illustrates the progression from current to forecast end-of-life conditions. There is a solution to the problem!

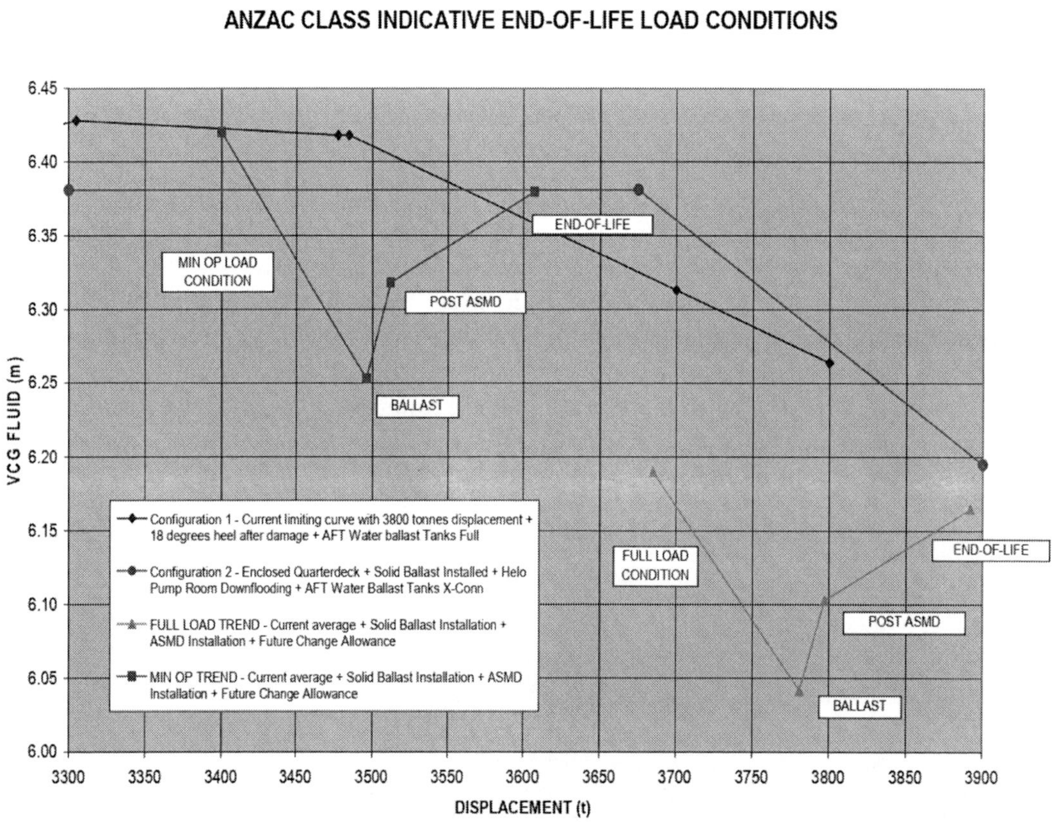

Figure 5: Full Load and Minimum Operating Load Condition to End-Of-Life

REAL ESTATE PLANNING

The ANZAC Frigate example of a gymnasium now residing in the Port Sonobuoy Stowage Compartment is an illustration of approved configuration change while the weight circuit machines residing in the Air Weapons Magazine is an illustration of crew initiative. Real estate planning requires quarantining as per the original, requirements based, design intent. These compartments, once lost, are difficult to reclaim however *corporate discipline* can overcome the *possession is nine-tenths of the law* approach to compartment allocation.

The Real Estate Planning System (REPS) is another component of the WOSM planning models which allows for real estate management, of high profile compartments, and is currently controlled through a CAD package of 3D drawings. The drawings depict the configured baseline allowing for bulkhead or deck space to be reserved for change tasks under early design consideration. REPS had its genesis in deconflicting aerials, radomes, close-in weapon system firing arcs and missile efflux above the hangar as Figure 6 depicts. The hangar itself has been recently incorporated into the suite to maintain helicopter clearance requirements. Photogrammetric depiction, or virtual vision, is an emerging application of photo realistic walk-through technology no longer just for computer game playing realism. Virtual Vision is currently being investigated at WOSM level following a pilot project of the hangar. Real estate can be more readily viewed in movie like quality and the print screen function allows for easy markup for text comments and space reservation. The technology is versatile and other organisations in the ANZAC Frigate community are interested in the same product for computer based training and condition monitoring.

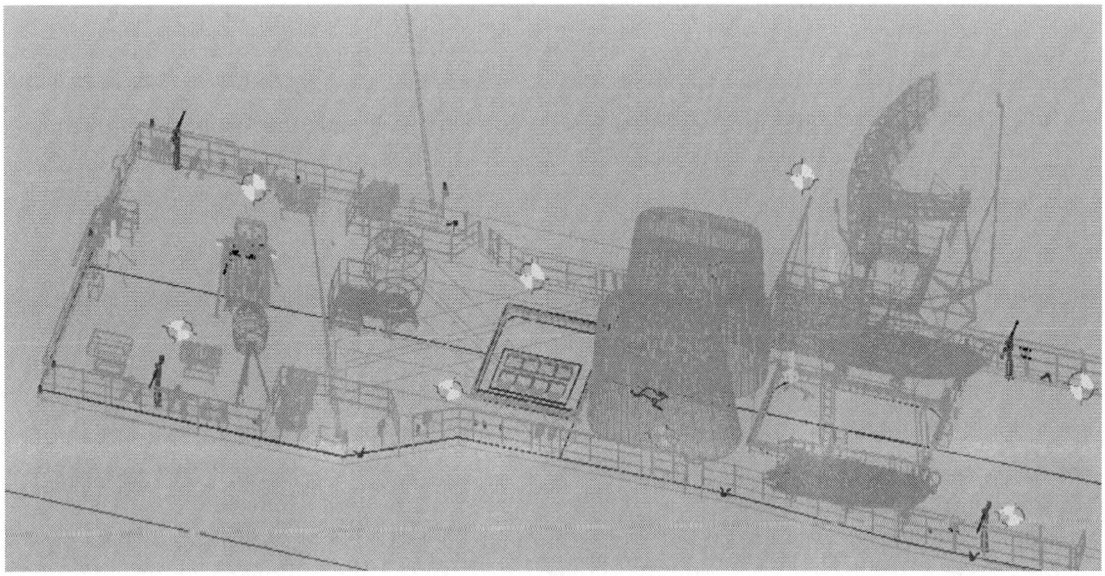

Figure 6: Screen Capture from the Real Estate Planning System

AVIATION FOCUS

Medium size helicopters are the versatile stalwarts of frigate aviation and the helicopters themselves evolve to meet future needs whilst the design margins for real estate, weight, electrical power, safe ammunition stowage and maintenance access are progressively consumed. This helicopter evolution tends towards heavier and larger designs that at worst case may no longer fit into the hangar or the handling system becomes inadequate. Envelope compliance monitoring is required as shown in Figure 7.

Aviation capability has been a recent focus for AIR 9000 Phase 8 – Future Naval Aviation Combat System. Waugh contends [12] the future proofing built into the ANZAC Frigate for deck strength, hangar real estate and handling system will be tested in integrating the chosen next generation helicopter. The handling system, as the interface between the ship and the helicopter, that remains tailored to handle one type of helicopter per ship is not future proofed. Handling system versatility is required as this leads to interoperability among partner Navies under joint operations with varying helicopter types being able to be hosted.

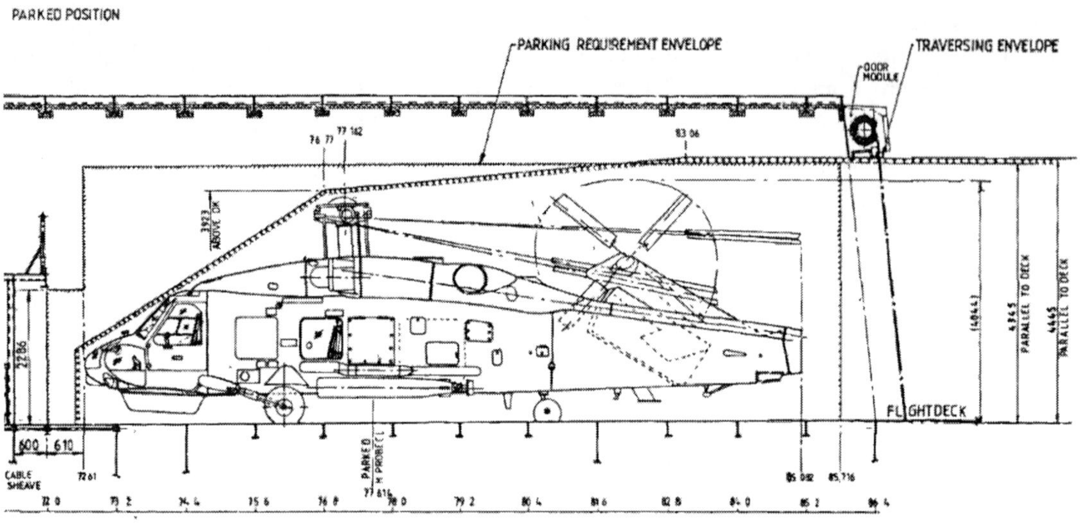

Figure 7: Baseline for Parking and Traversing Envelope Compliance Monitoring

ELECTRICAL POWER MARGIN MANAGEMENT

Electrical Power Generation and Distribution (EPG&D) is another important margin that requires management for the service life of the ship. However, once again, a specific percent of this margin is not reserved for the various warfighting capability areas. The EPG&D margin is typically ten percent for the service life of the ship [13]. However, this margin is associated with generating capacity, spareways in switchboards plus the cable support system and not converted or exotic supplies potentially resulting in shortcomings in service supply of 400Hz supplies and/or 24VDC supplies. Equally it is often overlooked that a new communication system, for example, is integrated at preliminary design given its power requirements however the uninterrupted power supply plus the self-contained air conditioner identified late in the design process triples the true power demand which can lead to a localised lack of power. Integration needs to be considered at WOSM level.

In ANZACs considerable effort is being extended to understand the actual load demand in various operational states so that an accurate picture of the actual margin capacity can be obtained. Design diversity factors given in standards such as BV 030 are very appropriate when planning an electrical system for a new ship, but of little use to the designer of a configuration change in service that will have an impact on electrical loads in a variety of operational states. Inclusion of the requirement and business rules for a minimum load state (in ANZACs this is called Vital Loads) is very important. In this way load shedding arrangements can be managed through life to ensure that in the event of a major failure or action damage, sufficient capacity is maintained to allow orderly power restoration.

MARGIN EFFECT ITERATIVE NATURE

Just as the design spiral is used to show the iterations towards design maturity the relationship in engineering change introduced in-service has a follow-on effect to all margins and ship performance characteristics. The WOSM level is often overlooked when introducing a capability that requires additional manning. Additional manning has an effect on the support services for heat load, real estate, power consumption, provisions capacity, fresh water consumption and sewage plant capacity. If not managed at WOSM level then systems can be found insufficient under increased manning. If systems require upgrading then this, in turn, effects ship displacement which affects the performance characteristics of speed, range, structural strength and shock.

The requirement for WOSM tools to manage at this level of iterative complexity has been identified by ANZAC SPO and a body of work has commenced.

FUTURE PROOFING FRIGATE DESIGNS

The ability to forecast the future requirements for the duration of the service life of a frigate is not possible. The warfighting requirements will change, that is certain, but to what capability? Given the relative short technology cycle to develop and manufacture some electronic systems it is possible the warfighting capability of a frigate in 20 years time remains to be invented. Therefore a combination of art and science is required for future proofing. This is no different to generic planning for future technology insertions through having both CUM and IGM available from acceptance into service as the platform typically has a service life of 25 years whilst most combat system components, with a shorter technology cycle, become obsolete or ineffective before 10 years has typically passed

Establishing realistic margins are part of the early design studies. Excess space and weight at ship acceptance can be considered a good characteristic. However, it is sometimes difficult to embody the lessons learnt from the past into future frigate designs. These may be spend-to-save initiatives such as commencing with a physically larger ship or having full redundancy in EPG&D. The effect of manning levels needs to be understood and a "margin" for personnel, and their effect on platform support systems, would be prudent.

CONCLUSIONS

The phrase coined by Brown of *innovate or die* should propel designers towards future proofing through a combination of art and science. Future proofing should take cognisance of interoperability and emerging technology. Excess space and weight at ship acceptance is a good characteristic and while acknowledging there is little room for innovation in the design of a monohull frigate the margins from the outset should be realistic. If there is little room for innovation then there should be low risk in the design decisions. This assumption can be a pitfall should there be little to no CMM and the requirements quickly change after contract signature.

This perception of low risk has a downside in low margin percentages being allocated at early design for both CUM and IGM which may lead to technology insertion or platform integration difficulties later in life. However, it is sometimes difficult to embody the lessons learnt from the past into future frigate designs. Spend-to-save initiatives such as commencing with a physically larger ship or having full redundancy in EPG&D should be researched under early cost-benefit analysis.

ANZAC Frigate Ship Performance Characteristics, Service Life Margins and Signatures are managed through the Technical Management Model, a high level plan within ANZAC SPO under a holistic approach to managing Whole Of Ship Characteristics. Models, databases and tools have been developed to enable management of these characteristics through to end-of-life. The iterative complexity of performance and margins has been identified by ANZAC SPO and a body of work has commenced to take the developed models, databases and tools to new functionalities.

DISCLAIMER

The opinions expressed are those solely of the authors and do not represent margin management philosophy embodied within the RAN Materiel Requirements Set.

BIOGRAPHY

Mal Waugh is currently the Stability Manager within the ANZAC Systems Program Office of the Defence Materiel Organisation, Australian Department of Defence. He is responsible for platform system margin management for the ANZAC Frigates of the Royal Australian Navy. Restoration of stability margins to end-of-life, such as installing solid ballast and enclosing the quarterdeck, plus preparing to integrate a replacement helicopter are the strategic focus points to his current employment.

Ziauddin Ahmed is currently the Whole of Ship Manager within ANZAC Systems Program Office of the Defence Materiel Organisation, Australian Department of Defence. His responsibility is to ensure that the Technical Integrity of the ANZAC Class is retained in the engineering activities conducted in the SPO. He was involved with design requirements and reviews of major capital programs of Royal Australian Navy. Before joining the ANZAC SPO, he was the Principal Naval Architect with the Air Warfare Destroyer Program.

REFERENCES

[1] Baker R. (1959), contribution to, *Some problems in the construction of warships today*, Transactions of the Institute of Naval Architects, Volume 101, London, UK.

[2] Brown D. K. (1991), *Innovate or Die*, Journal of Naval Engineering, Volume 33, Number 1, London, UK

[3] Friedman N. and Baker A. D. (2004), *US Destroyers: An Illustrated Design History*, Naval Institute Press, USA

[4] Heather R. G. (1990), *Preliminary Design Considerations for Fast Warships*, The Future of Surface Warships International Symposium, Paper 11, 4 June 1990, London, UK.

[5] Maritime Acquisition Publication (2007) Number 01-070 (MAP01-070), *Surface Ship and Submarine Margins Guidance*, Issue 02, UK.

[6] Materiel Requirements Set (MRS) Volume 2 Part 29, *Margin Requirements*, Issue 01, June 2008, Canberra, Australia.

[7] Brown D. K. (1989), *The Architecture of Frigates*, Journal of Naval Engineering, Volume 30, Number 3, London, UK.

[8] Acquisition Reform Office (1998), *Arsenal Ship Lessons Learned Report,* Defense Advanced Research Project Agency, VA, USA.

[9] Kehoe, Capt J.W, Graham, Cdr C, Brower, K.S. and Meier, H.A, (1981) *NATO and Soviet Naval Design Practice: Eight Frigates Compared*, Naval Engineers Journal Volume 93.

[10] Oliver Hazard Perry (FFG7) Design, at: http://www.globalsecurity.org/military/systems/ship/ffg-7-upgrade.htm

[11] Beck, D and Lord, J (1997), *Design and Production Of ANZAC Frigates For The RAN And RNZN: Progress Towards International Competitiveness,* Ship Production Symposium, New Orleans, USA

[12] Waugh M. K. (2009)., *Futureproofing Frigate Aviation*, RINA WARSHIP 2009 Conference, London, UK.

[13] Materiel Requirements Set (MRS) Volume 5 Part 8, *Electrical System Requirements*, Issue 04, November 2005, Canberra, Australia

Operating concepts for next generation submarines

Author(s) Name: Patrick Dematte, Christophe Pichon, Jean-Michel Bolot
Affiliations: DCNS - France
Email contact
patrick.dematte@dcnsgroup.com
christophe.pichon@dcnsgroup.com
jean-michel.bolot@dcnsgroup.com

The aim of this paper is to present DCNS R & D projects concerning the application of enhanced operating concepts of future submarines.

OVERVIEW

- DCNS presentation
- Problematic
  - General context
  - Submarine operating: current solutions
  - Status
- Solution studied for future usage
  - Action preparation tools - operating through a large touch screen
  - 3 D - Tactical situation operating
  - Progress
- Conclusion and way ahead

DCNS PRESENTATION

DCNS submarines designer and manufacturer

Since 1888, DCNS studies, designs, manufactures, supports and modernises submarines. At the present time, for "new" projects, DCNS is working on:

SSBN Le Triomphant type	(14 000 tons)	crew of 130	specific mission
SSN Barracuda type	(5 000 tons)	crew of 60	multi-missions
SSK Scorpene type	(1 800 tons)	crew of 32	multi-missions
SSK Andrasta type:	(900 tons)	crew of 21	littoral missions oriented.

These submarines are equipped with the DCNS generic combat system: the SUBTICS® which also is onboard modernized submarines.

PROBLEMATIC

General Context

To fulfil the evolution of operational needs, next generation submarines will have to lead a wide variety of missions, such as blue waters missions needing long range detection and long range weapons or green/brown waters missions, needing intelligence gathering sensors and accurate navigation means for heavy traffic areas or deep strike missions, needing ground representation for non real time mission preparation. Therefore, submarine combat systems will evolve to include the following capabilities to:

- communicate continuously : for instance, when diving, they will receive tactical data as well as strategic and environmental data, in real time. The submarine has also to perform emission in diving condition
- be integrated in naval forces
- launch and recover UUV or UAV
- integrate new sensors and new weapons.

Furthermore, in the future, the missions will be achieved with a limited number of submarines, each operated by a reduced crew.

Submarine operating : solutions used today

Over the last 40 years, the amount and performance of sensors have considerably increased, as well as the quantity of data received via the tactical data link, while the number of operators remained stable. Over time, several improvements have been implemented in order to manage the submarine systems.

At sensors level:

The extraction algorithms, extracting from raw data the synthetic tracks, have reduced the load for the operator. Concerning the Tactical Data Link data, simple filters based on geographic limitation or danger criteria have also contributed to significantly reduce the flow of data and the workload of operators.

Today, these algorithms have reached a good level of performances.

At combat system level:

In order to optimize the system operation, general and common MMI principles as well as a graphic charter have been defined and applied for all the sensors. Furthermore, the use of Common Multifunction consoles has become the standard.

A major gap is the change of the combat system architecture. Initially a simple aggregation of equipment, the combat system is now a real integrated system where all the significant data is collected, correlated and synthesised before being presented to the commanding crew. This change from a federated architecture to a centralized architecture s provides a unique tactical situation, which gives, at the commanding crew level, a unique vision of the operational theatre.

An over change is the use of a multipurpose network with segregation in the combat system to be compliant with the main operational tasks (mission preparation for the future action of the sub, nautical and tactical "real time" management, data forwarding for the sub crew...) and the security rules.

Status:

The combat system operating is carried out using standard screens with proven operating principles older than 30 years. When. However as far as new information must be processed, the only solutions today are either to add displays into the CIC (but due to physical constraints, it's quite impossible), or new images on the multi functions console and use of automation.

The fact is that operators today are overloaded in wartime situation: the data must be drastically synthesised and tactics are defined in a hurry and consequently not optimal.

In parallel, operators are from the Internet / iPhone / Gaming generation and use all day long touch screens and intuitive MMIs. Today, there are, more than 20 years of cultural gap between the civilian word and the military word on this domain. And, in the future, the operator workload will continue to increase, due to:

- new sensors such as UUVs
- new weapons such as cruise missiles, anti aircraft missiles or even a small-caliber gun
- use of real time and permanent communication capability
- integration in naval forces
- increase of brown water missions
- reduced crew.

So we have to totally rework the operating of our systems, and a real gap is necessary, using the latest technology available. If not, operators will be completely overloaded in wartime situation, therefore creating vulnerabilities.

Where to focus our efforts

At the sensors level and weapons level, DCNS now has decades of experience. Improvements are still feasible, but the available margin for enhancements is rather low.

On the other end, the tactical situation is becoming overloaded and operating is now possible only by limiting the number of processed data.

So, the most promising track is to focus our efforts on the tactical situation elaboration and tactical evaluation operating.

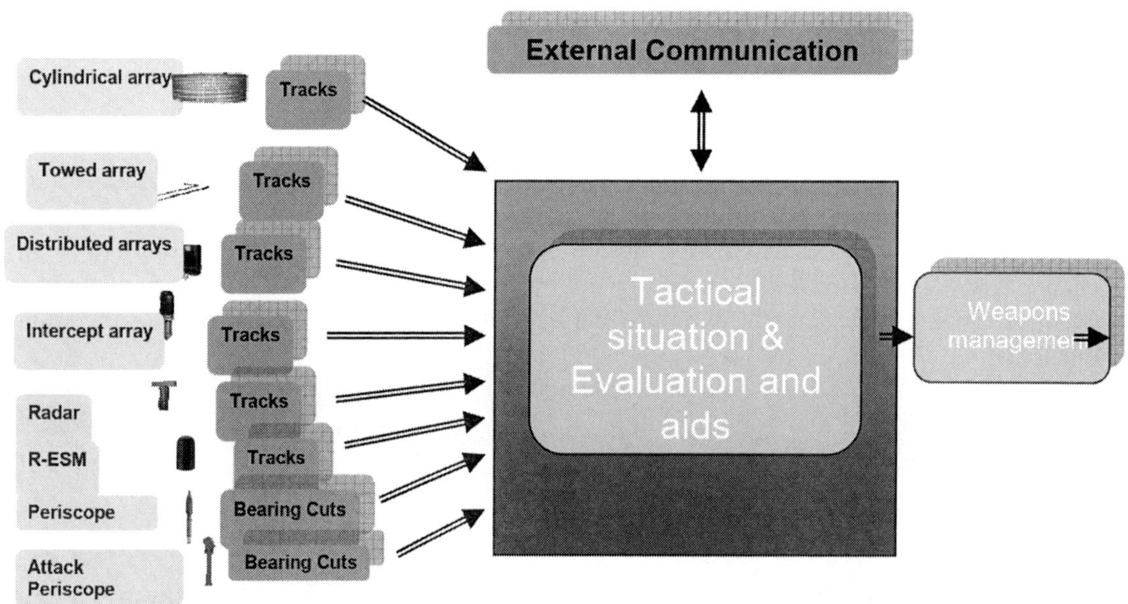

SOLUTIONS STUDIED FOR FUTURE USAGE

Studies in progress

Two studies being conducted by DCNS are presented today:

Study 1

Action Preparation: using a large touch screen display, the aim of this study is the development of tools which help the commanding crew to optimize the submarine tactics.
It is a short term study; a first version of the demonstrator is under evaluation.

Study 2

Tactical situation: new operating, through the use of 3D visualisation.
The objective is to use more intuitive and more effective operating.
It's a medium term study; a technology demonstrator has been implemented and evaluated.

Study 1: Action preparation

Methodology used

The study started in 2007. After a preliminary study, a demonstrator was specified and developed, base on a large touch screen.
In a first step, we started from our background:

- experience from systems in operation, mainly in the "surface" world (in the "surface" word, the usage of action preparation tools is more developed compared to what can be found in the "submarine" world)
- integration of our current developments (mainly for ASW and AAW frigates and new generation of SSN)
- existing algorithms for sensors performance prediction
- existing simulations to calculate the performance of submarine combat system in duel configuration (acoustic, electromagnetic, optic, infra red…)

Additionally, we collect and compile the views of a wide audience, mainly through:

- involvement of former naval officers(operational point of view)
- involvement of ergonomists (ergonomic point of view).

MMI principles:

The MMI principles are closed the ones used in the "iPHONE" and in "Windows" - Windows Icon Menu. The operator basically uses « drop-down » menus, the mouse and the touch screen.

Proposed solution

The solution is to propose one or several tactics (route, speed, immersion) considering:

- the current position of the submarine
- the environmental data, received in real time via communication equipment
- the use of geographic databases (bottom, currents, ...)
- tactical data exchanged via tactical data link (eg position underwater, surface or air for each friend and foe unit, future point of contact, launching zones…)
- the status of the platform (such as submarine autonomy, battery capacity level, radiated noise level in real time)
- the mission to be accomplished. A choice of typical pre-defined missions is proposed to the commanding crew.

Typical missions:

- Area sweeping
- Control of embargo
- Hunting
- Intelligence and reconnaissance in a defined area (including UUV or AUV launching and recovery)
- Convoy protection
- Sea Line of communication protection

- …

Operating – solution being evaluated

The demonstrator is based on a 2D sensitive screen (around 1 m * 1 m)

The basic view is the tactical situation (addition of the navigation map and the CMS tracks).

This view can be enriched by adding:

- external tracks received through the tactical data link or AIS
- radar tracks
- overlaying of data linked to a track as acoustical signature when required by the operator
- overlaying of raw data (range, bearing, contacts, zones or pictures, videos) coming from the radar, the optronic mast, the attack periscope, R-ESM and/ or C-ESM
- areas and zones (exercise area…)
- image inlay (e.g. typical landmark).

2D Table: tools

A lot of aids are available:

- standard tools for navigation and manoeuvre aid, including WECDIS
- route layout
- navigation alarm management function
- man overboard
- optimization tools: calculating routes such as hunting, road tracking, cinematic proposed for an control area
- opportunity to view the tracks history (back to the past) possibility of time accelerated future projections (possibility to play several scenarios, ...) (projection into the future).

Study 2 : 3D Tactical situation

Methodology used:

The study is led by a combat system architect (technical point of view) and an ergonomist (ergonomics point of view), taking into account:

- experience on systems in operation
- involvement of former navy officers (operational point of view)
- involvement of ergonomists (ergonomic point of view)
- use of simulations and models demonstrator
- operating constraints of future submarines in terms of crew / mission and onboard installation constraints
- new visualisation tools.

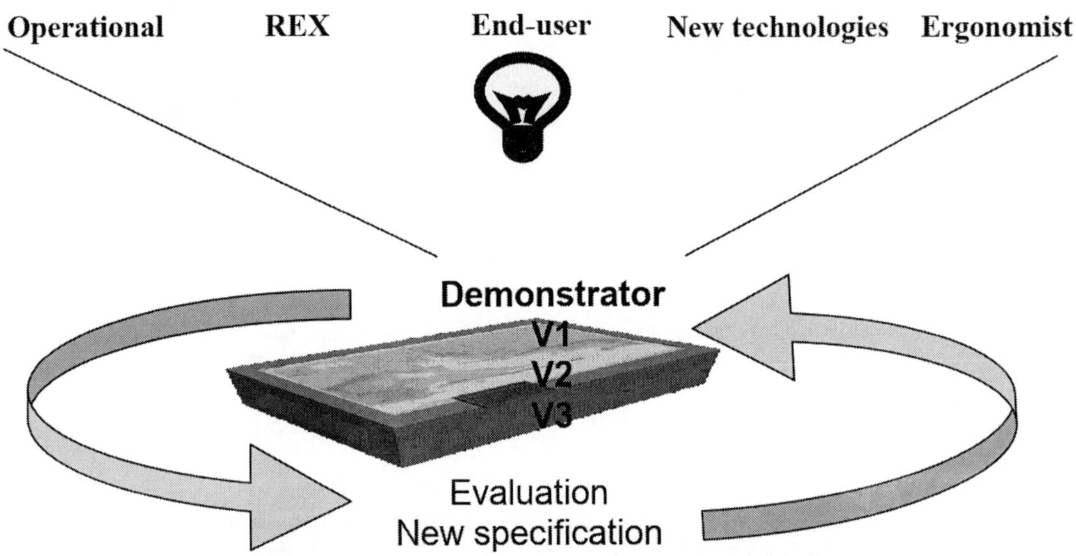

Model approach:

The model approach is a spiral development. The demonstrator is first used to check the technological feasibility, after that, an operational scenario is tested and in a following version, the MMI is tested.

V1: technology demonstrator (focusing on technology aspect)
V2: optimised demonstrator using operational scenario
V3: focusing on MMI demonstrator (shore and sea tests)
Then the functionalities and the hardware are integrated in our combat systems.

3D tactical situation - visualisation

The principle is an "Immersive" visualisation, using “active stereo glasses” and two screens of 120 * 80 cm. This represents a volume of approximately 1 m^3 (Holobench type solution - Workbench)
Two cameras allow the optical tracking of the active stereo glasses.

3D visualisation
Geographical area representation with raised topography- equivalent to a world in miniature - of the tactical situation
The speed and route of the tracks are visualised
The operator can choose between two representations for the tracks:
- NATO symbols
- near real symbols.

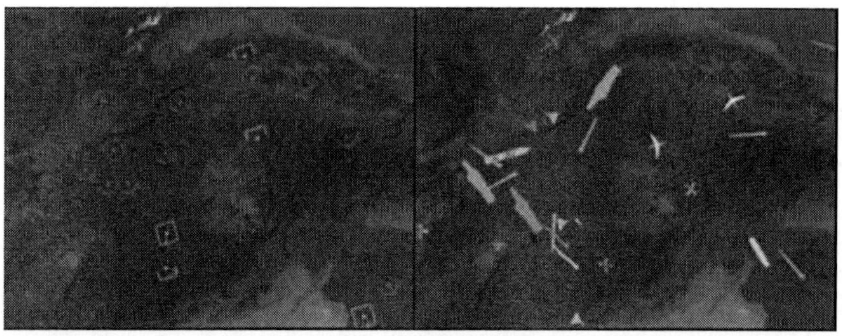

The shape corresponds to the type of mobile (aircraft carrier, frigate, submarine ...).
The colour identifies the track: friend / enemy / unidentified.
The speed and road are visualised by a 1 vector.

Furthermore, a 2D map (overview) is permanently displays, in addition to the tactical situation, allowing the operator to quickly situate himself within the theatre (mastering the loss of control of the context due to the zoom factor).

Slide 13 - Interaction
3D allows establishing a feedback between the head position (elevation) of the current operator and the projected view. Furthermore, several functionalities are proposed such as:
- the modification of the terrain elevation factor
- the scale factor can be modified (zoom)

- by finger pointing , the operator can select a track (via the 2-camera tracking, the software calculates the location pointed at the end of the gesture - like a laser beam
- the request of track data
- on a designated track, obtaining sensors detection range visualization / weapons detection range (data extracted from a local data base)
- additionally a touch screen zone allowing the implementation of widgets.

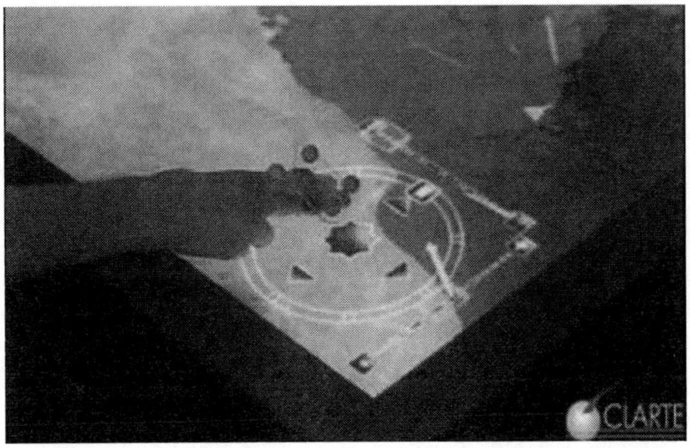

Progress

2D Tactical Table

The demonstrator: is today under test by the French Navy in the Navy Training Center Navy in Toulon – France.

3D Visualisation

Demonstrator evolutions:

- 2007 - 2008: first demonstrator developed
- 2008 2009: 3D Design Technology and REX. This demonstrator was presented in Laval/ France in 2009 in a virtual reality exhibition.
- 2009: review with the Navy.

CONCLUSIONS AND WAY AHEAD

The two studies presented are part of a global approach intended to totally reconsider the usage of our future submarines. DCNS believes this approach is essential if submarines of tomorrow are to be operated in an optimal manner.

2D table:

The initial feedback is positive. The goal now is a short term on board implementation in a submarine with progressive tools upgrades.

3D visualisation

The technology used is mature enough to test some operational scenario, test scheduled in 2010 (the demonstrator is planned to be exhibited in future exhibitions in Europe in the last quarter of 2010).

In 2011, the new versions of the MMI will be integrated and test with end-users, during full scale naval operations.

In 2012, DCNS intends to start systematic integration in the SUBTICS® class products.

General Operations

2010 - 2011: assessment and analysis of the impact on other operators: - redefinition of roles / tasks and new « complement scheme » for the combat system

The implementation targeted on a submarine with a reduced crew (about 20 including 8 for the combat system) within 5 years.

Operating submarine combat systems is becoming extremely complex. Both, “actions preparation tools” and “3D tactical situation operation”, experimentations are promising. Now that feasibility has been proven, DCNS considers it is a good way forward to implement them onboard SUBTICS® DCNS Combat System, in order to significantly improve wartime capacities and efficiency.

E-Craft, a Self-Transforming Vessel

Paul Rispin
Office of Naval Research, USA

paul.rispin@navy.mil

INTRODUCTION

The Program Executive Office, Ships of the United States Navy manages acquisition and complete life-cycle support for all U.S. Navy non-nuclear surface ships. The Naval Sea Systems Command, through its Naval Systems Engineering Division (SEA05), evaluates new ship concepts, formulates and develops design goals and requirements, and supervises the ship design process. The Office of Naval Research (ONR) coordinates, executes, and promotes the science and technology programs of the United States Navy and Marine Corps and is constantly looking for innovative scientific and technological solutions that can address current and future requirements. In the recent past ONR has built a number of technology demonstrators to explore ideas and technologies before they are passed on to NAVSEA for further development. Examples are SEA SLICE, SEA FIGHTER and SEA FLYER all in the area of high-speed multihull vessels.

As such it is often used by The Office of Naval Research has often used technology demonstrators, which are one step before a prototype and two steps ahead of a first of class construction, and are often an appropriate means to explore a technology for possible future use. In this case ONR is building a technology demonstrator to explore what it would take to design, build and operate a vessel capable of operating at high speed at sea, have good seakeeping characteristics while able to go close to shore in shallow water.

High speed is necessary to reduce the transit time that comes with the stand off distances forced by ever more capable surface to surface missiles. Seakeeping is to enable this high speed and facilitation of cargo handling in deeper waters while the shallow draft is to allow close approach to shore and, if possible, beaching. An ONR prototype, T-Craft, is exploring how to add a full amphibious capability using a different approach.

TECHNOLOGY

The primary goal of the E-Craft program is to examine one approach to the ship to shore connector – the use of a self-transforming ship with two modes, a SWATH mode for sea and a barge mode close to shore. The secondary goal was to provide for an afterlife for the vessel which would prevent it from becoming a funding liability to ONR and, at the same time, would provide the Navy with a stream of structural and performance data for a number of years.

The technology uses very large hydraulic cylinders to raise and lower a watertight car deck and, for the shallow water mode, to force it down into the water surface and thereby lift the rest of the vessel to provide the required shallow draft.

BACKGROUND

The origin of what is now E-Craft came from Lockheed Martin in 2002 with their Varicraft and Varicat concepts which came from early ideas for the Littoral Combat Ship competition. Lockheed tested models of the two concepts at the Offshore Model Basin in California where RADM Cohen, then CNR, became interested in the possibility of their being potential game changers for ship to shore connectors. Lockheed proposed to ONR a program to advance the Varicraft concept, but there wasn't enough funding to begin - promising though the idea was. Some funding was put into the 2004 budget while Lockheed formed an alliance with the Alaska borough of Matanuska-Susitna to get enough support to get moving. They used borough funding to begin preliminary concept design using Guido Perla Associates (GPA) in Seattle as design agent. By November 2004 Lockheed and GPA produced a preliminary concept design.

In April 2005 Lockheed's role in the program changed and the lead was assumed by Alaska Ship and Drydock (ASD) in Ketchikan AK, who were to build the vessel. This change in lead facilitated the design-build process we used for E-Craft. In August 2005 ONR entered into a tripartite Other Transaction Agreement (OTA) with ASD and the Borough, with Lockheed playing a lesser role.

E-CRAFT CONCEPT

The main elements of the design are a twin hull vessel that encloses a cargo deck that can transition between two modes. In the deep water mode (SWATH mode) the cargo deck is in a raised position and in the shallow water mode (barge mode) the watertight deck is lowered and forced into the water surface. The extra buoyancy gained by the deck forces the rest of the vessel upwards to retain equilibrium until the barge provides the controlling draft. This is the basic idea, shown in Figures 1 and 2, but taking the technology from a concept to an operating vessel proved to be challenging, time consuming and costly.

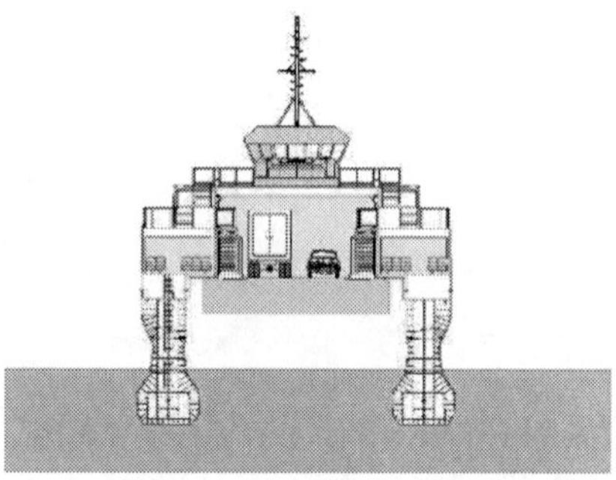

Figure 1 SWATH Mode

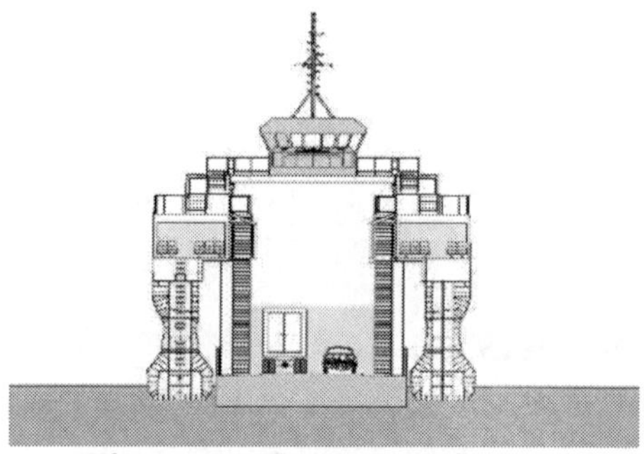

Figure 2 Barge Mode

E-Craft has a conventional four water jet configuration but, because of the change in draft, they cannot work in the barge mode. To power E-Craft in shallow water mode a set of thrusters was installed amidships whose nozzles were vectored for steering and to unbeach the vessel if necessary.

Figure 3 is an example of the ice conditions which will be faced by the vessel in its second life as a ferry on the Knik Arm of Cook Inlet. The area is tidal with a vertical

Figure 3 Cook Inlet in Winter

difference of 25 feet and a peak current of around 5 knots. To deal with this E-Craft was designed with an ice cutting bow, a strengthened steel hull (A0) and an "ice-knife" to protect the water jets during reversing. The narrow hulls necessary to keep the ratio of structure to deadweight low meant that machinery is very crowded below decks, almost as dense as a submarine. Finally, since the two hulls are connected structurally only by the bridges near the bow and near the stern, the ship is quite flexible.

DESIGN CONSTRAINTS

Because of the dual nature of the E-Craft requirements, suitability for Navy trials and for long term ferry operations in ice, a number of constraints were placed on the design. Classification was one of these. The High Speed Code was chosen to allow a design that would give a twin hull vessel even though the design speed was at the low end of the HSC range. The American Bureau of Shipping (ABS) was chosen to class the vessel to look at the effect of its rules on Naval type vessels and to permit use of E-Craft in commerce after receiving a US Coast Guard certificate of inspection. The US Coast Guard was involved in the process from the beginning

During the design phase we paid a lot of attention to the final ferry configuration while ensuring that we got the technology demonstrator we needed. Thus the passenger and cargo requirements (cars, trucks, equipment) should be as large as possible, constrained mainly by the cost of construction. Another driver (for both military and commercial reasons) was the desire to be able to operate from an unimproved landing area or even to beach the vessel if required. The third major driver was the requirement for the vessel to be able to operate in the tidal ice conditions near Anchorage AK in the winter months

DESIGN PROCESS

The design of the vessel underwent a major change early on. It was rebased from close to 1000 tons by reducing passenger numbers and cargo capacity, reducing hydraulic system weight by accepting slower vertical motion and by pruning all nice to have systems. Even so, the weight inched upwards again. We realized that we needed a higher standard of ice protection. This increased hull and scantling weight. Stability tests done in conjunction with the ice testing in St Johns required a 2 foot increase in beam. In addition, the growing weight of the lift system, the stability needs of the vessel and the minimum speed necessary to allow use of the High Speed Code gave a design spiral that was difficult to make converge.

Due to time constraints this was a design-build program from the beginning. After the rebasing, the design spiral finally converged at about 880 tons and detail design began. This revealed a large number of small components that also added weight. About six months into detail design many major components were ordered and six months later construction began. A year later, in December 2007, detail design was substantially complete. However, a dialog between the designers, ABS and the Coast Guard continued for a long time and it was not until May 2009 that the Contract Design Package was delivered. E-Craft turned out to be a more complex and difficult design than anyone expected.

Design of the Main Hulls

Figure 1 shows a cross-section of the vessel in the SWATH mode. The cargo deck is up and the passenger compartments are at either side. In the Barge mode hydraulic cylinders force the barge into the water and raise the rest of the vessel, giving the required shallow draft.

Figures 4 and 5 show general views of the two modes.

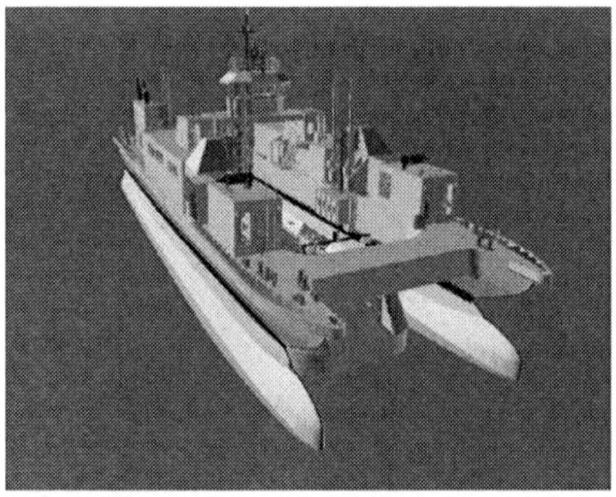

Figure 4 View with Barge Down

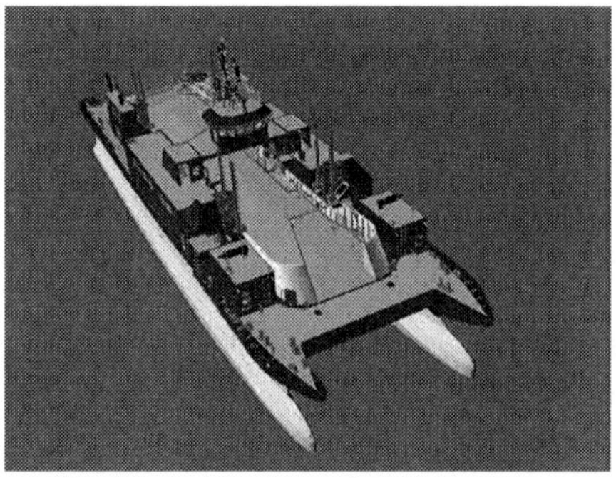

Figure 5 View with Barge Up

Design of the Bow

E-Craft is not designed as a true icebreaker. Instead, as an ice capable vessel, it can make its way through pan ice or cut through 1 foot thick fresh ice. To do this the vessel is flooded down to a displacement of 1058 tons so that the lower part of the bow is under the ice and cuts from below instead of from above. This accounts for the shape of the bow – a shape that was confirmed in the ice tank at St John's Newfoundland.

Design of the Barge

Since the barge was protected by the main hull from ice loads, it was possible to save a lot of weight by making it out of aluminium. It is basically a cargo deck on a watertight structural box, with a bow ramp for loading and unloading and fitted with oversized connections to the main vessel. This was the least complex part of the E-Craft design. The pilot house, its foundations and the passenger accommodations were also made of aluminium.

Design of the Lift System

The barge is moved vertically by four pairs of high pressure hydraulic cylinders shown in Figure 6. These are placed near the four corners of the barge. Each pair is inside a vertical cylinder welded to the barge. High on the inside of each hull are pairs of fittings to take the reactive loads. A high pressure hydraulic power unit and an extensive electronic control system ensure that the motions of the eight cylinders are synchronized and to prevent jamming of the barge at all times during its vertical motion. Extensive safety systems are in place to keep the barge level and at the commanded height. At either end of the barge motion four high strength steel locking 'pins' keep the barge in place without need for the hydraulic system to be running.

Figure 6 Hydraulic Cylinders

CONSTRUCTION OF E-CRAFT

Hull Construction

Construction of such a complex vessel with a need for high accuracy in joining its modules posed a number of problems. Each of 36 modules was built indoors and brought outdoors to be joined in pairs to form blocks, then superblocks and then the two hulls. Accuracy control at the module level was achieved by using a specially designed mandrel and rotating table so that as welding was completed for one section the assembly was rotated so that the great majority of welding was done downhand. The set up is shown in Figure 7. Openings were cut in each of the end bulkheads and special clamps held them in place. As work progressed and the outer skin was completed access was maintained through the hollow mandrel and the access openings shown in the figure. After removal of the module the openings in the bulkheads were welded shut and the module brought outside for painting and/or initial outfitting. It was at this stage that engines, thrusters, and gear boxes were installed and preserved for later integration. Figure 8 shows a module being moved outdoors. Figure 9 shows the building slipway and rail system.

Figure 7 Mandrel

Figure 8 Moving Module Outdoors

Figure 9 Construction Slipway

Figure 10 shows a block consisting of two modules vertically paired modules. The other figures show the bow, the stern and the layout of the blocks on the construction pad.

Figure 10 Block in Place

Figure 11 Stern Modules at Ground Level

Figure 12 Bow Superblock

Figure 13 Modules in Place

Figure 14 Welding Vessel

Figure 15 Water Jets

Figure 16 General View

Figure 17 After Painting

Barge Construction

The barge is aluminium and was made by a subcontractor in Washington state, along with the other aluminium modules. It was towed to Ketchikan for integration into E-Craft. There it will be fitted with a Lockheed designed ramp.

Figure 18 Half of Barge

Figure 18 Barge in Alaska

Lift System Construction

The most complex part of the vessel is the lift system. The design of the system and the fabrication of the hydraulic cylinders and the associated control and actuation systems were sub-contracted to Bosch-Rexroth. Figure 19 shows the reaction structure to hold the upper end of the cylinders. Figure 20 shows the addition of the vertical guideway that the barge fitting in Figure 21 encloses. After completion of the ship fittings, installation of the rest of the system began in November 2009

Figure 20 Ship Fittings for Cylinders

Figure 21 Ship Fittings

Figure 22 Barge Fittings

Figure 23 Final Design

DATA ACQUISITION SYSTEM

One of the Navy goals for this vessel is to collect structural, hydrodynamic and propulsion data for comparison with predictions and to form a basis for any future design of a ship of this sort. To do this ONR engaged the Full Scale Trials group at the Naval Surface Warfare Center, Carderock Division to plan and implement a suite of sensors and a Data Acquisition System to gather and store data from strain gages, accelerometers, torsion meters, wave height sensors and gyro sensors as well as from the waterjet, engine, lift system and navigation systems. The majority of the sensors are strain gages placed throughout the ship and the barge at locations that were determined by results of the finite element analysis of the hull response to various seaway and other loads such as ice impact. Accelerometers were added to enhance the response data stream. Propulsion performance will be derived from engine parameters and torsion meters. Maneuvering response, seakeeping data, thruster performance and especially water jet data will be recorded. To get data on cavitation performance of the Wartsila 810 jets, viewing ports were inserted into one of the water jets as well as pressure sensors.

All of these sensors are tied together by extensive signal wiring installed in various compartments. Most of the strain gages are in one hull with enough on the opposite one to identify non-symmetric loads when they occur. This wiring system ends in one of the passenger compartments where the central communication and signal conditioning unit will be installed.

An interesting opportunity is the planned autonomous data recording that will remain running for three years as the ship goes about its regular duties. The data from this will be stored and returned to Carderock at regular intervals for analysis as the ships age.

COMPLETION AND TRIALS

When this paper was written (November 2009) the structure of the ship was substantially complete as was most of the exterior and interior painting. The major outstanding work was the installation and testing of the lift system. This was to take place on land at the end of which the vessel will be launched using a dry dock that mates with the rails in the construction area. This is scheduled for late February 2010. Fitting out will continue for a few weeks, at the end of which builders and acceptance trials will take place, followed at the end of March, by turn over to the Borough who will crew the vessel from then on. Sea trials conducted by Carderock have been planned after which E-Craft will transit to Anchorage and enter into ferry service as MV Susitna.

APPENDIX E-CRAFT CHARACTERISTICS

Length Overall 195' 5"
Beam Overall 60' 0"

Design Draft 12'
Design Draft Full Load 970 LT

Ice Navigation Draft 13' 4"
Ice Navigation Displacement 1058 LT

Barge Mode Draft 4' 6"
Barge Mode Displacement 1013 LT

Maximum Speed 20 Kt
Cruise Speed 17 Kt
Speed in Ice 10 Kt
Barge Mode Speed 6 Kt

Vehicles 20 Standard or 14 Large
Tractor Trailer 1
Passengers 88

Propulsion Engines 4 x 2700 HP Diesel
Waterjets 4 x 2700 HP Axial Flow
Thrusters 2 x 1600 HP 360deg
Generators 2 x 300 KW Diesel

IMO HSC Code
ABS Certified +A1 Ice Class A0

Transitioning Towards Condition-Based Maintenance through Enhanced Decision Support Tools

L. Hammond
Defence Science and Technology Organisation, Australia
lloyd.hammond@dsto.defence.gov.au[1]
P. Kelly and J. Sikorska
CASWA Pty Ltd, Australia

ABSTRACT

The Navy's requirement to reduce cost of ownership and increase availability is the driving force behind the development of a data-driven, decision-support toolset for defence.

DSTO's implementation of a Condition Based Maintenance data environment necessitated a significant effort in obtaining and integrating system condition, failure, and history information to feed data-driven decision making processes that can ultimately provide Navy asset managers with an enhanced maintenance management capability for critical assets.

DSTO has developed and trialled a systems interface unit for accessing and transferring sensor and alarm data from Anzac Class ships to shore, as well as a web-based software tool, that can access, abstract, refine and present information from a large number of existing logistical data sets. The trials were undertaken with the support and cooperation of the ANZAC Systems Program Office, who provided both ship and shore-based trial environments. It can be demonstrated that these tools have the capability to reduce Navy maintenance costs and enhance platform system reliability.

1. INTRODUCTION

1.1 Background

The requirement to reduce the cost of ownership of the Australian Defence Organisation's (ADO's) assets is the key driver to the establishment of a Condition-Based Maintenance (CBM) program outlined in this and a previous paper [1].

The Navy, like many commercial organisations, has historically been guided by its vendors and Original Equipment Manufacturers (OEMs) with respect to maintenance planning. Such maintenance programs are typically expensive, overly conservative and require a restrictive regime of periodic planned maintenance and unacceptably high unavailability of assets. Conversely, many commercial organisations now embrace more flexible, CBM approaches to maintenance, integrated with sound risk management practices.

CBM is at its core, a Data-Driven Decision Making (DDDM) process since it is based on a knowledge of the actual condition of the asset. CBM adopts the premise that intervention will only occur when there is evidence that it will improve the availability of an item of equipment. As such, the efficient collection and subsequent management of real-time condition data is critical to the CBM paradigm, which views maintenance as a response to a changed asset state.

The tasks of maintenance engineers and asset managers would be enhanced with a predictive capability in order to optimise maintenance strategies. Answering questions like, "What happens to the probability of a failure if we postpone this oil change?" or, "How long before this unit fails?" or, "What is the remaining useful life of this asset" is the endgame for asset managers and maintenance planners.

Since CBM methodologies are based on the real condition of an asset rather than an assumed condition, they are inherently more effective for the maintenance of most assets. This is because CBM uses the all available data to determine the condition of an asset and whether maintenance is required.

DSTO has undertaken a four year program analysing the applicability of CBM and related maintenance techniques to ANZAC Class assets. In particular, the CBM approach has been trialled to assess its feasibility for maintenance of the Propulsion Diesel Engines (PDEs) on ANZAC Class frigates; the choice of RAN platform and system being largely determined by the ANZAC Systems Program Office (SPO). The ANZAC SPO has willingly engaged with the Defence Science and Technology Organisation (DSTO) and has enthusiastically participated in trialling new hardware and software.

The decision to implement CBM on the ANZAC Class frigates required improved access to existing data repositories as well as the measurement of physical parameters for the diesel engines, indicative of asset condition. To facilitate access to a wide range of disparate maintenance data sources, a software program called the CBM Toolkit was developed. The CBM Toolkit is a web-based tool that allows the user to (a) interact with a variety of data sources, (b) have the option of assessing the data quality and to undertake remediation if necessary; and, (c) present data in a format for analysis and/or post-processing depending on the user's job requirements.

To facilitate the collection of condition data, a concurrent DSTO project to develop hardware and software to extract sensor and alarm data from the existing ANZAC Class Control & Monitoring (C&M) system was undertaken. The hardware developed is referred to as the Systems Interface Unit (SIU) and has now been trialled on two ANZAC Class frigates with a concurrent trial of a third SIU at the Training Unit – ANZAC Ship Support Centre (TUASSC), HMAS Stirling, WA.

1.2 Overview of DSTO's CBM Program

The driver for this program, which also encompassed other related asset management objectives, was to reduce the through-life cost of ownership of RAN platforms without adversely impacting on their availability or reliability. The asset chosen as a test-bed for this study was the ANZAC Class V12 MTU Propulsion Diesel Engine (PDE), of which there are two on each frigate. As mentioned above, CBM is a DDDM process and consequently data is at its core. The data requirements for the CBM program necessitated access to failure and condition data histories for the PDEs. It was quickly discovered that there was a lack of online condition-monitoring data for the PDEs, although there were results from periodic spectrographic lube oil analyses and broad spectrum vibration analyses. Consequently and somewhat surprisingly, the biggest challenge of this program has been the recovery of reliable, real-time asset condition and failure data for use in CBM analyses.

DSTO's Maritime Platforms Division (MPD) commenced a technology development program in 2004, with the purpose of developing procedures that will provide RAN asset managers with improved decision support information for critical assets.

One of the tasks was to tackle the problem of a paucity of performance data. This solution was the development of hardware to stream platform C&M data from ship to shore. The second task was to develop procedures to abstract, refine and present this and other datasets in a form amenable to the user. The solution in this case was the development of a web-based program known as the CBM Toolkit incorporating these procedures. The CBM Toolkit solution to data acquisition issues was developed in the first instance to stimulate the CBM Program, but it was also undertaken to provide a new and rapid means of gaining access to numerous data sources to support the ANZAC Class in a variety of endeavours.

The solutions presented are intentionally generic in nature and they are intended to be applied more broadly across the Navy as the methodologies mature. DSTO's development of the SIU, the CBM Toolkit and their subsequent validation through a rigorous trialling program represents the basis of this paper.

1.3 CBM Fundamentals

CBM identifies the onset of a failure mode by detecting one or more measurable condition changes which are known to precede such failures. Changes are measured either in real-time with online sensors (e.g. temperature and pressure) or using offline analysis techniques (e.g. oil analysis, vibration analysis and acoustic emission). CBM is most effective when the detected condition change occurs with sufficient lead time before the failure occurs, allowing an opportunity to plan and undertake the required maintenance as illustrated in Figure 1.

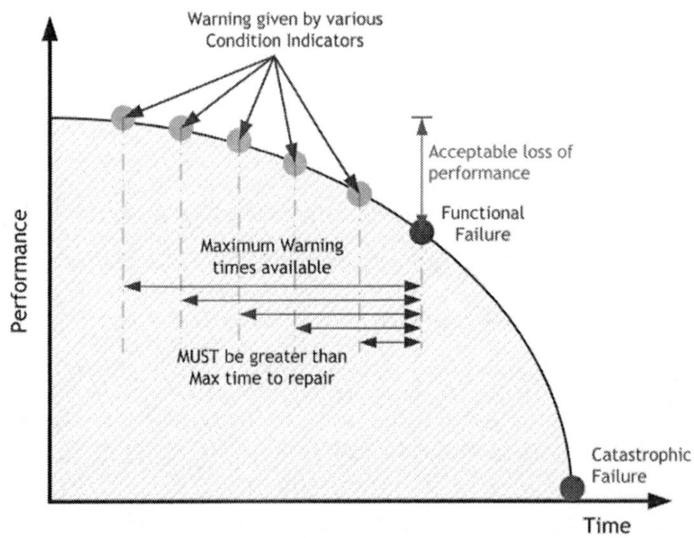

Figure 1 Performance deterioration versus time

Consequently, the choice of the Condition Monitoring (CM) technique is guided by its ability to provide an operator with the greatest lead time to an incipient failure as well as its sensitivity to the potential failure event. Complementary studies have also been undertaken by DSTO into the testing and evaluation of advanced sensors to provide more advanced warning of failures than is currently available [2]. Employing a CBM approach requires accurate information pertaining to an asset's:

- Actual failure modes;
- Actual and/or predicted consequences of failure; and,
- Trends of condition indicators.

2. DEVELOPMENT AND TESTING OF THE SYSTEMS INTERFACE UNIT

2.1 Introduction

The ANZAC Class C&M systems are connected to about 3500 onboard sensors and alarms, covering all major systems with the exclusion of the combat system. The data from these sensors represents a potentially significant data source for condition monitoring, prognostics and diagnostics yet until recently no C&M system data had been collected. The SIU was developed by DSTO to realise a capability of accessing and transmitting this valuable data back to a shore-based database.

2.2 Feasibility Study

In 2004 a feasibility study was commissioned by the ANZAC Systems Program Office (SPO) to investigate ways of extending the capability of the C&M system. The study proposed an enhanced system design which would be capable of capturing all C&M data and making this data available for remote interrogation. This solution (refer Figure 2), required the development of an SIU, which resides between the C&M system and the ship's Local Area Network (LAN), and uses web protocols to collect and buffer processed and raw sensor data.

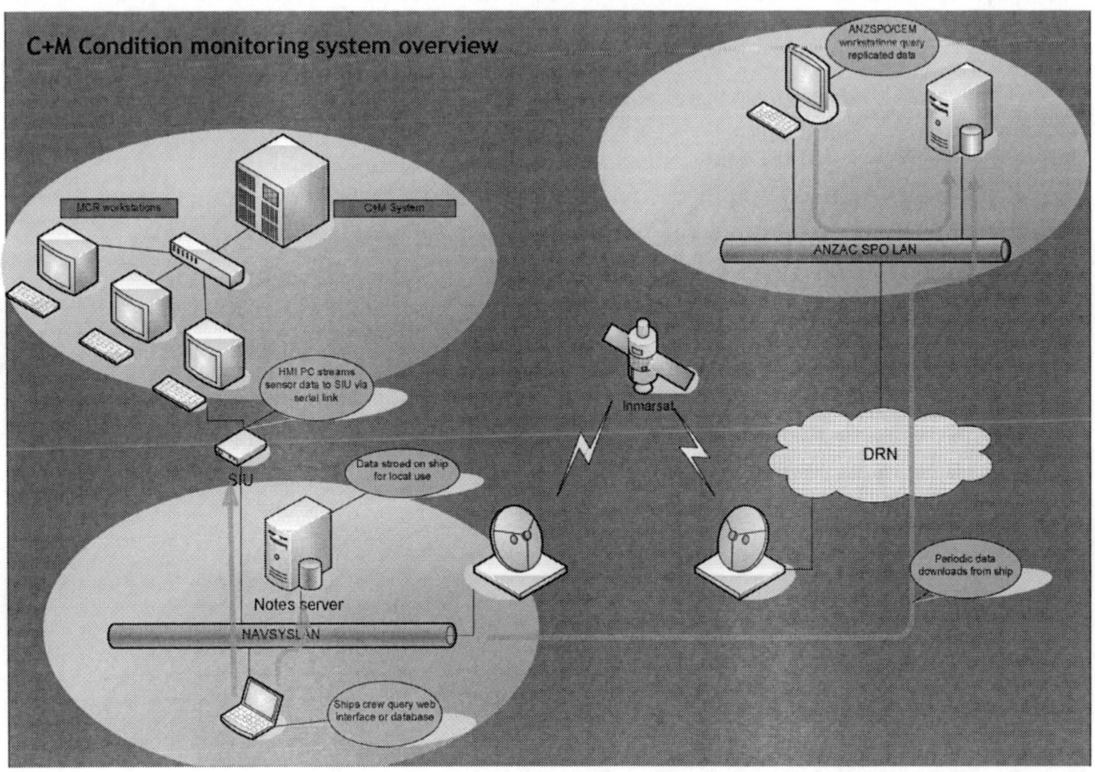

Figure 2 Overview of the ANZAC Class's C&M system, incorporating the SIU.

This required modifications to the existing C&M software to collect and store all sensor data from a serial port on one of the Machinery Control Room (MCR) workstations. The serial data stream could then be processed by the SIU which is connected to the ship's network. The SIU would store and then periodically forwards the C&M system data over the ships LAN, Inmarsat[1] and the Defence Restricted Network (DRN) to the CBM server at DSTO where it can be accessed by users via the CBM Toolkit.

2.2.1 Initial On-Shore Testing

Preliminary testing of an SIU utilised the C&M system training facility at TUASSC[2], at close proximity to DSTO at HMAS Stirling. The TUASSC tests were undertaken in order to minimise on-ship trialling. The initial testing verified the connectivity and the communications protocols of the SIU.

2.2.2 HMAS Ballarat Field Trial

In early 2009, a six months trial was successfully completed on the ANZAC Class HMAS Ballarat during her normal deployment. The objectives of the HMAS Ballarat trial were to:

- Evaluate the SIU's ability to access, buffer and transfer the C&M system data; and,
- Evaluate the SIU's internal software design to control the SIU's access to the C&M system and manage the transfer of the C&M data to shore-based users.

The trial was limited to the 74 C&M sensors associated with the two MTU V12 Propulsion Diesel Engines (PDEs). Once commissioning was completed, the SIU was shown to function according to its design intentions. C&M data for the two PDEs was transferred hourly to the onshore C&M database on a DSTO server. The SIU on HMAS Ballarat has continued to transmit C&M data from the two PDEs onboard HMAS Ballarat on an hourly basis. The C&M system data was validated by cross-checking of SIU collected data with manually recorded data observed on the C&M system's Local Operating Panels.

2.2.3 Extended Trial on HMAS Arunta

Following the success of the HMAS Ballarat trial, it was decided that more extensive trialling should be undertaken on another ANZAC Class frigate and at TUASSC, but with the objective of obtaining data from all C&M system sensors and alarms (approximately 3500).

Given the experience of the recently completed trial on HMAS Ballarat, the installations of the SIUs on HMAS Arunta and TUASSC were relatively straightforward. Connectivity was verified and communications were established. However, the SIU encountered problems when attempting to transmit the full data set to shore. This problem may be associated with the additional code integrated into the C&M system software to generate the required data for buffering and eventual transferral. A fix for this software problem is currently being sought.

[1] Inmarsat is a geosynchronous telecommunications satellite.
[2] TUASSC: Training Unit ANZAC Ship Support Centre, HMAS Stirling, Western Australia.

3. DEVELOPMENT OF THE CBM SOFTWARE

3.1 Introduction

To operate in a DDDM paradigm, there is an increased need to access data from a variety of independent databases. These databases contain condition, failure, repair and cost records, as well as the supporting technical documents. Table 1 shows a subset of the relevant RAN databases. Accessing numerous, disparate data sources has inherent problems. For example:

- The databases may be written in different languages;
- The databases may reside on a variety of hardware types with different operating systems;
- The database field nomenclature may vary between databases.

Table 1 Data sources for CBM Analysis

Data Sources	Data Contained
Asset Management Planning System	Failure history
Urgent Defect Notices	Failure history
Shipboard Control & Monitoring System	Condition data
Vibration Analysis – offline monitoring	Vibration analyses results
Spectrographic Oil Analysis - offline monitoring	Oil analyses results
Reliability-Centred Maintenance results	Potential failures
SRCO* - Maintenance Information System	Summaries of major works performed
Standard Defence Supply System	Spares costs
Work packs	Details of major overhauls

*: SRCO: Ship Repair Contract office

One way to achieve access to these data sources is to produce a centralized repository of data, which may necessitate the replacement or upgrade of existing maintenance management software tools, particularly as new data sources become available. These data warehousing projects have historically failed. A second approach is not to replace existing, independent systems' databases, but rather to produce an interface capable of:

- Obtaining data from the variety of relevant but disparate data sources;
- Validating the quality of the data abstracted; determining if it is fit for purpose; and if remedial action is required;
- Presenting the data in the most appropriate format for each user.

The latter option was adopted in this project because it is a more flexible approach and does not require significant alteration when new data sources become available. The CBM Toolkit is essentially a tool designed to facilitate the decision making process; its overriding philosophy is to free maintenance planners and engineers of the burden of mining data and allow them to concentrate on analysis and decision making. Three separate processes are incorporated into the CBM Toolkit. These are discussed below.

3.2 Data Source Abstraction

The data source abstractor is an information gathering layer that hovers over all of the data sources used by the CBM Toolkit and DSTO's other data driven projects. Security, connectivity and database language issues are managed by this system and it is presently being used for a number of related DSTO projects, in addition to the CBM Program.

3.3 Data Validation

The data required for CBM modelling must be validated as accurate in order to generate reliable outputs. The data validation process represents the most complex of all the CBM Toolkit functions and involves receiving data from the data source abstractor and inspecting each datum in its context to determine whether the information is:

- Valid;
- Should undergo automatic remediation; or,
- Requires manual intervention.

In the latter case, the data analyst would be notified of the information and the details of any discrepancies. The analyst could then decide whether there is sufficient corroborating information to attempt remediation or if the data should be returned to the originator for immediate correction. There is currently no known system that has the ability to perform data validation tasks reliably although DSTO has undertaken some recent research in this area as part of the CBM Toolkit development [3]. Consequently, this has been a primary driver for the development of the CBM Toolkit.

3.4 Data Presentation

The presentation layer, illustrated in Figure 3, receives information from the validation layer and displays it along side the equipment item to which it relates (i.e. the item being selected by the user from the equipment hierarchy).

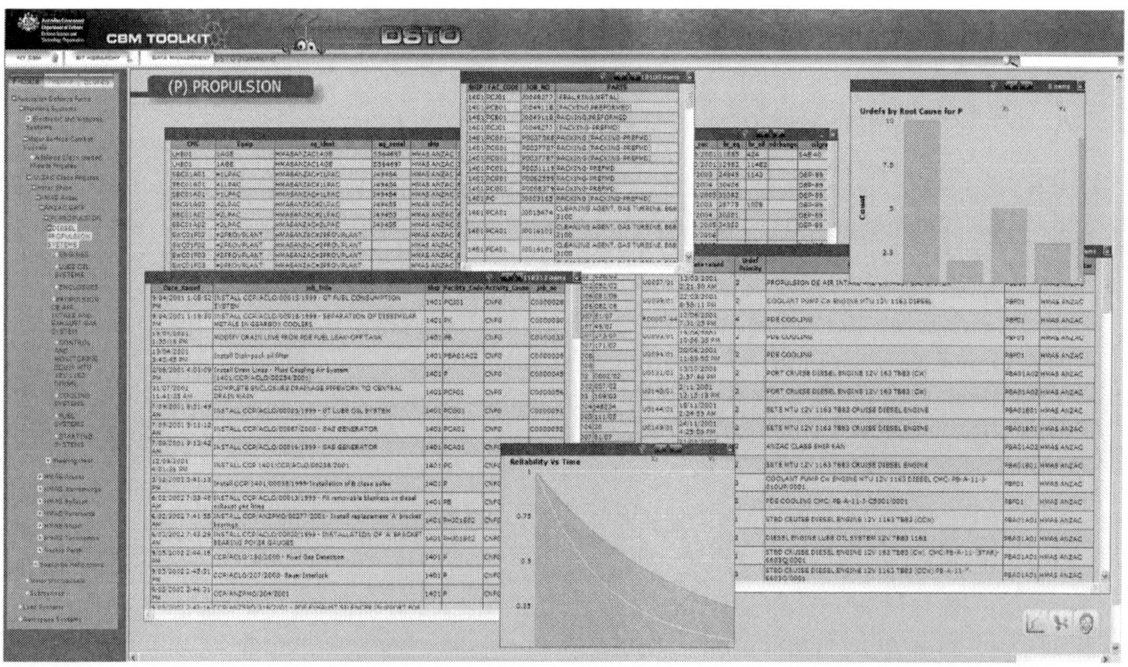

Figure 3 Example screen shot of a CBM Toolkit data presentation showing data extracted from data sources (intentionally cluttered to illustrate the numerous data sources available).

A number of user-defined hierarchies can be accommodated, enabling users to search data in their preferred context. Other features of the web-based user interface include:

- 'Happiness Indicators' (HIs) that can be created by individual users to present real time Key Performance Indicators (KPIs) relevant to their work area. HIs are typically queries that check the status of parameters from any combination of the available data sources. The HIs can be displayed with a choice of graphics (e.g. traffic light indicators);
- An advanced charting engine (The "Plot-o-Matic") to graphically present and interpret one or more datasets;
- An ability to download data to Microsoft Office applications for offline analyses;
- A functionality to create live-data reports in standard Microsoft applications;
- Customisable views allowing the user to personalise the CBM Toolkit interface, thereby simplifying access to the most pertinent information. This includes preferred bit hierarchies, data sources, reports, shortcuts and KPIs;
- Complete transparency of the underlying extraction code to allow users to use the same queries in other tools;
- A filter to evaluate/search data retrospectively if required; and,
- The ability to access a number of standard reporting templates, giving the user a push-button ability to generate asset management reports based on the latest available data.

In the 'Bit Hierarchy' screen shown, the user is able to select from the asset bit hierarchy in the left-hand pane and can then select the data sources that are of interest (via a right-mouse-click activated window). The data sources selected, which are presented in tabular form in individual windows, are displayed in the right-hand pane. They can then be individually filtered based on user criteria, then sorted and visually interrogated.

3.4.1 RCM Failure Mode Finder

A search tool, called 'RCM Failure Finder', has been developed [4,5] to find likely candidates for failure modes that correspond to a particular Urgent Defect[3] (URDEF) failure record, based largely on the information in the related URDEF signal. This set of potential failures was developed from RCM data. Searching is undertaken by:

1. The user selects the "URDEFS" data source and identifies an URDEF record of interest. The 'RCM Failure Mode Finder' screen then opens in a separate window with search results as shown in Figure 4. The URDEF data, shown at the top of the new window, is then compared against the list of potential failures from the RCM failure database.
2. Details of the Failure Modes and Effects Analysis (FMEA) record associated with any failure mode in the list can be viewed on the same screen by clicking the 'Noticeable Effects', 'Function', 'Functional Failure' and 'Observed Effects' links, also shown in Figure 4. Once satisfied with the available choice, the user can select the appropriate failure mode and return to the first screen.
3. If no suitable record can be found in the list, the user can either:
    a. amend the search text and repeat the search,
    b. explore the RCM failure database manually,
    c. flag the record as containing insufficient information for failure assignment,
    d. flag that no suitable RCM failure modes can be found for the URDEF record.

3 Urgent Defect notices are typically issued for system breakdowns requiring urgent attention.

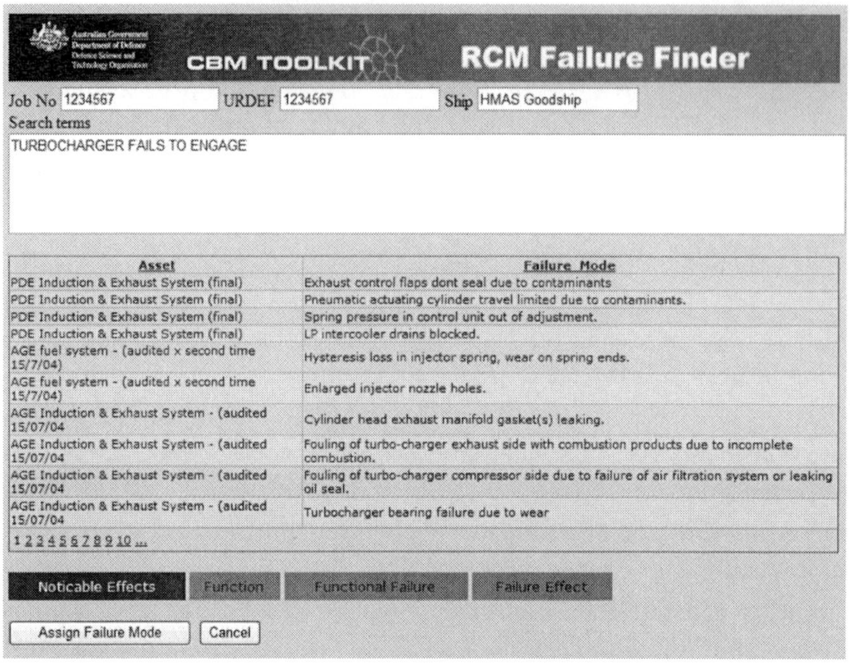

Figure 4 Example screenshot of the RCM Failure Finder using a fictional URDEF.

A new, separate database of 'clean' history records has been created and is updated following each new failure assignment since it is not permitted to write to the existing URDEF database. The RCM Failure Finder tool and process is still under development with proposed improvements to retrieval performance and functionality including, (a) ongoing enhancement of keywords to improve the likelihood of the most appropriate failure modes being at the top of the retrieved list; and, (b) adaptation of the tool for troubleshooting failures in the field (i.e. at time of failure).

3.5 CBM Toolkit Status

The CBM Toolkit software is operational and ready for field use although its capability should be extended through continual enhancement of the software and the related databases (e.g. cleansed failure histories as described in Section 3.4.1). Proposed enhancements include the addition of alternative search hierarchies, access to more data sources, the extension of reporting capabilities and the development of intuitive search and presentation functions.

Data quality has been a strong focus of this program due to the poor quality of historical data. Methodologies to profile and cleanse historical data, including the ability to select and apply data profiling metrics, have been incorporated into the CBM Toolkit. However, considerable further work is required to realise automated or semi-automated data cleansing algorithms, with the ultimate objective being to fully mine all pertinent information from existing data.

4. CONCLUSION

With the successful development of the SIU and the CBM Toolkit, the RAN now has a potentially powerful data-driven decision support capability for managing their maintenance needs. Once a minor software issue is resolved, it is anticipated that SIUs will be installed on all ANZAC Class ships in the near future, providing a plethora of performance data on critical systems; the data being collated into a C&M database for failure diagnostics and prognostics.

Furthermore, the SIU technology and the CBM Toolkit software have been developed as generic tools with uses beyond CBM analyses and with applicability to other ADO assets in addition to the ANZAC Class frigates. Many of the datasets required for CBM contain information that can be used for Life Cycle Costing and modelling.

The CBM Toolkit will provide the access to this growing database, as well as numerous other databases required for maintenance analyses. Further development is required to enhance the CBM Toolkit's data profiling and data cleansing capability. However, the concept of producing an interface capable of obtaining data from the variety of disparate data sources as opposed to accessing these data sources via a centralised repository of data has been realised. The completion of the CBM Toolkit trials program was delayed due to the inability to gain access to and therefore analyse the full C&M system datasets due to a C&M software related problem. This issue is being dealt with at the time of writing this paper.

The technology developed in this project represents a new paradigm for the RAN in its approach to maintenance management, one that can only be fully realised through organisational change and the dedication of staff to manage this new capability.

Asset management is increasingly being driven by competing forces, most notably the requirement to reduce operational budget costs whilst maintaining or enhancing platform reliability and increasing operational availability. The aim of the work presented in this paper is to encourage the optimisation of maintenance management processes by moving toward the DDDM paradigm. DDDM, in conjunction with innovative technologies, can be used in failure prognostics and diagnostics. It equips the asset manager with the knowledge to manage critical assets with limited capabilities, including being able to:

- Select optimal maintenance actions including the removal of unnecessary maintenance;
- Manage risk in different operational scenarios; and,
- Provide management with risk information in the face of competing considerations.

Data-Driven Decision Making is not presented as a panacea, but rather as an option of making more informed decisions based on real information rather than assumptions.

5. REFERENCES

1. Hammond, L., Kelly, P., (2008), "Improving Asset Management through Enhanced Decision Support", MAST2008 Conference Proceedings, Cadiz, Spain.
2. Sikorska, J., (June 2009), "Feasibility Report: AE Testing of RAN Diesel Engines", CASWA Pty Ltd.
3. Sikorska, J., (June 2009), "A Practical Framework for Evaluating Data Quality", CASWA Pty Ltd.
4. Sikorska, J., Hammond, L., and Kelly, P.J. (May, 2007), "Identifying Failure Modes Retrospectively using RCM Data", Conference: ICOMS 2007, Melbourne, IEAust.
5. Sikorska, J., Hammond L., (June 2009), "The Application of Case-Based Reasoning to Data Mining of Computerised Maintenance Management System", DSTO Technical Report, Department of Defence (submitted for publication).

Effect of Commonly Used Diesel Biocides on the Corrosion Behaviour of Marine Grade Aluminium Alloys

L.M. Pedrina & P.L. Mart
Defence Science and Technology Organisation Melbourne
Lenore.Pedrina@dsto.defence.gov.au

1. Abstract

A study was undertaken to investigate corrosion behaviour of marine grade AA5083-H321 and AA6082-T6 aluminium alloys in the presence of biocides used in distillate fuel. Condensation occurs within ship fuel tanks and the biocides easily leach out of the fuel phase into the water phase concentrating in the bottom of the fuel tanks.

Three biocides were evaluated, singularly and in combination, and two distillate fuels, F-76 and Automotive Diesel Fuel. No corrosion was observed due to biocides in the fuel phase, only in the water phase. DiEGME, premixed in F-76 distillate, did not cause corrosion. BC-250, either alone or in combination with one other biocide caused pitting corrosion. GrotaMar, either alone or in combination with one other biocide caused general corrosion. When all three biocides were present no corrosion was observed after 4 months exposure.

As DiEGME is premixed with F-76, it is recommended that its use continue. For Automotive Diesel Fuel, with no premixed biocide, addition of DiEGME is recommended in preference to BC-250 or GrotaMar. If the latter biocides are used, unpainted fuel tanks should be closely monitored with condensate drained off regularly, to avoid prolonged exposure.

2. Introduction

With the increased use of aluminium alloys in the shipbuilding industry a greater understanding of the effect of the various marine environments on these alloys is needed. In recent times marine grade aluminium alloys have been used to build increasingly larger vessels [1]. Aluminium ships are being utilised by customs, police, coast guard, other patrolling agencies, and as cruise vessels [1]. The non-magnetic property of aluminium also makes it ideal for minesweeping activities [2]. Marine grade aluminium alloys have been chosen to replace traditional steel structures because of their superior strength/weight ratio, reliability, and advancements in manufacturing and building techniques [1-3]. Research has concentrated on engineering changes, focusing on changing manufacturing/ship building techniques to ensure stress concentrations are as low as possible [1], and on the interaction of the external hull with the marine environment. However with the common practice of the hull also serving as the fuel tank, research is needed to understand the interaction of marine grade aluminium alloys with diesel/water environments that occur in the fuel tanks.

In the aircraft industry much research has gone into the effect of various environments on the corrosion resistance of aluminium alloys including the effects in aircraft fuel tanks. In particular, the interaction of these alloys, microorganisms and the DiEGME biocide in aircraft fuels has been investigated [4-6]. Condensation is commonplace in aircraft fuel tanks and in ship fuel tanks especially when operating in humid tropical environments where temperatures can vary markedly. When water is present in the fuel tanks biocide leaching occurs, with biocides concentrating in the water phase [6].

In aluminium ships the fuel tanks are largely unpainted exposing 'unprotected' aluminium to the fuel/water/biocide environment. Fuel and biocide use has not been assessed or adjusted with respect to the change in materials used for ship hulls. This paper investigates the behaviour of marine grade aluminium alloys when exposed to the fuel/water/biocide environment.

3. Experimental details

Samples of the aluminium alloys AA5083-H321 and AA6082-T6 were exposed to solutions of fuel/distilled water and biocides. AA5083 is an Al-Mg alloy and has a magnesium content typically 4.0-4.9%, while AA6082 is an Al-Mg-Si alloy with a magnesium content typically 0.6-1.2% and silicon 0.7-1.3%. The full compositions are shown in Table 1. The coupons were cut from as-received sheet alloy. The coupon dimensions were 25 mm x 100 mm with the AA5083 being 4.0 mm thick and the AA6082, 5.0 mm thick. Prior to exposure all the samples were ultrasonically cleaned in ethanol to remove any loosely adherent surface products.

Alloy	Si	Fe	Cu	Mn	Mg	Cr	Zn	Ti	Other	Al
AA5083	0.20%	0.35%	0.15%	0.15%	4.0-5.0%	0.15%	0.25%	0.10%	0.15%	Balance
AA6082	0.7-1.3%	0.50%	0.10%	0.40-0.10%	0.6-1.2%	0.25%	0.20%	0.10%	0.15%	Balance

Table 1 Typical Alloy composition of AA5083 and AA6082 [7].

The biocides used in this study were diethylene glycol monomethyl ether (DiEGME) commonly found in F-76 Naval Distillate; Diesel Fuel Biocide BC-250; and GrotaMar 71 Diesel additive. DiEGME is a pure chemical compound. BC-250 is a mixture of chemicals comprising >60% alkoxy alkanol, and smaller amounts (<10% each) of isothiazolones, magnesium chloride and magnesium nitrate. GrotaMar has as its majority component (>90%) the amine 5-methyloxazolidine. The manufacturers' Materials Safety Data Sheets should be consulted for specific information.

As well as the three biocides two different fuels were tested, F-76 Naval distillate (F-76) and BP Automotive Diesel Fuel (BP-ADF). Marine vessels are known to use both fuels. F-76 is premixed with the biocide DiEGME, BP-ADF has no pre-mixed biocide content.

As discussed in the introduction, water from condensation is common in the fuel tanks of platforms operating in a maritime environment. To simulate this situation, laboratory experiments were conducted in glass bottles with fuel floating above a water/biocide solution. Concentrations of biocides in the water portion siphoned from marine fuel tanks have been reported at concentrations as high as 20% for DiEGME [6], 20% BC-250 [8] and 2% for GrotaMar [8]. To encourage any possible corrosion the biocide levels for the experiments were set at these maximum levels. The glass bottles had a total volume of 1 litre. To create the fuel/water/biocide environments, 800 ml of fuel was added to 200 ml of the water/biocide solutions. To create the single biocide solutions of either 20% DiEGME or 20% BC-250 solution, 40 ml of biocide was taken and distilled water was added to make a 200 ml solution. For the 2% GrotaMar solution, 4 ml of biocide was taken and distilled water added to make a 200 ml solution. To mix the dual biocide solutions, 100 ml of each of the individual biocide solutions was added to 800 ml of fuel. To create the triple biocide solution, 100 ml of each of the individual biocide solutions, totalling 300 ml, was added to 700 ml of fuel.

Once the exposure solutions were mixed each bottle was shaken vigorously for 60 seconds to allow thorough mixing of the fuel and the water biocide solution. These were then allowed to rest and in all cases the fuel and water/biocide solution easily separated with the fuel floating above the water/biocide solution.

The alloy samples were placed in the test solution such that the metal crossed the water/fuel interface exposing the metal to both the fuel portion and the water/biocide portion of the mixture. Samples were exposed to the fuel/water/biocide mixtures for 4 months, held at 30°C in a water bath, and the progress of corrosion monitored. An example of an exposure bottle is shown in Figure 1. The red solution at the bottom is the water/biocide solution, which in this case is water/DiEGME/BC-250, and the yellow liquid is F-76 distillate.

Figure 1. AA5083 exposed to F-76/Distilled Water/ DiEGME/BC-250

After exposure the samples were removed and the surfaces analysed by optical microscopy and Scanning Electron Microscopy (SEM). Elemental analysis of the corrosion products was also undertaken by Energy Dispersive X-ray analysis (EDX). Pit depths were measured using a differential focussing technique on a metallurgical microscope.

4. Results & Discussion

The corrosion observed on samples was confined to the area exposed to the water/biocide solution and not the area exposed to the diesel fuel. Due to this, discussion of the results will concentrate on examination of the area of the samples exposed to the water/biocide portion of the mixtures.

4.1 Distilled water and single biocides

4.1.1 DiEGME

Both alloys exposed to the solution containing only DiEGME showed no signs of general corrosion or pitting corrosion after the 4 month exposure. There were no significant surface deposits and these samples did not differ in appearance from those exposed to the water/fuel mixture (no biocide) or samples that had not been exposed to any solutions at all.

4.1.2 BC-250

The samples exposed to the water/biocide solution containing only BC-250 showed pitting as well as a corrosion product of hydrated aluminium hydroxide complexes including sulphur and chlorides; see Figure 2 (a) & (b). The sulphur and chlorides have deposited from the BC-250 biocide which contains sulphur and chlorides whereas DiEGME and GrotaMar do not. The EDX spectrum of the corrosion product is shown in Figure 2 (c). The pitting of the alloys under these conditions was the deepest seen in this study, with average pit depths of 20 μm on AA5083 and 30 μm on the AA6082. Although slightly shallower, the number and density of pits on the AA5083 was greater than on the AA6082. Pits were counted and depths measured in 4, 1mm x 1mm areas of the surfaces. The pit density and the average pit depth were then determined. AA5083 had a pit density of 115 pits/mm^2 and AA6082 had a pit density of 68 pits/mm^2. Figure 2 (a) & (b) show examples of the pitting on the aluminium alloys exposed to the BC-250 biocide.

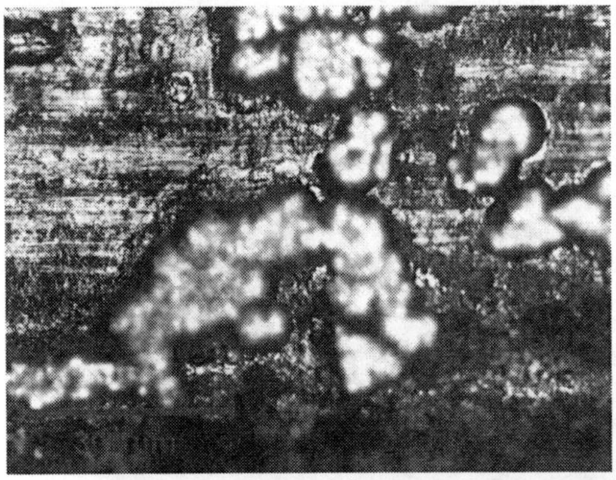

(a)

(b)

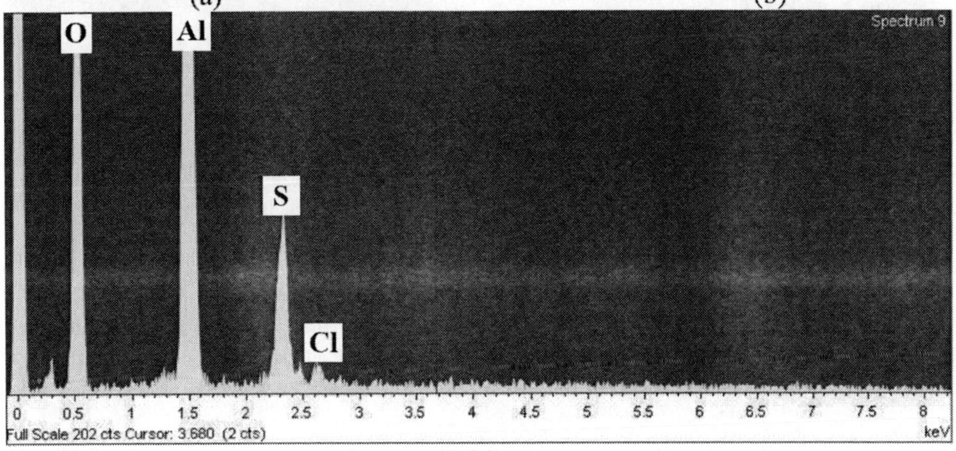

(c)

Figure 2. Pitting corrosion of marine aluminium alloys after 4 months exposure to BP-ADF/Distilled Water/ BC-250 solution. (a) AA5083; (b) AA6082; (c) EDX spectra of the corrosion product on AA5083.

4.1.3 GrotaMar

The alloys exposed to the water/biocide solution containing only GrotaMar had a thick surface deposit. Analysis of the AA5083 sample showed that the surface deposit was a loosely adherent shell-like coating on the alloy surface as shown in Figure 3 (a) & (b). EDX analysis of the surface deposit confirmed that this was a corrosion product layer for which the collected spectrum is shown in Figure 3 (c). The peak attributed to magnesium has two possible sources; it could arise from the substrate alloy as non-exposed AA5083 samples give a strong magnesium peak that is much larger than that seen in the EDX spectra in Figure 3, or it could arise from the hydrated aluminium/magnesium hydroxide corrosion product.

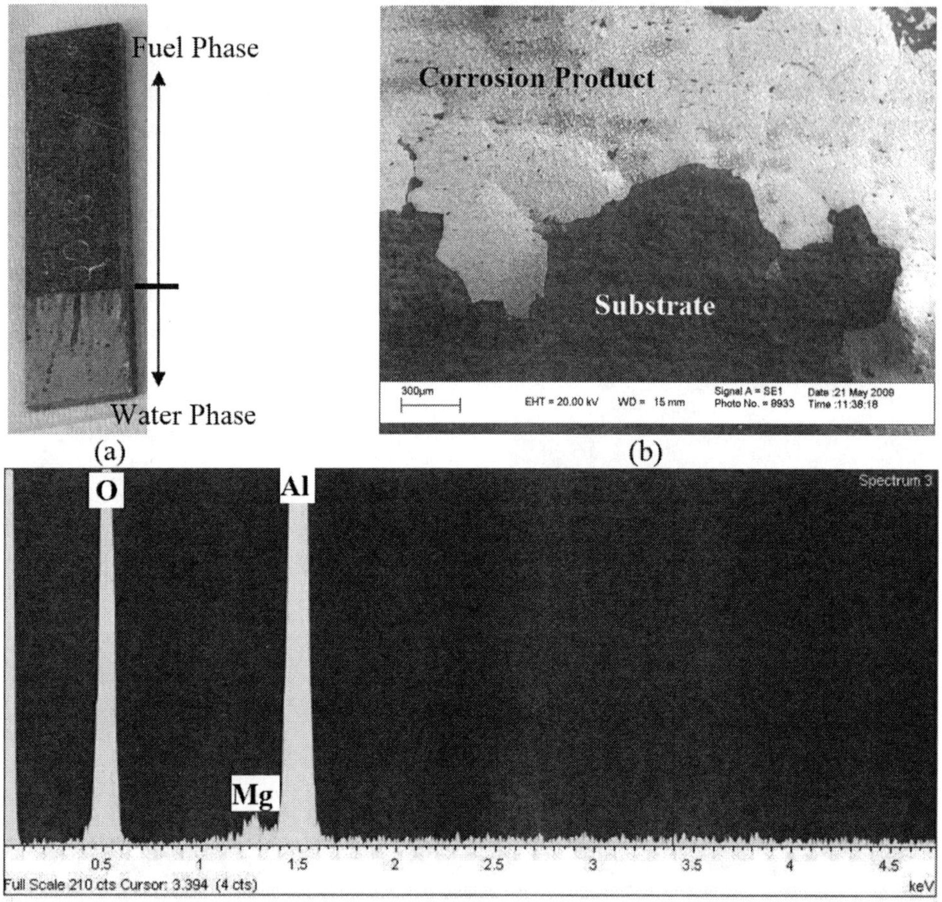

(c)

Figure 3. AA5083 after 4 months exposure to BP-ADF/Distilled Water/GrotaMar solution. (a) sample after removal from solution; (b) SEM micrograph of corrosion product on AA5083; (c) EDX spectrum of the corrosion product on AA5083.

AA6082 exposed to the 2% GrotaMar solution had a visibly thicker corrosion product layer than the AA5083 under SEM examination. Figure 4 (a) shows the AA6082 sample after 4 months exposure to the GrotaMar solution and Figure 4 (b) shows the thicker corrosion product layer that was not completely uniform. The EDX spectrum in Figure 4 (c) shows the aluminium hydroxide complex corrosion products however no detectable magnesium peak, consistent with the lower magnesium concentration of the alloy.

The corrosion product was removed from both the alloys by nitric acid (HNO_3) cleaning and the surface examined optically. The cleaning process involved the samples being immersed in a 1 M HNO_3 solution for 3 min at room temperature as described in ASTM Standard G1 [9]. Pitting corrosion was not observed, rather a general corrosion attack of the surfaces of both alloys had occurred.

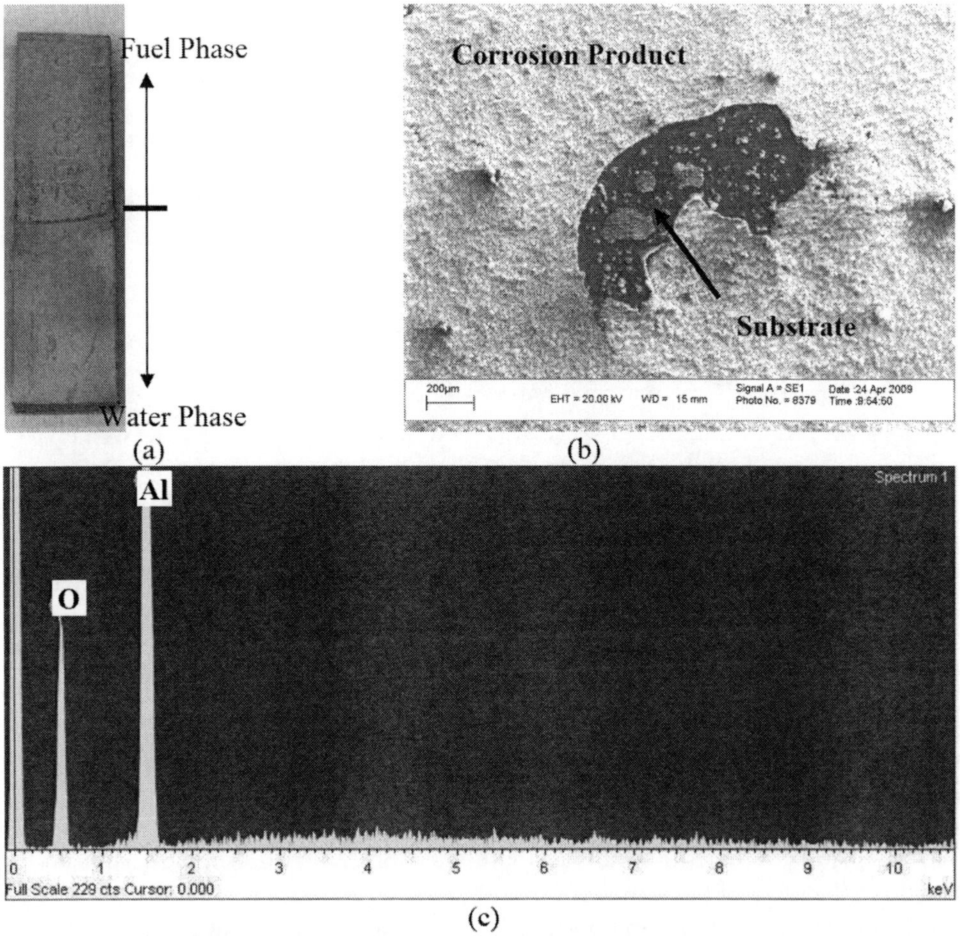

Figure 4. AA6082 after 4 months exposure to BP-ADF/Distilled Water/GrotaMar solution. (a) Sample after removal from solution; (b) SEM micrograph of corrosion product on AA6082; (c) EDX spectrum of the corrosion product on AA6082.

4.2 Distilled water and dual biocides

When fuel tanks of marine vessels are topped up, mixing of different fuels and more than one biocide can occur. This results in a mixture of biocides leaching out into the water fractions in the tanks. Therefore, combinations of the biocides were experimentally investigated to simulate the mixing of different fuel sources.

4.2.1 Distilled water/DiEGME/BC-250

The AA5083 exposed to F-76 distillate and the dual biocides DiEGME and BC-250 had a white corrosion product of hydrated aluminium hydroxide complexes including sulphur and chlorides from the BC-250 biocide, as shown by EDX analysis in Figure 5 (a). There was no detectable magnesium peak, which suggested that the AA5083 substrate was masked by the thickness of the corrosion product.

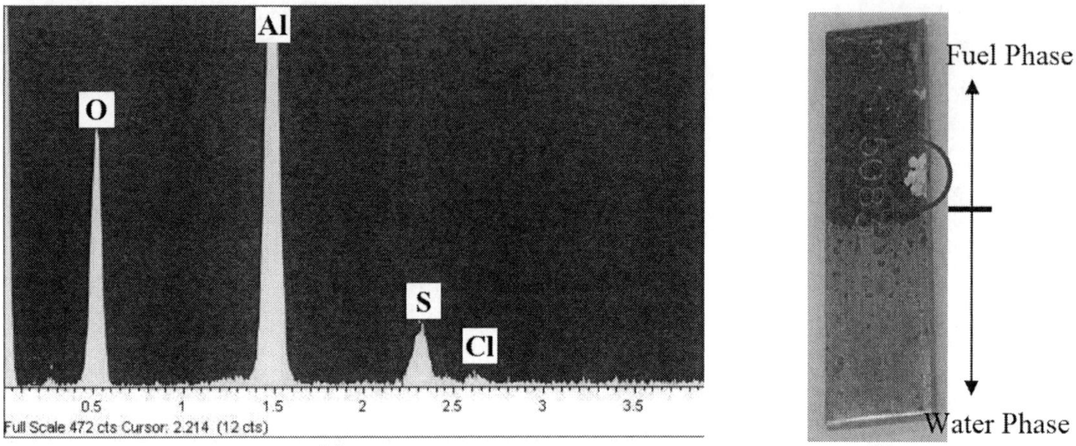

(a) **(b)**

Figure 5. AA5083 after 4 months exposure to F-76/Distilled Water/DiEGME/BC-250.
(a) EDX spectrum of the surface deposit on AA5083; (b) sample after removal from solution.

The small amount of sulphur and chlorine detected would have arisen from the BC-250 biocide which contains these elements. After exposure the area of the sample exposed to the F-76 distillate was largely free of corrosion products while the area exposed to the water/biocide solution had a deposit of white corrosion products. There was also an area on the surface close to the fuel/water interface where localised corrosion attack had occurred which appeared as a bright region, circled in Figure 5(b). As the sample was inclined this corrosion actually initiated at the fuel/water/biocide interface on the side of the sample. The corrosion migrated around the edge of the sample and onto the front surface as shown in Figure 5(b). Inspection by SEM showed there was substantial pitting corrosion in the bright area and along the edge of the sample. The SEM micrograph in Figure 6 (a) shows a pit on the surface of the sample in this region. EDX confirmed that the deposit around the pit is hydrated aluminium hydroxide corrosion products. Pitting corrosion occurred over the entire surface exposed to the water/biocide mixture but did not occur on the part of the sample exposed to the diesel fuel. Figure 6 (b) shows an example of this pitting corrosion.

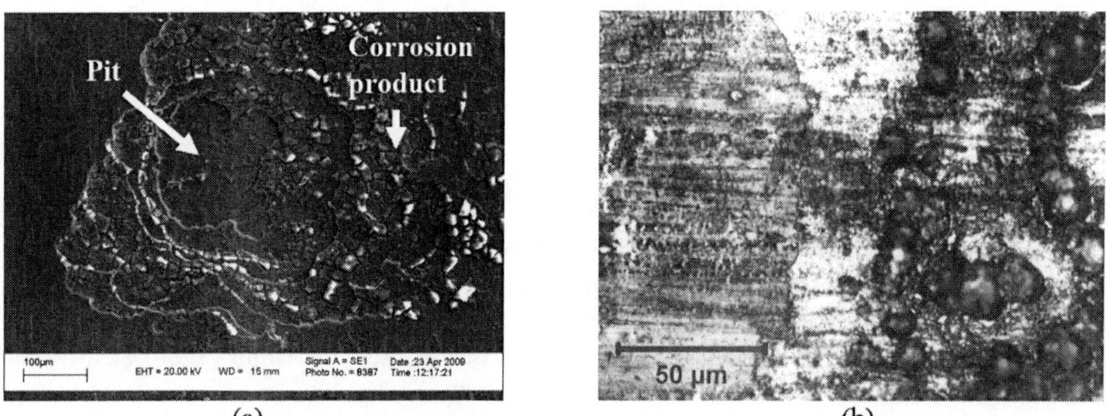

(a) (b)

Figure 6. AA5083 exposed to F-76/Distilled Water/DiEGME/BC-250.
(a) SEM micrograph of the localised pitting corrosion; (b) photo of the pitting corrosion.

The AA6082 exposed to the F-76/Distilled water/DiEGME/BC-250 also had a hydrated aluminium hydroxide corrosion product deposit on the surface, confirmed by EDX analysis, and surface pitting similar to that seen on the AA5083. Again this was confined to the part of the sample exposed to the water/biocide solution.

4.2.2 Distilled water/DiEGME/GrotaMar

Both of the alloy samples exposed to the DiEGME/GrotaMar solution showed very little evidence of corrosion. There was a thin corrosion product film on the AA6082 sample; however the AA5083 sample was indistinguishable from a non-exposed sample.

4.2.3 Distilled Water/BC-250/GrotaMar

A visual inspection of the two alloys exposed to the BC-250/GrotaMar solution revealed red staining on the surfaces due to the red dye in the BC-250 biocide. EDX analysis of the staining did not reveal any elemental composition of this deposit, as the aluminium signal from the substrate alloy dominated the spectrum. Optical microscopy as shown in Figure 7(a) & (b) revealed pitting of the surfaces, however this was different to the pitting seen on the samples exposed to BC-250 alone (shown in Figure 2). The pits in Figure 7(a) & (b) were closer together and more extensive across the surface of the metal. On the AA5083 sample (Figure 7(a)) there was extensive general surface corrosion as well as pitting occurring simultaneously. As a result it was impossible to determine a pitting density for AA5083 exposed to these conditions. Figure 7 (b) shows examples of the pitting attack on AA6082 exposed to these two biocides. This sample had a pit density of 93 pits/mm^2, higher than was observed with the BC-250 biocide alone.

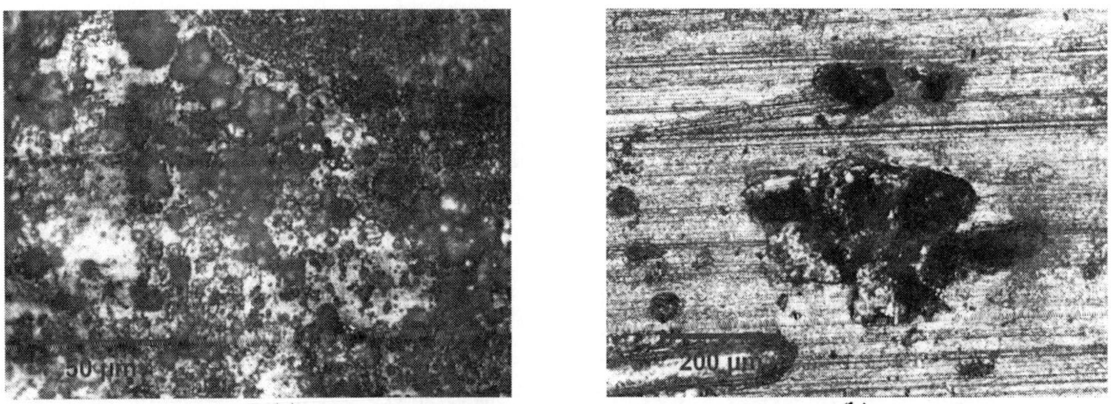

(a) (b)

Figure 7. Aluminium alloy surface pitting and general corrosion after 4 months exposure to BP-ADF/Distilled Water/BC-250/GrotaMar solution. (a) AA5083; (b)AA 6082.

4.2.4 Distilled Water/DiEGME/BC-250/GrotaMar

When all three biocides were present in the water mixture, prepared as outlined in the experimental details, no corrosion was seen and no surface deposits were detected. After exposure, the appearance of samples was identical to those which had not been exposed.

5. Conclusions

This study has shown that the fuel/biocide environment per se does not cause corrosion issues in marine grade aluminium tanks, however, when condensation occurs the biocides are able to leach out of the fuel and concentrate in the water portion. It is this environment that is particularly aggressive to marine aluminium alloys such as AA5083 and AA6082.

When DiEGME was the only biocide present no corrosion was observed after 4 months exposure. When the biocide BC-250 was present either alone or in combination with DiEGME or GrotaMar, pitting corrosion occurred. When GrotaMar was in the water mixture alone or in combination with DiEGME or BC-250, general corrosion of the exposed surfaces occurred. When either alloy was exposed to all three biocides in the ratios described no corrosion was seen after 4 months exposure.

When different fuel sources are mixed in ship tanks, mixing of biocides also occurs. This study has shown that the mixing of biocides creates corrosive environments resulting in higher general corrosion rates, and in the presence of BC-250 pitting corrosion occurs.

F-76 distillate is supplied pre-mixed with DiEGME biocide. This in itself does not create an aggressive environment as DiEGME on its own does not increase corrosion rates or cause pitting when concentrated in the water fraction within unpainted fuel tanks. However, the addition of commercially available biocides such as BC-250 and GrotaMar can cause significant corrosion. It is recommended that when F-76 distillate is the only fuel used that no additional biocide be added. If fuel is used that has no biocide premixed, then DiEGME is recommended. If BC-250 and GrotaMar are used, tanks need to be monitored closely and the condensate drained off as soon as practicable. This will prevent the unpainted aluminium alloy tanks having prolonged contact with concentrated water/biocide solutions.

6. References

1. Ferraris, S., (2005) "Aluminium Alloys in Third Millennium Shipbuilding: Materials, Technologies, Perspectives." Proceedings of The Fifth International Forum on Aluminium Ships 2005: Tokyo, Japan 11-13 October, p. 1-11.
2. Wang, X., (1997) "On Application of Aluminium in Ship Hulls." American Society of Mechanical Engineers, 3, p. 275-282.
3. Hoogovens Aluminium Walzprodukte, (1998) "Alustar Aluminium A New Generation Of Marine Grade Aluminium Alloys." HSB International, 47, p. 35-37.
4. Liang, Z.D., (1992) "Corrosion of Aluminium Alloy in Aircraft Fuel Tank Compartments Due to Condensed Water." Corrosion, 48(6), p. 514-517.
5. Marlowe, D.L. and Green, M.G., (2007) "Investigation into the cause and effects of widespread coating failures and corrosion on the internal surfaces of P-3 Orion wing integral fuel tanks." Corrosion Reviews, 25(5-6), p. 533-543.
6. McNamara, C.J., Perry IV, T.D., Leard, R., Bearce, K., Dante, J., Mitchell, R., (2005) "Corrosion of aluminium alloy 2024 by microorganisims isolated from aircraft fuel tanks." Biofouling, 21(5/6), p. 257-265.
7. ASM International, (1990) ASM Handbook Vol 2 Properties and Selection: Nonferrous Alloys and Special-Purpose Materials, Specific Metals and Alloys, Alloy and Temper Designation Systems for Aluminum and Aluminum Alloys. ed ASM Hanbook Committee, USA.
8. Hulston, C., (2007) DSTO Air Vehicles Division, Fuels and Lubricants Group, Personal Communication.
9. ASTM G1-90 (1999), "Standard Practice for Preparing, Cleaning, and Evaluating Corrosion Test Specimens - A1 Chemical Cleaning Procedures." ASTM International, West Conshocken, PA.

Flexible Composite Hydrofoils and Propeller Blades

N.L. Mulcahy[1], B.G. Prusty[2] and C.P. Gardiner[3]
[1] Pacific Engineering Systems International Pty. Ltd.,
Sydney, NSW, Australia.
lexm@esi.com.au
[2] The University of New South Wales, Sydney, NSW, Australia.
g.prusty@unsw.edu.au
[3] Maritime Platforms Division, Defence Science and Technology Organisation, Melbourne, Victoria, Australia.
Craig.Gardiner@dsto.defence.gov.au

INTRODUCTION

Shape-adaptive structures have properties that allow them to adapt to their working conditions, and have applications in a number of fields e.g. the INCOMPRO (Intelligent Composite Products) project designed and produced shape-adaptive structures as diverse as yacht booms, marine propellers, floor panels and pump impellers [1]. The design of shape-adaptive structures involving fluid-structure interaction is also called hydroelastic tailoring. A field of particular interest in the present day is that of flexible composite marine propellers and hydrofoils [2–14]. Propeller hydroelastic tailoring aims to achieve adaptive blade pitching whereby the pitch changes with loading in a specified manner, resulting in increased efficiency over a range of operating conditions. The use of composite materials makes it possible to design a laminate to achieve the required pitch changes.

There are several existing approaches to implementing hydroelastic tailoring of composite propellers:

1) The first approach sets performance goals for the flexible propeller at two operating conditions, and optimizes the composite laminate to achieve those goals (e.g. Lee and Lin [4], Lin et al. [13]). The first performance goal is that the flexible propeller should maintain the same performance in terms of torque, thrust and efficiency as a baseline rigid propeller at the designed operating condition. The second performance goal is that the pitch must reduce as much as possible as the inflow angle is reduced (i.e. inflow velocity reduced or rotational speed increased) to keep the torque within the narrowest possible range.
2) The second approach assumes that the propeller operates in a spatially varying wake, and designs the blades to adapt to the changing flow conditions encountered during each revolution (e.g. Chen et al. [5], Motley et al. [14]). This recognizes that while on average the propeller is operating at the design condition, an individual blade's incoming flow velocity varies throughout each revolution so that effectively the operating condition of the blade fluctuates about the design condition. To minimize efficiency losses and potentially benefit from the changing flow conditions, the efficiency of the flexible composite propeller should match that of its rigid counterpart at the design condition, and be higher at off-design conditions.
3) A third approach, developed by Blasques et al. [12], is applicable to controllable pitch flexible composite propellers. This method determines the optimal composite lay-up

and pitch angle to reduce the weighted average fuel consumption for two conditions (cruising and maximum speeds).

While the reasoning behind the first two approaches is different, the outcome is the same. The resulting flexible propeller will have the same shape and performance as a baseline rigid propeller at the design operating condition, the pitch reduces with increasing load, and the flexible propeller is more efficient than the baseline rigid propeller at off-design conditions. A common feature of the existing hydroelastic tailoring methods is that the flexible propeller is based on an existing rigid propeller.

The method to be outlined and applied in the present paper assumes that the flexible propeller is to adapt to spatially varying wake flow, but with the knowledge that this will also mean that the propeller will have improved off-design efficiency. It will be shown that for optimum efficiency, the efficiency curve of the flexible propeller must be tangential to that of the baseline rigid propeller at the design condition, and that therefore the design condition cannot be arbitrarily specified.

The remaining sections of the paper contain: (i) an outline of the proposed hydroelastic tailoring method, (ii) two design examples (hydrofoil and propeller), (iii) an outline of a method for estimating efficiency gain of a flexible propeller in a spatially varying wake, and (iv) conclusions. This paper does not explore the potential benefits that could be achieved from a composite flexible propeller in terms of reduced maintenance, weight, vibration and acoustic response, although these areas are the subject of current and planned work.

OUTLINE OF THE METHODOLOGY

While the method is primarily aimed at the design of flexible propeller blades, it can also be applied to the design of hydrofoil blades in certain situations. Consequently, the method will be outlined for propellers and the differences noted for hydrofoils.

Propeller characteristics are presented using the following coefficients: advance coefficient $J = V_A/nD$, thrust coefficient $K_T = T/(\rho n^2 D^4)$, torque coefficient $K_Q = Q/(\rho n^2 D^5)$, and efficiency $\eta = K_T J/(2\pi K_Q)$, where V_A is the mean axial inflow velocity, n is the rotational speed (revolutions per second, rps), D is the propeller diameter, ρ is the water density, T is the thrust, and Q is the torque. Curves of K_T, K_Q and η are usually plotted as functions of J.

Assume that a propeller is operating in the wake of a ship where the flow velocities are spatially varying. Incoming flow velocities at different points of the propeller plane can vary substantially from the average velocity. A propeller may be operating at the design advance coefficient, J_{design}, based on the average incoming flow velocity, but the advance coefficient of an individual blade effectively varies during each revolution as it passes through the spatially varying velocity field i.e. for part of the revolution $J < J_{design}$ and for the remainder $J > J_{design}$. The aim of hydroelastic tailoring is to benefit from the variation of J by designing a flexible blade that is more efficient than a similar rigid blade for a range of operating conditions near J_{design}.

Efficiency curves for a series of propellers are presented in Figure 1. The propeller series was created by taking a baseline rigid propeller and applying a series of pitch changes. A series of curves such as this is the starting point for the development of the proposed hydroelastic tailoring method for flexible propellers. The aim is to produce a propeller blade that will adapt to changes of the flow conditions relative to the design operating state, which is designated the design condition.

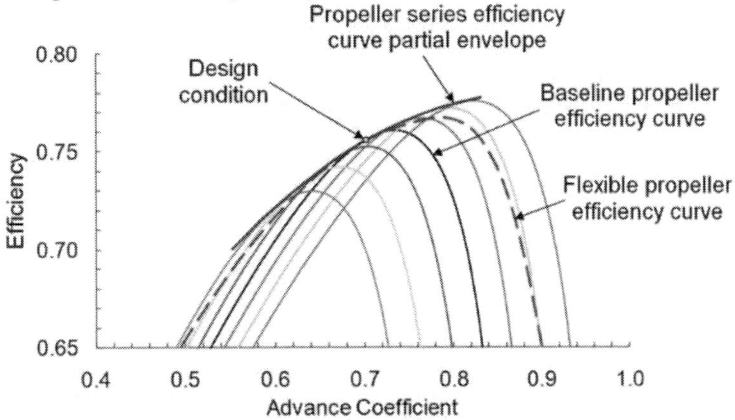

Figure 1: Propeller Series Efficiency Curves and Design Condition

The next step is to construct the partial envelope of the efficiency curves of the propeller series. To maximize the overall efficiency, the efficiency curve of the flexible propeller must be tangential to that of the baseline rigid propeller at the design condition J_{design} (Figure 1). If the two curves are not tangential at this point, the curve for the flexible propeller will cross that of the rigid propeller, so that for a range of J, either less than or greater than J_{design}, the flexible propeller will be less efficient than the baseline propeller. Given the appropriate material properties and unloaded shape, a flexible propeller may be designed with an efficiency curve corresponding approximately to the envelope of the propeller series. However, this objective may not be possible within the constraints provided by the strength and stiffness of available composite materials. It is important to note that points on the efficiency curve envelope at lower values of advance coefficient correspond to reduced pitch angle and higher propeller thrust, and therefore the pitch of a flexible shape-adaptive propeller must reduce as the load increases. So rather than aiming to achieve the envelope curve, the objective of the present method is to achieve the maximum reduction of pitch between the unloaded and loaded states, while limiting the stresses in the most highly loaded condition. The effect is to broaden the efficiency curve.

The outline of the proposed scheme for designing a shape-adaptive propeller is:

1) Determine the shape of a baseline rigid propeller to satisfy hydrodynamic requirements. Perturb the baseline shape to produce a series of propellers with slightly different pitches, noting that usually two perturbed shapes are sufficient.
2) Compute using steady flow analyses the efficiency curves of the propeller series and determine the operating condition corresponding to where the envelope of the curves is tangential to the efficiency curve of the baseline propeller i.e. the design condition. If the thrust at this condition does not satisfy the design requirements, the shape of the baseline propeller must be modified, and the procedure to this point repeated.
3) Determine the blade loads for the baseline propeller at the design condition.
4) Determine the allowable stresses or strains at the design condition, which may not be the most highly stressed condition within the operating range i.e. the allowable stresses or strains at the design condition will be less than those for the most highly loaded condition.
5) Using an optimisation technique, find the unloaded shape and composite lay-up of the propeller blade to maximise the average pitch angle change from the unloaded state to

the design condition, subject to the blade not being overstressed at the design condition. The unloaded flexible propeller is pre-twisted relative to the baseline propeller, and the direction of twist is such that the pitch reduces with increasing load.

6) Given the unloaded shape and composite lay-up, carry out a series of coupled hydrodynamic/structural analyses over a range of operating conditions to define the efficiency curve of the flexible propeller. The efficiency of the flexible propeller should be greater than or equal to that of the baseline rigid propeller over the full range of operating conditions.

For hydrofoils, curves of lift on drag (L/D) versus angle of attack at the root (i.e. root angle α_{root}) are considered in place of efficiency curves.

EXAMPLES

Design examples for a flexible hydrofoil and a flexible propeller are presented in the following sections. The finite volume computational fluid dynamics code CFD-ACE+ [15] is used for all hydrodynamic calculations, the finite element code for composites SYSPLY [16] for all structural calculations, and the general optimization code PAM-OPT [17] for the optimization calculations. Structural dynamic effects are not included in the analysis.

Hydrofoil

The proposed method can be demonstrated with a simpler structure than a propeller blade, namely a hydrofoil blade. The aim is to produce a flexible shape-adaptive foil that has improved lift on drag relative to a baseline rigid foil, for a range of angles of attack (0 to 10 degrees) at the specified flow velocity (8 m/s). It is assumed that the hydrofoil is to operate in disturbed flow conditions where the inflow direction of the water fluctuates about a mean value, and the foil is to adapt to these changing conditions.

The geometry of the baseline rigid foil is summarized in Figure 2. Both the chord and thickness taper linearly from the root to tip, and the tip is rounded. The angles of attack at the root and tip of the baseline rigid foil are identical. The angle of attack is varied by rotating the foil about the leading edge at the root. The simulation domain for the hydrodynamic analysis is 4.0 m × 4.0 m × 1.5 m, with the inlet 1.5 m upstream of the foil's leading edge at the root. The foil cantilevers from the mid-height of one of the exterior surfaces, and extends in the shorter direction of the domain. The mesh for the hydrodynamic analysis is fully structured with 2.85 million cells, while that for the structural analysis has approximately 3000 shell elements.

Create two perturbed foil shapes by imposing linearly varying twists so that the tip angle of attack is changed by +1 degree in one case, and -1 degree in the other. Figure 3 presents curves of L/D for the three foil shapes, and a tangent curve. The design condition is selected as the point where the tangent curve meets the baseline L/D curve (root angle of attack, $\alpha_{root} \approx 5.5$ degrees).

Material properties for a unidirectional carbon/epoxy ply are taken from Soden et al. [18] (AS4 Carbon/3501-6 Epoxy). Based on having two ply directions, a minimum ply angle difference of 60 degrees, and equal numbers of plies in each direction, the optimum ply angles are computed so as to maximise the average reduction of angle of attack over the length as loads are applied (from the unloaded state to the loaded state at the design condition). Maximum surface strains are constrained so that the factor of safety is greater than 2 at maximum load (corresponding to α_{root} = 10°). Final ply angles, relative to the tangent plane at the foil root (opposite to the flow direction), are 75° and 15°. Figure 4 presents (to scale) the unloaded shape relative to the loaded shape at the design condition. The tip angle of attack reduces by 2.2 degrees as loads are applied.

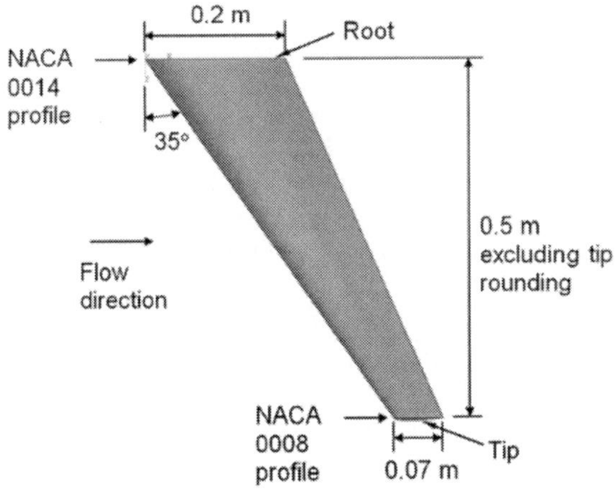

Figure 2: Hydrofoil (Plan View)

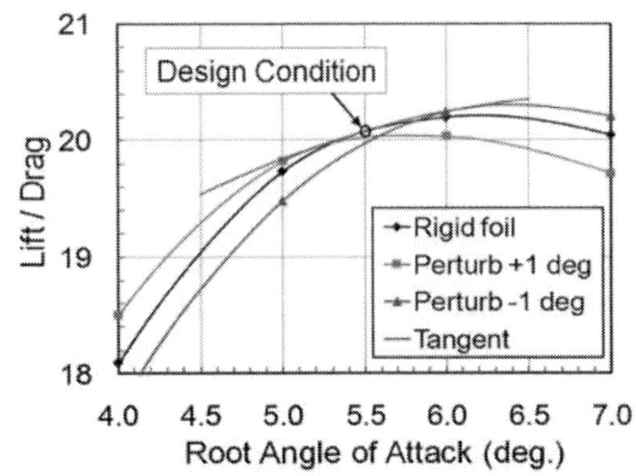

Figure 3: Hydrofoil Design Condition

Figure 5 presents the results of a series of steady coupled hydrodynamic/structural analyses. The computed L/D of the flexible foil is compared with that of the baseline rigid foil. As required, the flexible foil's L/D is greater than or equal to that of the baseline foil for the full range of root angles of attack from 0 to 10 degrees. The beneficial effect of hydroelasticity may have been greater if the geometry of the baseline foil was chosen to maximize flexibility through the selection of a any combination of a non-linear span-wise variation of thickness, a plan-form with larger chord length near the foil tip where the shape changes are greatest, or a composite material with lower elastic modulus and similar strength.

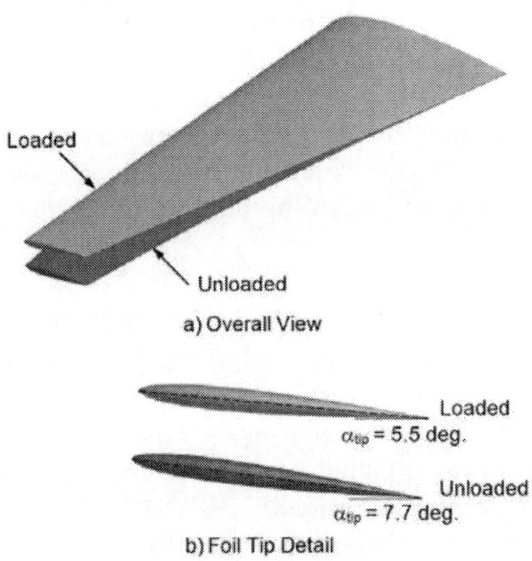

Figure 4: Unloaded and Loaded Shapes

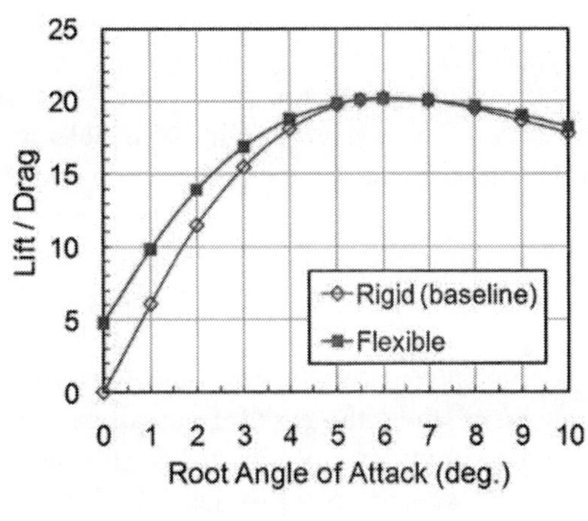

Figure 5: Comparison of Lift on Drag

Propeller

The aim is to design a flexible shape-adaptive composite propeller with efficiency that is greater than or equal to a baseline rigid propeller over its range of operating conditions. The baseline propeller is DTNSRDC Propeller 4384 [19,20], which has five blades and 108 degrees skew. The flexible propeller is to produce the same thrust, torque and efficiency as the rigid propeller at the design operating condition. The rotational speed is 7.8 rps for all calculations.

Hydrodynamic computations for a single blade are carried out using cyclic boundary conditions and a rotating frame of reference (Figure 6). The computational domain extends 0.45 m upstream, 1.53 m downstream and to a radial distance of 0.305 m. The mesh consists of approximately 3.9 million cells. The structural model of the blade consists of approximately 29 thousand shell elements on the mid-surface of the blade. The mesh density of the structure is greater than necessary for the stress calculations, but is required for proper functioning of the coupling of the structural and hydrodynamic models.

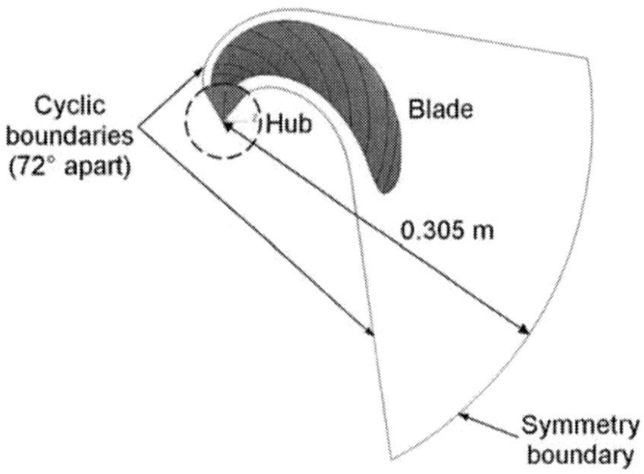

Figure 6: Propeller Computational Domain

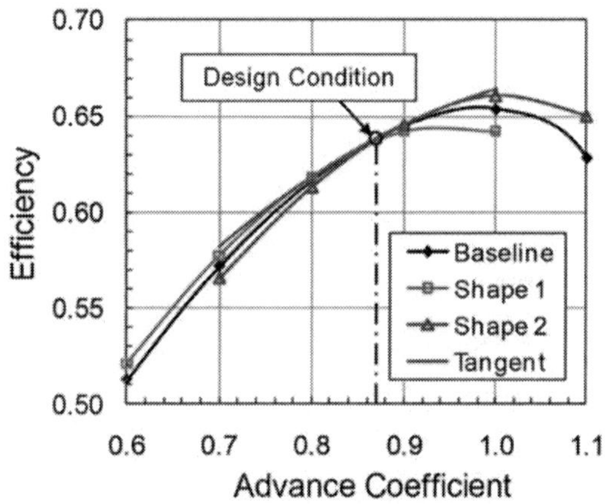

Figure 7: Propeller Design Condition

Two perturbed blade shapes are created from the baseline shape by applying the fluid loads for advance coefficient J = 0.9 to the blade with assumed material properties, and adding and subtracting the resulting displacements. The pitch angle change at radius 0.9R is approximately ±1.0 degree relative to the baseline shape. Figure 7 presents computed efficiency curves for the three shapes, plus a curve that is tangent to all three efficiency curves. The design condition is selected as the point where the tangent curve meets the baseline efficiency curve (advance coefficient $J \approx 0.87$).

The aim of the optimization is to find the composite lay-up that will result in the maximum reduction of pitch angle (averaged over the length) as the design condition loads are applied. In order to examine the effects of different degrees of flexibility, two different composite materials used and therefore two optimization analyses are required. For the present study, the cases are an all carbon/epoxy lay-up including strain limits, and an all E-glass/epoxy lay-up without strain limits. Properties for unidirectional AS4 Carbon/3501-6 Epoxy and Silenka E-Glass/MY750/HY917/DY063 Epoxy are taken from Soden et al. [18]. The adopted strategy is to have a symmetrical lay-up and four ply angles with half the plies placed at one angle, and the remaining plies evenly divided between the other three angles at -45°, +45° and +90° relative to the first angle. Ply orientations are defined relative to the element local x-direction, which is obtained for all elements from SYSPLY's geometric draping option with the starting point tangent to the trailing edge at the root. The resulting element local x-direction is approximately parallel to the blade centreline.

The optimum ply angle for the major direction of the carbon/epoxy lay-up is 34°. Rounding to 30° produces the ply sequence [120°,30°,75°,30°,-15°,30°]. Similarly, the E-glass/epoxy lay-up has an optimum major ply angle of 49° and resulting ply sequence [140°,50°,95°,50°,5°,50°]. Different unloaded blade shapes result from the two composite lay-ups although the baseline shape is the same. The E-glass/epoxy case is more flexible than the carbon/epoxy case. For example, at J = 0.87 the maximum blade tip deflection of the carbon/epoxy blade is 2.1 mm, and that of the E-glass/epoxy blade is 4.0 mm i.e. the E-glass/epoxy blade is approximately twice as flexible as the carbon/epoxy blade.

Comparisons of thrust coefficient, K_T, torque coefficient, K_Q, and efficiency, η, are presented in Figure 8 for the baseline rigid propeller and the two flexible propeller cases. These results are from a series of steady coupled hydrodynamic/structural analyses, for which the advance coefficient is varied from 0.4 to 1.3 while the rotational speed is held constant at 7.8 rps.

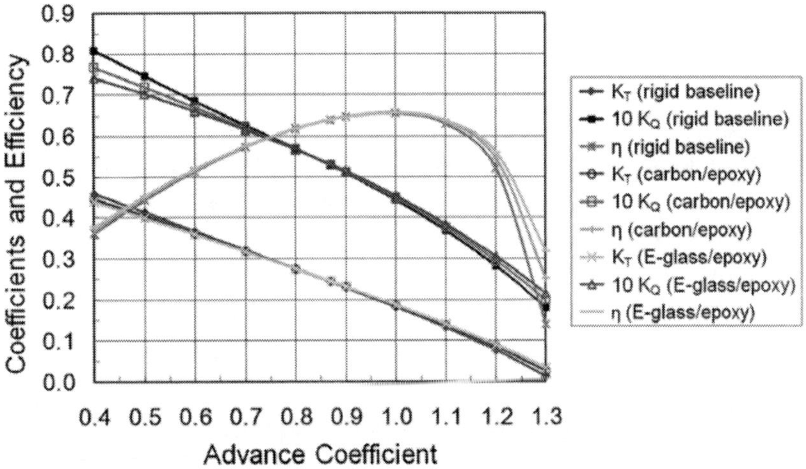

Figure 8: Comparison of Propeller Performance

Both the thrust coefficient and torque coefficient are reduced at small J and increased at higher J i.e. the absolute values of the slopes are reduced. The efficiency curves for the two flexible propeller cases are tangent to the rigid propeller curve at J ≈ 0.87, and show higher efficiencies away from this condition i.e. the effect of flexibility is to broaden the efficiency curve. Comparison of results for the glass/epoxy and carbon/epoxy laminates show that increased blade flexibility results in greater widening of the efficiency curve. Therefore, it is felt that maximizing flexibility, through selection of the optimum baseline propeller shape and composite material, will lead to improved performance.

FLEXIBLE PROPELLER EFFICIENCY GAIN

An expression to enable a quantitative assessment of the efficiency gain of a flexible propeller operating at the design condition in wake flow is developed in this section. Consider a propeller that is operating in the spatially varying wake of a single-screw ship. The propeller is operating at rotational speed n at the design condition, which is the same for the flexible propeller and its rigid counterpart. Assume that the inflow velocities at the propeller plane are symmetrical about the centreline of the ship and the propeller, and that the spatially varying flow velocity is:

$$V = V_A(1 + \sum_i A_i(r)\cos(i\theta)) \quad (1)$$

where i =1,... and the spatially varying component is a function of radius r and angle θ. Assume that the thrust coefficient K_T and torque coefficient K_Q vary linearly with J about their design condition values $K_{T(design)}$ and $K_{Q(design)}$ at J_{design}:

$$K_T = K_{T(design)} + s_T(J - J_{design}) = K_{T(design)} + s_T J_{design} \sum_i A_i \cos(i\theta) \quad (2)$$
$$K_Q = K_{Q(design)} + s_Q(J - J_{design}) = K_{Q(design)} + s_Q J_{design} \sum_i A_i \cos(i\theta) \quad (3)$$

where s_T and s_Q are the slopes of the thrust and torque curve, respectively, at the design condition. The K_T and K_Q curves for flexible and rigid propellers have different slopes e.g. thrust curves for flexible and rigid propellers have slopes $s_{T(flex)}$ and $s_{T(rigid)}$, respectively.

Average thrust power P_T and delivered power P_D over one revolution is:

$$P_T = \frac{1}{\pi(R^2-R_h^2)} \int_{R_h}^{R} \int_0^{2\pi} TV\, r\, d\theta\, dr \quad (4)$$
$$P_D = \frac{1}{\pi(R^2-R_h^2)} \int_{R_h}^{R} \int_0^{2\pi} 2\pi n Q\, r\, d\theta\, dr \quad (5)$$

where R_h is the hub radius and R is the propeller radius. The efficiency gain of a flexible propeller, over its rigid counterpart, when operating at the design condition is:

$$\Delta\eta = \eta_{flex} - \eta_{rigid} = P_{T(flex)}/P_{D(flex)} - P_{T(rigid)}/P_{D(rigid)}. \quad (6)$$

Substituting expressions for T, Q and V into P_T and P_D, and integrating results in:

$$\Delta\eta = c_f\, c_w\, \eta_{design} \Delta s_{T(design)} J_{design}/K_{T(design)} \quad (7)$$

where

$$\eta_{design} = K_{T(design)} J_{design}/2\pi K_{Q(design)} \quad (8)$$
$$\Delta s_{T(design)} = s_{T(flex)} - s_{T(rigid)} \quad (9)$$
$$c_w = \int_{R_h}^{R} r \sum_i A_i^2\, dr/(R^2-R_h^2) \quad (10)$$

c_f is a correlation factor that accounts for differences between the simplified quasi-static analysis, above, and a more accurate unsteady fluid-structure interaction analysis, c_w characterizes the spatial variability of the wake, and $\Delta s_{T(design)}$ is the difference of the slopes of the flexible and rigid propeller K_T vs. J curves measured at the design condition. Numerical values for c_w have been calculated as 0.036 for a particular ship's wake and 0.021 for a four cycle wake. Factor c_f has been estimated as 0.55 for data from Motley et al. [14], but the degree of variability of c_f is currently unknown.

Numerical values for a flexible propeller based on propeller 5474 from Motley et al. [14] are $\eta_{design} = 0.69$, $\Delta s_T = 0.13$, $J_{design} = 0.66$, and $K_{T(design)} = 0.17$. For this case the baseline propeller was aluminium. Adopting $c_w = 0.036$, and $c_f = 0.55$ results in an estimated efficiency gain $\Delta\eta = 0.0069$ in a ship's wake compared to the baseline propeller. Although such an efficiency gain at the design condition is modest, it could be expected to increase if compared to a stainless steel or nickel-aluminium-bronze (NAB) propeller which would have greater rigidity than an aluminium propeller. Furthermore, as shown for the propeller example above (Figure 8), and for propeller 5474 of Motley et al. [14], efficiency gains are higher at off-design conditions as a shape-adaptable design broadens the efficiency curve.

The efficiency gain expression (Eq. 7) indicates that there exists the possibility of increasing efficiency by increasing the blade flexibility, and therefore the thrust curve slope change $\Delta s_{T(design)}$, e.g. through the use of different composite materials.

CONCLUSIONS

A method has been presented for the hydroelastic tailoring of flexible composite shape-adaptive propellers, which is applicable to situations where the blades adapt to changes of flow conditions due to their rotation in a spatially varying wake. The shape-adaptive behaviour results in improved efficiency compared to a baseline rigid blade. The method is based on first determining the design condition by a technique taking account of perturbed blade shapes, and then using an optimization scheme to find the composite lay-up that maximizes the pitch reduction with load. Under certain circumstances, the method can be applied to the hydroelastic tailoring of composite hydrofoils. Potential benefits from reduced weight, maintenance, cavitation and vibration and acoustics have not been addressed in this paper. Reduction of propeller cavitation and noise is the subject of proposed research.

The method is applied to the design of a hydrofoil and a propeller. Both designs show small performance improvements relative to baseline designs. The degree of performance gain is proportional to the flexibility of the composite structure. The appropriate selection of the geometry of the baseline structure, upon which the flexible counterpart is based, is important e.g. flexible propeller blades should be thin and highly skewed [5]. As well, it appears that selection of composite materials to maximize flexibility, within strength constraints, should be investigated.

A closed-form expression for estimating the efficiency gain of a flexible propeller in spatially varying flow was developed. For the example given, the predicted efficiency gains at the design condition were small. However, in terms of efficiency the potential benefit of a flexible propeller is in broadening the efficiency curve and reducing the sensitivity of losses due to spatially varying wake. In application, even small gains in efficiency, across the range, can be very significant for a ship's operation.

ACKNOWLEDGEMENTS

Some of this work described herein was carried out as part of the research program P2.2.3 Composite Shape Adaptive Technology for Marine Propellers of the Co-operative Research Centre for Advanced Composite Structures (CRC-ACS). Their ongoing support, along with the support of the first author's employer, Pacific ESI, and the software vendor ESI Group, in particular the continued enthusiast support of Dr Argiris Kamoulakos and Dr Tom Kisielewicz, is most gratefully acknowledged.

REFERENCES

1. Kamoulakos A, (2002), "INtelligent COMposite PROducts", Final Report submitted to the European Community (Project BE-98-5505, Contract BRPR-CT98-9006), October 2002.
2. Gowing S, Coffin P, Dai C, (1998), "Hydrofoil Cavitation Improvements with Elastically Coupled Composite Materials", 25th American Towing Tank Conference, Iowa City, Iowa, 24-25 Sept. 1998, p. 4-13 – 4-17.
3. Lin HJ, Lin JJ, (1997), "Effects of Stacking Sequences on the Hydroelastic Behavior of Composite Propeller Blades", Proc. of ICCM-11, Volume I: Composites Applications and Design, July 1997, Gold Coast, Australia, pp.757-761.
4. Lee Y-J, Lin C-C, (2004), "Optimized Design of Composite Propeller", Mechanics of Advanced Materials and Structures, 11:17-30.
5. Chen BY-H, Neely S, Michael T, Gowing S, Szwerc RP, Buchler D, Schult R, (2006), "Design, Fabrication and Testing of Pitch-Adapting (Flexible) Composite Propellers", 2006 Propellers and Shafting Symposium, Sept. 12-13 2006, Fort Magruder, Virginia, USA. Society of Naval Architects and Marine Engineers, p.8/1–8/11.
6. Young YL, Michael TJ, Seaver M, Trickey ST, (2006), "Numerical and Experimental Investigation of Composite Marine Propellers", 26th Symposium on Naval Hydrodynamics, Sept. 17-22 2006, Rome, Italy.
7. Young YL, (2007), "Time-dependent hydroelastic analysis of cavitating propulsors", Journal of Fluids and Structures, 23:269–295.
8. Young YL, (2007), "Hydroelastic Behavior of Flexible Composite Propellers in Wake Inflow", 16th International Conference on Composite Materials, July 8-13 2007, Kyoto, Japan.
9. Young YL, (2008), "Fluid–structure interaction analysis of flexible composite marine propellers", Journal of Fluids and Structures, 24:799–818.
10. Liu Z, Young YL, (2007), "Utilization of Deformation Coupling in Self-Twisting Composite Propellers", 16th International Conference on Composite Materials, July 8-13 2007, Kyoto, Japan.
11. Plucinski MM, Young YL, Liu Z, (2007), "Optimization of a Self-Twisting Composite Marine Propeller Using Genetic Algorithms", 16th International Conference on Composite Materials, July 8-13, 2007, Kyoto, Japan.
12. Blasques JP, Berggreen C, Andersen P, (2008), "Hydro-Elastic Tailoring and Optimization of a Composite Marine Propeller", In: Proceedings of the 13th European Conference on Composite Materials, Composites for Sustainable Progress, ECCM-13, 2-5 June 2008, Stockholm, Sweden.
13. Lin C-C, Lee Y-J, Hung CS, (2009), "Optimization and experiment of composite marine propellers", Composite Structures, 89: 206-215.
14. Motley MR, Liu Z, Young YL, (2009), "Utilizing fluid–structure interactions to improve energy efficiency of composite marine propellers in spatially varying wake", Composite Structures, 90:304–313.
15. ESI-Group (2008), "CFD-ACE+ 2008 User Manual". ESI CFD Inc, Huntsville, AL, USA.
16. ESI-Group (2008), "SYSPLY 2008 User's Guide", Engineering Systems International, France.
17. ESI-Group (2008), "PAM-OPT 2008 Solver Reference Manual", Engineering Systems International, France.
18. Soden PD, Hinton MJ, Kaddour AS, (1998), "Lamina properties, lay-up configurations and loading conditions for a range of fibre-reinforced composite laminates", Composites Science and Technology, 58:1011-1022.

19. Boswell RJ, (1971), "Design, Cavitation Performance, and Open Water Performance of a Series of Research Skewed Propellers", Naval Ship Research and Development Center, Bethesda MD, Report 3339.
20. Kerwin JE, Lee C-S, (1978), "Prediction of Steady and Unsteady Marine Propeller Performance by Numerical Lifting-Surface Theory", Trans. SNAME, 86:218-253.

Marine Casualty Investigation, the new Australian Transport Safety Bureau

Author: Peter Foley, Director Surface Safety Investigations
Australian Transport Safety Bureau

peter.foley@atsb.gov.au

ABSTRACT

On 1 July 2009 the Australian Transport Safety Bureau (ATSB) became a statutory agency, completely independent of the Commonwealth Government's marine policy and safety regulatory agencies. This is one of many important government reforms affecting Australia's marine transport industry. The paper discusses the change in governance for the ATSB, the future directions of the agency and takes a look some investigations which were pivotal in instigating the change.

THE PAST

The ATSB was formed on 1 July 1999 from the amalgamation of the Bureau of Air Safety Investigation (BASI), the Marine Incident Investigation Unit (MIIU), and non-regulatory parts of the Federal Office of Road Safety (FORS). In addition, an embryonic Rail Safety investigation capability was established.

The staff of the ATSB are a small group of highly committed professionals, working together with a strong common purpose – to improve safety. The work of the ATSB is very much a team effort, on the part of aviation, marine, and rail investigators, as well as other ATSB staff. About 65 of the ATSB staff are transport safety investigators. They come from a range of professional backgrounds, including pilots, Licensed Aircraft Maintenance Engineers, air traffic controllers, master mariners, human factors specialists, and a range of engineering and technical disciplines.

In July 2003 the *Transport Safety Investigation Act* (TSI Act) came into force. For the first time Australia had comprehensive multi modal safety investigation legislation and a range of powers which enshrined the principle that the best outcome of any transport accident was the lesson in how to prevent it from occurring again. The Act remains the envy of similar organisations around the world.

LOCKHART RIVER

On 4 April 2007 the ATSB released the final report into the 15-fatality accident on 7 May 2005 involving a Fairchild SA227-DC Metro 23, registered VH-TFU, operated by Transair at Lockhart River. The aircraft flew into the side of a hill on approach to the airstrip in Instrument Meteorological Flight Conditions.

The investigation found several contributing safety factors in relation to the regulatory oversight of the aircraft operator and seven recommendations were made to the Civil Aviation Safety Authority (CASA).

An inquest into the accident was subsequently conducted by the Queensland Coroner. The inquest was adversarial, with the ATSB and CASA arguing the findings of the ATSB's investigation. In handing down his findings on 17 August 2007, the Coroner recommended that the Commonwealth Minister for Transport undertake a review to assess whether high level intervention was required, to ensure that there is a productive, collaborative focus on air safety between the ATSB and CASA.

The very nature of the role of the ATSB often places it in the position of reviewer of transport safety regulators, regulatory and other actions where there has been an accident or serious incident. There can be tensions arising from legitimate differences of opinion. Information about what happened in an accident or incident can be fragmented, allowing for a variety of hypotheses about what actually happened. The causes contributing to an accident or incident can often be diverse, leaving ample room for debate over what actually caused it. Differing professional judgements will inevitably lead to different views, often firmly held.

While tensions between safety regulators and the accident investigator should not be unexpected, the Coroner correctly observed that the tensions he had detected during the Lockhart River inquest went well beyond the constructive tension one would expect. Tension between safety regulators and accident investigators is not unusual – there are international examples of such tension and, in any case, some degree of constructive tension between the investigator and safety regulators should be expected given their respective roles. Although they are both important contributors to Australia's transport safety systems, and share the same long term goal (improving transport safety), they have quite different powers and functions. Creative tension can be a positive force. Professional disagreement, properly expressed, can lead to better outcomes overall as each party examines the views of the other and the expertise available to each is shared, debated and evaluated. A clearer picture can emerge and a better outcome may result.

MILLER REVIEW

Following the Lockhart River inquest and the Coroner's recommendation, the then Minister subsequently announced that a review would be conducted into the relationship between the CASA and the ATSB aimed at identifying potential areas for improved aviation safety outcomes through better cooperation and co-ordination.

In 2007 Mr Russell Miller AM was tasked by the then Government to review the relationship between the CASA and the ATSB. In finding there was room for improvement in the way the agencies interact, Mr Miller addressed the ATSB's governance structure. Mr Miller recommended that the Australian Government move to clarify the ATSB's independence as the national safety investigation agency.

The Australian Government, newly elected, accepted this key recommendation, which received strong support from industry, and addressed it in the National Aviation Policy Green Paper (2 December 2008). Investigations that are independent of the parties involved in an accident, transport regulators and government policy makers, are better positioned to avoid conflicts of interest and external interference. Independence for the accident investigation authority is consistent with international standards. Standard 5.4 of Annex 13 to the International Convention on Civil Aviation (the Chicago Convention) states:

The accident investigation authority shall have independence in the conduct of the investigation and have unrestricted authority over its conduct.

Various models were considered for the governance of the ATSB including that is could be created as a statutory authority with a Board of up to five Commissioners like New Zealand's TAIC and similar to the Dutch, Canadian, and United States models (Canada's TSB is used as a model in ICAO guidance). The powers of the Commission were to include most of those currently vested in the then Executive Director and the new Commissioners would delegate most of those powers to relevant ATSB staff for the day-to-day conduct of investigation and related activities. However, the power to approve and release investigation reports would remain with the Commission.

Independence for the ATSB was to be demonstrated in the following ways:

1. Independence from the regulator, operators, and the transport industry would enable the ATSB to investigate and comment impartially on any role these parties might have had or been able to play in preventing the accident.
2. Independence from the Police and regulatory authorities allows people to talk to the ATSB's investigators freely in the knowledge that the ATSB cannot use that information to punish someone or penalise an organisation.
3. Independence from the Police and regulatory authorities ensures that any prosecution brought by them does not compromise or delay the ATSB's own investigation and report on the accident or incident.
4. Independence from other agencies gives the ATSB full control over approving and publishing its reports and there is no perception of lack of independence.

Establishing the ATSB as a statutory authority in the form of a Commission would mean that the ATSB is separate from the Department and other portfolio agencies with its own Parliament-appropriated budget, and employs its own staff.

THE NEW ATSB

In 2009 the *Transport Safety Investigation Amendment Bill 2009* was introduced into the Australian Parliament to amend the TSI Act to establish the ATSB as a Statutory Agency within the meaning of the *Public Service Act 1999* (PS Act).

The Agency was established with a Commission structure on 1 July 2009. The position of the Executive Director of Transport Safety Investigation, appointed by the Secretary of the Department, no longer exists. Instead, Commissioners appointed by the Minister for five years, consisting of the Chief Commissioner and two part-time Commissioners (with the power to appoint more) are responsible for administering the functions of the TSI Act and exercising its investigation powers. The Minister has the power to provide notice of his or her views on the strategic direction for the ATSB, to which the ATSB must have regard. However, other than the ability for the Minister to require the ATSB to investigate a particular matter, the ATSB is not subject to a direction from anyone with respect to the exercise of its powers and performance of its functions.

The creation of a statutory agency also gives the ATSB responsibilities in its own right under the PS Act and the *Financial Management and Accountability Act 1997* and discretion with respect to the management of its staff and resources consistent with these Acts. The ATSB, therefore, has operational independence with respect to the exercise of its investigation powers and functional independence with respect to the administration of its resources.

MINISTER'S STATEMENT OF EXPECTATIONS (Draft)

The establishment of the Commission and the ATSB as an independent statutory authority underlines the Australian Government's commitment to ensuring the ATSB continues to be a global leader in transport safety investigation, research and analysis, and fostering public awareness of transport safety.

The strategic direction and guidance in this Statement is given in accordance with my powers under section 12AE of the Transport Safety Investigation Act 2003 (the Act) and is intended to clarify the broader Government policy framework in which the Commission and the ATSB operate.

Governance
The functions of the ATSB under the Act include:

- *receiving and assessing reports of transport safety matters, reportable matters, and other safety information that is prescribed by the regulations;*
- *independently investigating transport safety matters;*
- *identifying factors that:*
  - *contribute, or have contributed, to transport safety matters; or*
  - *affect, or might affect, transport safety;*
- *communicating those factors to relevant sectors of the transport industry and the public in any way, including in anyone or more of the following ways:*
  - *by making safety action statements;*
  - *by making safety recommendations;*
  - *by issuing safety advisory notices;*
- *reporting publicly on those investigations; and*
- *conducting public educational programs about matters relating to transport safety.*

With the ATSB's separation from the Department of Infrastructure, Transport, Regional Development and Local Government (the Department), the Commission has a responsibility to ensure the ATSB remains an active and effective participant in the transport policy and regulatory framework, working effectively with industry and other agencies while retaining its operational independence.

Government priorities
In releasing the Aviation Green Paper last year, I made clear that safety is paramount across all transport sectors and that high quality transport safety investigation and research into transport accidents and incidents are important elements of Australia's safety system. I expect the ATSB to continue to give priority to transport safety

investigations that have the potential to deliver the best safety outcomes for the travelling public. I also expect the ATSB to be prepared to provide occasional assistance to accident investigations in other countries, in accordance with international protocols. The ATSB will also need to prepare to meet the Council of Australian Governments' commitment for it to be the preferred investigator of rail accidents.

Portfolio and other relationships
While maintaining its independence, the ATSB should work with my office, the Department and other Australian Government agencies, including the Civil Aviation Safety Authority (CASA), Airservices Australia and the Australian Maritime Safety Authority (AMSA), towards the common goal of improving transport safety. I expect the ATSB will retain a close and constructive working relationship with the Department and ask the Commission to have particular regard to building strengthened working relationships with CASA and AMSA.

I also expect the ATSB to engage constructively in processes from time to time where it can provide information, assistance or advice on safety issues and any other government processes that may require ATSB's transport safety expertise. The interests of the travelling public are best served by cooperative working relationships between the various Commonwealth and State Government agencies with responsibility for safety matters.

Parliament
The ATSB has a responsibility to provide advice on its operations to me, the Parliament and, through the Parliament, the Australian public. Timely and accurate advice in response to requests for input to ministerial representations, parliamentary questions and other information and briefing should be given high priority. I note the ATSB has additional responsibilities in these respects flowing from its creation as a separate statutory authority.

Conclusion
I look forward to working with the Commission as you confront the challenging times ahead. I am confident I will receive your support and cooperation in achieving the goals outlined in this Statement. In this regard, I ask that you provide me with a statement of intentions within four months, outlining your program for meeting these expectations, including your performance milestones.

CONCLUSION

With the separation of the ATSB from the Department of Infrastructure and the formulation of the Commission to govern it, the ATSB has completed the transformation to a truly independent accident investigation agency which started when the Bureau was formed in 1999. The ATSB will continue to provide world class excellence in its independent safety investigations and research and now has a clearer mandate to focus on maximising the benefits for the transport industry by promulgating the lessons learned from its research and investigations through education.

Peter Foley 2009

SEA 21: Facilitating Global Maritime Partnerships

Rear Admiral James McManamon
Deputy Commander Surface Warfare (SEA 21)
Naval Sea Systems Command
United States Navy

PERSPECTIVE

Distinguishing characteristics of Pacific Rim nations are their vast coastlines bordering immense ocean space, creating unique challenges to and opportunities for maritime security. For naval forces and maritime communities at large, this geographic reality creates mutual imperatives for all of the delegates at the Pacific 2010 International Maritime Conference. Indeed, the key words are *mutual* and *community*, for none of the challenges confronting Pacific Rim states can be addressed adequately by one government alone, and no single Pacific Rim nation can fully embrace the tremendous opportunities afforded by proximity to the world's largest ocean.

Senior naval and maritime leaders recognize these imperatives. For example, in his March 2009 U.S. Naval Institute *Proceedings* article, "The Commanders Respond," Australia's Chief of Navy, Vice Admiral R.H. Crane, noted, "Australia's continued reliance on the maritime environment will dominate our thinking." Vice Admiral Crane also outlined the major future acquisitions the Royal Australian Navy (RAN) will make in air-warfare destroyers (modelled on the U.S. Navy's *Arleigh Burke-class* (DDG-51) Aegis destroyer) and amphibious assault ships that will help safeguard Australia's maritime interests. (1)

Later last year, U.S. Chief of Naval Operations Admiral Gary Roughead underscored this focus on the maritime environment in his remarks at the 19th Biennial International Seapower Symposium in Newport, Rhode Island. In addressing the mutual interests in these unique challenges and opportunities to the delegates of the 100+ nations represented at the September 2009 event, he noted:

> At this largest gathering of naval leaders in history, we here today represent our countries' efforts not only to defend our respective maritime interests and our shores, but also to secure the global maritime commons at a time of great challenge. The stakes are massive: our activities – individually and in partnership – are vital to the future generations. (2)

The U.S. Navy has an enormous stake in supporting and sustaining Global Maritime Partnerships (GMP), and the Surface Warfare Directorate (SEA 21) for the U.S. Naval Sea Systems Command (NAVSEA), is the Navy's principal agent for supporting our many coalition partners. We do this in numerous ways, principally by building both the *capability* to succeed in a multi-dimensional threat environment and the *capacity* to work together on the global commons.

The nations – as well as their navies, coast guards, and maritime industry partners – represented at the Pacific 2010 International Maritime Conference recognize the need to work together to address mutual threats to the global commons. From the SEA 21 perspective, this cooperation occurs both globally, under the auspices of the GMP, as well as regionally, in a growing number of maritime alliances, many led and sustained by governments represented at this conference.

SEA 21, as well as our national and international partners – including Australian's Defence Material Organization – has a keen eye on what it takes to support the GMP with mission-capable ships. As our nations and navies band together to address threats to our mutual interests, we must recognize that at the end of the day we will need to ensure our warships continue to meet the missions for which they are designed over their expected service lives.

THREATS TO MUTUAL INTERESTS

Clearly, the navies and coast guards of the Pacific Rim must work collectively to deal with security issues including terrorism and piracy, environmental violations, fisheries security and trafficking in illegal persons, weapons, or drugs. Indeed, navies and coast guards will be increasingly stressed to fulfil their myriad responsibilities while ensuring their ships and craft are well maintained and ready for tasking.

The need to work together to ensure continued economic prosperity and dissuade or defeat those who would use the oceans to do us harm is as critical now as it has ever been. Globalization has borne exponential growth in seaborne commerce, which has more than quadrupled in the past four decades. The tri-service U.S. maritime strategy, *A Cooperative Strategy for 21st Century Seapower*, put it in these terms:

> The oceans connect the nations of the world, even those countries that are landlocked. Because the maritime domain – the world's oceans, seas, bays, estuaries, islands, coastal areas, littorals, and the airspace above them – supports 90% of the world's trade, it carries the lifeblood of a global system that links every country on earth. Covering three-quarters of the planet, the oceans make neighbours of people around the world. They enable us to help friends in need and to confront and defeat aggression far from our shores. (3)

A critical element of this cooperation and collaboration will be a "whole of governments" approach to Maritime Domain Awareness (MDA). At its foundation, MDA supports a broad spectrum of operations and is a global process to understand changes in the maritime domain, how these changes might affect our vital interests, and how best to respond across this broad spectrum of traditional and trans-national threats to the maritime, or "contested" commons – and, by extension, to our homelands. (4)

Ultimately, the ability of the RAN to ensure Australian security and prosperity in accordance with the recent Australian Defence White Paper *Defending Australia in the Asia Pacific Century: Force 2030* will depend not just on the capability of that force, but on its capacity. (5) In seeking this capacity, Australia – as well as all the nations of the Pacific Rim – must depend not only on newly purchased ships but also on the current ships, submarines and craft already in their inventories. And like the RAN, the navies and coast guards of the other nations represented at the Pacific 2010 conference must also get more life and nautical miles out of their in-service ships and craft.

BUILDING PARTNERSHIP CAPABILITY *AND* CAPACITY

Today, globalization and the presence of a new generation of threats on the high seas, the littorals, and the near-shore land areas demand even closer cooperation among navies and coast guards everywhere. *Defending Australia in the Asia Pacific Century: Force 2030* explained that Australia's defence policy:

> ...entails the maintenance of alliances and international defence relationships that enhance our own security and allow us to work with others when we need to pool our resources...this defence policy means that we must have the capacity to lead military coalitions where we have shared strategic interests at stake with others...and make tailored contributions to military coalitions where we share wider strategic interests with others. (5)

This point of view is embraced by a growing number of nations, and commanders of many naval forces have become increasingly mindful of the centrality of partnership to the preservation of the safety and security of the world's oceans. (2) As the number of navies participating in these partnerships and their enthusiasm for this cooperation has grown, the results have been nothing short of stunning. Less than five years ago, about 70 navies and some 50 service chiefs attended the 17th Biennial International Seapower Symposium. By the 19th such symposium, more than 100 nations and 90 chiefs of service came together.

It is the *actions* these navies and coast guards are taking to work with their respective naval partners in other nations that are even more impressive. In addition to the partnering activities mentioned above, numerous navies and coast guards are taking the lead in partnering activities in their regions and are achieving real results. For example, Korea has stood up a Regional Maritime Information Exchange, and the Western Pacific Naval Symposium Maritime Security Information Exchange Seminars are helping to establish standard operating procedures for information sharing. In the Persian Gulf, more than 20 states attend the monthly Shared Awareness and De-confliction (SHADE) meetings in Bahrain, with the European Union, NATO, and maritime industry as active participants. The goal here is to develop and share unclassified common operating pictures and databases, and to synchronize combined maritime forces. In the Americas, navies have put in place the Inter-American Telecommunications Network, a secure, regional system for sharing information. The U.S. Southern Command has set up the Cooperating National Information Exchange System designed to improve maritime and air domain awareness by providing partner countries with real-time data on potential trafficking targets of interest. In the Mediterranean, Italy has taken the lead in establishing a Virtual-Regional Maritime Traffic Center supported by 29 navies and which has four separate sub-regional initiatives, including one with Lebanon and another with five North African littoral countries. In 2008, Italy also put in place a federated Trans-Regional Maritime Network — yet another regional piece of the global maritime partnership solution. (4)

Growing the capability *and* the capacity of naval and maritime power to prevail against the myriad threats is always challenging, but perhaps even more so today due to the worldwide economic downturn and the cost of new ships. That is why the U.S. Navy in general, and the Naval Sea Systems Command, in particular, are dedicated to ensuring that U.S. surface forces are ready for tasking in the most cost-effective manner, a dedication that extends to America's global maritime partners, as well.

MAINTAINING FIGHTING TRIM

As the U.S. Navy's "Bridge to the Fleet," the Deputy Commander for Surface Warfare (SEA 21) is NAVSEA's steward of fleet introduction, maintenance and modernization, as well as the inactivation and disposal of all U.S. Navy non-nuclear surface ships and craft. The men and women of SEA 21 – military, civilian, and contractors alike – along with our partners throughout NAVSEA and other organizations, such as our world-class Navy warfare centers and laboratories, are charged by the nation to ensure that today's and tomorrow's surface and expeditionary warships, and patrol craft are kept in "fighting trim" in the most cost-effective manner possible — to be ready for tasking in support of our regional Combatant Commanders as well as our allies and coalition partners.

In short, SEA 21's unique role is to ensure our surface ships keep pace with evolving threats and guarantee that they remain capable, relevant, and affordable warfighting assets for the nation. Getting more effective life and more nautical miles out of in-service ships is a formidable task, especially with ships' service lives running to 35-40 years, and when the time from lead ship delivery to the decommissioning of the last ship of a class can exceed 75 years.

As the history of warship service demonstrates, getting more life out of ships is not always easy or successful. For example, it was about three-quarters-of-a-century from the commissioning of the world's first "battleship," HMS *Dreadnaught*, in 1906 to the commissioning of the first Aegis guided-missile cruiser, USS *Ticonderoga* (CG-47), in 1983. Both warships introduced advanced and innovative capabilities that revolutionized naval warfare, but both were also retired much earlier than planned – *Dreadnaught* served only 14 years and *Ticonderoga* just 21 – because threat vectors, technology dynamics, and rising costs rendered these once top-of-the-line capital ships obsolete much too soon.

Today, we have learned valuable lessons from these and similar examples. What SEA 21 provides is efficient, effective, and affordable lifetime maintenance, upgrades, and modernization to sustain the right force of ships with the right readiness to satisfy global peacetime, crisis-response, and wartime operational requirements. We ensure that the Navy has the right technologies, systems, sensors, weapons, platforms, and people to guarantee our surface ships and craft are ready for tasking at a cost it can afford.

In carrying out these responsibilities, SEA 21 has become the focal point of the Navy's Surface Warfare Enterprise, supporting warfighting readiness while dealing with daunting fiscal realities, managing to the right metrics and synchronizing lifetime support efforts. Thus, SEA 21 serves as the flag-level entry point for fleet operators to make their needs known in Washington.

Moreover, SEA 21 stewardship and responsibilities go far beyond the U.S. Navy. The directorate also provides "layup-to-disposal" support to the U.S. Navy's inactive ships, ship transfers, and international fleets. This international program support is particularly important in helping our Navy build critical relationships among the world's navies and coast guards. The hundreds of U.S. Navy professionals in our engaged in this effort provide stewardship for an extremely active program of international fleet transfers and the requisite follow-on technical support to those receiving inactivated U.S. Navy ships.

Many of the naval and maritime forces represented at the Pacific 2010 Conference now own and operate these ships, and we remain your partner to help ensure their readiness and operability. And at the end of the day, we at SEA 21 know it's about building relationships, not just selling ships and weapons systems. And it is these relationships with our more than 50 international customers – from Australia to Albania, from the Philippines to Paraguay, and from Thailand to Tunisia – that enrich our work every day.

"UNDER THE HOOD" OF THE BRIDGE TO THE FLEET

As the U.S. Navy operates with other world navies and coast guards, having these former U.S. Navy ships in your battle force enhances interoperability in support of the Global Maritime Partnership. This initiative – and others– is just part of the trust building that strengthens our mutual efforts to achieve security on the global commons. To fully understand SEA 21's support to the GMP, a look "under the hood" at the SEA 21 portfolio – and especially the Aegis upgrade and modernization program – will provide valuable insight.

SEA 21's portfolio spans several product lines that support the U.S. fleet as well as Global Maritime Partnerships. SEA 21's span of work includes introducing surface ships delivered to the fleet last week, modernizing and sustaining ships delivered to the fleet decades ago, disposing of ships after their service to the nation is complete, and supporting our coalition partners via our International Fleet Support Program and follow-on technical support.

To do this effectively and efficiently, we have structured our operations in focused programs that are managed by senior, highly experienced, naval officers and supported by skilled military, civilian, and contractor teams. The SEA 21 programs that enable us to deliver life cycle stewardship for enhanced readiness include:

- In-service combatants (cruisers, destroyers, frigates, Littoral Combat Ship)
- Amphibious, Auxiliary, Mine Countermeasures, and Command Ships
- Inactive Ships and Ship Transfers
- International Fleet Support

All of these programs are important, and from a "best practices" point of view have learned lessons worth sharing with our partners. In that regard, a focus on the cruiser, destroyer and frigate enhanced readiness program and the Aegis cruiser and destroyer modernization program provides good insight into SEA 21 "best practices" as well as their implications for a growing Aegis international community.

This "Aegis Global Enterprise" – which Australia will soon join with its Hobart-*class* Aegis air-warfare destroyers – represents one of the most important international efforts in support of Global Maritime Partnerships. As increasing numbers of these capable Aegis ships are equipped with ballistic missile defense capability, we will see even more vital examples of collaboration. Indeed, when President Barak Obama announced he would put in place a sea-based missile-defense posture based on Aegis ballistic missile defense, it highlighted the critical role these warships will play in meeting the future threat environment.

AEGIS MODERNIZATION

SEA 21's In-service Combatants Program Office has overarching stewardship of our Aegis cruiser and destroyer modernization program. With a core mission to sustain, modernize and increase the war-fighting capabilities of surface combatants and a dedication to efficiency, readiness, evolution, and continuous process improvement, this office has embarked on the most comprehensive upgrade and modernization program in the history of the U.S. Navy.

The Navy's *Ticonderoga* (CG 47) cruisers and *Arleigh Burke* (DDG 51) destroyers are the backbone of the U.S. Navy surface force and comprise almost one-third of the entire fleet. But as capable as these warships are today, they must keep ahead of the evolving and proliferating threats from a broad spectrum of adversaries.

The U.S. Navy has put in place well-conceived and cost-effective programs to substantially upgrade and modernize our fleet of 22 *Ticonderoga-class* Aegis guided-missile cruisers and all of the 62 *Arleigh Burke-class* Aegis guided-missile destroyers. For DDG-51, this will mark the first time a ship program has begun a major upgrade effort while ships of the class are still being constructed. This plan will ensure these ships reach their 35-year service lives – well beyond what *Dreadnaught* and *Ticonderoga* achieved.

USS *Bunker Hill* (CG 52) is the first of 22 Navy cruisers to receive complete combat systems and hull, mechanical and electrical modernization. The oldest of our in-service cruisers, CG-52 began its shipyard availability in February 2008 and completed in the spring 2009. In 2012, we will commence a modernization program for all of the Flight I and II Aegis guided-missile destroyers, beginning with DDG 51.

The primary components of the Aegis cruiser and destroyer modernization program are:

- Combat systems upgrades for air dominance (e.g., Cooperative Engagement Capability and SPY radar upgrades), maritime force protection (e.g., Close-In Weapon System Block 1B, Evolved Sea Sparrow Missile, Nulka electronic warfare decoy system and SPQ-9B radar), and undersea warfare (e.g., SQQ-89A(V)15 anti-submarine system).
- Hull, mechanical and electrical (HM&E) upgrades and mission-life extension (e.g., SmartShip, all-electric auxiliaries, and space, weight and moment improvements).
- Optimized manpower and maintenance reductions as well as improved crew quality of life services (e.g., an expanded recreation center and space for a crew gym).

In remarks to the Engineering the Total Ship Symposium in September 2008, Admiral Roughead captured the importance of the Aegis cruiser and destroyer modernization program when he noted, "Cruiser modernization is key to us because we need those ships not just for their capability but for the capacity that we are going to need to have for the types of defense and offense that we are going to have to deal with in the future."

The primary objective of the Aegis modernization effort is to improve the ships' combat capabilities so the ships will remain mission-effective throughout the remainder of their service lives. The second is to make the ships less expensive to operate, maintain and modernize well into mid-century.

This decision to get full service life and more capabilities out of these Aegis warships was not made lightly, but with due diligence and reliance on sound engineering practices. What we have found through years of learning from best-practices is that as long as you maintain the combat systems' relevance, fully fund modernization, and do the right corrosion control and other maintenance, there is no reason why these ships cannot reach their expected service lives.

COST-EFFECTIVE STRATEGIES

From a requirements perspective, the Navy has taken great care in determining the warfighting requirements for our cruisers and destroyers in the years ahead. Our analysis has crossed mission areas, looking at everything from the submarine threat to the latest in anti-ship cruise missile and the ballistic missile threats. The result is a modernization plan that takes advantage of the promises of open architecture (OA) and will introduce new capabilities, in an evolutionary fashion, as technology and development bring them to maturity.

Decades of Aegis systems engineering discipline, built on the late Rear Admiral Wayne E. Meyer's (widely regarded as the "Father of Aegis") admonition to "build a little, test a little, learn a lot," has provided our Navy planners with the background, capabilities, and capacity to put in place an effective strategy to pace the Aegis cruiser and destroyer modernization program. Some of the upgrades are visible, such as replacing the pair of 5-inch/54-caliber gun mounts with the new 5-inch/62-caliber gun. But most of what is new is internal, or below the waterline.

For example, the Aegis Combat System (ACS) engineering program provides immediate and effective capability to counter the current *and* projected air, surface and sub-surface threats. Changes in the threat capability and advances in technology, such as fiber optics, local area networks and high-performance computing, are driving corresponding Aegis Weapon System (AWS) and ACS changes. This program provides the ACS engineering and weapon system development necessary for a continued increase in the capability of Aegis cruisers and destroyers. In addition to developing and integrating improvements to the AWS, this program integrates combat capabilities developed in other Navy R&D programs into the ACS. For example, modifications of AWS computer programs must be made to integrate these capabilities into the ACS so that battle effectiveness and ACS performance will be retained against the evolving threat.

There is more at stake than just upgrading two classes of warships. The 84 Aegis class guided-missile cruisers and destroyers currently in service or being built are the heart and soul of the Navy's surface fleet. These Aegis ships are the main battery for air, surface, and undersea warfare and are the ships that sail with Navy carrier strike groups (CSG) and expeditionary strike groups (ESG); they serve to protect these multi-billion dollar capital ships from multi-dimensional threats. In addition to protecting CSGs and ESGs, the 2007 tri-service maritime strategy, *A Cooperative Strategy for 21st Century Seapower*, calls for increasing emphasis on maritime stability operations and humanitarian/disaster relief. (3) Therefore, sending a single ship – often an Aegis cruiser or destroyer independently – to a remote littoral area means these ships are in harm's way if they do not sail with today's and tomorrow's technologies.

Additionally, these ships are a pillar of the U.S. ballistic missile defense efforts; through 2009, 19 Aegis cruisers and destroyers have been outfitted with the capability to track and engage ballistic missiles. The BMD modifications to the Aegis Weapon System, and the development and upgrade of the Standard Missile 3 (SM-3) with its hit-to-kill kinetic warhead, gives select Aegis cruisers and destroyers the capability to intercept short- and medium-range ballistic missiles in the ascent, midcourse and descent phases of their exo-atmospheric trajectories, as well as surveillance and tracking capability against long-range ballistic missile threats. These ships are often on station alone conducting this vital national mission. As a nation we recognize the need to upgrade more Aegis ships for this BMD mission, but these ships must also have the most modern capabilities across the board.

Our Aegis cruiser and destroyer modernization program is a cost-effective way for the U.S. Navy to deal with today's and tomorrow's threats – in an era of flat or potentially decreasing defense budgets. Ensuring these ships ably serve our fleet, our nation, and our allies for the ensuing decades is perhaps the most critical element in providing the Navy the tools it needs to execute its *Cooperative Strategy for 21st Century Seapower.* In addition, these powerful best-practices may be the best example that SEA 21 can share with our allies and coalition partners in what can now be seen as the beginnings of an "Aegis Global Enterprise."

A WAY FORWARD

While Global Maritime Partnerships are much more than surface warships and craft, they remain the most visible part of this important international undertaking. SEA 21 supports the partnerships "behind the scenes" with good stewardship of our naval vessels, including those of our friends and coalition partners, from cradle-to-grave. SEA 21 provides the most visible manifestation of the U.S. Navy's determination and fiscal discipline to make naval assets last longer through good stewardship.

The U.S. Navy is a great partner for building and modernizing ships. As our Navy's executive agent for upgrading and modernizing the overwhelming majority of ships and craft in the U.S. Navy, as well as acting as the Navy's "face" to more than 50 coalition partners and allies via our International Fleet Support Office, SEA 21 works closely with Australia's Defence Material Organization, and with the material organizations of our other partners and customers world wide to provide effective stewardship for surface ships and craft at prices we can afford.

References

(1) Crane, R.H. (2009), "The Commander's Respond," *U.S. Naval Institute Proceedings*, March 2009.

(2) Roughead, G., (2009) *Remarks at the 19th Biennial International Seapower Symposium*, Newport, Rhode Island, October 17, 2009.

(3) Conway, J.T., Roughead, G., Allen, T.W. (2007) *A Cooperative Strategy for 21st Century Seapower* (Washington, D.C., Department of the Navy, 2007), pp. 1-16.

(4) Stubbs, Bruce B., (2009) *Maritime Domain Awareness: Global Perspectives and Partnerships, Remarks at the Sixth Annual Coastal Surveillance Conference*, Singapore, November 2009, www.dodeaformda.navy.mil.

(5) *Defending Australia in the Asia Pacific Century: Force 2030* (2009) (Canberra, Australia, Australian Government Department of Defence, 2009), accessed at: www.defence.gov.au.

The Development of an 80m Multi-Role Trimaran Vessel, based on commercial and LCS2 design experience.

Tim Speer (tims@austal.com)
Austal Ships, Australia
Tony Armstrong (tonya@austal.com)
Austal Ships, Australia

OVERVIEW

1. A brief history of the commercial trimaran design development
2. The LCS program contracts, key events and the future acquisition plan
3. The LCS concept and the two solutions
4. The 102m trimaran platform
5. The 80m Austal Multi-Role Vessel; Design Maturity, Mission Capability, Key Design features
6. Conclusion

1. AUSTAL AND THE TRIMARAN DESIGN

Austal has been a successful builder of commercial high speed aluminium passenger and car ferries for more than 20 years. In the late 90s it was widely recognised in the commercial ferry industry that high speed vessels could only be successful if they offered a service to passengers which was recognised as enjoyable and value for money. On exposed routes, some high speed vessels were recognised as a fast mode of travel which regularly made many of the paying customers violently ill with motion sickness. Several services were nicknamed "vomit comets" which surely impacted their patronage. At this time, Austal commenced a significant research and development program to develop a high speed hull form which would dramatically out-performed the existing high speed monohull and catamaran hulls already in the market.

Catamaran vessels offer a significant resistance advantage over conventional displacement and planing monohull forms, making them significantly more economical to operate. The Achilles' heel of these catamaran designs is that unless they operate in relatively sheltered waters they can be susceptible to high lateral accelerations and also to slamming of the wet deck or cross-structure.

The trimaran concept was initially expected to provide a substantial improvement in passenger comfort over other high speed platforms available in the marketplace. Our initial work also suggested a possible speed improvement over an equivalent catamaran, and certainly over an equivalent monohull. Austal carried out model tests at small scale, and then at a large scale as the research and development program developed. An 11 metre instrumented and manned model was also built in order to assess the manoeuvrability and seakeeping ability.

The hull shape was optimised using SHIPFLOW, a numerical analysis tool from Flowtech, analysing over 300 variations of the hull and amah (sidehull) shapes before commencing a program of numerical analysis and model tank testing, including both resistance and seakeeping. Details of the design process and tank testing are given in Armstrong [1]. A

contract to build the 127 metre trimaran for an operator of fast ferries in the Canary Islands, Fred.Olsen SA (FOSA) was signed in June 2003.
This vessel christened "*Benchijigua Express*" has now been in operation for more than four years and remains the largest trimaran ever built with the same length overall as the US Navy Littoral Combat Ship (LCS) "*Independence*". While other high speed ships in the Canary Islands are forced to cancel sailings due to heavy weather, *Benchijigua Express* has never been forced to cancel a sailing due to weather in its four years of operation (Figure 1).

Figure 1: Benchijigua Express has been operating successfully in the Canary Islands for four years and has demonstrated the advantages of the trimaran design as a high speed platform

2. THE LITTORAL COMBAT SHIP PROGRAM

The LCS program was conceived by the US Navy under the Chief of Naval Operations, Admiral Vern Clark. The program was intended to leverage the latest commercial off-the-shelf technology to develop an agile and novel design capable of high speed operation in the littoral. The Focussed Mission High Speed Ship was an initial study of concepts undertaken by six competing teams in late 2002 and resulted in the award of three LCS Preliminary Design contracts in July 2003. The most significant feature of the Focused Mission Ships was their ability to employ modular mission packages or modules that could be easily added and removed from the ship to support specific missions.

The three preliminary designs (Figure 2) were awarded to:

1. The Lockheed Martin led team with a high powered steel monohull design based on the Fincantieri Destriero design.
2. The Raytheon led team with a FRP surface effect design based on the Norwegian Skjold Class design.
3. The General Dynamics led team with the Austal aluminium trimaran design based on the 127m *Benchijigua Express*.

Figure 2: Three LCS Preliminary Designs were developed during 2003, followed by the award of two Final Design contracts

Following completion of these preliminary designs, each team submitted a proposal for the final design and the subsequent detailed design and construction of their LCS. In May of 2004 the Lockheed Martin Team was awarded a seven month Final Design contract and the General Dynamics Team was awarded a seventeen month Final Design contract.

In January of 2005 the USN exercised the Lockheed Martin contract option for the Detailed Design and Construction of the LCS-1, which was later named "*Freedom*". In December 2005 (Fiscal Year 2006) the General Dynamics contract option for Detailed Design and Construction of LCS-2 was exercised. LCS-2 was later named "*Independence*".

Due to subsequent cost growth and budgetary constraints, a second Lockheed Martin and a second General Dynamics contract were awarded, terminated and re-awarded. These two vessels are currently under construction at Marinette Marine and Austal respectively.

Freedom completed preliminary acceptance trials in August 2008 and was delivered to the Navy in September 2008. *Independence* completed preliminary acceptance trials in November 2009 and is planned to be delivered to the Navy in December 2009. This equates to a total detailed design and construction period of 3 years and 7 months for *Freedom* and 4 years for *Independence.*

The two LCS teams are currently responding to a Request For Proposal for a further ten LCS ships to be awarded at a rate of two per year for the next five years. The award of the first two vessels in that contract are expected mid 2010. In Fiscal Year 2012, the Navy will reopen the competition to other shipyards to build additional vessels of the same design. The losing bidder in the initial LCS down select may still compete to build additional hulls in 2012 of the selected design, but other shipyards are also encouraged to compete as well.

3. THE TWO LCS SOLUTIONS

The design requirements for the two LCS designs are consistent; however the design arrangements of the two solutions are significantly different. The fundamental difference between the two hulls is that the trimaran configuration affords a significant increase in the internal volume and deck area when compared with the monohull. Although the trimaran is larger and has a greater overall beam, this multihull arrangement offers significantly lower resistance than the monohull due to the very slender hull configuration. The multihull also offers improved hull stability with significant reserve buoyancy in the outer hulls which become more deeply immersed as the ship heels.

The other significant difference between the two platforms is the hull construction material. The aluminium trimaran offers substantial savings in hull structural weight which could not be achieved with a steel hull. The structural weight of *Independence*, if it were constructed in steel, would be approximately double the structural weight of the current design. There were many challenges associated with the use of marine grade aluminium for primary hull structure of a surface combatant vessel for the USN. The design and construction process has undergone extensive review by both NAVSEA and ABS, and has been certified in accordance with the specified requirements.

Although *Freedom* has an aluminium superstructure, the hull is fabricated from steel. The overall loaded displacement of the two vessels has been reported as comparable even though *Independence* is a physically larger vessel. The installed power of *Freedom* is 64% more than *Independence* due to the difference in hull arrangement and the resulting power required to achieve the sprint speed of more than 40 knots. Table 1 illustrates some of the key differences in characteristics between the two platforms:

Table 1: Major differences in capability between the vessels in the Littoral Combat Ship Program

Characteristic	LMLCS	GDLCS	Comments on Differences
Hull form	Monohull	Trimaran	The trimaran hullform offers lower resistance and better seakeeping
Material	Steel hull, Aluminium superstructure	All Aluminium	
Length	115m (378')	127m (418')	10% longer
Beam	17.5m (57')	30.4m (100')	75% wider offering increased internal mission volume and flight deck area
Draft	4.1m (13.5')	4.5m (14.8')	10% deeper draft due to the slender multihull arrangement
USN Stability status [1]	2 (No stability margin)	1 (weight and VCG margin)	No Stability Margin on LMLCS
Operating Speed	40 knots	40 knots	
Range	3500nm @18 kts	4300 @18 kts	23% greater range
Gas Turbines	Rolls-Royce MT-30	GE LM 2500	
Power	2 @ 36MW	2 @ 22MW	GDLCS requires 40% less turbine power for maximum speed
Full Power Turbine Fuel Consumption	17.28	10.78	(t/hr estimated). 40% less fuel used
Diesels	FM Pielstick 16PA6B	MTU 8000	
Power	2 @ 6.48MW	2 @ 9.1MW	38% more diesel power allows higher cruising speeds with economy
Diesel Fuel Consumption	2.71	3.80	(t/hr estimated)
Total power	85MW	62MW	37% less power equates to less fuel consumption
Total Propulsion Fuel Consumption	19.99	14.58	(t/hr estimated) 27% less
Waterjets	Rolls Royce	Wartsila	
Gas turbine- 2 inboard	153SII- boost	LJD 160- steerable	300% more steering redundancy
Diesel- 2 outboard	153SII- steerable	LJD 150- steerable	
Transmission	Combining/splitting gears	Direct drive	The simpler shafting arrangement provides greater system redundancy
Generators	IF 1708T2	MTU 8V396	Proven MTU reliability
	4 @ 750KW	4 @ 800KW	7% more electrical power available
Thruster (manoeuvring and take-home)	None	Retractable, rotating	Improved austere port manoeuvre
Bridge window from water	11.7m	16.0m	37% higher for improved visibility
Weapon Modules	2	3	50% more
Flight Deck size (area)	Approximately 595m^2	Approx. 1030m^2	73% greater
H-60 helo spots	1	2	100% more
Height above water	6.7m	10.7m	60% higher results in less spray
Landing/retrieval system	Yes – Trigon	Not required	Platform more stable
Move stores to Mission Bay	Requires shore crane	Integral hangar elevator	Elevator provides flexibility to move stores, UAVs and maintenance parts
Elevator to Mission Bay	No (Hatches only)	Yes	Greater flexibility
Hangar size (area)	2 x H-60	2 x H-60 and multiple UAVs	
Hangar doors	1	2	More flexible arrangement
Mission bay size (area)	Compartmented	Approx.1400m^2 Open deck	
Side Ramp	Door only	Yes	Integral Ro-Ro capability
Boat Launch	Aft ramp + side door	Aft Twin Boom Crane	
Remote Mine-hunting v'h'le	Side launched	Aft Twin Boom Crane	
Boat handling area	Wet and floodable	Dry	
Berthing per room	8, 4, and 2 person	4 and 2 person	Improved habitability

[1] US Stability STATUS 1. An increase in weight and a rise of the ship's centre of gravity are acceptable. Added weight and moment resulting from changes will not require any compensation unless the magnitude of the additions is so large as to make the ship approach the stability limits.

US Stability STATUS 2. Neither an increase in weight nor a rise of a ship's centre of gravity can be accepted.

The significant difference in the installed power between the two designs has a direct impact on the LCS operating costs. The total ownership cost of the LCS class will depend very much on the operating and maintenance cost of the installed machinery and equipment. The LM2500 gas turbine is the most commonly used and proven propulsion engine in the US Navy fleet, with a documented history of reliability over 40 years. *"The U.S. Navy's use of the LM2500 power plant in more than 175 ships has proven it to be an extremely reliable, dense power unit. The U.S. Navy has accumulated over 13 million hours of service on our total LM2500 population since the beginning of the USS Spruance and USS Oliver Hazard Perry class applications."* - E. Alan Karpovitch, U.S. Navy Propulsion Program Manager [11].

The required manning of both LCS variants is comparable with a significant level of automation on both ships and with mission package crews only embarked on an as-required basis. The manning requirements and their impact on operating costs are therefore expected to be similar between the two LCS variants.

The estimated diesel fuel oil consumption for the two platforms at full power is illustrated in Figure 3 and demonstrates that if *Freedom* and *Independence* both require full power to cruise at a sprint speed of 40knots for 40 hours (1600 nm) then *Freedom* will consume approximately 200 tonnes more fuel than *Independence*.

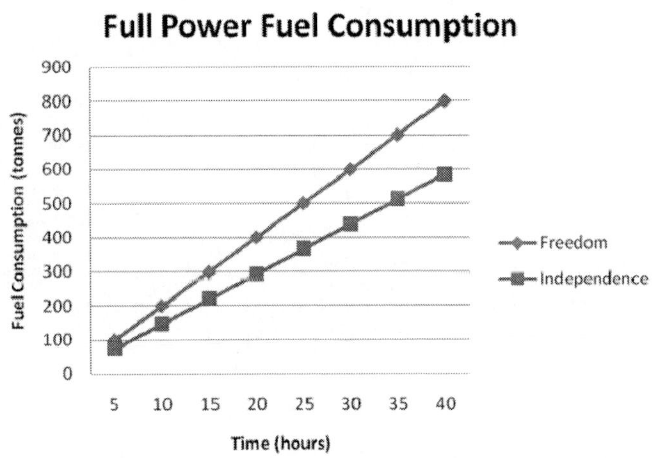

Figure 3: The fuel consumption of both LCS variants at full power is considerable, however Freedom's range and endurance reflects a power of 64% more than Independence

John P. Baggett, in his "Logistical Analysis of the Littoral Combat Ship (LCS) Operating Independently in the Pacific"[12], states that the GD LCS has a fuel capacity of 67,746 gallons (256,450 litres or 213 tonnes) more than the LM LCS. The paper goes on to say that *"The GD LCS is capable of travelling over twice the distance of the LM LCS at their respective economical speeds"*. This substantial advantage in both fuel capacity and fuel consumption means that *Independence* not only offers lower operating costs, but also offers significantly more range and endurance than *Freedom*.

One of the other real advantages of the trimaran hull is the improved motions offered by the platform in a range of sea conditions and ship headings. It is essential that the ships crew are not fatigued or suffering from motion sickness in order for them to effectively execute their duties. It is also important that ship motions are minimised to allow aviation activities, water craft launch and recovery and effective weapons operation. Austal has performed self-propelled model tests and full-scale trials to verify the accuracy of our seakeeping analysis. This analysis is performed with the aid of the VERES software developed by Marintek in Norway. The operability of various commercial platforms has been assessed in the past using the VERES tool. A 135m monohull has been compared with a 127m trimaran with both vessels operating in the same wave environment, at an operational speed of 39 knots and with a deadweight of 500 tonnes. The installed power on the monohull was 51MW and on the trimaran was 33MW. This analysis demonstrated that although the trimaran had a lightship displacement of approximately 1500tonnes compared with the much larger 2300tonnes for the monohull, the operability for the trimaran was comparable to the much larger monohull. The Motion Sickness Incidence (a measure of vertical acceleration in the passenger space) for the monohull was marginally better than the trimaran in head seas due to greater waterline length, however the performance of the trimaran in beam and following seas was similar between the two platforms. The key difference between the two platforms was apparent when comparing the lateral acceleration. The trimaran operability was more than 80% in beam seas while the monohull operability was only 54%. This illustrates the gentle roll characteristics offered by a trimaran platform when compared with a monohull.

The following pictures illustrate some of the key differences between the two LCS platforms:

Figure 4: The stern watercraft launch and recovery arrangement on Freedom is susceptible to flooding when compared with the twin boom crane arrangement on Independence

Figure 5: Both LCS variants offer large flight decks and hangars however Independence allows two MH-60 aircraft to operate simultaneously with a hangar door port and starboard

Figure 6: The bow wave and wake generated by the slender hulls of Independence compared with the significant wake produced by Freedom, illustrates the fundamental difference in the resistance of these two designs at high speed

4. THE 102 METRE TRIMARAN PLATFORM

The 102m trimaran (Figure 7) was designed to be the first "production" trimaran after the successful design and construction of the prototype 127 metre trimaran "*Benchijigua Express*", which was delivered in 2005 and has been successfully operating between the Canary Islands since then. *Benchijigua Express* was extensively instrumented and trialed over a period of many months at the shipyard and during daily operation, as befits a first-of-class design. The results of those trials were used to build an accurate manoeuvring and motions simulator which permitted us to investigate the effects of changes to the ship design to optimise the motion characteristics.

Figure 7: The 102metre Hull 270 is undergoing sea trials in January 2010 and is available for delivery in February, offering the next generation of proven trimaran technology

The trimaran is unique in ship design in that it easily permits the designer to choose the desired stability characteristics. The desired metacentric height GM_T is chosen along with the other principal dimensions such as length and overall beam, as further detailed in Armstrong [1]. The GM_T value determines the inertia of the waterplane area and hence the amah size. The shape of the righting lever (GZ) curve can then be easily manipulated to give the desired rolling motions by adjustment of the amah geometry above the waterline, which has an impact on the desired deck layout.

The stability of *Benchijigua Express* exceeds the requirements of the regulations by a considerable margin, even though it was deliberately designed with a low metacentric height (< 3m at full load), in order to create a long and slow roll, which would enhance the comfort for passengers and crew. A Ride Control system was included as an essential part of the design, comprising of a forward-mounted T-foil and anti-rolling fins fitted to the main hull aft of midships. These fins were not ideally placed, being positioned far aft in the engine rooms. The position of these fins made them effective in controlling roll motions, but created some yawing forces which had to be counteracted by the integrated rudder system. A tendency of the design to lie over at an angle of heel for long periods of time also caused anxiety amongst some passengers, fearful that the vessel may not come back upright! Subsequent changes to the ride control software have substantially overcome this issue, although it was felt that the GM_T value should be increased on the next generation of vessel.
For the design of the 102m trimaran, the GM_T was chosen by analysing the roll of the vessel in beam and following seas of a fixed wave height. The GM_T was varied by changing the size of the amahs. The resultant curves are illustrated in Figure 8, and the value chosen was based on the location where the roll angle was not excessive in stern quartering seas, i.e. GM_T > 5.2 meters approx.

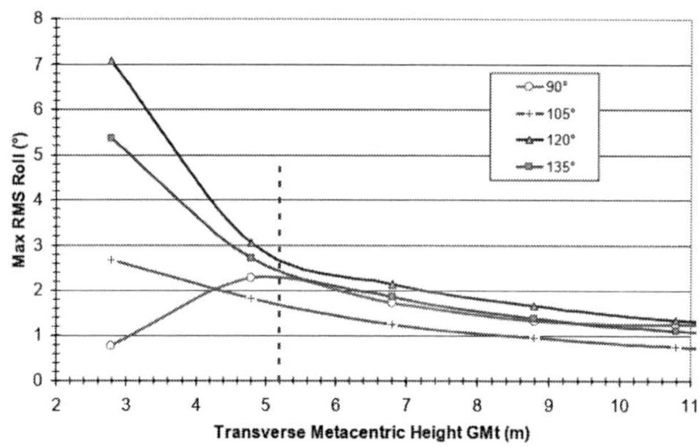

Figure 8: Maximum roll angles for various GM_T values in beam and stern quartering seas, 36 knots, $H_S = 2.0m$

Having taken a fairly conservative approach to the ride control on the prototype trimaran, based on existing technology, the ride control layout was amended for the 102m trimaran to obtain an improved performance. The roll fins were moved from the main hull to the transom of the amahs where they were redesigned as horizontal wings on long struts (T-foils). These are active fins operated by hydraulics and accelerometers. The software control algorithm has also been changed to better suit the motion characteristics of a trimaran, rather than use one developed for a catamaran.

The hull shape of the 102m trimaran was also modified from that used for *Benchijigua Express* because of the different weight and mass distribution, and to better suit operation at a different Froude number.

5. THE AUSTAL MULTI-ROLE VESSEL

Design Maturity

There are currently only five trimaran ships either in operation or under construction with a length greater than 70 metres. One is the 90 metre, 22 knot steel demonstrator craft "*Triton*", which has been converted for use by Australian Customs and Border Protection Service. The LCS-2, LCS-4 and *Benchijigua Express* have all been discussed in the previous paragraphs.

Austal has recently launched the 102m next generation trimaran design which is configured as a vehicle ferry. Austal are therefore at the forefront of design and understanding of trimaran craft.

Trimaran design work at Austal started in 2001 with a 149m long craft, and has been ongoing since that time. *Benchijigua Express* was completed in 2005, and designs at various other lengths have been developed, including 102 metres, 92 metres, 80 metres, and 55 metres. Design of a trimaran is a complex task as some of the physical characteristics do not easily scale to other sizes, as they do on conventional monohulls or catamarans, owing to the interference effects. In addition, regulatory intact stability is an important driver of the design, and stability is generally constant for all sizes of vessel.

The 92 metre and 80 metre designs were started in 2005, following on from the lessons learned on *Benchijigua Express*, and included numerical studies on the optimum amah shape and placement and main hull shape for maximum speed and minimum ship motions for passenger and crew comfort. These numerical studies used robust software that had been correlated against full scale trials on the 127m trimaran.

Numerical analysis supported by model tank testing was used to study the effect of the locations of the amahs. The transverse locations is principally driven by the regulatory stability requirements, together with the desired deck width, although studies were conducted to ascertain the interference effects from the proximity of the amahs with the main hull.

The longitudinal location of the amahs was a compromise between the competing performance requirements. The various studies found that the optimum location for high speed (well aft) agreed with the best location for course-keeping stability, but conflicted with the optimum locations for manoeuvrability (midships) and minimum loss of speed in waves (between midships and aft). It was found that the most suitable location for improved ship motions depended on the ships heading relative to the waves, with a good mean location for the amah LCB being at L/3 from the transom. A good compromise for the amah location was found for the relatively modest speeds of the 80 metre design with the transom of the main hull in line with the transom of the amahs, which also suited the main deck layout, ramp and stern-loading proposed arrangement.

The transverse location of the amahs has a major impact on the rolling motions and intact stability characteristics. The location was determined by the method of choosing a GM_T value, as described in Section 4.

A major driver of trimaran design is the ability to accurately estimate the design loads for the design of the ship structure. High-speed craft are generally designed to an acceleration level, but it is difficult to accurately estimate the accelerations resulting from operation in a seaway. Numerical studies were conducted over a range of sea states and headings to determine the ship motion and likely resultant accelerations. A major design driver is slamming at the bow, and the trimaran in this regard is an excellent platform, as in the forward third it is essentially a very thin monohull with no wet deck and the slamming loads are relatively small, especially when compared against those experienced by the wet deck of a catamaran.

Mission Capability

The extensive design development program described above has led to the development of the Austal Multi-Role Vessel (MRV) design. The MRV leverages many of the lessons learned from *Benchijigua Express*, *Independence* and the next generation 102m platform. The MRV is intended to exploit the advantages of using a single platform to conduct offshore security, mine countermeasure and hydrographic roles. The Austal MRV will employ modular concepts as demonstrated on the LCS program, that enable it to be employed in multiple roles to meet the needs of particular operations or strategic circumstances.

Figure 9 The Austal 80m MRV offers a substantial reconfigurable mission bay and a large flight deck which would not be achievable on a monohull design of similar length.

Some of the proposed mission capabilities are described below:

- Disaster relief and humanitarian assistance
- Troop transportation and austere port deployment
- Border protection and Customs duties
- Capacity for the launch and recovery of up to six 11 metre Rigid Hull Inflatable Boats
- Capacity for the launch and recovery of three 8metre Rigid Hull Inflatable Boats
- Magnetic dyad transportation, launch and recovery
- Decompression chambers in support of diving operations
- Expendable mines
- Remotely Operated Vehicle launch and recovery
- Mother ship / replenishment for patrol vessel fleet
- Submarine rescue equipment transportation
- At sea helicopter operational training

Key Design Features

Some of the key design features of the Austal MRV are provided below:

- Accommodations for 39 ships crew including mission crew and 50 seats and 48 additional berths for embarked personnel
- Separate engine rooms available for 2, 3 and 4 propulsion engine arrangements
- Large fuel capacity for a 5000 nm range
- Machinery and essential systems protected from surface attack by amah hulls
- Low Wash
- Low Radar signature
- Large stable flight and mission decks
- Flight deck for an 11 tonne helicopter (MRH90 or equal)

- Bi-folding and slewing aft ramp for stern or side loading
- Internal gantry crane for cargo deck equipment movements
- Helicopter hangar with refuelling capabilities
- Intact and damage stability compliance including compliance with high speed craft raking damage criteria (HSC2000)
- Aft RHIB launch and recovery ramp with overhead gantry crane providing access to the mission bay
- Advanced ride control that can stabilize the platform in a variety of wind and wave conditions and dynamically control vessel trim and heel
- Large stable platform for placement of weapons and sensors
- Aft crane for loading / unloading equipment from the main deck to the flight deck and ashore

7. CONCLUSION

The Austal MRV offers significant mission capability for a vessel of its length at only 80m. *Benchijigua Express* and *Independence* have demonstrated some of the real advantages offered by an aluminium trimaran platform including:

- Improved motions in a seaway
- Reduced resistance and operating fuel costs
- Superior aviation capability
- Limited draft for austere port operation
- Substantial mission bay capacity for reconfigurable mission packages

The MRV is an ideal platform to conduct offshore security, mine countermeasure and hydrographic roles in support of the Royal Australian Navy.

REFERENCES

1. Armstrong T & Holden K: "A New Generation of large Fast Ferry – from Concept to Contract Reality", 7th International Conference on Fast Sea Transportation (*FAST'03*), Ischia, Italy, October 2003.
2. Lockheed Martin LCS Team, www.lmlcsteam.com/solution.html
3. General Dynamics LCS Team, http://www.gdlcs.com/gd-lcs-solution
4. NAVSEA (SEA 05Z12), "COMNAVSEA SYSCOM WASHINGTON DC(uc) STABILITY STATUS OF IN-SERVICE SHIPS (UNCLAS)", April 09
5. Program Executive Office Ships, http://peoships.crane.navy.mil/LCS/factsheet.htm
6. Rolls Royce, "Fact Sheet, MT30 marine gas turbine", http://www1.rolls-royce.com/marine/downloads/gas_diesel/mt30_fact.pdf
7. Rolls Royce, "In-Depth", Issue 7, http://www1.rolls-royce.com/marine/downloads/article/id7_mt30.pdf
8. Ronald O'Rourke, "Navy Littoral Combat Ship (LCS) Program: Oversight Issues and Options for Congress", July 2007, http://www.dtic.mil/cgi-bin/GetTRDoc?AD=ADA471160&Location=U2&doc=GetTRDoc.pdf page CRS-35
9. Diana K. Massing, Lockheed Martin, http://www.lockheedmartin.com/news/press_releases/2008/060408_LCS1_GasTurbines_LightOff.html
10. CAPT Mike Good, Program Manager LCS Mission Modules, "LCS Program Overview for Northwest Florida Defense Coalition", May 2009, http://www.florida-edc.org/DSI/Littoral%20Combat%20Ship%20Program%20Overview.pdf
11. GE Marine, "The Wave Volume 4, Issue 3", Fall 2009
12. John P. Baggett, "Logistical Analysis of the Littoral Combat Ship (LCS) Operating Independently in the Pacific", Naval Postgraduate School, 2008

A Forensic Analysis of the Wreck of HMAS Sydney II — Setting the Scene

John Jeremy
Royal Institution of Naval Architects, Australian Division

ABSTRACT

Shortly after the discovery by the Finding Sydney Foundation of the wrecks of HMAS *Sydney* and HSK *Kormoran* off the coast of Western Australia the Chief of the Defence Force, ACM A. G. Houston AC AFC, established a Commission of Inquiry to inquire into and report on the circumstances associated with the loss of HMAS *Sydney* on 19 November 1941 after her encounter with the German raider *Kormoran*.

In May 2008 the Commission appointed the Defence Science and Technology Organisation in collaboration with the Australian Division of the Royal Institution of Naval Architects to provide expert advice and opinion on a wide range of technical issues relating to the ships and their loss and to determine the most likely cause for the loss of HMAS *Sydney* with all her ship's company.

HMAS *Sydney* was a Modified Leander-class light cruiser, one of three designed by the Admiralty for the Royal Navy in the early 1930s. Laid down as *Phaeton*, she was launched as *Sydney* and commissioned as HMAS *Sydney* on 24 September 1935. To aid the forensic analysis of the wreck of the ship the study team had as evidence many excellent photographs and video of the wreck taken by David Mearns. In setting the scene for the analysis of this material the paper describes the challenges faced by the Commission, DSTO and RINA in finding primary source material relating to the design and construction of *Sydney*.

The paper outlines the design origin of the Modified Leander-class cruisers and provides a description of HMAS *Sydney* as she was in November 1941.

THE INQUIRY

The wreck of HMAS *Sydney* was located by the Finding Sydney Foundation on 16 March 2008 about 112 n miles west of Steep Point, Western Australia, ending nearly 67 years of uncertainty about the location of the famous Australian cruiser [1]. The wreck and debris field was located at depths ranging from 2220 to 2480 m and, despite the challenges of the depth and prevailing weather conditions, David Mearns and the FSF team obtained many hundreds of very high-quality images and hours of video of the wreck.

Shortly after the location of the wreck, on 28 March the Chief of the Defence Force, Air Chief Marshall A. G. Houston appointed a Commission of Inquiry with the Hon. Terence Cole QC as President. The terms of reference of the Inquiry were broad:

"to inquire into and report on circumstances associated with the loss of HMAS *Sydney* (II) in November 1941 and consequent loss of life and related events subsequent thereto."

After discussions with both organisations, the Commission appointed the Defence Science and Technology Organisation (DSTO) assisted by the Australian Division of the Royal

Institution of Naval Architects (RINA) to provide expert technical advice and opinion to the Commission. Again, the instructions for the technical study were very broad. In addition to a comprehensive overview of the design and construction of both HMAS *Sydney* and KSK *Kormoran*, the Commission sought advice on the likely effect of the observed damage to both ships on their ability to float, move and fight. In particular, the Commission wished to have an opinion on the most likely cause of the loss of HMAS *Sydney* with all her crew.

Supplementary instructions expanded the scope of work to include a detailed description of HMAS *Sydney's* machinery and electrical services and equipment, fire-fighting arrangements, life saving equipment and communications.

DSTO and RINA established a team of twelve naval architects, engineers and scientists for the task, working in Melbourne, Sydney and Canberra.

TECHNICAL INFORMATION

Much has been written about the Leander-class and Modified Leander class cruisers and, in particular, HMAS *Sydney*. The technical description of the ship in some of the published works is very complete and of high quality, but for the DSTO/RINA investigation as many original source documents as possible were required to provide an accurate design and construction history of the ship. In order to facilitate the task of gathering this information, the Commission took steps to have all relevant archival material from major sources in Australia and the United Kingdom made available to the study team. Much came from the National Archives of Australia and the assistance provided by the NAA staff, particularly in Canberra and Melbourne, was outstanding.

An early task was to give the Commission guidance on the type of information needed for the study. In addition to as many construction and 'as fitted' drawings of HMAS *Sydney* as possible, ship specifications and original design information were sought. Most of the drawings of HMAS *Sydney* held by the NAA are in series MP551/1 in Canberra, and the original packing lists suggested that over 1000 drawings were likely to be available. At first the NAA could only locate the 'as fitted' hull drawings and a few examples of structural and other working plans which must have come from the bundles thought not to be held.

Essential information for the study included a lines plan and other design information which did not appear to be available in Australia. The Ship's Cover in the collection of the National Maritime Museum in Britain provided some relevant design material but it was the discovery of the missing working drawings in the NAA in Canberra which provided the lines plan and structural drawings. Copies of the drawings were provided by the NAA as high-resolution photographs. Before the plans could be photographed, they had to be conserved and flattened, a process which took time. Whilst priority lists were provided by the study team it was many months before the copying was completed and all the requested drawings were available.

More information came from the Australian War Memorial and the RAN Sea Power Centre. Private collections also contained much relevant information previously discarded by official sources. Missing almost entirely from collections in Australia and the UK are machinery and electrical drawings. However, ships built to Admiralty specifications in a particular period almost always all work the same way and gaps in original information were filled to some extent by standard Books of Reference (BR). Where these could not readily be found in libraries or archives an internet search often proved productive. Finally, only weeks before the study report was to be completed and presented to the Commission, the original tracings of HMAS *Sydney's* Machinery Information Book were found in the RAN's historical collection on Spectacle Island in Sydney. The discovery helped confirm the validity of some of the assumptions made about the ship and her systems.

Whilst HMAS *Sydney's* 'as fitted' drawings had, in some cases, been modified to include changes to the ship since completion, the Ship's Book (in the NAA Canberra) and other files (in the NAA Melbourne) provided the history of her refits. As some changes were made hastily during the first two years of World War II, for which drawings may not even have existed, photographs from the Australian War Memorial's collection also proved invaluable.

The search for original technical information was at times frustrating. That which was found was often of very high quality — the original 'as fitted' drawings were in ink on linen and in good condition, although some were slightly water damaged. One wonders how much of today's predominantly-electronic information will be available for researchers with a similar task in seventy-five years' time.

THE MODIFIED LEANDER-CLASS CRUISER HMAS SYDNEY [2]

Figure 1 — HMAS *Sydney* in Sydney Harbour on 10 February 1941
(Australian National Maritime Museum, Sam Hood Collection, Neg. No. 20357-1)

HMAS *Sydney* was one of three Modified Leander-class light cruisers originally ordered for the Royal Navy. *Phaeton*, *Amphion* and *Apollo* were all to serve in the RAN as HMA Ships *Sydney*, *Perth* and *Hobart*. *Phaeton* was the first to be completed, as HMAS *Sydney*. The cruisers were designed by the Admiralty under the supervision of the Director of Naval Construction, Sir Arthur Johns KCB CBE RCNC. The Constructor responsible for the design was the distinguished naval architect Charles Lillicrap who was head of the cruiser section at the time. Lillicrap was also responsible for the design of the highly-regarded County-class cruisers (including the RAN heavy cruisers *Australia* and *Canberra*) and the ships of which the Modified Leanders were a development — the Leander-class light cruisers.

British cruiser design and construction during the 1920s and 1930s was dominated by the obligations imposed by the Washington Treaty of 1924 which limited cruisers to a standard displacement of 10 000 tons and guns not exceeding 8-inch calibre. The County-class cruisers were designed to these limits, but they were expensive ships and the Royal Navy could not afford to build sufficient ships of this type to satisfy the need for cruisers, estimated to be about 70 ships. A smaller ship (HMS *York*) was ordered in 1927, fitted with six rather than eight 8-inch guns and a sister ship (HMS *Exeter*) was ordered a year later [3].

Further designs of slower and smaller ships fitted with 8-inch or 6-inch guns were considered, and in 1927 the Admiralty approved the design of a cruiser fitted with eight 6-inch guns on a standard displacement of 7,154 tons. This ship became HMS *Leander*, ordered in the 1929 program.

Charles Lillicrap was a great believer in the use of welding for ship construction, and the new design incorporated welding where he believed prudent in order to reduce weight. *Leander* turned out over-weight but the four subsequent ships of the class — *Achilles*, *Neptune*, *Orion* and *Ajax* — benefited from considerable weight savings and welding was used extensively in subsequent cruiser designs [4].

In October 1932, during the detailed design and early construction of the Leander-class cruisers, the high price of the ships prompted the Third Sea Lord and Controller of the Navy to order a review of the design and he proposed number of changes 'in these times of financial stringency'. These included reducing the gun fire-control equipment, elimination of magazine cooling and even a reduction of internal telephones [5].

The changes to the fire-control equipment included the elimination of the after director control tower (DCT). Recognising that, if the forward DCT was put out of action or was temporarily blanked, control would devolve on one of the turret officers, a non-revolving control position fitted out with a minimum of necessary equipment was proposed to be fitted on the after superstructure [6]. These changes, which were approved by the Admiralty Board on 10 November 1932 [7], were the source (in part, at least) of later dissatisfaction in the arrangements in the Leander-class and Modified Leander-class cruisers as expressed by Captain J. W. A. Waller of HMAS *Sydney* in October 1938 [8] and the Vice Admiral, Light Forces, Mediterranean in December 1940 [9].

Meanwhile, in July 1932, an investigation had been completed into a revised machinery arrangement for the later ships of the Leander Class. In the ships then ordered, the propelling machinery comprised six boilers and four sets of turbines, two sets of 24 000 SHP driving the outer shafts and two sets of 12 000 SHP driving the inner shafts for a total of 72 000 SHP. The boilers were in three boiler rooms ahead of a forward engine room, a gearing room and an after engine room. In the last three ships, it was proposed to separate the boiler and engine rooms to reduce the vulnerability of the machinery plant to damage. The installation would comprise four boilers in two boiler rooms separated by the forward engine room, with four sets of turbines of equal power — 18 000 SHP on each shaft [10]. The modified arrangement is shown in Figure 2. The new arrangement required the length of the ship to be increased by 8 feet between perpendiculars and the length of side armour to the upper deck level extended to 141 feet instead of 84 feet in *Leander*.

Two of the four turbine-driven generators were replaced with two diesel-driven generators arranged in wing compartments outboard of B boiler room. These were not intended to act as emergency generators but to enable the steam plant to be shut down in port. Emergency generators were not fitted in Australian cruisers until 1943–44.

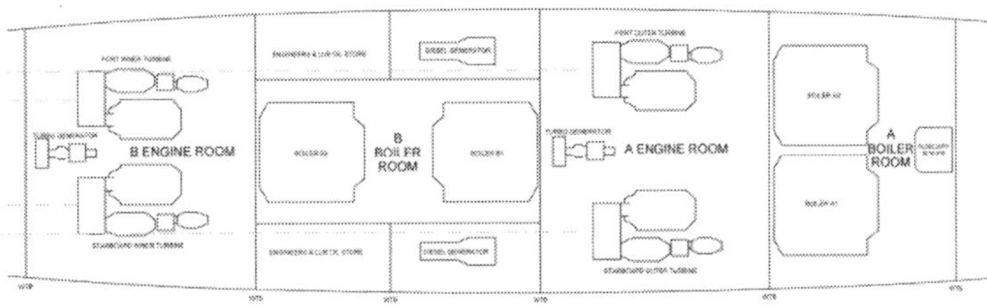

Figure 2 — Arrangement of machinery in the Modified Leander-class cruisers [11]

Despite the proposed changes resulting in even further congestion in an already very crowded ship, they were adopted for the three light cruisers ordered under the 1932 programme, with the ships becoming known as the Modified Leander Class. *Amphion* was ordered from HM Dockyard, Portsmouth, *Apollo* was ordered from HM Dockyard, Devonport, and *Phaeton* was ordered from Swan Hunter & Wigham Richardson Limited at Newcastle upon Tyne.

In early 1934, HMS *Phaeton* was acquired by the Australian Government as a replacement for the old cruiser HMAS *Brisbane*. She was renamed HMAS *Sydney* [12]. HMS *Amphion* and HMS *Apollo* were completed for the Royal Navy and were also acquired by Australia in 1938 becoming HMAS *Perth* and HMAS *Hobart*.

Description of HMAS *Sydney*

HMAS *Sydney* was laid down on 8 July 1933, launched on 22 September 1934 and completed on 24 September 1935. Her general particulars as completed are shown in Table 1 [13, 14]:

Displacement, light	6701 tons
Displacement, half oil	8056 tons
Displacement, full load	8940 tons
Displacement, standard	7198 tons
Length overall	562’ $3\frac{7}{8}$”
Length between perpendiculars	530’ $0\frac{3}{8}$”
Breadth extreme	56’ 8½”
Breadth moulded	56’ 0”
Depth, from under side of keel to underside of Upper Deck amidships	32’ 0” at side
Draught, at standard displacement	
Forward	15’ 3”
Aft	17’ 3”
Oil fuel capacity	1,800 tons
Machinery	Parsons single-reduction geared steam turbines
Boilers	Four
Shafts	Four
Power	72 000 SHP
Maximum Speed (designed)	32½ knots
Range	7000 n miles at 16 kn

Table 1 — General Particulars of HMAS *Sydney* as built

General Arrangement

The external profile and plan of HMAS *Sydney* is shown in Figure 3, a reproduction of the wireless telegraphy rig ‘as fitted’ drawing.

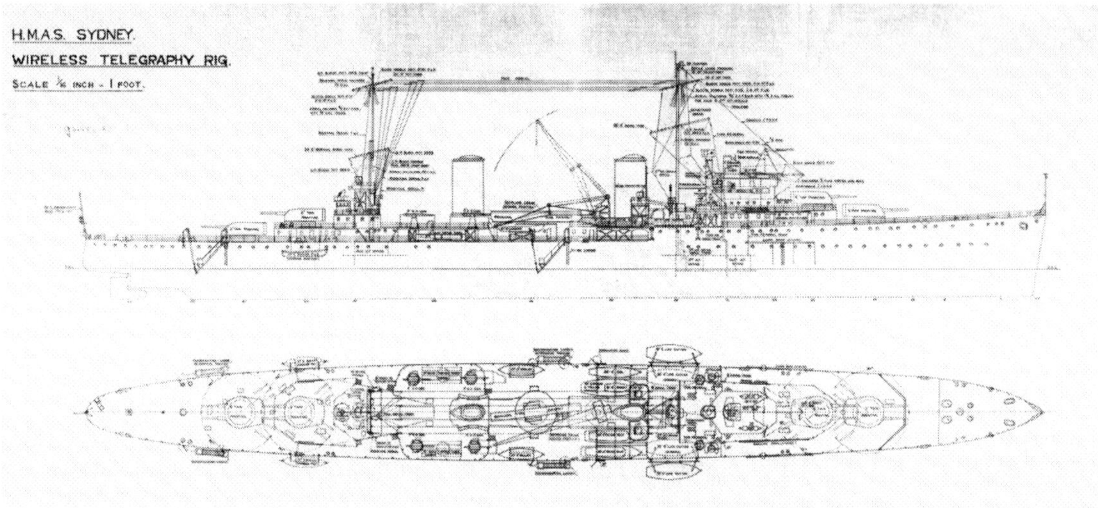

Figure 3 — External profile and plan of HMAS *Sydney* [15]

HMAS *Sydney* was arranged conventionally for a British cruiser of the time, with two superimposed 6-inch mountings (A & B) forward of the bridge and two (X & Y) aft. The aircraft catapult was fitted amidships and the secondary armament of 4-inch guns arranged abeam of the after funnel. The ship's boats were arranged amidships and were generally launched by the aircraft crane, except for the two cutters which were intended as sea boats and could be launched by davits. Her two 27 foot whalers could also be launched from davits aft when in port. The arrangement of the ship's boats when the ship was completed is shown in Figure 4. The ship also carried a number of Carley floats. The number carried varied from the time the ship was commissioned in 1935 to her loss in 1941. The arrangement of Carley floats in November 1941 is shown in Figure 5.

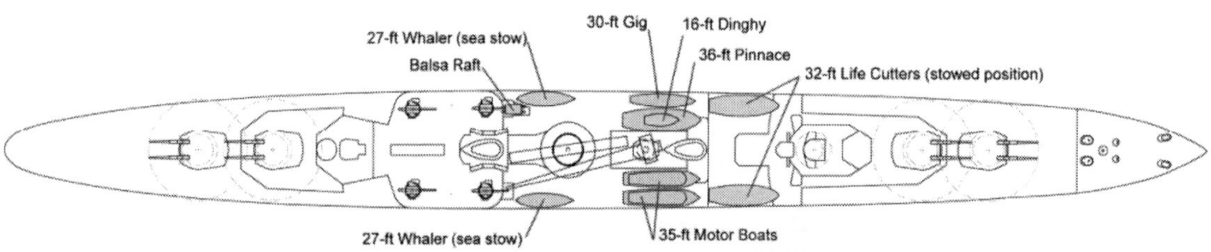

Figure 4 — Arrangement of boats in HMAS *Sydney* as completed

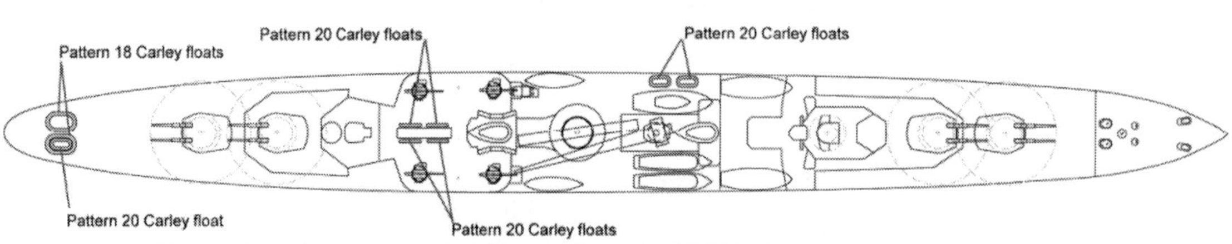

Figure 5 — Arrangement of Carley floats in HMAS *Sydney* in November 1941

Armament

The main armament of HMAS *Sydney* was eight 6-inch BL Mk XXIII guns in four Mk XXI twin mountings. The gun fired a 112-lb shell. It was first introduced into Royal Navy service in HMS *Leander*, and was fitted to all subsequent RN cruisers, except HM Ships *Tiger*, *Lion* and *Blake* which were completed to a modified design well after the war. The range of the gun was 25,800 yards at an elevation of 45 degrees.

The Mk XXI mounting was also introduced in HMS *Leander* and was fitted to all Leander-class, Modified Leander-class and Arethusa-class cruisers. The mounting had hand ramming but was powered by a self-contained hydraulic system.

The secondary armament of HMAS *Sydney* included four 4-inch high angle Mk V guns on Mk IV mountings. The mounting weighed 7 tons and was manually operated. The guns could be used against surface targets or aircraft. They could fire a 31-lb shell to a maximum height of 28,750 feet. Against surface targets the maximum range was 16,300 yards. Ready-use ammunition for the 4-inch guns was carried in 12 ready-use lockers on the 4-inch gun deck, which containing 20 rounds. The main 4-inch magazine was located aft of B Shell Room.

For close-range engagements *Sydney* was fitted with twelve 0.5-inch Vickers machine guns in three quadruple Mk II mountings, one fitted on top of the After Control Position and two on each side of the Flag Deck. They were supplemented by a number of 0.303-inch machine guns.

As built, HMAS *Sydney* was also fitted with four 3-pounder saluting guns on the after deckhouse, but they had been removed by 1941.

Construction and Subdivision

HMAS *Sydney* was constructed of steel and transversely framed and was predominantly of riveted construction. However, in keeping with her designer's advocacy of welding in ship construction, a considerable amount of the structure was welded, including decks and bulkheads, particularly towards the ends of the ship where stresses were lower.

The hull of HMAS *Sydney* was subdivided into a large number of watertight compartments. There were 23 transverse watertight bulkheads below the Platform Deck, 16 of which extended to at least the Upper Deck. Longitudinal bulkheads extended from the watertight transverse bulkhead (WTB) at frame 25 to the WTB at frame 86 and between the WTBs at frames 131 and 191. These bulkheads, which were below the Platform Deck, formed the boundary of important compartments like magazines and the Transmitting Station.

The introduction of the divided machinery spaces described above (see Figure 2), a practice adopted for all subsequent British cruiser designs, has proved to be somewhat controversial due to the introduction of longitudinal bulkheads (in *Sydney* between transverse WTB 116 and WTB 135) well off the centreline extending from the inner bottom to the Upper Deck.

The late David K. Brown RCNC, a past Deputy Chief Naval Architect in the UK MOD(N) was critical of this aspect of cruiser design pointing out that, in the event of flooding of one wing space as well as both engine rooms and the after boiler room, stability would be greatly reduced and rapid capsize was likely. This type of flooding contributed to the capsize of five ships. HMS *Cleopatra* was torpedoed in this area and went to a large angle of heel (about 15°) but was saved by rapid counter flooding [16]. When this layout of machinery spaces was adopted the Constructors at the Admiralty did not have the benefit of World War II experience and perhaps did not regard the extent of damage which would cause such flooding as likely. It should be noted that this type of flooding was not a factor in the loss of HMAS *Sydney*.

In addition to the subdivision provided by the longitudinal and transverse watertight bulkheads, HMAS *Sydney* was provided with a double bottom between WTB 76 and WTB 151 which extended from the keel to the Upper Deck between WTB 86 and WTB 151.

Compartments within the boundaries of decks and watertight bulkheads were formed by the erection of minor bulkheads. Minor bulkheads enclosing wet spaces like washplaces and heads and spaces like switchboard rooms were constructed of swedged 5 lb steel plate welded to the deck and deckhead. Bulkheads to storerooms and similar spaces were constructed of 3/16 inch thick steel-faced plywood. The plywood bulkheads were generally stopped 1 foot below the deckhead and the gap filled with wire mesh. Longitudinal cabin bulkheads were constructed of 5 lbs steel — transverse bulkheads of plywood — all stopped 1 foot below the deckhead with wire mesh above [17].

Armour and Protective Plating

The original Staff Requirement for the Leander-class cruisers specified that the ships should be protected so that:

- The ships side was immune to 6-inch shell fire above 10,000 yards,
- Crowns to be immune to 6-inch shell fire below 16,000 yards,
- The sides were to be immune to 4.7-inch shell fire above 4,700 yards, and
- Crowns to be immune to 4.7-inch shell fire at all ranges [18].

The ship's side armour was intended to protect the machinery spaces. It did not, for example, extent sufficiently far forward to protect the Main Switchboard Room.

The armour, as fitted to HMAS *Sydney*, is shown In Table 2 [19]. NC is non-cemented armour and D1 is high-tensile steel.

Machinery Spaces	**Sides:** 4 in, comprising 3 in NC and 1 in D1 **Crown:** $1\frac{1}{4}$ in D1 **Ends:** $1\frac{1}{2}$ in D1
Magazines, B Shell Room	**Sides:** $3\frac{7}{8}$ in, comprising $3\frac{1}{2}$ in NC and $\frac{3}{8}$ in D1 **Crown:** $2\frac{3}{8}$ in, comprising 2 in NC and $\frac{3}{8}$ in D1 **Ends:** $2\frac{7}{8}$ in, comprising $2\frac{1}{2}$ in NC and $\frac{3}{8}$ in D1
Transmitting Station, Low Power Switch Room & Gyro Compass Compt.	**Fwd, Sides:** 1 in D1 **Crown:** $1\frac{1}{2}$ in D1 **Ends:** $1\frac{1}{2}$ - $2\frac{7}{8}$ in D1
A, X and Y Shell Rooms and Thrust Block Recesses	**Sides:** 1 in D1 **Crown:** 1 in D1 **Ends:** 1 in D1
Steering Gear Compartment	**Sides:** $1\frac{1}{2}$ in D1 **Crown:** $1\frac{1}{4}$ in D1 **Ends:** $1\frac{1}{2}$ in D1
Turrets	**Faces:** 1 inch D1 **Sides:** 1 inch D1 **Rears:** 1 inch D1 **Crowns:** 1 inch D1
Turret Trunks and Ammunition Lobbies	1 in D1
D.C.T.	**Sides:** 1 in D1 **Crown:** 1 in D1
Compass Platform, Remote Control Office and Plotting Room	**Sides:** $\frac{3}{8}$ in D1 **Crown:** $\frac{1}{2}$ in D1
Upper Bridge	**Sides:** $\frac{3}{8}$ in D1
After Control	**Sides:** $\frac{3}{8}$ in D1 **Crown:** $\frac{1}{2}$ in D1

Table 2 — HMAS *Sydney* Armour and Protective plating

Machinery

HMAS *Sydney* was fitted with four Admiralty 3-drum watertube boilers in two boiler rooms. A Boiler Room, the forward boiler room, was also fitted with an auxiliary boiler to supply saturated steam for domestic purposes and to supply the steam-driven forced-draft fans when flashing up the main boilers. The main boilers were arranged abreast (A1 starboard and A2 port) with superheated steam supplied to A Engine Room through a main stop valve on the centreline at WTB 116. Under normal conditions, A1 and A2 supplied the steam for the propulsion turbines in A Engine Room only, but an emergency stop valve was fitted which enabled steam to be directed to B Engine Room, or vice versa [20]. A Boiler Room was pressurised and was entered by airlocks, port and starboard, at the forward end.

Auxiliary steam was supplied to A Engine Room through two connections port and starboard. Auxiliary steam could also be supplied to B Engine Room, and vice versa. The auxiliary steam drove turbo generators, fuel and feed pumps, air ejectors, circulating water pumps and evaporators [21].

A Engine Room was fitted with two sets of Parsons single-reduction geared turbines, one driving the port-outer propeller shaft and the other the starboard. All turbines drove three-bladed fixed-pitch propellers. Each 18 000 SHP turbine set comprised a cruising turbine, a high-pressure turbine and a low-pressure turbine mounted on top of the condenser and connected to a reduction gear box. Apart from auxiliaries the engine room also contained a 300 kW turbo generator.

B Boiler Room was fitted with two boilers, arranged fore and aft. Like the forward space, B Boiler Room was pressurised by the forced draft fans and was entered through two air locks, port and starboard.

B Engine Room was similar to A engine Room with two sets of propulsion turbines driving the port and starboard inner propeller shafts. Another 300 kW generator was fitted in this space along with the usual auxiliaries and four evaporators for the production of feed water and fresh water for domestic use.

Electrical power was generated and distributed at 220 volts DC and the distribution system was configured as a ring main which could be divided into sections in action conditions.

Both engine rooms were ventilated by electrically driven fans and in each the manoeuvring valves were located at the forward end of the space.

External Communications

HMAS *Sydney* had three W/T offices, the main aft below the main mast and the secondary and auxiliary offices were located forward on the platform deck. The communications sets which were fitted, Type 48, 49, 45 and 43a, were capable of wireless telegraphy only.

Modifications after Completion and Proposed Modernisation

Whilst some modifications were made to HMAS *Sydney* during refits in Australia, most were minor and did not significantly change her configuration. During the war she was fitted with degaussing, some sidelights were plated over to reduce her vulnerability to damage and some

20 lb steel plates were fitted to the guardrails around the 4-inch gun deck to provide some limited splinter protection for the gun crews.

A modernisation was scheduled for December 1939 to March 1940 in Sydney when it was planned to replace the single 4-inch guns with twin 4-inch Mk XIX high-angle guns as had been fitted to her sisters *Hobart* and *Perth* in 1938 when they were both still in Royal Navy service. The modernisation refit was deferred indefinitely in October 1939 [22].

Despite the inevitable compromises made during the design process, HMAS *Sydney* was a fine, well-built modern cruiser constructed to the highest British Admiralty standards.

REFERENCES

1. Finding Sydney Foundation, 2009, *Summary of Pre-search and Search Activities — Search for HMAS* Sydney *(II)*, Submission to the Commission of Inquiry (COI), COI document FSF.013.0001.
2. This section is largely drawn from the DSTO/RINA report DSTO-GD-0559 *Report on Technical Aspects of the Sinking of HMAS* Sydney *and HSK* Kormoran. The report can be accessed at www.defence.gov.au/sydneyii as Exhibit 106.
3. Brown, D. K., *A Century of Naval Construction*, Conway Maritime Press, London, 1983, pp. 136–137.
4. Brown, D. K., *A Century of Naval Construction*, pp. 146–147.
5. Controller to DNC and others, 7 October 1932, HMAS *Sydney*, Ship's Cover, National Maritime Museum, UK, COI document UKAA.015.0001.
6. Director of Naval Ordnance and Director of Training and Staff Duties to ACNS, Controller and First Sea Lord, 20 October 1932, *Proposals for the Reduction in Fire Control Equipment of New Construction 6" Cruisers on Account of the Urgent Need for Economy*, HMAS *Sydney*, Ship's Cover, National Maritime Museum, UK.
7. Minute 3014, Admiralty Board Minutes, 10 November 1932, HMAS *Sydney*, Ship's Cover, National Maritime Museum, UK, Document UKAA.015.0001.
8. Commanding Officer HMAS *Sydney* to RADM Commanding HM Australian Squadron ref. 75/3599, 21 October 1938, *HMAS* Sydney, *Fighting Efficiency — Protection of Fire Control System and Alternative Controls*, NAA: MP 1049/5, 2026/3/294.
9. Vice Admiral, Light Forces Mediterranean in HMS *Orion* to Commander-in-Chief Mediterranean, copied to Australian Commonwealth Naval Board, 28 December 1940, *Armament of Orion Class*, NAA: MP1049/5, 2026/3/397
10. DNC to Controller, 25 July 1932, HMAS *Sydney*, Ship's Cover, National Maritime Museum, UK, COI document UKAA.015.0001.
11. Drawn from HMAS *Sydney*, Platform Deck (As Fitted), NAA: MP551/1, 136AF/75.
12. Controller Office Acquaint No. 100 of 28 May 1934, HMAS *Sydney*, Ship's Cover, National Maritime Museum, UK, COI document UKAA.015.0001.
13. HMAS *Sydney*, Form D240, 27 September 1934, HMAS *Sydney* Ship's Book, COI document NAA.090.0001
14. Raven, A. and Roberts, J., *British Cruisers of World War II*, Arms and Armour Press, London and Melbourne, 1980, p. 416.
15. HMAS *Sydney*, Wireless Telegraphy Rig (As Fitted), NAA: MP551/1, 136/59AF.
16. Brown, D. K., 2000, *Nelson to Vanguard, Warship Development 1923–1945*, Chatham Publishing, London, p. 73 and pp. 164–166.

17. *Specification for the Building of a Cruiser of the 1932 Programme — "Leander" Class (Modified), Part 1 Hull, A. Structural*, 1932, HMSO, London.
18. Raven, A. and Roberts, J., 1980, *British Cruisers of World War II*, Arms and Armour Press, London and Melbourne, p. 143.
19. HMAS *Sydney*, Armour & Protective Plating (As Fitted), NAA: MP551/1, 54AF.
20. HMAS *Sydney*, *Machinery Information Book*, Folio 1, Main Steam Arrangement in Engine & Boiler Rooms.
21. HMAS *Sydney*, *Machinery Information Book*, Folio 2, Auxiliary Steam Arrangement.

The interpretation of the evidence and visualisation of the sinking of HMAS Sydney II

L. de Yong, R. Neill and S.M. Cannon,
Defence Science and Technology Organisation, Maritime Platforms Division, Melbourne, Victoria, Australia

ABSTRACT

On the 19th of November 1941, the Royal Australian Navy Modified Leander Class light cruiser HMAS *Sydney II*, en route to Fremantle, intercepted the disguised German raider, the HSK *Kormoran*, about 100 nautical miles west of Steep Point off the coast of Western Australia. In the ensuing battle, *Sydney* was sunk with the loss of the entire crew of 645 men. *Kormoran* was subsequently scuttled with the loss of 81 men.

Following the discovery of the wrecks of *Sydney* and *Kormoran* in March 2008, the Defence Science and Technology Organisation (DSTO) in collaboration with the Australian Division of the Royal Institution of Naval Architects (RINA) were appointed to provide expert advice and opinion into the sinking of *Sydney*.

The evidence from the wreck site of *Sydney* and the debris field on the ocean floor was extensively evaluated and interpreted using several advanced computer analysis techniques. Historical records and documents were also studied and evaluated to complement the computer analysis to determine the underlying cause of the sinking of *Sydney*. The analysis provided a scientific insight into the battle sequence, what it would have been like to be on board and why there were no survivors. DSTO also produced a highly realistic animation of the battle in order to show the probable sequence of events and to visualise the scientific findings. In essence, DSTO and RINA analysed the survivability of *Sydney* against the *Kormoran*.

This paper takes the evidence of the weapons damage to *Sydney* and presents the interpretation of that evidence. It explains how and why *Sydney* sank, why there were no survivors and presents a visualisation of the engagement.

INTRODUCTION

The sinking of HMAS *Sydney* II on November 19th 1941 off the coast of Western Australia, after a battle with the German raider HSK *Kormoran,* resulted in the greatest loss of life in Australian naval history. *Sydney* sank with the loss of the entire crew of 645 and *Kormoran* was scuttled by its own crew with the loss of 81 men. A degree of mystery has always surrounded the loss of *Sydney* as the location of the battle site was unclear. There was also disbelief that the "pride of the fleet" could have been sunk by a German raider with the loss of all hands and a reluctance by many to accept the official story which was based only on the accounts of the survivors of *Kormoran*.

In March 2008, the wrecks of *Sydney* and *Kormoran* were found by the SV *Geosounder*, under charter to the Finding Sydney Foundation, approximately 207 km off the coast of Western Australia. Both ships were lying in approximately 2500 m of water. Given the possible wealth of new evidence about the engagement that could be gleaned from the wreck site, the Chief of the Australian Defence Force established a Commission of Inquiry (COI) into the circumstances surrounding the sinking of *Sydney* [1]. In support of this enquiry, the Australian Defence Science and Technology Organisation (DSTO) in collaboration with the Australian Division of the Royal Institution of Naval Architects (RINA), was requested to provide expert advice and opinion to the COI in several scientific areas related to understanding why *Sydney* sank and why there were no survivors. This included scientific analysis of the design and weapons systems of the ship, its structural integrity, its vulnerability to weapons, flooding and seakeeping, a fire assessment, analysis of the damage control organisation and procedures, lifesaving equipment and survival at sea and the search and rescue process. The analysis provided a scientific insight into the battle sequence, what it would have been like to be on board and why there were no survivors. DSTO also produced a highly realistic visualisation of the battle in order to show the probable sequence of events and to visualise the scientific findings. In essence, DSTO and RINA analysed the survivability of *Sydney* against the *Kormoran*. This paper reports on the outcomes of this work [2].

BATTLE DAMAGE

The damage sustained by *Sydney* was unique in that it occurred to both port and starboard sides over an extended period of time. *Sydney* received a torpedo strike to the bow, was subjected to 15 cm gun bombardment as well as an ongoing attack from 3.7 cm and 2 cm weapons [2]. Another paper in this session presents the details of the battle damage to *Sydney* due to the 15 cm shells and the torpedo. This showed that it was hit by 87 individual 15 cm shells, 41 on the port side and 46 on the starboard side. Of these hits, 37 were contact detonations, 35 penetrated the structure they hit and the remainder did not penetrate the structure. This damage caused flooding, fires and incapacitated many of the personnel on board.

PROGRESSIVE FLOODING

The holes in the side of *Sydney* due to the weapons strikes and the damage to the bow caused by the torpedo strike resulted in a degree of flooding which was analysed using the Cooperative Research Navies (CRN) computer code, FREDYN [3]. The PARAMARINE [4] code was used to generate the compartment model of *Sydney* which is shown in Figure 1. The compartments in yellow are those that potentially could be flooded due to the damage sustained from the 15 cm weapons and the hole created from the torpedo. As a verification of the accuracy of the computer model, a comparison of the draught and trim of *Sydney* against the original ship design data was carried out. The calculated data was subsequently found to be very close to the data in the original ships log.

FREDYN is a temporal code that predicts the motions of a ship in moderate to extreme seaways including small and large amplitude motions, broaching and capsize. It may also be used to predict the progressive flooding of a damaged ship and the implications this flooding has on the overall motions and stability of the vessel in the encountered seaway. Estimates of time to sink or capsize can also be determined. Utilizing the predicted damage from the vulnerability analysis and that which was observed with the video footage, the ingress of flood water and subsequent flooding throughout the ship was modelled. The consequences of this flooding were then determined by modelling the ship motions.

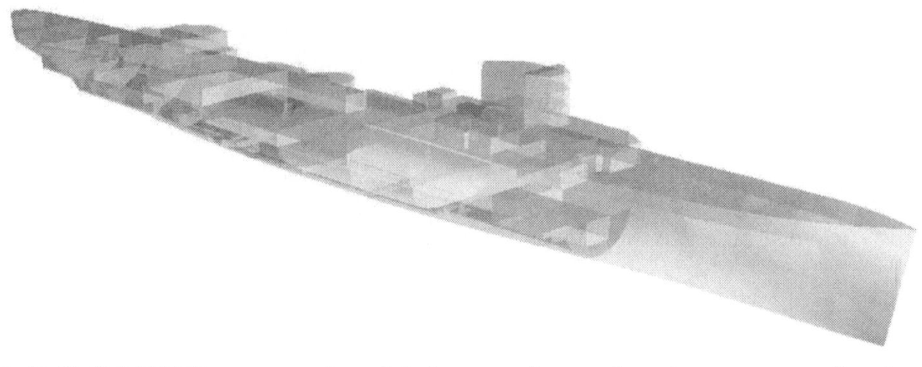

Figure 1: PARAMARINE generated model showing the predicted progressive flooding on Sydney [2]

The flooding analysis showed that as the sea state deteriorated in the hours following the battle (the German survivors indicated that the sea state increased from SS3 to SS4), more of the 15 cm shell holes became submerged below the waterline and the flooding increased [2]. As a consequence, *Sydney* began to roll to a greater extent (rolling from 15° up to 40°) and as time progressed, *Sydney* experienced significant rolling motions such that the edge of the deck became immersed. Eventually *Sydney* rolled to an angle beyond that from which she could recover, lost buoyancy and sank.

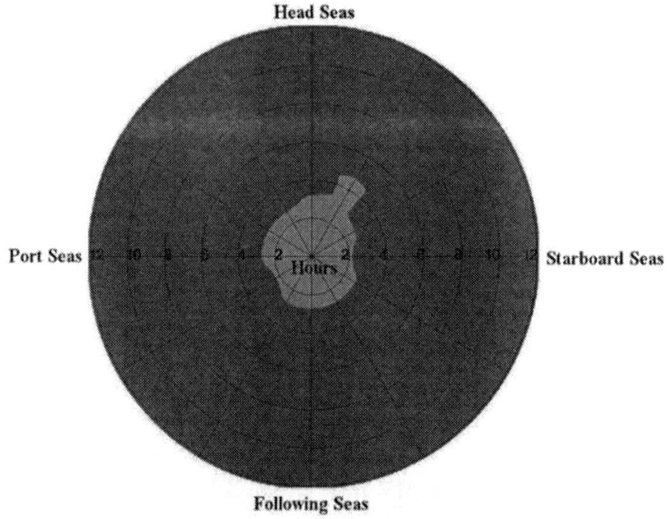

Figure 2: Plot indicating the time Sydney remained afloat in sea state 4 considering flooding through hull penetrations, internal damage and torpedo damage. Green indicates that the ship is afloat, and red indicates that the vessel has sunk [2]

Figure 2 provides one of the analytical outputs of the flooding analyses. The plot is in the form of a modified polar plot for sea state 4. The concentric circles represent time in hours, the centre being the time of the battle and the outer circle representing 12 hours after the battle. The direction that *Sydney* is travelling with respect to the primary sea swell is given by the sectors. Directly up the page is the head seas case where *Sydney* is travelling into the oncoming waves, whereas down the page represents *Sydney* and the waves travelling in the same direction. The green areas indicate when *Sydney* is still afloat whereas the red indicates that *Sydney* has foundered. Based on the German accounts, after the engagement *Sydney* sailed away in starboard beam seas. As can be seen from Figure 2, it is predicted that *Sydney* would have remained afloat for about 3 hours after the battle provided the sea state was constant at sea state 4 and the speed was constant; any variation in these parameters would have an effect on this time.

STRUCTURAL INTEGRITY

It has been asserted that the catastrophic event that caused the *Sydney* to sink was the separation of the bow from the rest of the ship. The German account did not indicate this was an issue and it is speculation following the discovery of the wrecks. To evaluate this, the structural integrity of *Sydney* in the forward part of the hull where the torpedo struck and the ultimate strength of the damaged section were evaluated against the structural loads imparted on *Sydney* by the sea. The Cooperative Research Ships (CRS) software code, PRECAL [5] was used to determine the magnitude of the structural loads experienced by *Sydney*. PRECAL is a 3-dimensional linear structural response and seakeeping code, which uses an advanced three dimensional panel method. This provided an analysis of the loads on the bow section of *Sydney* where the torpedo struck.

The software code, ULTSTR [6] was then used to calculate the ductile collapse (ultimate) strength of the forward section of *Sydney*. It is based on a variety of empirically determined strength of material solutions for the most probable failure modes for unstiffened and stiffened plate structures. By comparing the results calculated using these two computer codes, it was confirmed that the *Sydney* was intact after the battle and her bow did not break off on the surface but most likely broke off during the sinking process. Figure 3 shows an original drawing and the corresponding ULSTR model for frame 27 and Figure 4 shows a plot from the analysis [2]. Frame 27 was chosen as it was the closest frame to where the torpedo hit and where the major structural damage was and, fortuitously, it was a frame for which detailed drawings and structural data was available.

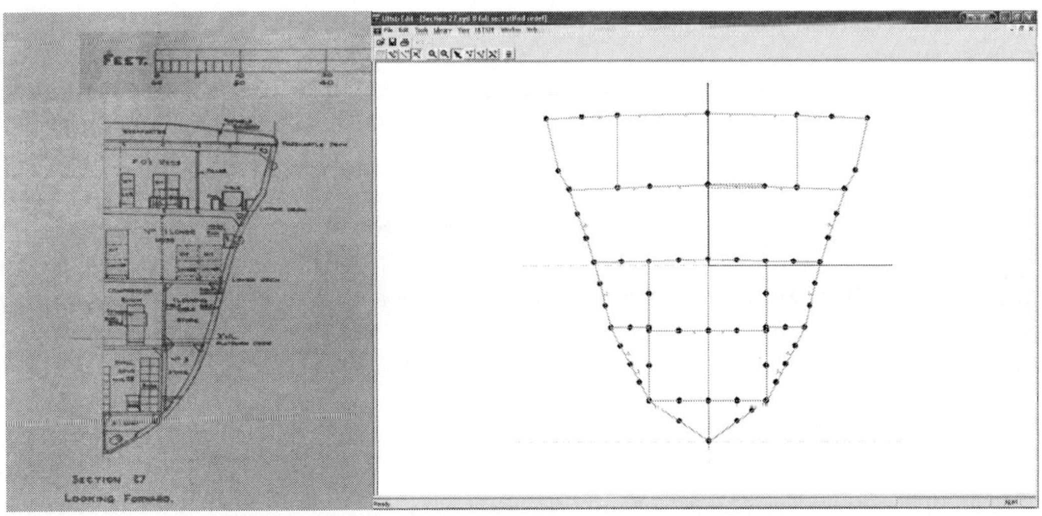

Figure 3: Diagram showing an original drawing for the structure of Sydney at Frame 27 and the corresponding ULTSTR model [2]

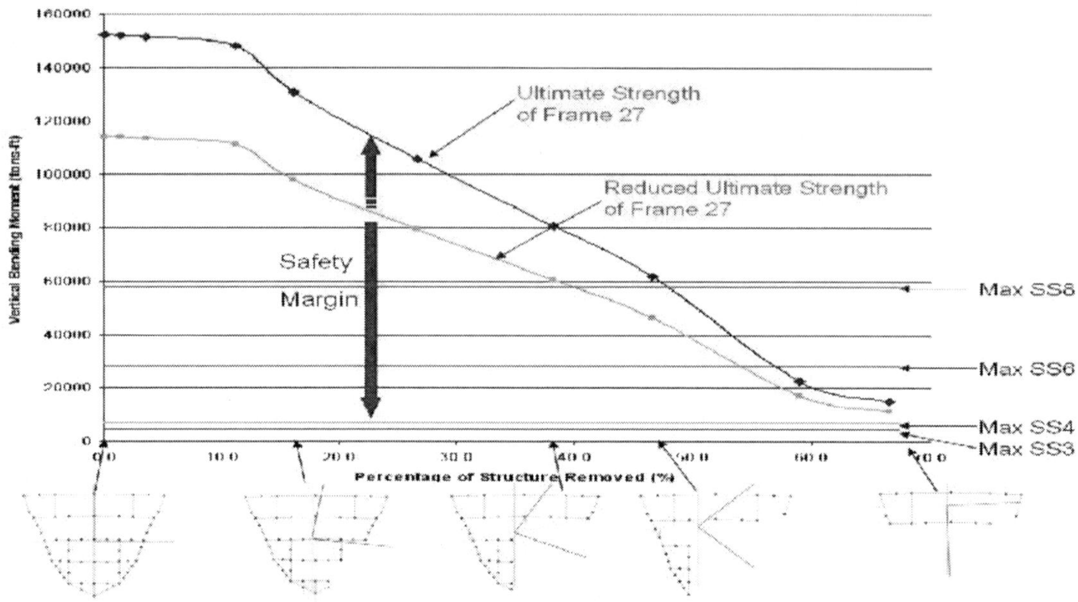

Figure 4: Comparison of ultimate strength bending moments due to sea loads at Frame 27 for asymmetric damage [2]

The horizontal lines on the graph show the loads imposed on frame 27 for various sea states. The capacity of the structure or the residual strength is also shown on the graph for varying degrees of damage. In all cases the capability far exceeds the loads and therefore it is highly unlikely that the bow broke off on the surface even with significant torpedo damage. It is therefore considered that the bow separated after *Sydney* sank below the surface.

FIRE DAMAGE AND DAMAGE CONTROL

The extent of fire damage on Sydney was determined by an analysis of the weapons damage, analysis of the many photographs recorded by the Remotely Operated Vehicle (ROV) (Fig. 5) and video footage as well as examination of fire damage from other naval incidents over the last 70 years [2]. DSTO has extensive experience in the analysis of fire spread on naval platforms from accidents and from field trials on decommissioned RAN ships. Details of the fire fighting equipment on board as well as the makeup of the damage control (DC) crew assigned to fight fires on *Sydney* were limited so other WWII ships' equipment, DC organisation and responses to fire or torpedo damage were studied to evaluate the extent to which the fires on *Sydney* could be controlled during and after the battle.

This analysis showed that damage control and the operation and control of the ship would have been severely hampered by extensive fires, smoke, internal damage and loss of electrical power. The main fire in the forward section would have grown to a large conflagration after the battle. Smoke from this fire would have been drawn into the forward machinery space due to the location of the ventilation inlets. Shell hits on the aft superstructure during the battle near the Turret would have ignited more fires. On the Platform Deck, the forward switchboard room and breaker rooms were hit with multiple weapons fragments. Loss of power would have resulted in reduced lighting, limited ventilation for smoke extraction, and lack of pumping ability for controlling flooding and supply of water for fire fighting. On the Lower Deck Damage Control stations 1 and 2 were hit by multiple shells and weapons fragments leading to a large number of casualties from the crew involved in repairing the ship and controlling the fires and flooding.

An accurate estimate of the number of casualties resulting from the battle damage was extremely difficult to determine. Limited information was available on the location of the crew at Action Stations and the dynamic nature of the battle, over a timeframe of 30 minutes meant that the exact location of the crew at any time will never be known. However, based on the available data, a likely scenario of the casualties that can be directly related to the battle engagement showed up to 70% of the crew would have been incapacitated by the effects of the battle weapons and being trapped in spaces due to fires. It would be expected that the remaining crew would be below decks in the aft end of *Sydney*, trying to control the ship from the aft machinery space and the steering compartment, managing the wounded and possibly trying to conduct damage control operations. The rolling of the ship as the sea state increased would have created a situation where physically moving around the ship would have become dangerous if not impossible.

Figure 5: Fire damage to officers galley (left) and recreation space/heads (right) [2]

Weapons damage would have caused considerable difficulty for the crew to move about the ship to escape or to conduct damage control activities. This would have significantly restricted the ability of the crew to shore leaking hatches and doors and the forward bulkheads. The crew would have been attempting to stop the ingress of water and remove flood water from the ship. But without electrical power, their ability to successfully achieve this would have been limited.

There was also damage to the firemain in the forward part of the ship. The firemain was a single pipe which was likely to have been severed in several places early in the engagement. This would have limited the fire fighting options in the forward part of the ship and any fires in this part of the ship would have been able to spread below decks. The openings in the hull from the weapons would also have provided oxygen to fuel the fires.

LIFESAVING AND SEARCH AND RESCUE

This battle was virtually unique as not only was the crew subjected to a constant barrage of 15 cm shells, they were continually being subjected to attack by 2 cm and 3.7 cm shells. It is estimated that the detonating 15 cm shells generated a combined total of over 200,000 steel fragments impacting on *Sydney*. Furthermore, the crew had to control the ship which had been holed by a torpedo. Over this time frame numerous fires were burning, spreading plumes of smoke and toxic gases over and throughout the ship.

An investigation of the lifesaving equipment on *Sydney* was carried out by detailed analysis of photographs of *Sydney* from the National Archives, the Australian War Memorial, numerous books, the RAN Sea Power Centre and private photographs obtained by the COI [2]. Original drawings of *Sydney* identified locations for life boats (there were 10 lifeboats of varying sizes providing a capacity for 375 crew) and 9 Carley floats (life rafts which provided capacity for another 274 crew). Searching through archival documents provided details of recorded changes, written and photographic, to equipment at various times. Photographs of the wreckage in the debris field from the ROV were used to determine damage to the life boats and the type and number of life boats that went down with the ship. The evidence of these images was, in large measure, consistent with descriptions provided by the German survivors of the battle. In short, none of the ship's boats could be launched following the battle due to the damage to the ships crane; the crane was required to lift all bar the two Cutters into the water. The Cutters were on davits and the weapons damage to the hull around the davit holders (Figure 6) suggests that these boats were likely blown off during the initial encounter and probably suffered significant damage rendering them useless for lifesaving. The Carley floats were less robust than the lifeboats and most would have been easily damaged during the engagement (most were in the midships area) or were blown off. This partially answered the question of why there were no survivors.

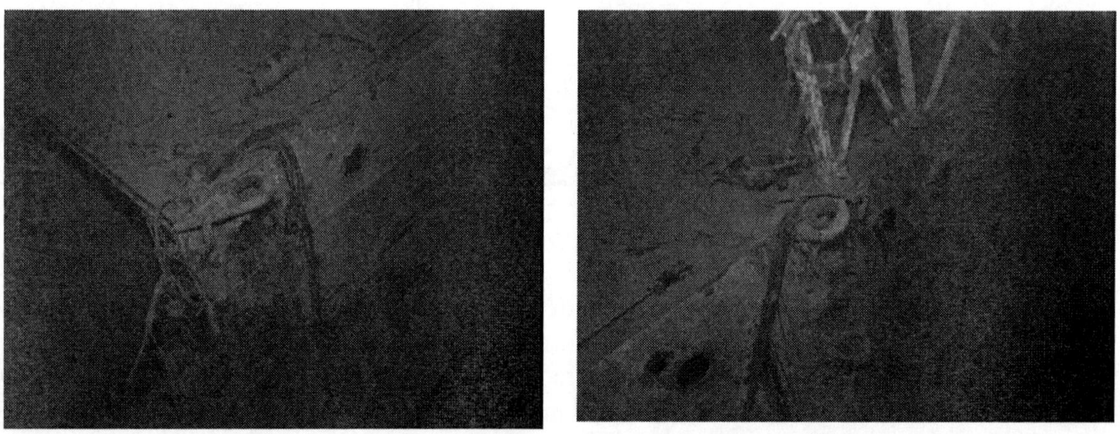

(a) *(b)*

Figure 6: Locating holes for davits for starboard Cutter (a) aft (b) forward [2]

The other part of the answer was provided through an analysis of the search and rescue process carried out by the aircraft and ships and the factors influencing the survival at sea of individuals. Again, this was based on an analysis of archival documents, RAAF reports, and various books. This was complemented with the use of modern search and rescue analysis to determine the likelihood that the search process conducted 67 years ago would have actually found any survivors. The results showed that it was highly unlikely that any search and rescue process would have spotted any individuals in the water (the sorties were conducted at a height that was appropriate to locate a damaged or disabled vessel but too high to possibly detect individuals in the water) and probably unlikely to have sighted any Carley floats. Interestingly, several small items were found floating in the ocean along with three Carley floats (one RAN type, two foreign) but all were sighted by the searching merchant ships and not by the searching aircraft. A Carley float was picked up by *Heros* but it was not sighted by any aircraft even after the location had been searched multiple times over a period of 3 days by multiple aircraft [7].

There was no evidence of any systematic order to abandon ship or to evacuate *Sydney*. There may have been some crew who fell into the water, were blown into the water due to weapons effects or who simply jumped into the water. They would have relied on their personal life belts for survival. These would have only provided a lifesaving capability for hours at the best [7]. Royal Navy data from WW II showed that a person in a Carley float would typically have survived for 3-5 days [8]. Modern prediction tools show that for a person in water at 20° C, survival time would have been around 20 hours [9]. The water temperature at the time of the encounter was 23 – 24° C [10] which increases the predicted survival time to around 45 – 60 hours which is still less than the time taken to start the search process.

The search and rescue process was delayed for 4 days and it was initially conducted to locate *Sydney* rather than bodies or lifeboats; this suggests that the air search was conducted at a height that was far too high to have detected bodies in the water or even life boats or Carley floats. Over the subsequent days, the air search was conducted at 1500 ft which was unlikely to have been low enough to detect bodies in the water and helps explain why no survivors were found.

Any bodies that drowned would have sunk probably to the seabed. Over time, the human body decomposes and usually reaches a point where it become buoyant and then floats back to the surface. In this situation, the bodies may have been as deep at 2500 m and their buoyancy is unlikely to have overcome the weight of water above them and they probably never floated back to the surface. It must also be remembered that 60 of the crew from *Kormoran* were thrown into the water during their evacuation and none of their bodies were found.

VISUALISATION AND THE ENCOUNTER

DSTO also produced a three dimensional visualisation model of *Sydney*. The aim was for the model to be used (a) to image particular locations on the ship, (b) to undertake 'forensic visualisations' of the vessel (e.g. investigating the interaction of potential shell trajectories with the vessel) and (c) to generate realistic animated visualisations of critical phases of the engagement between the two vessels. The original ship design drawings and Finding Sydney Foundation photographs were used extensively to develop this model.

The three-dimensional visualisation package 'Blender' was selected for use to generate the high-fidelity, computer-based model of the ship [11]. This is open-source software which benefits from having a very active user and developer base. While Blender is not an engineering drawing or development package, it can support development of three-dimensional models which are both precise and of high-fidelity. This was particularly important as "a picture is worth a thousand words" and it enabled one to see what happened to *Sydney,* how the battle sequence progressed and what the consequences were. Figure 7 shows part of the visualisation sequence and Figure 8 shows the use of the visualisation in forensic analysis. One of the ship's boats showed clear evidence that a shell had passed through it without exploding. By locating the point of impact of the shell on the ship, and inputting entrance, exit and impact locations into the visualiser, it was possible to determine the trajectory of the shell. This was consistent with a close-range engagement in which the ships were virtually abreast of each other.

Figure 7: Visualisation of Sydney showing rolling motion and deck edge immersion due to flooding [2]

Figure 8: Recreation of a shell hit on the port side Whaler which shows the very shallow angle of the trajectory consistent with the close range of the overall engagement [2]

SUMMARY

This paper has outlined some of the scientific tools and visualisation techniques that were applied to the investigation to support the Commission of Inquiry into the loss of HMAS *Sydney*. The paper presents the results of the analysis which goes towards the findings of why *Sydney* sank and why there were no survivors from *Sydney*.

Acknowledgements

The authors of this paper would like to acknowledge the following colleagues who completed much of the scientific analysis outlined above: Michael Buckland, Grant Gamble, John Jeremy, Tim Lyon, Pat McCarthy, Michael Skeen, Brett Morris, Brigitta Suendermann, and Terry Turner. Additionally the authors would like to acknowledge Jim Nicholls and Anthony Travers for their valuable assistance with the visualisations.
The authors would also like to acknowledge that the photographs in the paper were provided by the Finding Sydney Foundation.

References

1. HMAS Sydney II Commission of Inquiry - Department of Defence, http://www.defence.gov.au/sydneyii/tor.htm, March 2008
2. Buckland et al. (2009), HMAS Sydney II Commission of Inquiry. Report on the Technical Aspects of the sinking of HMAS Sydney and the HSK Kormoran. DSTO-GD-0559
3. *FREDYN Version 9.9*, (2008) [Computer software] Maritime Research Institute Netherlands
4. *PARAMARINE Version 6.0* (2008) [Computer software], Gosport, Hampshire, GRC Limited.
5. van't Veer, A. P. (2008) *PRECAL V6.4 Theory manual*, MARIN Report No. 17926-2-CPS, Maritime Research Institute Netherlands
6. Adamchak, J. C. (1997) *ULTSTR* (1997): *the revised program for estimating the collapse of ship hulls or hull components under longitudinal bending or axial compression,* Report No. NSWCCD-TR-65-97/21, West Bethesda, MD, Naval Surface Warfare Centre, Carderock Division
7. Olsen, W. (2000) *Bitter Victory – the death of HMAS Sydney*, Nedlands, WA, University of West Australia Press
8. Critchley, M. (1943) *Shipwreck survivors: a medical study,* London, J&A Chrurchill
9. Tikuisis, P. (1997) Prediction of survival times at sea based on observed body cooling rates, *Aviation, Space and Environmental Medicine,* 68 (5) 441-448
10. *NODC (Levitus) World Ocean Atlas* (1998), Boulder, CO, NOAA-CIRES Climate Diagnostic Centre
11. Blender Foundation (2008) *Blender Version 2.48A* [Computer software], Amsterdam, Amsterdam Blender Institute

The identification of the battle damage to HMAS Sydney II

M. Buckland, G. Gamble, M. Skeen, L. de Yong
Defence Science and Technology Organisation, Maritime Platforms Division, Melbourne, Victoria, Australia
Michael.Buckland@dsto.defence.gov.au

ABSTRACT

On the 19th of November 1941, the Royal Australian Navy Modified Leander Class light cruiser HMAS *Sydney II*, en route to Fremantle, intercepted the disguised German raider, the HSK *Kormoran*, about 100 nautical miles west of Steep Point off the coast of Western Australia. In the ensuing battle, *Sydney* was sunk with the loss of the entire crew of 645 men. *Kormoran* was subsequently scuttled with the loss of 81 men.

The wreck site of *Kormoran* and *Sydney* and the debris field on the ocean floor were surveyed by *Geosounder* and many hours of video footage of both wrecks was produced. This footage was analysed by the DSTO and the RINA to assess the extent and type of damage to both *Kormoran* and *Sydney*. This paper looks at the identification of the significant damage observed from the survey of the *Sydney* wreckage. This analysis involved examining and identifying the structural and equipment damage to both ships. Observations of the footage of the wreck site of *Sydney* identified 87 individual 15 cm shell hits and extensive fire damage. Most of the ships boats were found in various degrees of decay on the seabed.

INTRODUCTION

Following the discovery of the wrecks of *Sydney* and *Kormoran* in March 2008, the Defence Science and Technology Organisation (DSTO) in collaboration with the Australian Division of the Royal Institution of Naval Architects (RINA) were appointed to provide expert advice and opinion into the sinking of *Sydney*.

The only information available to identify the probable cause of the battle and sinking of *Sydney* were the images of the wreck site gathered by the *Geosounder*, the historical records of *Sydney* and the German eyewitness battle accounts. Based on this information personnel from DSTO and RINA surveyed, identified and recorded the likely causes of the damage evident from the wreck site. This data is recorded in the Report on Technical Aspects of the Sinking of HMAS *Sydney* and HSK *Kormoran* [1].

The initial discovery of the *Sydney* wreck site revealed a 20 m section of the bow had separated from the hull and now lies inverted in the debris field. The hull section of Sydney is sitting in a favourable position for identification of major damage, however during the 67 years it has spent on the seabed silt and corrosion has masked sections of the ship making it impossible to positively identify all of the damage sustained during the battle.

A significant part of the process of identifying the cause of the loss of *Sydney* was in the battle damage assessment of *Sydney*. This paper looks at the damage resulting directly from the battle including the impact damage of the shells on the structure of *Sydney*, the effects of blast and fragment damage of the shells and the structural damage resulting from the torpedo detonation. This damage needed to be distinguished from damage resulting from the sinking processes and impact with the sea bed. The other important information was the extent of damage resulting from the fires onboard *Sydney* as German accounts state that *Sydney* was ablaze after the action.

BATTLE DAMAGE ASSESSMENT

Torpedo Damage

The bow of *Sydney* lies inverted on the ocean floor on its starboard deck edge (Fig. 1a). The bow section shows significant damage due to it tearing apart from the main hull section and its final impact with the ocean floor. However the plating along the port side of the bow appears largely undamaged from these effects and from this section it is possible to conclude that the bow has broken off at frame 19.

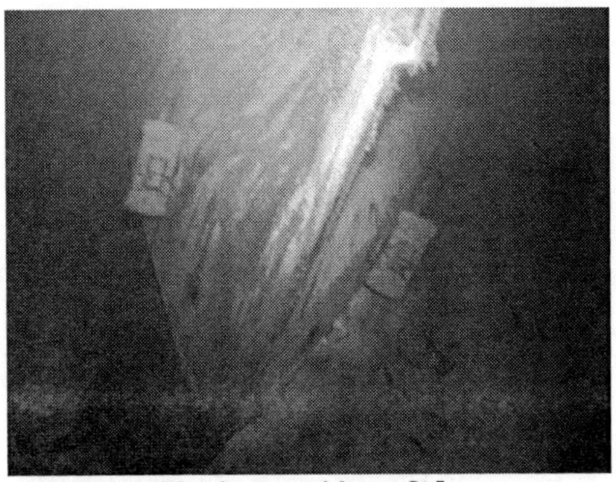

Figure 1a The inverted bow [1]

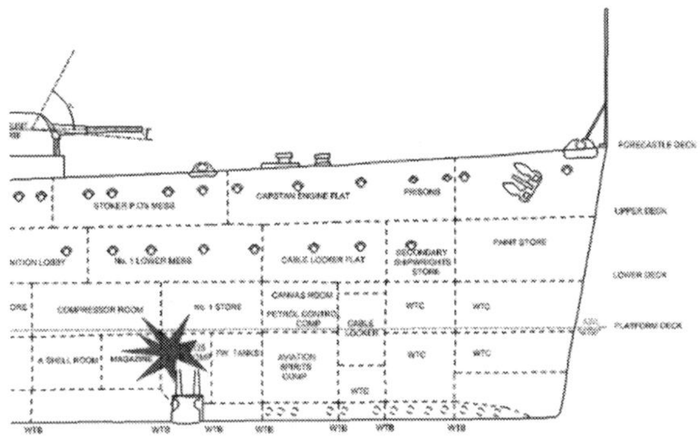

Figure 1b The probable impact point of the torpedo [1]

Damage close to the port side of the bow's keel shows the typical concave indentation of an explosion on this surface consistent with torpedo damage (Figure 2a). This damage indicates that *Sydney* was hit by a torpedo on the port side. As the hole edge is at frame 19, the detonation occurred aft of this frame, at a location between frame 19 and frame 30. The probable detonation location from the evidence would be in the vicinity of Frame 25, around the Type 125 Sonar Dome (Figure 1b).

The flaring on the starboard side panels shown in Figure 2b indicates that the damage from the torpedo detonation on the port side has progressed and either caused a hole or bulging on the starboard side of the hull. The fresh water tanks in the vicinity of the detonation would have been one mechanism that would have transferred the explosive energy from the port to starboard side resulting in this flaring.

The area of the bow failure on the main section of the wreckage of *Sydney* shows that significant secondary damage has occurred from a number of mechanisms; the torpedo explosion, the ripping of the bow from the main hull section during sinking, the force of the water flowing over the ship during sinking, and the impact of the main hull section on the ocean floor.

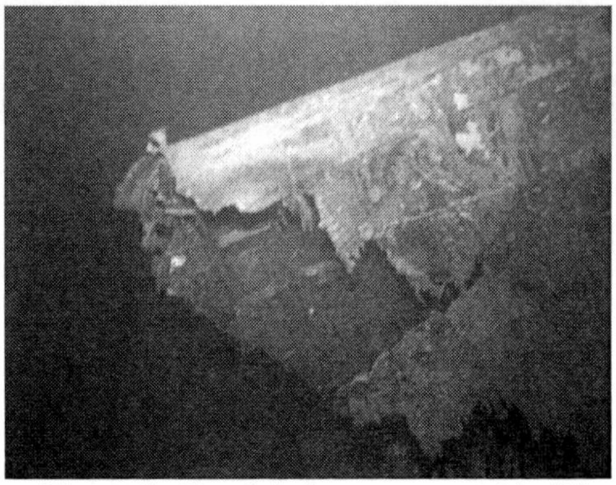

Figure 2a Break in bow on port side [1]

Figure 2b Break in bow starboard side [1]

Examination of the main hull section on the starboard side shows a large section of hull plate that has been torn and bent right past the 'A' Turret to frame 35. This damage may have been attributed to a detonation in the forward magazine, but this theory has been discounted by considering video footage of the ship which shows that the bulkheads surrounding the major magazine in this region to be intact.

The extent of damage to the port and starboard side of the ship due to the combination of torpedo and sinking damage makes it very difficult to definitively determine the size of the torpedo damage.

Without being able to define an accurate damage radius and detonation location it is difficult to analytically evaluate the size of the charge used in the torpedo, but the damage sustained is consistent with that expected from a single torpedo with a 280 kg warhead detonating close to the port side of the hull.

Shell and Fragment Damage

From historical records and German accounts, it was known that the *Kormoran* was armed with a number of 2 cm and 3.7 cm guns along with six 15 cm guns. The 15 cm guns were capable of firing a range of ammunition both point detonated and base fused. The survey showed that *Sydney* has been extensively damaged by shell hits on both her port and starboard sides.

Shell and fragment damage was identified by a number of distinguishing features and estimates were provided for the size of damage to *Sydney*. This was done using a CAD package to overlay a grid over the ROV images and use this grid to estimate the size of features in the photos (Figure 3). It is expected that these measurements have an accuracy of ±50 mm.

The measurement of the size of the hole and damage zones was important to identify the likely size of shell used to cause the damage. The location and likely compartments that the shells hit and entered were also matched with the General Arrangement drawings to identify likely damage to essential equipment.

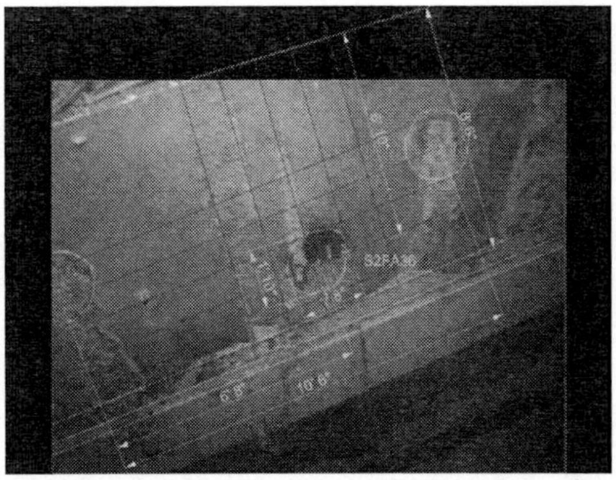

Figure 3 Image of *Sydney* showing CAD overlay used to take measurements [1]

Typical examples of damage resulting from shell hits to the turrets are shown in Figure 4. Figure 4a shows a weapon hit on 'X' turret where the shell has penetrated into the turret, the hole size is estimated to be 15 cm and has cleanly passed into the turret without signs of detonation on the surface. 'A' Turret received considerable damage from hits to the starboard side. Figure 4b shows the petalling damage resulting from a weapon which has detonated on impact with the starboard side armour.

Figure 4a Weapon damage to 'X' Turret [1]

Figure 4b Weapon hole in 'A' turret plate [1]

Critical hits for the *Sydney* were identified to both the Director Control Tower (DCT) and High Angle Control Station (HACS) as shown in Figures 5a and 5b. Both of these structures were critical for accurately firing *Sydney*'s guns. On the DCT there are at least one or possibly two hits to the support structure and another hit penetrating the main DCT. If any function in the DCT had remained operational after the hit to the port side, which is highly unlikely, the additional hits on the starboard side would have definitely put the DCT firing control permanently out of operation. The hit to the HACS would have permanently disabled its function.

Figure 5a Damage to DCT [1]

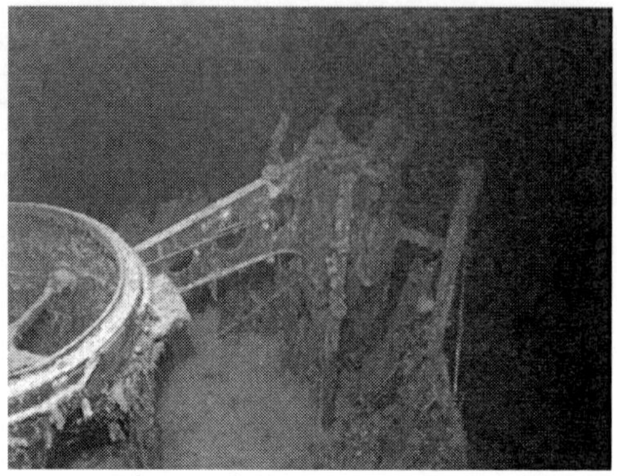

Figure 5b Damage to HACS [1]

The starboard side of *Sydney* has also received multiple hits along the hull, particularly around the armour plating protecting the engine room. The value of the armour plating around the engine space is demonstrated by a number of these hits where there is obvious shell impact but the shells have not penetrated the hull (Fig. 6a). A number of hits have penetrated the armour plating and Figure 6b shows a hit that has probably penetrated the armour and possibly detonated inside *Sydney*.

Figure 7a and Figure 7b show hits that have probably detonated on the outside of *Sydney* and the fragments from the detonation have still been able to penetrate the armour. The weapon hits shown in Figure 7a and **Error! Reference source not found.**7b would have caused some internal damage to *Sydney* but this damage would be much less than if the shell had detonated on the inside of the ship.

It is also interesting to note that a number of the hits on the armour plating appear to have impacted at an angle. For example, the hit shown in Figure 7b is likely to have been fired from a position somewhere forward of *Sydney*. This is evident from the direction of the fragment markings on the hull. It is not possible from the evidence available to predict the angle at which this shell hit *Sydney* with any accuracy.

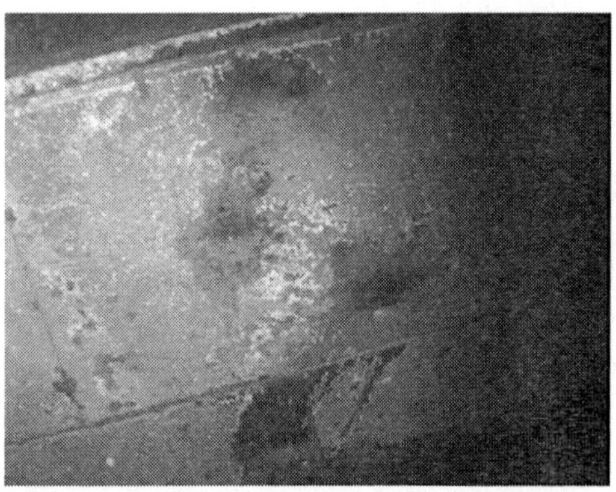

Figure 6a Non-penetrating shell damage to the armour plating[1]

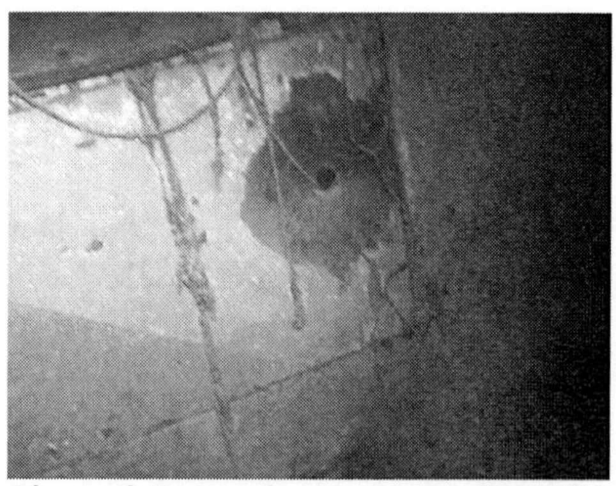

Figure 6b Penetration to armour plating [1]

Figure 7a Example of a shell that has probably hit *Sydney* on an angle [1]

Figure 7b Example of shell that has hit the armour with fragments causing the penetration [1]

Fire Damage

Evidence of fire is based on paint damage from the ROV video and still images. Often the areas affected by fire are accompanied by corrosion which will also occur where metal has been damaged from fragments. Corrosion can grow into rusticles due to the action of micro-organisms in the sea and these are often used as evidence of fire damage. In some locations it is difficult to distinguish corrosion due to organisms from that due to fire damage. However, large areas of paint damage on the ship where structural damage is relatively low indicate the location of fires which are likely to have caused damage to systems and disruption to ship operations. Based on observations of fire damage to other ships, a survey of fire damage to *Sydney* was undertaken. Examples of fire damage are shown in Figure 8.

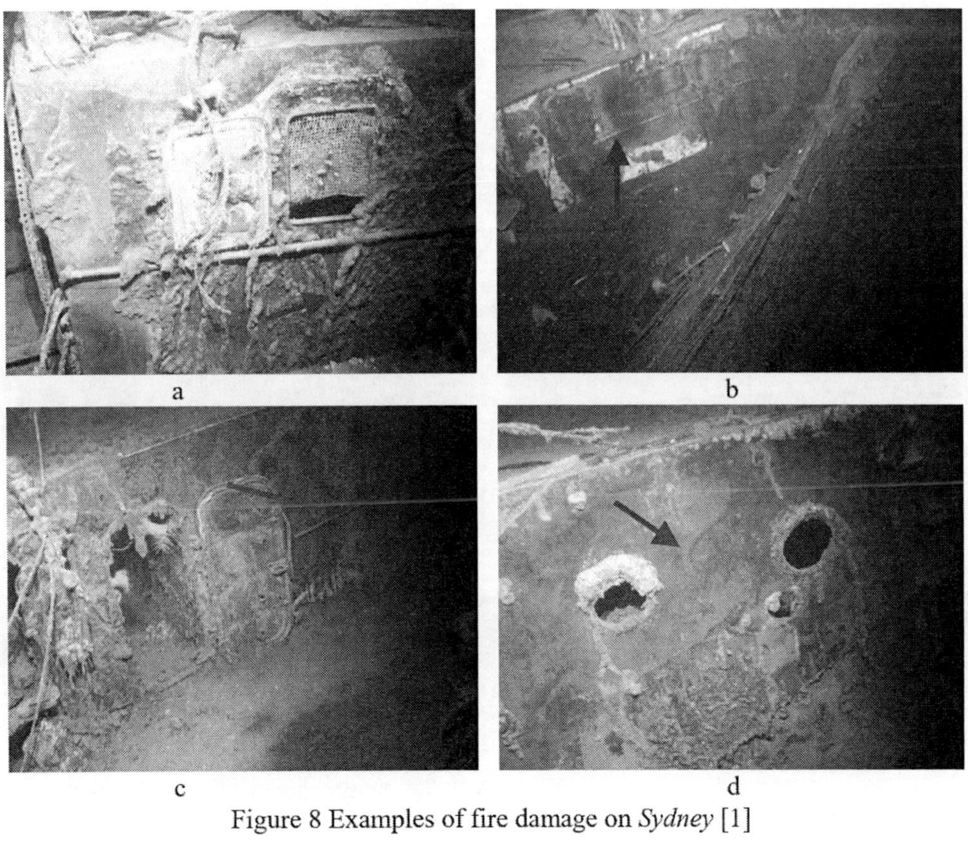

a b c d

Figure 8 Examples of fire damage on *Sydney* [1]

The extent of fire damage identified by the survey is illustrated in Figure 9. It shows the fire extended over the entire Bridge structure and across the breadth of the Forecastle Deck, and down the Lower Deck hull on the port side, and down to the Upper Deck hull on the starboard side. Fire is evident in the Midships below the aircraft platform and across the Upper Deck. Fire is also evident on the aft superstructure and Upper Deck on the port side and aft superstructure on the starboard side, below the aft control position. The Upper Deck is also burnt below 'X' turret. The starboard hull (Lower Deck) is burnt around a shell impact/penetration.

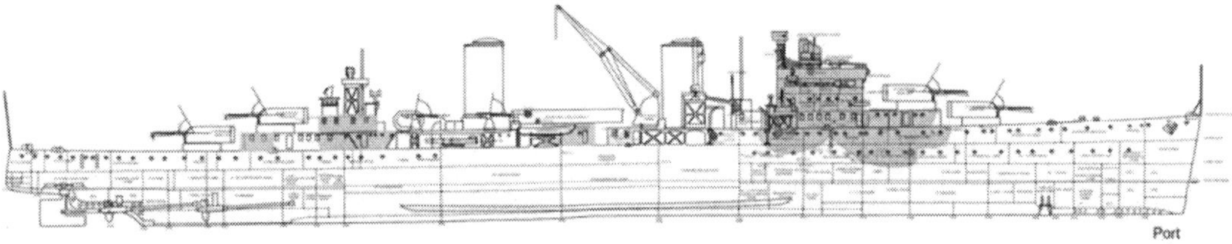

Figure 9 The areas of *Sydney* that were affected by fire damage are shaded in orange [1]

Other Structural Damage

The photographic images and video footage of *Sydney*'s wreck site contains evidence of structural damage that was due to the foundering process and/or the impact of the ship with the seabed. This section identifies the locations of this type of structural damage and provides brief explanations of the likely causes. It is important to identify this damage and disassociate it from the damage sustained during the engagement.

Separation of the Bow

At the start of this section it was noted that the bow of *Sydney* lies separately from the rest of the ship. The initiating cause was the hole associated with the torpedo damage. Upon reviewing the wreck images it is apparent that a significant part of the ship's forward structure is missing. The break in the Forecastle Deck can be identified and its approximate location is shown in Figure 10. The break lies approximately where the timber deck planking ended at frame 20. The bow section lies in the debris field upside down, but there is a considerable amount of other hull plating and deck structure also lying in the debris field.

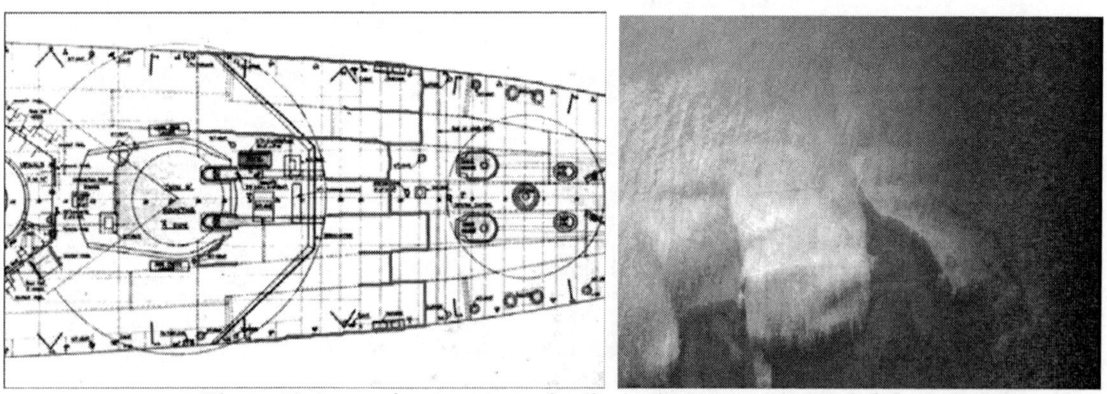

Figure 10 Approximate extent of splits in the Forecastle Deck [1]

Implosion Damage

Implosion damage is caused when an airtight container has pressure applied externally to a level at which the container collapses. Implosion damage has been evidenced on many ships during sinking, particularly those which sink rapidly before all spaces within the ship have filled with water. For example, the wreck of the cruiser USS *Quincy*, lost during the battle of Savo Island, shows very similar implosion damage to that evident on the wreck of *Sydney*. Figure 11a shows the aft section of the ship's Upper Deck on the starboard side. The Upper Deck has collapsed downwards and the side shell bent inwards over an extensive length, possibly as far forward as the watertight bulkhead at frame 154.

Figure 11a: *Sydney*'s Upper Deck on the starboard side [1]

Figure 11b. Quarterdeck bollards on the port side of the Upper Deck aft [1]

The quarterdeck bollards on the port side of the Upper Deck at frame 196 aft are shown in Figure 11b. The deck has been forced down several feet, exposing the capstan shaft on the ship's centreline and bending the ship's side slightly inwards.

The extent of this buckling and collapsing damage therefore extends from the stern through to about frame 154. The damage is evident on both sides of the ship. Figure 12 shows a plan of the Upper Deck marked with the extent of the collapsed region. Also shown in the diagram are two thick red lines indicating where there are splits between the Upper Deck and the 'Y' Turret barbette.

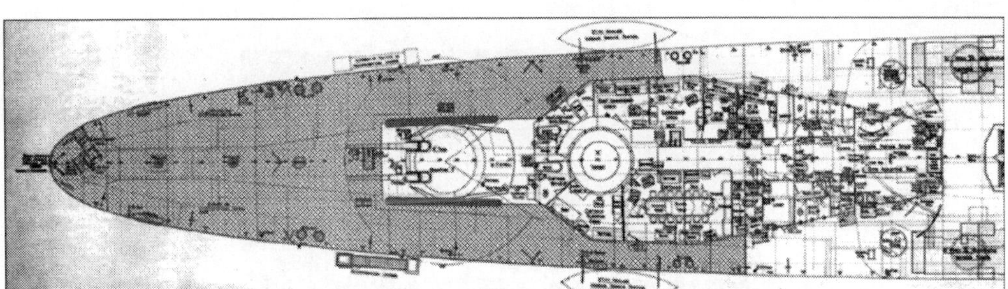

Figure 12 Upper Deck plan showing the extent of the collapsed region [1]

Impact with the Seabed

The main hull of *Sydney* shows some evidence of damage caused when the ship impacted with the seabed. The stern section shows some signs of collapse where it has come to rest on the seabed, shown in Figure 13a, although some of this damage may have been caused by implosion.

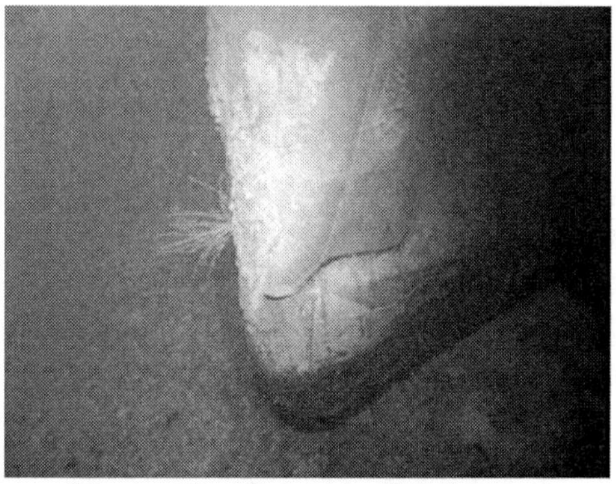
Figure 13a The stern of *Sydney* on the seabed [1]

Figure 13b The ship's port propellers on the seabed [1]

Moving further forward along the seabed there is evidence that the impact with the seabed caused major damage to the propellers, shafts and the associated supporting structure. One of three-bladed highly-pitched propellers is shown in Figure 13b. This is the port inner propeller and the shaft has been partly withdrawn by about 8–10'. The shaft bracket has separated from the hull and the fractured top arm is also visible. The propeller is partly buried in the sand or silt on the seabed.

Collision Damage

The final type of damage that is evident from the wreck site is that associated with collision damage. As *Sydney* was sinking several large pieces of wreckage broke away from the hull, including the masts and funnels, and collided with other parts of the wreck. The image in Figure 14 shows the forward screen of the superstructure near the Captain's Sea Cabin and the Bridge.

Figure 14 The forward screen of *Sydney*'s Bridge superstructure [1]

The front of the Bridge structure is collapsed and crushed above the Captain's Sea Cabin and the Navigating Officer's Cabin. It is evident that the Bridge has been crushed. The starboard side of the Bridge has been torn away and the port side bent in. This severe damage almost certainly occurred as the ship sank, since the bullet-proof roof over the compass platform on the Upper Bridge Deck is lying intact in the debris field resting against the DCT. It is unlikely to have survived intact if it had not left the ship before the damage to the Bridge and the Upper Bridge Deck occurred. It is possible that some large piece of wreckage hit the Bridge as the ship sank. A number of collisions of this nature during the sinking would account for some of the damage to topside structure of the ship.

Ships Boats

Examination of the *Sydney* wreck site shows that there is no evidence of any of the Carley floats in the debris field or any evidence of any still attached to the wreck. However a number of the ship boats still remain in the debris field (Figure 15). The boats are in varying degrees of decay. The boat shown in Figure 15 appears to be in very good condition with only limited decay of the wooden planking and the frame. The equipment inside is mostly missing and has presumably fallen out during the sinking. The paint on the outside of the boats is in good condition and again there is little evidence of any fire damage.

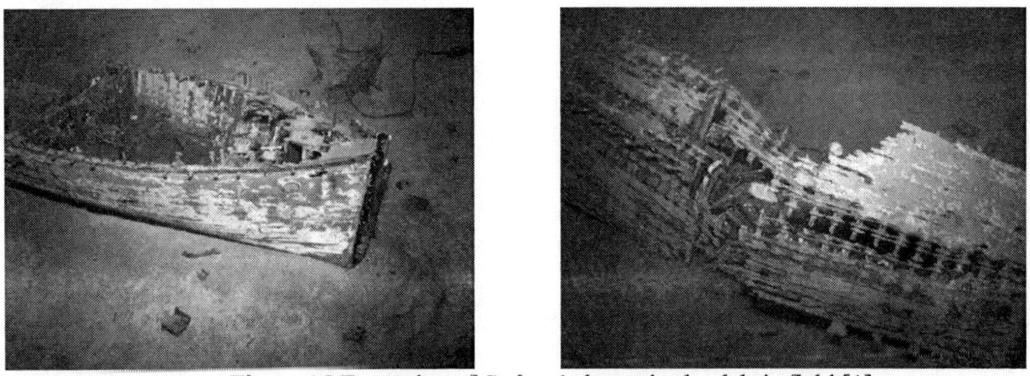

Figure 15 Examples of *Sydney*'s boats in the debris field [1]

CONCLUSIONS

This paper has outlined some of the observed damage identified by DSTO and RINA in surveying the footage of the *Sydney* wreck site that was gathered by *Geosounder* in March 2008. The information gathered was used as a basis for a detailed analysis of the causes of the loss of *Sydney* [1].

ACKNOWLEDGEMENTS

The authors of this paper would like to acknowledge the extensive analysis of the wreckage carried out by Stuart Cannon, John Jeremy, Brigitta Suendermann and Patrick McCarthy. Additionally the authors would like to acknowledge the Finding Sydney Foundation for the use of the photographs.

REFERENCES

[1] M.Buckland, S.M. Cannon, L. de Yong, G. Gamble, J.C. Jeremy, T. Lyon, P.McCarthy, B. Morris, R.A. Neill, M. B. Skeen, B. Suendermann and T. Turner. Report on Technical Aspects of the Sinking of HMAS *Sydney* and HSK *Kormoran*. DSTO-GD-0559, Jan 2009.

HSK Kormoran

M. Buckland and S.M. Cannon
Defence Science and Technology Organisation, Maritime Platforms Division, Melbourne, Victoria, Australia
Michael.Buckland@dsto.defence.gov.au

ABSTRACT

Following the discovery of the wrecks of HMAS *Sydney II* and HSK *Kormoran* in March 2008, the Defence Science and Technology Organisation (DSTO) in collaboration with the Australian Division of the Royal Institution of Naval Architects (RINA) were appointed to provide expert advice and opinion to assist in the Commission of Inquiry.

The aim of this paper is to describe *Kormoran's* construction and in particular to provide estimates of the time to de-camouflage and fire her weapons. The wreckage of *Kormoran* was analysed to assess the extent and type of damage sustained during the battle. It is concluded that *Kormoran* sustained extensive damage which is largely attributed to the consequence of storing mines onboard and to that sustained during the scuttling process.

DESIGN

Kormoran (Ship 41) (shown in Figure 1) began life as a cargo liner built for the Hamburg-Amerika Line, also known as Hapag, and had been launched under the name of Steiermark from the Deutsche Werft at Kiel. Steiermark was a new type of passenger ship intended for the East Asia run. Steiermark had completed ship trials before the war, the outbreak of which had prevented her being taken into service.

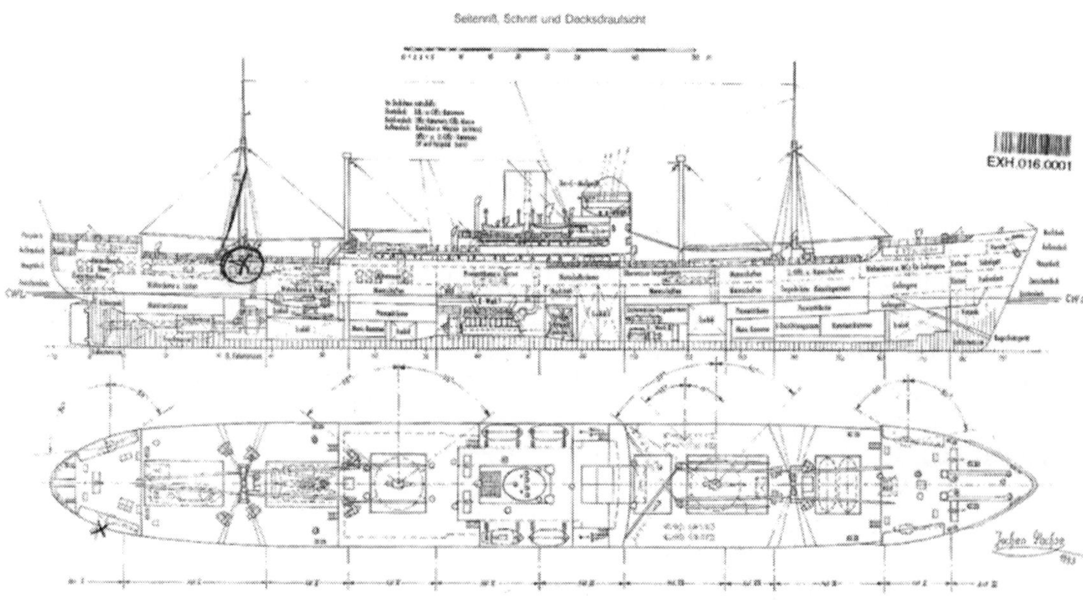

Figure 1: General Arrangement of Kormoran [1]

The principal particulars of *Kormoran* were:

Displacement:	19,900 tons
Dimensions:	167.5 m LOA, 157.0 m LWL x 20.2 m beam x 8.5 m draught
Machinery:	Four Krupp-Germania 9-cylinder four-stroke diesel engines delivering 3,600 BHP Two propulsion electric motors rated at 6,370 SHP each Total SHP 12,740 for 17.5 knots (full load, clean hull)
Bunkers and Radius:	Diesel fuel 5,200 tons; range 84,500 miles at 10 knots; 74,000 miles at 13 knots; 50,000 miles at 17 knots
Protection	Splinter protection for chart-house, the helm and engine-room telegraph posts
Armament:	Six 15 cm guns in single mounts, two 3.7 cm A.A. in single mounts, five 2 cm cannon in single mounts, several 7.92 mm machine guns; six 21" torpedo tubes in two twin-tube above-water mounts and two submerged tubes.; two aircraft and one light fast motor boat
Complement:	400

MAIN ARMAMENT

Kormoran's main armament consisted of six 15 cm guns. The Nos. 1 and 2 guns were fitted under the Forecastle Deck, No. 1 on the starboard side and No. 2 on the port side. The Nos. 3 and 4 guns were in Holds 2 and 4 respectively. The Nos. 5 and 6 guns were fitted under the quarterdeck, No. 5 on the starboard side and No. 6 on the port side. With this arrangement *Kormoran* was limited to a full broadside of four guns rather than six. The location and firing arcs of the six 15 cm guns are shown in Figure 1 and Figure 2.

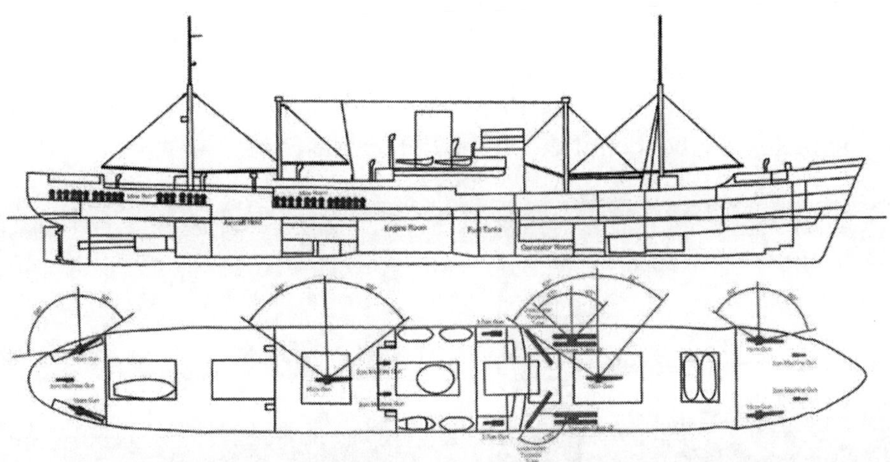

Figure 2: Weapons identified within the wreckage are marked as green and match those presented in reference material [1].

No.1 and No.2 Gun

The two 15 cm guns mounted beneath the Forecastle were concealed behind hinged, counterweighted steel plates that opened upwards. Figure 3 shows the probable design of the camouflage mechanism based on the description of *Kormoran* crew members [2], the underwater photographs of *Kormoran* and photographs of other German auxiliary cruisers.

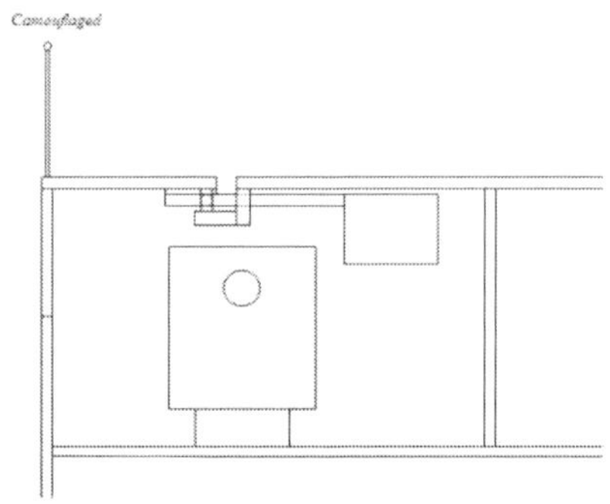

Figure3 a: No 1 gun in the Camouflaged condition [2]

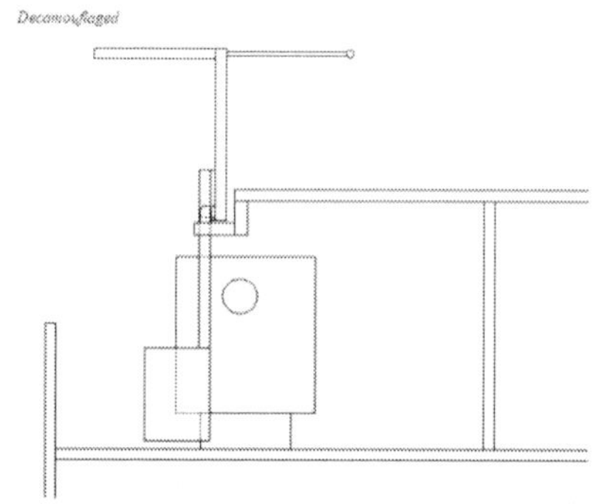

Figure3b: No 1 gun in the De-camouflaged condition [2]

The key feature of the camouflage equipment for the forward guns was the large counter-balanced weights. These enabled the guns, which would have been stowed pointing in the forward direction to be to be exposed within two seconds. In the estimate of the time taken to fire it was assumed that there was sufficient room to begin training the gun at the same time the de-camouflaging process began. Therefore the estimate of the time taken from the order to de-camouflage and fire was between 15 and 18 seconds for the forward 15 cm gun.

Figure 4 and Figure 5 show *Kormoran's* No.1 and No. 2 guns. The damage seen is most likely due to damage from the collapse of the masts and/or the sinking process. There are no signs of weapons damage sustained from *Sydney's* shells. The hinges for the camouflaging plates can be identified on the Forecastle Deck.

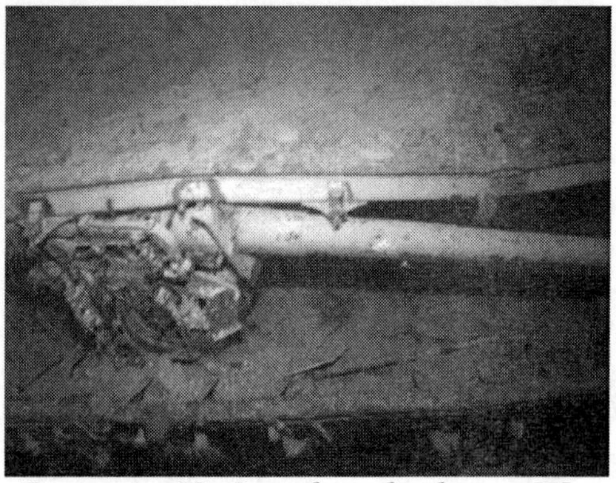

Figure 4: No.1 starboard side gun [2]

Figure 5. No.2 port side gun [2]

No. 3 and No. 4 Gun

The two 15 cm guns were mounted on the centreline of the ship in Nos. 2 and 4 cargo holds, and were concealed behind collapsible hatch coamings. It can be seen from both Figure 1 and underwater photographic evidence that these guns were fitted inside pits or tubs that were lower than the surrounding deck. This was done to ensure that the guns did not protrude above the standard height hatch coamings. It did, however, restrict how far the guns could be depressed.

Figure 6 is a photograph of *Kormoran's* swimming pool concealed inside a hatch (most probably the hatch over Hold No. 6 which also contained the Leichte Schnellboote (LS3) motor boat. It can be seen that a light canvas tarpaulin was used to cover the hatch and conceal the LS3 and swimming pool underneath. It can be reasonably assumed that a similar light canvas tarpaulin was used to cover the hatches over Nos. 2 and 4 cargoholds and conceal the 15 cm guns inside.

Figure 7 shows the probable design of the camouflage system.

Figure 6: Kormoran — Swimming Pool [3]

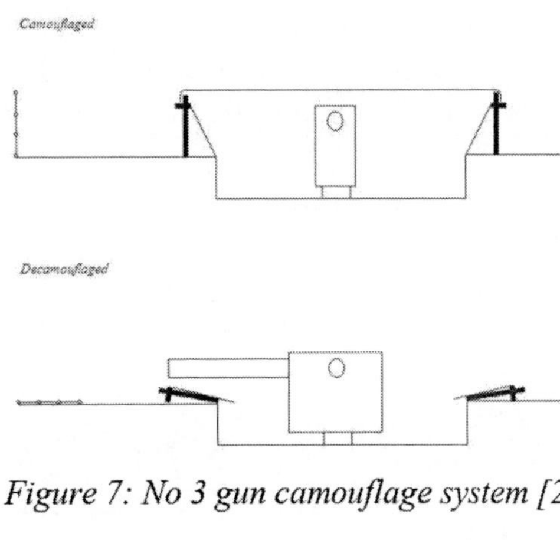

Figure 7: No 3 gun camouflage system [2]

The major difference between the de-camouflaging equipment for these guns is that the canvas tarpaulin covering the hatch needed to be removed and the collapsible hatch coamings needed to be dropped (see Figure 7). This was estimated to have taken five and two seconds respectively. The guardrails also needed to be dropped, however it was not essential as the guns could be fired with these in situ.

There was sufficient room within the cargo holds for the guns to be trained to 45° to starboard prior to the de-camouflage command. Therefore, these guns would have only had to train 45° to bear upon the target abeam. Using the same rates for training this equated to 6 seconds for the best rate and 8 seconds for the conservative rate. In total it was therefore estimated that the time taken to de-camouflage and fire the guns within the holds was between 16 and 18 seconds.

Figure 8: Kormoran's No. 3 15 cm gun in hold [2]

No.5 and No.6 Gun

The two 15 cm guns mounted at the stern beneath the poop deck were concealed behind hinged, counterweighted steel plates that opened upwards in a similar manner as for the No. 1 and No. 2 guns.

The major difference between the aft starboard gun and the forward starboard gun is the gun's orientation prior to the de-camouflage order. In this case the gun would have a 25° degree bearing when stowed and would only need to be trained 65° to bear on a target abeam. Therefore the total time from the order to de-camouflage and open fire would be shortened to 12 to 14 seconds depending upon the training rate used.

3.7 cm Anti-aircraft Guns and 2 cm Cannon

Two 3.7 cm army anti-aircraft guns were installed on port and starboard sides by the bridge and five 2 cm cannon were fitted, two forward on the Forecastle, two amidships (at the aft end of the superstructure), and one aft on the quarterdeck. The locations of these guns are shown in Figure 2.

The five 2 cm cannon were concealed below the deck and raised hydraulically. Figure 9 shows the forward port side 2 cm gun, with the camouflaging cover open next to it. The starboard figure shows the gun platform is separate from the deck allowing for the gun to be raised from below the deck.

Figure 9a: 2 cm gun fwd port side [2]

Figure9b 2 cm gun fwd starboard gun [2]

TORPEDOES

Kormoran was equipped with six torpedo tubes. Two sets of twin tubes were mounted port and starboard above the waterline and forward of the bridge structure. The tubes could be trained outboard using a cogwheel mechanism fitted with ratchets. The crew had to manhandle the tubes into the correct notch and then pull the ratchet forward. The other two torpedo tubes were mounted below the waterline in the No. 3 hold immediately forward of the bridge. The location and firing arcs of the six torpedo tubes are shown in Figure 2.

Deck mounted Torpedo Tubes

The deck mounted tubes were concealed behind steel plates and had to be pushed into the firing position after the covering plates had been raised. Figure 10 shows the probable design of the camouflage mechanism.

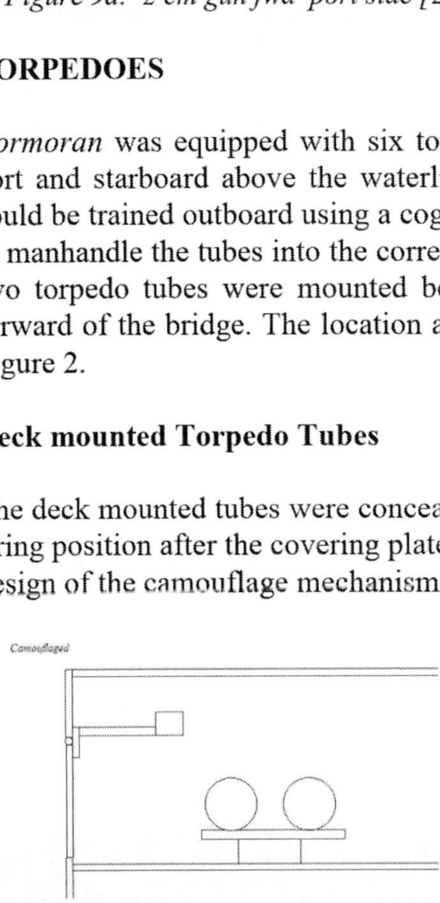

Figure 10a: Deck mounted torpedo tubes: Camouflaged [2]

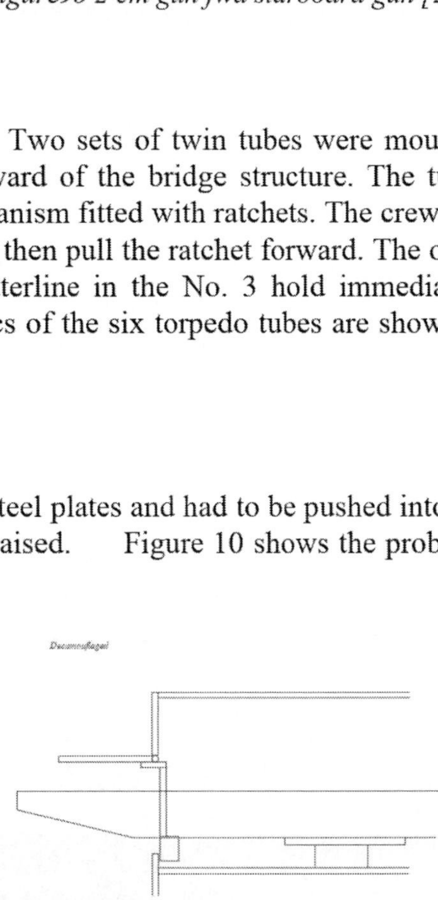

Figure 10b. Deck mounted torpedo tubes De-Camouflaged [2]

Figure 11 shows the flaps used to camouflage the starboard and port side above water torpedo tubes. Unfortunately the ROV was unable to view into the space below the flap to observe the torpedo tubes.

As with the guns it is assumed that the torpedoes were loaded and set with the depth and speed settings prior to the firing order being given. It is also assumed that the bearing of the target was known.

Figure 11a: Above water torpedo flap
Starboard side [2]

Figure 11b: Above water torpedo flap
Port side [2]

The time taken to raise the camouflage concealment flap was estimated to be about two seconds. It was assumed that the training of the torpedo tubes was completed with manual labour. As the torpedo tubes were trained forward prior to the de-camouflage order they would have needed to be trained 90° which was likely to take 30 seconds. Therefore the total time required to fire a torpedo was 32 seconds. This time would be larger if *Kormoran* was required to alter course to bring on a more favourable bearing. A change of about 10° may have taken about 10 seconds.

Fixed Underwater Torpedoes

Kormoran was fitted with a fixed underwater tube to port and starboard, mounted at an angle of 35° abaft the beam (see Figure 12). These tubes were not fitted with modern extension sleeves, with the result that the vessel's speed affected the torpedo as soon as it left the tube. There was therefore a danger that it might be forced off course. Further, they could not be used at a ship speed of greater than three knots.

All activities relating to the firing of a fixed underwater torpedo could be carried out prior to the de-camouflage order. Therefore these could be fired instantly, assuming *Kormoran's* speed was low enough and the target was on the correct bearing.

Figure 12a: Below water torpedo tube
Starboard side [2]

Figure 12b: Below water torpedo tube
Port side [2]

Torpedoes Used

Kormoran carried fifteen torpedoes. These torpedoes were most likely 53.3 cm G7a T1s. The 53.3 cm G7a T1 torpedoes entered service about 1938 and had a 280 kg Hexanite explosive charge. They had a range of 6,000 m at 44 knots, 8,000 m at 40 knots and 14,000 m at 30 knots.

It is unlikely that *Kormoran* had any fire control system for the torpedoes (such as fitted on *Sydney's* Bridge). It is probable that the Torpedo Officer had a simple sight on the Bridge and telephone communication with the tubes.

Mines

Kormoran carried 360 (EMC) moored type contact mines (see Figure 13) and 30 (TMB) magnetic ground mines. The former could be deployed in coastal waters up to the 200 fathom line, but the latter, which rested on the seabed, could only be laid in water with a maximum depth of 10 fathoms.

The mines were stowed on the mine deck — a specially constructed deck which ran aft from No. 4 hold to the quarterdeck. The mine deck, situated one deck below the Upper Deck, ran for almost a third of the ship's length. The mines were laid through two concealed doors in the stern. The mine deck and concealed stern doors are shown in Figure 1.

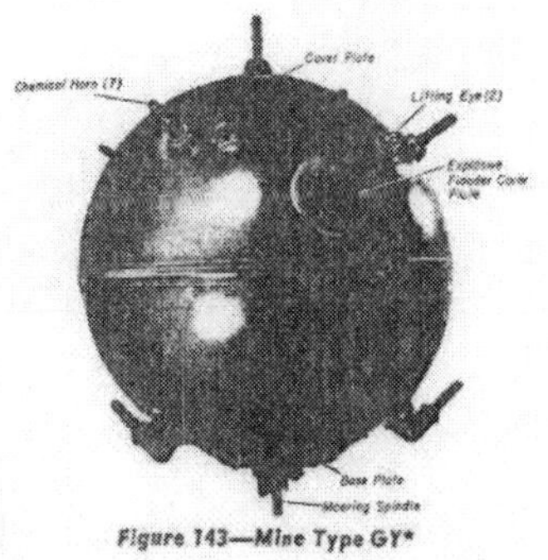

Figure 13: German EMC mine [4]

HSK Kormoran Wreck Site

The *Kormoran* wreck site consists of two main features; the first being a large section of the *Kormoran* hull which remains intact on the seabed. This section is approximately 90 m to 95 m of the *Kormoran* hull and coincides with the un-hatched section shown in Figure 14. The bow is intact up to the start of the superstructure.

The bow section shows obvious damage from the sinking process and the effects of its masts falling onto the upper decks. However, no obvious signs of weapons damage can be identified along the hull section which remains intact.

The second feature consists of the large pieces of the aft section of the hull and superstructure which are scattered in a tangled mess within the debris field. No shell weapon damage can be derived from this section of the hull and no individual weapon event can be distinguished from the wreckage except that the damage was from a extremely large detonation.

The *Kormoran* wreck does provide sufficient information to enable the confirmation of the identity of the German Merchant raider, its weapons and camouflaging arrangements.

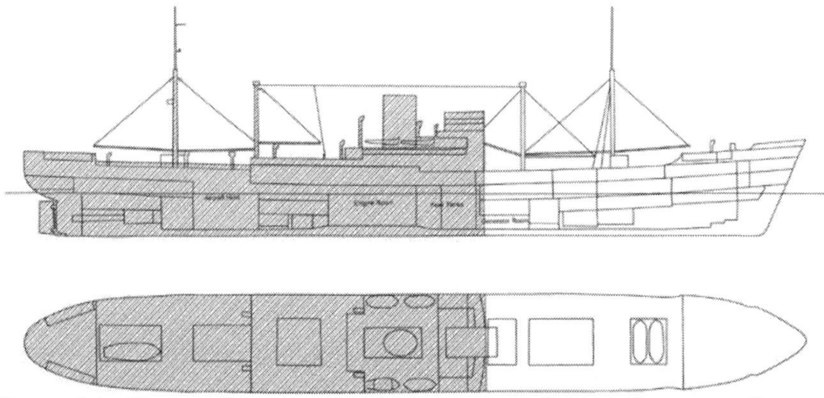

Figure 14: Extent of damage to Kormoran as it lies on the ocean floor [2]

HSK Kormoran's Vulnerability to Shell Fire

Kormoran was built as a merchant vessel and no information is available of any protective armouring of the structure or magazines. *Kormoran* was well armed and its distribution of weapons over the entire length of the ship meant *Kormoran's* war fighting capability was extremely difficult to knock out with the use of 4" or 6" shells. To stop this firepower would require a minimum of four accurately placed shells extending the length of the ship to disable the four 15 cm guns able to fire broadside. *Kormoran* would then still have had the torpedo tubes both above and below the waterline to use as its major offensive weapon.

The lack of armour to the engine rooms made these positions vulnerable to Armour Piercing (AP) shell hits penetrating into the engine room and disabling the mobility of the ship.

Kormoran's greatest vulnerability to an High Explosive (HE) shell was the large stowage of mines which the ship could carry. *Kormoran* could stow 360 EMC mines each having an explosive weight of 290 kg, and these were stored above the waterline in a large compartment space extending along most of the aft structure. The location of these mines above the waterline made *Kormoran* vulnerable to above water weapon attack.

Detonation of one of these mines would result in a mass detonation of over 100 tons of explosive in this area and totally destroy the rear of the ship. Figure 15 illustrates the extent of the damage as predicted using the DSTO vulnerability assessment code XVAM.

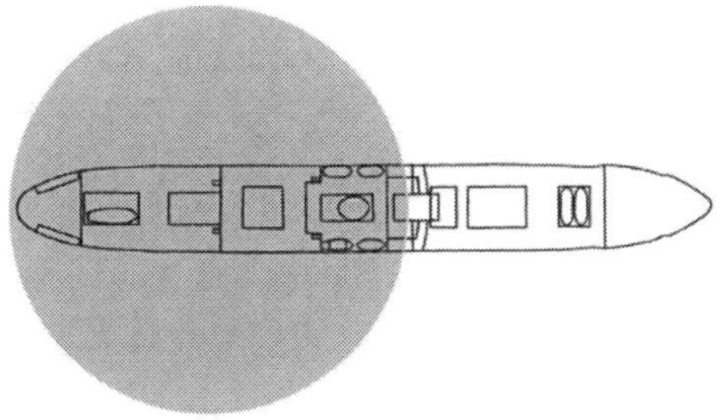

Figure 15: Extent of damage resulting from mass detonation of mines [2]

CONCLUSIONS

This paper has outlined the design features of *Kormoran* and provided an estimate of the time to de-camouflage her weapons and fire at *Sydney*. It must be appreciated that this is an estimate and the use of a highly trained, experienced crew with the element of surprise would have added to the efficiency with which *Kormoran* operated.

Evidence from the wreck site suggests that little damage was inflicted on *Kormoran* by *Sydney* during the battle. It was highly likely that the damage experienced by *Kormoran* was due to a limited number of well placed shots fired by *Sydney* that eventually lead to the German crew scuttling *Kormoran* by detonating the mines carried by *Kormoran*. The consequences of this were devastating.

ACKNOWLEDGEMENTS

The authors of this paper would like to acknowledge the extensive research on German armed raiders and the analysis that was carried out by Tim Lyons to predict the de-camouflage times. Additionally the authors would like to acknowledge the Finding Sydney Foundation for the use of the photographs.

REFERENCES

[1] HSK Kormoran showing armament, EXH.016.0001, [Drawing] Sydney, NSW, HMAS Sydney II Commission of Inquiry.

[2] M.Buckland, S.M. Cannon, L. de Yong, G. Gamble, J.C. Jeremy, T. Lyon, P.McCarthy, B. Morris, R.A. Neill, M. B. Skeen, B. Suendermann and T. Turner. Report on Technical Aspects of the Sinking of HMAS Sydney and HSK Kormoran. DSTO-GD-0559, Jan 2009.

[3] Schmalenbach, P. (1979) German raiders: a history of auxiliary cruisers of the German Navy, 1895-1945, Annapolis, Md, Naval Institute Press.

[4] US Navy Department t (1945) Mine identification manual (Second revision), Ordnance pamphlet 898, US Navy Department.

Analysis of the Economy of Fuel Cells Used in Merchant Ships

Lars Erik Mangset, Tore Longva, Tomas Tronstad, Det Norske Veritas

DNV Research & Innovation, Program for Maritime Transport, Veritasveien 1, N-1322 Høvik, Norway. Tel. +47 6757 9496, tomas.tronstad@dnv.com

Abstract

A shift towards more environmentally sustainable shipping is likely to open up for new technology in the maritime sector that is economically competitive to the current incumbent technology represented by diesel engines/generator sets. Marine fuel cell concepts are a promising new technology, which can offer several technical and operational advantages. This study assesses the competitiveness of marine fuel cell applications operated on Liquefied Natural Gas (LNG) relative to diesel generator sets, with focus on economic and environmental performance. With regard to the latter, with a carbon containing fuel such as LNG, fuel cells reduce CO_2 emissions by about 30% at maximum continuous rating and even more at partial load levels. NO_x and SO_x emissions are reduced up to 100%. As to the economics, the current higher investment costs of fuel cells are compensated by lower fuel costs (LNG). The modelled results show that with today's high fuel oil price (2008), an energy efficient technology such as fuel cells, will be economically competitive even with capital expenditures double that of the incumbent technology. CO_2 taxation is expected to have less influence on the profitability.

1. Introduction

There is a great focus on making shipping more environmentally sustainable, while maintaining cost competitiveness. The context of climate change has increased the political scrutiny towards the maritime sector (EU, 2002; EU, 2007; EPA, 2007) which forms expectations towards a stricter environmental tax regime, either through direct taxation or a quota based trade mechanism. Although emission abatement technologies applied in combination with traditional diesel power exists, particularly with respect to mitigating NO_x and SO_x emissions, the market pull for long term environmental sustainable solutions is likely to entail new technological paradigms, which increase energy efficiency and emit less CO_2.

A possible move toward new technological paradigm will require any emerging technological rival to be competitive with regard to economics, performance and operability. This paper assesses the competitiveness of fuel cell concepts with electric drive systems installed in three case ships; offshore supply vessel, dry cargo ship and container ship. The fuel cell concept comprises fuel cells running on liquefied natural gas (LNG) which is assessed relative to what is termed the incumbent technology – diesel generators running on medium distillate oil (MDO). Competitiveness is assessed based on economical and environmental performance. Operability issues are commented on a qualitatively level only.

The study includes quantification of some of the main future uncertainties, the first being a future environmental tax regime and/or incentive mechanisms to spur emission reductions. Secondly, the development in future fuel prices, which currently represents between 20 and 60 percent of the overall cost in shipping (Corbett, 2006), is assumed and modelled. Thirdly, future technological development for fuel cells gives expectations for increased commercial viability, and the study quantifies price and operational costs in this regard.

1.1. Marine fuel cell applications

In short, a fuel cell is similar to a battery, except that a fuel cell produces electricity for as long as it is fed with an oxidant, i.e. a fuel. Its working principle includes an electrochemical process which converts the chemical energy stored in a fuel directly to electrical energy. Fuel cells became best known to the public through the high-cost applications onboard the Apollo space expeditions in the 1960's (Karin et al., 1999). However, the research and development activities undertaken since the 1960's has now matured to a level where some fuel cell types are on the brink of commercialisation. On this basis, several pilot and demonstration projects have been undertaken and promising developments in this regard make it attractive to consider marine fuel cell applications. The technology is continuously improving, where the different fuel cell types are heading for different market segments. The low temperature cells, of which most known is the PEM cell operated on pure hydrogen, is directed towards transport use, mainly in the automotive sector. The direct methanol fuel cell, not so different from the PEM cell, is being developed mainly for portable electronic devices, such as cell phones and laptops. For the high temperature cells, capable of operating on a multiple of fuels including hydrogen, natural gas, methanol, landfill gas, biogas, LPG etc, application developments tend to focus on stationary power use, such as base load power production.

The promising advances in the many development and demonstration projects make it attractive to consider fuel cells for marine applications. Today, there are a number of demonstration projects globally with several fuel cell types being of particular interest for marine use. This study is based on results from the FellowSHIP project (FellowSHIP 2008), which builds on leading Norwegian maritime industrial competence in combination with state-of-art fuel cell technology with the goal to develop clean and efficient power packs for use in ships. Economical and environmental consequences regarding fuel cell in ships have also been addressed in earlier studies (Tronstad and Endresen, 2006; Sødal, 2003).
The future development in technology and production capacity will lead to cost reductions in manufacturing, and hence reduced selling prices.

Marine fuel cell applications can be utilised in two ways; as main propulsion power or as auxiliary power production. Fuel cell applications are inherently more energy efficient compared to conventional technology, and especially so at partial loads. Due to absence of moving parts, fuel cells are in practice vibration-free and have excellent noise characteristics (Karin et al., 1999). The technology is modular in design, and can be fitted into unconventional machinery spaces. In addition, the technology includes zero - or ultra low - emissions to air.

The use of LNG is a proven alternative for use in ships, already adopted as main fuel in several offshore supply vessels as well as in large ferries operating in Norway.

The emission reduction of using fuel cells fuelled on LNG compared to diesel generators running on MDO is a consequence of several effects (Renaas, 2005):

- LNG has a relative higher energy content which reduces the specific fuel consumption, and thus formation of CO_2 emissions per used kWh (Table 4).
- LNG has a lower carbon content which reduces the amount of CO_2 per ton fuel consumed (Table 4).
- The fuel cell has a higher efficiency than diesel engine, reducing CO_2 emissions per used kWh (Table 4).
- LNG contains no sulphur, thus eliminating formation of sulphur-oxides (SO_x) during use as fuel (however, this study focuses on CO_2 emissions and SO_x emission gains are omitted).
- Due to the absence of a combustion process, fuel cells produce no nitrous oxides (NO_x) or particular matters (PM) (however, this study focuses on CO_2 emissions, NO_x and PM emission gains are omitted).

2. Modelling approach

Methodology for calculating ship fuel consumption and emissions are reported by several studies (e.g. Endresen et al., 2003; 2007; 2008; EMEP, 2002; Corbett, 2003; Eyring et al., 2005a; 2005b).

This study estimates the deviation (Δ) in environmental and economic performance of fuel cells versus diesel generators. The modelling approach is based on (Tronstad and Endresen, 2006; Longva et al., 2008). In particular, the following approach is applied:

a) Exogenous parameters (those that are not changed in the model) for operational profile and ship characteristics for three case ships are drawn **(Equations 1 – 8).**
b) Endogenous (those changed in the model) values are assumed for future fuel prices and emission taxes and/or other emission reducing initiatives **(Future scenarios).**
c) Economic and environmental results are compiled for each year of operation and used to assess the competitiveness of fuel cell relative to the incumbent technology. **(Equations 9-10).**

The detailed modelling approach is presented in the following.

2.1. Calculated costs and emissions

2.1.1. Calculating environmental performance of diesel generators vs. marine fuel cells

The emission (*E* – tonnes/year) is calculated based on average specific emissions per engine output (*ASE* – g/kWh), installed power (*p* – kW), days per year (*d* – days) and engine load (*l*). Similarly, fuel consumption (*F* – tonnes/year) is calculated through specific fuel consumption (SFC – g/kWh). The calculation is shown in Eq.1a and 1b.

$$E = \frac{ASE \cdot p \cdot 24 \cdot d \cdot l}{10^6} \tag{1a}$$

$$F = \frac{SFC \cdot p \cdot 24 \cdot d \cdot l}{10^6} \tag{1b}$$

Several methods exist for calculating average specific emissions based on emission amount per tonne fuel. The approach used herein is to determine emissions based on engine load. In this study an average emission and fuel consumption rate is assumed, representing an average engine load.

The emission level (E_A – tonnes/year) for the ship when using fuel cells is calculated based on the emission reduction effect (*R* -%), as shown in Eq. 2:

$$E_A = E \cdot (1 - R) \tag{2}$$

The emission reduction (ΔE - tonnes/year) when installing fuel cells is (Eq. 3):

$$\Delta E = E - E_A = E \cdot R \tag{3}$$

2.1.2. Calculating economic performance of diesel generators vs. marine fuel cells

Economic performance is assessed by summing total cost differences (ΔC_{Total} – €) for the fuel cells compared with diesel generators. Total cost is a function of capital expenditures (ΔC_{CapEx} – €), operational costs (ΔC_{OpEx} – €) and changes in fuel costs (ΔC_{Fuel} – €) less the savings of reduced environmental tax exposure (ΔS_{Em} – €), converted to a Present Value (PV) which is the present value of future cash flows (i.e. future cost levels in this study) (Ross et al., 2005) (Eq. 4).

$$\Delta C_{Total} = \Delta C_{CapEx} + \Delta C_{OpEx} + \Delta C_{Fuel} - \Delta S_{Em} \tag{4}$$

Capital expenditure (ΔC_{CapEx}) is the total investment sum being the purchase cost and the installation cost. Additional cost associated with fitting a LNG tank and the LNG onboard fuel system is not taken into account. The investment cost (ΔP_C - €/kW) is a present value and may be converted to a lump sum directly (Eq. 5).

$$\Delta C_{CapEx} = \Delta P_C \cdot p \tag{5}$$

Operational costs include maintenance costs and costs related to spare parts. These costs incur as the engine is operating and is a function of installed power (*p* - kW), engine load (*l* - %), days at sea per year (*d* - days) and expected life time (years) (*n*). Operational unit cost (ΔP_O – €/MWh) is defined as hourly operational cost per MWh. A discount rate for capital costs (r) is used to calculate the present value of future costs. (Eq. 6).

$$\Delta C_{OpEx} = \frac{d \cdot p \cdot l \cdot 24}{10^3} \cdot \sum_{i=0}^{n-1} \frac{\Delta P_O}{(1+r)^i} \tag{6}$$

Fuel cost difference (ΔC_{Fuel}) is a function of the price of fuel ($\Delta P_{Fuel}, P^i_{Fuel}$ – €/tonne) multiplied by fuel consumption (*F* – tonnes/year) and is calculated as the price difference between MDO and

LNG for a given year. In addition to change in price, increased fuel consumption (ΔF – tonnes/year) may lead to extra costs. The total difference is calculated as shown in Eq. 7.

$$\Delta C_{Fuel} = \sum_{i=0}^{n-1} \frac{F^i \cdot \Delta P_{Fuel} + \Delta F \cdot P_{Fuel}^i}{(1+r)^i} \qquad (7)$$

The CO_2 tax savings (ΔS_{Em}) are based on a tax per tonne emitted (P_T – €/tonnes). The tax may also represent a quota price for one tonne emitted, as it would be similar to reducing the cost of operating the ship. This is shown in Eq. 8.

$$\Delta S_{Em} = \Delta E \cdot \sum_{i=0}^{n-1} \frac{P_T^i}{(1+r)^i} \qquad (8)$$

Economic results are compared using Internal Rate of Return (*IRR*) and Profitability Index (*PI*). PI is the ratio of present value of the future expected cash flows *after* the initial investment divided by the initial investment (Ross et al., 2005). The PI shows the profitability of the project also taking into account capital costs. IRR is the discount rate when the net present value=0, thus showing total rent of the investment alternative in the absence of capital costs. (Eq. 9 and 10).

$$PI = \frac{PV - \Delta C_{CapEx}}{\Delta C_{CapEx}} \qquad (9)$$

$$NPV = \sum_{i=1}^{T} \frac{\Delta C_{Total,i}}{(1+IRR)^i} - \Delta C_{Total,0} = 0 \qquad (10)$$

If PI>1, then the project delivers value above the initial investment cost, taking into account discounted future cost savings and capital expenditures. If PI<1, then the investor simply loses money in the project. If IRR>0, then the investment has a positive rentability in the absence of a capital cost (discount rate). As such, if IRR<discount rate, then the project fails to deliver the minimum required rent. In this study, the applied discount rate is 8%, an IRR below this threshold indicates that fuel cells are outperformed by the incumbent technology.

2.2. Case ships and assumptions

In order to assess the use of fuel cells, three typical case ships are designed; supply vessel, dry cargo and container. The individual ship profile is based on typical ship type parameters such as deadweight tonnes (Dwt), cruise speed, days at sea, engine loads and fuel consumption. Assumptions for all case ships are shown in table 2.

The supply vessel is set up with a diesel electric propulsion system with four diesel generators. In the supply vessel case, one diesel generator is replaced with a fuel cell. For the dry cargo ship case the auxiliary generator is replaced, and in the container ship case one of the four auxiliary generators is replaced. Using equations 1 – 8, the economic and environmental performance of fuel cell in each ship type is determined. The replaced diesel generator for the supply vessel case is assumed to operate 150 days per year (3600 hours). For the dry cargo case the replaced auxiliary generator is assumed to operate 150 days per year (3600 hours), identical to the days on hire (350 days) considering only the days the ship is not at sea. For the container ship the generator is assumed to run 250 days per year (6000 hours) as these power the reefer container on board. Replacing a diesel generator with a fuel cell in this study does not lead to a change in operational mode for that engine (including operating hours).

Ship type	Supply	Dry Cargo	Container	
Deadweight	5000	6500	85000	Tonnes
Speed	15,0	12,5	25,0	Knots
Typical engine load profiles (*l*)	85%	85%	85%	%
Installed power: Main engines	4*2000	2800	55000	kW
Installed power: Aux. engines	-	300	4*3000	kW
Specific fuel consumption (*SFC*)	199	199	199	g/kWh
Specific CO_2 emissions (*ASE*)	630	630	630	g/kWh
Power converted to fuel cell (*p*)	2000	300	3000	kW
Engine operation days (*d*)	150	150	250	Days
Yearly kWh per installed fuel cell	300000	45000	750000	kWh

Table 1: Characteristics and operational profile of the three case ships

2.2.1. Setting endogenous cost levels of fuel price and environmental taxation (scenarios)

During the life span of a ship, fuel price of MDO and LNG will vary. From 1985 – 2006 the price development of LNG and MDO were tightly correlated (0.94), albeit with different annual growth rates (1.5% and 3.9% respectively) (BP, 2006; Bunkerworld, 2008).

An emerging uncertainty concerning price development of marine diesel is prospected regulation on sulphur content which in reality effectively bans the use of heavy fuel oil. This will lead to increased demand of distillate fuels such as MDO. In contrast to oil, LNG is stipulated to account for a growing share in the global energy and electricity mix (Shell, 2001). Historically LNG has been regionally priced but technological advancement opens for a move towards a more a global LNG market, and hence also price (Jensen, 2004). Currently, Japan is the dominating importer of LNG with 39% of global LNG import in 2006 (BP, 2006), and the Japan import price will function as a baseline target for a global LNG price in this study. The fuel prices assumed for LNG and MDO in this study are 225 € and 401 € per tonne respectively (P^0_{Fuel}) reflecting current prices. Extrapolated future price growth is developed based on historical data from (Bunkerworld, 2008; BP, 2006) and is summarised in Table 3.

Tax exposure for emitting SO_x, NO_x and CO_2 can represent a significant cost component for the ship operator. However, a recent agreement was reached in the Marine Environment Protection Committee (MEPC) on an amendment to Marpol Annex VI restricting the SO_x and NO_x emission to below 10% and 20% respectively (IMO, 2008)[1]. Although these limits will apply on ships constructed after 2016 any taxes on emitting SO_x and NO_x is not modelled in this study. Tax exposure relating to emitting CO_2 is today absent in the maritime sector, but expectations are that a tax regime may be introduced in the future. Such a possibility is modelled herein. Fuel cell concepts which emit less CO_2 than the incumbent technology therefore represent a lower environmental tax exposure.

Taxation of CO_2 can be done through voluntary schemes and through regulated (enforced) schemes. In this study, it is assumed a taxation (or price) for CO_2 without constructing an overall scheme for how such taxation would be enforced (including setting limit for free allowance of CO_2 emission and the possibility to sell emission quotas). Two projections are made; a price level reflecting today's price of emitted CO_2 of about 20 € per tonne and a significant higher price of 100 € per tonne, thus incorporating expectations that future trading schemes will be stricter in price terms (Norpool, 2008) (Table 3).

Future change in fuel price and the introduction of a CO_2 taxation scheme constitute the drawn future scenarios. Because of the judgementally sat values applied in the forecasting, the absolute values must be interpreted with care. The purpose of constructing such scenarios in this study is not to predict the price of fuel or tax in a given year, but to show the sensitivity of performance given changes in these variables. Table 3 highlight the scenarios.

---

[1] A sulphur cap on fuel will gradually be lowered towards 0.5% in 2020 outside the emission control area (ECA) and 0.1% from 2015 inside the ECA. NOx limits will be 14.4 g/kWh and 3.4 g/kWh outside and inside the ECAs respectively (IMO 2008)

Name scenario	Description	Assumptions – values
Historical development continues, no CO_2 taxation	Price of MDO and LNG continue to grow at average annual growth rate from 1985-2006	Annual price growth MDO = 4% Annual price growth LNG = 1.5%
Historical development continues, medium CO_2 taxation	Price of MDO and LNG follow historical development, price of CO_2 emission the same as current price level	Tax per tonne emitted CO_2 = 20 € per tonne
Historical development continues, high CO_2 taxation	Price of MDO and LNG follow historical development, price of CO_2 are higher than today's Kyoto level	Tax per tonne emitted CO_2 = 100 € per tonne
High LNG price growth, no C02 taxation	Price of LNG grow at the double rate as MDO, conversely from the historical development	Annual price growth LNG = 8% Annual price growth MDO= 4%

Table 2: Description for applied scenarios which constitute the endogenous cost variables

2.2.2. Setting level for investment and operational costs

The applied capital expenditures of diesel generators is based on USCG (1999), adjusted according to industry inputs. The capital expenditures for marine fuel cells are based on CFC (2008). Table 4 list capital expenditures levels in € per kW.

Type	Price
Diesel generators	2000 €/kW
Fuel cells	4500 €/kW
Difference (ΔC_{CapEx})	2500 €/kW

Table 3: Modelled capital expenditures (ΔC_{CapEx}) for the incumbent technology and fuel cells. Marine fuel cells incurs 2500 €/kW added capital expenditures.

Maintenance costs are assumed alike for diesel generators and fuel cells, except for stack replacement cost. Fuel cells currently have a significant cost component in replacing worn-out stacks, which can be regarded as an expensive spare part. The price is set to 1500 €/kW per exchange, which has been assumed in this study to occur every 30 000 hours, taken to be a compromise between typical stack life time under a steady-state base load and a somewhat cyclic operational profile. Estimating three replacements during a lifetime of 120 000 hours (20 years life time), the operational cost is set to:

$$\frac{1500\,€/kW}{\left(\frac{120000h}{3}\right)} = 37.5\,€/MWh$$

With the future expected requirements to NO_x emission it is relevant to compare the fuel cell machinery with a diesel engine equipped with a selective catalytic reduction (SCR) plant. The SCR has an operational cost of about 7 to 8 €/MWh (ENTEC, 2005) reducing the cost difference to 30 €/MWh for fuel cell (C_{OpEx}). The capital expenditures for the SCR is not taken into consideration in the diesel alternative as it represents only about 3 % of the cost of the fuel cell.

2.2.3. Other assumptions

All prices are denominated in Euro (current prices), and inflation or currency movements are not taken into consideration. EU recommends a discount rate of 4-6 % (EU, 2008), but this applies to public investments. For a ship owner the required rate of return is higher. In addition there should be a risk premium to cover the uncertainties of the investment. This paper uses a discount rate (r) of 8 per cent.

The CO_2 emission calculations are based on carbon content of the fuel, the fuel energy content and engine efficiency (see Table 1). The total CO_2 reduction (R – see equation 2) for fuel cell is calculated to be 29% (*R*), as outlined by the calculation shown below (the value 3.6 represent the conversion from kWh to MJ).

$$\frac{2.75 \cdot 3.6}{45 \cdot 0.49} = 449 \text{ g } CO_2 \text{ per kWh for fuel cells}$$

$$\frac{3.17 \cdot 3.6}{43.1 \cdot 0.42} = 630 \text{ g } CO_2 \text{ per kWh for diesel engines}$$

	Diesel generator (MDO fuel)	**Fuel Cell (LNG)**	
Energy content	43.1	45	MJ/kg fuel
Carbon content	3.17	2.75	Grams CO_2 per gram fuel
Engine efficiency	42	49	%

Table 4: Energy content, carbon content and engine efficiency for diesel generator and marine fuel cells (1kWh = 3.6 MJ) (Renaas, 2005).

3. The modelling results and discussion

Results are shown in table 5 summarising relative performance of the case ships in four scenarios incorporating a fixed value for the installation of fuel cells. This value is set to 2500 €/kW (Table 5).

Case ship	**Reduced CO_2 emission per year (ΔE)**	**Scenario**	**Internal Rate of Return (IRR)**	**Profitability Index (PI)**
Container	2 771 tonnes	Historical development continues, no CO_2 taxation	2 %	0.87
		Historical development continues, medium CO_2 taxation	7 %	0.95
		Historical development continues, high CO_2 taxation	11 %	1.26
		High LNG price growth, no C02 taxation	-	0.16
Dry Cargo	168 tonnes	Historical development continues, no CO_2 taxation	2 %	0.53
		Historical development continues, medium CO_2 taxation	3 %	0.57
		Historical development continues, high CO_2 taxation	5 %	0.76
		High LNG price growth, no C02 taxation	-	0.09
Supply	1 312 tonnes	Historical development continues, no CO_2 taxation	2 %	0.56
		Historical development continues, medium CO_2 taxation	3 %	0.62
		Historical development continues, high CO_2 taxation	6 %	0.84
		High LNG price growth, no CO_2 taxation	-	0.16

Table 5: Summary of economic and environmental performance for all scenarios. Applied additional investment cost sat to 2500 € per kWh for marine fuel cells compared to diesel generators.

With historical development in fuel prices and high CO_2 taxation, the results show fuel cells as economically competitive replacing one of the four auxiliary generators in the container ship. In all the remaining scenarios fuel cells incur higher cost levels than diesels (i.e. PI<1). A general pattern is that fuel cell installed in the container case ship is most economically competitive. In the dry cargo ship case and supply vessel case, fuel cells perform relatively equal.

As a general conclusion, the relative profitability of a high cost investment such as fuel cells is very dependant on the number of running hours. The supply vessel has more installed power than the dry cargo ship; however the same amount of hours in operation for the fuel cells leads to relatively equal profitability among these two ships. This is partly due to assumed linear relationship between cost and size of the fuel cell application, i.e. the unit cost (€/kW) does not decline with increased size of the unit installed. This in turn underlines the importance of fuel price to the economic competitiveness of fuel cells. CO_2 taxation does not have a significant impact on the profitability, but could tip the balance from a non-profit to a break even situation, as the example of container ship show for the high CO_2 taxation scenario.

In addition to the economic impact, environmental benign solutions such as fuel cells will allow for more flexible route planning, owing to the bypass of any environmental control area restrictions. Reducing emission of CO_2 and the zero emissions of NO_x and SO_x moreover provide value in alleviating the environmental impacts from ships which is beneficial for society at large.

3.1. Results with variable investment costs

As the fuel cell technology develops the price is expected to fall. Diesel generator solutions are far more mature and are not expected to see significantly price reductions. This makes profitability analysis as a function of price difference interesting.

Figures 1, 2 and 3 depict the profitability for each case ship in all applied scenarios as a function of additional needed capital expenditures for fuel cells relative to the incumbent technology (i.e. price difference between fuel cells and diesel generators). The operational cost (stack replacement) is adjusted accordingly, with the same percentage rate.

With price of fuel following historical growth trends, added capital expenditures of about 2000 €/kW appear to be a break even level in order for fuel cells to be economically competitive against the incumbent technology, provided a high degree of utilization of the fuel cell in terms of operating hours. In this case study, such a profile is depicted in the container case ship (Figure 1). With a lower degree of utilization of the fuel cell with the same fuel price scenario, added capital expenditures of about 1500 €/kW appear as the maximum price in order for fuel cell to be economical competitive. This is illustrated in the supply vessel case and the dry cargo ship case (Figure 1 and 2).

In the event of a reverse trend in the growth of LNG and MDO prices (and no CO_2 taxation), with higher price for LNG than MDO, the higher capital expenditures of fuel cells must be no more than about 1200 €/kW for high utilisation and about 900 €/kW for low utilisation.

The CO_2 taxation has a small effect on the profitability for the relatively high fuel cell investment costs. As the investment cost decreases, the impact of CO_2 taxation becomes more important, but in such cases, fuel cells will already be profitable.

Hence, since fuel cells running on LNG fundamentally incur lower fuel and taxation costs, higher capital expenditures can be justified without compromising profitability.

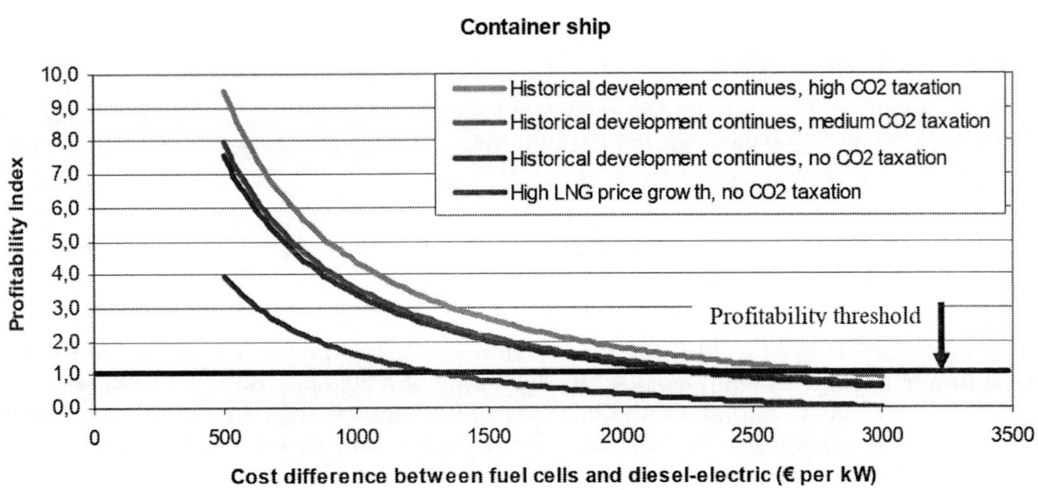

Figure 1: The profitability of container ship case in all applied scenarios as a function of added capital expenditures

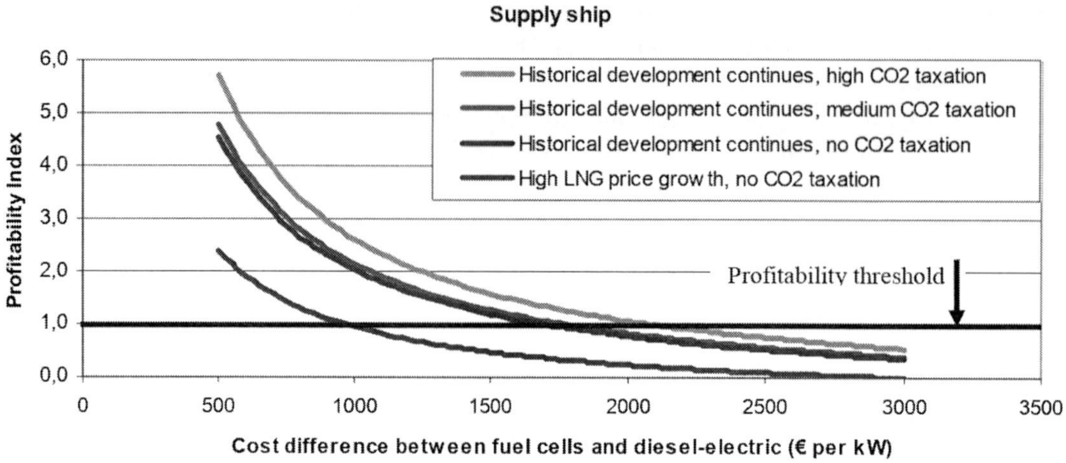

Figure 2: The profitability of dry cargo ship case in all applied scenarios as a function of added capital expenditures

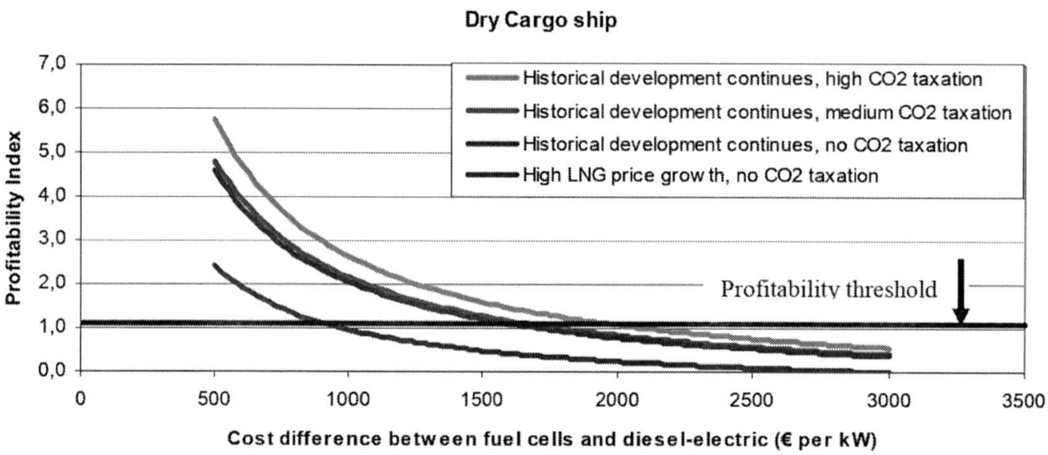

Figure 3: The profitability of supply vessel case in all applied scenarios as a function of added capital expenditures

3.2. *Discussion*

Based on the reference studies the data input should provide a rather good indication of the relative performance and air emission data for marine fuel cells. Larger uncertainties concerning the absolute price/cost levels will have to be expected.

Environmental performance will be enhanced using fuel cells regardless of externalities. Economic competitiveness is to a large extent determined by future fuel price development and the absolute price of fuel cells.

This study has assumed LNG to be the fuel of choice for marine fuel cells. Other alternatives include hydrogen, methanol, and bio-gas to name some. The phasing in of fuel cells in the world fleet is to a high degree dependant on the availability of a suitable fuels, the switch to such fuels as main fuel for marine operators and other non-technical variables such as the competence of machinists. These factors, as well as other more general uncertainty factors owing to the limited marine experience of fuel cells, suggest that a higher risk should be factored in the analysis. This could be manifested in a higher discount rate for the investment of fuel cells compared to diesel generators, imposing stricter economic requirements for fuel cells compared to diesels.

The numbers of fuel cell operating hours used in this study are derived from the typical operational profile of auxiliary generator sets in the chosen case ships. This may not be representative for how fuel cells will be phased in for ship applications in practice as the results indicate that the higher the number of operating hours, the better the competitiveness of marine fuel cells.

Moreover, the somehow in-tangible values of fuel cells, specifically with regard to any reputation effects utilisation of a greener technology might impose for the owner, potential regulative restrictions on the use of diesel fuel oils, as well as the general social benefit of lower CO_2 emissions suggest that the competitiveness of fuel cells are influenced by variables outside the realm of (at least) traditional economic analysis.

4. Conclusion

This study has modelled the difference in fuel consumption and emission profiles of three case ships running on two different technologies; diesel generator run on MDO, and fuel cells running on LNG. Due to the better energy efficiency of fuel cells and current favourable fuel prices (LNG vs. MDO) (2008), the modelled results show that fuel cells run on LNG will be economically competitive with a capital expenditures level twice that of the incumbent technology. Current fuel cell prices may be approaching this level, as indicated in this study.

Thus reduced capital expenditures levels and high fuel prices of MDO relative to LNG will further increase the competitiveness of fuel cells. CO_2 taxation is expected to have less influence on the profitability, unless reaching 100-200 € per tonnes CO_2. Table 6 summarizes the pro's and con's for fuel cells compared with diesel generators.

	Financial	**Non-financial**
Quantifiable	+ Reduced fuel cost + Reduced environmental taxation exposure - High up front capital expenditures	+ Reduced CO_2 emission by 29%, even more for partial load operation. + No emission of SO_X + No emission of NO_X + No emission of particular matters
Non-quantifiable	+ Flexibility in voyage route planning, due to absence of environmental control area restrictions - High uncertainty in cost level due to novel technology	+ Enhanced environmental stewardship and positive reputation effects for the ship-owner + Less noise and vibrations, improved onboard working conditions + Complies with future regulations regarding NO_X and SO_X. - Uncertainty related to storage of novel fuels (exception for LNG which is proven for ship use) - Uncertainty regarding fuel infrastructure - Uncertainty regarding operability issues

Table 6: Conclusion of overall competitiveness of fuel cells operated on LNG compared with diesel generators. This study has focused on quantifiable performance indicators.

5. Further research

The most important limitations of this study concern the limited availability of marine fuel cell performance data, and uncertainties relating to the general diffusion of fuel cells and utilisation of LNG as fuel on a worldwide basis. These aspects could provide more complete baseline for assessing the competitiveness of marine fuel cells. The cost of LNG fuel systems including storage tanks was omitted in this study, and future work will have to include updated cost figures from the industry on this issue. Generally speaking, the cost figures related to both diesel generator sets as well as fuel cell systems used in the study should be perceived indicative only and future work should emphasise on increasing the reliability of these figures.

The future introduction of fuel cells in ships could take different paths; from replacing one or more of the auxiliary diesel generators to a switch to an all electric ship powered by hybrid arrangements of fuel cells and gas engines. Future work should prepare comparisons of different alternatives, as this will also increase the usefulness of the studies. This analysis has used an average engine load representing an average of a typical operation profile for the case ships. A more detailed study of the actual load levels would improve the study results, with dedicated efficiencies for partial load operations to name some.

Reference list

Behrens, H.L., Ø. Endresen, A. Mjelde & C. Garmann. (2002) Environmental Accounting System for Norwegian Shipping – EASNoS Phase 1, Documentation of Principles, Methodology & Structure Report No. 2002-1645.

Bunkerworld (2008): www.bunkerworld.com, March 2008.

CFC 2008. Personal Communication with CFC Solutions, Germany, Mr. Michael Poulain, March/April 2008.

Corbett, J. 2006. "Environmental Impact of ocean Shipping in a Freight Context" A report to OECD, Paris, November.

Douglas-Westwood 2006. "The World marine propulsion report 2006 -2010". DW Report no 357 -06

Endresen, Ø., E. Sørgård, J. K. Sundet, S. B. Dalsøren, I. S. A. Isaksen, T. F. Berglen, and G. Gravir. (2003). "Emission from international sea transportation and environmental impact". Journal of Geophysical Research, 108 (D17), 4560, doi:10.1029/2002JD002898.

Endresen, Ø., Mjelde A., Sverud T., and Sørgård E. (1999). "Data and models for quantification of Ship pollution" Det Norske Veritas, Report no. 98-2059, web site: http://research.dnv.com/marmil/

Entec UK Limited (2005), *Service Contract on Ship Emissions: Assignment, Abatement and Market-based Instruments, Task 2b- NO_x Abatement*, European Commission Directorate General Environment.

European Union (EU). (2002). "Communication from the Commission to the European Parliament and the Council: A European Union strategy to reduce atmospheric emissions from seagoing ships", COM (2002) 595 final. Volume I, November 2002. Available at http://europa.eu.int/eur-lex/en/com/pdf/2002/act0595en01/1.pdf

European Union (EU). (2007). "Policy issues GHG emissions and shipping", Presentation by Mark Major, Europan Commision. Available at http://regserver.unfccc.int/seors/file_storage/86um9rfgntlaajx.pdf

FellowSHIP (2008). FellowSHIP phase 1 and phase 2 (2003-2008) project development results (restricted information)

International Maritime Organization (IMO). (2008). Available at: http://www.imo.org/Newsroom/mainframe.asp?topic_id=1709&doc_id=9123, April, 2008

Jensen. K. T. (2004). "The Development of a Global LNG Market: Is it Likely? If so, when?". Oxford Institute for Energy Studies, Alden Press, Oxford.

Karin, Zvi ; Leavitt, N. ; Costa, T. ; Grijalva, R. (1999). "Marine Fuel Cell Market Analysis. U.S Coast Guard Research and Development Center". Available at: http://handle.dtic.mil/100.2/ADA373706

Longva, T., Eide, S.M. and Nyhus, E. (2008). "Achieving Emission Reductions: Analysing Cost and Consequences towards 2020". Det Norske Veritas Report no. 2007-1214.

Maritim Miljøsatsing (Maritime Environmental Effort), the Norwegian Research Counsils program MARITIM, web site, http://research.dnv.com/marmil/, 2002.

Norpool. (2008). Trading of EU allowances at Nord Pool, http://www.nordpool.no/marketinfo/co2-allowances/allowances.cgi

Reenaas, Marte. (2005). "Solid Oxide Fuel Cell Combined with Gas Turbine versus Diesel Engine as Auxiliary Power Producing Unit onboard a Passenger Ferry". NTNU Master Thesis 2005. Available at http://www.indecol.ntnu.no/indecolwebnew/publications/mastertheses/Marte%20Reenaas%20var05/diplomoppgreenaas_v05_pdf.pdf

Ross, A. R., Westerfield, R. W. and Jaffe. J., F. (2005). "Corporate Finance, 7th ed". McGraw Hill, New York.

Shell (2001). "Dynamics as Usual". Available at: http://www.pecc.org/PECC2003Brunei/papers/plen-session3/hilton.ppt#665,2,Slide 2

Søldal, S. (2003). "Fuel cells in shipping: higher capital costs and reduced flexibility". SNF Report NO 23/03.

Tronstad, T. and Endresen, Ø. (2006). Modelling Economic Break-Even Conditions for Fuel Cells in Merchant Ships. WHEC, 13th-16th of June 2006, Lyon, France.

United States Environmental Protection Agency (EPA) 2007. "U.S. Proposal to Cut International Ship Emissions Finds Key Support". 28th of June 2007.

United States Coast Guard (USCG) (1999). Life cycle cost model, version 1.1. 26 October 1999. United States Coast Guard.

AUTHOR INDEX

AUTHOR INDEX